Cabinetwork and Joinery

Comprising Designs and Details of Construction With
2021 Working Drawings and Twelve Plates

Paul N. Hasluck

SOTHIS PRESS

ISBN 978-1-0365-0303-1

This edition is published by
Sothis Press © 2023. The company
is registered in England and Wales.
The Company Number is 14900389.

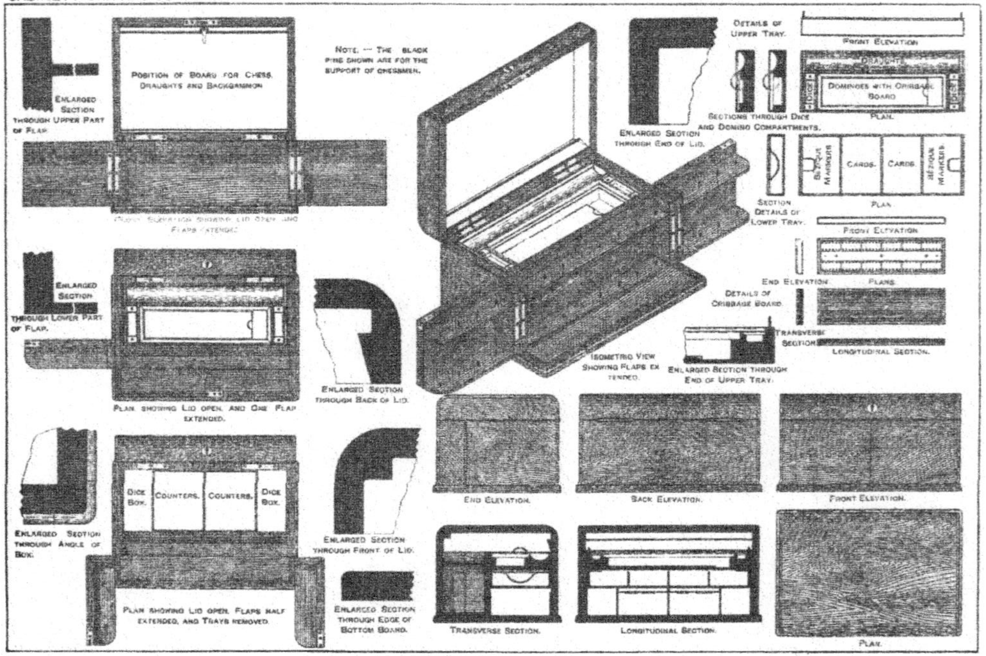

1

GAMES COMPENDIUM.

SOTHIS PRESS

Sothis Press aims to completely reimagine the public domain book with an emphasis on universal access to captivatingly designed print editions. We believe that technology will enable every single book ever published to be perpetually accessible in beautiful print form. Sothis is beauty in ubiquity.

www.sothis.press
1 Waterhouse Square, London, ECIN 2ST

CASSELL'S

CABINETWORK

AND

JOINERY

COMPRISING DESIGNS AND DETAILS OF CONSTRUCTION
WITH 2,021 WORKING DRAWINGS
AND TWELVE COLOURED PLATES DRAWN TO SCALE

EDITED BY

PAUL N. HASLUCK

CASSELL AND COMPANY, LIMITED
LONDON, NEW YORK, TORONTO AND MELBOURNE
1912

PUBLISHERS' NOTE

CABINETWORK AND JOINERY is a natural outcome of, and sequel to, CASSELL'S CARPENTRY AND JOINERY, with which work it is uniform in style and price. Whilst the chief object of that work was to explain constructive principles adopted in the related crafts of carpentry and joinery, the present purpose is to give some hundreds of examples showing further how these principles are applied in everyday practice.

The reader is here assumed to be acquainted with hand tools and appliances—their shapes, care, and uses; with timber—its qualities, varieties, and selection; with the different forms of joints and their adaptability to various conditions; with the setting out of work—including the preparation of rods; and with the principles of construction in woodwork; all these matters are fully dealt with in WOODWORKING and in CARPENTRY AND JOINERY, two other volumes in this series. The present book devotes but little space to the rudiments of cabinetwork and joinery, but makes a direct and immediate appeal to the constructive instinct of the craftsman by presenting him with an extensive and varied range of designs of completed articles, accompanied by full explanatory notes. No less than 250 different designs with details are included in this book, the illustrations numbering 2,021 in all.

In almost every case the objects here illustrated have been made by their designers, who also contributed the original drawings and descriptions either to WORK or to BUILDING WORLD, and it is from the columns of those two weekly journals that this volume has been compiled. The thoroughly practical character of the work is thus assured, the contributors including the foremost master-craftsmen of the day.

Emphasis is laid on the fact that the book is concerned with actual practice only. In every case the designs are workmanlike, and the host of detail illustrations—of which this book certainly contains more than any other of its kind, if any other exists—will be welcomed by all craftsmen who seek for thoroughness and sound constructional practice.

CONTENTS

LIST OF COLOURED PLATES

CABINETWORK AND JOINERY

DESIGNS AND DRAWINGS

TABLES

Introduction.

THIS book is uniform in style, size, and price with "Cassell's Carpentry and Joinery," and whilst the chief object of that work was to describe constructive principles, the present purpose is to provide a wide range of designs and working drawings. Information on first principles, tools, materials, and processes is quite outside the present scope; for these subjects the reader should refer to a companion volume, "Woodworking" (Cassell & Co.), which fully describes the manipulation of tools and materials, and devotes much space to the discussion of elementary exercises and simple examples; and he should consult, also, "Cassell's Carpentry and Joinery," already mentioned, which constitutes a reliable treatise on the technology of the two related crafts. The present purpose being chiefly to illustrate and describe typical examples of work, all other matters are but minor considerations here.

Kitchen Table with Turned Legs.

With regard to kitchen tables, a turned leg generally gives more satisfaction than the ordinary plain tapered leg, common to cheap tables, and costs but little more. Fig. 1 is a section through part of a kitchen table, showing one of the legs as well as the general construction. Into this table is introduced a slide for a drawing-board, or it may be utilised for a baking-board. There is also a drawer below. Fig. 2 shows the appearance of the front of the table. Part of the front rail is cut away to allow for the thickness of the board, and then stretchers are dovetailed down to it as shown in Fig. 3. The size of the drawer is next decided on, the necessary amount of stuff is taken out, and rebated stretchers are inserted between front and back rails. The drawer (Fig. 4) has pieces screwed to the sides to travel along the rebate in the stretcher, as clearly shown in Fig. 1. The bottom of the drawer is checked in square, and blocked with square fillets. This kitchen table may be made of good yellow pine throughout, with pins inserted through the legs into the tenons. Blocks may also be glued in at the back of the leg and rail. Bottom stretchers may be dispensed with when the rails and legs are well bound together.

Rectangular Side Table.

The next table to be considered is a side or hall table. Fig. 5 is a half elevation of the front of such a table. The legs are checked to receive the rails, which are carried all round, mitered at the corners, and screwed from the back through the legs. They are also blocked at the back, similarly to those in the previous example. A fielded channel is cut down the rail, finishing in a diamond pattern at the centre. A narrow rail is also carried

Fig. 1.—Vertical Section through Part of Kitchen Table with Turned Legs.

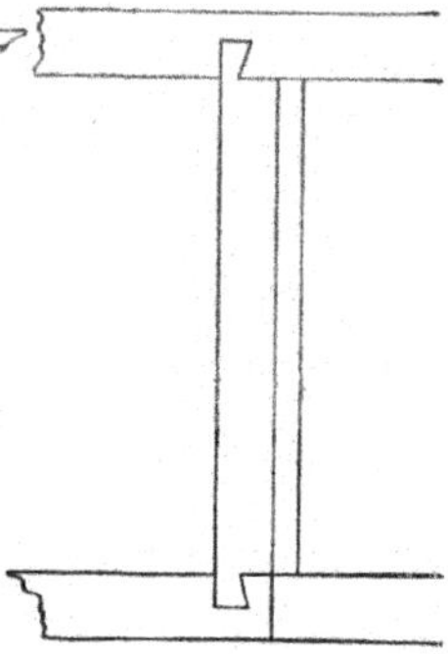

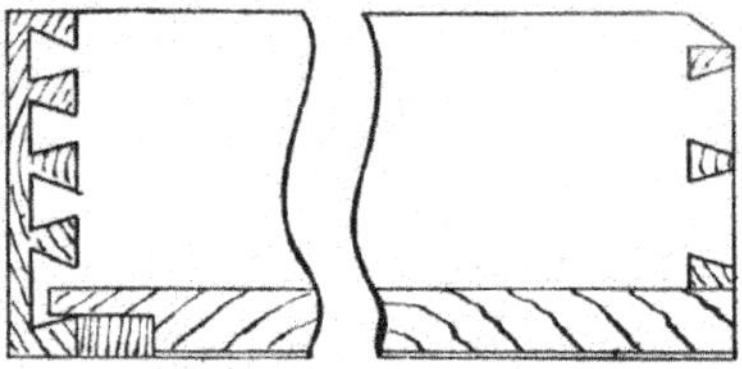

Fig. 4.—Section through Drawer.

Fig. 3.—Plan of Drawer Stretcher.

Fig. 2.—Part of Front Rail of Kitchen Table.

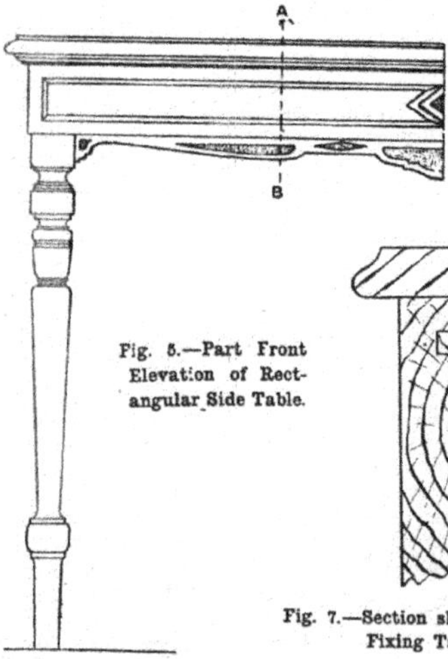

Fig. 5.—Part Front Elevation of Rectangular Side Table.

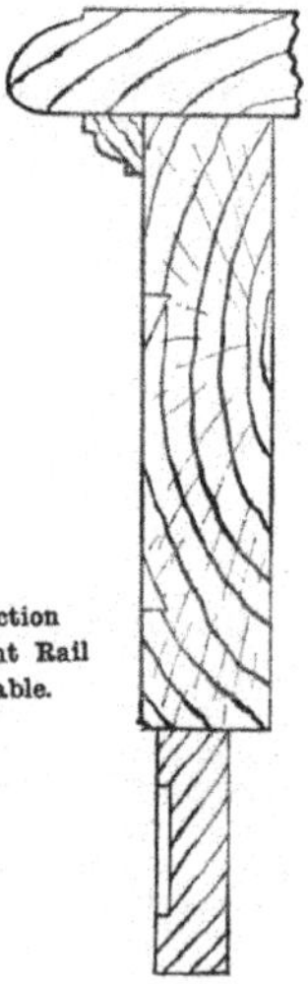

Fig. 6.—Section through Front Rail of Side Table.

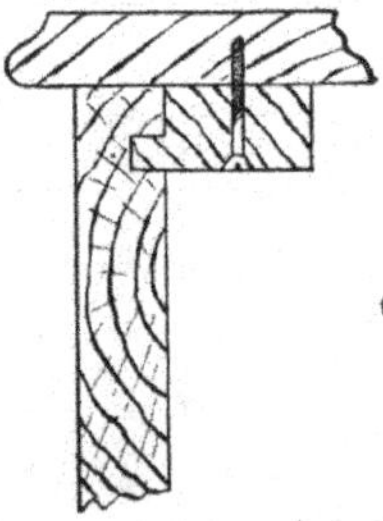

Fig. 7.—Section showing Method of Fixing Table Top.

along the front, and channelled out in a similar manner. A small moulding (see Fig. 6) is carried along the rails immediately under the table top, and this top is secured as shown in Fig. 7. Small mortices are cut in at the back of the rails, and blocks, having a small tenon to correspond with the mortice, are screwed to the under side of the top. This allows the top to shrink and swell without any danger of splitting. Fig. 8 is a horizontal section through a corner of the table, showing the leg rebated to receive the rails, which have a projection of about $\frac{3}{16}$ in. Fig. 9 shows a half design for a suitable leg for the table.

Fig. 9.—Half Design for Table Leg.

Fig. 10.—Square-frame Table with Elliptical Top.

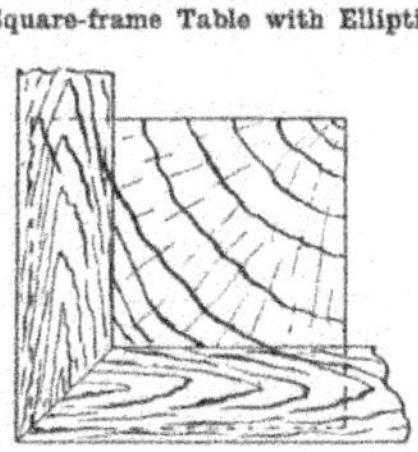

Fig. 8.—Horizontal Section through Corner of Side Table.

Square-frame Table with Elliptical Top.

A parlour table will next be considered. A design for such a table is shown at Fig. 10. The part plan (Fig. 11) shows the top to be an ellipse, but the frame is kept square. The dotted lines show diagonal stretchers, which carry a shaped shelf with a small fore edge planted down on top of it. The legs are turned, and the outside corners of the square are rounded off. The stretchers are half checked where they cross each other, and are tenoned to the legs. The shelf is shaped and

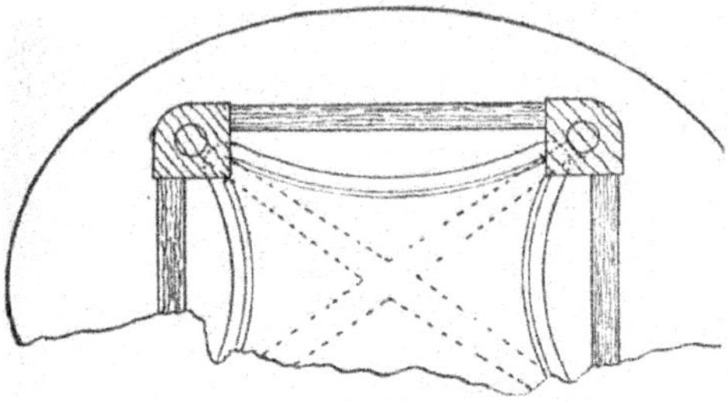

Fig. 11.—Underneath Plan of Square-frame Table with Elliptical Top.

bradded down on top of them, the fore edges being fixed to the shelf. A drawer is placed in the middle of the rail; it has a moulding run on the top and bottom edges of the front. This moulding is also

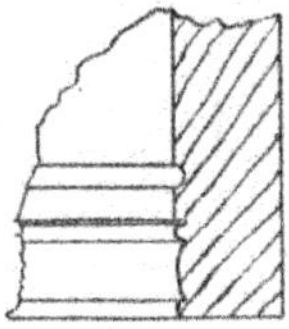

Fig. 13.

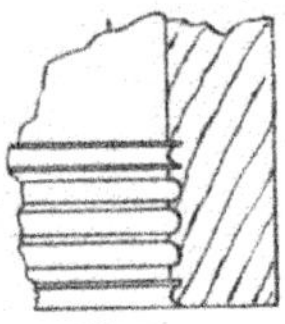

Fig. 14.

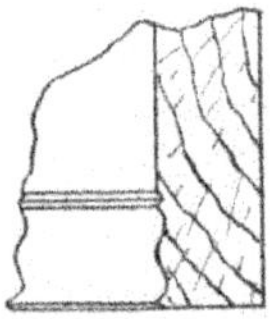

Fig. 15.

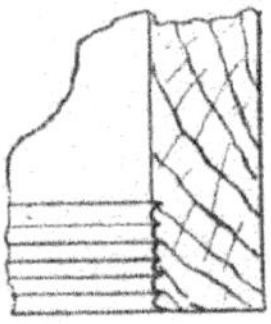

Fig. 16.

Figs. 13 to 16.—Sections of Various Mouldings.

run on the three other rails to carry out the same effect. Fig. 12 is a section through the drawer, showing the moulding and also the drawer bottom, which is fielded in this case instead of being square. Figs. 13 to 16 show different arrangements of mouldings. The edge of

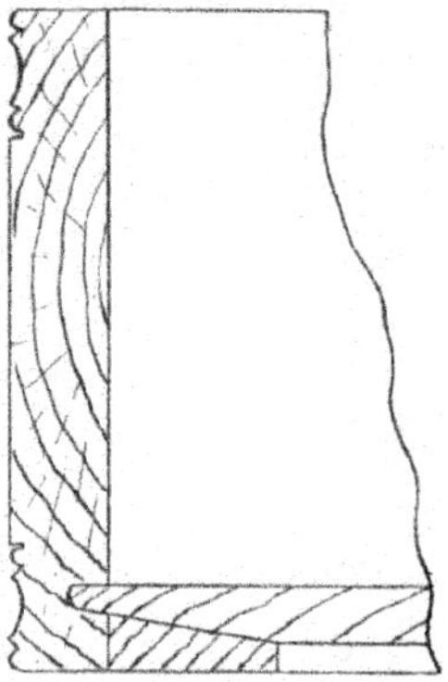

Fig. 12.—Section through Table Drawer.

the shelf is also moulded (see Fig. 17). The top is fixed in the same way as that of the rectangular side table (see Fig. 7, p. 2).

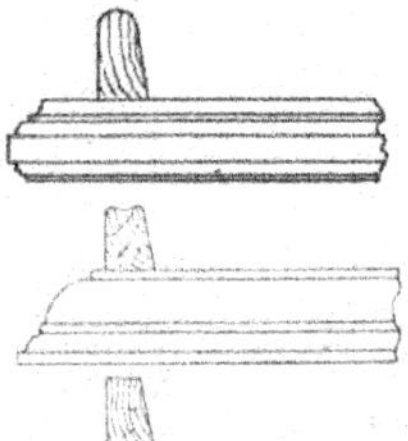

Fig. 17.—Alternative Mouldings for Table Shelf.

Extending Kitchen Table.

The material for making the extending table (Figs. 18 and 19) is good red or white deal, or the best pine. The sizes of the

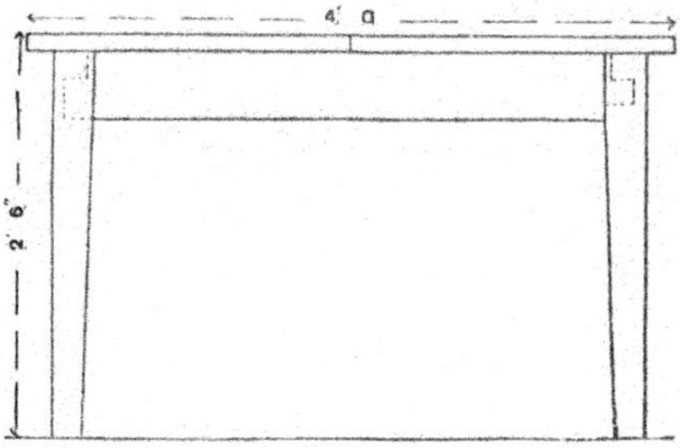

Fig. 18.—Side Elevation of Extending Kitchen Table Closed, without Leaf.

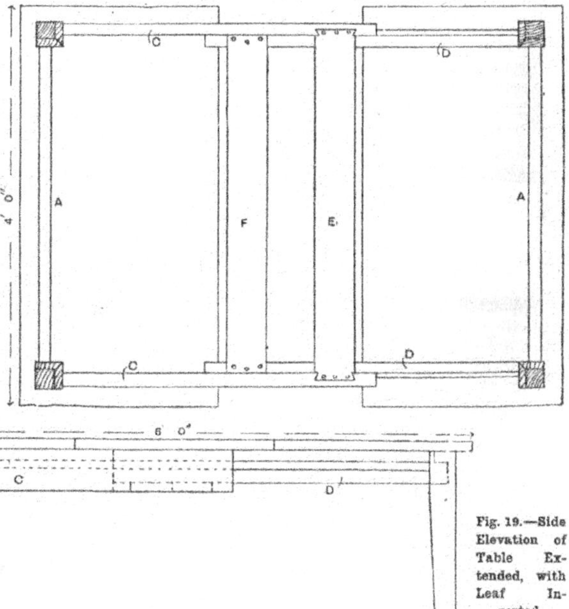

Fig. 20.—Underneath Plan of Extending Kitchen Table.

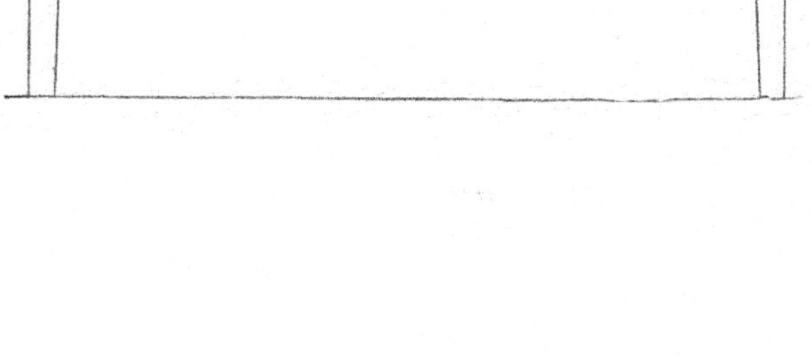

Fig. 19.—Side Elevation of Table Extended, with Leaf Inserted.

several pieces are as follow :—**Four legs, 2 ft. 7 in. by 3 in. by 3 in.; two end rails A (Fig. 20), 3 ft. 8 in. by 5½ in. by 1½ in.; two side rails C (Figs. 19 and 20), 3** ft. 5 in. by 5½ in. by 1½ in.; two **inner rails** D (Figs. 19 and 20), 3 ft. 5 in. by 4½ in. by 1½ in.; and two cross rails E and F (Fig. 20), 3 ft. 8 in. by 5½ in. by 1¼ in. These are cutting-out sizes, and allow for waste. The top is formed of three leaves (two fixed and one movable), each being made of three jointed boards, ploughed and tongued, or dowelled and glued. The square legs look better if tapered to about 2¼ in. at the bottom as shown. The joints connecting the rails and legs are shown in Fig. 21; the mortices are at right angles, and meet each other, thus allowing the tenons to be mitered as at G (Fig. 21). The parts should be firmly glued in, the side rails ploughed, each from the top edge, and the tongue, of some hardwood, inserted and glued to the inner rails D. To hold together the outer rails C a cross rail E is dovetailed in (see Fig. 20), and as there is no great outward strain on the inner rails, the cross rails F can be screwed to them. The two fixed flaps should be secured by screwing into them obliquely through the rails. Should it be desired to extend the table to two or more distances to suit flaps of different widths, a handy contrivance for holding the rails firmly in position is shown in

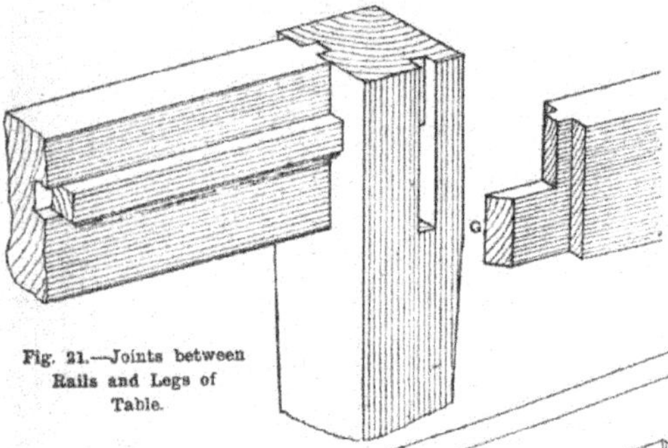

Fig. 21.—Joints between Rails and Legs of Table.

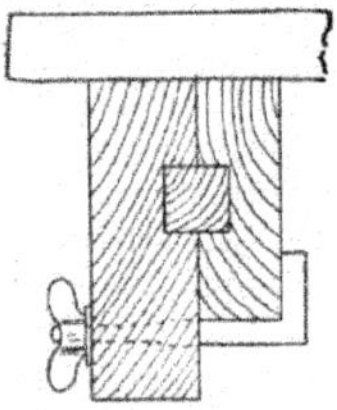

Fig. 23.—Section through Side Rails of Table, showing Method of Securing Them.

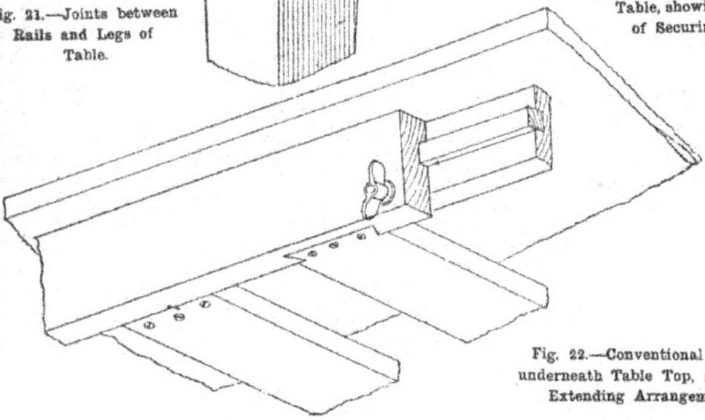

Fig. 22.—Conventional View underneath Table Top, showing Extending Arrangement.

Figs. 22 to 24. It is of square iron, bent to a right angle and rounded at the outer end, which is screwed for a nut. Fig. 23 shows that by tightening the nut the two rails are gripped together.

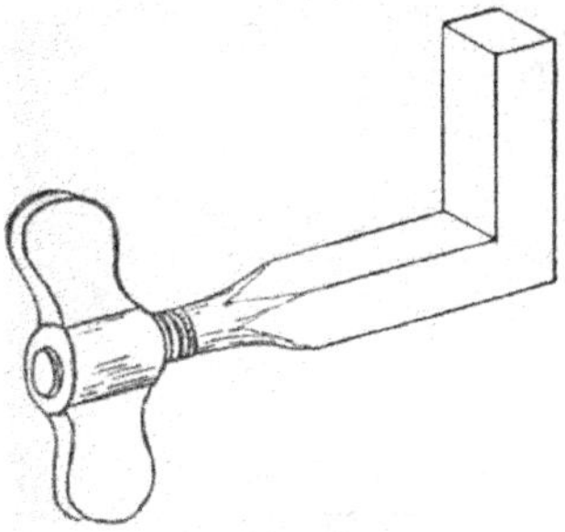

Fig. 24.—Clamp and Nut.

Extending 6-ft. to 9-ft. Dining Table with Screw.

The simple extending dining table shown by Figs. 25 to 28 should be made of mahogany, oak, or walnut. In Fig. 28, A is a half-plan from underneath, and B a half-plan with top removed. The legs are turned from stuff about 5 in. square. The outer rails for the framework may be solid, or the outside portion may be of ½-in. stuff glued to a backing as shown. The inner rails or slides and the two cross rails should be of hardwood, such as beech or birch. The special screw can be obtained from almost any furnishing ironmonger. In constructing the framework, plane the stuff to the sizes given, and set out the mortices of the legs and the tenons of the rails (see Fig. 29). The mortices made for the inner sliding rails are farther from the front edge of the legs than those for the outer rails, as shown in Figs. 28, 30, and 31. Plough the slides from their top edges, 1 in. wide and ½ in. deep. This groove can be made with a ½-in. plough-iron, or with a rebate plane by fixing a strip of wood at the right distance parallel to the top edge. A piece of hardwood should be planed so as to fit nicely in the grooves, and should then be firmly glued into the groove of the inner rail. The moulding on the bottom of the outer rails O (Figs. 31 and 32) is next fixed with glue and screws. The cross rails A and B (Figs. 30 and 31) should be dovetailed to the sliding rails—B to the inner rails, and A into the projecting moulding, as shown in Figs. 28 and 33. Make these dovetails carefully, or the rails, through not being parallel, will prevent proper working. When the joints are properly fitted, those between the legs and the rails, and between the cross rails and the rails, should be glued together, keeping the legs and rails square. The cross rails should also have a couple of screws inserted, as shown in Fig. 33.

Fig. 25.—Extending 6-ft. to 9-ft. Dining Table with Screw.

Top of Extending Table.—Well-seasoned material for the top can be obtained in widths about 1 ft. 6 in., and each half will then require only one joint. The leaves are also 1 ft. 6 in. wide. The top should be dowelled and glued, and the under sides of the top and leaves trued up. Next join together the two portions of the permanent top and the two leaves, and dowel them with hardwood pins about $\frac{3}{8}$ in. in diameter, projecting about $\frac{5}{8}$ in. (see Fig. 34). The whole top should then be turned bottom side up, the framework stretched out to its full length (see Fig. 30), put on and fastened to the two permanent parts of the top by screws

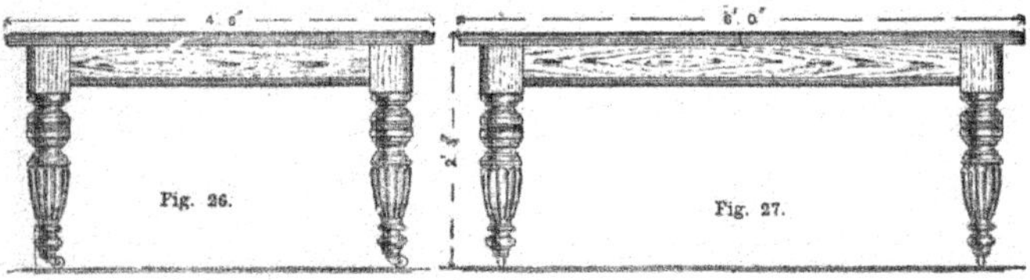

Figs. 26 and 27.—End and Side Elevations of Extending Dining Table.

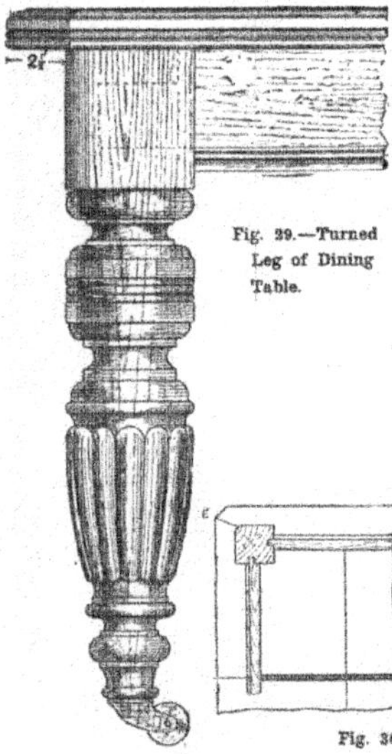

Fig. 29.—Turned Leg of Dining Table.

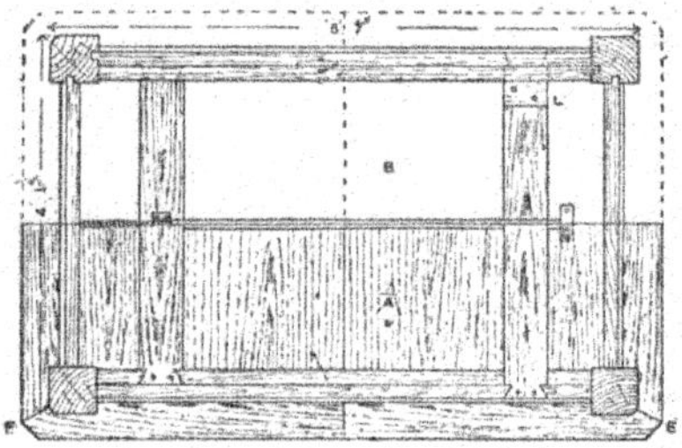

Fig. 28.—Half Plan and Half Underneath View of Extending Dining Table.

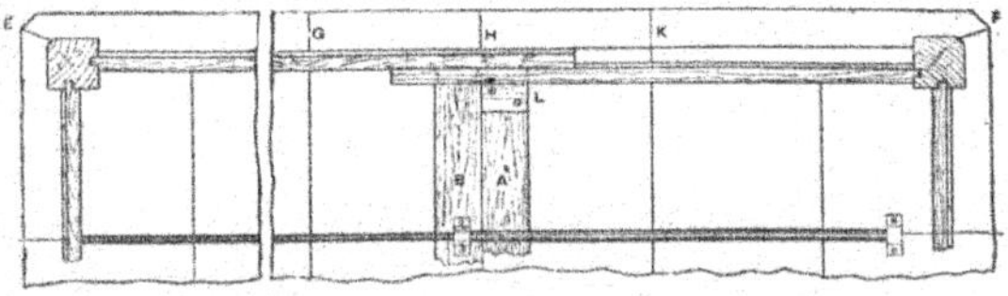

Fig. 30.—Part Plan of Framework of Table, Extended.

illserted obliquely, as clearly shown in an inustration (Fig. 32) given below.

Fixing the Screw and Barrel.—The screw and barrel should now be fixed. Secure the handle end of the screw to the end rail of the table. To the cross rail B (Figs. 30 and 31) fix the box in which the screw works, and which holds one end of the barrel; fix the other end of the barrel to the under side of the top, a wood block probably being necessary for this purpose. Slightly tighten the screw so as to hold the top firmly together, plane the top and leaves, and work the moulding round the edges.

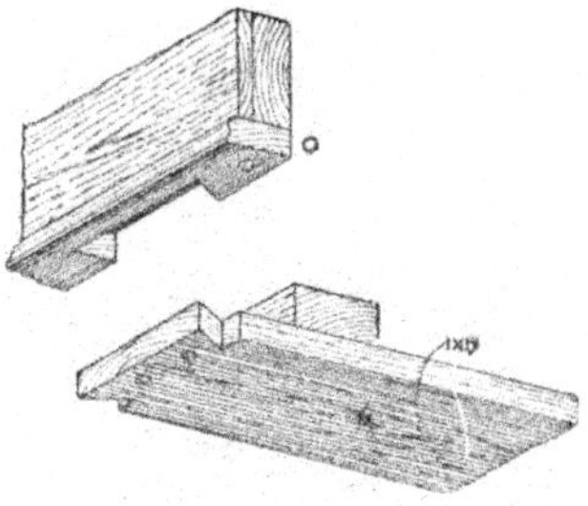

Fig. 33.—Joint between Cross and Side Rails of Table.

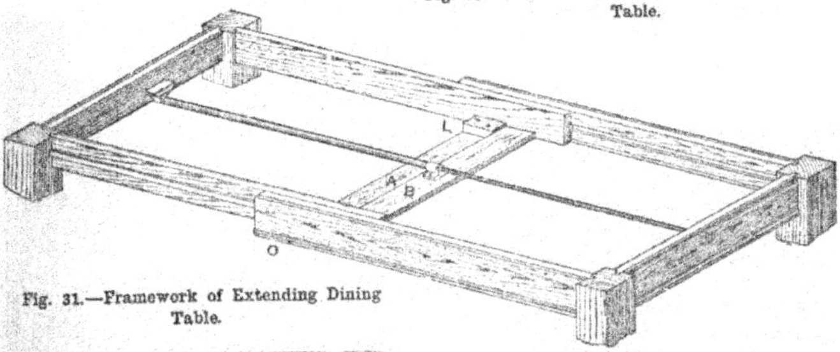

Fig. 31.—Framework of Extending Dining Table.

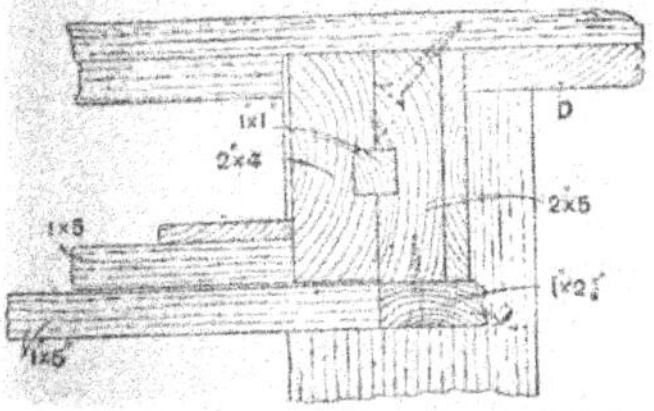

Fig. 32.—Section through Sliding Rails of Table.

Completing the Extending Table.—The thicknessing fillet shown in section by D (Fig. 32), having a smal bead worked on one edge and the other rounded, should be mitered at the angles E and F (Figs. 28 and 30), and fixed with glue and screws. Have the two side pieces long enough to reach from end to end, thus taking in the two leaves; cut with a fine saw where the joints of the leaves should occur, as shown by G, H, and K (Fig. 30). A stop L prevents the framework moving too far. The table when closed is 6 ft. long, and will extend to 9 ft. with two 1-ft. 6-in. leaves. It would be firmer if it extended to 8 ft. 6 in. only, with leaves 1 ft. 3 in. wide. This would give the sliding rails a lap of 2 ft. instead of 1 ft. 6 in.

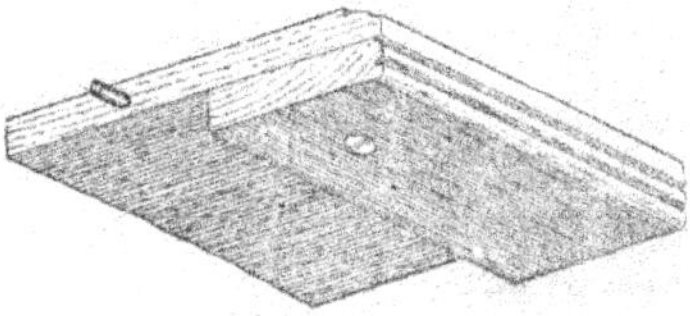

Fig. 34.—Corner of Table Leaf.

Fig. 35.—Falling-leaf Gate-leg Table with Turned Legs and Rails.

Falling-leaf Gate-leg Table with Turned Legs and Rails.

Fig. 35 is the general view of a table with a flap supported by gate legs. All the legs and lower rails are turned. Mahogany, oak, walnut, pitchpine, and yellow pine are suitable woods. The sizes of the various pieces may be varied to suit requirements. Having cut the necessary pieces to the several lengths, plane them up to the proper sizes. If desired, the legs and rails may be turned before being planed, but this is not always so satisfactory as planing up material true beforehand. Next set out the legs and rails for mortice-and-tenon joints. Fig. 36 shows all the framework. The ends of the turned rails have the tenons mitered at the extremities (see Fig. 37); this allows the cross mortices in the leg to be made so that the tenons nearly meet. The long turned rails to which the rails of the movable legs are attached are set further back from the face of the legs; therefore the tenon must be made nearer the front of the rail,

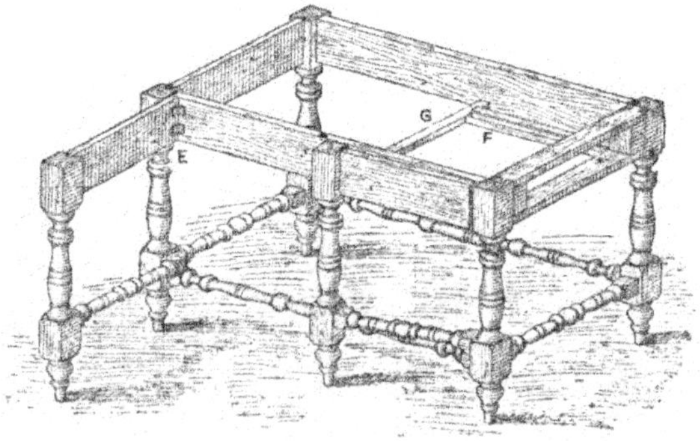

Fig. 36.—Framework of Falling-leaf Gate-leg Table.

and a thicker piece taken from the back; see A (Fig. 37). The upper rails at the sides and one end have a shoulder on the outside only with a haunch as at B (Fig. 37). The lower rail for the drawers is stub-mortised and tenoned together (see C), and the upper rail is dovetailed into the leg as shown at D. The turned rails to the movable legs are tenoned; they have square shoulders, and are at right angles to the

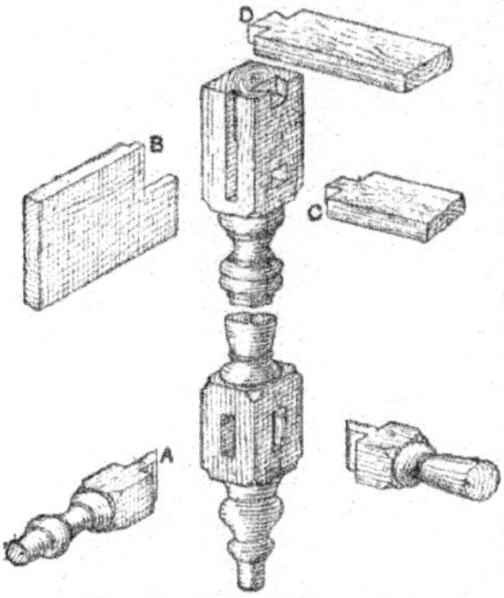

Fig. 37.—Joints in Framework of Falling-leaf Table.

legs, the upper rail being a little on the slope, as shown at Fig. 38. This is to allow of the upper rail being connected to the upper main rail by means of back flaps, as shown at E (Fig. 36). After the joints have been made and fitted they should be glued together. Then the top of the legs and rail should be planed off level with each other, and the movable legs and rails secured in position. A good

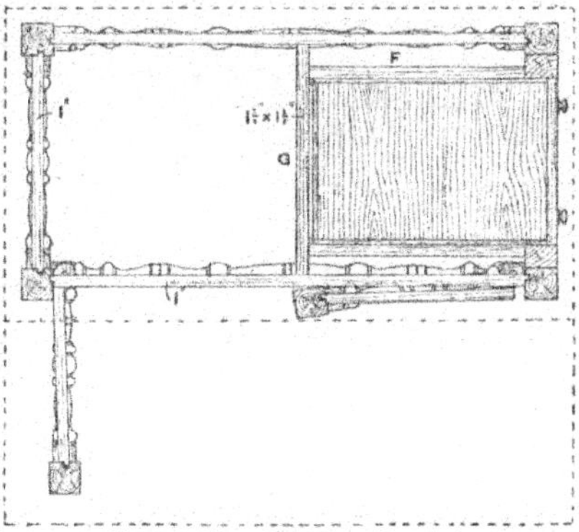

Fig. 38.—Plan of Framework of Falling-leaf Table.

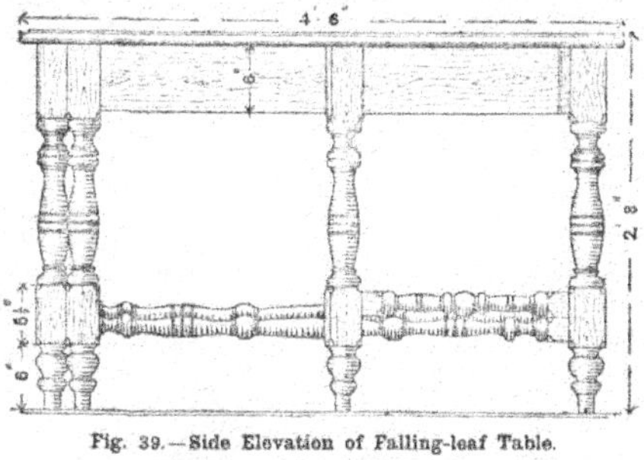

Fig. 39.—Side Elevation of Falling-leaf Table.

method of connecting the rail and movable legs to the main rail is by inserting a flat-headed bolt secured on the under side with a nut; but a stout screw answers. The drawer, fitted at one end, may be the same width as the distances between the legs, or it may be narrowed, as shown, by inserting a block at each side. In the latter case the runners F (Figs. 36 and 38) should be fixed into the lower front rail, and another rail G at the back. If the drawer occupies the whole width, the runner can be fixed to the broad side rails. The drawer is of the ordinary dovetail construction. Next prepare the top. If hardwood is used, the best plan will be to join the pieces by dowelling and gluing them together. If pine or similar soft wood is used, the joint should be ploughed, cross-tongued, and glued together. Figs. 35, 39, and 40 show the edges of the top moulded, and the joint between the top and the flap would have a much better appearance if the inside edge of the flap were hollowed, so that when the flap is down the moulded edge would be in the form of a rule joint, as shown at Fig. 41.

Small Table with Round Top.

The table shown by Fig. 42 is supported on three shaped legs made from stuff

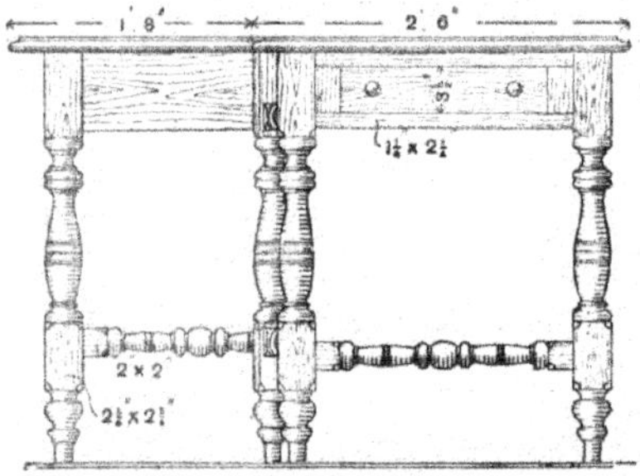

Fig. 40.—End Elevation of Falling-leaf Table.

$\frac{3}{4}$ in. thick when planed; they are 2 ft. 4 in. high, and cut from board 11 in. wide to the dimensions given in Fig. 43. The table top is 1 ft. 6 in. in diameter by 1 in. thick. If it is made from two or more boards, their edges should be trued up

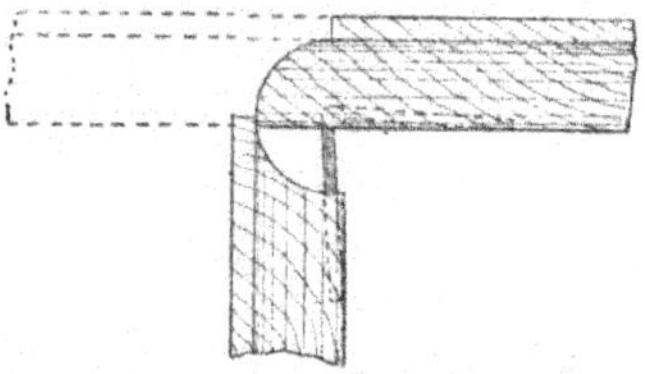

Fig. 41.—Joint between Table Top and Flap.

square and parallel, and the boards dowelled, glued, and cramped till set; then the circle can be struck and cut with a bow-saw, and the moulding on the edge formed. Fig. 44 is an underneath plan of the table top, showing the position of the legs and

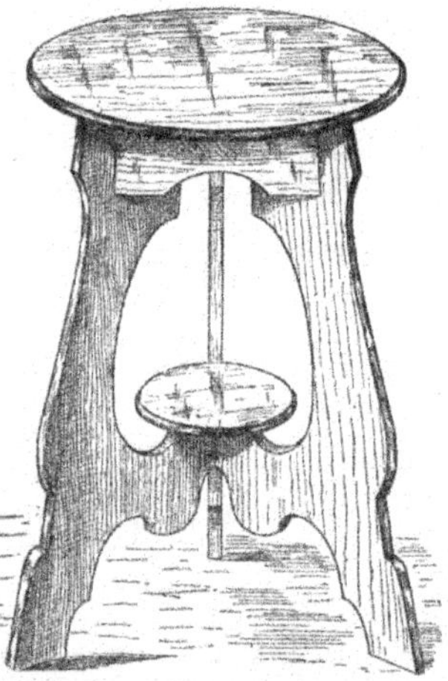

Fig. 42.—Small Table with Round Top.

brackets. The legs are stub-tenoned to the top as shown in Fig. 45, the joints of the legs immediately under the lower shelf, and the stub-tenons that fit in the mortices in the shelf being shown in Fig.

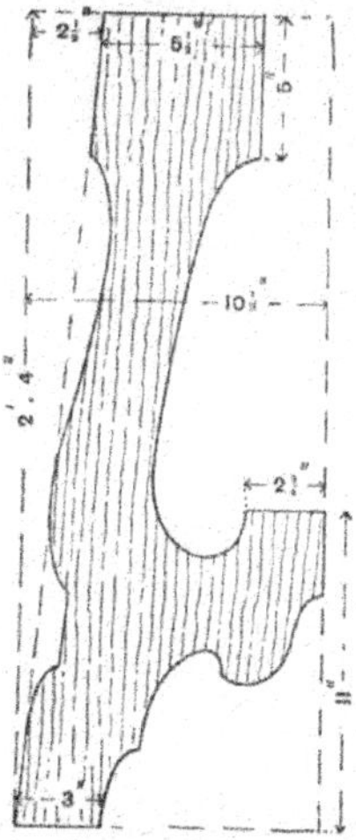

Fig. 43.—Shaped Leg of Small Table.

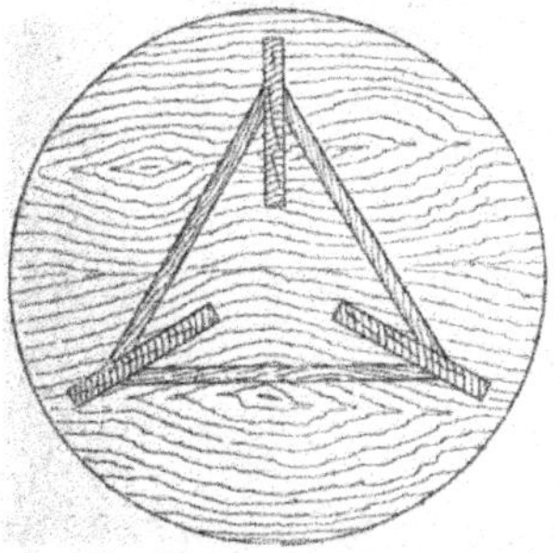

Fig. 44.—Underneath Plan of Small Table with Round Top.

46. The shelf is 8 in. in diameter by $\frac{7}{8}$ in. thick, and a moulding is worked round its edge similar to that on the table top. The legs should be shaped at the centre joint as shown, and firmly brought together while marking at right angles across each joint the position of the grooves. Take the legs apart, and work out the grooves $\frac{3}{8}$ in. deep and $\frac{1}{8}$ in.

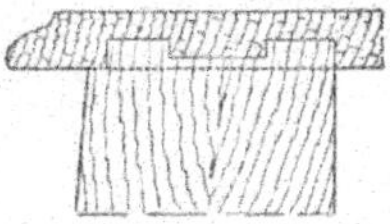

Fig. 45.—Connection of Table Legs to Top.

wide. The joints are then ready for gluing. They are brought together, and oak or mahogany cross-grained feathers are slipped into the grooves. The shelf and table top are glued on over the tenons, and weighted to keep them firmly in place while the glue is setting. When the glue joints have properly hardened, fit the brackets to their respective places. If the table is to be enamelled, the brackets may be bradded on; but if it is desired to stain and polish the table, they should be glued and blocked.

Square-top Table with Fretted Brackets.

The table shown by Fig. 47 has a square top with moulded edges, square tapered legs, and fretted brackets, and also has

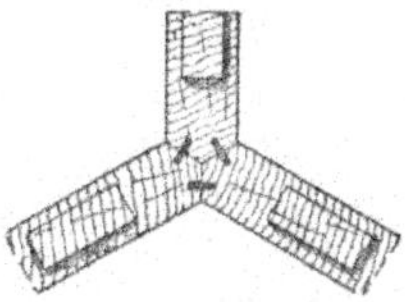

Fig. 46.—Joints of Table Legs under Lower Shelf.

a shaped shelf underneath. The top is 1 ft. 6 in. square by 1 in. thick, Fig. 48 being a half top and half under-side plan. It may be in one or more pieces. The legs are 2 ft. 3 in. long by $1\frac{3}{4}$ in. square at the top and $1\frac{1}{4}$ in. at the foot. The legs are splayed to bring the feet in the same vertical line as the edges of the table-top. The correct splay or batter for the brackets and shoulders may be obtained by making on a board a full-size dimensioned drawing of one side of the table. The brackets are prepared from $\frac{5}{8}$-in. stuff, fretted as shown in Fig. 49, and are haunch-tenoned to the legs; they are also grooved on the inside for the tongued blocks which are used to secure the table top in position.

Fig. 47.—Square-top Table with Fretted Brackets.

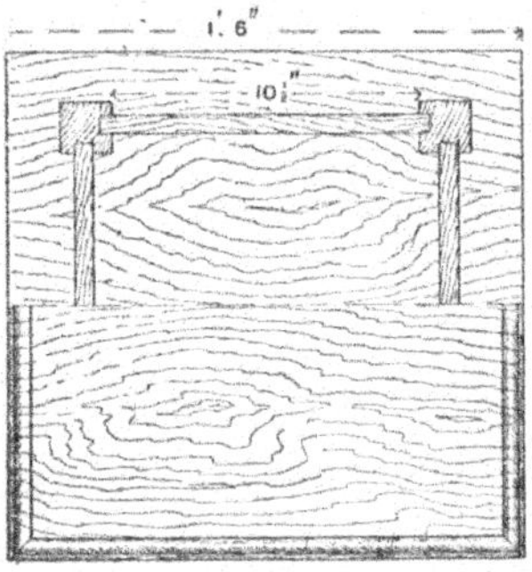

Fig. 48.—Half Plan and Half Underneath View of Square-top Table.

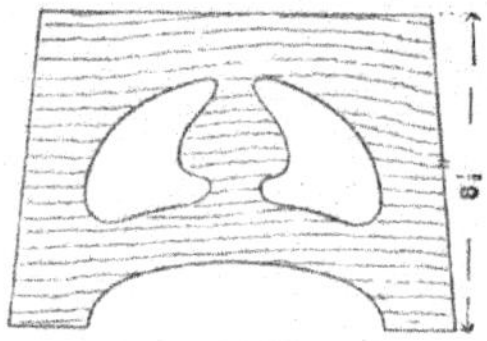

Fig. 49.—Fretted Bracket.

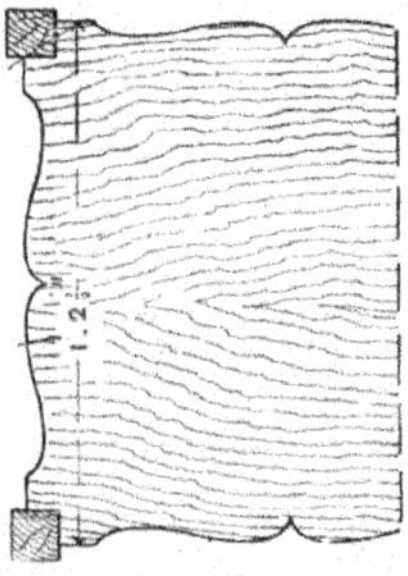

Fig. 50.—Part Plan of Table Shelf.

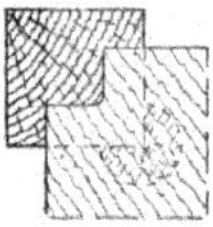

Fig. 51.—Shelf Housed to Leg of Table.

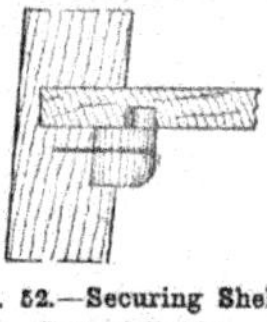

Fig. 52.—Securing Shelf to Leg of Table.

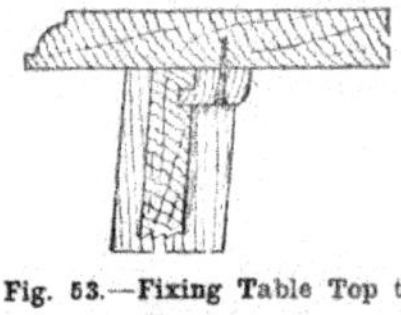

Fig. 53.—Fixing Table Top to Brackets.

Put the brackets and legs together temporarily, and see that they are quite home to the shoulders; then mark a distance of 1 ft. up on each leg for the groove in which the shelf is to fit, and take the dimensions between the legs for setting out the shelf, which is shown in part plan by Fig. 50. The shelf is housed to the legs as shown in Fig. 51, and further secured from spreading by a small block stub-tenoned to the under side of the shelf and bradded to the legs (see dotted lines in Fig. 51, and the side view, Fig. 52).

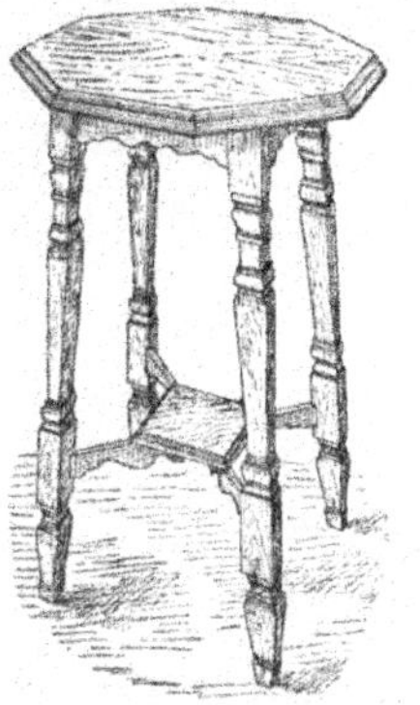

Fig. 54.—Octagonal Occasional Table.

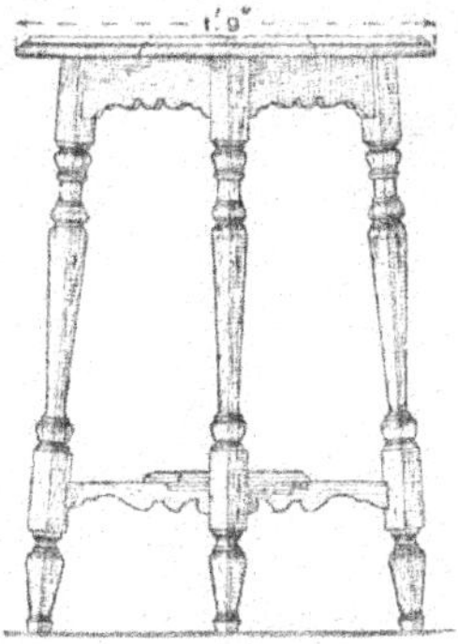

Fig. 55.—Elevation of Octagonal Occasional Table.

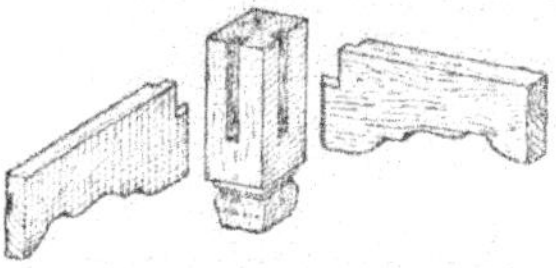

Fig. 56.—Joint between Top Rails and Leg.

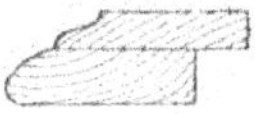

Fig. 58.—Section of Table Top.

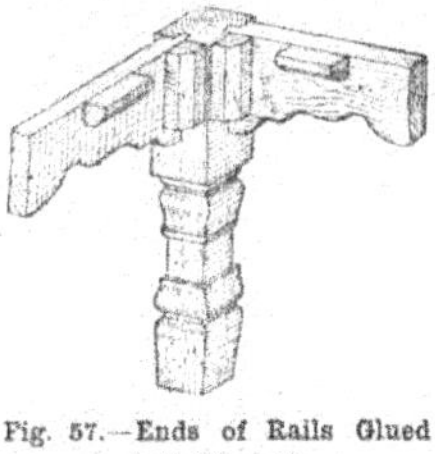

Fig. 57.—Ends of Rails Glued and Blocked.

The method of fixing the table top is shown by Fig. 53. The work should be glass-papered off and given a coat of size and whiting. When thoroughly dry, it should be rubbed down smooth and given a second coat; then again rubbed down, dusted, and given a coat or two of enamel paint. Alternatively, the table may be stained, sized, and varnished.

Octagonal Occasional Table.

Fig. 54 is a perspective view of an octagonal table. Fig. 55 is an elevation. The table is 2 ft. 6 in. high. The top is octagonal, 1 ft. 9 in. wide across the flat. The legs are square turned. First set out the legs for the haunched mortices at the upper ends, and for the stub-mortices towards the lower ends. The rails are 3 in. by $\frac{5}{8}$ in., and the upper ones have haunched tenons which mitre together in the legs as shown at Fig. 56. The tops of the legs are $1\frac{5}{8}$ in. square. The two lower rails cross each other with a halved joint. The ends of these lower rails are stub-tenoned to fit the mortices in the legs. When the legs and rails have been worked and properly fitted, they should be cleaned off and glued together; also glue blocks in the angles of the upper joints as shown at Fig. 57.

Fig. 58 is a section of the octagonal top, which is built up of two thicknesses, the top piece being $\frac{7}{16}$ in. thick and the under strips $\frac{5}{8}$ in. thick by $2\frac{1}{4}$ in. wide. In making this part, care should be taken to get good mitres between the various pieces forming the lower part of the top. This done, they should all be glued to the upper part, and then the edges may be moulded. The top is fixed to the top rails by small blocks glued to the rails, and also to the under side of the top. These blocks should be planed so as accurately to fit the angle formed by the two parts. On the lower rails is a small shelf which has a moulded edge. The shelf is fixed by means of glued blocks. As illustrated, the shelf is square, but an octagonal shape might be more suitable.

Lady's Work-Table with Sliding Body.

The work-table shown in elevation and section by Fig. 59 has a top 3 ft. by 1 ft. 4 in. Under the frame is a sliding body or well whose interior is fitted as a lady's work-box, with additional accommodation for small pieces of unfinished work. A tray, furnished with the usual assorted compartments for needles, cotton, etc., rests in the top of the well (see Fig. 60), and may be removed bodily, but access to the interior of the well is obtained ordinarily through the central compartment, which is bottomless, but which may, if preferred, be covered with a lid, stuffed outside with cotton-wool to form a pincushion. Immediately over the sliding body is a

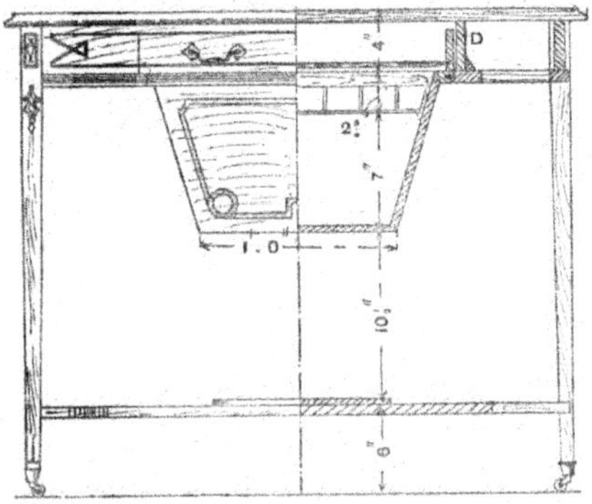

Fig. 59.

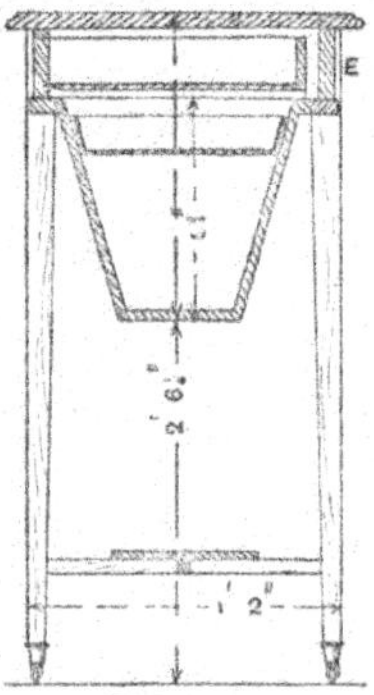

Fig. 60.

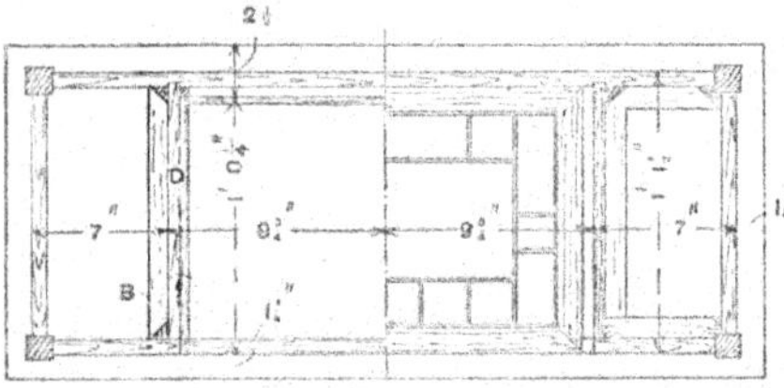

Fig. 61.

Figs. 59 to 61.—Elevations, Vertical Sections, and Horizontal Sections of Lady's Work-table with Sliding Body.

drawer working between solid guides D (Fig. 59) framed into the rails of the table as shown in Figs. 61 and 62. A shaped stretcher is framed between the legs, and carries a small oval shelf, as shown in the half-plan, Fig. 63. The top and drawer rails, and the rails of the framing, are of $\frac{3}{4}$-in. stuff, and the legs are $1\frac{1}{4}$ in., tapering to $\frac{3}{4}$ in., the joints connecting them to the rails being shown in elevation in Figs. 62 and 64, and in plan in Fig. 65. The well is of $\frac{1}{2}$-in. pine, square jointed and bradded, and afterwards veneered with wood to match the remainder, which might be either mahogany or walnut, the marquetry inlays being holly, sycamore, or satinwood. Fig. 66 illustrates the sliding body. A wide rail is framed between the sides of the table, and a hardwood tongue, oak for preference, is grooved and glued in it; this is stopped back 1 in. from the front edge. A similar but slightly thinner rail is tongued and grooved to the well, as shown in Fig. 62; a plough groove is made in its edge, in which works the hardwood tongue. This must be fitted accurately so that there is no side play. It is best to fit it first rather tightly, and then to rub powdered French chalk over the tongue. The rim of the well should be dowelled at the mitres, as shown in Fig. 67. The tray, which is 2 in. deep, is made of $\frac{5}{16}$-in. stuff bradded together and glued, and lined with silk. It is advisable to glue a piece of green baize on the bottom to prevent scratches on the table top when the tray is taken out. To provide an opening for the drawer, the front rail of the table is cut through from the bottom edge to within $\frac{1}{2}$ in. of the top, and the cross guides D (Fig. 61) are kept flush with the ends of the opening. These guides must be well fitted and fixed with glued angle blocks, and

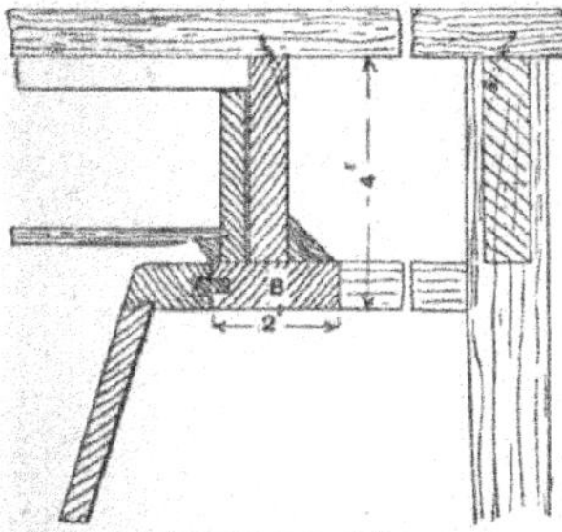

Fig. 62.—Part Section of Tray and Drawer of Lady's Work-table.

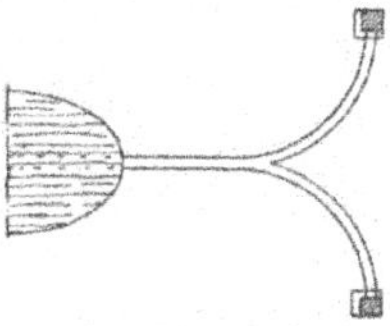

Fig. 63.—Half Plan of Stretcher.

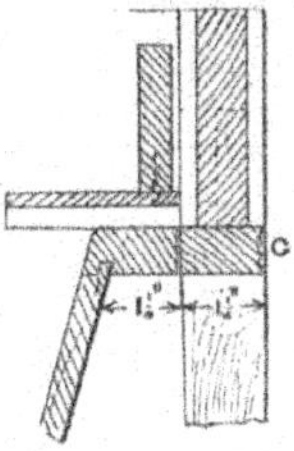

Fig. 64.—Part Section of Tray and Drawer.

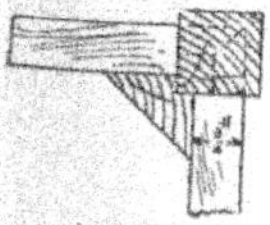

Fig. 65.—Joint in Work-table Legs.

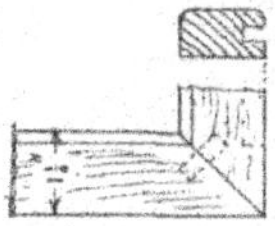

Fig. 67.—Joint in Rim.

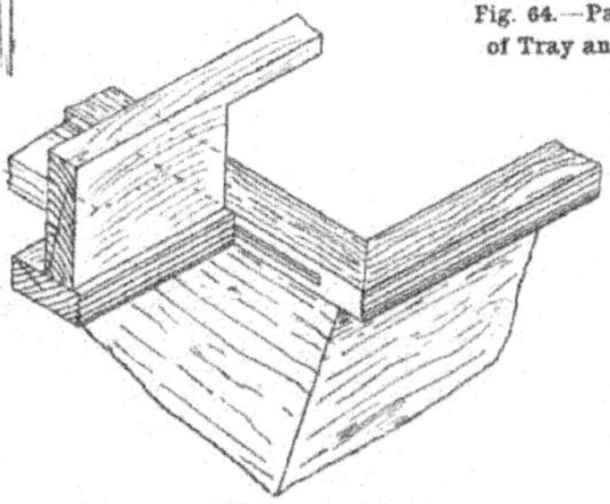

Fig. 66.—Well, Partly Withdrawn.

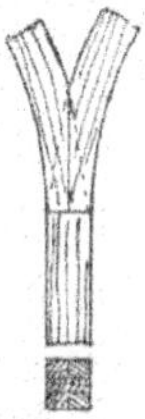

Fig. 68.—Joint in Stretcher.

screwed to the top as shown in Fig. 62, the bearers B (Figs. 61 and 62) being screwed to them. The table top, not being very wide, may be solid, and screwed to the framing as shown in Fig. 62. The moulded rails C (Fig. 64) should be cut in tight between the legs, and fixed after the framework is glued up. It will be noticed that there is no opening at the back, neither the drawer nor the sliding body passing out on that side, and the groove in the sliding rim must be taken through at the back, but stopped in the front to prevent disfigurement of the moulded rail (see Fig. 66). The joint in the shaped stretcher is shown in Fig. 68. The two curved ends are first jointed together square, and then grooved diagonally through the middle; they are then glued up and bradded, thus forming a tapering mortice into which

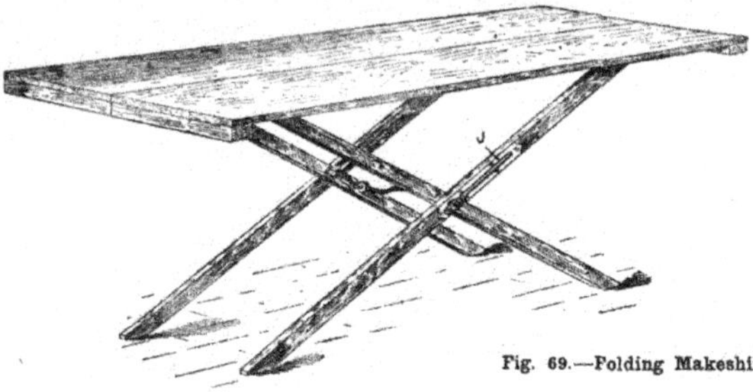

Fig. 69.—Folding Makeshift Table.

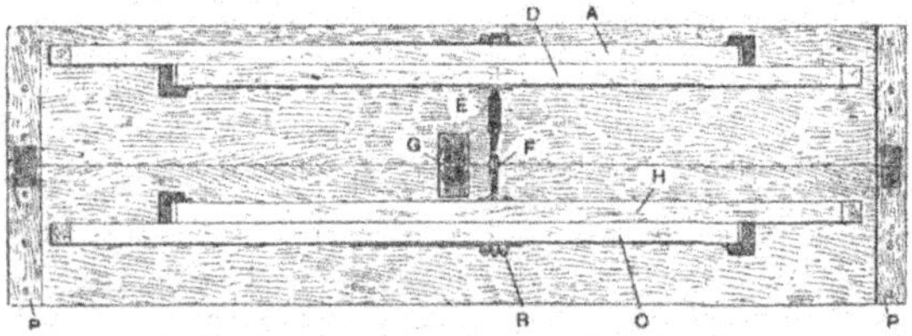

Fig. 70.—Underneath View of Table, with Legs Folded.

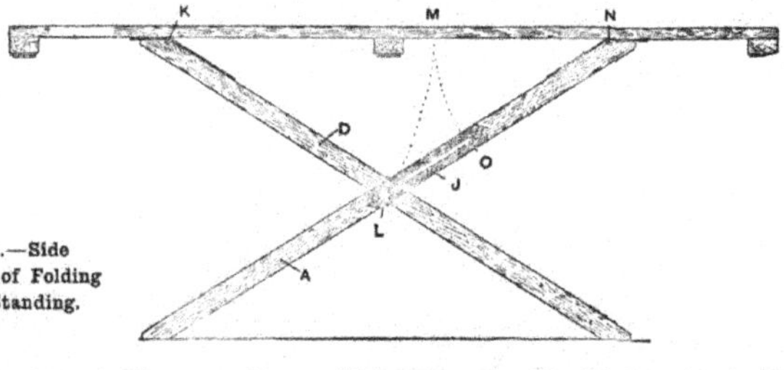

Fig. 71.—Side Elevation of Folding Table, Standing.

the tenon on the straight rail is fitted, the opposite ends of the curved portions being tenoned into the legs.

Folding Makeshift Table.

Fig. 69 shows a table that is very handy when a large table is required out of doors, or for extra accommodation indoors, and which can also be used as a paperhanger's table. Most folding tables are somewhat small, but this one is 5 ft. 10 in. long by 2 ft. wide. When folded it is 5 ft. 10 in. by 1 ft. by $3\frac{1}{2}$ in., and as all the parts are connected there is no fear of any portion being mislaid when wanted. A suitable wood for construction is deal, painted green or stone colour. Figs. 70 and 71 show the construction. The leg A (Figs. 70 and 71) stands in $1\frac{1}{4}$ in. from the edge of the top. The wing nuts B (Fig. 70) should be as short as possible, say $\frac{1}{2}$ in. long, because the shorter they are the nearer the leg C can be to the outer edge of the top. This leg when folded comes as near D as the nut will allow it. On this also depends the distance apart of the legs when touching the floor; the farther they are apart the firmer the table will stand. The iron stretcher must have its centre F exactly opposite the centre of the hinge G, and the part E must be sunk to allow the legs C and H to close over it (see Fig. 76). To obtain the length of the slot in the plate J (Fig. 71), draw the arc L M from centre K, and then M O from N. Closing the leg D to the top moves the stretcher and nut up the slot. When the legs are closed the two halves of the top can be folded together as in Fig. 72.

The sizes of wood are: For top, two boards 5 ft. 10 in. by 1 ft.; for the ends, four battens P (Fig. 70), 1 ft. by 2 in. by $1\frac{1}{8}$ in. thick; two blocks G, 3 in. by 2 in. by $1\frac{1}{8}$ in. thick, and four legs 4 ft. 7 in. by $1\frac{5}{8}$ in. wide and $1\frac{1}{8}$ in. thick, with just the sharp edges planed off. The battens are screwed to the top, and $1\frac{1}{2}$-in. iron butt hinges are used for joining the two halves forming the top. The tops and bottoms of the legs are bevelled as in Fig. 71, and the two inside legs D and H are bored to take the $\frac{3}{8}$-in. round iron stretcher, the two outer legs A and C being slotted for the same purpose. The plates J (Figs. 71 and 73) prevent the nuts bruising the wood; they are $1\frac{1}{2}$ in. wide and of $\frac{1}{8}$ in. sheet iron. The hinges for the top ends of the legs (see Figs. 74 and 75) are about 6 in. long by $1\frac{1}{8}$ in. wide at the narrow portion, and the wide part is 3 in. by 2 in. long; the plates are countersunk at opposite sides, as shown in Fig. 75. In fixing the hinges, the narrow ends are screwed to the inside edges of the legs, and then the wide ends are screwed to the top. The stretcher is shown in Fig. 76. The straight threaded ends go through the inside legs, and to prevent turning, half the diameter is let into the side of the leg, and the plate R (Figs. 76, 77, and 78) on the leg prevents the stretcher pulling through when the nuts are slackened if the table is being closed. When the table is open it is 2 ft. 5 in. high, and the nuts are screwed tight to prevent the legs closing. The dotted lines in Fig. 76 show the position of the legs C and H when closed about F over the flattened part of the stretcher.

Fig. 72.—Folding Table, Closed.

Fig. 73.—Slotted Plate.

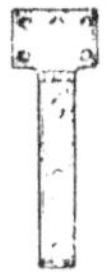

Fig. 74.—Hinge for Legs.

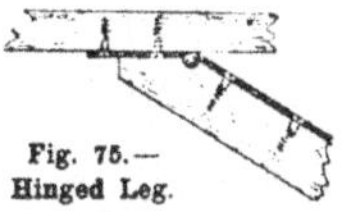

Fig. 75.—Hinged Leg.

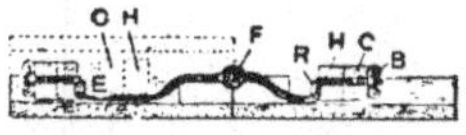

Fig. 76.—Elevation of Stretcher.

Fig. 77.

Fig. 78.

Figs. 77 and 78.—Stretcher Clamp.

Strong Portable Folding Table.

One of the chief points to be observed in making a folding table of the kind shown by Fig. 79 is that the legs, which fold up against each other, should, when housed, be flush with, or a little way below, the flush of the rails, otherwise they are apt to be in the way. The dimensions given are suitable for a table which may be roughly used, but, for a light serviceable table, the scantlings may be reduced. The length of the top is 4 ft. 9 in., the height 2 ft. 6 in.; the width may vary from about 2 ft. upwards, according to requirements. The width and length of the top should be settled in order to get the exact size of the frame, to which it is screwed, and which is 2 in. smaller all round. This frame is dovetailed together at the corners and glued. To test the frame for squareness, place a wood rod diagonally from corner to corner and mark it. Try it on the reverse corners, and, if the mark coincides, the frame is square. This is a better method than using a try square, as the long rails might be bent somewhat, and this would lead to error. When the glue has set, clean off the sides and ends, and make the edges fair. Mark off, on the top edge, the position of the screw holes, and bore them. The angle and position of the screw are shown in Fig. 79, where a channel is seen cut out with a gouge, leaving a square shoulder for the head of the screw. After the top is screwed down to the frame the legs may be proceeded with. These are 3 in. by 3 in., and are made tapered on the inside edges from the bottom to about 1 in. below the frame. To allow the legs to fold up properly, a dead piece is screwed to the under side of the top at one of the ends. The thickness of this piece is deducted from the length of the legs. Rails are mortised and tenoned into the legs at the top, and narrow spars are fixed in the same manner at the bottom. Flap hinges are used, and are screwed to the legs and top in the one case, and to the legs and fixed piece in the other. To hold the legs firm when down, small flush slip bolts should be let into them, the plate to receive the bolts being sunk in the rail. Another pair of slip bolts should be let into the outside edges of the right-hand pair of legs, to keep them in their place when packed up. The dotted lines in Fig. 79 show the position of the legs when they are folded up. The scale of Fig. 79 is 1 in. to 1 ft.

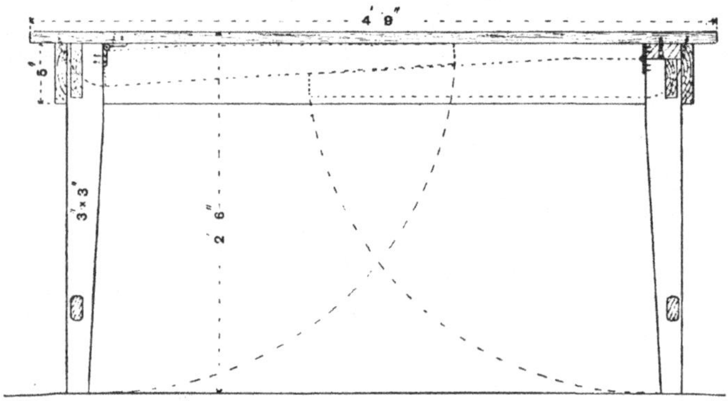

Fig. 79.—Strong Portable Folding Table.

CHAIRS.

Dining-room Armchair.

OAK is the best material to use for chairs of the description shown by Fig. 80. Fig. 81 is a sectional elevation of the frame and Fig. 82 an elevation of the back. A thin wood mould for the back legs should first be made to the dimensions given in Fig. 81. After lining out, cut the legs out of $1\frac{1}{2}$-in. wood, and dress them up to the shape of the mould. From the bottom edge of the seat rail, taper the legs, the inner side only, to 1 in. thick at the foot (see Fig. 82). Mortise the legs for the seat rail and stay rail, which should be tenoned as shown at Fig. 83, and rebate the inside edges of the upper part of the legs $\frac{1}{4}$ in. deep, leaving a belt $\frac{7}{8}$ in. wide, which should be rounded to form a bead (see section, Fig. 84). The shaped top rail is finished to the same width as the legs, and is rounded in the same way, being secured to the top end of the legs with dowels. Prepare the front legs as shown in Fig. 81. The part above the seat rail, forming the pillar for the arms, is turned, while below the seat rail the leg is tapered, and shaped at the foot as shown. This done, cramp the legs and rail together. Draw a full-size plan of the seat, to give the bevels and shoulders of the side rails. The part plan (Fig. 85) shows the positions of the tenons. The short tenon on the back rail allows the tenon of the side rail to pass it, thus giving strength to the side rails where it is most needed. Make a thin mould for the arms (Fig. 86), which are $1\frac{1}{4}$ in. thick and flat-rounded on the edges. After being fitted to the back legs, each is secured with glue and a screw, which is driven through from the back and sunk below the face, the hole being filled with a wooden plug (see Fig. 86). In the front of the arm is bored a hole $\frac{3}{4}$ in. in diameter for the pin turned on the leg pillar.

Fig. 80.—Dining-room Armchair.

Small Chair to Match Dining-room Armchair.

The small chair shown at Fig. 87 would go well with the armchair last described. A sectional elevation of the frame is given at Fig. 88, and an elevation of the back at Fig. 89. The back leg is $1\frac{3}{8}$ in. thick at the seat rail, tapering down to $\frac{7}{8}$ in., and terminating in a bulb at the foot. From 1 in. above the seat rail the legs are reduced to 1 in. thick. The back legs slope from 1 ft. 4 in. apart at the top to 1 ft. at the floor. This necessitates the rails it is connected to the rails with short tenons at the top and bottom, as shown in section at Fig. 88. The top rail C (Fig. 89), $\frac{7}{8}$ in. thick, is ornamented with some simple carving; it is mortised to the leg $\frac{1}{8}$ in. in from the front, the same as the stay rail. The moulded caps (Fig. 91)

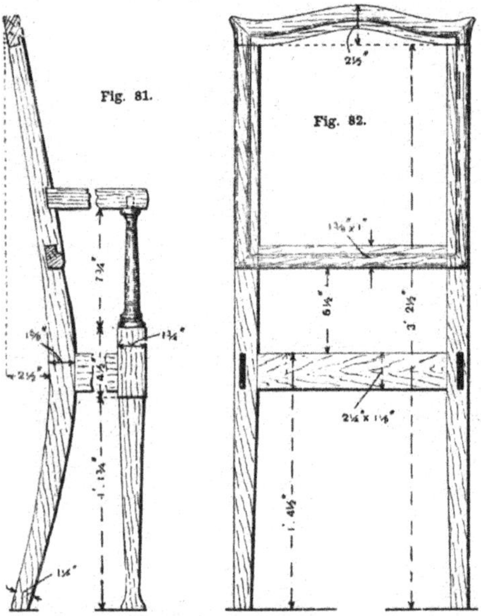

Figs. 81 and 82.—Sectional and Back Elevations of Dining-room Armchair.

Fig. 83.—Tenon on Armchair Rail.

Fig. 84.—Section of Upper Part of Back Leg.

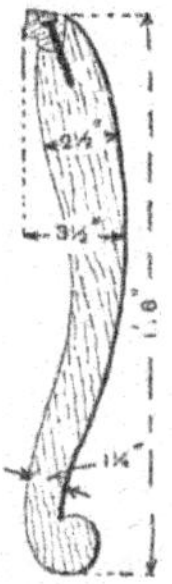

Fig. 86.—Arm of Armchair.

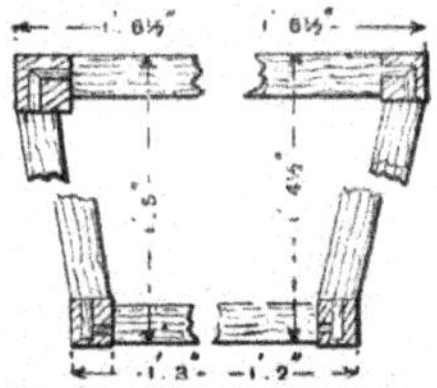

Fig. 85.—Plan of Seat Rails of Armchair and Small Chair.

being bevelled to the rake, which can be obtained by drawing a full-size half plan of the back. The seat rails may be made of birch, and clamped with oak 1 in. deep by $\frac{1}{4}$ in. thick, to form a rebate for the stuffing (see section Fig. 90). The stay rail A (Fig. 89) is $\frac{7}{8}$ in. thick, and is mortised to the leg $\frac{1}{8}$ in. in from the front. The fretted slat B is made of $\frac{1}{2}$-in. stuff, and is kept flush with the rails at the front; are fixed to the top of the legs with glue and two fine brads. Wood $2\frac{1}{2}$ in. in section is required for shaping the front legs. After being cut to the shape shown in Fig. 88, the leg is cut a second time to the same shape on the front, thus producing a corner leg. The corners are rounded with a spokeshave gradually from the top of the leg down to the foot, where the section becomes circular. The stump of the legs should be left projecting ½ in. above the top edges of the rails (see Fig. 88), to give greater strength and to allow a deeper tenon on the rails. The small bracket D should be glued in place after the chair is cramped together, and should be shaped to match the leg. Fig. 85 shows the part plan of the seat rails for the small chair, and gives the sizes. The directions given with Fig. 85 for obtaining the bevels for the side rails, tenons, etc., of the armchair apply equally to this case. The clamps on the seat rails (Fig. 90) are kept flush with the squares on the front legs, and the rebates carried across the legs, after the chair is cramped up.

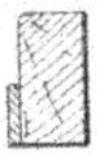

Fig. 90.—Section of Seat Rail.

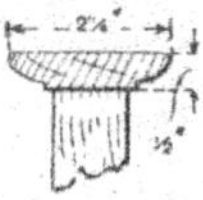

Fig. 91.—Capping Piece.

Fig. 87.—Dining-room Small Chair.

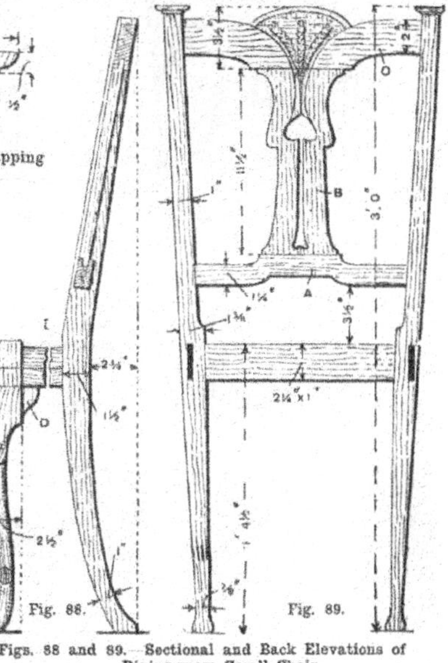

Figs. 88 and 89.—Sectional and Back Elevations of Dining-room Small Chair.

"Quaint" Easy Chair.

The "quaint" easy chair (Fig. 92) is not strictly a "stuff-over" chair, although the only wood parts seen are the legs. In stuff-over work the arms and head are stuffed up to form round bolsters, but the "quaint" easy chair is finished up square. Birch or beech may be used for the different members, the visible parts of the legs being veneered with some choice wood, or worked from the solid to harmonise with surrounding furniture. The back framing (Fig. 93) should be taken in hand first. The legs are 2⅝ in. square, cut as shown to a 4-in. sweep at the foot, which is chamfered off to 1½ in. from the inside, as shown in Fig. 93. The head rail is 2 in. by 2½ in., with a sweep as shown in the crown of 2 in. The stuffing rail is 1½ in. by 1¾ in., and the seat rail 2⅝ in. by 2½ in. These rails are joined to the legs by mortice and stub-tenons, which are afterwards draw-bored and pegged. This completes the back frame, which is 3 ft. 4 in. high to the top of the legs, and 2 ft. 3 in. wide (see Figs. 93 and 94). The front legs are 2½ in. square, tapering to 1½ in. at the toe, the top portion being

Fig. 92.—"Quaint" Easy Chair.

cut down the centre to a depth of 1 ft., as shown in Fig. 95. Fig. 96 is a half-plan of the seat frame. The front seat rail is tenoned into the legs 9 in. from the bottom, and measures over all 2 ft. 3 in. The back and front are now ready for joining together. The seat rails are jointed with

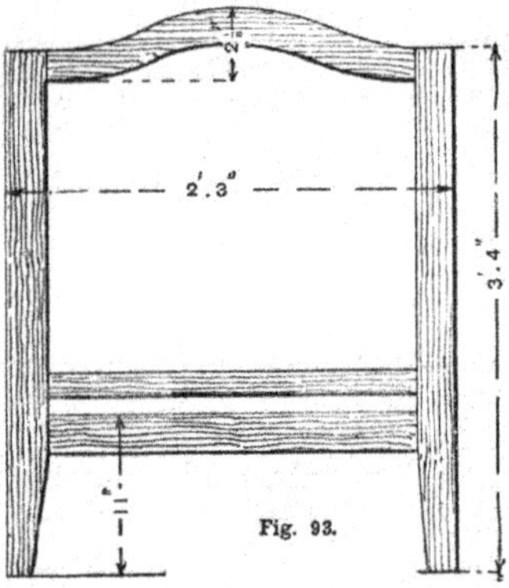

Fig. 93.

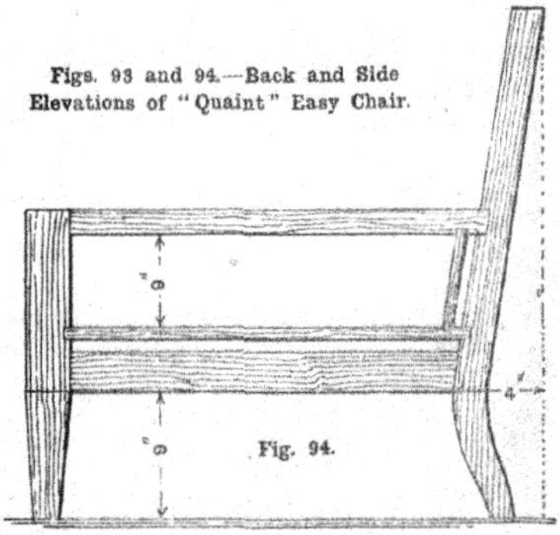

Fig. 94.

Figs. 93 and 94.—Back and Side Elevations of "Quaint" Easy Chair.

mortice and tenons, the rake of the back being set before the measurements of the arm rails are taken. The arm rails are housed into the legs, and glued and nailed. Let the glue set, and then level off, allowing the back to stand slightly lower than the front. Rasp or shave off all sharp corners that are to be covered with the stuffing. The legs are fitted with castors, having plates, not sockets.

Upholstering "Quaint" Easy Chair.

To upholster a chair of this description is hardly a job for the woodworker, but it is convenient to understand the process. Turn the frame bottom up, and web the bottom. No. 12 English grey webbing should be used from back to front, with six lengths of webbing, and from side to side with eight, using good ⅝-in. tacks, and doubling over the ends of the webbing. The insides of the arms are webbed with six lengths from top to rail, and two lengthwise. The inside of the back will require six lengths up and nine across. The springs for the seat (see Fig. 97) should be No. 9 hard 8 in., and the back swell springs No. 7 soft 7 in. Sixteen springs are put in the seat and seven in the back swell (four in the bottom row and three in the top). These are sewn to the webbing and covered with best quality hessian spring canvas. Fig. 97 also shows how the canvas is fixed over the back swell springs, being sewn to the webbing about

1 ft. from the top. The coverings are tacked on the back of the stuffing and seat rails. The insides of the arms are stuffed firm, the top edge being stitched

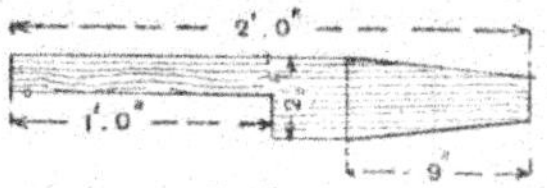

Fig. 95.—Front Leg of "Quaint" Easy Chair.

up square (see Fig. 98). The front edge of the seat, and the top and sides of the back as far as the arms, are also stitched up square. The chair should be stuffed with horsehair, but cocoa-fibre or alva will be quite suitable for stitching up edges and first stuffing. Finish the work right out in calico, and then lay on

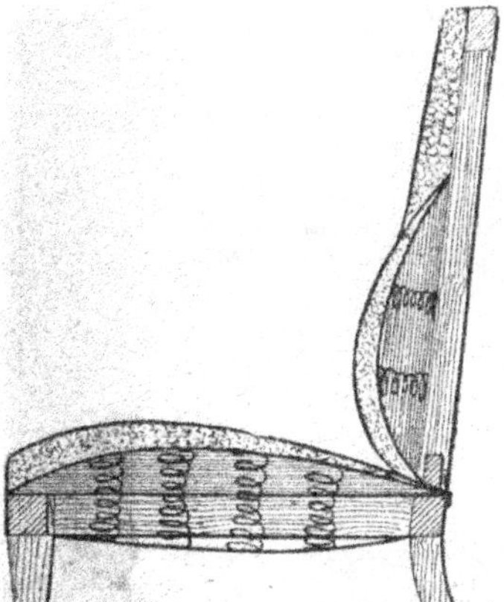

Fig. 97.—Vertical Section of Easy Chair showing Springs and Upholstery.

the outer covering of cotton or linen tapestry, which can be procured 50 in. wide. If care is taken in cutting out, 3 yd. will be sufficient for both coverings and outside linings, which are of the same material. The outer side of the back and arms should be webbed with a cheap cotton webbing to prevent the linings sagging. The tack-

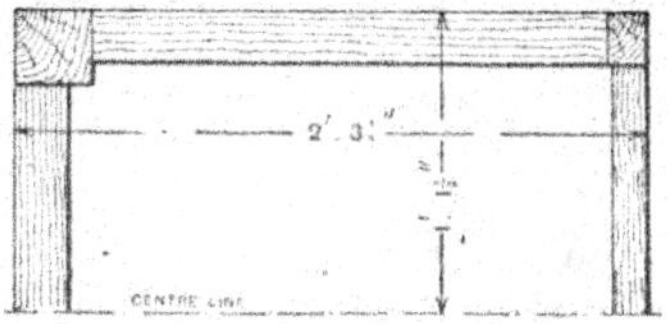

Fig. 96.—Half Plan of Easy Chair Seat Frame.

ing lines are hidden with a $1\frac{1}{4}$-in. chair braid, secured and finished off with $\frac{7}{8}$-in. copper-headed nails; the braid and nails can be procured from any upholsterer's warehouseman. The nails are inserted $1\frac{1}{2}$ in. apart; no advantage is gained by putting them closer, while the effect may be entirely spoiled. To keep out dust the seat is underlined with a piece of black forfar.

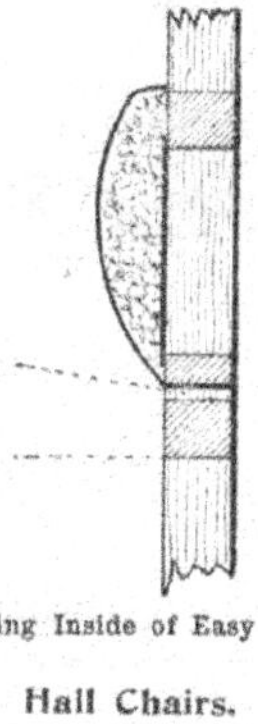

Fig. 98.—Stuffing Inside of Easy Chair Arms.

Hall Chairs.

Hall chairs are generally made of mahogany, oak, or walnut, the selection being governed by the style of the other hall furniture. Figs. 99 and 100 show a suitable design, and later figures will give alternative designs for the back. The measurements about to be given are for a chair suitable for a narrow hall or passage. The construction is the same in all the designs, as also is the thickness of the wood. The backs are of 1-in. stuff; elevation as in Fig. 99. First draw the two horizontal lines of the seat, making the top of the seat 1 ft. 6 in. from the floor; next the outsides of legs, 1 ft. 3½ in. apart,

Fig. 99

Fig. 100.

Figs. 99 and 100.—Front and Side Elevations of Hall Chair.

Fig. 102.—Section of Hall Chair Leg, Tenons and Mortices.

Fig. 103.—Section showing Wood Plug and Screw.

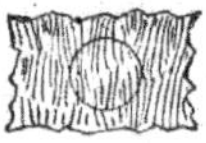

Fig. 104.—Elevation of Part of Chair Back, showing Plug.

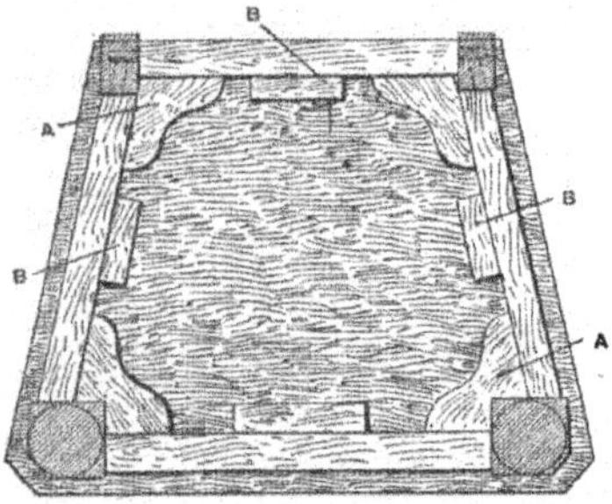

Fig. 101.—Underneath View of Hall Chair Seat.

Fig. 105.—Section showing Turned Button and Screw.

Fig. 106.—Alternative Seat Mouldings.

front legs, 2¼ in. square, this being the largest diameter of the turning; back legs, 1¾ in. by 1⅛ in. at the top, and 1¼ in. by 1⅛ in. at the bottom; seat rails, 1¾ in. wide by 1⅛ in. thick; and seat of ¾-in. stuff. Begin by setting out the front the seat projecting ½ in. at front and sides, which makes it 1 ft. 4½ in. across the front; then the back and back legs, 12 in. across. To simplify the copying of the backs, the right-hand side of each design is spaced out in 1-in. squares. The side elevation

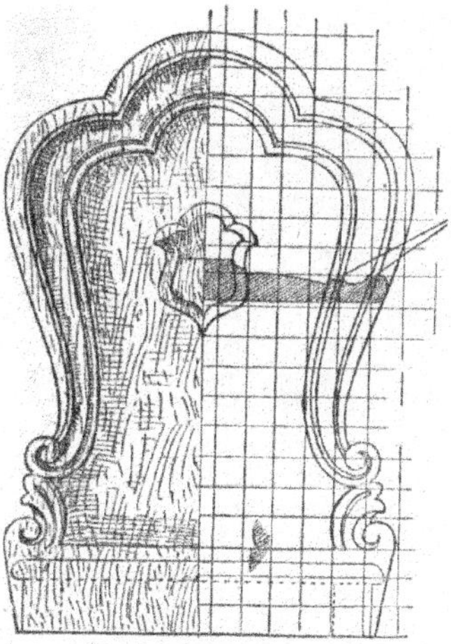

Fig. 107.—Hall Chair Back

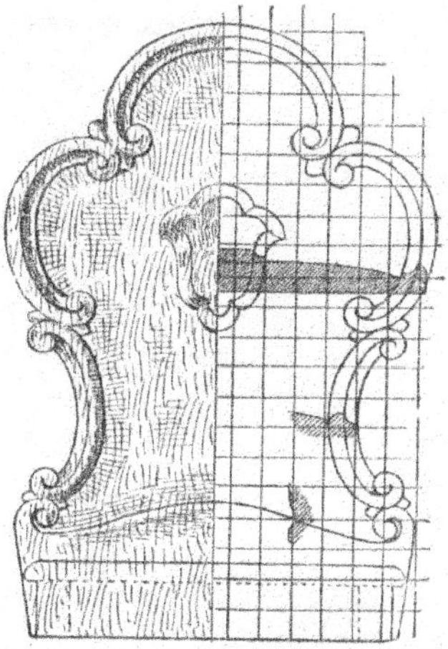

Fig 109.—Second Alternative Design for Hall Chair Back.

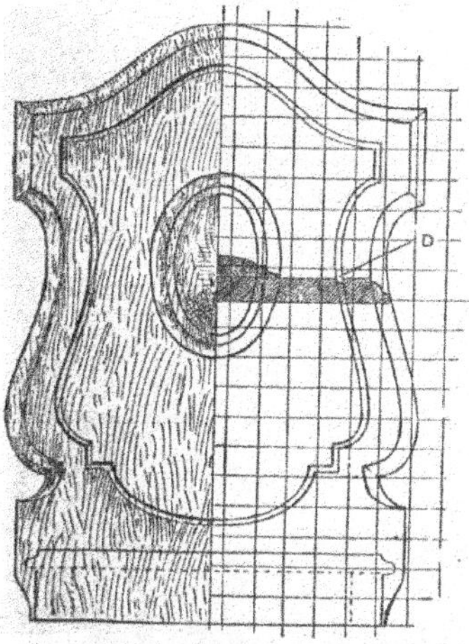

Fig. 108.—First Alternative Design for Hall Chair Back.

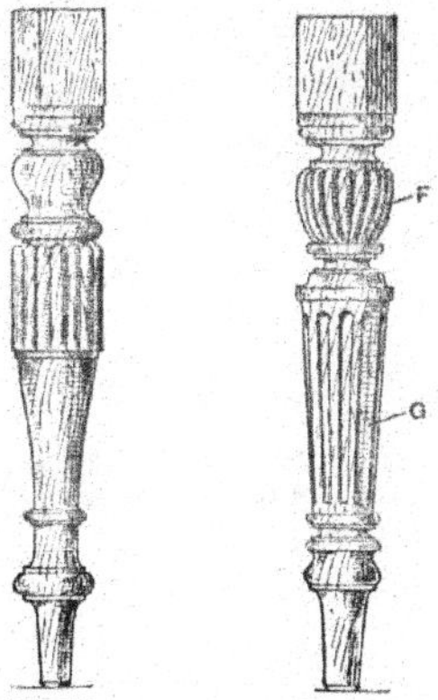

Fig. 110. Fig. 111.

Figs. 110 and 111.—Designs for Hall Chair Legs.

Fig. 112.

Fig. 113.

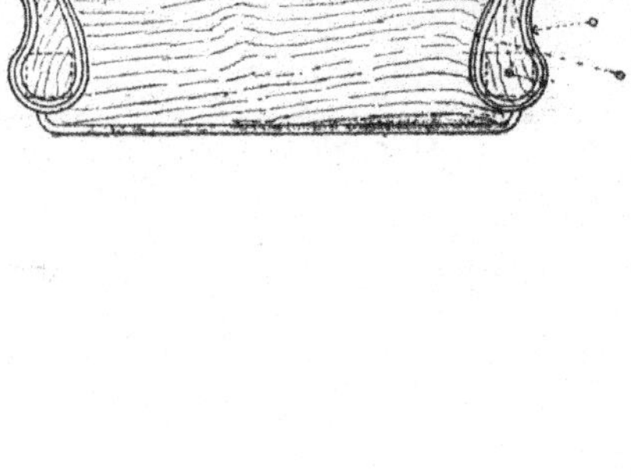

Fig. 114

Figs. 112 to 114.—Side and Front Elevations and Horizontal Section of Carved Oak Hall Chair.

Fig. 116.—Part of Top of Hall Chair Back (see A, Fig. 113).

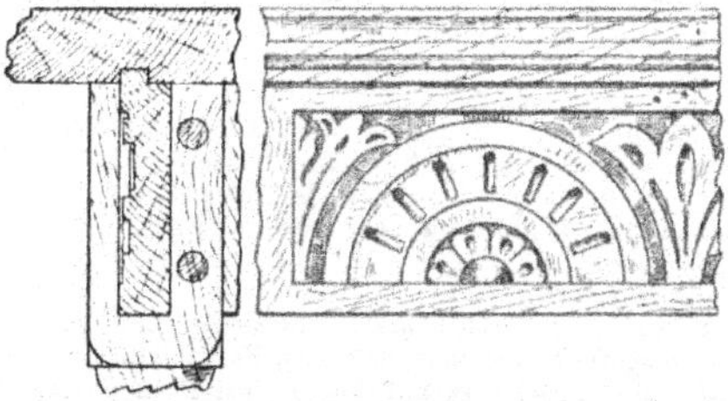

Fig. 117.—Cross Section and Details of Carving at Front of Hall Chair Seat (see B, Fig. 113).

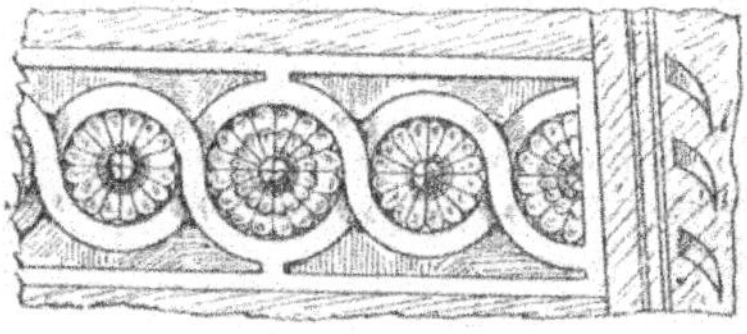

Fig. 118.—Carving on Hall Chair Back (see C, Fig. 113).

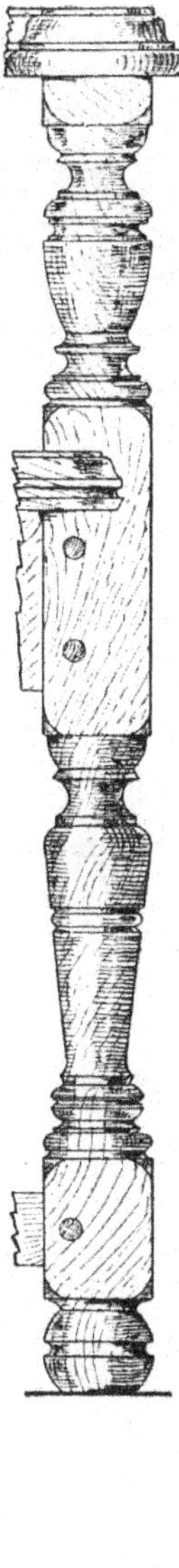

Fig. 115.—Front Leg of Carved Hall Chair.

is shown in Fig. 100. The pitch of the back and back legs is obtained by drawing a perpendicular line from the bottom of the leg, and then allowing the leg at the seat to stand back 3 in., the back 2⅞ in., and the top of back 1 in. The length from the outsides of the front and back legs is 1 ft. 0¾ in. Next set out the plan of the seat as in Fig. 101, the rails standing back ⅛ in. from the face of the front legs and ¼ in. from the back of the back legs. From the plan get the shoulder bevels of the side rails. To obtain the fullest length of tenon, the side rails should be mitered where they meet, as shown in Fig. 102. To strengthen the frame, braces 1⅜ in. thick are glued and screwed to the rails (see A, Fig. 101). When fixed, they are planed level with the top edges of the rails, and the seat is glued to them. The blocks B are glued in for further security of the seat. The back legs are cut away near the top to allow the chair-back to fit close against the back seat-rail (see C, Fig. 100). The chair-back is fixed with screws and glued, a centre-bit hole being first bored ⅛ in. deep, and then the hole is countersunk to receive the screws. The top hole is plugged to match the same way of the grain as the chair back (see Figs. 103 and 104); or a turned button may be used to fill the hole, as in Fig. 105. The corners of the front legs and the front of the seat are bevelled (see Fig. 101), sections of alternative patterns for the seat moulding being shown in Fig. 106. The backs are cut to shape with a fret-saw or bow-saw and then carved. Fig. 108 is the simplest in form, having merely an ogee moulding worked on its outer edges and a hollow D forming a marginal line (see section on right-hand side). The oval patera in the centre is made up of two gluings of ¼-in. stuff, the lower having a hollow on its edge and the upper being flatly rounded. If preferred, the patera may be carved out of a single piece ½ in. thick. The back will require bevelling at the bottom to give it the necessary pitch (see Fig. 100). The back shown in Fig. 107 is more elaborate, and requires greater skill in carving. The quirks E are first cut with a parting tool—that is, a tool of a V-shape—and the round and hollow sections afterwards carved with gouges and chisels, the centre shield, ⅜ in. thick, being glued on as in Fig. 108. The design given in Fig. 109 is executed in the same manner as that in Fig. 107, the shield in this also being ⅜ in. thick. The enlarged patterns of legs shown by Figs. 110 and 111 are suitable for any of the backs; the turned members F (Fig. 111) may be carved, and the shafts fluted or reeded as at G. The chairs should be finished off with french polish.

Carved Oak Hall Chair.

A hall chair in oak is shown in side and front elevation by Figs. 112 and 113, and in horizontal section on the line X X (Fig. 113) by Fig. 114. An enlarged detail of one front leg fitted to the arm of the chair is presented by Fig. 115. Details at A B C (Fig. 113) are shown on a larger scale by Figs. 116, 117, and 118. Figs. 112 to 114 are drawn to a scale of approximately 1¼ in. = 1 ft., and Figs. 115 to 118 to a scale of approximately 3 in. = 1 ft. In working from Figs. 112 to 114, first construct an accurate scale, noting that the distance from the ground line to the top line in Fig. 112 measures exactly 47 in.; from this all other dimensions can be obtained.

SIDEBOARDS.

Plain Sideboard with Pedestal and Full-length Shelf.

The sideboard design (Figs. 119 to 121) affords opportunity for variation in the treatment of details to suit individual requirements, and would look well if executed in wainscot oak, birch, or Italian walnut. If made according to the directions about to be given, a very substantial and handsome piece of furniture will result. The back is made movable for convenience of packing. The principal dimensions are: Height over all, 8 ft. 5½ in.; width of body, 4 ft. 4 in.; depth, 1 ft. 9 in.; top, with flaps, 6 ft. 3½ in. by 1 ft. 10¼ in.; shelf, 4 ft. 6 in. by 10¼ in.; pedestal, 3 ft. 6¾ in. high. Fig. 119 shows the front elevation, Fig. 120 the end elevation, and Fig. 121 the general plan; the half A being above the top, and showing the flap raised, and the half B below the top, being sectional. Fig. 122 shows a horizontal section in two heights drawn to a larger scale, the half section A being taken through the lockers, and the half-section B being taken through the drawers. Figs. 123 and 124 show the complete vertical section, broken, however, in order to economise space; as all dimensions are marked, no difficulty will be experienced in setting out a full-sized drawing.

Variations.—The top is shown wrought solid, chamfered, and carved in low relief, with a tongue moulding; this necessitates a special and somewhat difficult joint for the flaps, so that the appearance of the top may be the same whether the flaps are up or down; a joint easier to make, though not so well in keeping with the design, would be the common rule joint, with a half-round worked on the edge of the top. The top and flaps, for the sake of economy, might be built up with a ½-in. top, glued and blocked to ⅝-in. by 3-in. marginal pieces mitered at the angles. The brackets for the flaps also might be hinged with brass butts in place of the wood hinge to be described. The panel under the shelf might be replaced either by painted tiles or by silvered glass, in which case the framing would have to be rebated instead of being ploughed as shown. The joints of the door panels might be placed diagonally instead of vertically, and flat chamfers might be substituted for the hollows on the standards, if these are found too difficult to work.

Working Drawings.—Begin by making full-size drawings of the sections shown by Figs. 122, 123, and 124, of course not employing broken lines. It will be found best to make two separate horizontal sections, repeating the drawings on each side of the centre line shown in Fig. 122. The dotted line in the half-plan marked A indicates the top front rail shown in section at M (Fig. 123). The dotted lines in B half are the drawer runners and division rails (see also Fig. 123). Figs. 123 and 124 will be drawn in line with each other, Fig. 123 above Fig. 124, at the proper distance apart, according to dimensions given; and it will be advisable to make horizontal sections through the frieze rail E E, the mirror back F F, and the framed panel between the shelf and cupboard top G G. This done, take off the quantities of stuff required; the cutting list of these on p. 33 will probably be found useful, as indicating the necessary allowances for preparing; the dimensions given are the rough sizes, the finished sizes being obtained from the drawings, and carefully worked to in planing. It will be found, in some instances, that the rough size is very nearly the finished size; this occurs in unimportant places, where a shaving more or less is of no consequence, as in the back panels, etc., where to use the next size of stuff would require much labour in reducing it to dimensions.

Cutting List.—The following is a list of the stuff required to construct the sideboard as measured from the setting out,

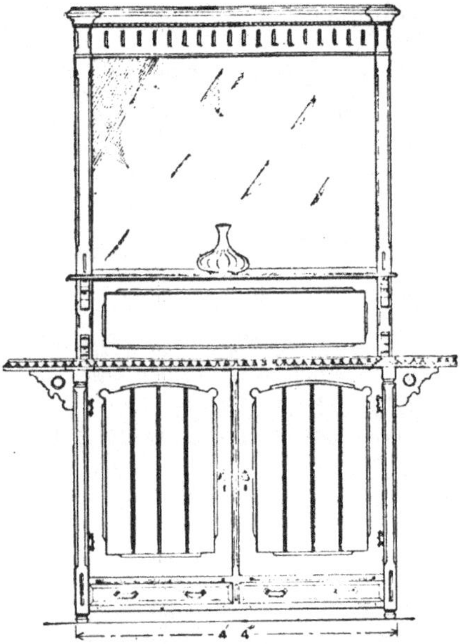

Fig. 119.

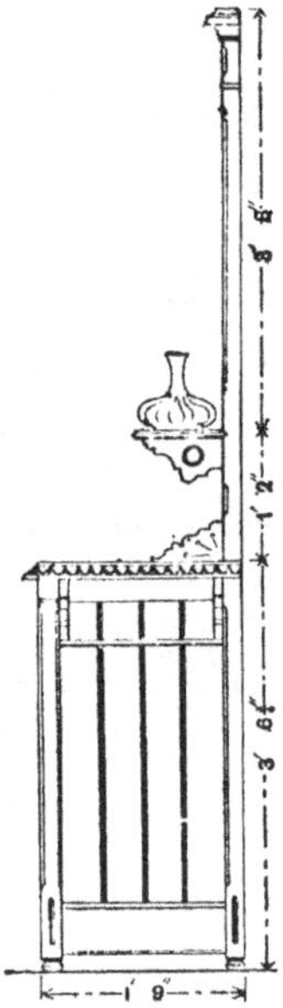

Fig. 120.

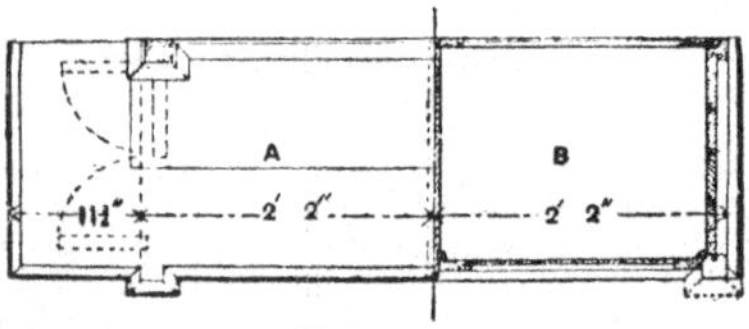

Fig. 121.

Figs. 119 to 121.—Front and End Elevations and Plan of Plain Sideboard with Pedestal and Full-length Shelf.

sufficient substance being allowed for cleaning up to the finished sizes:—

Description.	No. of Pieces.	Lgth. ft.	in.	Brdth. ft.	in.	Thickness inches.	Wood.
Back—							
Cornice	1	4	2	0	2½	1½	Wainscot
Do. breaks	1	1	8	0	2½	1¼	Do.
Do. backing	1	4	2	0	2¼	1	Pine
Standards ..	2	4	11	0	1¾	1¾	Wainscot
Frieze	1	4	2	0	5½	¾	Do.
Neck moulding	1	5	0	0	1¾	¾	Do.
Shelf	1	4	7	0	10⅝	¾	Do.
Guard bead ..	1	4	1	0	1⅜	⅜	Do.
Mirror Back—							
Rails	2	4	3	0	3⅜	⅞	Deal
Stiles	2	3	2½	0	2½	⅞	Do.
Muntins ..	2	3	2	0	2½	⅞	Do.
Panels	3	3	0	1	2¼	3/16	Do.
Glass fillet ..	2	4	2	0	0⅜	3/16	Do.
Shelf Panel— ..							
Stiles	2	1	3	0	2¼	¾	Wainscot
Rails	2	4	3	0	2¼	¾	Do.
Panel	1	3	9½	0	10¾	5/16	Do.
Brackets ..	4	0	9	0	5	1½	Do.
Mirror	1	4	1¼	3	0¼	¼	Brit. plate
Table Tops—							
Centre	1	4	5	1	10½	1¼	Wainscot
Flaps	2	1	0¼	1	9⅛	1¼	Do.
Brackets ..	4	0	8	0	7	¾	Do.
Do. fillets	2	1	6	0	1	¾	Do.
Pedestal—							
Standards ..	2	3	6	0	1¾	1¾	Wainscot
Do. ..	2	3	6	0	2	2	Do.
Top rails ..	2	1	9	0	2½	1	Do.
Bottom do. ..	2	1	9	0	5¼	1¾	Do.
Top partition	1	2	11	0	2½	¾	Do.
Do.	1	2	11	1	6	¾	Pine
Bottom do. ..	1	0	3¾	0	2¼	¾	Oak
Do. ..	1	0	3¾	1	6	¾	Pine
End panels ..	8	2	8½	0	4½	¾	Wainscot
Front rail ..	1	4	2¾	0	2	¾	Do.
Doors—							
Stiles	4	2	10½	0	2½	⅞	Wainscot
Top rails ..	2	2	0	0	4	⅞	Do.
Bottom do. ..	2	2	0	0	3½	⅞	Do.
Panels	8	2	4¼	0	4⅝	½	Do.
Stops	4	2	10	0	0½	¼	Do.
Back— ..							
Stiles	2	3	4	0	3	¾	Deal
Top rail ..	1	4	3	0	3¼	¾	Do.
Bottom do. ..	1	4	3	0	7	¾	Do.
Muntin ..	1	2	10	0	5	¾	Do.
Panels	2	2	6	1	8	½	Do.

Cutting List (*continued*):—

Description.	No. of Pieces.	Lgth. ft.	in.	Brdth. ft.	in.	Thickness inches.	Wood.
Carcase—							
Solid bottom ..	1	4	2¾	0	2½	⅞	Oak
Do.	1	4	1½	1	6	⅞	Deal
Framed do. ..	1	4	2¾	0	1½	⅞	Do.
Do. ..	1	4	2¾	0	2½	⅞	Oak
Drawer runners	2	1	5	0	1¾	⅞	Do.
Do.	1	1	5	0	3¼	⅞	Do.
Dust boards ..	2	1	10	1	5	⅜	Deal
Drawer front	2	2	0½	0	3⅝	¾	Oak
Do. sides ..	4	1	7½	0	3⅝	¼	Bass wood
Do. backs	2	2	0½	0	3	¼	Do.
Do. bottoms	2	2	0	1	7	¼	Do.
Do. blocking	1	7	6	0	0⅜	¼	Oak
Buttons ..	16	0	1½	0	1½	¾	Do.

Fittings.—Two 2¼-in. brass lever locks; four pairs of 2½-in. brass butts and plates; two door pulls, mediæval, to pattern; four drawer handles to pattern.

Preparing Stuff.—The stuff being cut out, plane up the best sides and edges, straight, square, and out of winding. When all is faced, set gauges to the various thicknesses and widths, and reduce each piece to the requisite finished sizes, marking each, as finished, with its appropriate name. If the stuff has to be left unfinished at any time, carefully pack the pieces together and weight them, or fasten them down with handscrews, to prevent warping. Pick the stuff for figure, try to balance the grain, and keep points of figure upwards. Joint up the partitions, dust boards, drawer bottoms, panels, etc., ploughing and tonguing the joints where possible.

Setting Out.—Assuming the stuff all prepared, begin to set out the carcase. Take a front standard or leg and lay it on the rod upon the vertical section drawn from Figs. 123 and 124, in the position it will occupy when framed. Square up the top line, floor line, width of end rail, thickness of front rail M, the two drawer divisions, and the bottom end rail, remembering that the end and front lines go upon the adjacent inside faces. Square up also the chamfer stops and the V-chases at top and bottom. Mark over the mortices for the rails—these will be kept ⅜ in. within the sight lines, and ⅞ in. wide for top rail—and a 3-in. mortice in centre of width

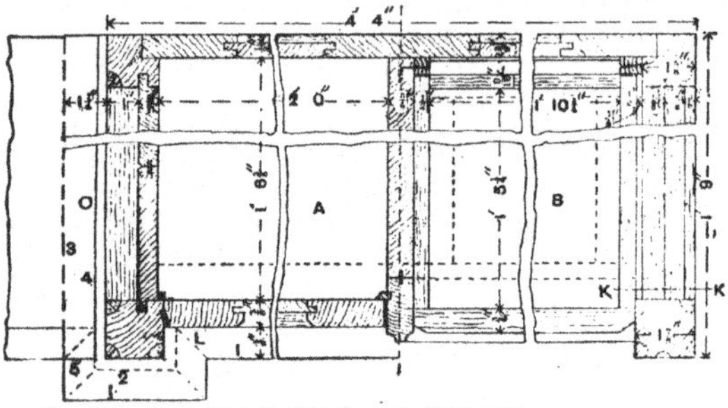

Fig. 122.—Horizontal Section through Sideboard.

of rail for the bottom one. It will be noticed that the top rails are only ⅞ in. thick, while the bottom rails are 1⅝ in., the same thickness as back standards; this is to provide room at the top for the flap brackets to fold back out of sight, therefore different gauging will be required. A $\frac{5}{16}$-in. tenon should be used at top. Gauge from the outer or face side so that the face of the rail stands full ¾ in. from the face of the standard. A ½-in. tenon kept in the centre of the standard can

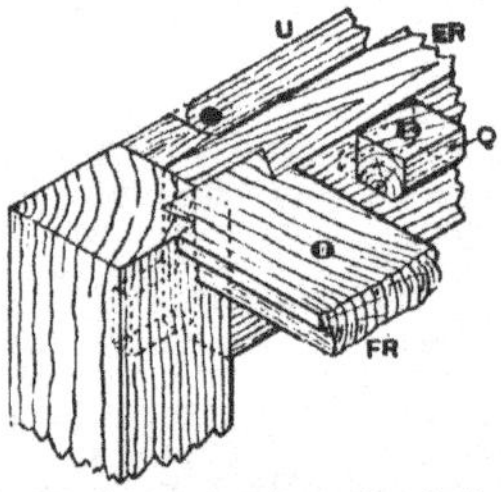

Fig. 125.—Top End of Front Standard of Sideboard.

be used for the bottom rail. Cut in the mortices in the back standards, which may be 1½ in. deep, and may go through and be wedged at the back. The top front rail will be dovetailed in after the ends are framed up, as shown in isometric

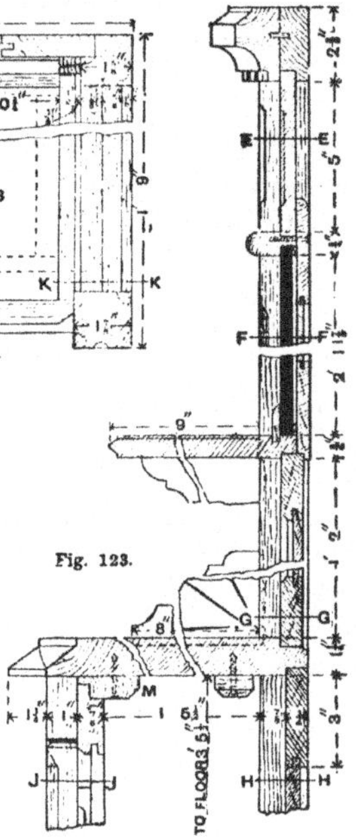

Fig. 123.

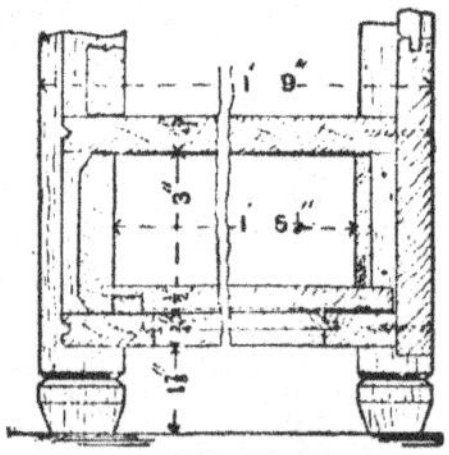

Fig. 124

Figs. 123 and 124.—Vertical Section of Sideboard, showing Locker, Back, Drawer, etc.

view by Fig. 125. The lettering signifies: —E R, end rail; F R, front rail; Q, button; U bracket. The two division rails at bottom will have $\frac{3}{8}$-in. mortices in centre of thickness of standard 1 in. deep; these mortices should taper so that the tenon tightens as it is driven in. This first standard being now completely set out, pair the others with it and strike the lines over where required—namely, face and end lines on the other front standard, and end lines only on the two back standards; all the lines should be struck over in pencil. Make a wood gauge, and line in the chamfers

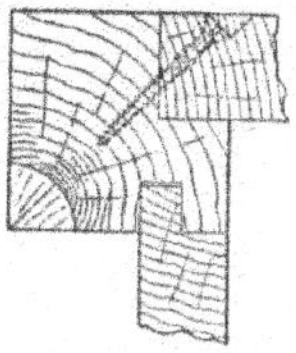

Fig. 126.

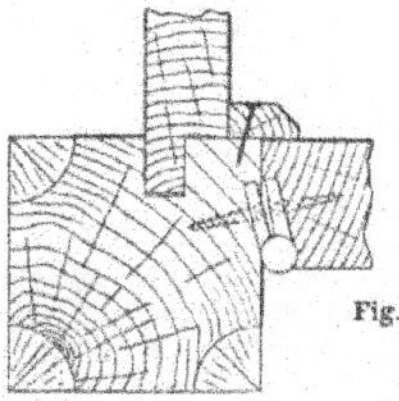

Fig. 127.

Figs. 126 and 127.—Sections through Sideboard Back and Front Standards respectively, on Lines H H and J J (Fig. 123, p. 34).

as shown by details (Figs. 126, 127, and 128); then set a fine tooth gauge and run it down the face of the panel groove; this will be 1 in. from end faces. Gauge $\frac{3}{4}$-in. by $\frac{1}{2}$-in. rebates on the back standards to receive the framed back, stopping the rebate $1\frac{3}{4}$ in. from the floor line. Set out the end rails from the section (Fig. 122), allowing for $1\frac{1}{2}$-in. tenons at the front ends, and $1\frac{3}{4}$-in. at the back. These tenons should have square shoulders. The top rail should be set back $\frac{3}{4}$ in. full from the face of the standard, so use a $\frac{3}{4}$-in. slip with the gauge. Gauge $\frac{1}{4}$-in. by $\frac{1}{4}$-in. grooves for the panels $\frac{3}{4}$ in. on for the buttons (see Fig. 125). Chamfer the bottom rails $\frac{5}{8}$ in. by $\frac{3}{8}$ in.; gauge a $\frac{1}{4}$-in. by $\frac{1}{8}$-in. rebate on the top inside edge, and a $\frac{1}{4}$-in. by $\frac{1}{8}$-in. groove at the bottom edge to receive the divisions. See Fig. 129, which is a cross-section through the bottom

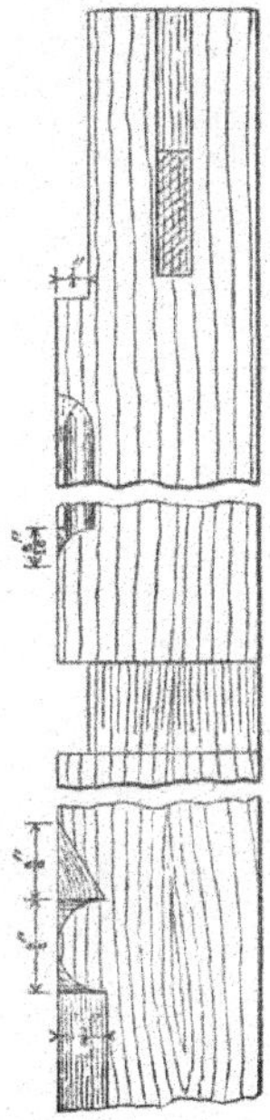

Fig. 128.—Details of Sideboard Top Standard.

rail, as seen from the back of the case, the section being taken on the line K K (Fig. 122); Fig. 129 is one-quarter full size. Having shot the front edge of the top partition straight, lay it on a front standard, and square over the sight lines of the top rail and the bottom of the cupboard. Allow $\frac{3}{16}$ in. at the bottom end for the housing (see Fig. 130, which is an isometric view of the drawer rails and partitions, one-quarter full size), and ¾ in. full at the top end in order to finish flush with the top side of the top rail into which the standard will be jointed with ¾-in. tenons. Gauge a ¼-in. by ⅛-in. groove, 1⅜ in. from the front edge on each side for the door-stops, line in the chamfers, and square over the length. The lower partition requires simply gauging to width, and squaring over to the sight lines between the two bottoms, an allowance of $\frac{3}{16}$ in. being made at each end for the housing. A small flute is worked on the front

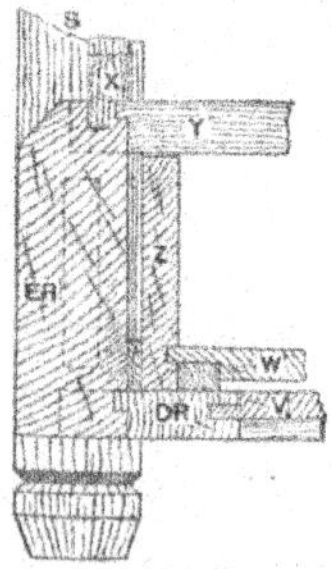

Fig. 129.—Section on Line K K (Fig. 122, p. 34) showing Foot of Front Standard, Rail, Framed Bottom, etc., seen from Back of Sideboard.

edge. as shown in Fig. 119. The grain should run with that of the top partition.

Cupboard Doors.—The doors will next claim attention. Set out the stiles from the vertical section (Fig. 123), working from the sight lines of the rails. Mark over two lines for the top rail, one for the springing, and one for the crown, the mortice being kept in line with this, to avoid the sunk ring in the corner; a ¼-in. mortice will be made, ⅜ in. from the face; this will allow for the panels a $\frac{3}{16}$-in. groove without stopping. The width of the top mortice should be 1¼ in., and that of the bottom one 1¾ in. Set off the chamfers, pair the stiles, and strike over the remainder of the lines. The lengths of the rails will be found from the horizontal section (Fig. 122). All the shoulders should be square, as the chamfers are stopped; gauge the tenons and the face lines of the ploughed grooves; the top rails cannot be so gauged at this stage, as they are not yet shaped. After the tenons and shoulders are cut and fitted, the sweep may be struck on the bench by means of a rod and bradawl. First set a radius of 2 ft. 3 in., then strike intersecting arcs from the corners of the rail, and, from the point of intersection as a centre, describe the curve; work the edges, and plough the ring, but do not cut it until after wedging up. The panels should be matched and chamfered (the two outside boards being left square), then glued up and set out from the framing, and the tongue worked all round. A piece ½ in. by 2 in. long will have to be glued on each top corner, and the board left square at the top end to serve as a bottom for the corner sinking, and the ploughed groove will be made correspondingly deeper.

Pedestal Back, Centre Partition, etc.—The pedestal back will be set out in a similar manner. The vertical pieces should be mortised, and the horizontal rails tenoned through them; the muntin being stub-tenoned into the rails. In this case the best side will be inside. The panels should be flush inside, and rebated and chamfered all round. Set the rails out rather full, so as to have enough stuff to make a tight fit after the carcase is glued up. The bottom of the cupboard should be laid face side up on the rod, the insides of the front standards squared up, and the centre partition marked. It will be noted in the table of quantities (p. 33) that the oak edging is longer than the deal; this is to allow for 1-in. tenons in the standards, the deal being rebated into the rails ⅛ in., as shown in Fig. 129. Make due allowance for the rail, setting back ¼ in. (see Fig. 122). Stop the housings ⅜ in. from tho front edge, as shown in Fig. 130. In gauging the tenons, use a ½-in. slip, as the bottom sets back that distance. Set out the framed bottom from this, the cross-rails or runners from the section (Fig. 124), and the dust-boards from the framing. The rails need not be tenoned longer than ⅜ in.—the depth of the ploughed groove. The middle runner is a double one, and is grooved to receive the partition.

Setting Out Drawers.—This may now be dealt with, though it would be advisable in actual work to leave the setting out until the carcase is put together. Shoot the fronts $\frac{1}{16}$ in. wider than the finished size; square them to length between the partition and the standard; set back the thickness of the sides, and square the lines over on to the worse side. Run a $\frac{1}{2}$-in. cutting gauge on the ends; pair the back, and square over. Lay one of the sides on the rod, and square up inside the back and front. The back should be kept $\frac{1}{2}$ in. clear of the back of the pedestal in order to provide room for a stop. Allow $\frac{1}{2}$ in. on the front end and $\frac{3}{8}$ in. on the back end Y (Fig. 129) for dovetails, and square over. Gauge a $\frac{1}{4}$-in. groove $\frac{3}{8}$ in. up from the bottom edges. In setting out the bottom, allow $\frac{3}{16}$ in. extra at each end for a tongue into the sides, and $\frac{1}{4}$ in. at the front. Gauge the width so as to overhang the back $\frac{3}{8}$ in., and run a rebate round three sides wide enough to receive the blocking, which must be glued to the sides only, and not to the bottom.

Back.—It is not necessary to repeat the instructions for setting out the framing, as those that were given for the pedestal will again apply. Bear in mind, however, that the upright pieces will be mortised, and the horizontal ones tenoned. Keep the panel of the mirror back flush on the inside. The shelf panel, which is framed of $\frac{3}{4}$-in. stuff with a $\frac{5}{16}$-in. panel, is set back $\frac{3}{8}$ in. from the face, and has stopped chamfers wrought all round. This panel need not be rebated, but may be bevelled as shown in Fig. 123. The lettering in Fig. 129 signifies :—S, standard; E R, end rail; X, panel; Y, solid bottom; V, framed bottom; D R, drawer runner; W, drawer bottom; Z, drawer side.

Top Standards.—Lay one of the top standards, face upwards, on the height rod, with the lower end projecting $1\frac{1}{4}$ in. beyond the sight line of the top of the pedestal, and square up on the inside edge the sight lines of the top and cornice, also those of the shelf, the groove for the neck moulding, and the stops for the chamfers. Square over on the face the sight lines of the brackets, and a line $\frac{1}{4}$ in. within each to form a stop for the grooves. Square also on the face the sinking for the shelf; the sinkings for the necking and cornice should be squared across the face, and also outside. Pair the other standard with this, and square the lines over. The different sections at the various heights are shown in Figs. 131, 132, and 133, half full size, and they must be gauged accordingly from the face side, sinking the plough grooves and rebates $\frac{3}{8}$ in. deep. The portion between the necking and

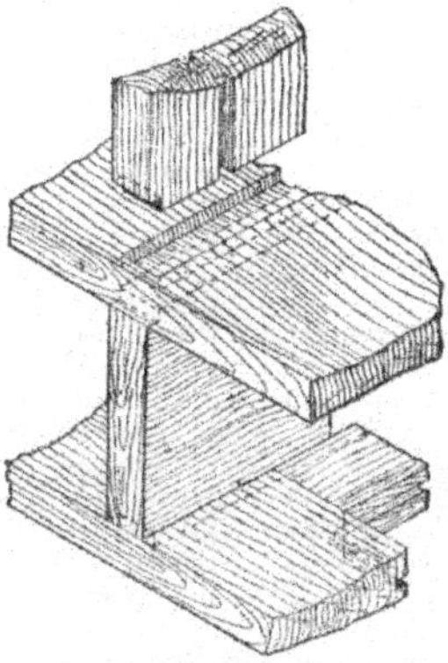

Fig. 130.—Part of Sideboard Drawer Rails and Partitions.

the cornice must be ploughed for the frieze panel, that between the necking and shelf rebated for the mirror, and that between the shelf and the top ploughed with a $\frac{3}{8}$-in. groove for the shelf panel. A mortice should be cut in the top end in line with the ploughed groove to receive a tenon on the cornice backing, which can be continued right through and wedged, as it will be covered by the return cornice. The bottom end of the standard will be rebated back half its thickness, and screwed into the table top. Set out the cornice backing (see Fig. 134, which is half full size) from the plan, square up the shoulders from the sight lines of the standards, and gauge the tenons from the back, also gauge the rebate shown in Fig. 134. This will complete the setting

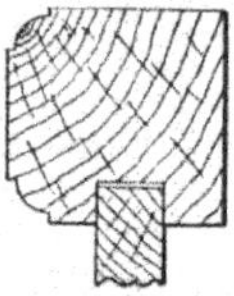

Fig. 131.—Section of Sideboard Top Standard on Line E E (Fig. 123, p. 34).

out, the remaining portions being fitted as the work proceeds. All the framed pieces should be glued up first, cleaned off, and set aside till wanted. Glass-paper should not be used inside the cupboards or drawers; the work should be left smooth from the plane. Next get ready the carcase, prepare the end panels to size, frame the rails together, work the chamfers, and glue up the two ends. Clean off inside, fit the buttons and the front top rail, and glue up. When the work is quite dry, fit on the top, fit in the drawers, and screw in the back. Fit the top and bottom panels in, groove the brackets into the standards, the lower one by a dovetailed groove as shown at Fig. 133, the upper one by screws from the back. When the brackets are in place and the panels in, stand the back on the table and mark the position of the brackets.

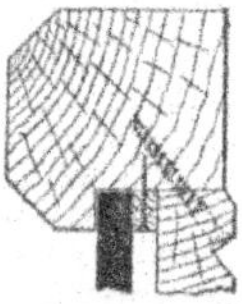

Fig. 132.—Section of Sideboard Top Standard on Line F F (Fig. 123, p. 34).

Take them out, and form a dovetailed groove in the table top ⅜ in. deep. Glue up the standards and brackets, slide them into position on the top, and screw up the standards to the same. The mouldings, shelf, glass bead, and back can now be fitted, the top buttoned on, the flaps, doors, etc., fitted and hung, the locks and furniture put on, the work cleaned off with fine paper, and taken to pieces for polishing.

Preparing Carcase.—Mortise the standards. The front ones are 1½ in. deep; the back ones are carried right through, and wedged. All mortices that do not go through should be tapered to the bottom, about $\frac{1}{16}$ in. at each end, so that the tenon will drive in tight. The two fronts will be mortised on the inside face edges with ⅝-in. mortices 1 in. deep. Plough the panel grooves on the inside faces, stopping them at the rail lines; also rebate the back standards on the back side to receive the framing. Cut the tenons on the rails, and plough both; make due allowance for the difference in thickness. Rebate and chamfer

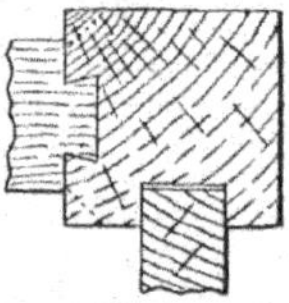

Fig. 133.—Section of Sideboard Top Standard on Line G G (Fig. 123, p. 34).

the top edge of the bottom rails, and plough the lower edge inside, as shown in Fig. 119. Cut the shoulders, and fit the work together; fit in the panels (which should have been glued up after being chamfered), alternating the grain of the wood for the sake of effect. Glue up the framed bottom with the dust panel flush on the top side, cut the tenons on the front rail, and rebate the ends, leaving a ¼-in. tongue on the top side. House in the partition, stopping the housing ⅜ in. from the front edge, as shown in Fig. 130; work it $\frac{3}{16}$ in. deep with a router. Work the cupboard bottom in the same manner, except that it will require housing on both sides. Fit in the partitions, and mark them where fitted; then fit the bottoms into the grooves and mortices in the framed ends. Mark the line of the back rebate, and plane off to width (see Fig. 124). The two partitions should be reduced to exactly the same width; the lower one will require nothing else to be done to it. The

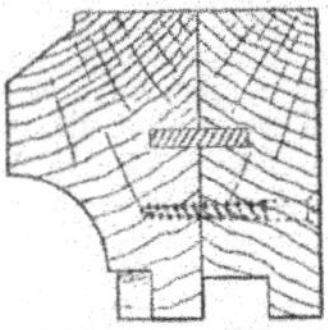

Fig. 134.—Section of Sideboard Cornice.

upper one will want ploughing on each side ¼ in. by ⅛ in. for the door stops, 1$\frac{3}{16}$ in. from front edge. Cut a ⅜-in. tenon at the top end ½ in. from the face to go into the top rail. Do not fit the top front rail in until the carcase is glued up, as a dovetail is required that will be partly in the rails and partly in the standards. Square a line over on each side of the partition level with the shoulder, and on these set out three 1½-in. by ¼-in. mortices for the buttons (shown in Figs. 123 and 125, p. 34). When the shoulders are all up and the carcase is true in both directions, knock it to pieces and work the chamfers; and when these are finished, cut the stops to the lines with sharp chisels. The top stop is a plain chamfer with the hollow butting square against its bottom; the lower one is a triangular pyramid with the hollow dying down upon it on each side A cardboard template should be cut to the shape of the foot, and applied all round, the V being cut with a chisel, and the bottom bevelled off with a tenon saw. Scratch in the flute at the bottom, and this will complete the standard. Work the chamfers on the partition, and the flutes on the bottom and division. Clean all off, and glasspaper the insides of the standards. If they are to be wax-polished they can now be glued up; if to be french-polished, the face edges of the standards, the edges of the rails, the end panels, and the sight margins of the partitions and divisions should be polished before being glued up. The work being ready for gluing up, fit the end panels into the rails, glue the tenons and mortices, enter them, and knock up. Lay the work on the bench out of winding, and cramp it up; wedge the back standards, and turn a 1¼-in. screw into the front tenons from the inside. Having glued up both ends, clean off the inside, glue and nail in the drawer partition, glue the end tenons and tongues, enter them in their places, and cramp up. Try with a rod for squareness, and brace the work in position; leave the cramps on until dry. Next fit in the top rail, keeping it 1 in. back from the face of the leg, insert the partition in the housing of the bottom. Glue the tenon and the dovetails, and drive on the top rail; nail it down at the ends, and wedge the tenon in the centre; put screws through the back edge of the bottom anglewise into the back standards. When the work is dry, clean off and level the top ready to receive the table.

Table Tops and Flaps.—The top T (Fig. 136) may be prepared in one piece, the breaks being cut and the moulded edges returned in the solid, but the appearance would be nearly as good and the work would be much easier if the breaks were formed by gluing on separate pieces after the main top was worked and moulded, the joint being made in line with the margin of the moulding, and the internal angle being mitered as shown at Fig. 124. This method having been decided upon, plane up the top, shoot the back edge, and lay the top on the carcase. Mark a 1-in.

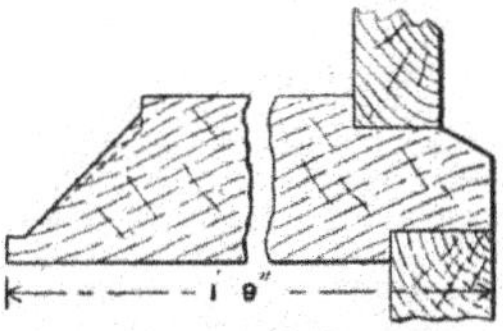

Fig. 135.—Section through Top of Sideboard.

margin to the face of the standards in front, and a $\frac{5}{16}$-in. margin at the ends; then cut and shoot to size. Set a cutting gauge to 1 in., and gauge the front edge both ways for the moulding (see section, Fig. 135); run plough grooves ⅛ in. deep, and finish the chamfer with a rebate plane. Work a quarter-circle hollow on each end, as shown in Fig. 136, for the flaps to work in; cut out the pieces to form the breaks exactly 2½ in. longer than the thickness of the standard, joint them on, and dowel them in position. Work the chamfer round as on the top, and glue on and cut the outside, back to the line of the top, as shown at 5 (Fig. 122). The full line in Fig. 121. Form the joint as shown in the section (Fig. 136), and hinge the flaps with a pair of table butts. Keep the centre of the knuckles in line with the joints M and N (Fig. 136), the hinges being sunk as required. Be careful to form a clear way for them, as indicated by the dotted line in Fig. 136. Figs. 135 and 136 are half full size. Having hinged the flaps and cleaned the top off, set out the toothed mould with a cardboard template, spacing the teeth to get a half-

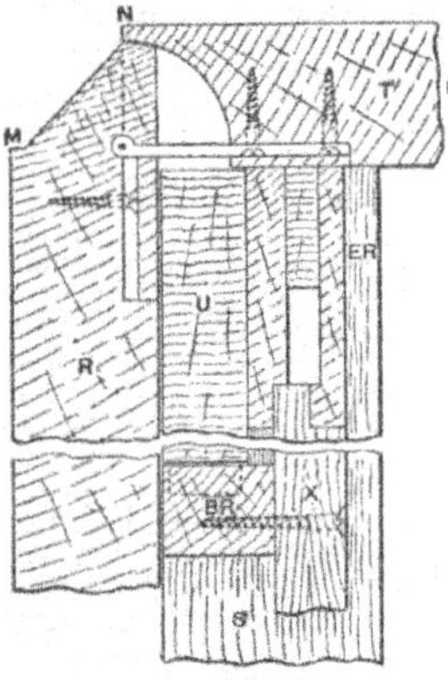

Fig. 136.—Sectional View showing Hingeing of Sideboard Flap.

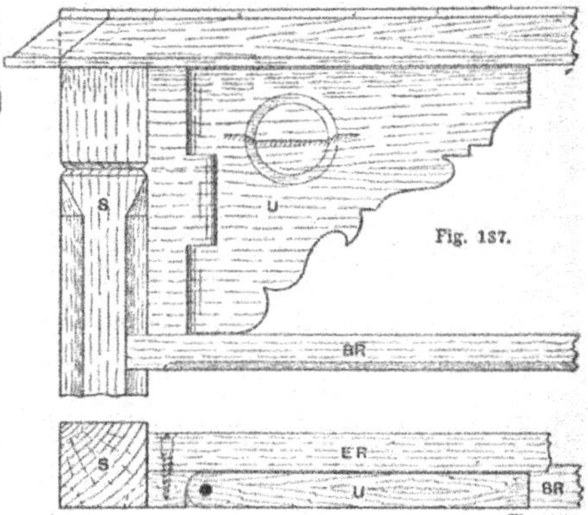

Figs. 137 and 138.—Elevation and Plan of Sideboard Bracket. Fig. 138.

in the section indicates the finished line at the top, and the dotted line at 3 the apparent edge of the top when the flap is down, which is, however, the joint edge of the latter, as will be seen by the section (Fig. 136), where that line is shown at M; see also Fig. 122, where O represents the joint edge of the flap. Lay the top face-down on the bench, turn the carcase out on it, button the top to the carcase, and screw the front rail. Next prepare flaps R (Fig. 136), the front edge and end being moulded in pairs. Stop the front chamfer at the joint ends for 1¼ in. back; this will be returned in the solid when the flap is hung in order to form a mitre, as shown tooth on each side of the mitre. Sink them $\frac{1}{16}$ in. deep at the top, and diminish to nothing at the bottom. Lift up the flaps, scribe the outline of the moulding, take the flaps off, and work the return; then, when the flaps are up, the top will appear as in the plan (Fig. 121); when down, as in Fig. 122; lines 1, 1, 3 being the outline of the top.

Brackets.—The brackets U should be got out of ¾-in. stuff to the shape shown in the elevation (Fig. 137), the grain being horizontal. Form the back ends into a half round, bore a ⅛-in. hole through the centre of the round, divide the width into three equal parts, and cut out the centre part.

Shape pieces ¾ in. by 1½ in. by 6½ in. to fit the backs of the brackets accurately, thus forming a hinge joint; bore the hole through the centre, and insert pieces

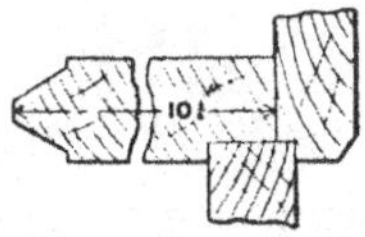

Fig. 139.—Section through Sideboard Shelf.

of iron wire for pivots. Cut a ¾-in. by 1-in. bracket rail B R between the standards, as shown in Figs. 120 and 136. Stub-tenon the bracket back into it, and glue it in position, screwing from the inside. Fig. 137 shows an elevation, and Fig. 138 a plan of the bracket, one-quarter full size, with the top and flap removed.

Fitting the Back.—Work the rebates, and plough grooves in the standards; form the sinking for the cornice necking and the seat of the table top. Groove in the brackets $\frac{5}{16}$ in. deep, the lower one with a dovetail groove; work the chamfers and stops, and prepare the panelled frame under the shelf and the frieze panel. Place them in position, and glue up the standards, first, however, grooving the cornice rail to receive the tongue of the cornice. Fit the shelf (Fig. 139 being the section) tight between the standards, the ends running over, and returning in the solid on the outer face of the standard. Between the standards the shelf will run back to the rebate line to form a seat for the glass and a rebate for the back. Plough a

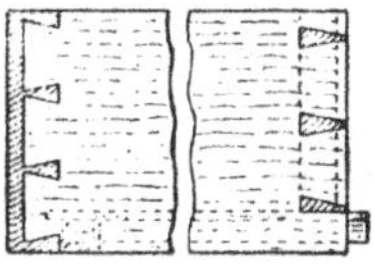

Fig. 140.—Side Elevation of Sideboard Drawer.

groove in the shelf to receive the guard bead, which can be put in after the shelf is fixed. Dowel the shelf on the brackets, house the guard bead in the standards ⅛ in. from the back, and put in the glass and its back last of all. When all is fitted up, mitre the cornice between the standards, and glue and screw from the back (see Fig. 134). The standards being sunk ¼ in., and the cornice upon them projecting ¾ in. in front of the general line of the cornice, the break will require to be 1¾ in. thick. Mitre these up, and glue and fix them with handscrews. Next place the back on the table, and mark the position of the rebate and the bracket grooves; form the latter slightly tapering, so that they will be easy to release. A screw can be turned in from underneath; one can also be put in from the back end of

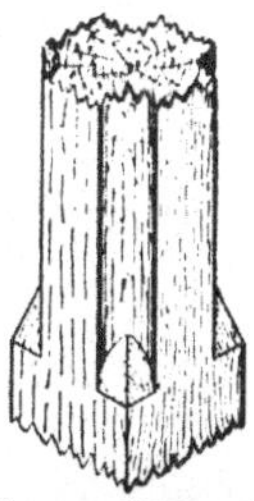

Fig. 141.—Stops on Sideboard Standard.

the standard, which should run down to the rebate in the top.

Preparing and Fitting Drawers.—Fit the fronts tight in the opening, and square off the backs to the same length. Square the sides 1 ft. 5¼ in. between the sight lines; allow ½ in. at the front end and ⅝ in. at the back for the dovetails, which should be set out as shown in Fig. 140. Pair the sides, handscrew them together, cut the sockets with a dovetail saw, and remove the core. Plough the sides ⅝ in. up from the bottom edge, the groove being ¼ in. wide and $\frac{3}{16}$ in. deep; the grooves in the fronts should be ¼ in. deep. Fix one of the fronts in a bench screw, take one of the sides, and run a ½-in. cutting gauge on each. Put a ¼-in. slip in the ploughed groove, drop the side upon it, keep the end to the gauge line, hold it firmly with the left hand, and draw the

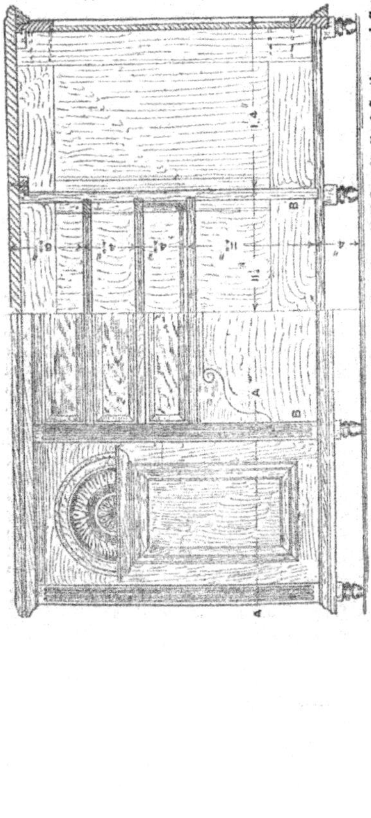

Fig. 142

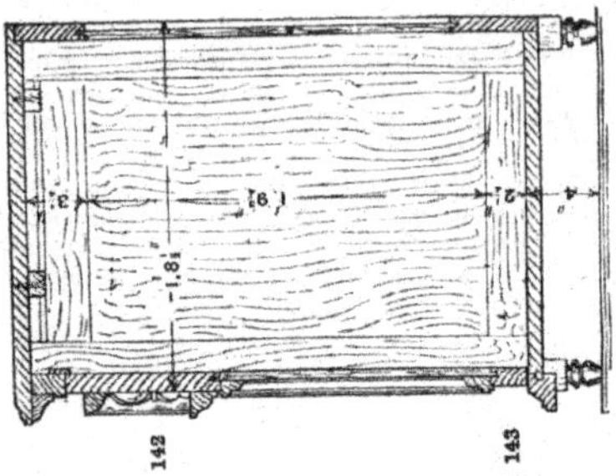

Fig. 143

Figs. 142 and 143.—Half Front Elevation, Half Longitudinal Section, and Cross Section of Sideboard Pedestal.

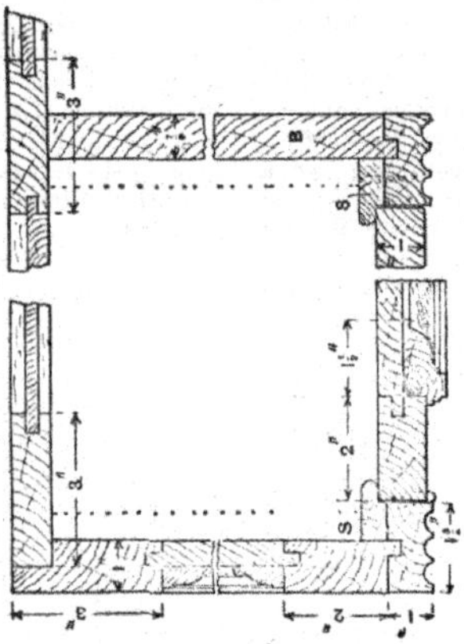

Fig. 144.—Section of Cupboard of Sideboard Pedestal.

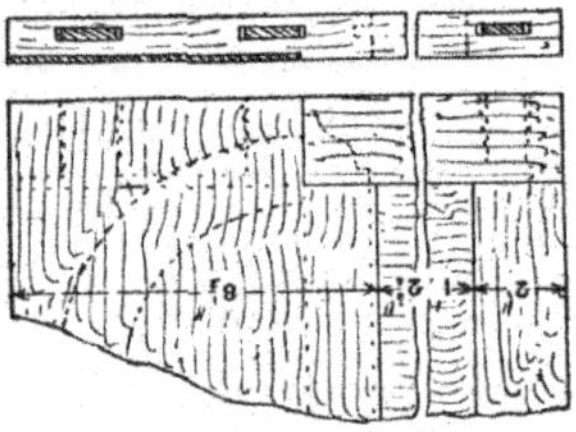

Fig. 145.—Detail of Door of Sideboard Pedestal.

dovetail saw through the cuts with the right, grasping the saw about the middle. Turn the front round, square down upon the inside the marks off the pins just made, and cut down with a dovetail saw, leaving the lines showing so that the pins may fit tight. Mark each end in a distinctive manner so as to avoid confusion, and repeat the process at the other end. For the back, put the slip in the side groove, keep it pressed tight against the back and also to the gauge line, and run the saw through the cuts in the same way. Cut down the pins as before, and remove the core with a bow saw and chisel. Having cleaned out the pins and sockets, take a shaving off the insides of the drawers, then glue and knock together. Cut the bottoms to size, rebate wide enough to receive the blocking, slot the back for the screw, slip the bottom into the groove, and glue in the blocking, the drawer being first carefully squared. When the work is dry, clean off the ends of the pins and try the drawers in the openings; they should run easy, yet without any play. A slip ¼ in. by 1 in. will be required at each end to bring the rail up to the thickness of the standard and form a guide for the drawer (see z, Fig. 129). Place stops in the ½-in. space at the back. Keep the drawer fronts in position, and glue and brad these stops to the standard. Screw in the case-back, work the chamfer round the drawer front, sink in the handles, and clean off ready for polishing.

Fitting Doors.—The doors can next be fitted in. Cut rods to the size each way of the openings, transfer these sizes on to the doors, and cut off to the lines. Shoot all the edges, after which the door should fit exactly; but if it is too tight, ease it a little. Rebate the top rail ½ in., as shown at Fig. 123; let in the butts, the knuckles being allowed to project the thickness of the ornamental plates, which afterwards are screwed on with round-headed screws (see Fig. 127). Place one of the doors in position mark the position of the butts on the standards, set a pair of dividers to the distance of the butt edge from the inside of the door, and scribe down against the stops of the standard (shown in Fig. 141). Sink the butts into this line, tapering up to nothing at the knuckle, and screw them in. Fit the locks and handles, insert the mirror, brad in the slips, screw up the back, and the sideboard is complete.

Sideboard Pedestal.

A sideboard pedestal may be constructed in solid wood, as illustrated in Figs. 142 and 143. Generally the same methods might be adopted for veneered work, except that the doors would then be framed up in a manner similar to the end frames, and veneered over all. Briefly described, the construction of the pedestal is as follows. The ends are panelled and moulded frames of 1-in. stuff, mortised and tenoned together, the panels being flush inside; the back stiles, 3 in. wide, are rebated on the edge to receive the back framing, and the front stiles, 2⅜ in. wide, are tongued on the edge to fit the fluted pilasters, as shown in Fig. 144. The moulding should be fixed with screws from the inside. The pilasters are worked, glued on the edges of the ends, and cut in flush between the top and the plinth mould. The divisions B (Figs. 142 and 144) are made of 1-in. deal, tongued to fit the pilasters, and are housed ⅛ in. in the bottom, the top edge having three mortices for buttons. The drawer divisions are housed ⅛ in. into the upright divisions, the housings being covered by the pilaster. The framed drawer divisions should have their side rails in oak or other hardwood, the front rails being of wood to match the remainder. The solid division may be of deal, edged with hardwood. The bottom is of ¾-in. deal, tongued into the plinth mould and also into the end frames. The divisions are grooved into this, and are glued and nailed through the plinth mould, which is of 2-in. by 1½-in. hardwood, glued and blocked to the bottom. Its ends run across the pilasters, and mitre with the return mould, which is glued and screwed to the sides as shown in section in Fig 142. The top is of 1-in. hardwood, rebated for the back, and overhanging the front and ends by 1½ in. It is solid moulded on the top side, and has a planted mould 1 in. by

Fig. 146.—Sideboard with Carved Panels and Bevelled Mirrors.

Fig. 147.

Fig. 148.

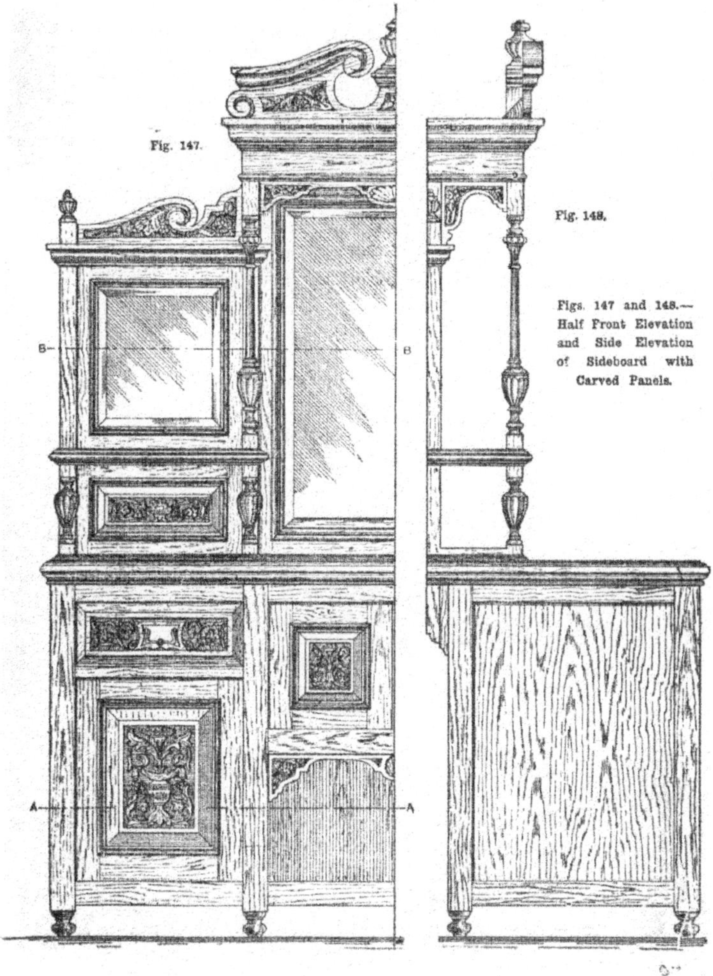

Figs. 147 and 148.—Half Front Elevation and Side Elevation of Sideboard with Carved Panels.

$1\frac{1}{2}$ in., glued underneath to add to its apparent thickness. The end pieces of the planted mould are better if put on in short lengths across the grain, to prevent the top splitting when shrinking, and to make the top appear solid. This mould is better if rebated in the framing,

Fig. 149.—Vertical Section on Centre Line of Sideboard with Carved Panels.

Fig. 150.—Part Horizontal Section of Sideboard on Line A A (Fig. 147, p. 45).

Fig. 151.—Part Horizontal Section through Mirrors, etc., of Sideboard on Line B B (Fig. 147, p. 45).

because as it is not glued to the framing it may curl off. A top rail crossing the divisions is notched into them and dovetailed into the ends. The door, having a carved and moulded pediment, requires a wide top rail as a backing, and the best construction is shown in Fig. 145, in which the top rail is mortised and tenoned to the stile, which is rebated on the face, and the rail is lipped across it, thus hiding the joints, which would otherwise interfere with the design. The shoulder of the rebate in the stile is made level with the top edge of the cornice mould, that and the carved pediment being screwed from the back. An easier but inferior method of making the door would be to dispense with the tenons, and simply halve the rail and stile together, gluing them, and fixing with a handscrew until dry; then add a few panel pins on the inside.

The door is bolection moulded, with a flush panel inside; when hung, it is sunk $\frac{1}{8}$ in. below the face of the front, as shown in Fig. 144, which is a section on A A (Fig. 142), and the stops S (Fig. 144) are rebated to receive it. The back is a $\frac{3}{4}$-in.

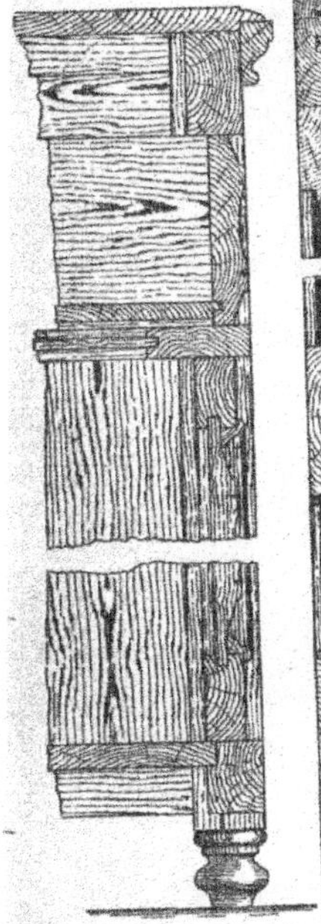

Fig. 152.—Part Vertical Section of Lower Part of Sideboard through Cupboard and Drawer.

Fig. 154.

Fig. 153.

Figs. 153 and 154.—Part Vertical Sections through Upper Part of Sideboard.

panelled frame, with 3-in. stiles and muntins. Two methods of panelling are shown in Fig. 144. The drawers are dovetailed together in the usual way, and have a small moulding planted on the fronts. The turned feet have square shanks, and are glued and screwed to the bottom.

Sideboard with Carved Panels and Bevelled Mirrors.

A general view of a good class sideboard is shown by Fig. 146, elevations and sections being presented by Figs. 147 to 149. Figs. 150 and 151 are hori-

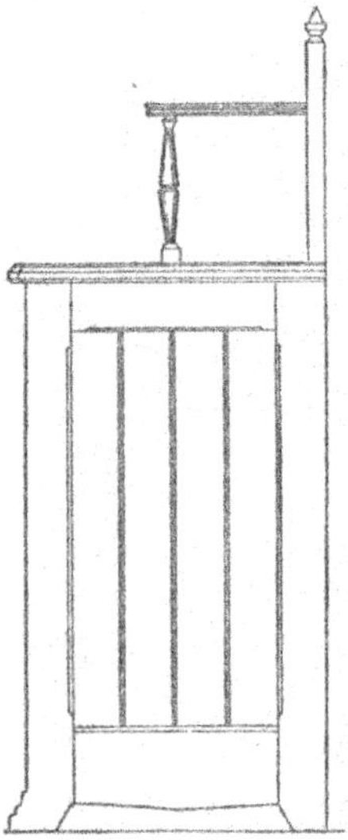

Fig 155.

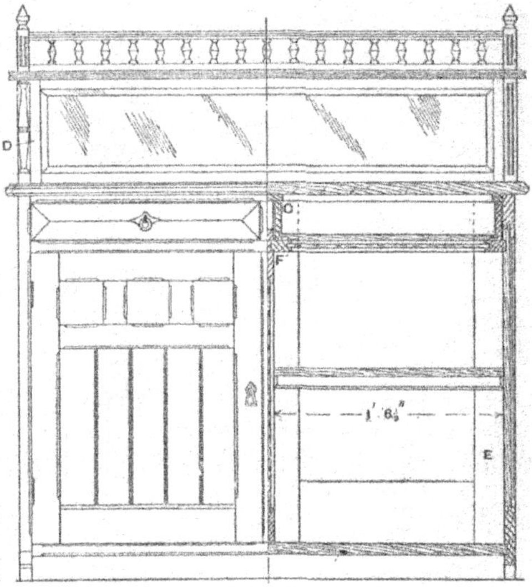

Fig. 156.

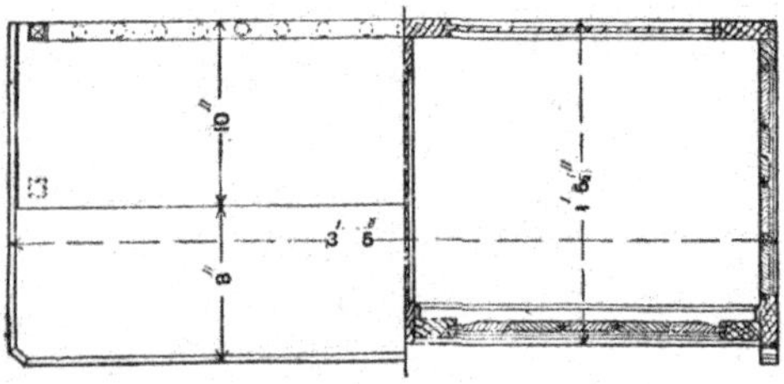

Fig. 157.

Figs. 155 to 157.—End Elevation, Half Front Elevation and Half Longitudinal Section, and Half Plan and Half Horizontal Section (at E, Fig. 156) of Early English Sideboard.

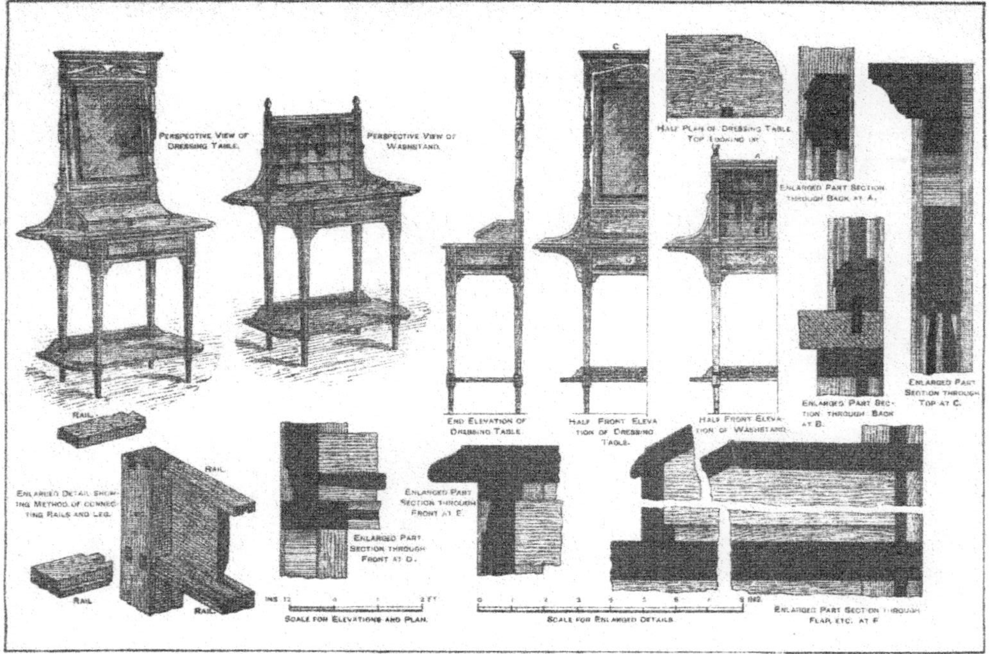

BIJOU DRESSING TABLE AND WASHSTAND.

zontal sections taken respectively on the lines A A and B B (Fig. 147); the second figure being drawn to a larger scale. Fig.

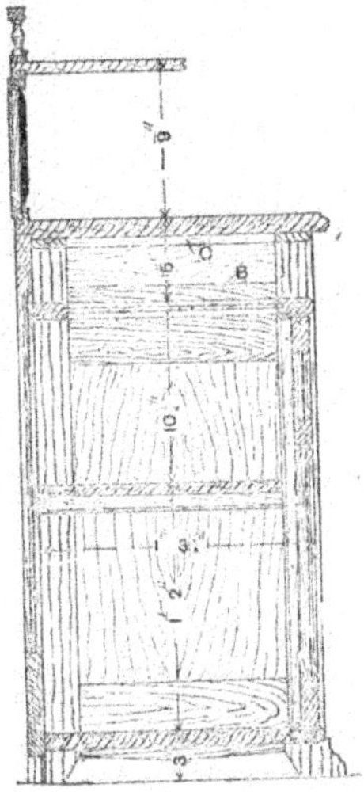

Fig. 158.—Vertical Cross Section of Early English Sideboard.

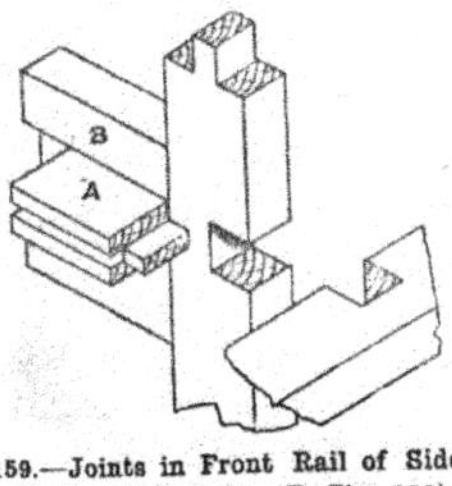

Fig. 159.—Joints in Front Rail of Sideboard Cupboard Top (see F, Fig. 156).

152 is a part vertical section of the lower part through the side cupboard and drawer, whilst Figs. 153 and 154 are part vertical sections through the upper portion, showing

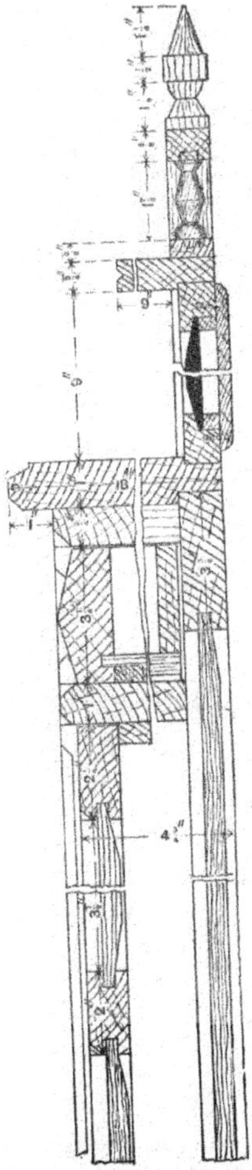

Fig. 160.—Enlarged Vertical Section (Broken) of Early English Sideboard.

side mirror, panel, and cornice, as well as a detail of the framing at C (Fig. 149).

Early English Sideboard.

The design shown by Figs. 155 to 158 would look well if executed in oak, and either stained brown or fumigated, the chamfers being left in the natural colour; or, if mahogany is chosen, the chamfers should be stained a deep red. Figs. 155 and 156

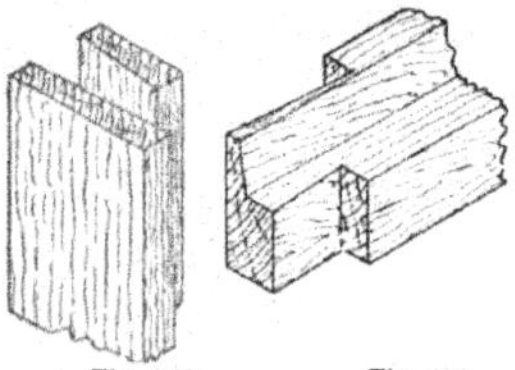

Fig. 161. Fig. 162.

Figs. 161 and 162.—Joint for Mirror Frame of Sideboard.

represent elevations, and from these and the plan (Fig. 157) a general idea of the construction can be gained. Fig. 158 shows a vertical cross section. As will be seen, with the exception of the top and bottom and the two shelves, which are solid, all the carcase is composed of framing. This method of construction, whilst entailing slightly more labour, yields much more satisfactory results in economy, strength, and lightness; but there is no objection to is required; but care should be taken to select stout and fairly true stuff, and, in the preliminary preparation, to plane off only just sufficient of the best surface to ensure the parts being straight and out

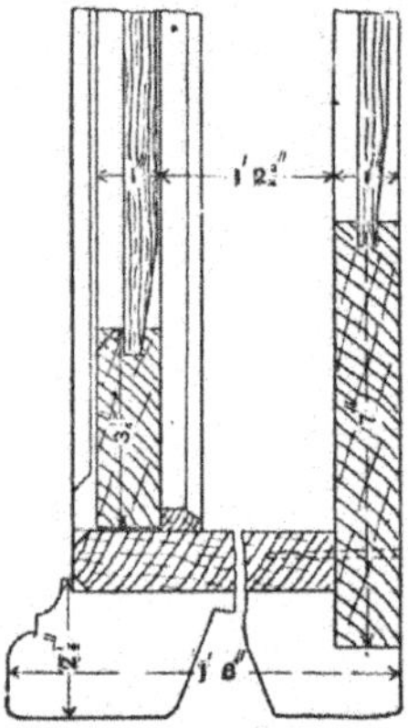

Fig. 164.—Enlarged Vertical Section through Bottom Part of Sideboard.

of winding, as the whole will have to be gone over again. Thicknessing is not absolutely necessary, unless the original thickness of the stuff varies considerably. The top, if possible, should be got out of one piece, but if jointing is found necessary,

Fig 163.—Enlarged Horizontal Section through Drawers of Sideboard.

substituting solid divisions to the cupboard and drawer compartment. With the exception of the drawer fronts, which are 1¼ in. thick, no stuff thicker than 1 in. dowelled joints are more suitable than tongued; any joints in the bottom and shelf should be ploughed and tongued. The V-jointed panels in the doors and ends should preferably be made up in narrow widths with tongued joints, but, if desired, may be made in one width, and the V-grooves worked with a small rebate plane. The case bottom, cupboard top, and division are housed into each division, and have a tenon cut on their ends as shown, which fits into the panel groove in the front and back rails. The corresponding rails above the drawers are mortised to receive a ⅜-in. tenon cut on the ends of the division stiles, which run

Fig. 165.

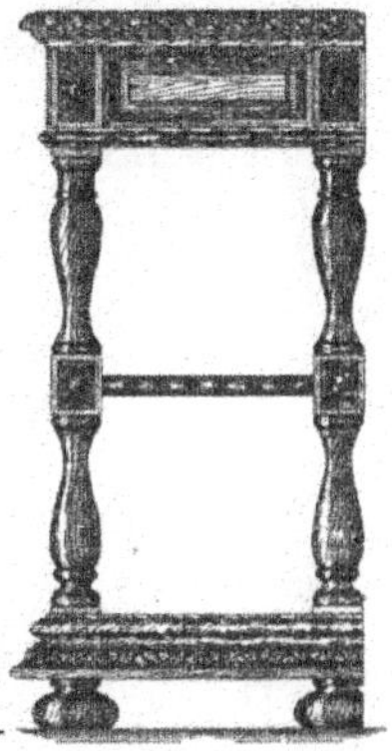

Fig. 166.

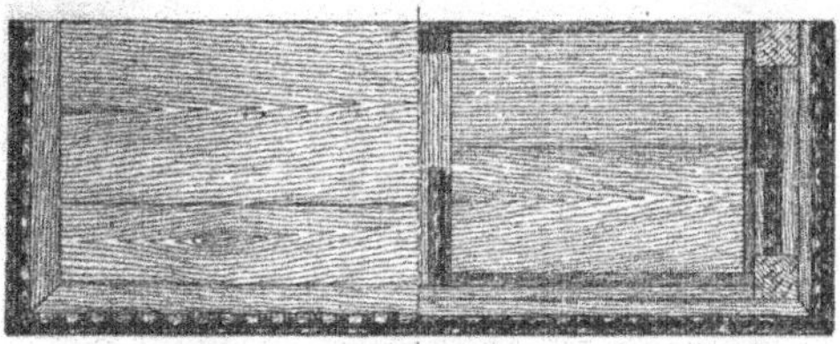

Figs. 165 to 167.—Front Elevation, End Elevation, and Half Plans (with and without Top) of Dinner Waggon.

Fig. 167.

other and into the sides $\frac{3}{16}$ in. deep. The method of making the joint in the front rail of the cupboard top is shown in Fig. 159. The front rail and the stile of the division are each notched half-way through on opposite edges, and driven tightly together; the drawer runners A are grooved into the top rail of the right up for this purpose. The top division B is made to stand 1 in. above the runners, to act as a guide for the drawers, and a tilting piece C (Fig. 156) is screwed to the under side of the top to prevent the drawers tilting up when being drawn out. Similar pieces are glued and screwed flush with back and front rails, upon the two ends, as shown in the section (Fig. 156), and to these the top is fixed by means of screws passing through slots to allow for shrinkage. The back is square-framed of 1-in. stuff with ⅜-in. panels, nailing flat on the edge to the bottom, and setting in rebates square, are tenoned through the top, and are notched to receive the shelf. The rails of the gallery, which finish respectively ⅜ in. and ⅝ in. thick, are stub-tenoned in. The face of each standard has a sunk ovolo with double chamfer

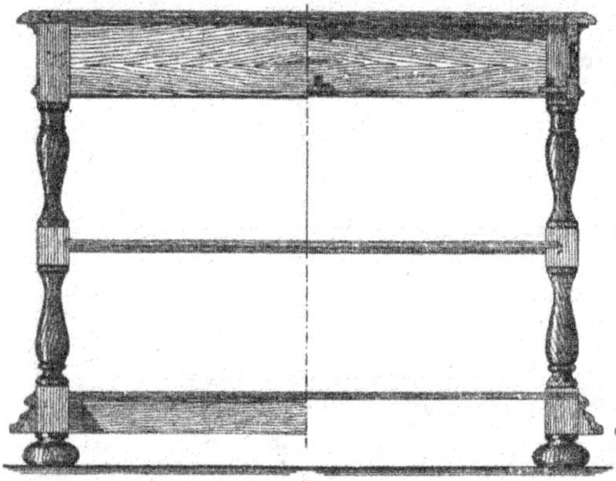

Fig. 168.

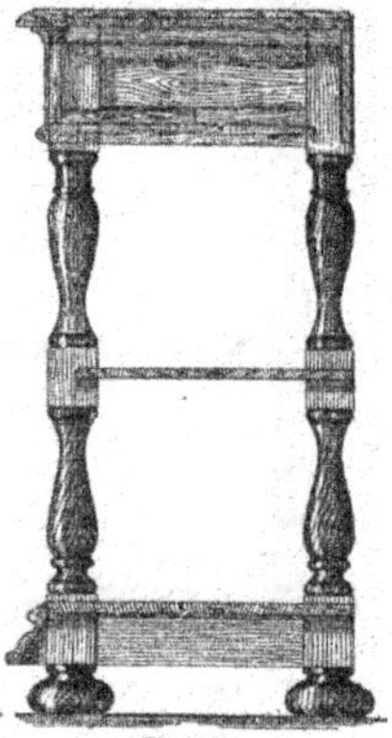

Fig. 169.

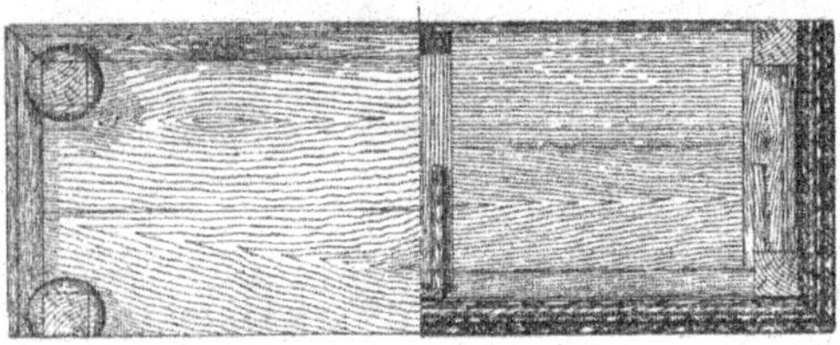

Fig. 170.

Figs. 168 to 170.—Half Back Elevation and Half Longitudinal Section, Cross Section, and Half Underneath Plan and Half Horizontal Section through Drawer Level of Dinner Waggon.

in the sides and top. The mirror-frame below the shelf is dovetailed at the angles as illustrated by Figs. 161 and 162, and fitted tight between the end standards, and sunk into ⅛-in. rebates in the shelf and top. The frame is out of ¾-in. stuff, and stands ⅛ in. below the standards; these are 1 in. scratched in, and the ends are moulded into square finials (see detail, Fig. 160). The shelf is V-moulded on the edge and ends, and is supported by two 1-in. turned columns.

Doors.—The doors are hung with a pair of 2-in. by ¾-in. brass butts, and fitted with 2-in. brass cupboard locks, and brass mediæval drop handles; they are kept ⅜ in. below the flush of the framing, and

Fig. 171.—Half Horizontal Section of Dinner Waggon through Shelf Level.

are stopped against ⅜-in. by ¾-in. chamfered slips.

Drawers.—The drawers are dovetailed and grooved in the usual manner, and stopped against ½-in. square blocks at the back (see Fig. 163).

Cutting List.—The following cutting list shows approximately the quantity of stuff required, the actual sizes being obtained with exactness from the rod when the work is set out full size:—Carcase: One top, 3 ft. 6½ in. by 1 ft. 6½ in. by 1 in.; one bottom, 3 ft. 3 in. by 1 ft. 4½ in. by 1 in.; two shelves, 1 ft. 7 in. by 1 ft. 2½ in. by ¾ in. Cupboard top: Two rails, 3 ft. 3 in. by 2 in. by 1 in.; four runners, 12½ in. by 1¾ in. by 1 in.; two panels, 1 ft. 4½ in. by 1 ft. 1 in. by ⅜ in.; two top rails, 3 ft. 3 in. by 2 in. by ¾ in.; one cross rail, 1 ft. 1 in. by 2½ in. by ¾ in.; two end ditto, 1 ft. 1 in. by 1 in. by ¾ in. Division: Two stiles, 2 ft. 5 in. by 2 in. by 1 in.; one rail, 1 ft. 3½ in. by 4½ in. by 1 in.; one ditto, 1 ft. 3½ in.

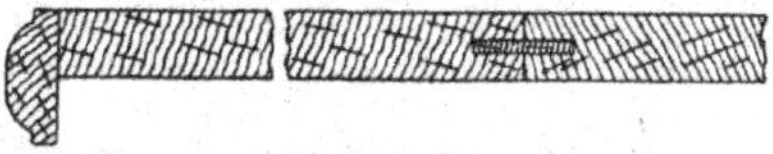

Fig. 172.—Section through Dinner Waggon Shelf.

by 3 in. by 1 in.; one panel, 1 ft. 7½ in. by 1 ft. 1 in. by ⅜ in. Ends: Four stiles, 2 ft. 9 in. by 2⅞ in. by 1 in.; two rails, 1 ft. 4 in. by 2⅞ in. by 1 in.; two ditto, 1 ft. 4 in. by 5 in. by 1 in. Panels: Eight pieces, 1 ft. 11½ in. by 3 in. by ⅜ in. Doors:

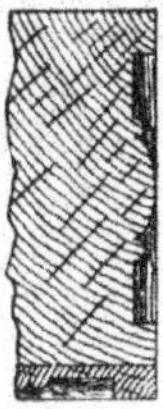

Fig. 173.—Section through Top Square of Dinner Waggon Leg.

Four stiles, 2 ft. 1 in. by 2½ in. by 1 in.; two rails, 1 ft. 7 in. by 2½ in. by 1 in.; two ditto, 1 ft. 7 in. by 2⅛ in. by 1 in.; two ditto, 1 ft. 7 in. by 3⅛ in. by 1 in.; six panels, 4 in. by 4 in. by ½ in.; ten pieces, 1 ft. 1¼ in. by 3 in. by ½ in. Drawers: Two fronts, 1 ft. 6⅝ in. by 3¼ in. by 1¼ in.; two backs, 1 ft. 6⅝ in. by 2¾ in. by ½ in.;

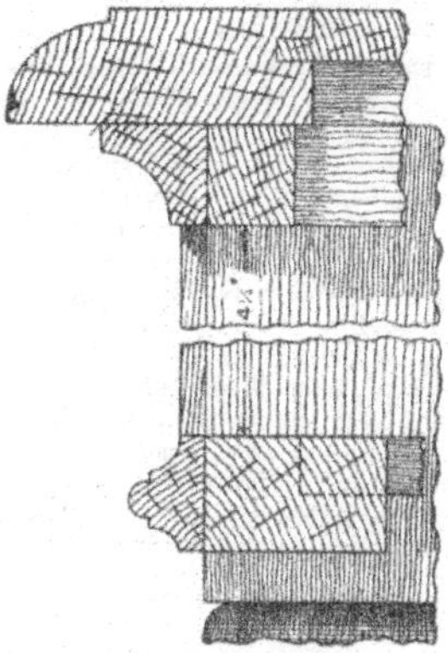

Fig. 174.—Section through Part of Dinner Waggon Top.

four sides, 1 ft. 3 in. by 3½ in. by ½ in.; two bottoms, 1 ft. 6 in. by 1 ft. 4 in. by ½ in. Back: Two stiles, 2 ft. 7 in. by

3 in. by 1 in.; one muntin, 3 ft. by 5 in. by 1 in.; one rail, 3 ft. 3 in. by 3¼ in. by 1 in.; one ditto, 3 ft. 3 in. by 7 in. by 1 in.; two panels, 1 ft. 9 in. by 1 ft. 3 in. by ½ in. Mirror frame: Two rails, 3 ft. 2½ in. by 1⅛ in. by ¾ in.; two stiles, 9 in. by 1⅛ in. by ¾ in.; one back, 3 ft. 3 in. by 9 in. by ⅜ in.; one shelf, 3 ft. 5 in. by 10 in. by ¾ in.; two standards, 1 ft. 3 in. by 1 in. by 1 in.; two columns, turned, 9 in. by 1 in. by 1 in. Gallery: One rail, 3 ft. 3 in. by 1 in. by ⅝ in.; one ditto, 3 ft. 3 in. by 1 in. by ½ in.; eighteen balusters to pattern out of ¾ in. by ¾ in. by 2½ in.; sundry strips

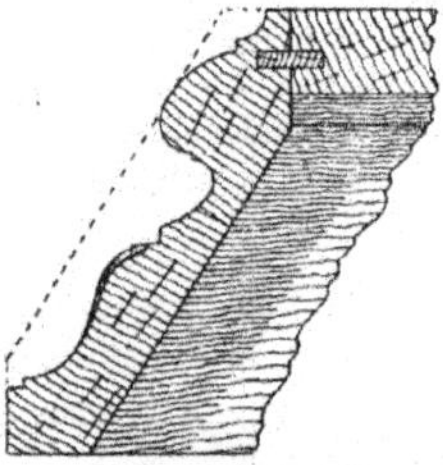

Fig. 175.—Enlarged Section through Part of Dinner Waggon Bottom.

for fillets, etc. The interior and the backs may be of deal; the bottom is edged with a 2-in. slip of hardwood; the foot pieces, 2½ in. by 1 in., are glued on to the edges of the ends. Figs. 155 to 158 are printed 1 in. = 1 ft.; Figs. 160 and 163 are 3 in. = 1 ft.; Figs 159 and 162 are 2 in. = 1 ft.; and Figs. 161 and 162 are half full size (approximate). So many exact dimensions are given in the illustrations that it is an easy matter to construct accurate scales. Fig. 164 is an enlarged vertical section through the bottom part of the sideboard.

Dinner Waggon.

Views of a dinner waggon are presented by Figs. 165 to 169. A half plan underneath and a half horizontal section are shown in Fig 170; a half horizontal

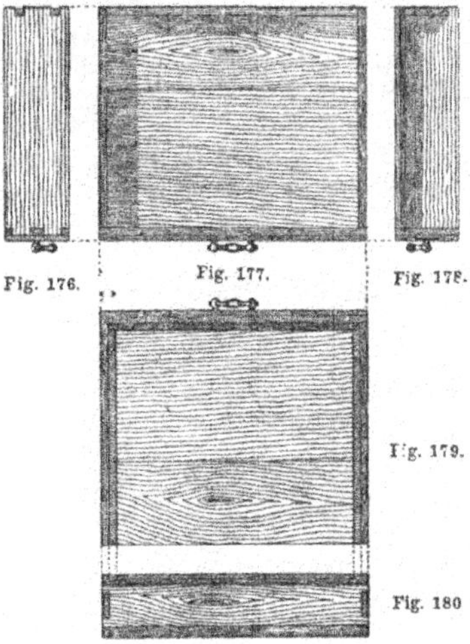

Fig. 176. Fig. 177. Fig. 178. Fig. 179. Fig. 180

Figs. 176 to 180.—End Elevation, Plan, Cross Section, Underneath Plan, and Back Elevation of Dinner Waggon Drawer.

section through the shelf level is shown by Fig. 171; details of construction are illustrated in the sectional views (Figs. 172 to 175); whilst full particulars of the drawers are given in Figs. 176 to 180. Figs. 165 to 171 are reproduced to a scale of 1 in. = 1 ft., as are also the views of the drawer above.

OVERMANTELS AND CHIMNEY-PIECES.

Dining-room Overmantel.

The overmantel shown in elevation by Figs. 181 and 182 should be made of oak or of walnut. It has three bevel-edged mirrors, two semicircular brackets shelf, and backing could be made of bass-wood, and the remainder of American satin walnut. This bass-wood, or whitewood as it is often called, will require two or more applications of stain to bring it to the same tone as the satin walnut, and

Fig. 181.

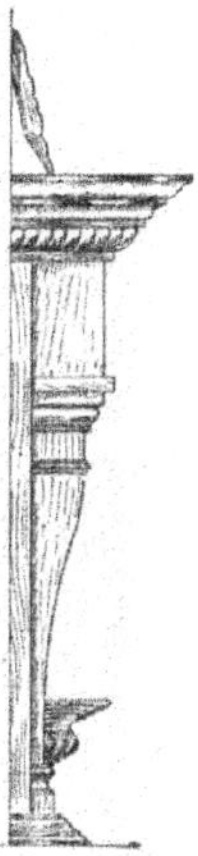

Fig. 182.

Figs. 181 and 182.—Front and End Elevations of Dining-room Overmantel.

being fixed immediately below the side mirrors, and above the glass a narrow shelf is carried the full length of the overmantel and supported by four carved and fluted pilasters. The cornice projects considerably, the top forming a wide shelf for pottery, etc. Thoroughly well-seasoned timber should be used, and if a less expensive material is desired the frame (Fig. 183), if this is carefully attended to, the difference between the two when polished will be scarcely perceptible. Both these woods bruise rather easily, so that care must be exercised to guard against injury while cramping up the work, etc. In Fig. 183, which is one half the back view, dimensions are given from the centre line. The four stiles are 3½ in. by 1⅛ in., while the rails are respectively 2½ in. by 1⅛ in., 2 in. by 1⅛ in., and 1½ in. by 1⅛ in., all mortised and tenoned together, rebated for the mirrors, and grooved for the wood panels and backing. The details of construction are clearly indicated in the sections (Figs. 184, 185, and 186), Fig. 185 being taken on A B (Fig. 184), and Fig. 186 on C D. The pilasters are attached by screws driven from the back of the stiles, and the shelf is then fixed to the pilasters and also screwed to the rail from the back. Next secure the four straight brackets. Figs. 187 and 188 show sections of the fluting and beads on the pilasters and brackets. The tops of the brackets are covered by a board 4½ in. by ½ in. by 3 ft. 8 in. long, and from this the cornice springs. The top shelf is sunk

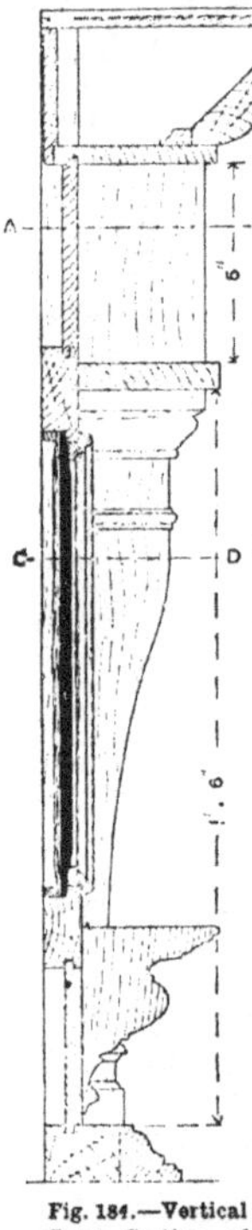

Fig. 184.—Vertical Cross Section of Overmantel through Side Mirror.

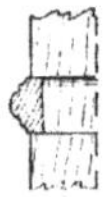

Fig. 190.

Fig. 191.

Figs. 190 and 191.—Fixing Loose Bead on Overmantel Pilasters.

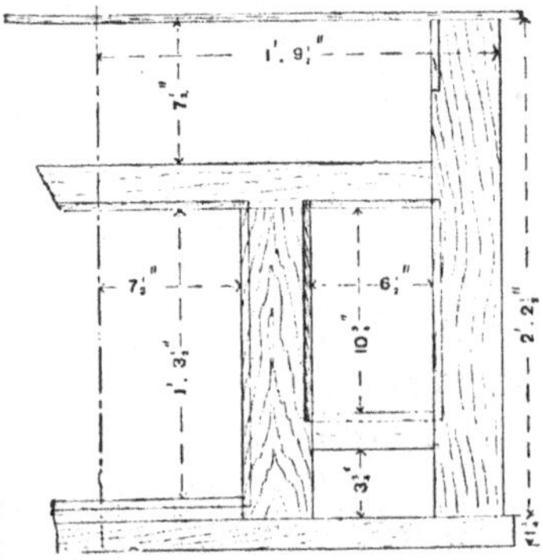

Fig. 183.—Half Elevation of Overmantel Framework.

Fig. 187.—Fluting on Pilasters.

Fig. 188.—Beading on Brackets.

Fig. 189.—Carving on Cornice.

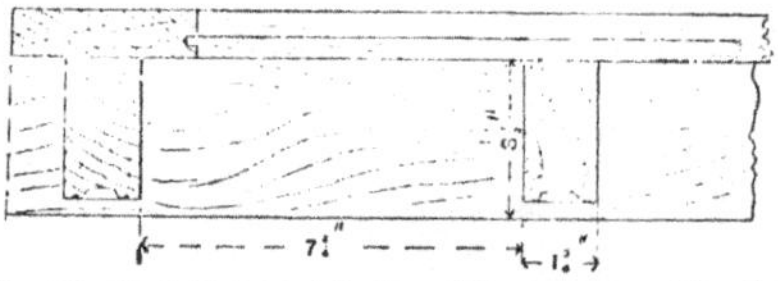

Fig. 185.—Part Horizontal Section of Overmantel on Line A B (Fig. 184).

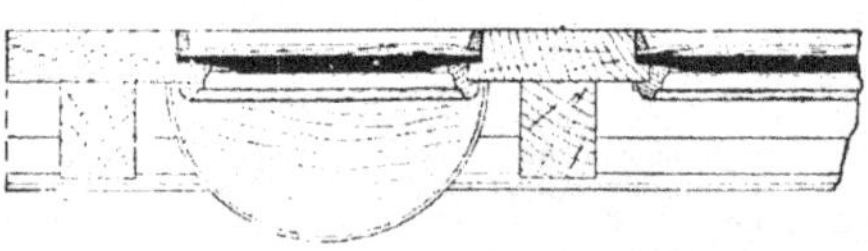

Fig. 186.—Part Horizontal Section of Overmantel on Line C D (Fig. 184).

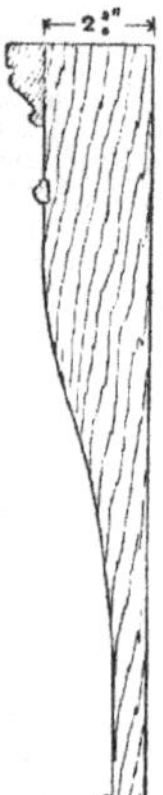

Fig. 192.—Section of Pilaster for Overmantel.

in a rebate in the top of the cornice, and ½ in. below the top edge, which serves as a stop to prevent articles sliding off. On the quarter-round ovolo part of the cornice moulding suitable ornamentation may be introduced to relieve the uniformity of so much straight work in the deep cornice, the effect being shown in Fig. 182. A suitable section is shown in Fig. 189. The pattern is made with hand-carving tools, and machine-carved egg-and-dart moulding, or dentils, may be introduced with good effect. Figs. 190 to 192 show the method of securing the small beadings that are mitered round the pilasters. The semicircular brackets are turned from one piece, which is afterwards sawn through lengthways, the ornamentation being formed

Fig. 193.—Front Elevation of Overmantel with Shelves and Turned Pillars.

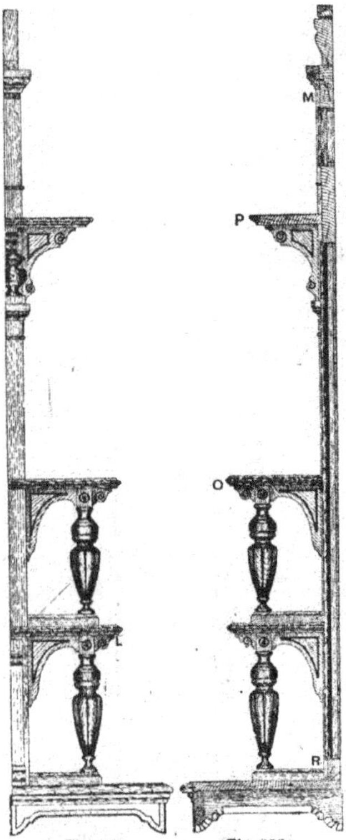

Fig. 194. Fig. 195.

Figs. 194 and 195.—Side Elevation and Cross Section of Overmantel.

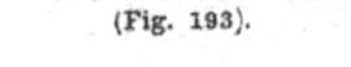

Fig. 201.—Horizontal Section of Upper Part of Overmantel on Line D D (Fig. 193).

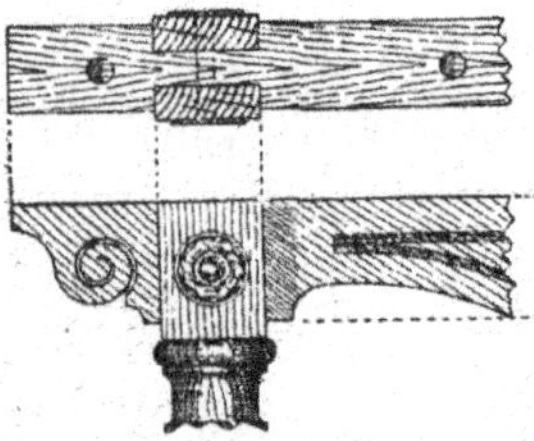

Fig. 196.—End of Overmantel Shelf supported on Pillars.

Fig. 200.—Cross Section of Overmantel on Line B B (Fig. 193).

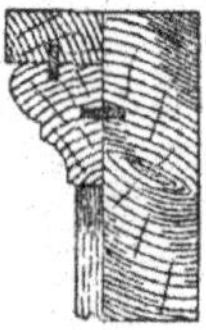

Fig. 202.—Detail of Moulding on Overmantel at M (Fig. 195).

Fig. 197.—Elevation and Half Cross Section of Pillar for Overmantel.

Fig. 198. Fig. 199.

Figs. 198 and 199.—Cross Section of Overmantel Shelves at O and P (Fig. 195) respectively.

Fig. 203.—Horizontal Section through Overmantel on Line C C (Fig. 193).

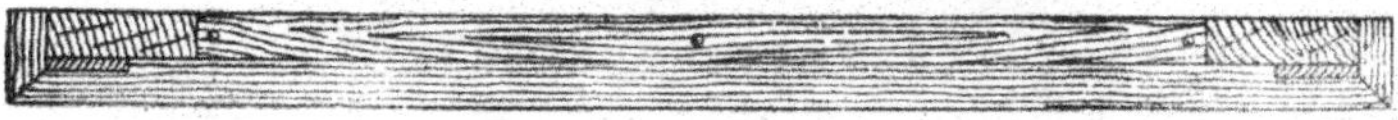

Fig. 204.—Plan of Overmantel Top with Pediment Removed.

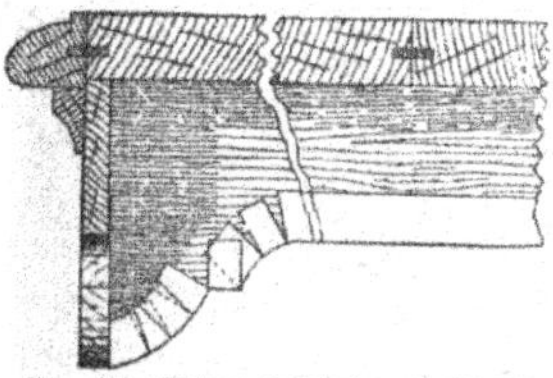

Fig. 205.—Enlarged Section of Overmantel Base on Line G G (Fig. 203).

Fig. 207.—Cross Section and Elevation of Overmantel Side Finial (see K, Fig. 193).

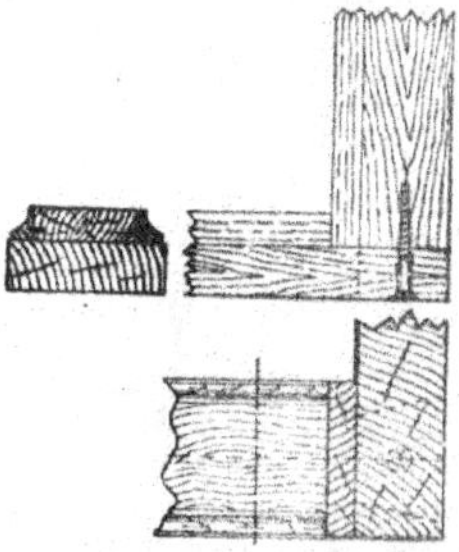

Fig. 208.—Part of Pillar Support and of Bottom of Overmantel Back (see R, Fig. 195).

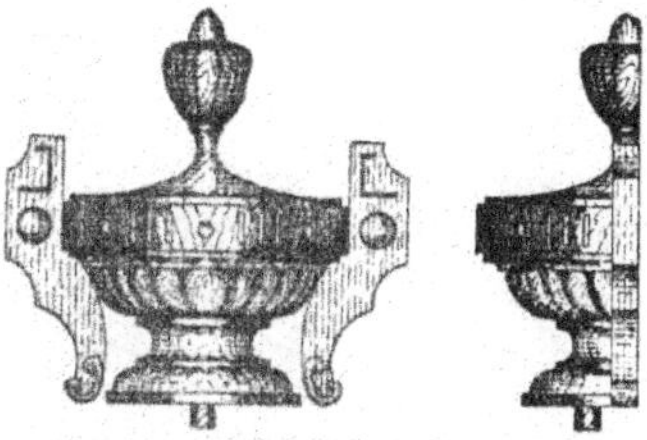

Fig. 206.—Front and Side Elevations of Overmantel Centre Finial (not shown in Fig. 193).

Fig. 210.—Detail of Moulding on Overmantel at N (Fig. 200).

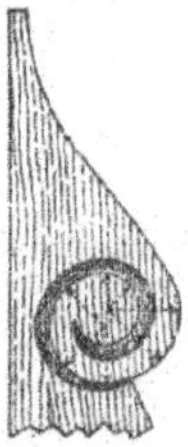

Fig. 209.—Scroll of Overmantel at Q (Fig. 193).

Fig. 211.

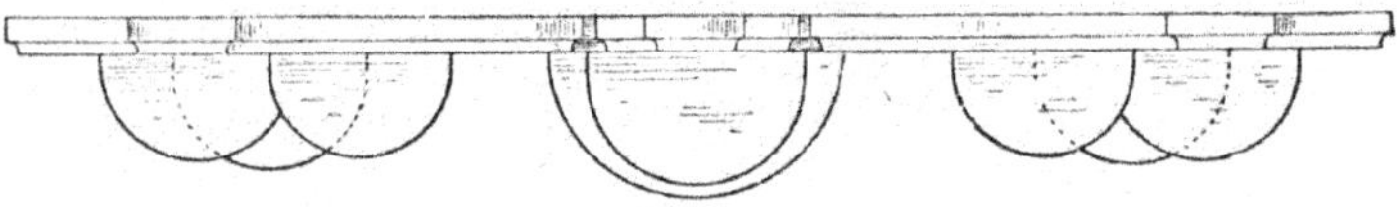

Fig. 212.

Figs. 211 and 212.—Front Elevation and Plan of Hanging Overmantel with Circular Mirror.

with the hand-carving tools. Finally, a small bolection moulding is mitered round the framing for the mirrors, while the backing and glass are retained by slips bradded on.

Overmantel with Shelves and Turned Pillars.

Figs. 193 to 195 are views of a handsome overmantel, the chief feature being the turned pillars which support the shelves. These views and the detail figures (Figs. 196 to 210) show clearly the whole of the construction. Beneath each of the detail figures (Figs. 196 to 210) is a descriptive title, and further reference to them in the text would be superfluous. It may be said that the overmantel has an extremely good effect if made in oak with dark wood mouldings on the edges of the shelves.

Hanging Overmantel with Circular Mirror.

The overmantel shown by Fig. 211 may be made of pine, painted and enamelled white or a pale shade of green, with the outer edges, ornamental lines, and edges of the shelves gilded. Fig. 213 shows the end elevation, and Fig. 214 a section through the mirror. The extreme width is 5 ft., and the height 5 ft. $1\frac{1}{2}$ in. A full-size drawing should first be made on sheets of brown or white lining paper pasted together. To secure the exact shape, make a tracing of the left-hand half of Fig. 211, and draw lines at right angles to each other to form squares of about $\frac{1}{8}$ in.; then, on the full-size drawing, space out the same number of squares to occupy 2 ft. 6 in., this being half the width of the overmantel, and get the height in the same way. The lines in each corresponding square are then copied. Should a smaller size of overmantel be preferred, decide on the width required, and space out with the same number of squares as in the tracing. The wood should be about $1\frac{1}{8}$ in. thick, the sizes of the various pieces being obtained from the full-size drawing. The top centre portion A (Fig. 211) is tenoned into the sides B as shown by the dotted lines on the right-hand side, the lower

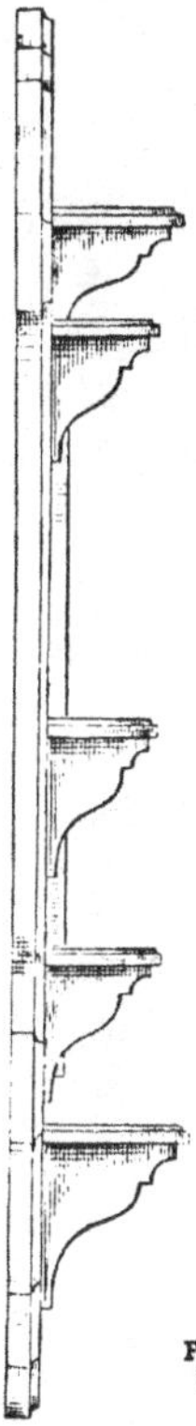

Fig. 213.—End Elevation of Hanging Overmantel.

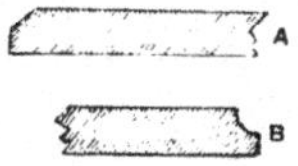

Fig. 215.—Sections of Shelves for Hanging Overmantel.

Fig. 216.—Section, showing Shape of Diagonal Lines on Overmantel.

Fig. 217.—Alternative Moulding for Overmantel Mirror.

Fig. 214.—Vertical Section through Hanging Overmantel.

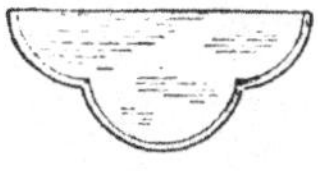

Fig. 218.

Fig. 219.

Figs. 218 and 219.—Plans of Overmantel Shelves.

centre part C being treated in the same way. To make up the corners D, separate pieces are fitted. To receive these, the sides B should have grooves about $\frac{1}{2}$ in. deep and the same width as the mortices worked on their edges. The corner pieces are then provided with a tongue to fit the grooves, the joint against the upper and lower parts being simply glued. The pieces D may be put in roughly as regards shape, and the circular opening cut to shape with a bow-saw when the framing is glued together. Care should be taken to select well-seasoned wood, or it may warp in its wider parts. A saving of material may be effected by gluing the prominent parts of the sides, top, and bottom at the joints, as shown by the dotted lines E, F, and G (Fig. 211). To accomplish this, a template or mould of thin wood or cardboard should be made from the working drawing; then, by placing the pattern on the timber, the shaped pieces are cut out of the board as desired. The prominent portions of the sides must be glued on after the sides have been mortised. The whole frame is then cramped together. After levelling the face and back of the frame, the outside shape should be marked on and then cut with a bow-saw, cleaning up with a spokeshave, file, and glasspaper. The outer edge may be bevelled as in A (Fig. 215), or hollowed with a gouge as in B (Fig. 215). The

Fig. 220.—Front Elevation of Chimney-piece Ornamented with Mouldings.

marginal lines H (Fig. 211) and the diagonals (see Fig. 216) are formed with a small gouge or parting tool. The moulding which surrounds the mirror may now be glued on, and further fixed with screws driven through from the back; it should be about $1\frac{1}{2}$ in. wide and $\frac{7}{8}$ in. thick, and should project to form a $\frac{3}{8}$-in. rebate for the glass (see I, Fig. 214). The moulding is got out in curved sections, each section being jointed and butted against the next, and when all are glued on it is turned or carved. If carving is not desired, a bevelled edge (see Fig. 217) formed with a spokeshave may be substituted. The shelves and bracket are made of $\frac{3}{4}$-in. stuff. Plans of these are shown in Fig. 218, and a side elevation in Fig. 213. Figs. 218 and 219 are alternative patterns for the shelves. The brackets are screwed from the back of the frame.

Circular Mirror.—A plain glass mirror may be used for the centre; but a bevelled one is much more effective, and to get the full benefit of the bevel, which should be $1\frac{1}{4}$ in., the glass should measure only $\frac{3}{8}$ in. more than the opening, thus taking up $\frac{3}{16}$ in. of the bevel all round. The mirror, which should be coated at the edges with lampblack, is then fixed in position with small triangular blocks K (Fig. 214) about $1\frac{1}{2}$ in. long. The blocks should be of such a thickness as to form supports

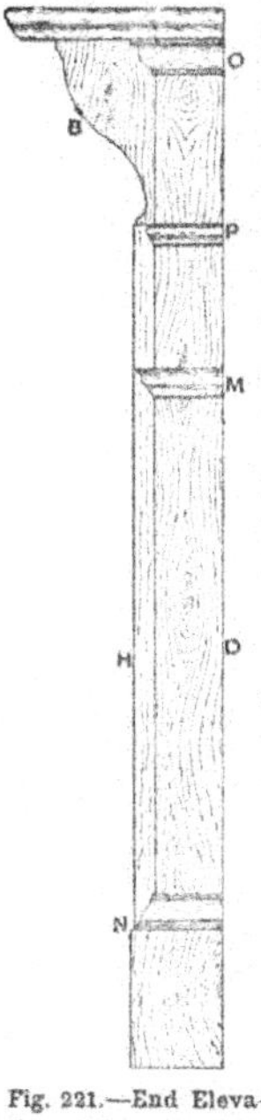

Fig. 221.—End Elevation of Chimney-piece.

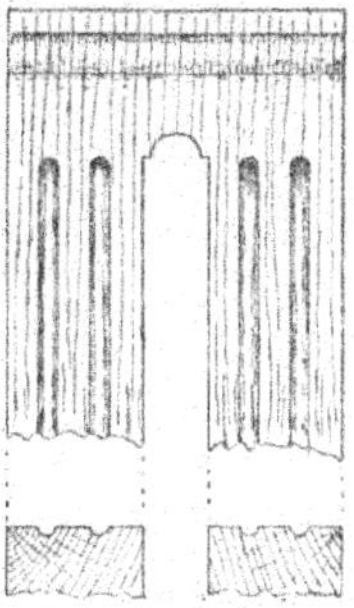

Fig. 225.—Chimney-piece Bracket.

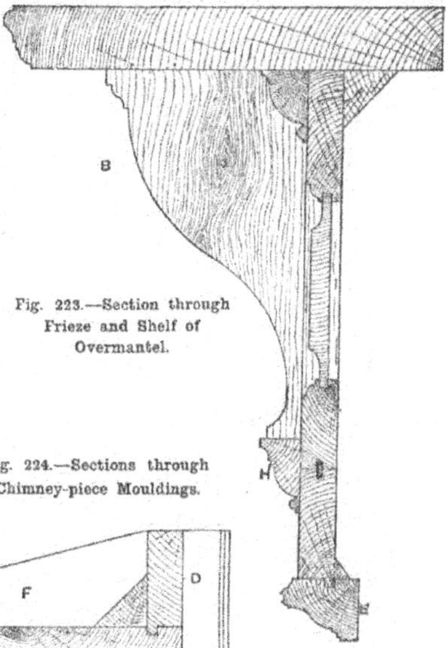

Fig. 223.—Section through Frieze and Shelf of Overmantel.

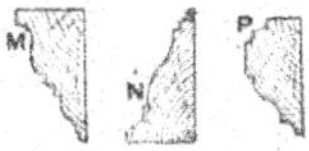

Fig. 224.—Sections through Chimney-piece Mouldings.

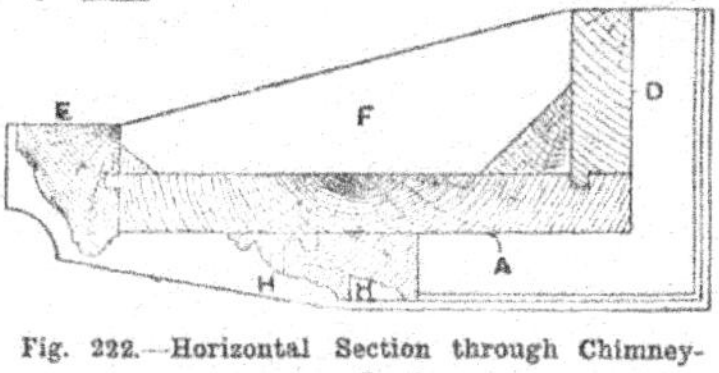

Fig. 222.—Horizontal Section through Chimney-piece Jamb.

for the ½-in. wood back L, which is secured with thin screws driven in a slanting direction into the frame. The overmantel is fixed to the wall by means of brass plates screwed to wooden plugs.

Fig. 226.—Front Elevation of Chimney-piece with Fret Ornament.

Chimney-piece Ornamented with Mouldings.

The chimney-piece shown by Fig. 220 is not complicated, and it has an effective and substantial appearance. The dimensions are: height, 4 ft. 6 in. from hearth to top of shelf; width, 4 ft. 8 in. over the jambs; width of frieze and shelf together, 1 ft. 4 in.; the shelf is 5 ft. 10 in. long, 11 in. wide, and 1½ in. thick; the jambs are 9 in. wide over all, and the opening is 3 ft. 2 in. by 3 ft. 2 in. Fig. 220 shows the front elevation, and Fig. 221 the end elevation. Two boards A A (Fig. 220, and shown in section in Fig. 222) form the jambs; these are 7½ in. broad (without the tongue), and they extend from the top of the base to the under side of the shelf. The panelled frieze (shown in section in Fig. 223) is made to fit between the jambs, and is neatly jointed and fixed with dowels and glue; there is only 2 in. of the joint seen (C, Fig. 220), the rest of it being hidden behind the bracket B. The panelled frame shows a margin of 1 in. all round; the top rail is wider by 2 in. than the margin shows, to accommodate the neck moulding (see section Fig. 223), and the bottom rail requires an extra width of form the ends of the chimney-piece (see section, Fig. 222, and side elevation, Fig. 221). The plinth or base of the jambs is a solid piece of wood, represented by the outer lines of Fig. 222, 10½ in.

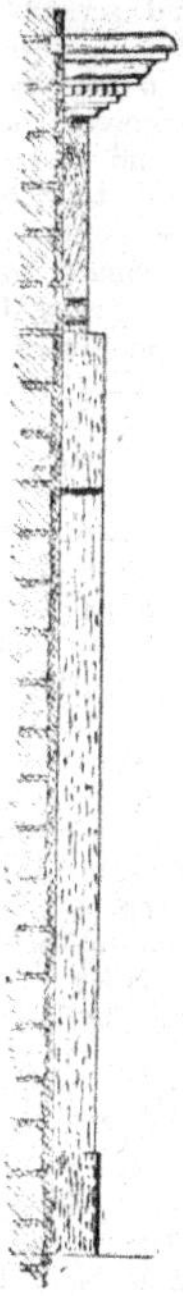

Fig. 227.—Side Elevation of Chimney-piece with Fret Ornament.

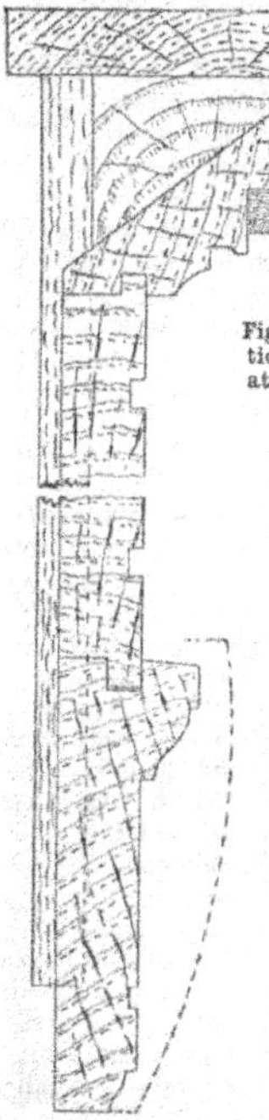

Fig. 228.—Section of Chimney-piece on Line A B (Fig. 226).

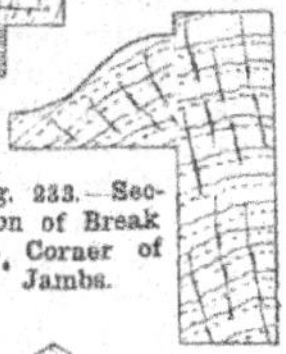

Fig. 233.—Section of Break at Corner of Jambs.

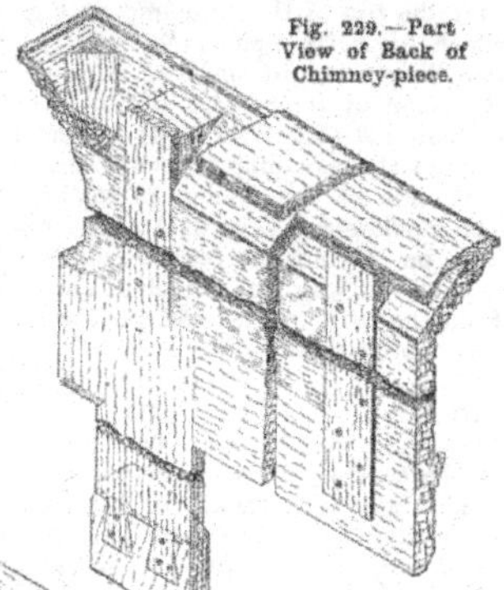

Fig. 229.—Part View of Back of Chimney-piece.

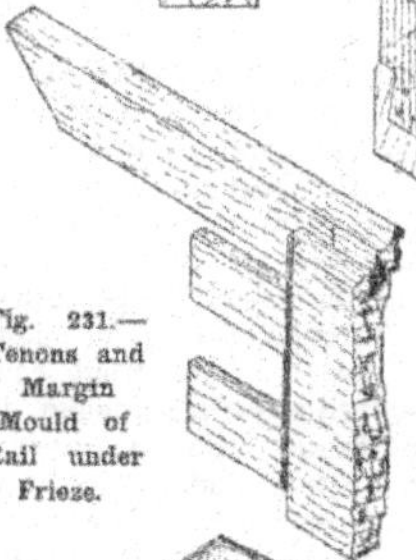

Fig. 231.—Tenons and Margin Mould of Rail under Frieze.

Fig. 230.—Return End of Mantelshelf.

Fig. 232.—Top Corner of Jambs.

3½ in., and is tongued into the moulding E; this rail can be made in two pieces if more convenient, as shown in Fig. 223, and the stiles are also kept the extra width required to pass behind the brackets B. Two pieces D, 2½ in. broad, the same length as the jambs, are tongued into them to broad, 7 in. high, and 4½ in. thick. For fixing this a stump should be allowed to project up behind the jambs, filling the space F (Fig. 222).

The Mouldings.—The moulding E is planted round, mitered at the angles, and stopped against the block or base; it must be well glued and angle-blocked behind, as shown in section. The moulding H is planted upon the frieze and jambs, showing a 2-in. margin between it and the moulding E. A break of 1 in. is made over the top of the moulding M (Fig. 224), which helps the appearance greatly. This moulding is glued on, and screwed from the back of jambs and frieze. The base moulding N (Fig. 224) butts on the square edge of the moulding H, and returns round each side of the jambs, as N (Fig. 221). The mouldings M (Fig. 224) and P (Fig. 224) butt on H, and are returned on the ends in the same way; all are glued and screwed from the back of the jambs. The brackets B supporting the shelf are 9½ in. long, 5 in. broad, and 5 in. thick (Fig. 225 shows part enlarged elevation and a section of the edge); they have an open space of 1 in. in the centre, and need not be solid. The easiest way is to make two brackets, 2 in. in thickness, and glue a 1-in. strip, shaped as shown, between them, at the top only; the part plan of the edge shows the fluting. These brackets are fitted between the shelf and the moulding H, and are glued and screwed from the back of the frieze; a 2½-in. screw is also put through the open space in the centre of the bracket into the shelf; the shelf is also well screwed and angle-blocked from the inside. The neck moulding O, shown in section and elevation, butts on the brackets on each side, and returns on the sides of the jambs. The sizes of the mouldings are as follows:—Neck moulding O, 2 in. by 1¼ in.; moulding P, 1¾ in. by ¾ in.; moulding M, 2 in. by ⅞ in.; base moulding N, 2 in. by 1 in.; moulding H, 2 in. by 1 in.; moulding E, 2 in. by 1½ in.

Chimney-piece with Fret Ornament.

The chimney-piece shown in front elevation by Fig. 226 and in side elevation by Fig. 227 is suitable for a large room furnished in oak in the Classic style. The fret ornament sunk in square in the frieze breaks up the large plain surface of the deep frieze-board, and harmonises with the stopped sinking running round the interior edge of the under portion. The centre piece, shaded dark, is preferably inlaid with ebony or a rich-coloured walnut, according to taste. The plinth blocks are round-faced, as indicated by the dotted line in Fig. 228, the jambs being double-dovetailed, housed in solid, except the front edge, and well glued and screwed to blocks. The cornice above the frieze is dentilled, and finished with a mantel-shelf as shown. Fig. 228 represents a section on line A B (Fig. 226), and shows how the frieze-board is tongued to the under portion and also into the cornice; the cornice at the top being tongued to the mantelshelf. Fig. 229 is a conventional view from the back, showing the details of the plinth blocks, and also showing how the jambs are cut and continued up till they reach the under side of the mantel-shelf, into which they are tenoned about ⅝ in., the cornice being glue-blocked to this extension as shown. Between the two extensions two other pieces are partly housed in, and are well screwed to the frieze-board and to the under portion. These also tenon into the shelf, and are blocked in the same way. The cornice is mitered at the corners and well secured, the corner block being glued in as shown. The mantelshelf is half rounded on the front edge, and ploughed for the cornice tongue, the ends having tongued to them a return piece, which is blind-nailed and glued to the main shelf (see detail Fig. 230). The under portion of the chimney-piece is double-tenoned, and the margin mould is worked in the solid as shown in Fig. 231, the ends of this mould running across the tops of the jambs to mitre with the returns that break out from the upright moulds on the jambs as detailed in Fig. 232. The moulds on the jambs are preferably worked solid, but can be planted on—that is, glued, and screwed from the back. The break at the corner, shown enlarged in Fig. 233, is worked in the solid, or built up in two pieces, to conceal the end grain of the overhang of the jamb underneath (see Fig. 229). At the top, also, a piece must either be planted or left on the bottom external corners of the frieze-board (see Fig. 229). The fitting together should be done with extreme care, and the screwing, gluing, and blocking should be thoroughly workmanlike. The mantelpiece may be screwed to the cornice with brass screws filed flush, or may be blind-nailed. Half of the fret ornament should be drawn full size on stiff tracing paper, and reversed to mark the other half. Where the fret band crosses on the diagonal lines, the sinking should be somewhat deepened. Alternatively, the design might be executed in whitewood, enamelled white or cream, with the sinkings and centrepiece finished in gold leaf. The scale of Figs. 226, 227, and 229 is ¾ in. to 1 ft.; that of Fig. 228 is 3 in. to 1 ft.; and that of Fig. 233 is half full size; while Figs. 230, 231, and 232 are reproduced to the scale of 1½ in. to 1 ft. The above scales are approximate.

Fig. 234. Fig. 235.

Figs. 234 and 235.—Front and End Elevations of Chimney-piece and Overmantel.

Chimney-piece and Overmantel.

The chimney-piece and overmantel shown by Figs. 234 and 235 was designed and executed for a study. The material is wainscot oak of selected figure, and fumigated to match the furniture. The total height is 9 ft. 11½ in., and the width 5ft. 1 in., exclusive of projection. Full details are shown by Figs. 236 to 242. The mantel-shelf runs level with the top of the dado rail fixed round the room, the section of the moulding on the edge of the mantel-shelf corresponding with the moulding on the upper part of the dado rail. The walls of the study are covered with an ingrain paper of dark apple-green tint; the picture rail is of oak; the cornice round the ceiling is tinted to match the paper and woodwork.

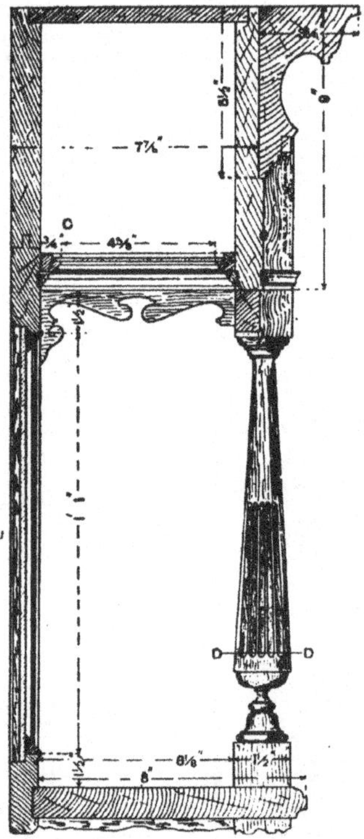

Fig. 236.—Section of Upper Part of Chimney-piece and Overmantel.

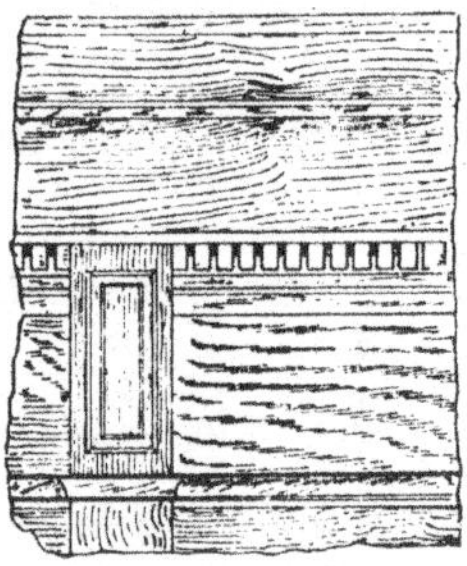

Fig. 237.—Enlarged Part Front Elevation of Overmantel at A (Fig. 234).

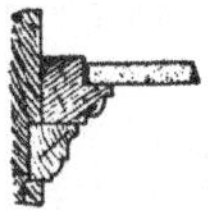

Fig. 238.—Enlarged Detail of Overmantel at C (Fig. 236).

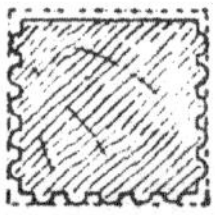

Fig. 239.—Enlarged Section of Overmantel Pillar at D D (Fig. 236).

Fig. 240.—Plan of Overmantel Shelf.

Chimney-piece.—The chimney-piece is constructed of 1¼-in. wainscot, with moulded base and twin trusses; these are shaped as shown, with sunk moulded panels in the shaped part. A small astragal moulding, ⅜ in. wide, is housed into and mitered

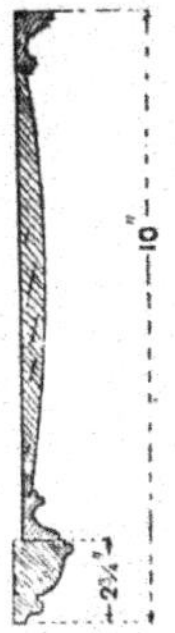

Fig. 241.—**Part Vertical Section of Overmantel Top at B (Fig. 234).**

round the truss at the springing of the shaped part, a space of 1 in. being left between this and the neck moulding. The moulding is 1½ in. wide, and is housed and mitered round the truss. A space of 4½ in. is left between the truss and the cap, the latter being formed with a moulding 2¾ in. wide, prepared for and carved into an egg-and-dart moulding supporting the mantel-shelf, which is 1¼ in. thick. The space between the neck and cap moulding on the truss is ornamented by five ⅜-in. reedings with ¼-in. projection, the space of 3 in. being divided equally. The lower moulding or plinth forms the base. The frieze between the mantel-shelf and marginal moulding is planted on the face of the frame; this is swelled and returned at the ends, the returns showing the same as the face; it stands on the top edge of the marginal moulding fixed round the opening to fireplace. Sienna marble slips are fixed between the oak moulding and the stove. The moulding on the mantel-shelf is formed partly on the shelf itself, the 2-in. thickness being made up by an additional moulding tongued to it; this makes the edge bold, and also acts as a clamp at the ends to prevent the shelf warping.

Overmantel.—The overmantel is 5 ft. 10 in. high by 5 ft. wide, and is fitted with six shelves, three on each side. These six shelves are carried by scrolled brackets and square-turned and reeded intermediate pillars, and are shaped and moulded on the edges, the diminished end finishing on the muntin. A plain 1-in. by ½-in. fillet projecting ¼ in. is fixed into a groove in the muntin. The fillet is cut away where the shelves come, to allow them to fit up to the frame, which is put together and formed into open panels, the dimensions being as given in the front elevation. A bolection moulding is mitered round the panel, and fixed to receive the glass, which is of plain polished silvered plate;

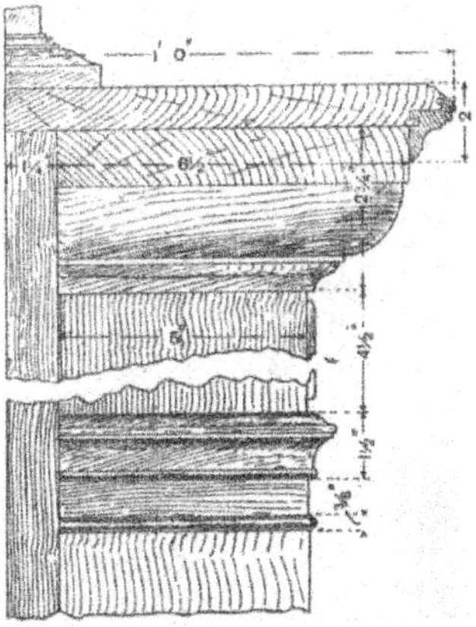

Fig. 242.—**Enlarged Detail of Mantelpiece and Pilaster (see E, Fig. 235).**

bevelling was objected to on account of the prismatic colours which frequently show. The frieze and cornice are built up as shown; the lower part of the frieze has a small moulding as a necking, the plain edge being relieved by small scrolled aprons

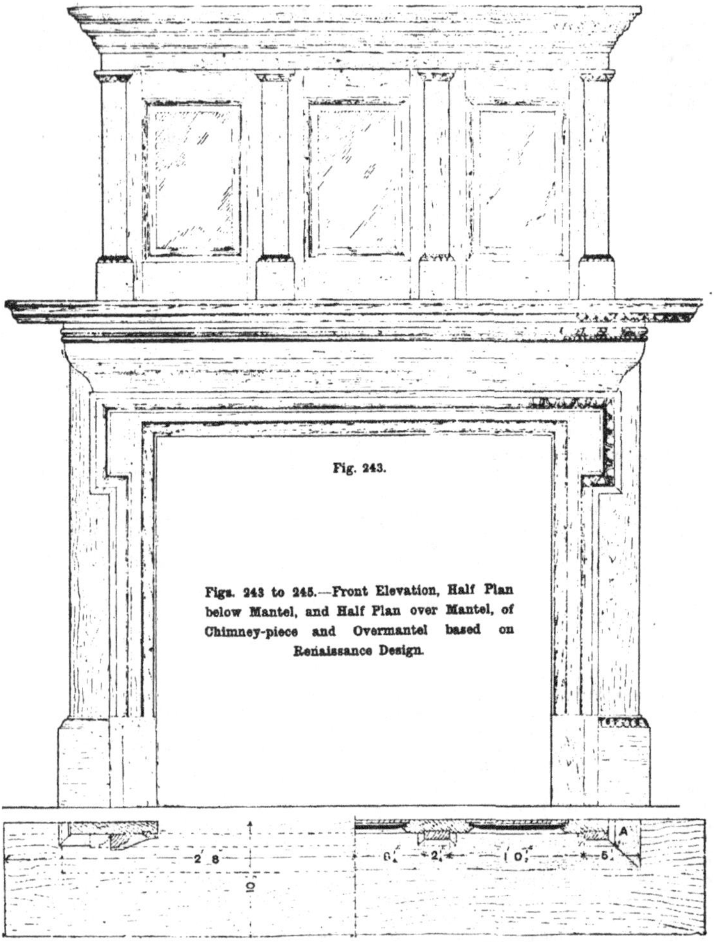

Fig. 243.

Figs. 243 to 245.—Front Elevation, Half Plan below Mantel, and Half Plan over Mantel, of Chimney-piece and Overmantel based on Renaissance Design.

Fig. 244. Fig. 245.

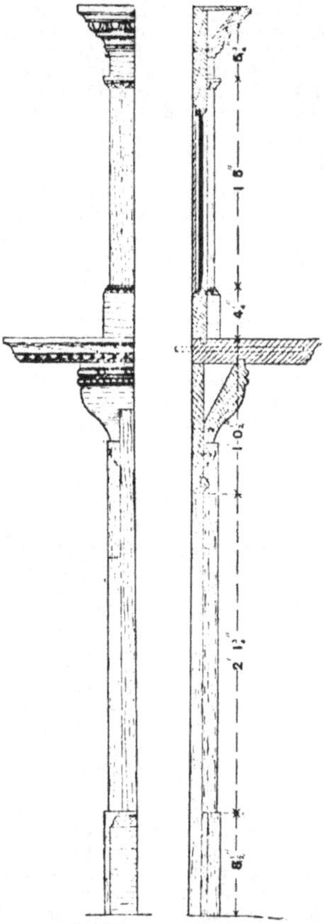

Fig. 246. Fig. 247.

Figs. 246 and 247.—End Elevation and Cross Central Section of Chimney-piece and Overmantel.

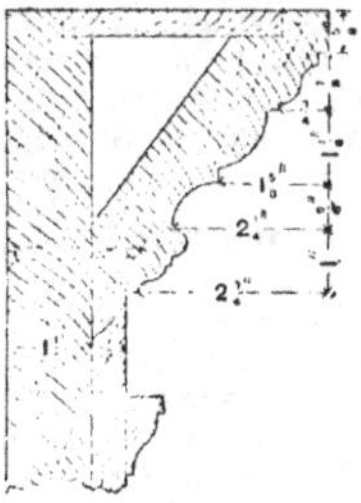

Fig. 248.—Section of Overmantel Cornice.

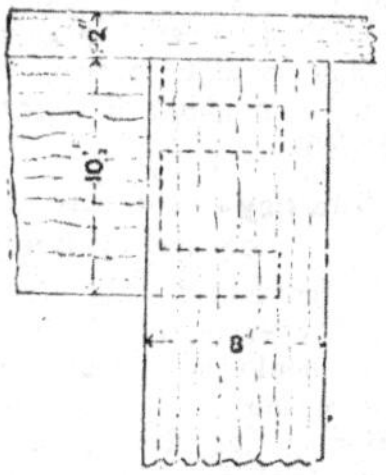

Fig. 250.—Joint of Chimney-piece Jambs.

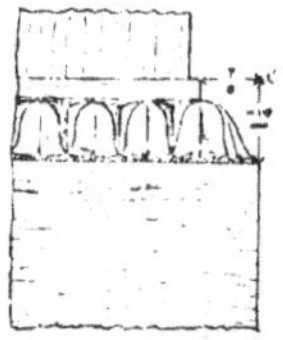

Fig. 251.—Detail of Chimney-piece Plinth.

Fig. 252.—Enlarged Detail of Overmantel at A (Fig. 245).

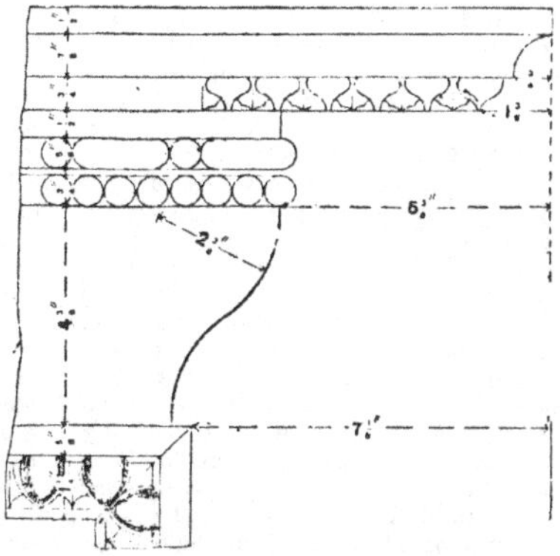

Fig. 249.—Detail of Mantel Mouldings.

fixed to it. The soffit is formed with a piece of silvered plate-glass fitted into a small moulded frame, which is supported on a moulded fillet grooved into it as shown.

Chimney-piece and Overmantel based on Renaissance Design.

Figs. 243 to 252 show a chimney-piece and overmantel of Renaissance character. It should be executed in dark mahogany or walnut. Fig. 243 shows a front elevation; Fig. 244 a half-plan below the mantelpiece; Fig. 245 a half-plan above the mantelpiece; Fig. 246 an end elevation; Fig. 247 a vertical section through the centre; Fig. 248 an enlarged section of the cornice, etc.; Fig. 249 an enlarged elevation of the mantelpiece and details of the mouldings; Fig. 250 an elevation of the head and jamb of the chimney-piece showing the method of making the joint; Fig. 251 an enlarged detail of the plinth of the chimney-piece; and Fig. 252 an enlarged section at A (Fig. 245). The following is the required cutting list:—

Cutting List.—Overmantel: Mantel back, two stiles, 2 ft. 4½ in. by 4⅝ in. by 1¼ in.; two muntins, 2 ft. 4½ in. by 7 in. by 1¼ in.; three bottom rails, 1 ft. 2 in. by 4¾ in. by 1¼ in.; three top rails, 1 ft. 2 in. by 4¾ in. by 1¼ in.; three pieces of deal to joint on these, 1 ft. 2 in. by 4½ in. by 1¼ in.; three pieces of silvered plate with ⅝-in. bevelled edges, 1 ft. 3 in. by 8½ in. by ⅜ in.; one frieze, 3 ft. 11 in. by 1⅞ in. by ⅝ in.; four pilasters, 1 ft. 9½ in. by 2¼ in. by ⅝ in.; one cornice, 5 ft. 4 in. by 5 in. by 1½ in.; one cover-board, deal, 4 ft. 4 in. by 3½ in. by ⅜ in.; one plinth piece, 2 ft. 2 in. by 4½ in. by ⅝ in.; one piece for necking, 2 ft. 2 in. by 1¼ in. by ¾ in. Chimney-piece: One mantel-board, 5 ft. 4½ in. by 11 in. by 2 in.; one head-piece, 4 ft. by 4½ in. by 1 in.; piece of deal to joint on 6¼ in. wide; two plinth pieces, 7 in. by 8½ in. by 1¼ in.; 10-ft. run of 2 in. by 1¼-in. echinus moulding; two jambs 3 ft. 9 in. by 8½ in. by 1 in.; one bed-mould, 5 ft. 3 in. by 8 in. by 2 in.; two plinth blocks, 8½ in. by 4½ in. by 1½ in.; 7 ft. run of 1⅛-in. by ¾-in. double ogee moulding.

Construction of Chimney-piece and Overmantel.—The jambs and head are framed together first as shown in Fig. 250, a pair of $\frac{5}{16}$-in. stub tenons being used. These are well glued and screwed from the back. The ogee border moulding is rebated as shown in Fig. 247 and the frame grooved to receive it. This is fitted tight and glued in. The plinth blocks are glued and screwed from the backs, as is also the carved ovolo moulding. The mantel-piece has the end mouldings returned in the solid, and is stiffened with three ⅜-in. iron bolts as shown in the section (Fig. 247). These may be left projecting 3 in., and may be cemented into the wall. The piece is secured to the head with screws counter-sunk from the top. A ⅝-in. groove should be made in the under side to receive the tongue of the bed-mould, and this must be stopped 6 in. from each end. The bed-mould is fixed to the back first with screws, and the mantel dropped on it. The return ends of the bed-mould are mitered on, and a cross tongue should be grooved into the joints; at the back end the moulding finishes partly against the face and partly running over the edge; alternatively the carved beads may be got out separately and sunk into grooves ¼ in. deep in the bed-mould. The back of the overmantel is framed up in one piece, the inner stiles being shams; they are slot-mortised over the rails. All the tenons are stopped and screwed from the back. The framing is double-checked, once for glass and once for the wood panel, as shown in the detail Fig. 252. The pilasters are all sunk ⅛ in. into the back, and glued in; the two outside ones should be rebated as shown, and a good joint made at the outside before fixing them. The plinth and necking should be mitered round and fixed. The plinths should be sunk in ⅛ in., and glued on, the fronts first, and allowed to dry, then the end pieces fitted and glued to them. The outside pilasters require their neckings and plinths to be carried on flush with the back side of the framing. The frieze is next fitted. This is simply glued on the face of the framing on the top of the pilasters. The cornice is worked out of a parallel piece of stuff

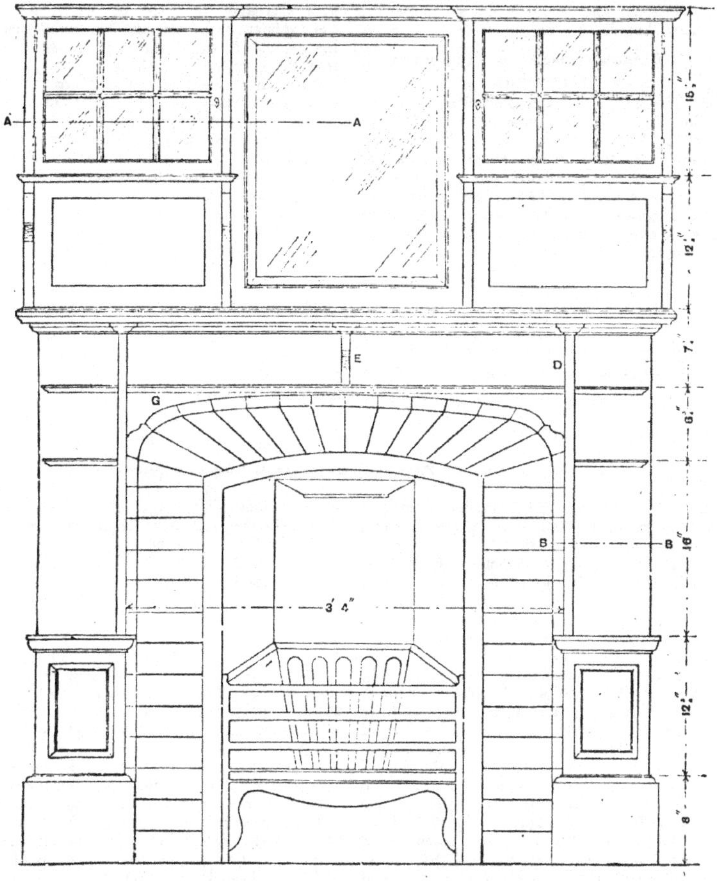

Fig 253.—Elevation Plan of Chimney-piece and Cupboard Overmantel.

as shown in Fig. 248, and fixed with screws from the back. It is lipped $\frac{3}{16}$ in. over the frieze, and rebated out $\frac{3}{8}$ in. for the cover-board.

Chimney-piece and Cupboard Overmantel.

The design for a mantel fitment shown by Fig. 253 comprises a chimney-piece with framed jambs supporting elliptic shelves in the corners, and an ogee central shelf below the mantel-board. The overmantel contains a bevel-edged mirror and a panelled back, the wings being fitted with cupboards having glazed doors; the cupboards surmount shelves which rest shaped ends of the cupboards are housed $\frac{1}{4}$ in. into the mantel-board, as shown in Figs. 254 and 256; the bottom shelf of the cupboard is housed into the standards, the moulded edge running across the front and mitering with a return piece planted on the face of the standard. The tops run over the standards, and are moulded in the solid, the standards being housed into them and nailed. The mirror is framed into the back as shown in Figs. 255 and 258, being fixed with sprigged fillets. Should the wall be at all damp, it would be advisable to brad on an additional deal back, but in ordinary cases painting is sufficient. Fig. 259 shows a

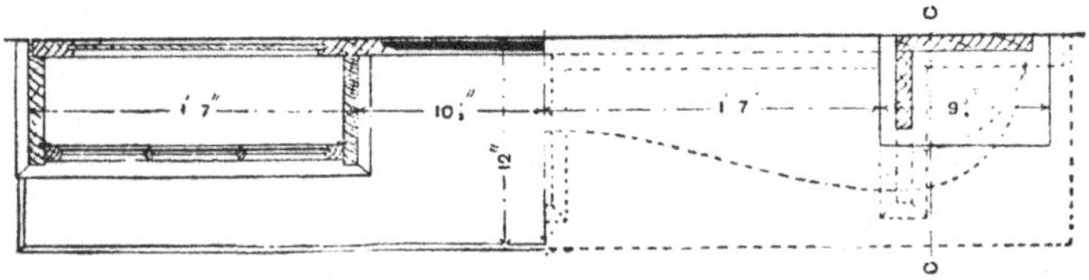

Fig. 254.—Half Sectional Plans of Chimney-piece and Cupboard Overmantel.

on the cupboard ends, which are shaped into brackets. The size of the opening for the fireplace is 3 ft. 4 in. by 3 ft. 6 in., and the outside dimensions of the fitment are—Height, 6 ft. 6½ in.; width, 4 ft. 11 in.; greatest projection, 12 in. The design would look equally effective in fumigated oak, or yellow pine stained and polished. Fig. 254 represents a plan showing on the left a half section through A A (Fig. 253), and on the right a half section through B B (Fig. 253); the dotted lines indicate shelves and mouldings above the line of section. Fig. 255 is a vertical section on line C C (Fig. 254), and shows the general construction. Fig. 256 shows an enlarged detail section through the cupboard and parts above mantel-board. Fig. 257 shows a horizontal section of the same parts. The doors, which are made of 1-in. stuff, dovetailed at the angles, are rebated and moulded with a $\frac{3}{8}$-in. by $\frac{3}{16}$-in. lamb's-tongue. The back framing is of 1-in. stuff, with $\frac{1}{2}$-in. sunk square panels. The vertical section through the base and surbase of the jamb; the surbase is formed with a hollow boxing, having $\frac{3}{4}$-in. panelled framing in front, 1-in. plain ends, and the 1-in. jamb at the back. A $\frac{1}{2}$-in. board forms the top, oversailing and forming part of the ovolo moulding planted round the front and ends. A $\frac{3}{4}$-in. by 8-in. ovolo moulded plinth forms the base, and this is screwed and blocked to the framing. Fig. 260 represents a horizontal section through the surbase, and shows two methods of construction, that on the left being suitable for painted work, that on the right for polished hard woods. Fig. 258 gives a sectional elevation (to a somewhat smaller scale than the other details) of the parts immediately below the mantel-board, E indicating the central bracket, F the shaped side of the jamb, and G the head lining. Fig. 261 illustrates the method of fastening the head lining to the jambs by a slot dovetail; the shelf H (Fig. 258), being housed both in the back and ends, must be notched back at the front edge sufficiently to let it come forward and clear the back lining while the latter is being driven into place; afterwards the shelf is knocked back into the groove and nailed through the jambs, the corner shelf hiding the nail holes. The jamb F and the bracket E should be dovetail grooved into the mantel-board to keep it from casting; and the stuff for this board should be specially selected, and cut radial to the annual rings. Figs. 253, 254, and 255 are reproduced to a scale of 1 in. to 1 ft., the details to a scale of 3 in. to 1 ft., with the exception of Fig. 258, which is to the scale of 2 in. to 1 ft.

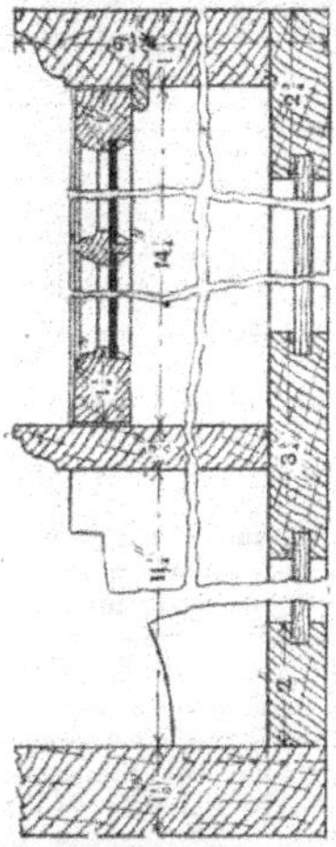

Fig. 256.—Detail of Overmantel Cupboard.

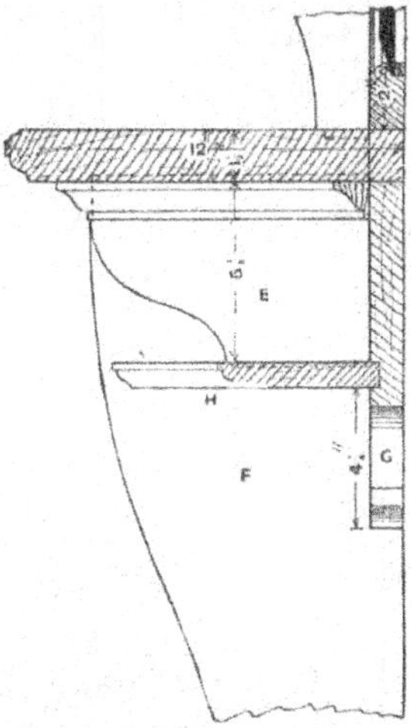

Fig. 258.—Section through Mantel-board and Shelf.

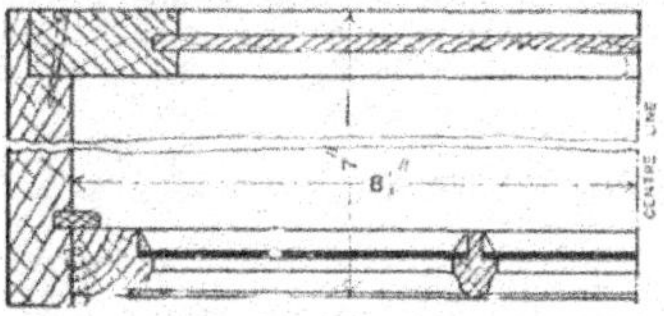

Fig. 257.—Half Plan of Overmantel Cupboard.

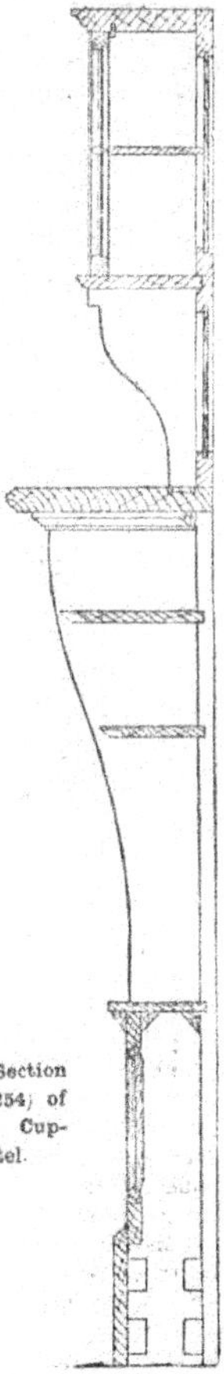

Fig. 255.—Vertical Section on Line C C (Fig. 254) of Chimney-piece and Cupboard Overmantel.

9-ft. Chimney-piece for Drawing-room.

Figs. 262 and 263 show a design which is suitable for the drawing-room of a large house, and is intended to be executed in dark-coloured mahogany or walnut. The extreme dimensions are: Width, 9 ft.; height, 7 ft. 6 in. Fig. 262 shows the front elevation; Fig. 263, an end elevation; Fig. 264, a half-sectional plan through the cupboard above the mantel-board; Fig. 265, a similar half-plan through the lower cove; Fig. 266, a half-plan of the central portion; Fig. 267, a section through one of the wings; Figs. 268 and 269, enlarged detail section through centre of cupboard, etc.; Fig. 270, a detail section through overmantel and chimney frame; Fig. 271, an elevation of the top corner of the cupboard door, showing its construction and the method of mitering the bars; Fig. 272, a part horizontal section of the cupboard door; A (Fig. 273), a vertical section through the side of the cupboard and mantel-board, showing the method of connecting them; and B (Fig. 273), the plan of the same parts, the side of cupboard being omitted, to show the grooves. A and B (Fig. 274) represent respectively the elevation and plan of the opposite side, showing the break in the mantel-board. Fig. 275 shows a section of the moulded edge of the lower brackets, and Fig. 276 the moulded edge of the shaped shelves. The panelled backs to the wings are made in one piece on each side from floor to mantel, the lower panels being moulded in the solid and the upper panels square sunk. The chimney frame is made up separately, and tongued to the winged framing as shown in Fig. 266. The overmantel back is framed up in one piece (see Fig. 264), the middle portion being double rebated for glass and back board, as shown in Fig. 270, the moulding around the mirror being planted on the face. The cornice is built up in three pieces (the lowest member being worked upon the edge of the soffit), and is carried round the square columns, which stand diagonally, as shown in the plan (Fig. 264). This position causes the salient mitre to

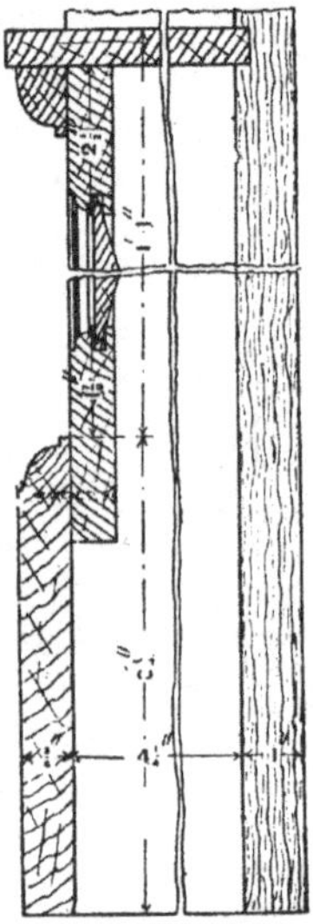

Fig. 259.—Base of Chimney-piece Jamb.

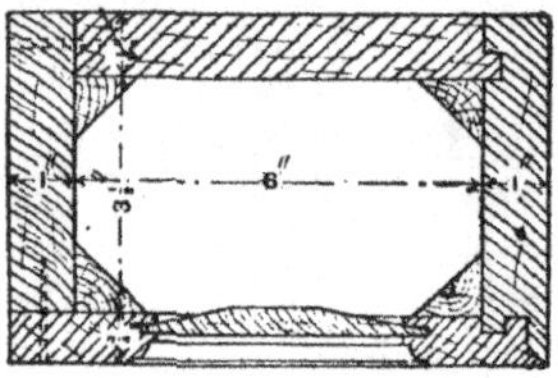

Fig. 260.—Section of Chimney-piece Surbase.

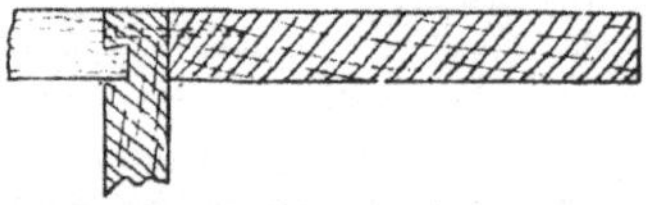

Fig. 261.—Detail of Joint in Chimney-piece (see D, Fig. 253).

appear as a straight line in the elevations. The columns are carried through the cornice, finishing flush with the under side of the cover board. They are bored to receive a $\frac{3}{4}$-in. dowel, turned on the spherical finials. The sides of the cupboards are housed into the soffit of the cornice, to which they are nailed as shown in Fig. 271; a dovetail tongue, worked on their lower ends, fitting into a similar groove

Fig. 262. Fig. 263.

Figs. 262 and 263.—Front and End Elevations of Drawing-room Chimney-piece.

Fig. 264. Fig. 265.

Figs. 264 and 265.—Half Horizontal Sections of Chimney-piece on Lines B B and C C (Fig. 262).

Fig. 266.—Half Horizontal Section of Chimney-piece on Line D D (Fig. 262).

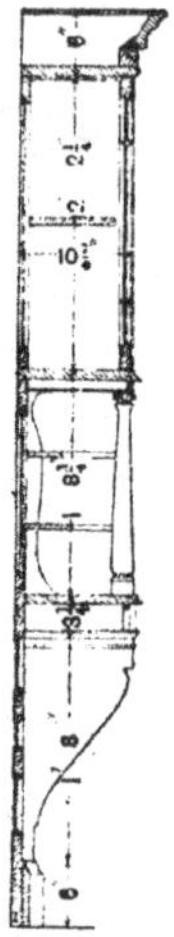

Fig. 267.—Cross Sectio of Chimney-piece on Line A A (Fig. 264).

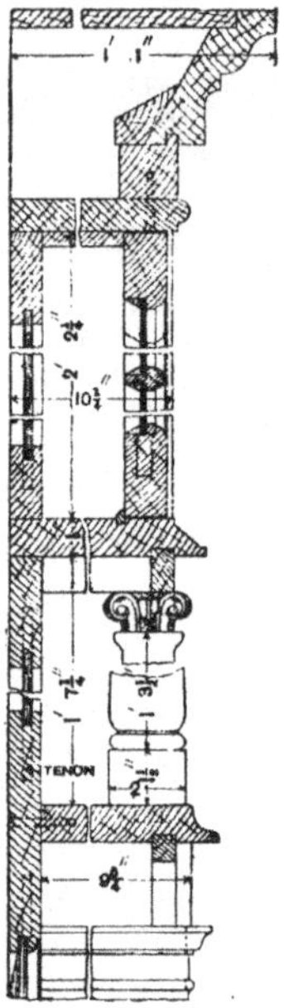

Fig. 268.

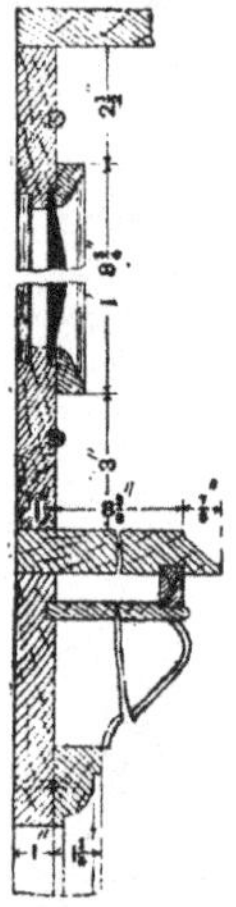

Fig. 270.—Central Vertical Section of Chimney-piece.

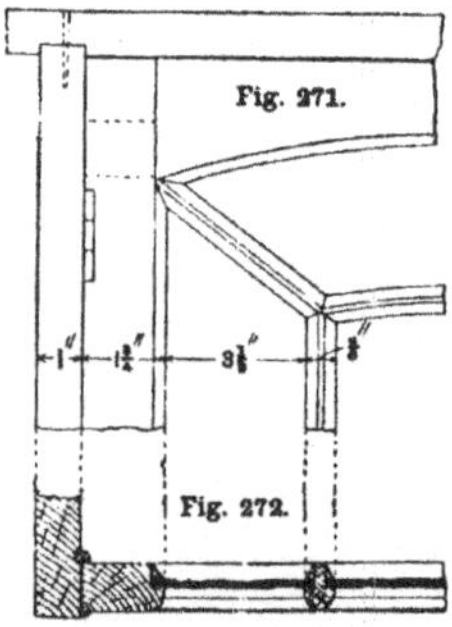

Figs. 271 and 272.—Enlarged Details of Corner of Chimney-piece Cupboard Door.

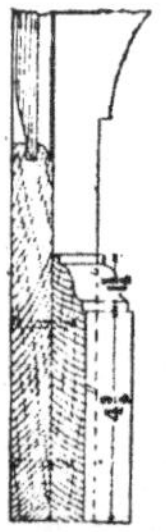

Fig. 269.

Figs. 268 and 269.—Enlarged Vertical Cross Sections through Chimney-piece.

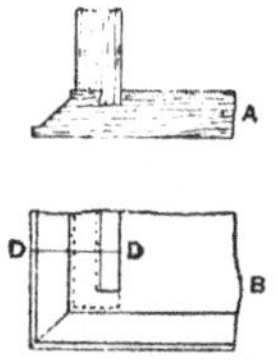

Fig. 273.—Section and Plan of Chimney-piece Cupboard Side.

in the mantel-board, serves to secure them to this. The central portion of the mantel-board is lined out, to increase its apparent thickness, as shown in Fig. 270; the wings

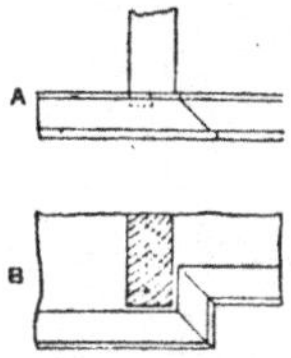

Fig. 274.—Elevation and Plan of Chimney-piece Cupboard Side.

are fitted with shaped apron pieces, blocked solid over the dwarf columns. These columns are dowelled at each end, are turned square, and have volutes carved on each angle of the cap. It will be seen that the brackets supporting the lower shelf are housed into the latter, and they are secured to the back with screws, their lower ends running down to the floor, where the plinth breaks around them as shown in Figs. 265 and 269. The upper shelves are shaped elliptical, as shown in Fig. 266, and are housed into the columns at back and front. The cupboard doors are sunk ⅛ in. below the sides, and are

Fig. 275.—Moulding on Edge of Chimney-piece Brackets.

Fig. 276.—Moulding on Edge of Chimney-piece Shelves.

hung with the knuckle of the hinge flush with the edge of the side, so that the door will open back clear of the edge. The marginal bars in the door are dovetailed at the angles, the moulding cut away to the mitre line, and the angle bars saddled over the square, as indicated by the dotted line in Fig. 271. Figs. 262 to 267 are reproduced to ½ in. to 1 ft., and Figs. 268 to 275 2 in. to 1 ft. (approximately). Fig. 276 is one-third full size.

BOOKCASES.

Simple Bookcase.

FIGS. 277 and 278 illustrate a simple bookcase with a cupboard underneath. The height from the floor line to the top is 6 ft., the width outside the ends is 3 ft., and the depth from back to front is 11 in. The lower part is enclosed by a pair of panelled doors, and the upper part, to receive the books, is enclosed by a pair of folding doors with glass panels, the two parts being divided by a table shelf, having a projecting moulding along the front and returned round the ends. The skirting or plinth is also mitered along the front and ends. The cornice is shaped from a piece of stuff 3 in. by ¾ in., and is also mitered round. The doors, both upper and lower, are made the full width of the case, and are hinged to the side as shown, and the inside of the cupboard is fitted with two shelves which are movable, being cut in clear of the sides; they rest on small fillets, which are screwed to the sides. The inside of the bookcase is fitted with three shelves; these also are movable, and rest on similar fillets. The shelves, when fitted in this manner, are easily raised or lowered as required for variation in the size of the books.

Cutting List.—The materials required for the simple bookcase are as follows:—Two lengths, 6 ft. by 10½ in. by ⅞ in., for sides; one, 2 ft. 11 in. by 11 in. by ⅝ in., for bottom; one, 2 ft. 11 in. by 11 in. by ⅝ in., for top; one, 3 ft. 1½ in. by 1 ft. 0½ in. by 1 in., for table shelf; five lengths, 2 ft. 11 in. by 9½ in. by ⅝ in., for shelves; one, 6 ft. by 3 ft. by ½ in., for matchboard back; one, 5 ft. by 3 in. by ⅝ in., for skirting; one, 5 ft. 8 in. by 3 in. by ¾ in., for cornice; four pieces, 2 ft. 3 in. by 2⅝ in. by 1 in., for door stiles; four pieces, 1 ft. 6½ in. by 2⅝ in. by 1 in., for door rails; two pieces, 1 ft. 11 in. by 1 ft. 2 in. by ½ in., for door panels; four pieces, 3 ft. 7 in. by 2 in. by 1 in., for door stiles; two pieces, 1 ft. 6½ in. by 2⅛ in. by 1 in., and two pieces, 1 ft. 6½ in. by 1⅞ in. by 1 in., for door rails; four pieces, 3 ft. 4 in. by ⅜ in. by ¼ in., and four pieces, 1 ft. 3 in. by ⅜ in. by ¼ in., for glazing bead; two pieces, 3 ft. 2¾ in. by 1 ft. 3 in., of clear selected 21-oz. sheet glass; four 1¼-in. brass cupboard knobs; two 2¼-in. brass straight cupboard locks; two brass thread escutcheons; two 3-in. iron-necked bolts; four pairs of 2½-in. brass butt hinges; two dozen 1-in. iron screws; four dozen ¾-in. No. 7 screws; and ¼ lb. of ¾-in. panel pins.

Construction of Simple Bookcase.—The sides are prepared and set out in pairs, the edges shot quite straight, and lines squared across to the dimensions. The back inside edge of each side is rebated ½ in. deep to receive the back. The top ends are dovetailed or rebated to receive the top, and the bottom ends are grooved on the inside 3 in. up to receive the bottom, and also 2 ft. 6 in. up to receive the table shelf. The top and bottom shelves are prepared both alike, level with the edge of the sides, and the table shelf is moulded on the front edge, and cut long enough to mitre at each end to receive the two return mouldings as shown in Fig. 279. The cornice moulding (Fig. 280) is worked, or a piece of ordinary architrave moulding may be used and mitered round. The doors are mortised and tenoned together, and

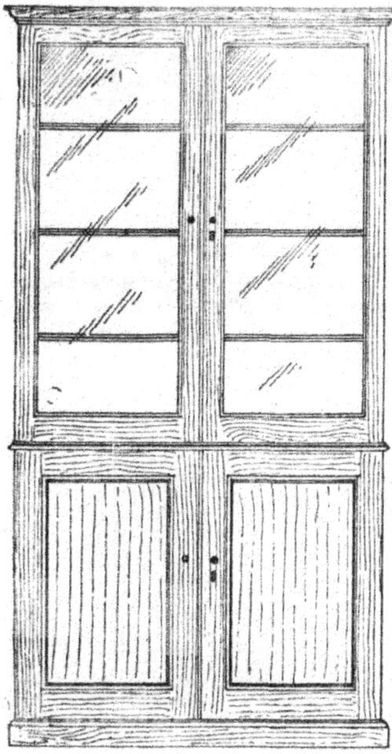

Fig. 277.

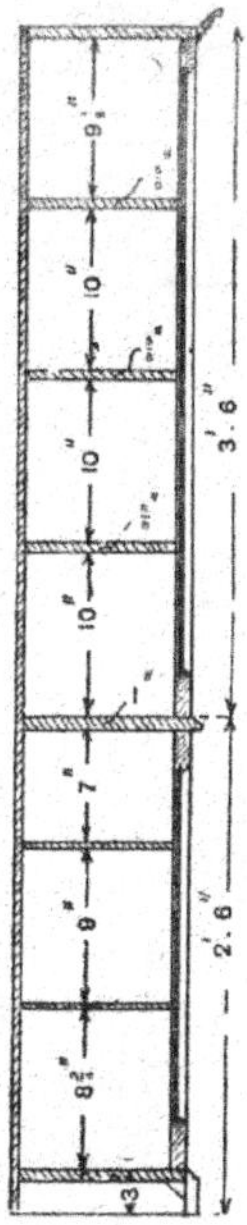

Fig. 278.

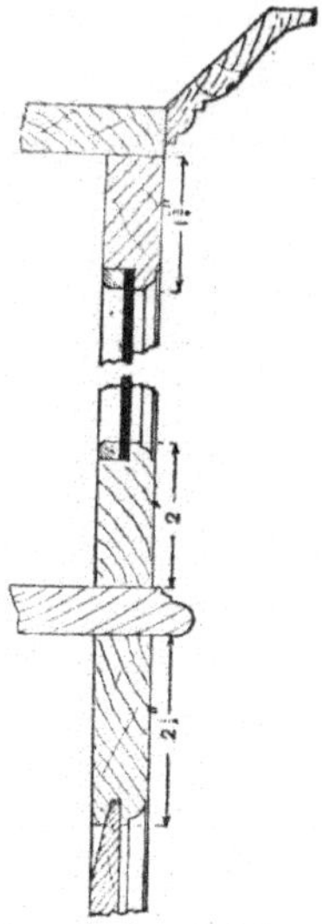

Fig. 280.—Vertical Section of Bookcase Cornice and Top Part.

Figs. 277 and 278.—Front Elevation and Vertical Section of Simple Bookcase.

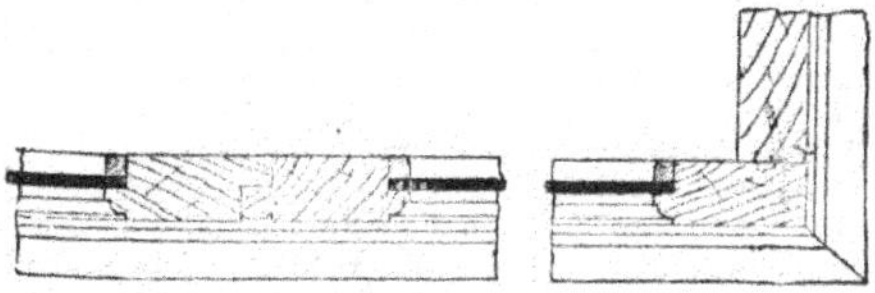

Fig. 279.—Part Horizontal Section of Simple Bookcase.

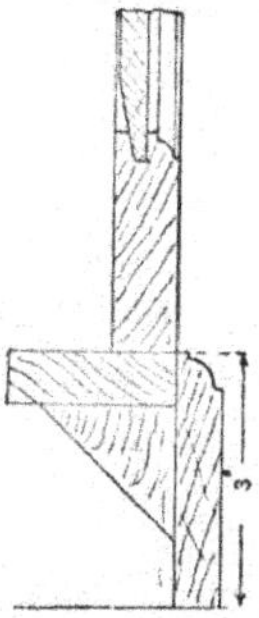

Fig. 281.—Section of Bookcase Plinth, etc.

moulded and grooved to receive the panels. The upper doors are rebated to receive the glass panels, the glass being secured by small beads or fillets mitered round and fixed with panel pins. Fig. 281 is a section of the plinth.

in the ends (see Fig. 283). The doors are framed together in single panels, the glass being fixed with beads. Each door has two hooks fixed to it, level with the top and flush with each end rail (see Fig. 284). These hooks hang on rollers A, and are

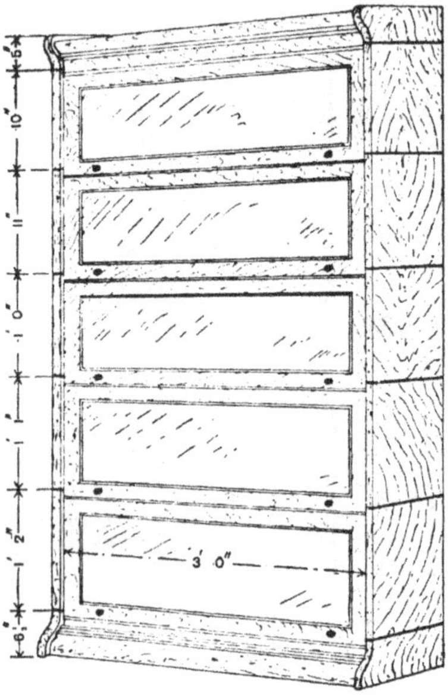

Fig. 282.—General View of Sectional Bookcase.

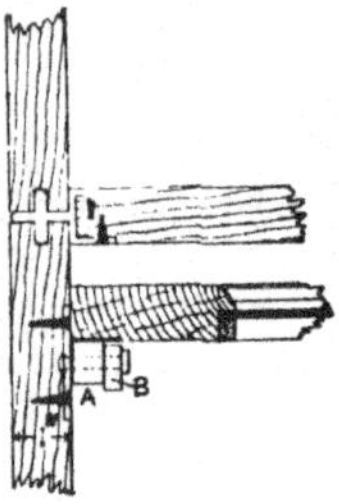

Fig. 283.—Method of Fixing Shelves and Roller Bearings of Sectional Bookcase.

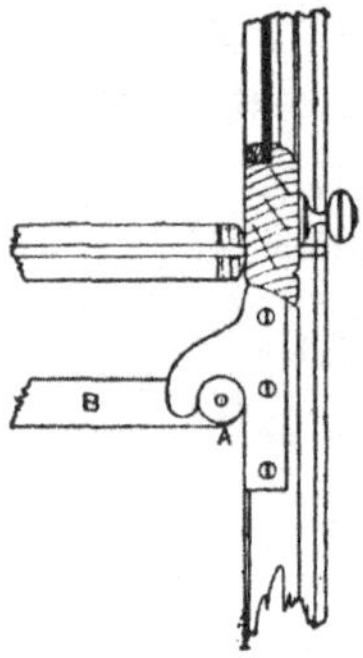

Fig. 284.—Enlarged Section through Intermediate Part of Sectional Bookcase.

Sectional Bookcase.

The bookcase shown by Fig. 282 is constructed in sections, each part being a case in itself, so that more or fewer sections may be used as required. The chief feature of this method of construction is that the parts, such as the doors, ends, shelves, and backs, are interchangeable. No nails, screws, or dowels are used; the shelves have malleable iron ends, with tongues at top and bottom, which fit into the grooves

easily removed by lifting off the roller. To open a door in order to reach a book from the case, the door is raised by the knobs until horizontal; it is then pushed into the case, sliding along on the roller and steel guide B. When the case is not in use, the door is withdrawn and allowed to fall into its original position. Fig. 283 shows the door lying on the roller; also the method of fixing the end of the steel guide. The fixing of the opposite end is seen in Fig. 285. The backs are dovetailed to the ends so as to slide in or out as desired. An enlarged section through the bottom section is presented by Fig. 286.

Cabinet Bookcase.

A cabinet bookcase is shown by Fig. 287. The finished sizes of timbers are: Top,

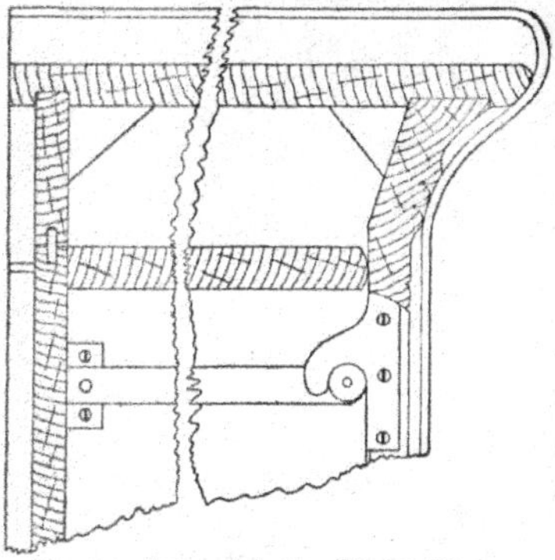

Fig. 285.—Enlarged Section through Upper Part of Sectional Bookcase.

3 ft. 3 in. by 1 ft. 4 in.; outside ends, 3 ft. 11 in. by 1 ft. 2 in.; two shelves and bottom, 2 ft. 11¾ in. by 1 ft. 1½ in.; two inside ends on each side of door, 1 ft. 6¼ in. by 1 ft. 1½ in.; and the shaped span-rail beneath the bottom, 2 ft. 11¾ in. by 6 in. wide. All are of 1-in. stuff, which should finish about 15/16 in. thick. Two bottoms beneath the drawers are 10¾ in. long by 1 ft. 1½ in. deep, of ¾-in. stuff. The door stiles and rails are of 1¼-in. stuff, and the panel is of ½-in. stuff. The shaped back above the top is of 1-in. stuff, and 3 ft. 1 in. long by 1 ft. 5 in. high. The shelf, 2 ft. 11 in. by 6 in., and the brackets beneath, 9 in. by 6 in., are also of 1-in. stuff. The drawer fronts are 1 in. thick, and the sides and bottoms are of ½-in. stuff. The back of the carcase consists of three muntins, 3 ft. 6 in. by 3½ in., of 1-in. stuff, and the backs between are 3 ft. 6 in. long and about 1 ft. 2 in. wide and ⅜ in. thick. First mark out half the front elevation and the end, full size. The height of the door compartment is 1 ft. 6 in., and the recesses along the side are 10½ in. wide, and the drawer fronts are 4 in. high. The top projects 1 in. all round. The wood must first be planed to thickness, length, and width. The top has an ovolo moulding A (Fig. 288) worked on the front and end edges, or, if this is inconvenient, a bevelled edge B (Fig. 288) will be suitable. Next groove the ends ⅓ in. deep for the two shelves and bottom, but take care that the grooves stop ¼ in. short of the front edges of the ends, so that they will not show on the front. The back edges of the ends A (Fig. 289) also require rebating to receive the side muntins. In the two long shelves plough ¼-in. grooves, ¼ in. from the front edges, to receive strips E (Fig. 290), on which are glued the leather valances which hang over the books. The shaped span-rail beneath the bottom is also grooved or housed in ⅛ in. deep and stands ⅛ in. from the face of the ends. In putting together, first fix the two long shelves and bottom with nails or thin 2¼-in. screws;

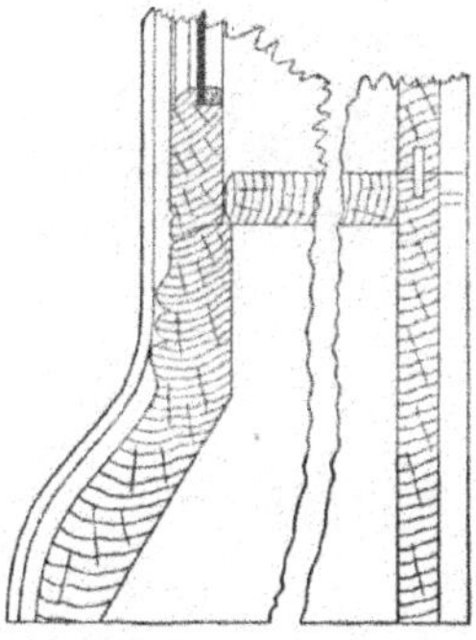

Fig. 286.—Enlarged Section through Lower Part of Sectional Bookcase.

next the inside ends and bottoms under the drawers; and finally the top, nailing it through into the ends from the top side. The shaped span-rails under the drawers are fitted between the ends and then secured

Fig. 287.—General View of Cabinet Bookcase.

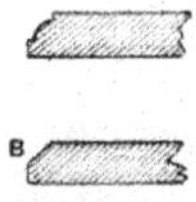

Fig. 288.—Cross Sections through Cabinet Bookcase Top.

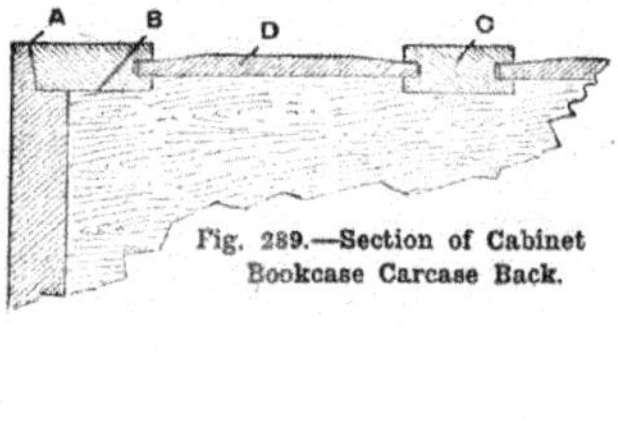

Fig. 289.—Section of Cabinet Bookcase Carcase Back.

Fig. 290.—Strip for Leather Valance of Cabinet Bookcase

with nails. Let them stand back $\frac{1}{16}$ in. The muntins of the back may now be fixed, the shelves being cut away a little at B (Fig. 289), and for the centre muntin C. The latter and the sides are grooved to receive the thin backs D, which slide up from below and are screwed where they come over the shelves and bottom. The sides of the drawers are lap-dovetailed to the fronts as in Fig. 291, and, instead of weakening the sides by grooving for bottoms, $\frac{3}{8}$-in. slips F (Figs. 291 and 292) are grooved and the top edge G is half-rounded. The drawer bottoms are pushed along the grooves and into the groove H (Fig. 291) of the drawer fronts. The door stiles and rails are $1\frac{1}{2}$ in. wide without moulding, and are mortised and tenoned together. The lines J (Fig. 293) on the panel are cut with a gouge, and the oval fan-like pattern is carved to the section shown. Figs. 294 to 298 show the method of marking the

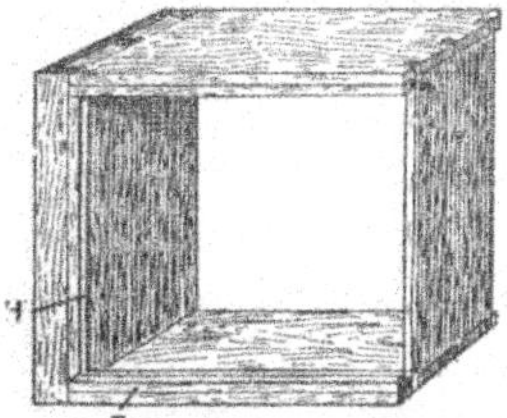

Fig. 291.—Part of Cabinet Bookcase Drawer.

Fig. 292.—Detail of Cabinet Bookcase Drawer.

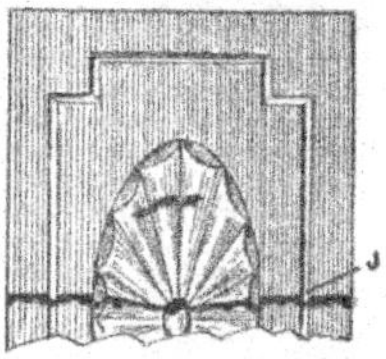

Fig. 293.—Door Panel of Cabinet Bookcase.

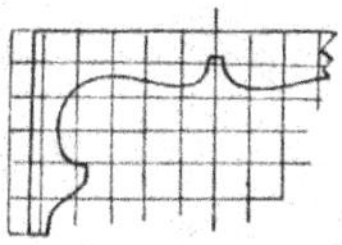

Fig. 295.—Pattern for Cabinet Bookcase Bracket.

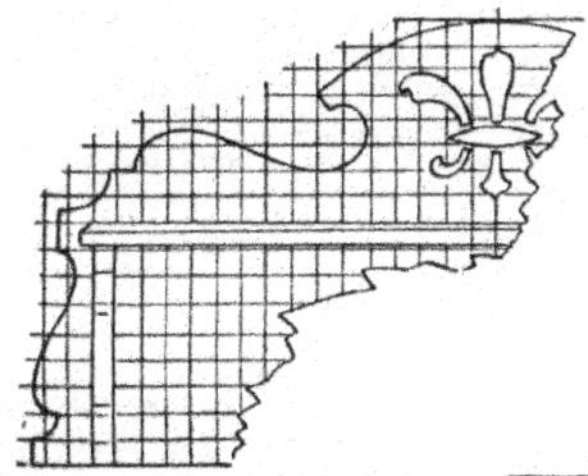

Fig. 294.—Pattern for Shaped Back of Cabinet Bookcase.

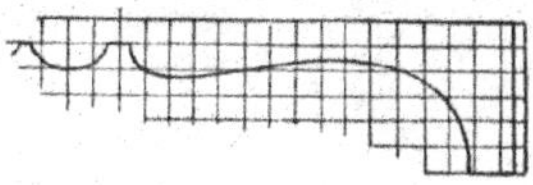

Fig. 296.—Pattern for Cabinet Bookcase Span-rail.

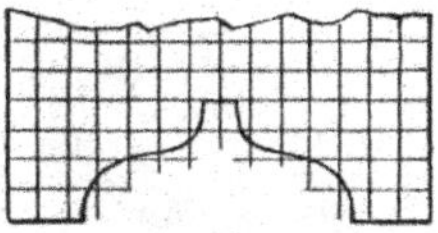

Fig. 297.

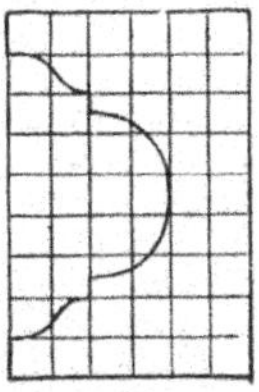

Fig. 298.

Figs. 297 and 298.—Patterns for Cabinet Bookcase Brackets.

Fig. 299.—Section of Cabinet Bookcase Door.

designs for brackets and span-rails. If the mortising and tenoning for a door as Fig. 299 cannot be done, make the door of a piece of 1-in. stuff, with two battens behind it to prevent it warping, and work a hollow to take the place of moulding. Or a curtain of some soft fabric may be substituted for the door. The design in the centre of the back is cut through. The

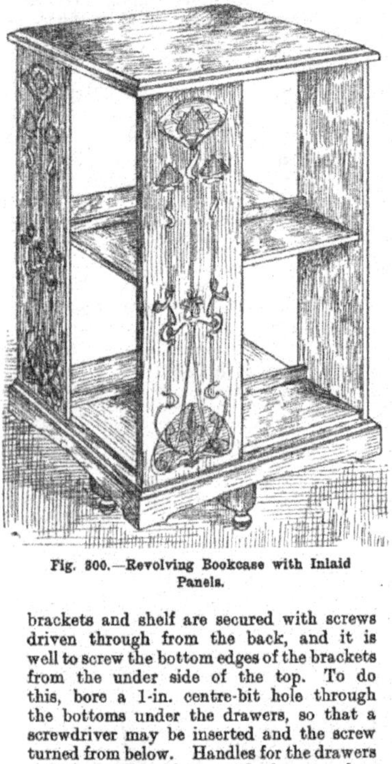

Fig. 300.—Revolving Bookcase with Inlaid Panels.

brackets and shelf are secured with screws driven through from the back, and it is well to screw the bottom edges of the brackets from the under side of the top. To do this, bore a 1-in. centre-bit hole through the bottoms under the drawers, so that a screwdriver may be inserted and the screw turned from below. Handles for the drawers and door, also hinges and three or four coats of paint, will complete the job. The cabinet would also look very well if made of mahogany, stained dark, and french-polished. In this case a silvered mirror should be framed in under the shelf of the upper part. Of course, the construction would be somewhat different, as nails could not then be used.

Revolving Bookcase with Inlaid Panels.

The revolving bookcase shown by Figs. 300 to 303 might be constructed in English oak, but almost any hardwood would look well. Details are illustrated fully by Figs. 304 to 308. It consists chiefly of two parts, the revolving case and the support or stand. The wood for the centre box of the case should be grooved and tongued together as shown in the horizontal section. Before fitting together, the insides should be rebated at the top to receive the collar, which is a piece of wood ⅝ in. thick, with a hole bored through the centre a trifle larger than the diameter of the pole (see Fig. 308). Its object is to keep the case running truly round the centre pole. A number of blocks should be glued in the internal angles, the centre ones being pushed into position first with the aid of a stick. The box should be set out for the housings into which the shelves are fitted. The shelves are made up of four pieces, mortised and tenoned together as shown in Fig. 305; two shelves only are illustrated, but others may be added to suit requirements. The bottom shelf can be glued before fastening to the box, but the shelves in the centre must be glued into the housings round the box. Laths are screwed round the box directly above and beneath the shelves, to form additional support and to prevent the books being pushed into another compartment. The laths on the upper side should be placed first, and a screw or two driven into them from the under sides of the shelves. The top and bottom ends of the centre box should fit nicely, and should be fixed into position with screws. A hole is bored through the bottom for the centre pole. A pin and plate (Fig. 306) must be procured; the plate should be let into the under side of the top, and should be firmly bedded to the wood. The top is square, with a moulding worked round the top edges to the section given. Four pieces about 7 in. long should be screwed to the under side of the top in the positions indicated, the outside faces of these being directly over the edges of the shelves below. The strips should be chamfered along one edge and returned in the ends. To hold the top firmly, a few screws should be put diagonally through the top of the box and into the bookcase top. The plinth (see Fig. 304) is made of two pieces glued and blocked together. Both pieces are

Fig. 301.

Fig. 302.

Fig. 303.

Figs. 301 to 303.—Front Elevation, Vertical Section (on Line Y Y), and Horizontal Section (on Line X X) of Revolving Bookcase.

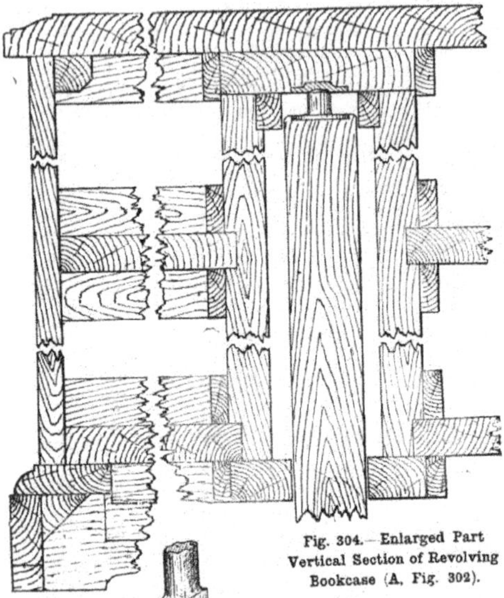

Fig. 304.—Enlarged Part Vertical Section of Revolving Bookcase (A, Fig. 302).

Fig. 306.—Pin and Plate for Revolving Bookcase.

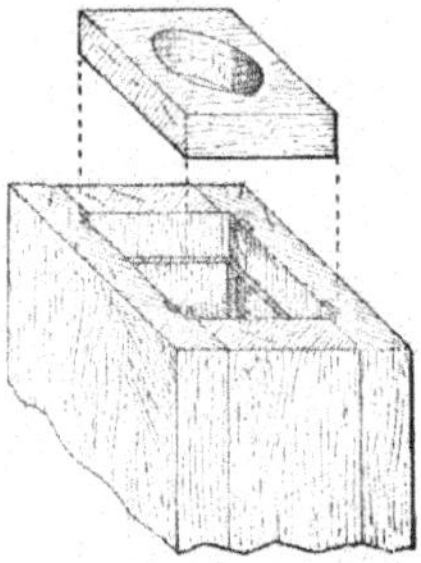

Fig. 308.—Top of Revolving Bookcase's Pillar Casing (C, Fig. 302), with Collar shown separately.

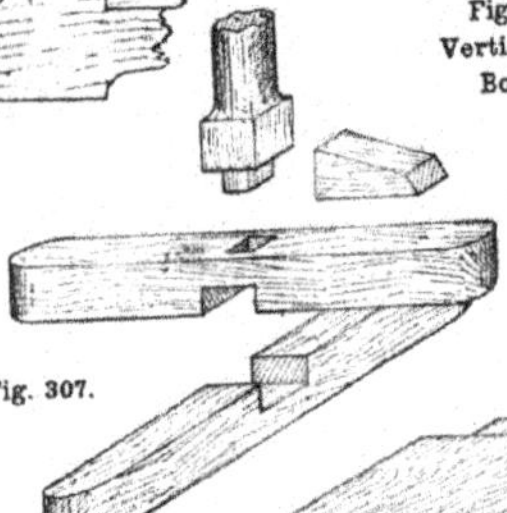

Fig. 307.

Fig. 307.—Foot and Bottom of Revolving Bookcase Pillar (B, Fig. 302).

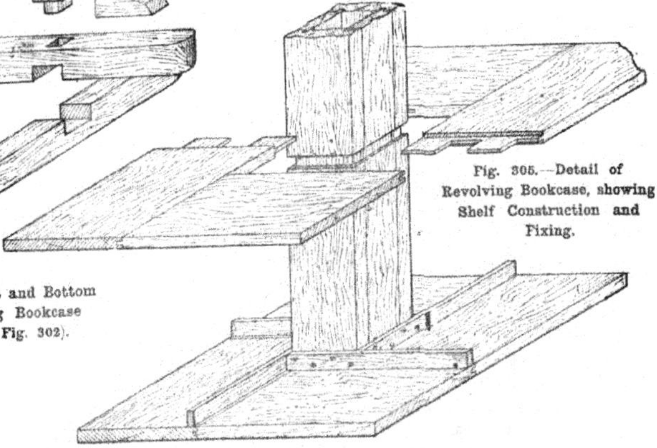

Fig. 305.—Detail of Revolving Bookcase, showing Shelf Construction and Fixing.

mitered together at the corners; the top one, having a thumb moulding worked along the edge, is screwed to the bottom shelf. The sides or faces of the plinth may be rectangular if desired, or may be cut to the shape given in Fig. 301. The panels or laths round the sides should be screwed to the bottom and intermediate shelves, also the strips under the top, using round-headed screws. As an alternative to the panel, five laths are shown in the horizontal section (Fig. 303), and these may be reeded on the face.

Stand for Revolving Bookcase.—The construction of the stand is clearly shown in the sectional elevation (Fig. 302) and the enlarged detail (Fig. 307). The foot consists of two bearers halved together at the centre, a mortice being made through the top one for the reception of the stub tenon on the end of the post (see Fig. 307). The centre post is turned to 1½ in. in diameter, and should be left square at the bottom as shown. Four blocks of the shape shown should be well glued and screwed to the foot of the post and the bearers. The ends of the bearers are rounded as shown, and should have ball castors fixed under the ends, and one in the centre if desired. The pin should be screwed into the top of the post, care being taken to get it in the centre. The bookcase is slipped over the post, in which position it will remain.

Finishing Revolving Bookcase.—The design on the panels lends itself to various ways of treatment, and careful consideration in arranging the coloured woods is desirable. The grain of the wood should follow the main and radial lines of the design as far as possible, and the horizontal and vertical lines should be kept in some of the larger masses, this procedure imparting strength to the design. Another method of making the panels would be to stain the designs by the aid of a stencil process. Should this method be adopted, the design would require altering slightly to obtain the ties in the stencil plate. The bookcase should be finished by polishing, etc., and this will vary according to the wood used. If desired, brass handles may be fixed to the panels for revolving the bookcase.

Reading Table with Revolving Bookshelf.

A reading table with revolving bookshelf is illustrated by Figs. 309 and 310. The table consists of a circular top 1 ft. 7 in. in diameter and 1 in. thick, with an ogee moulding worked on the edge; this stands 2 ft. 6 in. high on a 2⅝-in. turned pillar, supported by four shaped claws 1¼ in. thick. The table is provided with a revolving candle-holder working immediately under the top and just clear of its edge, so that the whole surface of the top is available for papers, etc.; lower down, and at such a height as just to clear the knees of a person sitting at the table, is a revolving bookshelf having three arms radiating from the centre, each 11 in. long by 6 in. wide, and provided with side galleries 3 in. in height. Fig. 310 shows on the left hand a half-plan of the top (the dotted outlines at A indicating the position of the claws), and on the right a half-plan just above the bookshelves, the dotted lines indicating the candle-holder and arm. Fig. 311 is a section through the centre of the pillar, showing the iron stem to which the table-top is attached and on which revolve the candle-arm and the plate to which the bookshelves are secured. These parts are shown separately in plan and elevation in Figs. 312 and 313 respectively, Fig. 314 being the plan of the candle-arm.

Pillar of Revolving Table.—Referring to Fig. 311, the iron stem, which is 17¾ in. long and ⅝ in. in diameter, is welded to a plate $\frac{5}{16}$ in. thick and 5 in. in diameter, with two 6-in. by 1½-in. by ¼-in. arms welded on as shown in Fig. 312. These arms are placed at right angles to the direction of the grain of the table-top, and are sunk flush with its under side, the circular portion being screwed on the surface. The pillar of the table is made in two parts, each being bored through to take the stem; the hole should be of such size that the stem will pass through with slight pressure; it should not be too tight for subsequent removal, or so loose that there is side play. A small washer shrunk on the stem as at A (Fig. 312) would prevent wear of the end of the pillar by the revolving arm,

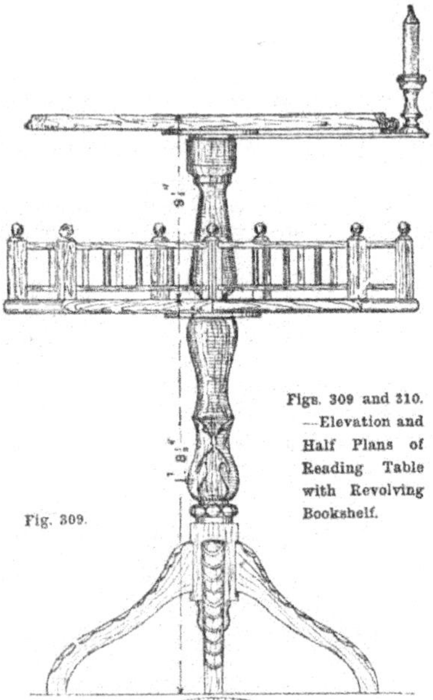

Fig. 309.

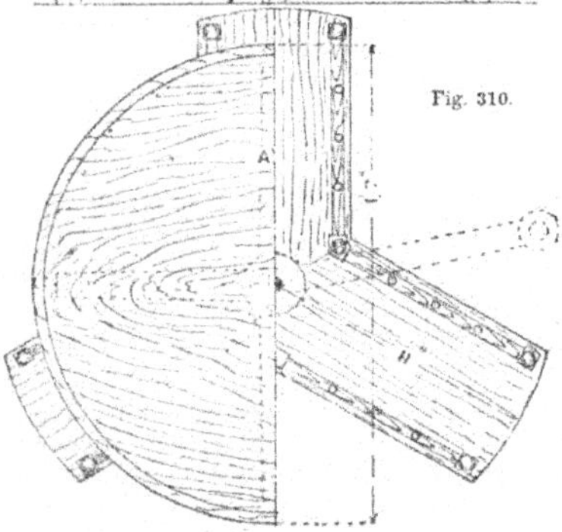

Fig. 310.

Figs. 309 and 310.—Elevation and Half Plans of Reading Table with Revolving Bookshelf.

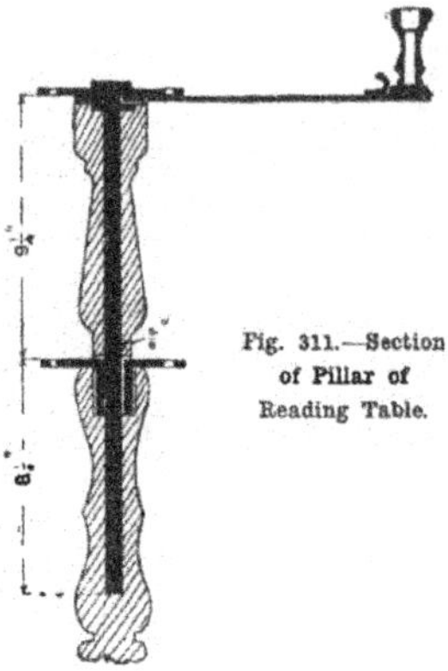

Fig. 311.—Section of Pillar of Reading Table.

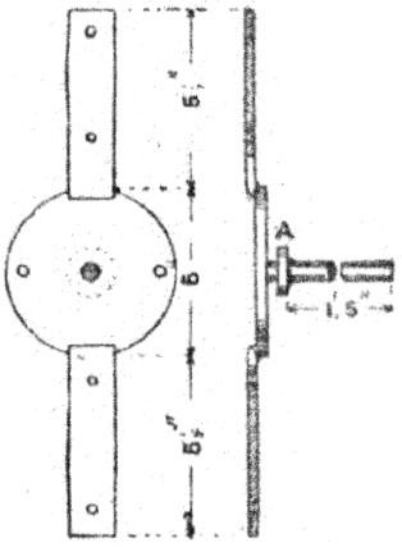

Fig. 312.—Table-plate and Stem for Reading Table.

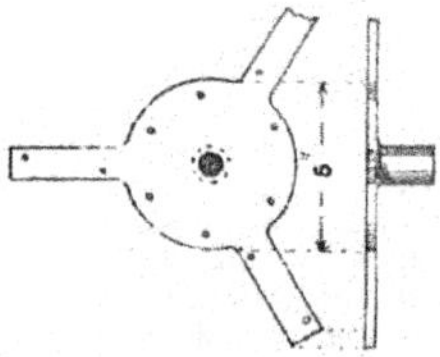

Fig. 313.—Shelf-plate and Sleeve for Reading Talbe

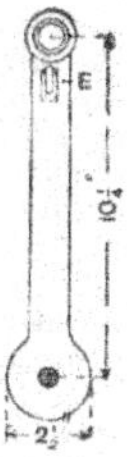

Fig. 314.—Candle-arm for Reading Table.

Fig. 316.

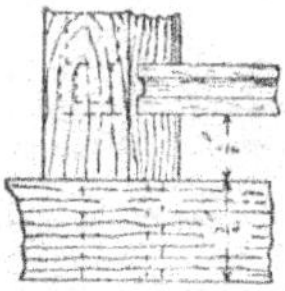

Fig. 317.—Gallery Post, etc., of Reading Table.

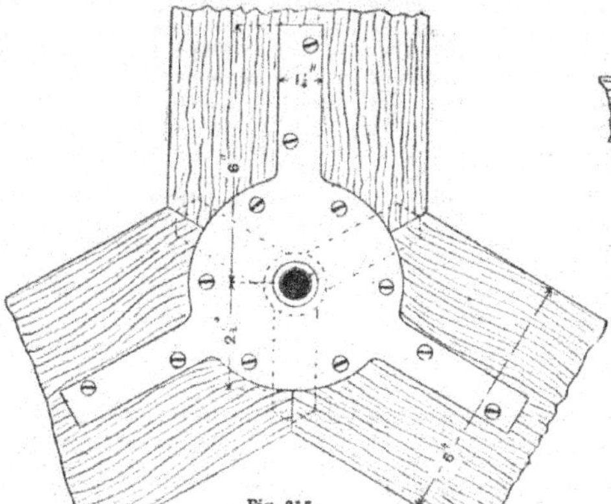

Fig. 315.

Figs. 315 and 316.—Underneath View and Cross Section of Reading Table's Bookshelf Centre.

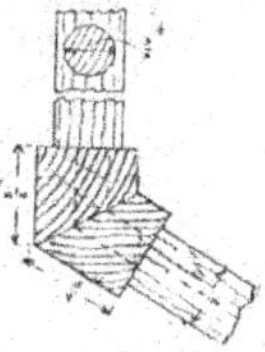

Fig. 318.—Plan of Gallery Post and Rails of Reading Table.

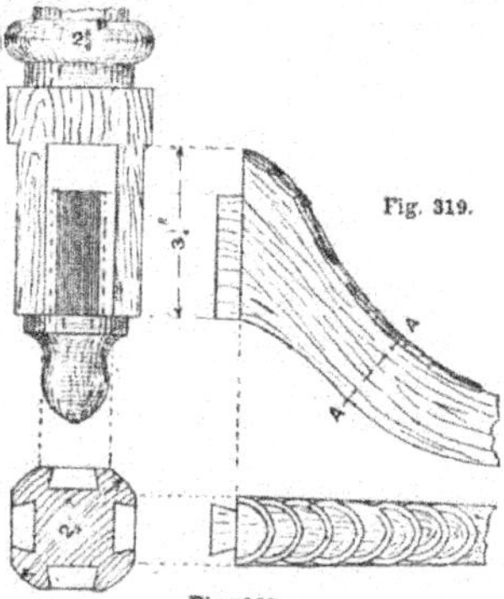

Fig. 319.

Fig. 320.

Figs. 319 and 320.—Foot of Reading Table Pillar.

Fig. 321.—Cross Section of Reading Table Claw.

but is not absolutely necessary, and, if it is used, the candle-arm must be placed in position before the washer is fixed. The lower end of the upper half of the pillar is turned down to 1¼ in. in diameter where it passes through the bookshelf, and the upper end of the lower part of the pillar is turned out to receive the sleeve attached to the bookshelf plate (see Fig. 313). This hole should be the exact depth of the sleeve, and of such a size that the plate can revolve freely.

Bookshelf of Revolving Table.—Fig. 315 is an enlargement of the central portion of the bookshelves, the under side being shown with the metal plate attached; this need not be sunk in, as it cannot be seen when in position. The ends of the shelves are mitered and cross-tongued together as indicated by the dotted lines in Fig. 315 and shown in the section (Fig. 316). The blacked portion in Fig. 315 shows the aperture in the plate for the stem, and the dotted circle the one in the woodwork for the shank of the pillar. The side galleries of the shelves are formed of ½-in. by ⅜-in. rails with fluted edges (see Fig. 317). They are tenoned into ¾-in. by ¾-in. standards with ball-and-cushion ends; these are tenoned through the shelves and wedged as shown in Figs. 317 and 318, the standards at the junctions being mitered and glued; the spindles between the rails are ⅜-in. plain rods dowelled in ¼ in. at each end.

Claw Foot of Pillar.—Figs. 319 and 320 show the method of fixing the claws in dovetail slots in the pillar on four squared faces; the dovetails are notched down 1 in. from the top, and should fit tightly, and may advantageously be made with slightly tapering sides, so that they will tighten as they are driven in place. Fig. 321 is a section of the claws on the line A A (Fig. 319). The wood for the claws should be selected of strong, even grain, and the pattern for cutting should be placed on the stuff, so that the grain may run as long as possible through the curved parts.

Finishing Revolving Table.—Cut and fit up all the parts before any carving is done. When fixing the iron stem in the pillar, the lower half should be coated with Brunswick black or varnish, which will adhere both to the wood and iron and make the parts secure. It may be desirable at some future time to remove the upper portion passing through the pillar, and it should therefore be coated with grease and blacklead; all the working parts should be similarly treated. These metal fittings can usually be obtained of manufacturing ironmongers in dozen sets only; but any skilful blacksmith could make them if supplied with full-size drawings of each part separately and a scale sketch similar to Fig. 311. The candlestick could be of brass, screwed to the arm, and a small brass hook as shown at E (Fig. 314) should be brazed on to clip the edge of the table and prevent the arm dropping. Figs. 309 to 314 are drawn to a scale of 1½ in. to 1 ft., and Figs. 315 to 321 are 3 in. to 1 ft., with the exception of Figs. 317 and 318, which are half full size.

CABINETS.

Small Museum Cabinet.

A SMALL museum cabinet for the display of curiosities is shown to a scale of ½ in. to the foot by Figs. 322 to 325; in addition to which figures, details one-quarter size are shown as follows: Fig. 326, detail plan of part of case; Fig. 327, detail vertical section through front; and Fig. 328, detail section of plinth to base. The outside dimensions are—4 ft. wide by 1 ft. 6 in. deep by 6 ft. 3 in. high. The case is constructed of Austrian wainscot oak for all outside parts, which are french-polished; while the parts that are hidden are of best yellow pine. All the materials must be perfectly seasoned. The lower part of the case, comprising the base, is panelled and moulded, and is made independent of the upper part or case proper. This latter is constructed on the air-tight principle, the opening sashes or doors having hook joints on the meeting stile and air-tight beads to the hanging stile. The frame has air-tight fillets at top and bottom, the ends being framed to match the front, and glazed with ¼-in. British polished plate glass. The case is lined inside on the back and bottom with velvet plush; the back of ⅜-in. matchboard is papered before being covered. The case is fitted with plate-glass shelves, shaped as shown, and supported on bronzed iron or brass shelf brackets, fastened with set screws to vertical standard bars, which are tapped at intervals of 1½ in. to 2 in. for convenience in raising or lowering the position of the shelves. The doors are hung on brass arrow butts, three to each door. The left-hand door is fitted with a brass bolt at top and at bottom; the right-hand one, with an eccentric handle and catch, and a small sash lock.

Materials for Small Museum Cabinet.—The materials required are:—For the base: Wainscot plinths, one 4 ft. 2 in. by 4 in. by 1 in.; two 1 ft. 8 in. by 4 in. by 1 in. Pine rails, one 4 ft. 2 in. by 4 in. by ⅞ in.; two 1 ft. 8 in. by 4 in. by ⅞ in. Wainscot rails, one 4 ft. 1 in. by 2 in. by 1 in.; one 4 ft. 1 in. by 1½ in. by 1 in.; two 1 ft. 6 in. by 1½ in. by 1 in.; two 1 ft. 6 in. by 2 in. by 1 in. Wainscot stiles, six 1 ft. 3 in. by 3 in. by 1 in. Wainscot muntins, two 1 ft. by 3 in. by 1 in. Wainscot panels, two 1 ft. 8 in. by 8 in. by ½ in.; two 1 ft. 1 in. by 8 in. by ½ in. Wainscot moulding, two 4 ft. 6 in. by 1¼ in. by ½ in.; two 3 ft. 2 in. by 1¼ in. by ½ in. Wainscot base capping, one 7 ft. 3 in. by 2 in. by 1 in. Pine top, one 3 ft. 10 in. by 1 ft. 5 in. by ⅞ in. Deal back, one 4 ft. by 1 ft. 3 in. by ⅞ in. Deal bearer, one 1 ft. 6 in. by 3 in. by 1¼ in. Twelve deal angle blocks, 3 in. long; 3½ doz. ⅝-in. No. 9 screws for panel moulding; 1 doz. 1¼-in. No. 10 screws for fixing plinth; 14 ft. run ⅜-in. feather cross-tongue. For the case: Wainscot stiles, two 4 ft. 8 in. by 1⅞ in. by 1 in.; two 4 ft. 8 in. by 1½ in. by 1 in.; two 5 ft. 2 in. by 1⅞ in. by 1 in.; two 5 ft. 2 in. by 1¼ in. by 1 in.; two 5 ft. 2 in. by 2 in. by 1 in. Wainscot rails, one 4 ft. by 1⅛ in. by 1 in.; two 1 ft. 6 in. by 2⅜ in. by 1 in.; one 4 ft. by 1⅞ in. by 1 in.; two 1 ft. 6 in. by 2⅜ in. by 1 in. Pine rails, one 4 ft. by 3½ in. by ⅞ in.; two 1 ft. 6 in. by 3½ in. by ⅞ in. Wainscot rails, four 2 ft. by 1½ in. by 1 in. Wainscot cornice, one 8 ft. by 4½ in. by 2½ in. Pine lining.

one 4 ft. by 3½ in. by ⅜ in. Wainscot fillet, two 4 ft. by 1¼ in. by ⅝ in. Pine fillet, one 4 ft. by 2 in. by ¾ in. Pine top, one 4 ft. by 1 ft. 6 in. by ¾ in. Pine bottom, one 4 ft. by 1 ft. 5 in. by ¾ in. Deal matched back, one 5 ft. by 4 ft. by ¾ in. Feather tongue, one 7 ft. run ¾ in. Wain-

Fig. 322.

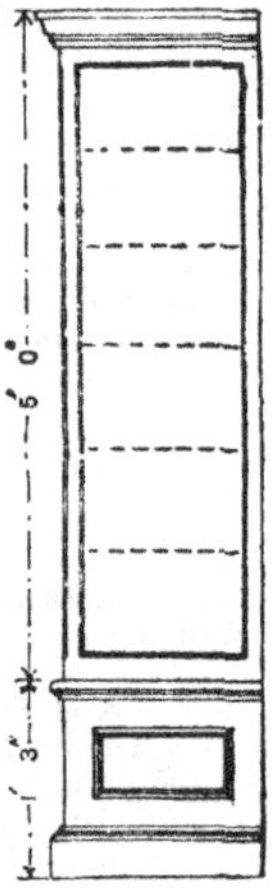

Fig. 323.

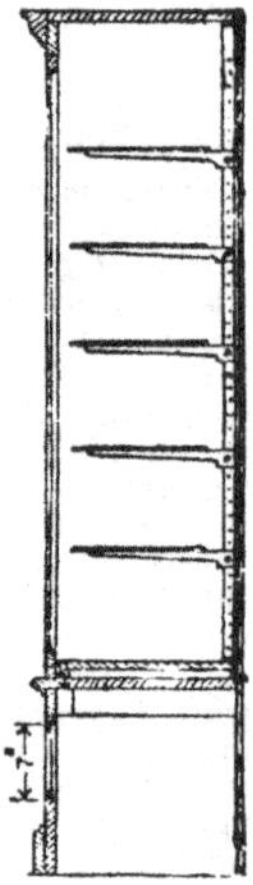

Fig. 324

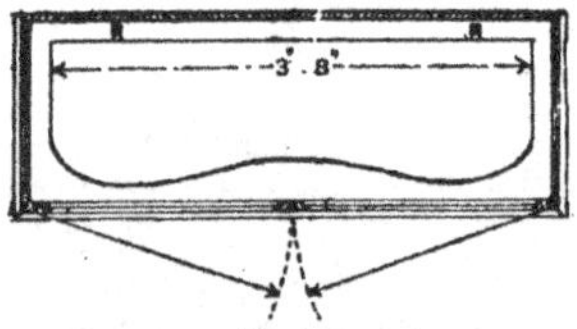

Fig. 325.

Figs. 322 to 325.—Front Elevation, Side Elevation, Vertical Cross Section, and Plan of Small Museum Cabinet.

scot glazing bead, 50 ft. run ½ in. by ¼ in. Plate glass, two 4 ft. 3¾ in. by 1 ft. 8⅜ in. by ¼ in.; two 4 ft. 3¾ in. by 1 ft. 3 in. by ¼ in. Standard bars, two 4 ft. 9¾ in. long, with base and top plate drilled and countersunk for screws; ten 12-in. shelf brackets and set screws; five glass shelves, 3 ft. 8 in. long by 1 ft. 1 in. wide, cut to shape, edges ground and polished; 5 yds. white lining paper; 24 ft. super. velvet plush; two brass-necked bolts; three pairs brass butts; one eccentric handle; one brass sash lock.

China Cabinet.

Figs. 329 to 331 show a china cabinet, in making which a pair of sash doors has been utilised. The following materials will be required: Deal bottom, 3 ft. 5 in. by 1 ft. 6 in. by 1 in.; two deal shelves, 3 ft. 5 in. by 1 ft. 5½ in. by ¾ in.; deal division, 2 ft. 10½ in. by 1 ft. 5½ in. by ¾ in.; deal matchboard back, 3 ft. 5½ in. wide by 3 ft. 1 in. high by ½ in. thick; two mahogany ends, 3 ft. 6 in. by 1 ft. 7 in. by 1 in.; mahogany top, 3 ft. 9 in. by 1 ft.

8 in. by 1 in.; mahogany skirting or plinth, 7 ft. 2 in. by 3 in. by ¾ in.; deal fillet under top, 3 ft. 4½ in. by ¾ in. by ¼ in.; mahogany rebated stop on doors, 2 ft. 10 in. by 1¼ in. by ½ in.; brass cut 2-in. cupboard lock and screws; two brass flush bolts, 4 in. by ⅝ in., and screws; two pairs of 3-in. brass machine-made butt hinges. First set out the work to the dimensions given in the illustrations. Figs. 329 and 330 are front and side elevations, and Fig. 331 is a horizontal section. The doors being each 2 ft. 9¾ in. by 1 ft. 8½ in., the carcase must be framed accordingly. The ends are prepared 1 ft. 7 in. wide; and if the mahogany cannot be obtained in one width, they must be put together with a tongued and glued joint, and afterwards faced on the outer side and thicknessed. The exact length to which they should be cut is 3 ft. 1 in., which allows ¼ in. to go into the groove in the top and sufficient to run down to the floor level. From the lower end on the inner face a groove is prepared, 3 in. up, to receive the deal bottom (see Fig. 332). The mahogany top must be prepared in the same manner as the ends, and cut 3 ft. 8 in. long by 1 ft. 8 in. wide. Grooves are prepared at each end on the under side to take the ends (see also Fig. 332), the outside of the groove being ⅞ in. from the end; they should be stopped 1 in.

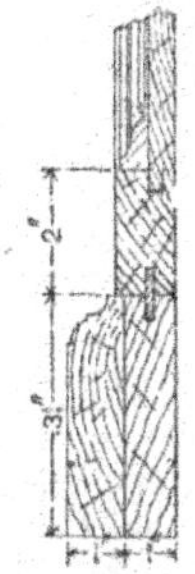

Fig. 328.—Plinth and Base of Museum Cabinet.

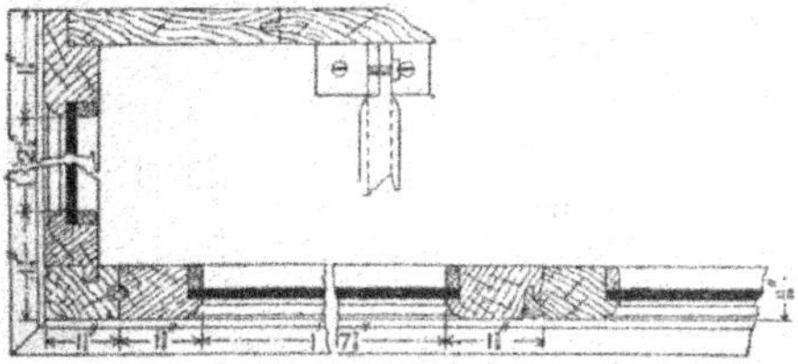

Fig. 326.—Part Horizontal Section of Museum Cabinet.

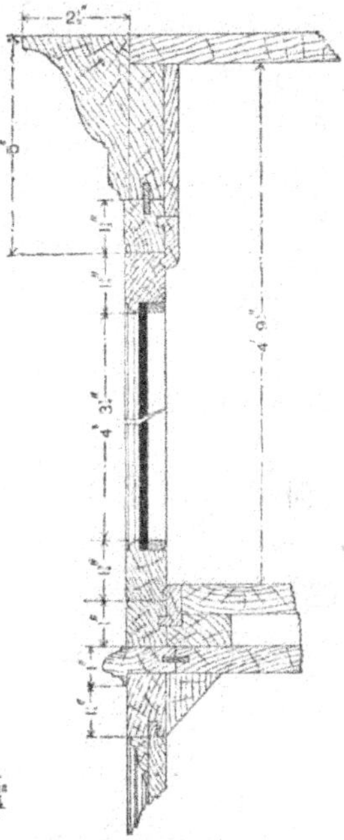

Fig. 327.—Vertical Section through Front of Museum Cabinet.

from the face edge (see Fig. 333), the amount of the projection. Rebates must be formed in the back edges of the top shown. The top should then be moulded along the front edge and return ends (see Figs. 332 and 333). The bottom is of deal,

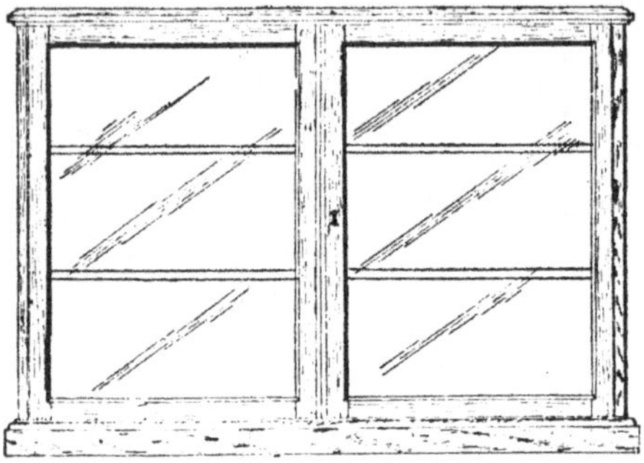

Fig. 329.

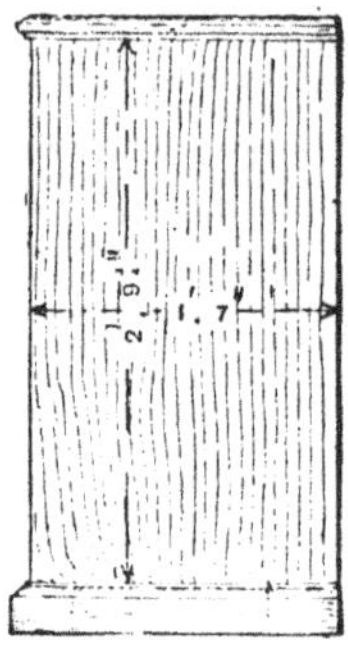

Fig. 330.

Figs. 329 to 331.—Front and Side Elevations and Horizontal Section of China Cabinet.

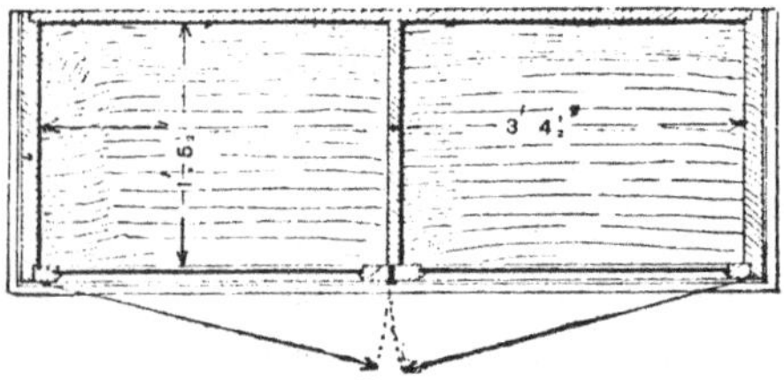

Fig. 331.

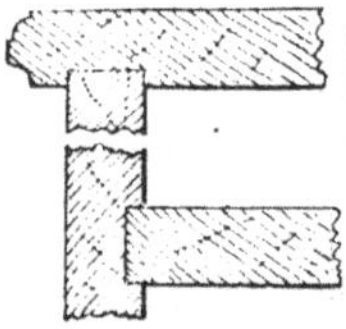

Fig. 332.—Framing of China Cabinet.

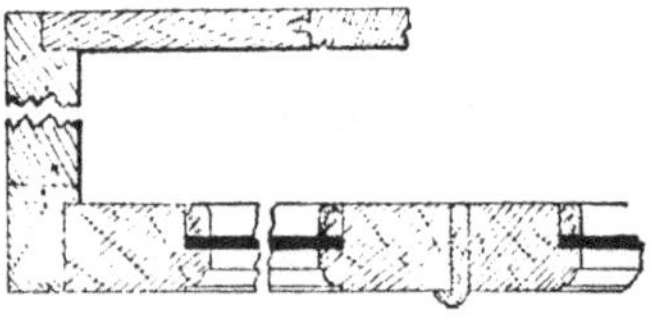

Fig. 334.—Enlarged Part Horizontal Section of China Cabinet.

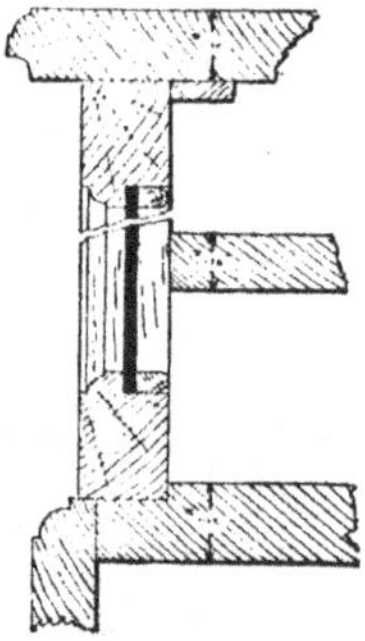

Fig. 333.—Enlarged Part Vertical Section of China Cabinet.

and ends to receive the $\frac{1}{2}$-in. back (see Fig. 334). The rebate in the ends to receive the doors must be prepared the exact thickness of the doors and sunk $\frac{1}{16}$ in., and a $\frac{1}{8}$-in. bead glued on the front edge as prepared from two 9-in. boards glued together; it is rebated on the front edge only to receive the doors, and grooved in the centre to receive the vertical division. A groove is similarly prepared on the under

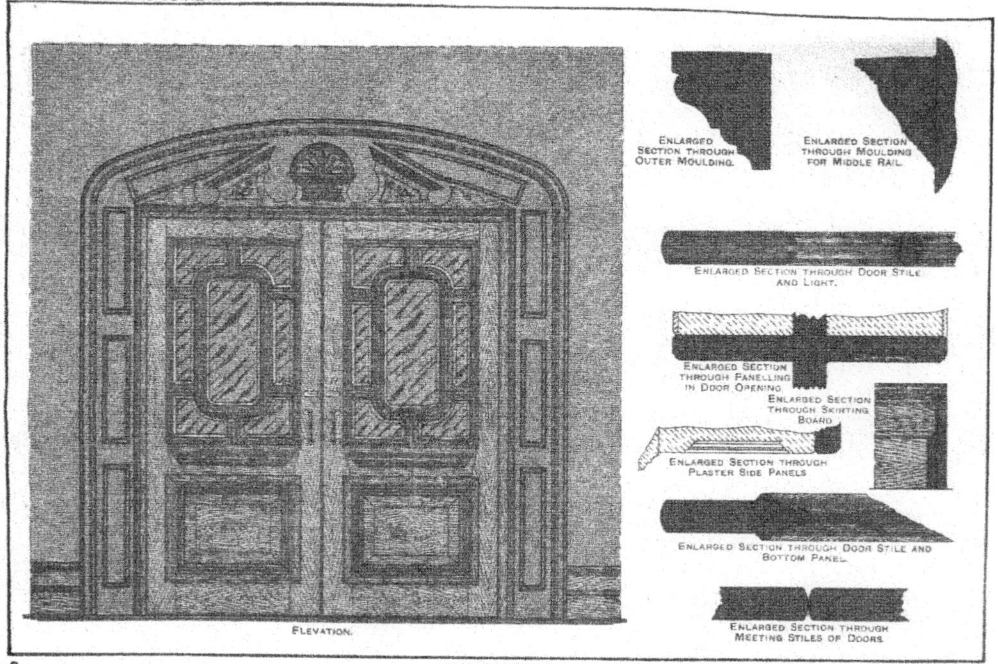

3

SALOON DOORS.

side of the mahogany top, but this is stopped, whilst that in the bottom runs right through. The exact cutting length of the bottom is ½ in. over the clear dimension between the grooves in the top, 3 ft. 4½ in.; the extra ½ in. goes ¼ in. each side into the grooves in the ends. The vertical division is cut 2 ft. 10¼ in., and is 1 ft. 5½ in. wide. The shelves are cut clear between the ends and the division, and rest on deal fillets screwed to the ends and the division. The skirting, which is moulded, and mitered at the angles, should be glued to the edge of the bottom along the front, and further strengthened by angle-blocks glued on the inside. The returns are fixed by means of screws through the ends, which continue down to the floor level. The joint at the meeting stiles of the doors is covered by a moulded and rebated stop (see Fig. 334). The dimensions given are calculated to allow the doors to finish 1 in. thick.

Sheraton Corner Cabinet.

Sheraton, the celebrated cabinet-maker, constructed furniture which was very light and graceful in appearance, but which was so skilfully put together that many examples, after the wear of a century, are practically as perfect as when they left the workshop. Rosewood and Spanish mahogany, both solid and in veneer, were his favourite woods; and these were generally inlaid with sandal, or satinwood and ebony. Sheraton also largely employed stained and shaded woods in inlaying for decorative effect, festoons and running scrolls being much employed, and the bars in the cabinets were convoluted and extremely fragile in appearance. The Sheraton cabinet shown in elevation in Fig. 335 and in section by Figs. 336 and 337 is intended to stand in a right-angle corner, and is of 1 ft. 10 in. side, out to out, with 6-in. returns. It stands 7 ft. 1 in. high, with a 1-ft. 10-in. front. Fig. 337 is a section at X X (Fig. 335). Make the cabinet of dark mahogany. The fine legs run through both compartments are got from 1¼-in. stuff. The shelves are cut around and sunk into grooves in the legs to a depth of ⅛ in., as in Figs. 338 and 339. The lower shelf is shown by Fig. 343. The ends of the legs are tenoned through the top, and nailed. The cornice, of ¾-in. stuff, is glued and bradded on the edge of the top, the ⅜-in. cover-board being nailed on top and fitted tightly, but not glued to the over-cornice. This is glued and blocked to the top of the cabinet with shaped return pieces, as shown in Figs. 335 and 336. The cornice and over-cornice are both inlaid, as shown in Fig. 340, the lines being of black walnut or ebony, or stained to represent the latter; the lozenges are satinwood; the enclosed shaped panels are rosewood or beefwood; and the tendril ornament either satinwood shaded or green ash. The various panels and bottom shelf can also be inlaid with a similar pattern of running tendrils or line-work. Figs. 339 and 341 show alternative methods of inserting the panelling in the framing. Fig. 339 has the panels sprigged into rebates or checks; this is the cabinet-maker's method, and it is easier but weaker than the joiner's method shown in Fig. 341, which is a section at A (Fig. 337). The glazed door in the upper part should be mortised and tenoned together, the dotted lines in Fig. 338 indicating the size and position of the tenons. The shoulders are made square, the rebate for the glass being formed by an inserted slip bead, which is mitered round after the frame is glued and cleaned off. This bead should be gauged to width and glued in. The bars, ½ in. by 13/16 in., are cut square against the stiles and rails, but are mitered to each other as shown in Fig. 342. The circular bars may be made in various ways, the best method being to bend the bar in the solid round a shaped drum, splicing the ends as shown in Fig. 342, and working to section after gluing up. An easier method is shown in the upper part of Fig. 342. Two rings for the bead and for the tongue are turned in the lathe; then a small groove is turned in the back of the bead, and the tongue ring inserted into this, crossing the grain of the two rings as much as possible. A third method is to work each bar in four pieces in the solid, cut them out, and dowel them together with butt joints. The door of the lower cupboard, shown in section in Fig. 341, is mortised and tenoned together and solid moulded with a ½-in. Grecian ovolo, the

Fig. 335.

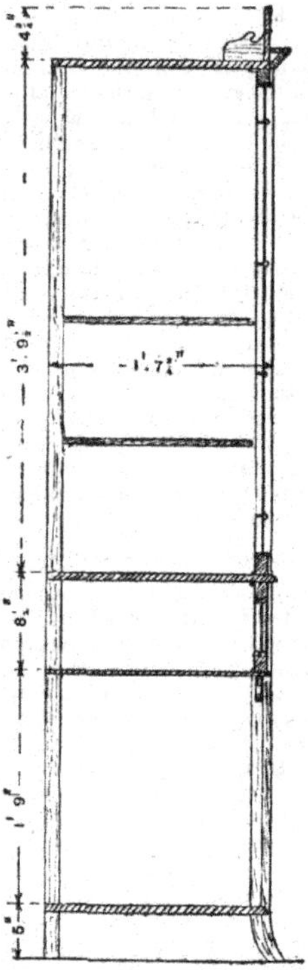

Fig. 336.

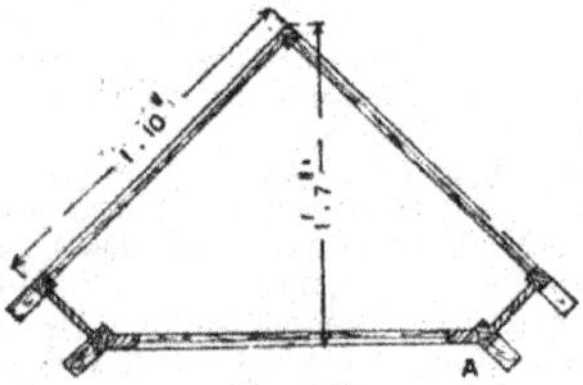

Fig. 337.

Figs. 335 to 337.—Front Elevation, Vertical Section, and Horizontal Section of Sheraton Corner Cabinet.

Fig. 343.—Part Elevation of Sheraton Cabinet's Lower Shelf.

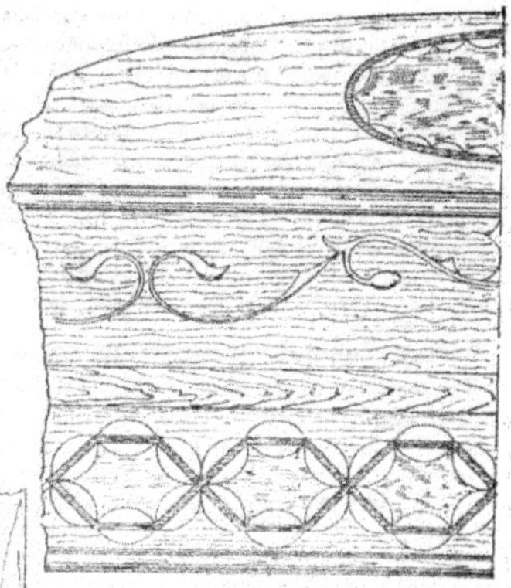

Fig. 340.—Part Elevation of Cornice of Sheraton Cabinet.

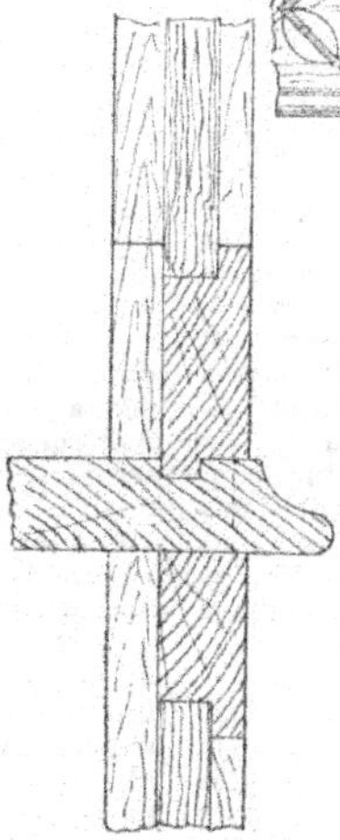

Fig. 339.—Section of Shelf, etc., of Sheraton Cabinet.

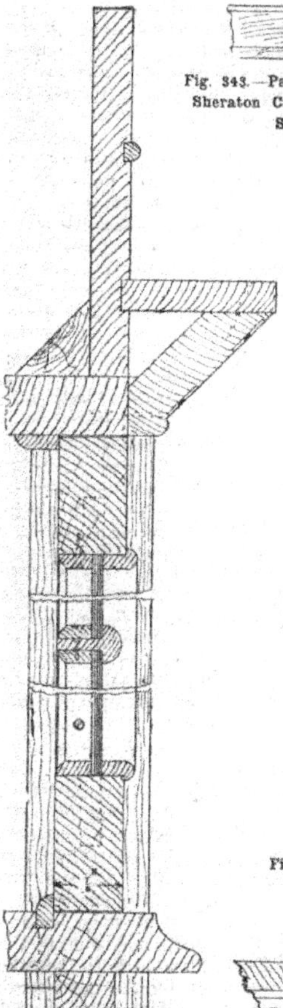

Fig. 338.—Section of Door and Cornice of Sheraton Cabinet.

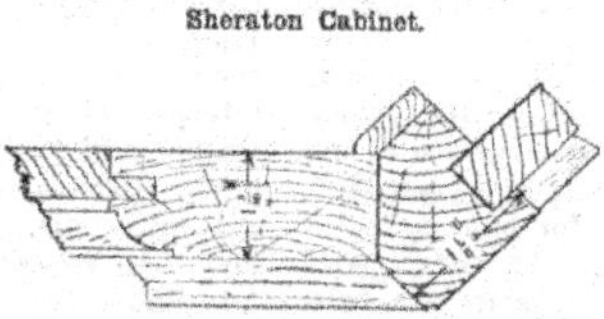

Fig. 341.—Horizontal Section through Corner of Sheraton Cabinet.

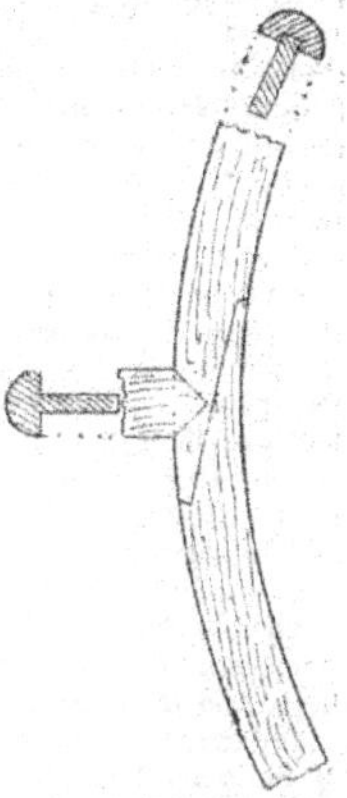

Fig. 342.—Method of Joining Bars of Sheraton Cabinet.

panel being rebated in, and finishing flush inside. The upper portions of the two front legs should be square and parallel, as shown in Fig. 341, and ½-in. slips are glued on for door-stops. The apron rails beneath the cupboard should be framed into the legs, and the cupboard bottom is glued and blocked to them. The back of the glazed cabinet is made of ⅜-in. pine boards, grooved in, and should be covered with purple plush or cloth. The bottom part is left open.

Corner Cabinet for Drawing-room.

The cabinet illustrated at Fig. 344 looks well if made of pine or deal, painted and enamelled white or pale green; or it may be made of hardwood, such as mahogany or walnut, and french-polished. The cupboard consists of two parts, the lower extending from B to A and the upper from A to C. The upper part is intended for the display of bric-à-brac, the door panel being of clear glass, and underneath the cupboard are two plate-glass mirrors. Before starting the construction, make a full-size plan of the lower part, as in Fig. 345, the top, of course, being omitted. The sides against the wall are 1 ft. 7 in. long, and the breaks D to E 8 in. long, the front being about 1 ft. 4 in.—the exact length can be obtained from the plan. The legs are of 1-in. stuff, 3 ft. long by 2¾ in. wide. The back leg F shows the thickness of one leg narrower by its width than the others, to make the joint as shown. The two large shelves, having been got out to the proper size and shape, should be cut away at the angles to receive the legs. Next get out the top, which should be an inch larger all round than the plan. The height from the floor of the bottom shelf is 6 in., and that of the shelf above 1 ft. 8 in. Mark the positions of these shelves on the legs, and the widths of the shaped span-rails and arches as well. The method of setting out these rails and arches is shown at Figs. 346 to 349. They must stand back ¼ in. from the face of the legs, and must be got out ½ in. longer than sight measure, to allow for housing them ¼ in. deep into the legs. The next parts to be prepared are the backs from A to G (Fig. 344); these are of the same thickness as the legs, running the same way of the grain, and glued and jointed to the legs. The shelf at G is supported at the back and sides by being screwed from underneath into the back. The inside ends of the curtained recess, indicated by dotted lines in Fig. 345, are glued and jointed to the front legs and backs. Before the legs are finally put together, they must be shaped on the outer edges to the form shown enlarged in Fig. 350. The top is nailed to the legs and shaped span-rails. The top is made to project 1 in. at the back in order to allow the upper part of the cabinet to get close to the wall, otherwise it would not do so, owing to the skirting board that usually runs round the bottom of the wall. Two brass screw eyelets are screwed behind the top centre span-rail, as a support for the wire rod on which the soft silk curtains are hung. The small shelves at H (Fig. 344) are ⅝ in. thick, and are fixed with nails driven through the backs and inside ends.

Upper Part of Corner Cabinet.—For the upper part of the cabinet, make a full-size plan as shown in Fig. 351, allowing the sides to be 1 ft. 8 in. long and fully ¾ in. thick. The left-hand side of the plan shows a section through the door, and the right-hand side a section through the mirror. The door posts J are 1 ft. 1 in. from the corner, and are got out of 1¼-in. stuff, bevelled to shape as shown. The extreme height from A to C (Fig. 344) is 3 ft. 6 in.; height from A to the shelf below the door, 11 in.; height from A to the top of the cupboard, 2 ft. 7 in.; door stiles and rails, 2 in. wide, including moulding; height of small corner shelf from the top of the cupboard, 7 in. The small shelves near the mirrors are halves of a 5-in. disc, and are fixed 6 in. from the top A; both shelf and bracket are ⅝ in. thick. Enlarged drawings of the shaped parts of the backs are given at Figs. 352 and 353. Each back is jointed to make one piece of the necessary width. To allow the backs to intersect at the corner, one is made the thickness of the stuff less in width than the other. The openings for the mirrors are cut out, a margin of 1½ in. being allowed for what appear as stiles and rails (see Fig. 351). To form a rebate for the mirrors, which should be of bevelled glass, half-round mouldings are glued and nailed to the face,

Fig. 344.—Drawing-room Corner Cabinet.

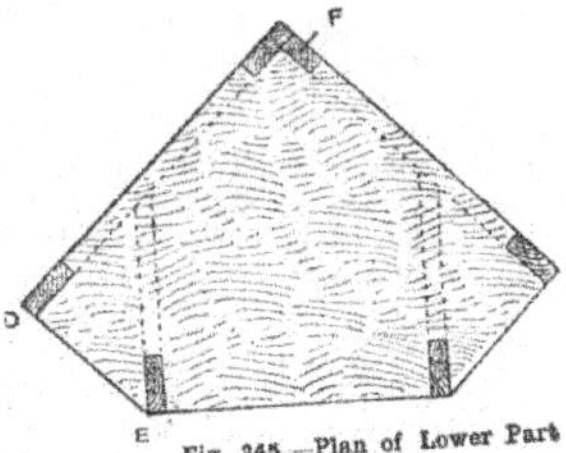

Fig. 345.—Plan of Lower Part of Cabinet.

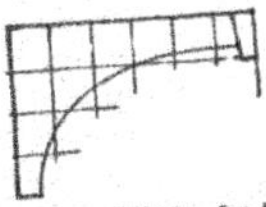

Fig. 346.—Pattern for Top Span-rail of Cabinet.

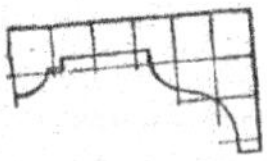

Fig. 347.—Pattern for Middle Span-rail of Cabinet.

Fig. 348.—Pattern for Bottom Front Span-rail of Cabinet.

Fig. 349.—Pattern for Bottom Side Span-rail of Cabinet.

as shown in the enlarged section at Fig. 354. To protect the glass, a ½-in. back K is inserted. The shelf below the door is ⅝ in. thick, projecting ½ in. from the door and posts. The top of the cupboard is ⅞ in. thick, projecting 1 in. from the door and posts (see full lines in Fig. 351). The door posts having been fixed to the backs, the shelf and top may be secured with nails driven through the backs. The door stiles and rails are mortised and tenoned in the usual way. The tracery pattern in the door is made of very thin stuff, such as can be obtained from any dealer in fretwork materials. An enlargement is shown at Fig. 355, half the design being set out in squares for copying. The door is hung with a pair of 2-in. brass butts; a lock may be fitted, or a brass handle as shown. The inside of the cupboard will look well if lined with an art shade of velveteen. As it may be desired to have a door instead of the curtains in the lower part, a design for this is given at Fig. 356. The two sections of the cabinet are fastened together with screws driven from the under side of the top A.

Centre Cabinet for Drawing-room.

The centre cabinet illustrated by Fig. 357 may be made of mahogany, inlaid with satinwood. The various parts should be kept as light as is consistent with strength. Each side is the same in appearance, but one is constructed to open as a door. Such a cabinet may be made of any size to suit requirements; the dimensions of the one shown here are: Height to top shelf, 4 ft. 10 in.; height to cabinet top, 4 ft.; and 1 ft. 2 in. to the top edge of the moulding which rests on the legs. The sides of the glazed cabinet are 1 ft. 9 in. wide, and are made independent of the top and lower framing. The first part to be taken in hand is the lower framing. The legs are each 1 ft. 2 in. long, and 1⅝ in. square at the top, tapering to 1 in. square at the bottom. The span rails A (Fig. 358) are of 1¼-in. stuff, 3⅝ in. wide, and are tenoned into the legs; the tenons should be made as long as possible by mitering the ends, as shown in the sectional plan, Fig. 359. The moulding B (Fig. 358) is ¾ in. wide, and projects ½ in. from the face of the legs and cabinet; it is glued on the face of the rails and legs, and mitered at the corners. Before finally gluing together, the satinwood stringing on the outside faces of the legs, and the fan pattern at the ends of the rails, must be inlaid.

Top of Centre Cabinet.—The top of the cabinet projects 1½ in. all round the carcase. It is of 1-in. stuff, and underneath are strips 3 in. wide by ⅝ in. thick, mitered at the corners, these forming the lower member of the moulding, as at C (Fig. 360). These strips are well screwed to the under side of the top. On the upper face of the top is a line of stringing 1½ in. from the edge, breaking inwards 1½ in. at the corners (see Fig. 361). In the centre is a fan-shaped patera 6 in. in diameter. This ornament may be obtained from inlayers, or it may be omitted. In inlaying pateras and corner pieces as in the lower rails, the general method is to veneer the surface, after fitting the inlaid portions to the veneer. The inlays are first secured by gluing paper on the face; then the veneer is glued to the face of the wood and held by clamps and a heated caul. As an alternative method, the inlaid portions may be sunk into the solid wood by cutting away the surface to receive the inlay.

Top Shelf and Supports.—The top shelf (see plan, Fig. 362) is of 1-in. stuff, and is 10 in. square; the moulding worked round its edges is shown by Fig. 363. Lines of satinwood stringing are inlaid on the top face of the shelf, standing in from the edges 1¼ in., and breaking inwards at the corners 1¼ in. The shaped pieces (Fig. 364) under the corners of the shelf must be set out full size, so as to get the right shape. To do this, make a full-size plan, as in Fig. 362. Draw on the plan a 1¼-in. square representing the bottom end of the shaped piece, the outside lines being level with the cabinet carcase; and the same of the top end, 1 in. square, standing in ½ in. from the edge of the shelf. Now, allowing 9¼ in. for the height of the shaped pieces, set out a side elevation as in Fig. 364; in this way the exact contour is obtained. To avoid spoiling good wood, it would be well first to make one roughly in pine for a

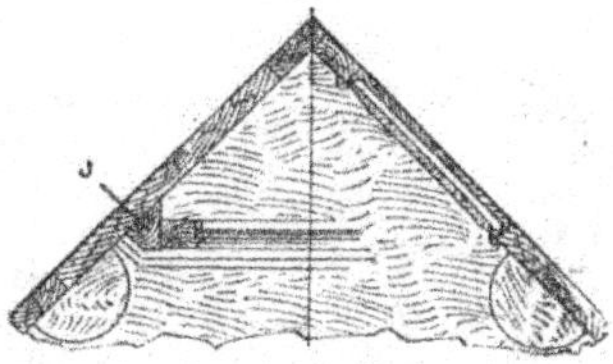

Fig. 351.—Plan of Upper Part of Cabinet.

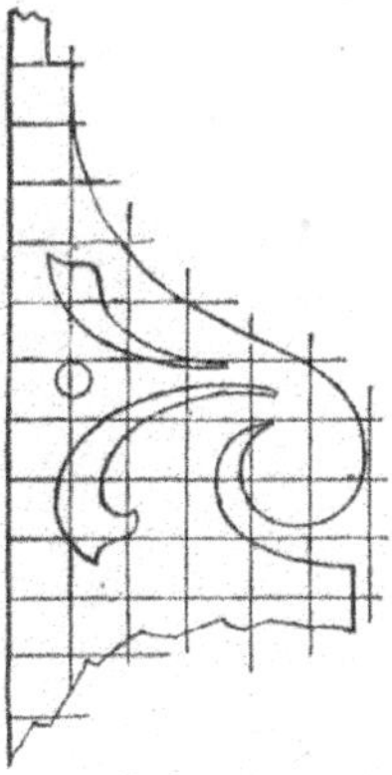

Fig. 352.—Pattern for Part of Cabinet Back.

Fig. 350.—Part of Leg of Corner Cabinet.

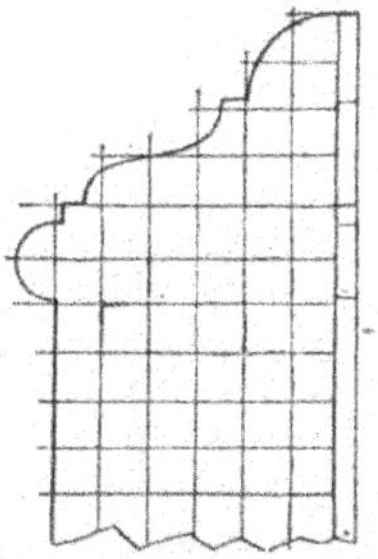

Fig. 353.—Pattern for Top of Cabinet Back.

Fig. 354.—Detail of Cabinet Mirror.

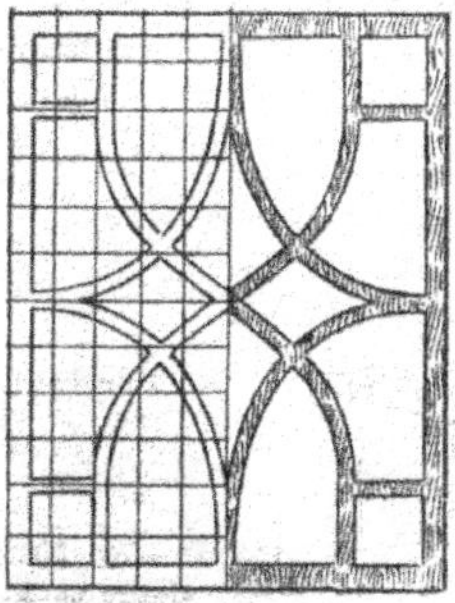

Fig. 355.—Tracery of Door of Cabinet.

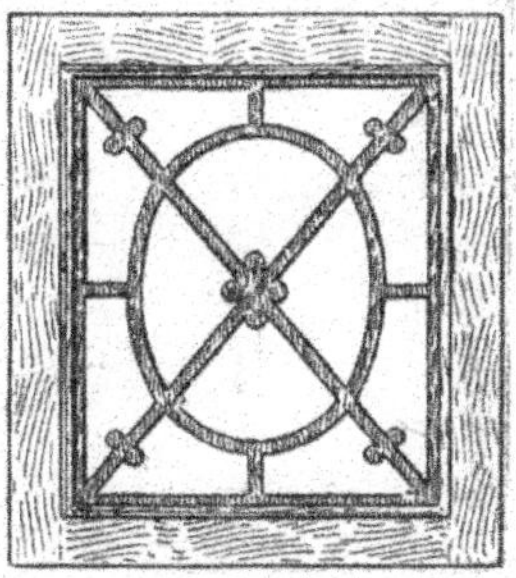

Fig. 356.—Door instead of Curtains in Lower Part of Cabinet.

Fig. 357.—Centre Cabinet for Drawing-room.

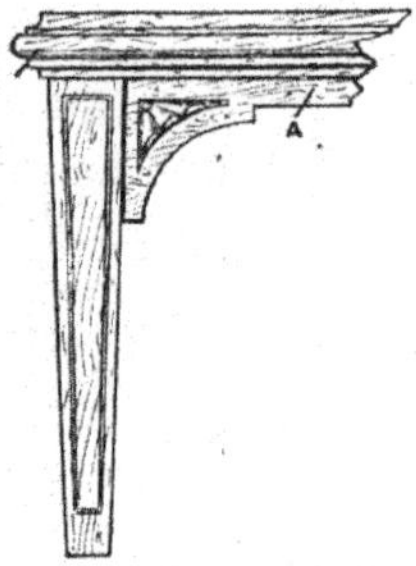

Fig. 358.—Leg, etc., of Centre Cabinet.

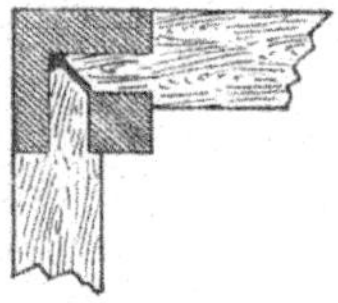

Fig. 359.—Centre Cabinet Span-rails tenoned to Leg.

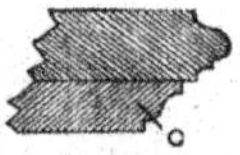

Fig. 360.—Top Moulding for Centre Cabinet.

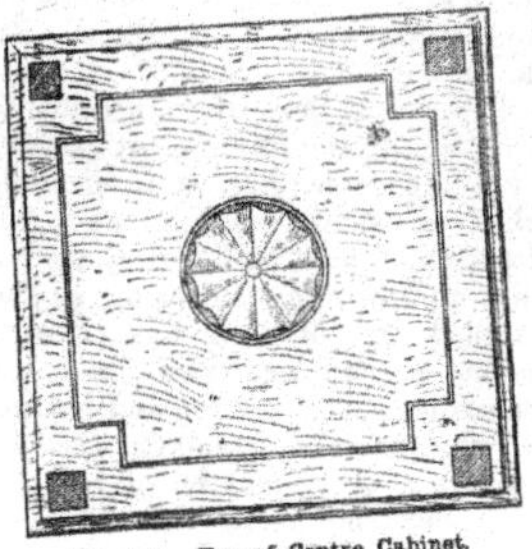

Fig. 361.—Top of Centre Cabinet.

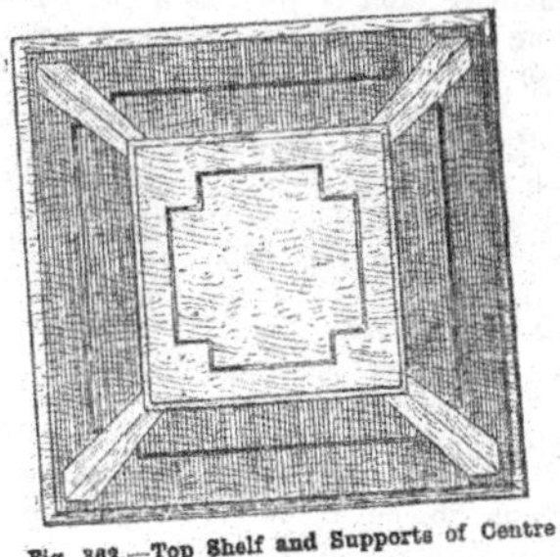

Fig. 362.—Top Shelf and Supports of Centre Cabinet.

Fig. 363.—Top Shelf Moulding of Centre Cabinet.

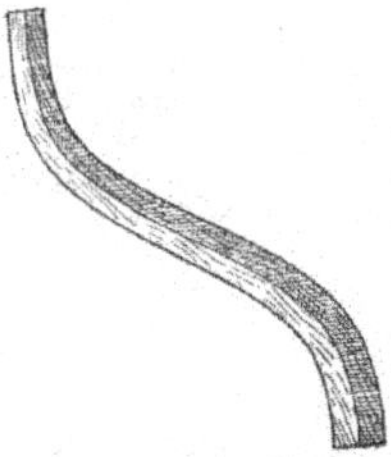

Fig. 364.—Top Shelf Support for Centre Cabinet.

Fig. 366.

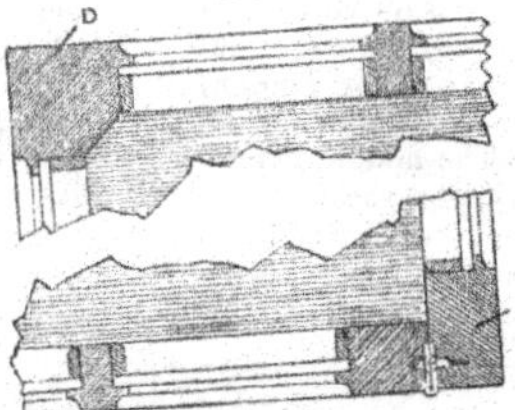

Fig. 367.

Figs. 366 and 367.—Sections of Glazed Framing of Centre Cabinet.

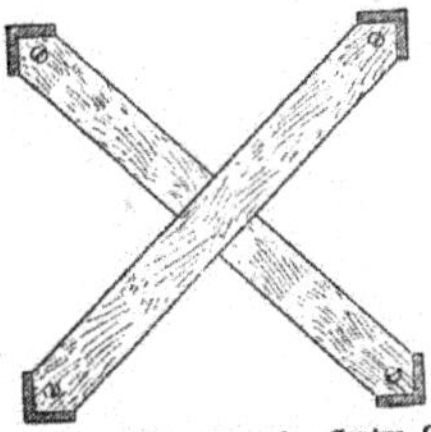

Fig. 365.—Plan of Rails under Centre Cabinet's Top Shelf.

Fig. 368. Fig. 369.

Figs. 368 and 369.—Part Section and Plan showing Centre Cabinet's Top connected to Bottom.

trial. This may then be used as a pattern for marking out on the mahogany board previous to cutting out with the bow or band-saw. To fix the shaped corner pieces to the shelf, two strips about 1 in. wide by ⅜ in. thick are halved together, and their ends sunk into the shaped supports and screwed as shown in Fig. 365. The lower ends of the shaped pieces are screwed from the under side of the top. To simplify the work, the upper shelf and the corner supports may be omitted without impairing the appearance of the cabinet.

Carcase of Centre Cabinet.—The carcase now remains to be made. The putting together by mortising and tenoning of the framework requires careful and neat workmanship. The upright corner posts on the opposite side to the door are 1½ in. square, with the inside corners bevelled off, as at D (Fig. 366). Those right and left of the door are 1½ in. by ⅞ in., as at E (Fig. 367); and the door stiles are 1 in. wide by ⅞ in. thick. The middle bars and cross bars are ¾ in. wide on the face by ⅞ in. thick. The top and bottom rails are 1½ in. wide by ⅞ in. thick, and are rebated to receive a pine top and bottom, each ⅞ in. thick. Fig. 368 is a section, and Fig. 369 a part plan, showing the method of connecting the top and bottom to the framing with screws. To receive the plate glass a $\frac{3}{16}$-in. rebate is worked on the cross bars and framing; and the edges next to the glass are hollowed with a quarter circle moulding, as shown in Figs. 366 and 367. A line of satinwood stringing is inlaid in the middle of the cross bars, corner posts, and top and bottom rails. Two shelves are required opposite the cross bars; these may be of pine, covered with a suitable shade of velveteen, the bottom of the carcase being covered with the same material. The shelves are fixed by slanting screws at each corner. The glass may be fixed with putty, or with beads, as shown in Figs. 366 and 367. Three small butt hinges and a very narrow lock are required for the door. The usual french-polishing will complete the cabinet.

COUCHES, SOFAS, AND SETTEES.

Sofa.

THE parlour sofa shown at Fig. 370 is good and substantial, and by no means difficult to make. The whole of the woodwork may be yellow pine; or, if preferred, exposed parts may be walnut or mahogany. The bottom frame (see half plans, Figs. 371 and 372) is made of $2\frac{1}{2}$-in. by 2-in. stuff (see section of front rail, Fig. 373), which should be free from twist, the corners being halved together. The braces are housed into the front and back rails, and take the centre dowel of the legs; these are turned from blocks 10 in. by 5 in. square (see Fig. 374), the toe being cut to take either pin or socket castors. The position of the legs is shown in Fig. 371; they are fixed to the braces by means of dowels, and well glued. The plinth (see Fig. 373) runs right along the front rail and along each end, the corners being mitered. The arm scrolls, which should be set out as shown in Fig. 375, are cut from $\frac{7}{8}$-in. stuff. The narrow outline is the show-wood or outer scroll, and the wider one the stuffing scroll. The stuffing scrolls are framed up with housed braces to the exact width of the bottom frame (see Fig. 376), and are jointed to the bottom frame by three dowels at the bottom of each scroll. The outer front scrolls are extended at the bottom to meet the plinth on the sofa front, as shown in Fig. 370, and are rounded or beaded on the edge; they are then glued to the front stuffing scrolls. The back is made of $\frac{7}{8}$-in. stuff. A half elevation of this is shown at Fig. 377, and a section through the top rail at Fig. 378. It should be framed together with dry hardwood dowels. Before the ends are trimmed, the back frame should be screwed in position; then the ends can be marked off to the sweep of the stuffing scrolls and dressed down to the lines. A half-round mould is fixed to the top edge of the back, two curved

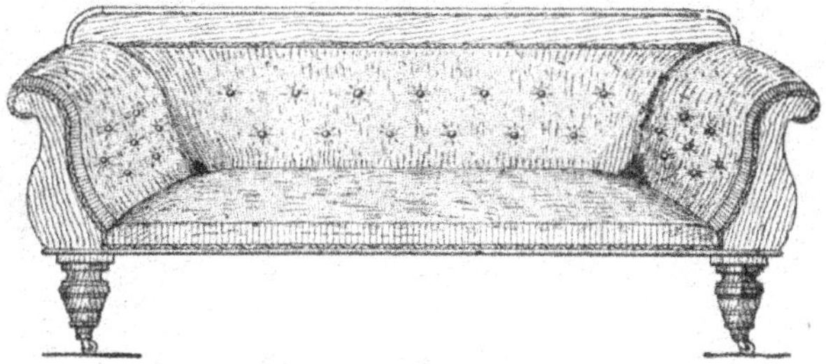

Fig. 370.—Front Elevation of Sofa.

pieces being worked round the corners. A **V**-groove is made with a parting tool or scratch beader, 1 in. from the lower edge, to serve as a tacking line for the stuffing. Before stuffing, rasp or shave off all sharp edges on the woodwork. A section of a stuffed arm and spring seat is shown at Fig. 379. The seat will require two dozen springs, placed in eight rows of three each, of a variety known as 9-in. hard furniture springs. The front edge of the seat and the front sweep of the arms are well stitched up with three rows of stitches. A section of the stuffed back is shown at Fig. 380; the top edge is formed into a roll and stitched up. The stuffed portions having been finished in canvas or calico, the woodwork can be stained and polished, and touched up when the cover and gimping are placed in position. A good quality of American leather, or hair seating, will make a suitable covering for a sofa of this kind.

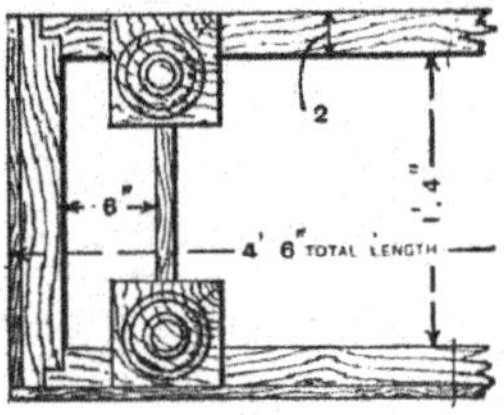

Fig. 371.

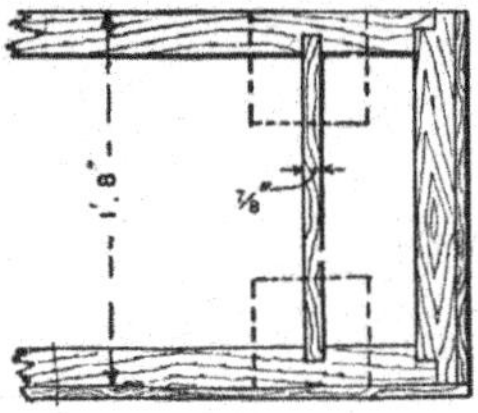

Fig. 372.

Figs. 371 and 372.—Half Plans of Sofa's Bottom Frame.

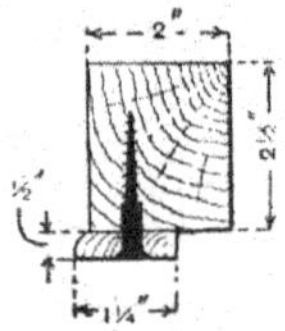

Fig. 373.—Section of Sofa's Front Rail.

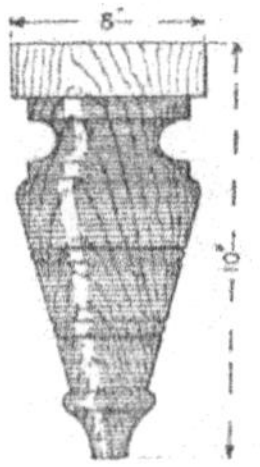

Fig. 374.—Leg of Sofa.

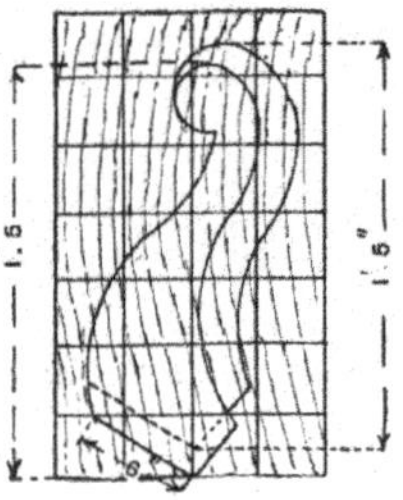

Fig. 375.—Pattern for Sofa Scrolls.

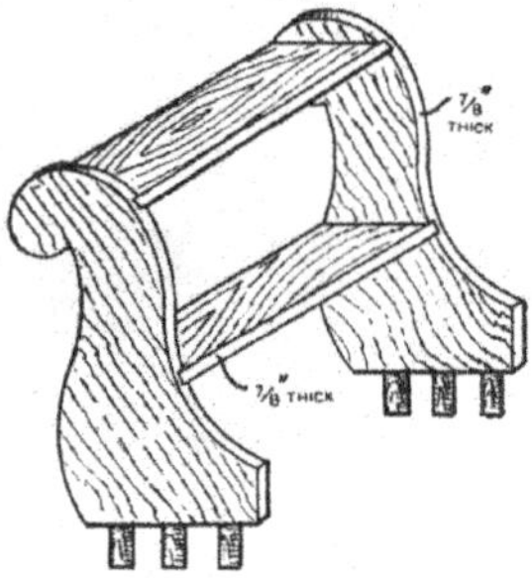

Fig. 376.—End Frame of Sofa.

Cromwell Couch.

Of the Cromwell couch (Fig. 381), the first part to be made is the bottom frame (Fig. 382), for which 14 ft. 6 in. of pine 3 in. by 1¾ in. will be required. Plane up the stuff and form a rectangular frame 5 ft. 6 in. long by 1 ft. 8 in. wide, with ordinary halved joints at the corners. The spring rails are 1 ft. 6 in. long, 3 in. wide, and ⅞ in. thick, after being dressed. Two pieces, each 1 ft. 6 in. long by 3 in. wide, are needed for the leg stays, the one at the head being let into the side rails at a distance of 6 in. from the end (inside measurement); the stay at the foot is attached in the same way at a distance of 3 in. from the rail. The spring rails are let into the side rails for a depth of 2½ in. from the top, and are secured from the front and back with nails. The curve at the foot is made out of a piece of stuff 2 in. by 3 in.; leave it of the full

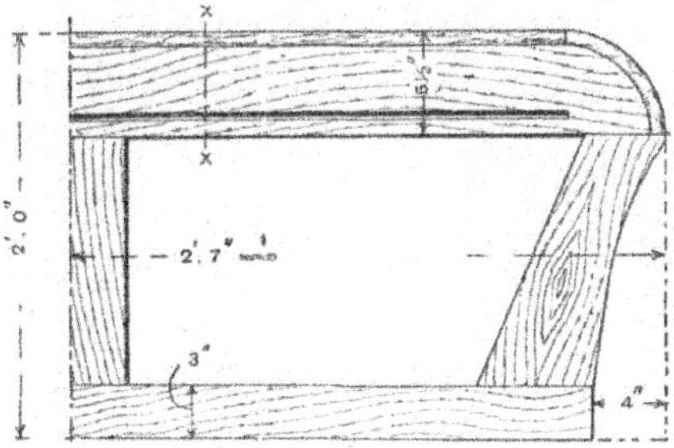

Fig. 377.—Half Elevation of Back Frame of Sofa.

width in the centre and work out the curve to the ends, making it 1 ft. 8 in. over all. The curve is nailed on the foot rail. The four legs are each 9 in. long (exclusive of the castors) and 5½ in. wide, and are each secured to the frame with three dowels, two of the dowels passing into the side rails

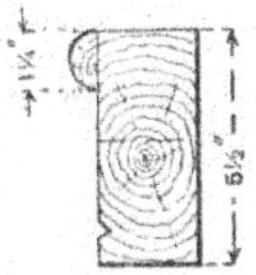

Fig. 378.—Top Back Rail of Sofa.

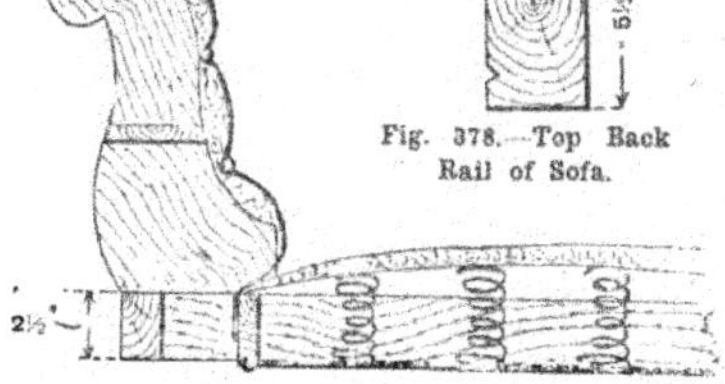

Fig. 379.—Section of Sofa's Spring Seat and Stuffed Arm.

and one into the leg stays. Make the dowels a clean driving fit, glue them in, and allow the work to dry. The foregoing refers to a couch having a spring seat; if a cushion seat is desired, the spring rails can be substituted by a solid boarded bottom, or a cross-webbed bottom covered with Hessian cloth. The stuffing scrolls

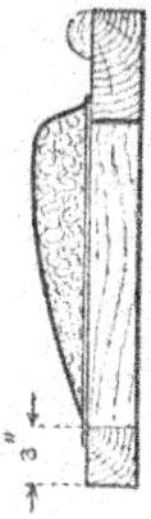

Fig. 380.—Section of Stuffed Back of Sofa.

are made from 1-in. stuff, and are sawn out with a band saw, a jig saw, or a compass saw, to the pattern shown by Fig. 383 (reproduced to scale of 1½ in. to the foot, approximately). Make a full-size pattern or template of stout paper or cardboard, place it on the wood, and mark off the positions. After sawing, dress the edges of the curve with a spokeshave, and shoot the bottoms with a trying plane. Frame up with two cross-pieces, fitting in grooves 2 in. from the top and 6 in. from the bottom of each frame. The outside measurement of the scroll frame must be of the exact width of the bottom frame, measured across the end. Gauge from the outside for the dowels in the centre and along the bottom frame, bore the holes with a centre-bit, glue, and drive the dowels in, being careful to make a good fit. The outside scroll, moulding, and back, with the exception of the stuffing rails and the bottom rail, should preferably be made of hardwood, such as mahogany, walnut, birch, etc. The moulding is of a simple quarter-circle pattern, and runs along the whole length of the front, round the end, and up the back to the terminal. If any difficulty is experienced in forming the bend at the end, make a few saw kerfs half-way through the moulding from the back bottom edge. Make lap joints at the corners, and finish off with a file. The outside scroll is secured with screws from the inside of the stuffing scroll.

Back of Cromwell Couch.—The back can now be made. Saw out the terminal to the pattern given at Fig. 384, which is reproduced one-eighth full size. This terminal

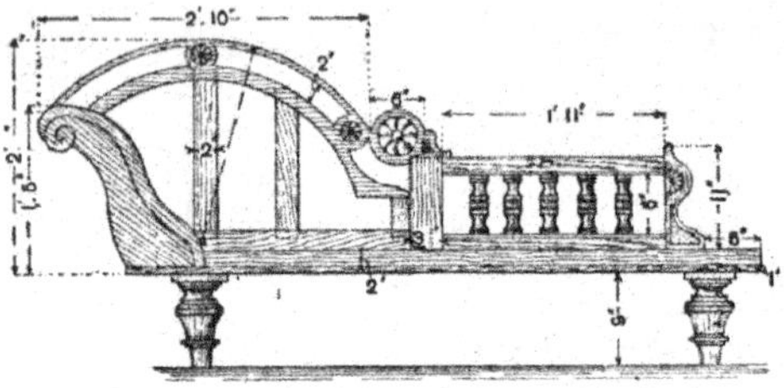

Fig. 381.—Front Elevation of Cromwell Couch.

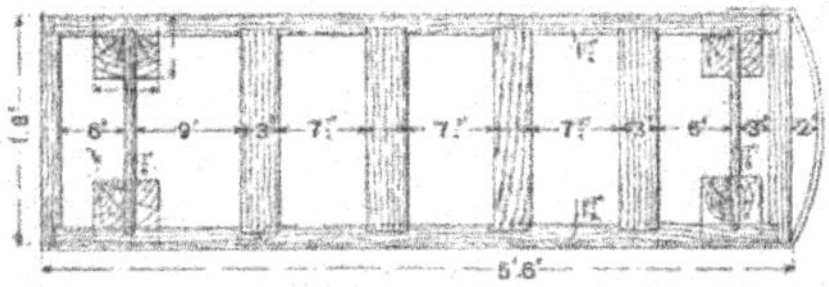

Fig. 382.—Plan of Bottom Frame of Cromwell Couch.

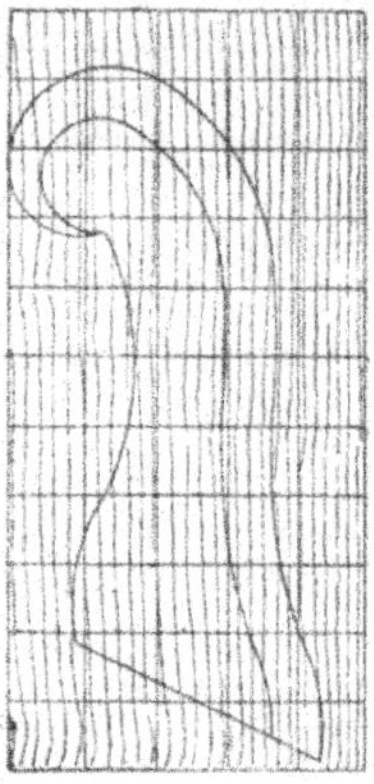

Fig. 383.—Pattern for Scroll of Cromwell Couch.

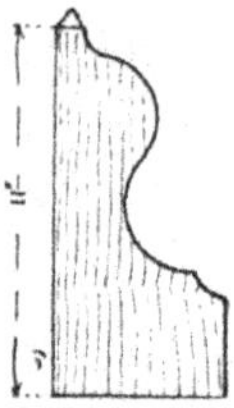

Fig. 384.—Pattern for Terminal of Couch.

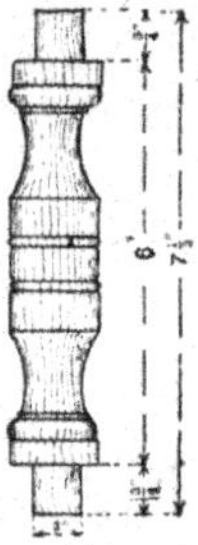

Fig. 386.—Spindle of Couch.

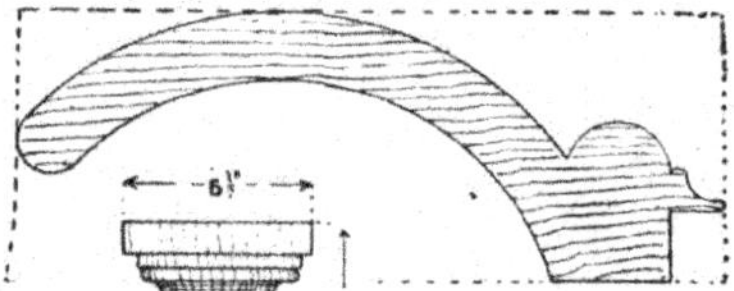

Fig. 387.—Pattern for Couch Back.

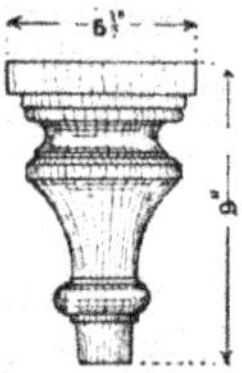

Fig. 388.—Leg of Couch.

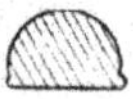

Fig. 385.—Section of Couch Handrail.

The edges of both scrolls are flush with each other under the couch head, the front of the stuffing scroll being raised about 2 in. on the front, which, when stuffed, will give a height of 3 in. above the outside scroll—thus making what is known as a German arm, which does not require a pillar.

should be made 5 in. wide at the bottom, and $1\frac{1}{2}$ in. thick, and the curved edge, after being sawn and dressed, is reeded with four $\frac{1}{4}$-in. half-round reeds, and the top pointed to a pyramid 1 in. long. The two spindle rails are $1\frac{1}{2}$ in. thick, and are beaded on the inside edge; they are stump-tenoned into the terminal and middle rail, leaving a space 6 in. wide for the spindles. The handrail at the top of the spindle frame is of the shape shown at Fig. 385. The spindles (Fig. 386) are 6 in. long, exclusive of the rail, and with a screw through the foot ornament. The supporting rails of the curve are made from pine 2 in. wide and 1 in. thick, and the upright rails are stump-tenoned into the curved back and bottom

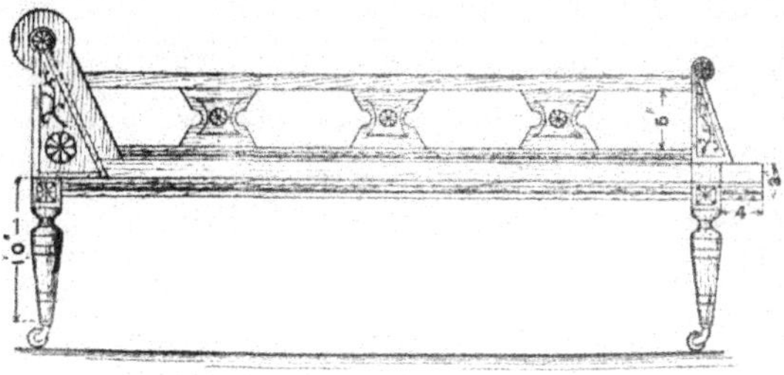

Fig. 389.—Front Elevation of Drum-head Couch.

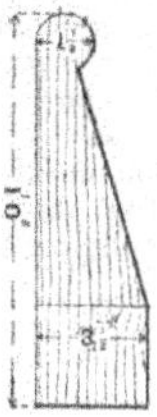

Fig. 394.—Terminal of Drum-head Couch.

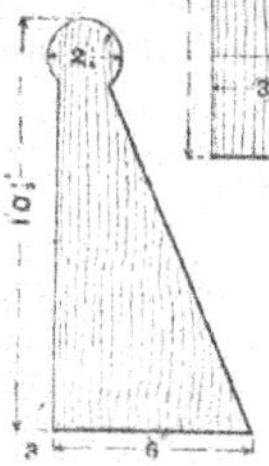

Fig. 392.—Outside Scroll of Couch.

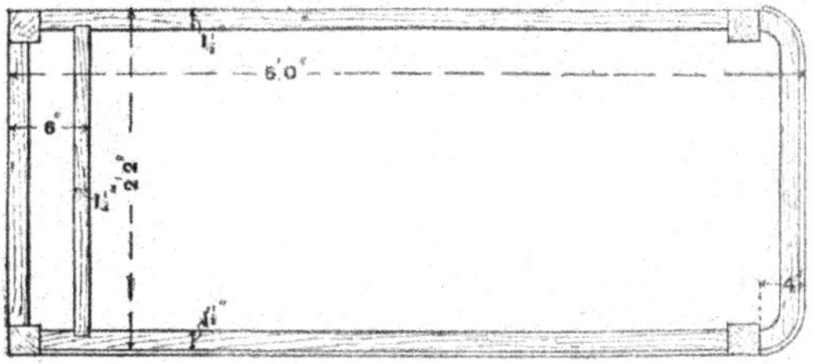

Fig. 390.—Bottom Frame of Drum-head Couch.

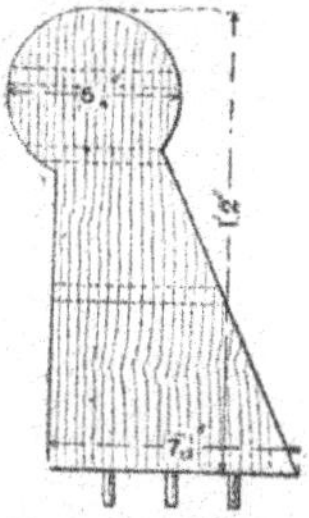

Fig. 393.—Stuffing Scroll of Couch.

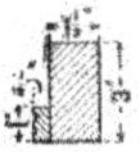

Fig. 391.—Section of Couch Rail.

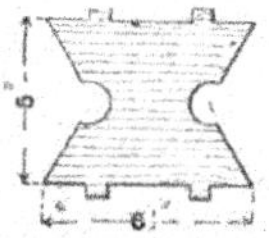

Fig. 395.—Baluster of Couch.

tenon at each end, and 1¼ in. thick. Care must be taken to see that they are quite upright before being glued. The curved back is made from 1½-in. stuff, and is sawn out to the pattern shown at Fig. 387. The back is made full in order to leave sufficient material for dressing, etc., and is fastened with dowels to the inner side of the middle back rail. When all the joints have been made and squared, glue up, and cramp until set. The back is secured to the carcase with two 3-in. screws in the terminal, two similar screws in the middle rail, and four 2-in. screws in the bottom back rail; two 3-in. screws are put through the end of the curved back, passing into the back stuffing scroll. The carving is simple incised work, and is worked as follows:—Mark out the design on the wood, follow the lines with a V-tool or veiner, then go over it with a ¼-in. spade tool, finally cleaning out with a regulator. The small rosettes can be made with the V-chisel alone, and the long lines on the back can be marked out with a scratch gauge. After brass socket castors have been fitted on the couch legs (Fig. 388 shows one of the legs) the frame is complete and is ready for upholstering.

Drum-head Couch.

Fig. 389 shows in front elevation the frame of a drum-head couch; for it, stained moulding, if worked independently of the rail (see Fig. 391), will require kerfing to the sweep of the corners. The bottom frame is stiffened by a rail housed into the side rails as shown in Fig. 390, and also by an iron rod 1 in. wide by ¼ in. thick, screwed flush on the under side of the frame.

Head and Back of Drum-head Couch.—After the bottom frame is glued and cramped the couch head can be set out. Fig. 392 shows an outside scroll. First, from 1¼-in. stuff, with a compass saw or band saw cut a pair of stuffing scrolls (Fig. 393); these are framed up by cross rails housed into their inner faces as shown by the dotted lines in Fig. 393. The head, when complete, should measure 2 ft. 2 in. across, and is

Fig. 396.—Box Ottoman.

birch or mahogany or walnut to harmonise with surrounding furniture may be used. The style of decoration lends itself readily to incised carving, and the rosettes may be worked in low relief, when brown oak would be a suitable wood. First make the bottom frame (Fig. 390); the side and end rails are each 3 in. by 1½ in. thick, and are dovetail-mortised into the leg stumps, which are 2¼ in. square by 1 ft. long without castors. The curved foot-rail is dowelled into the leg stump; this rail, if worked from the solid, will require rebating on the outside curve to a depth of 2 in., leaving a projection ½ in. wide by 1 in. deep for working a plain centre beading to match the one which is stuck on the front. The secured to the bottom frame by three dowels at each side. The ornamental scroll is fixed by gluing and screwing through the inside of the stuffing scroll, and should lie flush with the beaded moulding as shown in Fig. 389. The couch back is made of two 1¼-in. rails, each 2 in. wide and 4 ft. 3 in. long, the bottom rail being centre beaded to match the front moulding. These rails are secured by mortice joints to the terminal (Fig. 394), and before the rails are secured the three balusters (Fig. 395) must be cut and fitted. The terminal is half jointed 3 in. from the bottom and secured to the side rail by three screws. The two rails are cut halfway through where they meet the back stuffing scroll, and are screwed into the back. In this pattern of couch the bolster arm and head are firmly stuffed, and the seat and head swell are sprung; but before beginning stuffing rasp all the edges over which the covers will be laid.

Box Ottoman.

A box ottoman (see Fig. 396) is often quite roughly nailed together, as the inside is lined with print or calico. A useful size is 3 ft. 6 in. by 1 ft. 7 in. and 1 ft. 2 in. deep, all outside measurements, with the lid 3 ft. 7 in. by 1 ft. 8 in., this allowing ½-in. projection all round the box. The cross battens on the lid (see Fig. 397) are 4 in. wide by 1 in. thick, and a little shorter than the inside width of the box. For the covering of the outside, cretonne or damask will look well, with a slightly different shade for the festoons and tails. When the box is made, the first thing is to put on the pleated sides and ends. These should be tacked on ½ in. lower than the top edge to prevent all the tacking parts coming over each other. Four castors, about 2 in. high, should be screwed to the under side of the box. The pleated sides and ends must reach to within ¾ in. of the floor, the method of pleating being shown in Fig. 398. Next cut out the festoons, which, when finished, must measure 5 in. at the widest part, and 1 in. at the ends below the bottom edge of the fringe. As shown in Fig. 399, the ends are sloped away, so that when gathered there will not be too much bulk of stuff for the tacking. The festoons are tacked on pleats, underneath the fringe. Before cutting out the festoons, practise on a waste piece of material, which may be used as a pattern to mark out the finished material. The corner and side tails may now be added, and should be set out on paper to Figs. 400 and 401. For Fig. 400 the centre line B E is 12¾ in. long, B F 1¾ in., F D 2¾ in., D to H 4¾ in., B A 5⅛ in., D C 8¾ in., and H G 5⅛ in. The sweep of line C G E is obtained by the intersection of lines A G and G H. The segments are about equal, and are, say, 1 in. wide at the top and 2 in. at the bottom. Set out Fig. 401 from the following measurements:—B E 12 in., B F 1 in., F D 3¾ in., D H 4⅞ in., B A 3¼ in., D C 6¼ in., and H G 3¼ in. The segments are about ⅝ in. wide at the top and about 1⅜ in. wide at the bottom. A narrow fringe is sewn to the bottom edges of the tails. The plan of

Fig. 397.—Box Ottoman before Covering.

Fig. 398.—Folding of Pleats for Box Ottoman.

folding is shown in Figs. 402 and 403. In tacking on the festoons and tails, the distance from the top edge of the box is regulated by the headings of the fringe, for if the fringe is of an open trellis pattern the festoons and tails would be seen through the fringe. This is fixed on by suitable nails or gimp pins. Next cover the inside of the box with the print, calico, or glazed holland, letting it lap on the top edges of the box. If preferred, the box may be lined first, the lining being brought to the outsides of the box before the pleated parts are fastened. The lid may have a flat stuffing of hair or flocks. The outer covering is brought to the under side of the lid, and then the lining is added, the battens being first covered. The lid is now connected to the box with three brass butt hinges; and to prevent the lid falling back and wrenching off the hinges, tapes are connected with each end to inside the box. When the lid is closed, the tapes fall inside.

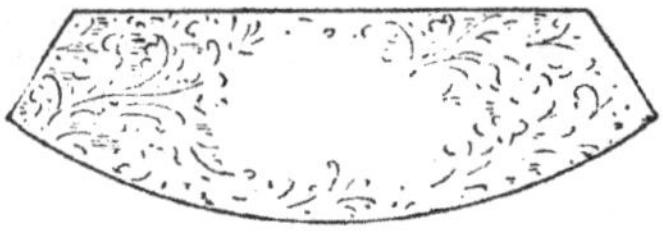

Fig. 399.—Festoon Piece for Box Ottoman.

Fig. 402. Fig. 403.

Figs. 402 and 403.—Folding Tails for Box Ottoman.

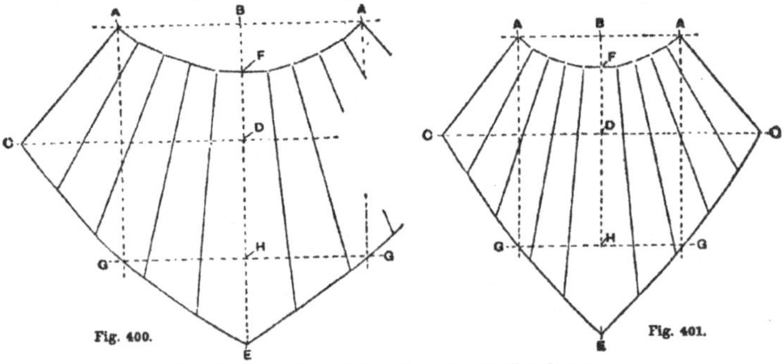

Fig. 400. Fig. 401.

Figs. 400 and 401.—Development of Tails for Box Ottoman.

Box Ottoman Settee.

The box ottoman settee shown in Fig. 404 has a framework nearly all the parts of which can be got out of 1-in. by 11-in. boards, a size which is stocked in yellow pine by timber dealers. Begin by making the box or settee body (Figs. 405 and 406). For dovetail joints at the corners the dimensions given will apply, but when cross-cutting the boards for plain lap joints, the two end pieces should be 1 ft. 7 in. long, the other dimensions remaining the same. To get the depth, it will be necessary to join two pieces of stuff, and with care a full width and a half of the 11-in. boards will, after jointing and shooting, come out the correct width for each side. Before the carcase is put together, the rebates for the bottom should be ploughed out as shown in Fig. 406. The bottom is strengthened by three cross-braces, 2¼ in. wide by 1 in. thick, let in flush with the bottom boards (see Fig. 405).

Fig. 404.—Box Ottoman Settee.

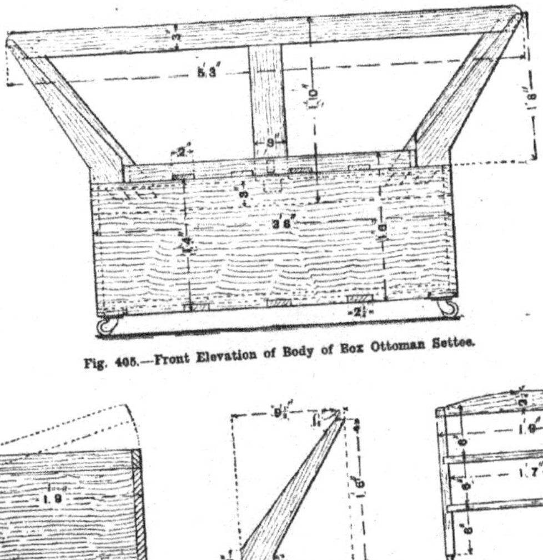

Fig. 405.—Front Elevation of Body of Box Ottoman Settee.

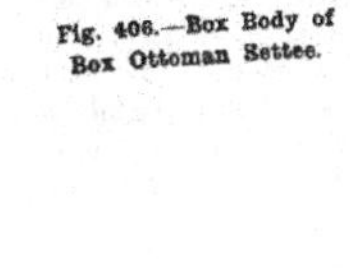

Fig. 406.—Box Body of Box Ottoman Settee.

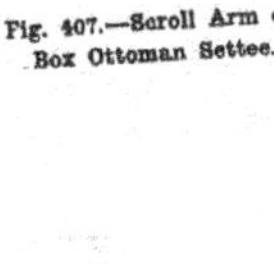

Fig. 407.—Scroll Arm of Box Ottoman Settee.

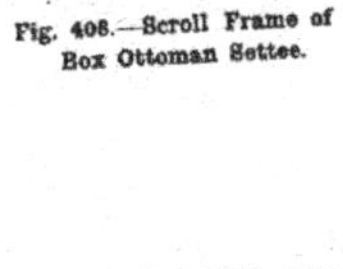

Fig. 408.—Scroll Frame of Box Ottoman Settee.

At each corner, glue and screw a castor block, 3 in. square by 1 in. thick. Next get out four scroll arms as Fig. 407, dressing all the angles to the same sweep. The bottom edges which fit on the box sides are shot square and true with the trying plane; frame should correspond with the width of the box, and when placed in position the two faces should be a good fit. The frames are secured to the box by dowel joints, two ½-in. dowels in each scroll being sufficient if the work is properly done. Use

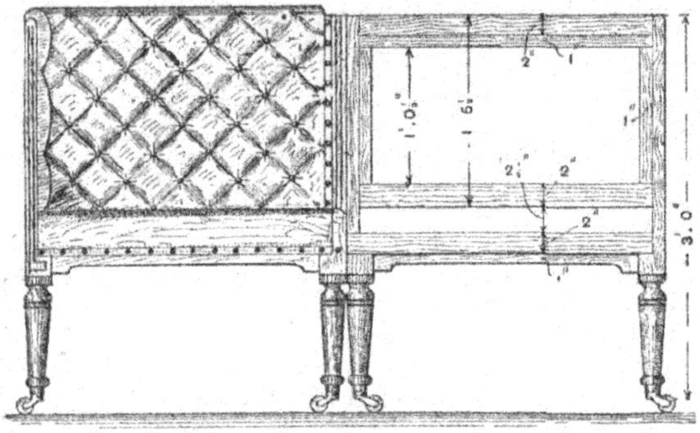

Fig. 409.—Half Front Elevations of Dividing Settee and Framework.

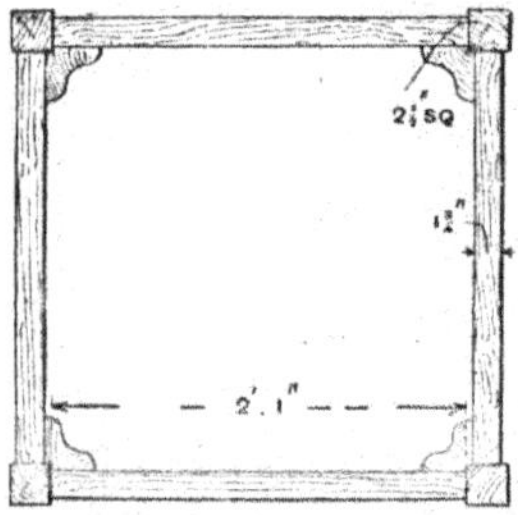

Fig. 410.—Seat Frame of Dividing Settee.

Fig. 411.—Pair of Turned Legs before Cutting in Two.

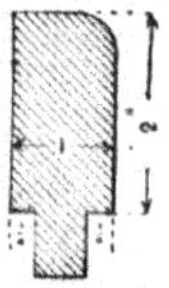

Fig. 412.—Top Rail of Dividing Settee.

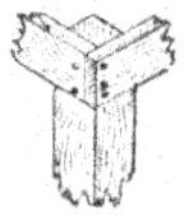

Fig. 413.—Joint of Settee's Top Rail and Corner Leg.

this is important, the correct fitting of the settee head to the body depending on it. Frame the scrolls in pairs, as shown in Fig. 408, by housing two cross rails to a depth of ½ in. into the inner sides of the scrolls, and by a head-rail, which is cut to a sweep of 2¼ in. in the centre to 1 in. at each end. When put together, the outside width of the dry hardwood dowels and fresh hot glue. The back is 5 ft. 3 in. long at the top, but it would be advisable to check this measurement before cross-cutting, as more or less rake may have been given in fitting the two arm frames. The top and bottom rails are jointed by a centre piece 3 in. wide, and by two shaped end pieces, which are marked and cut to the outline of the back scrolls after the back is made up. The dimensions given in Fig. 405 are face measurements only, and if it is intended to mortice-joint the back, allowance must be made for tenons in cutting the stuff. The top corners are rounded, and the back is fixed by screwing to the box back and the back scrolls. The seat frame is 3 ft. long by 1 ft. 9 in. wide and 2 in. thick, and the corners are dovetailed and four cross rails are let in flush with the bottom edges (see Fig. 405). For a loose cushion seat, the seat frame need only be made on the flat from 1-in. stuff, the loose cushion lying on the top. Special attention should be paid to castoring, as the settee, when the box portion is filled with goods, will have considerable weight; use 2-in. plate castors with solid brass runners. Before beginning to stuff, shave or rasp off all the sharp edges over which the covers will pass. The back is removed and stuffed on the bench. The lower part of the back and the insides of the arms are tufted and buttoned. The seat is sprung with eight 6-in. chair springs, fixed two to each cross-rail. A cotton tapestry or cretonne, at about 1s. a yard, will be a suitable material for covering the settee, and the whole of the outsides of the box and scrolls should be covered by pasting and tacking, the edges being finished with coloured furniture cord, slip-stitched on. The inside of the box should be either stained and varnished or painted light blue.

Dividing Settee.

A settee constructed as shown in Fig. 409 is found very convenient where room is a consideration. Fig. 409 shows the combination as used to form a settee, but if the parts were placed back to back, the result would take the place to a certain extent of the useful though cumbersome centre ottoman. Another position, gained by partly dividing the front, and allowing the back corners to touch, would result in each sitter being independent of his neighbour, or the settee can be properly divided and used as two corner chairs. Full dimensions for one-half only are given, as these will apply exactly to the other half, the only difference in the halves being that they are right- and left-handed. Fig. 410 shows the seat frame. To begin with the construction of the settee, the legs and stumps are got from 2½-in. square stuff. Two legs can be set out and turned from a piece 3 ft. 2 in. long, the turning being done before sawing out the legs (see Fig. 411). Four legs, as shown in Fig. 409, will be required, and also two stump feet, each 1 ft. by 2½ in., and two back corner legs, which are left 2½ in. square for the full length, the other legs being cut down to avoid undue heaviness. To get the two seats close together, the rails are sunk ⅜ in. behind the level of the feet; and if the edges of the upholstered seats are properly stitched up, they will slightly overhang and close the breach. The rail can be either stump-tenoned or dowelled into the legs; and if the former method is adopted, allow for tenons when cutting off the stuff. The seat frame will be greatly strengthened by wood dogs glued and screwed to the rails, as in Fig. 410. The top rail is shaped and tongued as shown in section at Fig. 412, and the method of attaching this to the back corner legs is explained by Fig. 413, the top of the leg being cut away to the thickness of the rails. Stuffing rails are fitted all round the seat frame, and two such rails are added to complete the panel rebates in the back. The decoration of the show-wood portions can be of inlaid stringing or reeding. The settee is mounted on eight socket castors, and in order to prevent motion when in use as a settee, four brass catch hooks, two underneath and two behind, engage with the brass eye-screws, and keep the two halves together. The dimensions given are suitable for pine or whitewood; but for hardwood, such as mahogany, walnut, oak, etc., the various members can be diminished by about one-fifth.

COAL VASES AND CABINETS.

Coal Vase.

The coal vase shown by Figs. 414 to 417 should be made in oak. The two sides must be got out first, these being of ¾-in. full board when finished. They are cut to the shape shown by Fig. 415, the extreme measurement being 1 ft. 5 in. long by 1 ft. 1 in. high. They are then grooved half-dovetail and rebated on the back edge, as shown by Fig. 417. The back and bottom are of ½-in. board, and the rebate and bottom groove should be made accordingly. The top is of ¾-in. board, which should be not more than 2 in. below the extreme top point of the sides. It is cut 1 ft. ¾ in. long by 10 in. wide, ⅜ in. at each end being dovetailed on the under side to fit the grooves. The front edge is planed to a bevel; the back edge comes to the corner of the rebate. The bottom, of ½-in. board, is 1 ft. ¾ in. long by 1 ft. 3 in.

Fig. 414.—Coal Vase.

wide, the length being measured along the grain; it is then treated the same as the top. When this is done, the pieces are slid in place in the grooves, using glue, which must be hot and not too thick. See that the bottom is not out of square; then stand it upside down and glue four blocks at the corners as at Fig. 416.

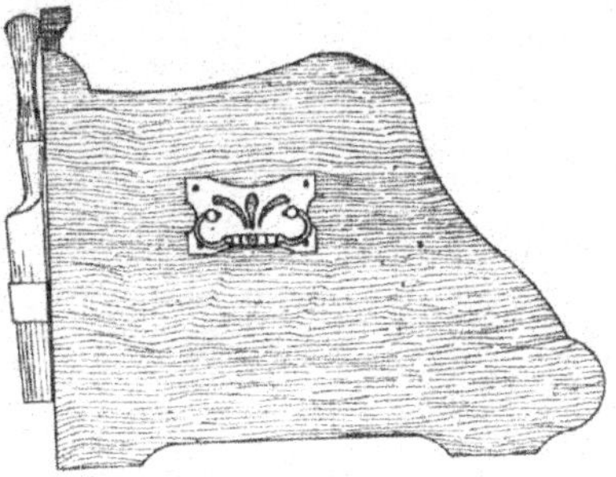

Fig. 415.—Side Elevation of Coal Vase.

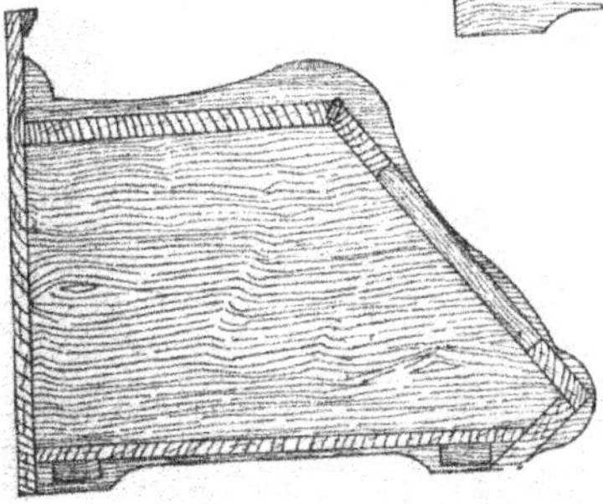

Fig. 416.—Cross Section of Coal Vase.

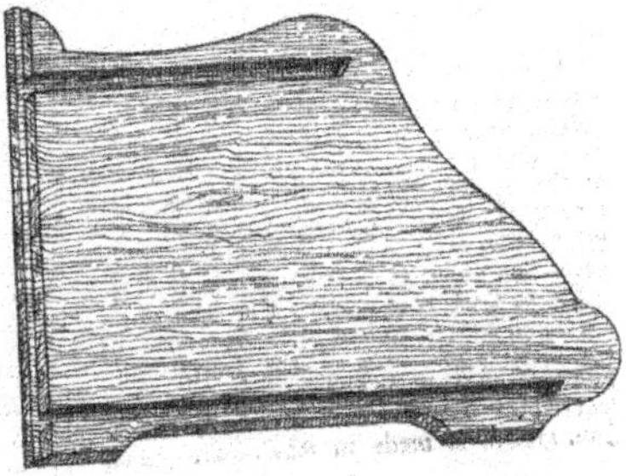

Fig 417.—Inner View of Side Piece of Coal Vase.

Back of Coal Vase.—The back, with pediment attached, is 1 ft. ¾ in. long by 1 ft. 3 in. high by ½ in. thick, the lower edge being cut out like the sides; the upper part, which forms the pediment, is cut to the design shown in Fig. 414. Five flutes are then worked on the front with a ⅜-in. gouge, and a piece of 1-in. by ⅝-in. curved ogee moulding is put on top, as shown in section by Fig. 418. The back is then fixed in place with screws to the sides, bottom, and top. The foot piece at the front measures 1 ft. long by 3¼ in. wide by ¾ in. thick. It is neatly fitted and secured by two sprigs through each of the sides, and to the front edge of the bottom, glue being also used. The sprigs should be punched below the surface, the holes being filled with a stopping to match the wood.

Fig. 418.—Section of Coal Vase Pediment.

Fig. 420.—Strap Hinge for Coal Vase.

Fig. 419.—Section of Coal Vase Lid.

Fig. 421.

Fig. 422.

Figs. 421 and 422.—Shovel Holder for Coal Vase.

Lid of Coal Vase.—The lid is next taken in hand. Its frame should be made of 2½-in. by ¾-in. material, the joints being mortised and tenoned. An ovolo mould is worked on the inner front edges, and a ¼-in. groove is made to receive the panel, which is added when the frame is put together (see Fig. 419). A flat ogee mould is worked round the edge of the face side of the panel, which is of, say, ½-in. board. A nicely figured piece should be chosen for this purpose if the panel is left plain as shown, but of course it may be carved or inlaid according to the worker's taste or ability, in which case the plainness of the material used does not matter. The lid is hinged to the front edge of the top, after being planed to fit, and special hinges are to be had for the purpose, of the same make

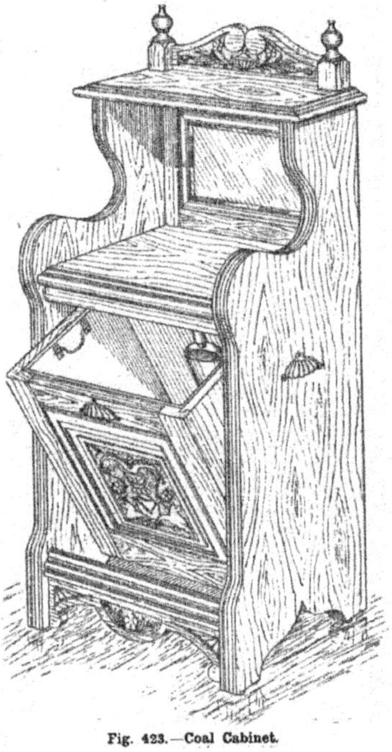

Fig. 423.—Coal Cabinet.

as those used for piano falls, but shorter. Strap hinges in brass or copper may be used (see Fig. 420), and are much easier to put on, though they are generally used to give ornament to plain lids, which are simply of board, the grain running vertically, with a narrow clamp at each end to prevent warping. The coal vase is now ready for polishing. When this is done, add the fittings, including two handles for the sides, the hinges, and the brass knob towards the lower edge of the lid.

Coal Shovel and Holder.—A small brass shovel can be obtained to match, on which may be put a wood handle of the same material as the box. Figs. 421 and 422 show the shovel holder, which is made from $1\frac{1}{2}$-in. by $\frac{1}{16}$-in. thick strip brass. This is easily hammered to the shape shown, and, when properly polished and lacquered, is screwed to the back, 5 in. from the floor. This holder keeps the shovel handy, though out of sight (see Fig. 415). A lining of galvanised sheet-iron should be made; but this, of course, is certainly a job for the sheet-metal worker.

Fig. 424.

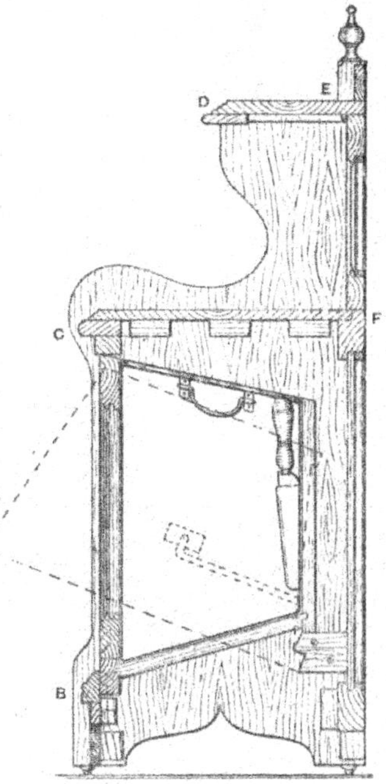

Fig. 425.

Figs. 424 and 425.—Front Elevation and Cross Section of Coal Cabinet.

Fig. 428.—Section of Bottom Rail, Span-rail, etc., of Coal Cabinet (B, Fig. 425).

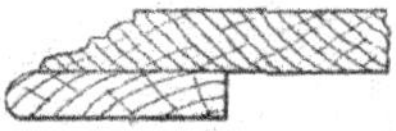

Fig. 430.

Fig. 430.—Top Shelf of Coal Cabinet (D, Fig. 425).

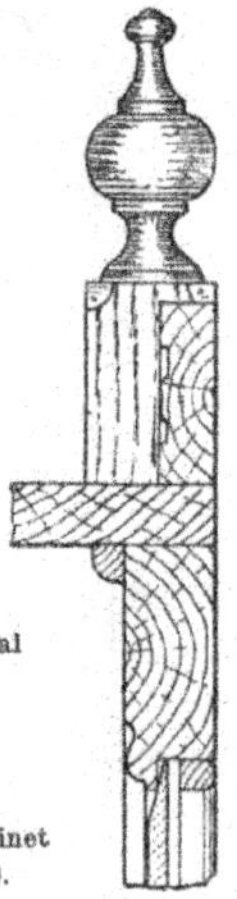

Fig. 431.—Detail of Coal Cabinet Pediment, etc. (E, Fig. 425).

Fig. 431.

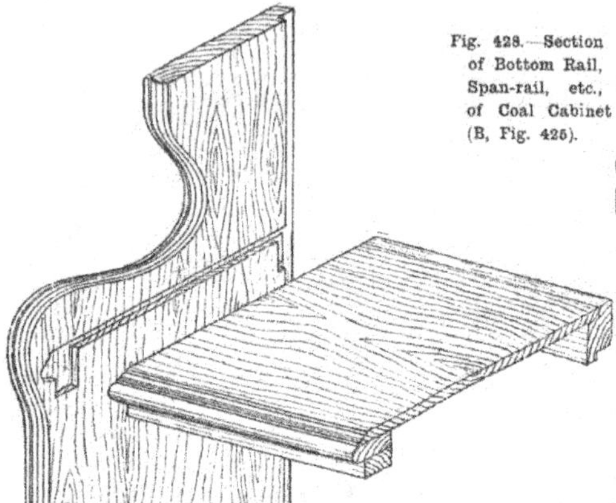

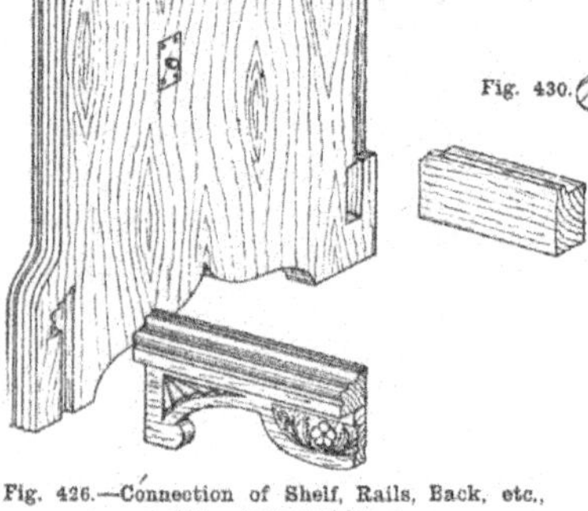

Fig. 426.—Connection of Shelf, Rails, Back, etc., to Side of Coal Cabinet.

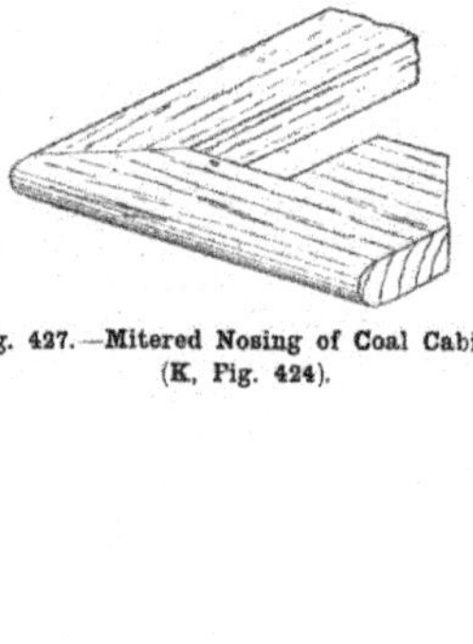

Fig. 427.—Mitered Nosing of Coal Cabinet (K, Fig. 424).

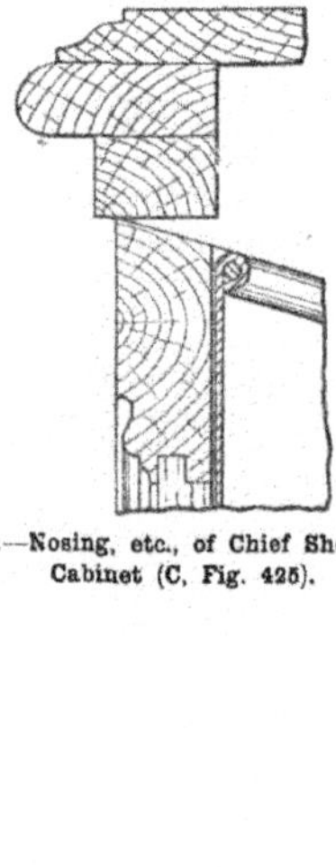

Fig. 429.—Nosing, etc., of Chief Shelf of Coal Cabinet (C, Fig. 425).

Coal Cabinet.

The coal cabinet shown by Fig. 423 (p. 120) is ornamental as well as useful. The coal box falls forward when required, the coal and coal shovel being quite hidden from sight when the box is pushed back into place. An illustration of the pivot and grooved piece by means of which the rotating movement of the coal box is obtained will be given later. Fig. 424 is a front elevation. The construction is fully explained in the vertical cross section (on line A A, Fig. 424) shown by Fig. 425. Some instructive details are presented by Fig. 426; this illustration shows the method of grooving, housing, mortising and rebating the sides to receive the shelf, rails, back, etc. Fig. 427 is a detail explaining the method of mitering the nosing at K (Fig. 424). Enlarged details at B, C, D, E and F respectively are shown by Figs. 428 to 432. An enlarged part horizontal section on the line G (Fig. 424) is presented by Fig. 433. The arrangement (already referred to) of supporting the coal box on pivots is illustrated by Fig. 434, the actual coal box of iron, without its wooden container, being shown by Fig. 435.

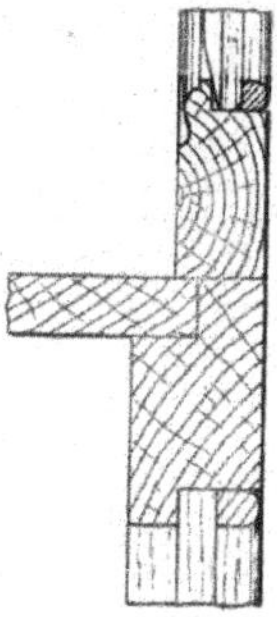

Fig. 432.—Detail of Back of Coal Cabinet (F, Fig. 425).

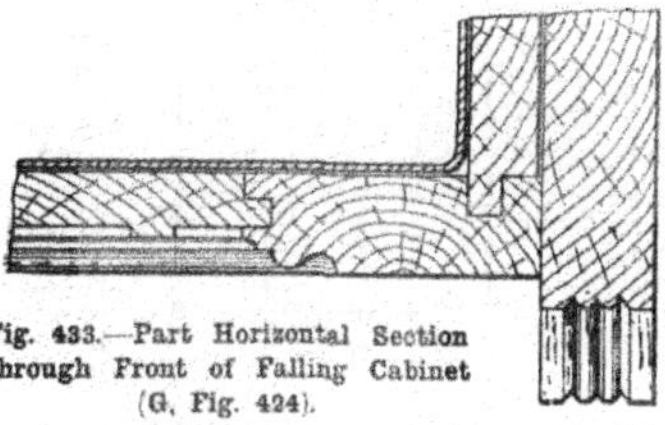

Fig. 433.—Part Horizontal Section through Front of Falling Cabinet (G, Fig. 424).

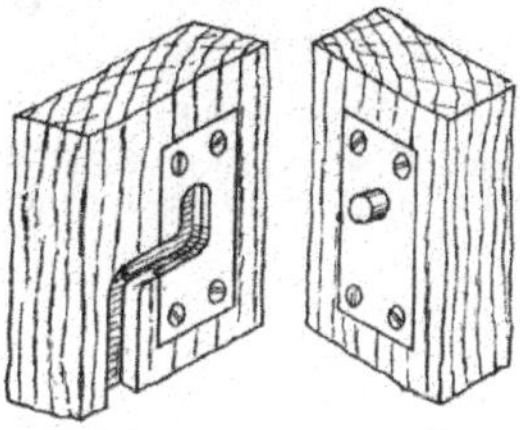

Fig. 434.—Plates and Groove for Rotating Coal Box.

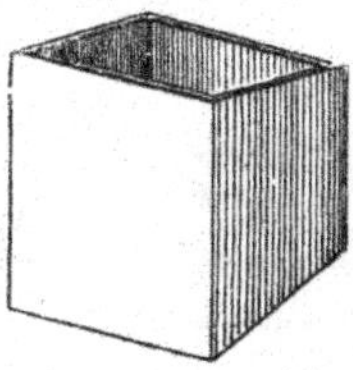

Fig. 435.—Sheet Metal Coal Box for Cabinet.

MUSIC FURNITURE.

Music Cabinet.

THE music cabinet shown in Figs. 436 to 439 is 4 ft. 9½ in. high by 2 ft. 2⅜ in. wide and 1 ft. 6 in. deep, excluding the projection of the moulding on the top. It is arranged with flaps and sliding trays, instead of the usual open shelves and drawers, the arrange-

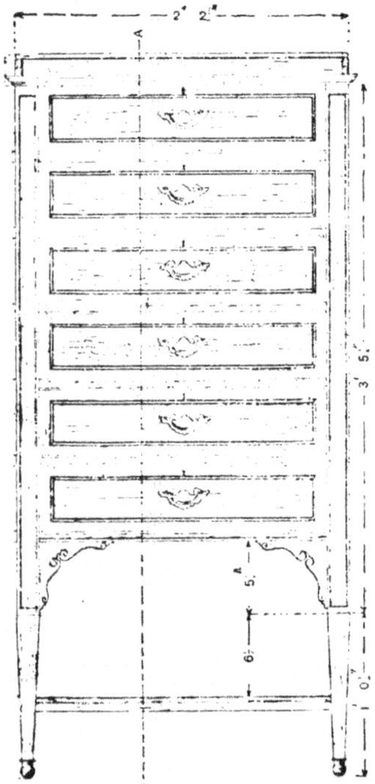

Fig. 436.

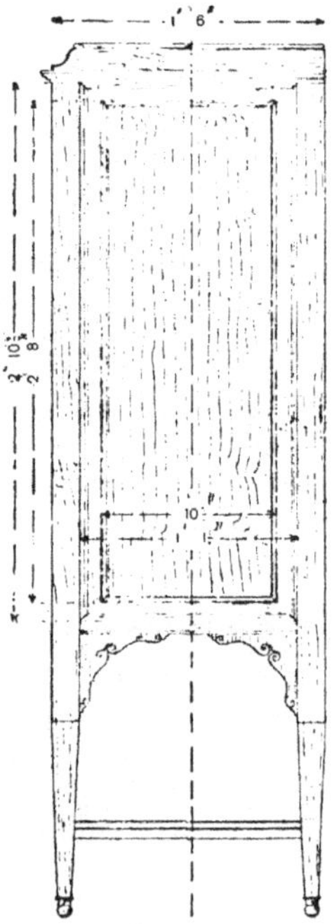

Fig. 437.

Figs. 436 and 437.—Front and Side Elevations of Music Cabinet.

ment giving easy access to sheet and other music. The flaps may all be down at the same time if necessary, and be held level with the division on which the tray slides by the brass quadrant stays. These stays are flaps are secured, when closed, by spring catches fixed on the inside, and can be opened only by a key. For some purposes, it will be found more convenient to use flush spring catches on the outside, without a key.

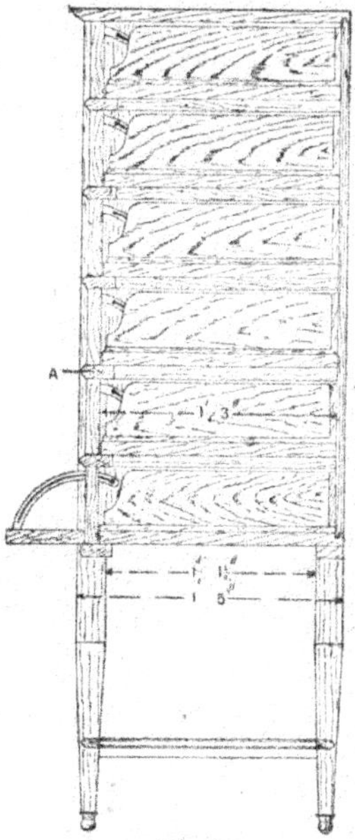

Fig. 438.

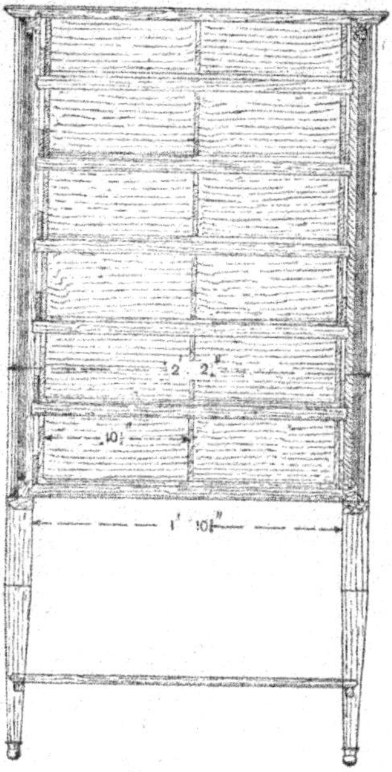

Fig. 439.

Figs. 438 and 439.—Cross Section and Longitudinal Section of Music Cabinet.

made of $\frac{1}{16}$-in. sheet brass turned up at one end and fixed to the flap and slotted in the centre to work on the screw. The tray is drawn out by a finger being placed through a hole in the division, the latter being in the centre of the tray. The flap when down forms a table on which the tray rests when partly drawn. It is unnecessary to draw the tray out the whole distance, except to place it on a table or counter, and the

Material for Music Cabinet.—The walnut required will be as follows: Four legs, each 4 ft. 9 in. by 2 in. by 2 in.; six front rails, each 2 ft. 2 in. by 2¾ in. by 1 in.; two end rails, each 1 ft. 4 in. by 2¾ in. by 1 in.; one division, 1 ft. 11 in. by 1 ft. 4½ in. by 1 in.; five divisions, each 1 ft. 11 in. by 1 ft. 3½ in. by 1 in.; one back rail, 2 ft. 2 in. by 2¾ in. by 1 in.; one top, 2 ft. 5 in. by 1 ft. 6 in. by 1¼ in.; six flaps, each

1 ft. 11 in. by $5\frac{1}{4}$ in. by $\frac{7}{8}$ in.; one bottom shelf, 2 ft. 2 in. by 1 ft. 5 in. by $\frac{3}{4}$ in.; two rails, each 2 ft. by 1 in. by $\frac{7}{8}$ in.; two rails, each 1 ft. 5 in. by 1 in. by $\frac{7}{8}$ in.; one skirting, 2 ft. 3 in. by $2\frac{3}{4}$ in. by $\frac{3}{4}$ in.; two skirtings, each 1 ft. 6 in. by $2\frac{3}{4}$ in. by $\frac{3}{4}$ in.; six tray bottoms, each 1 ft. 11 in. by 1 ft. $3\frac{1}{2}$ in. by $\frac{1}{2}$ in.; twelve tray sides, each 1 ft. $3\frac{1}{2}$ in. by 4 in. by $\frac{1}{4}$ in.; six tray backs, each 1 ft. 11 in. by 4 in. by $\frac{1}{2}$ in.; six tray divisions, each 1 ft. $3\frac{1}{2}$ in. by 4 in. by $\frac{1}{2}$ in.; one back of cabinet, 2 ft. 11 in. by 2 ft. by $\frac{3}{4}$ in.; four end stiles, each 3 ft. by $1\frac{3}{4}$ in. by 1 in.; four end rails, each 1 ft. $2\frac{1}{2}$ in. by $1\frac{3}{4}$ in. by 1 in.; two end panels, each 2 ft. 9 in. by $11\frac{3}{4}$ in. by $\frac{1}{2}$ in.; four mouldings, each 2 ft. 11 in. by 1 in. by 1 in.; four mouldings, each 1 ft. 2 in. by 1 in. by 1 in.; four mouldings, each 2 ft. 9 in. by 1 in. by $\frac{1}{2}$ in.; four mouldings, each 11 in. by 1 in. by $\frac{1}{2}$ in.; six astragal mouldings, each 4 ft. by $\frac{1}{2}$ in. by $\frac{3}{8}$ in.; six brackets, each 6 in. by 6 in. by 1 in.; and twelve guides, each 1 ft. 3 in. by 2 in. by $\frac{1}{4}$ in. The following will also be required: Four $1\frac{1}{4}$-in. Acme ball castors and screws; six fancy brass drop drawer handles and screws; six 2-in. brass spring locks and screws; six pairs of 2-in. brass arrow butts and screws; twelve 7-in. by $\frac{3}{4}$-in. by $\frac{1}{16}$-in. brass quadrants; twelve No. 12 screws; and forty-eight $\frac{5}{8}$-in. No. 10 iron screws. The flaps are $5\frac{1}{16}$ in. deep, and are parted by $\frac{13}{16}$-in. rails. The bottom ledge is $6\frac{5}{8}$ in. from the ground, and the rail round the top is $2\frac{1}{2}$ in. deep.

Fig. 440.—Enlarged Detail of Top of Music Cabinet.

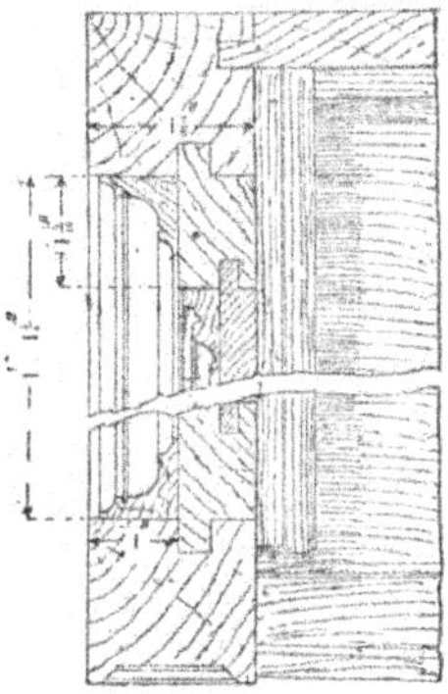

Fig. 441.—Part Horizontal Section of Music Cabinet.

Fig. 442.—Castor for Music Cabinet.

Legs of Cabinet.—To set out the various parts, first face the four legs straight, and square one edge and gauge each side to $1\frac{7}{8}$ in. thick; select for the front the two legs having the best figure, then place the four legs on the bench and put a square line across them 1 in. from the end selected to be the top, and from this line measure off 3 ft. $5\frac{3}{4}$ in., this being the distance to the commencement of the diminish of the lower end of the leg. From the second line measure 1 ft. $0\frac{1}{8}$ in. to determine the extreme length. Next set out for the sinking of the panels on the face side of the two front legs, the length of the panels being 3 ft. $3\frac{1}{2}$ in., leaving $1\frac{1}{8}$ in. from the underside of the top and from the line of the diminish

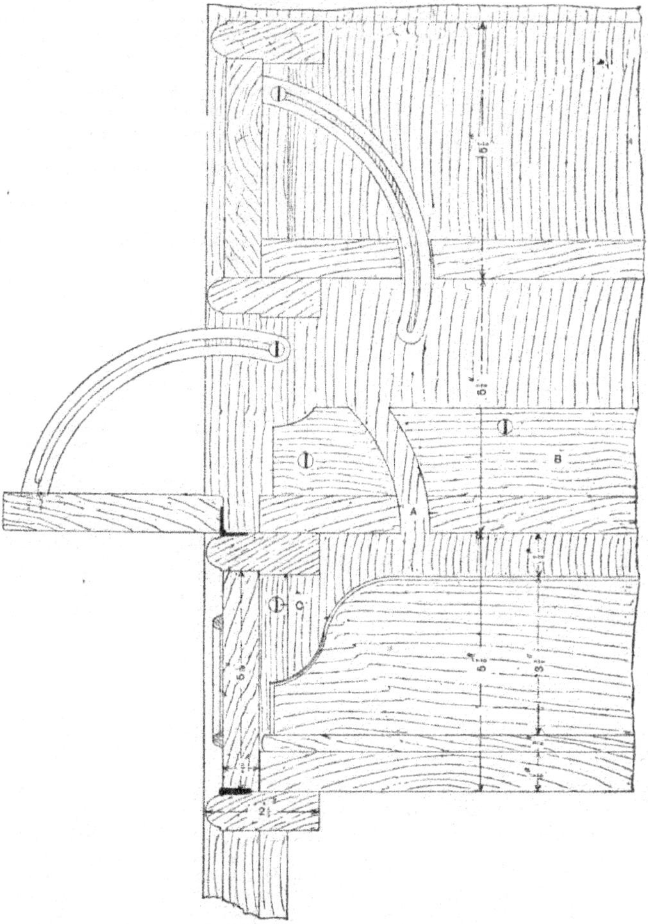

Fig. 443.—View showing Details of Music Cabinet Flaps.

(see Figs. 440 and 441). If more convenient to the maker, the moulding may be worked separately and fixed in with glue and needle points after the panel has been polished. From the line of diminish cut off the legs

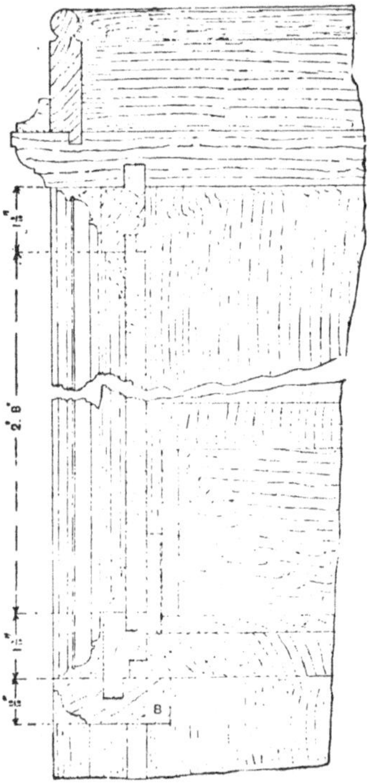

Fig. 444.—Part Vertical Section of End of Music Cabinet.

to 1 ft., and reduce them on each side, leaving them $1\frac{1}{4}$ in. square at the bottom, this being the size of the inside of the socket of the castor (see Fig. 442). The two back legs may now be prepared, but without the panel.

Rails.—Again take the front legs, and set out the mortices for the rails A (Fig. 438); these rails are $2\frac{1}{2}$ in. wide by $\frac{13}{16}$ in. finished thickness at the rebate and $\frac{7}{8}$ in. at the back part. These rails are framed into the legs. Set off from the line of diminish a distance of $6\frac{1}{2}$ in., and prepare a thin slip of wood as a gauge $5\frac{1}{16}$ in. long by $\frac{13}{16}$ in. wide. The length will be the exact width of the flap, and the width the thickness of the rails. From the $6\frac{1}{2}$-in. line mark on the thickness of the rail, and again the length for the width of the flap, and repeat this process until six spaces are marked. This setting out will of course be on the inside face of the front legs. Now prepare a double mortice gauge, and set it for the mortices, the first mortice being $\frac{1}{2}$ in. from the face of the leg; allow a $\frac{3}{8}$-in. mortice, $\frac{3}{8}$-in. intermediate space, and a $\frac{3}{8}$-in. mortice again, leaving $\frac{1}{4}$ in. on the inner face of the leg. The mortices are 1 in. deep, and the chisel should not go beyond the gauge line in the width, otherwise when the tenon is fitted a gaping joint will show. The front rails which fit these mortices will be 2 ft. $1\frac{1}{2}$ in. long by $2\frac{1}{2}$ in. wide and $\frac{7}{8}$ in. thick, finished sizes. After they are gauged to size, set out the shoulder lines with 1 ft. $10\frac{5}{8}$ in. between them, and gauge the ends for the tenons, allowing the front edge, which is a moulding, to stand back from the face of the leg $\frac{1}{16}$ in. The underside of the rail is then rebated back $\frac{1}{16}$ in. deep to form a stop for the flap (see Fig. 443). The moulding is then worked on the front edge, the square of which projects $\frac{1}{16}$ in. beyond the face of the flap. The legs may now be prepared for the side rails B (Fig. 444), and in this case a double tenon is not necessary, so that a mortice $\frac{5}{8}$ in. wide is made and set back $\frac{3}{4}$ in. from the outer face of the leg, this being repeated on the inner face of the back leg. Take the two side rails and prepare them as before, cutting the tenons on the ends and leaving 1 ft. $1\frac{1}{4}$ in. between the shoulders. Next groove the two front legs on the back face to receive the tongue on the side panelled framing, as in Fig. 441. The two back legs will also be prepared in the same manner, with an additional groove on the inner face to receive the tongue on the back board. Stop all the grooves at the line of the mortice for the lowest rail. Then face up the four end stiles and square the edges, and gauge to $1\frac{5}{8}$ in. full by $\frac{7}{8}$ in. thick. The top and bottom rails must also be prepared to the same size. On the face edge of the four stiles set out 3 ft. $1\frac{7}{8}$ in. and $1\frac{5}{8}$ in. from each end. Take the rails 1 ft. 2 in. long, and set out on the face edge $10\frac{5}{8}$ in. between the sight lines, and in each end of the stiles prepare a mortice, $1\frac{1}{8}$ in. wide from sight and $\frac{3}{8}$ in. thick; these mortices go right through the stiles, and the remaining $\frac{1}{2}$ in. beyond the mortice forms the haunching. Prepare the rails with a tenon on each end groove in the bottom rail of the side framing. The length of the bottom will be 1 ft. $10\frac{5}{8}$ in. between the shoulders, allowing in addition the $\frac{3}{8}$-in. tongue on each end, and the width will be 1 ft. $3\frac{3}{4}$ in. On the top side, along the back edge, is made a groove for the tongue on the lower end of the back; the latter is $\frac{3}{8}$ in. thick, so that the face of the groove will be $\frac{5}{8}$ in. from the back edge. The grooves for the remainder of the divisions may then be set out, and will be $5\frac{7}{8}$ in. from the under side in each case, the finished thickness of the divisions being $\frac{7}{8}$ in. Sink them $\frac{3}{16}$ in. deep, and cut the

Fig. 447.—Shaped End of Music Cabinet Division.

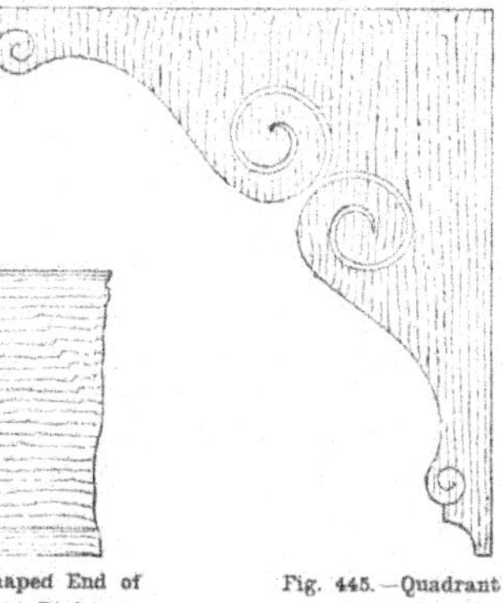

Fig. 445.—Quadrant Bracket of Music Cabinet.

Fig. 446.—End of Music Cabinet Tray.

to fit, and the two panels 2 ft. $8\frac{3}{4}$ in. long by $11\frac{5}{8}$ in. wide by $\frac{3}{8}$ in. thick finished. Prepare on each a tongue $\frac{3}{8}$ in. deep by $\frac{3}{16}$ in. thick to fit the grooves in the stiles and rails. When the ends or side framings are glued up, the back shoulders must be bevelled off, and the tongue fitted into the legs so that these are quite flush with the inside face of the legs.

Tray Divisions.—The grooves to receive the divisions on which the trays slide must now be set out and sunk. The under side of the first division or bottom will be level with the top side of the rail, and on each end of the bottom is prepared a tongue $\frac{3}{8}$ in. square (see Fig. 444), to be fitted into a divisions to exactly 1 ft. 11 in. long by 1 ft. $3\frac{5}{8}$ in. wide, fitting them tight up to the back. Now from the front edge of each division set out the $\frac{3}{4}$-in. housing at each end for the quadrant. Cut in $\frac{1}{2}$ in. from the end as at A (Fig. 443); the guides B for the trays will be 2 in. wide by $\frac{5}{8}$ in. thick, and are cut and fixed with $\frac{5}{8}$-in. screws, the countersunk heads being driven just below the surface.

Top of Cabinet.—The top is 2 ft. 4 in. long by 1 ft. $5\frac{7}{8}$ in. wide by $1\frac{1}{4}$ in. thick, and a moulding is worked on the face and return edges, while on the under side grooves are prepared to receive the tongues on the side framings and back (see Fig. 444).

The upper face is also grooved to receive the tongue on the lower edge of the skirting, but the exact position of these grooves must be obtained when the carcase has been put together temporarily.

Bottom Shelf.—The bottom shelf, fixed about 6 in. from the floor line, will be 2 ft. $0\frac{3}{4}$ in. long by 1 ft. $3\frac{1}{2}$ in. wide by $\frac{3}{4}$ in. thick, and is moulded all round. A small rail, 1 in. by $\frac{7}{8}$ in., is prepared and mortised into the legs to support the shelf, which is cut clean between the legs without housing. Next face the six pieces, 1 ft. 11 in. long by $5\frac{1}{4}$ in. wide by $\frac{7}{8}$ in. thick, gauge them to $\frac{3}{4}$ in.,

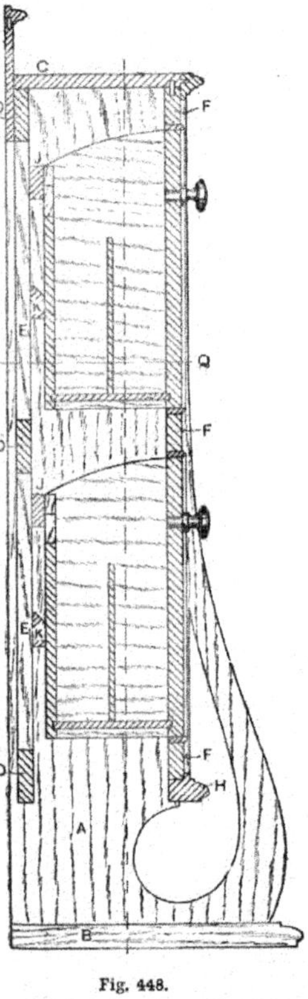

Fig. 448.

Fig. 449.

Figs. 448 and 449.—Cross Section and Half Front Elevation and Half Longitudinal Section of Music Cabinet with Swinging Drawers.

and then fit them into the spaces provided. Mark each one as fitted on the bottom edge with a small chisel, hang each at the bottom edge with a pair of 2-in. arrow butts, and fit on the quadrants (Fig. 445). The top screw c (Fig. 443) is a No. 10 round-headed brass screw, ¾ in. long, fixed into the leg (see also Fig. 446).

Spring Catches, Trays, etc.—The spring catches or locks should next be fitted to the flaps, and the striking plates let into the rails. The astragal moulding fixed on the face of the flap should be prepared and polished, but not mitered round until the polishing of the flap is done. For the bracket shown in Fig. 445 the legs and rails should be slightly recessed, so that no gaping joint shall be seen. For the trays, cut the six ⅜-in. bottoms to 1 ft. 9⅞ in. long by 1 ft. 3 in. wide, with the front edge slightly rounded. The six backs are 1 ft. 9⅝ in. long by 3¾ in. wide by ⅜ in. thick, the twelve sides 1 ft. 2⅞ in. long by 3¾ in. wide, and the six divisions 1 ft. 2⅞ in. long by 3¾ in. wide by ⅜ in. The sides are dovetailed to the back at the angles in the ordinary manner, the pins being cut on the back. Also the front ends of the sides are cut to shape and the top edges are slightly rounded. The centre division is tenoned at the back, three small mortices being prepared in the back, and are finally glued and wedged. The front end of this division is cut to the shape shown in Fig. 447, and may have a hole by which the tray is withdrawn. The bottom is fixed to the back, sides, and divisions by screws inserted from the under side.

Music Cabinet with Swinging Drawers.

The music cabinet shown by Figs. 448 and 449 is convenient of access, and occupies very little floor space, the width over mouldings being 1 ft. 11 in., the depth 1 ft. 2 in., and the height 3 ft. 9 in. Fig. 448 is a cross section through the middle of the cabinet, Fig. 449 showing a half front elevation and a half sectional elevation. The thick lines on the drawer fronts in the half front elevation indicate strips of inlay. Two side pieces A (Fig. 448) are shaped as shown, grooved and tongued into the feet B. The upper ends of the side pieces are connected by the top piece C, which is tongued and rebated at the ends (see Fig. 449), and tongued along the front edge (see Fig. 448). The tongues, of course, are for securing the moulding attached to the ends and the front edge. The side pieces are further connected by two frames, one at the front and the other at the back. The three rails of the back frame are shown at D in Figs. 448 to 451, and the stiles of the same frame at E. The rails of the front frame are shown at F, and the stiles at G. Both frames are attached to the sides by means of

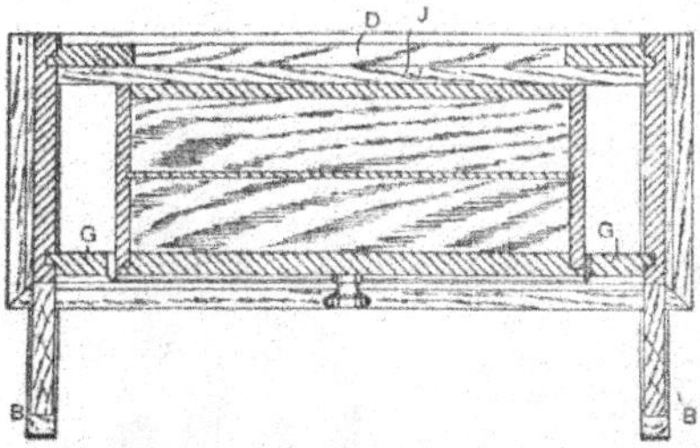

Fig. 450.—Horizontal Section through Music Cabinet with Swinging Drawers.

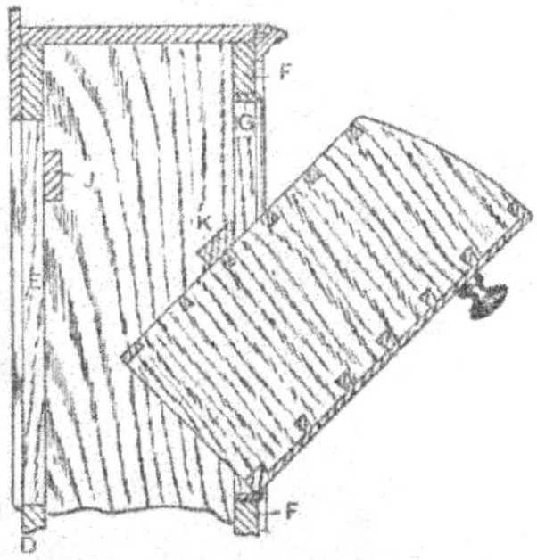

Fig. 451.—Dodged Section through Upper Part of Music Cabinet showing Drawer Extended.

tongues fitting into stopped grooves, as shown in Fig. 450, which is a horizontal section taken at P Q (Fig. 448). Angle blocks (not illustrated) are glued on at intervals to strengthen the framework. To the bottom of the front frame a moulding H (Figs. 448 and 449) is fastened, and is returned across the outsides of the end pieces A. Around the edges of the two openings in the front frame small beads are fastened.

Swinging Drawers.—The receptacles for the music sheets are two swinging drawers, each with a partition down the middle, and hinged to the front frame. The hinges are not shown in the illustrations, but a comparison of Figs. 448 and 451 will make the arrangement clear. In Fig. 451, it will be seen, the drawer is shown swung round upon its bottom forward edge. Hence the centre lines of the hinge pins must be made to

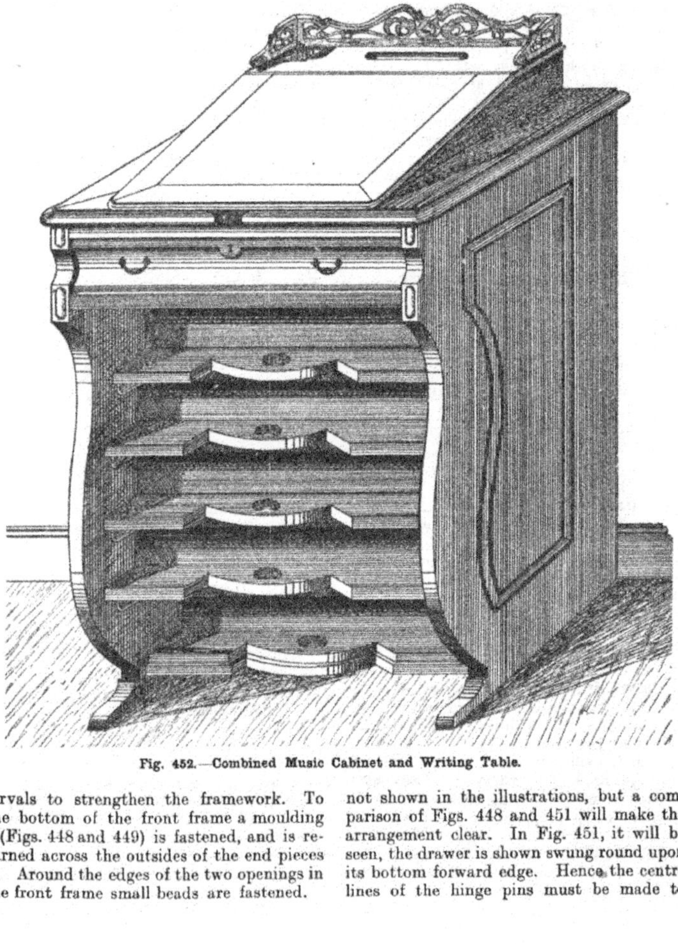

Fig. 452.—Combined Music Cabinet and Writing Table.

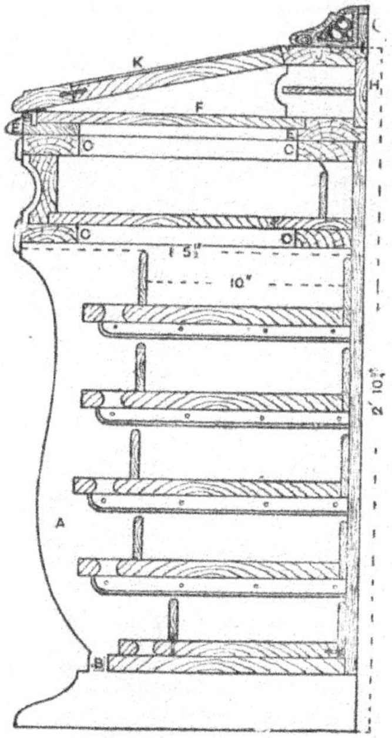

Fig. 453.—Vertical Section through Combined Cabinet and Writing Table.

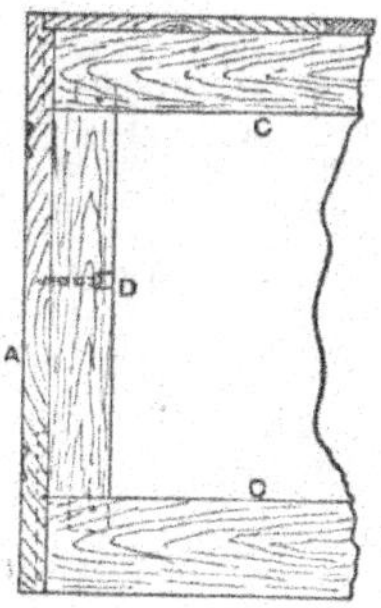

Fig. 454.—Part Horizontal Section of Combined Cabinet and Writing Table.

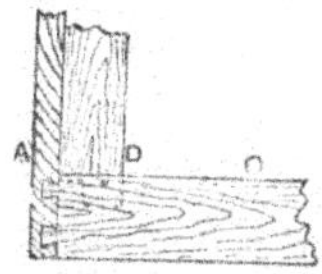

Fig. 455.—Top Drawer Divisions of Combined Cabinet and Table Dovetailed to Gables.

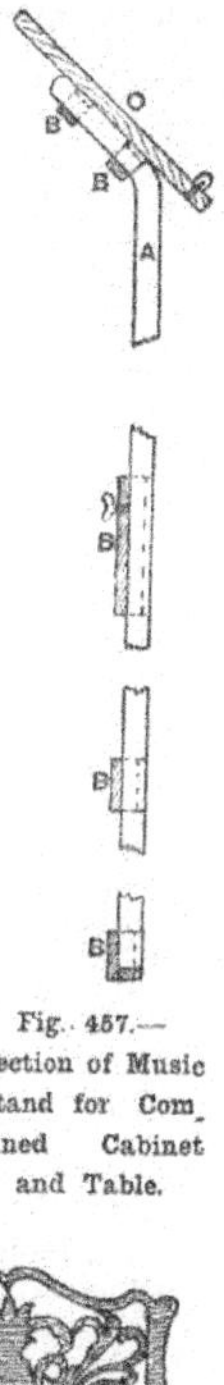

Fig. 457.—Section of Music Stand for Combined Cabinet and Table.

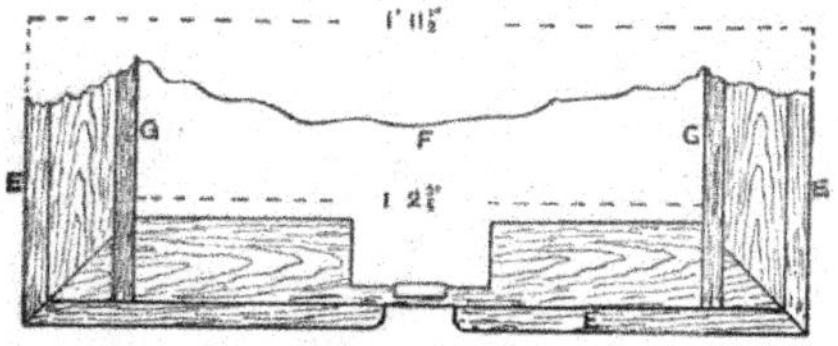

Fig. 456.—Part Plan of Combined Cabinet and Table, with Desk Lid Removed.

Fig. 458.—Music Easel.

coincide with that edge. When the body of the cabinet is built up, and the drawers have been trimmed to fit nicely in place, they may be removed, and the hinges laid in their proper positions upon the bottoms of the drawers and scribed off, and recesses cut out. Then, replacing the drawers, before the hinges are screwed into place, a scriber is drawn along the edges of the recesses already made, in order to mark off those which have to be cut in the frame. When being screwed into place, the hinges must be first attached to the drawers. In making the drawers, common dovetails should be used for the back, and lap dovetails for the front. In order to avoid undue strain upon the hinges, the drawers should rest, when closed, against battens J, which are screwed to the back frame. To prevent the drawers falling too far forward, each has a batten K screwed to it; this, when the drawer is fully open, rests upon the front frame (see Fig. 451). It will be found necessary to fasten these battens on from the back, after the hingeing of the drawers has been completed. It is obvious that if the drawers are to remain open without being held or fastened, the centre of gravity of the drawer and its contents must be outside the centre line of the hinge. This may be readily secured if the drawer from back to front does not exceed 5 in. The ends of the drawers are curved, so that they may swing about the hinges without fouling the frame; but the ends may be straight, if the line of the edge be the chord of the arc instead of the arc itself. Not only the drawer, but the music sheets also, must swing clear of the frame; hence the depth of the drawer measured at its shallowest part—that is, at the back—must not be less than 11 in. At the front it should be about 1 ft. 1 in. Oak, mahogany, or black walnut may be used for the construction of the visible parts. The frame at the back, and the whole of the drawers with the exception of the fronts, may be of yellow pine or deal or basswood.

Combined Music Cabinet and Writing-table.

The music cabinet and writing-table (Fig. 452) has solid gables A (Fig. 453), these being $\frac{7}{8}$ in. thick; the bottom shelf B, and the drawer divisions C, with the exception of the two upper ones, are raggle-dovetailed into them. E (Fig. 453) shows the moulded top; F, desk bottom; H, desk back; J, fixed part of desk top; and K, desk lid. This is shown in Fig. 454, in which A shows the gables; C C, drawer divisions; and D, drawer slide. The top drawer stretchers are dovetailed down on the top end of the gables as shown in Fig. 455, in which A shows the gables; C, drawer division; and D, drawer slide. Drawer slides D are tenoned into the front and back drawer divisions and dowelled. A screw is also put through the middle of the drawer slide to draw the gable up and keep it straight. The sunk bead on the gables is cut in with a cutting gauge and finished with the chisel, rebate plane, and glass-paper. The gables could be left plain if so desired, but this is a very simple method of breaking up the plain surface. The top part E, with the lamb's-tongue moulding, as shown in Figs. 453 and 456, is $3\frac{1}{2}$ in. broad and 1 in. thick. It is carried along the front and ends, and mitered at the corners. The back part is butted between the two ends. The inside edge is rebated $\frac{3}{8}$ in. on to receive a piece of yellow pine F, $\frac{5}{8}$ in. thick. This closes the open space above the drawer, and forms the button for the desk. The desk sides G are screwed up from the under side. The back H is kept $\frac{1}{2}$-in. above the flush of the sides, and is lap-dovetailed to them. The fixed part J, to which the desk flap is hinged, is rebated to fit down on the top edge of the back, and is screwed down to it as well as to the sides. These screws are placed in such a position that they are covered with the frets at the back of the desk. These frets are pinned down to the top of the desk, and the corners are mitered and bradded. The lid K is made up of three outside pieces with a moulding on them, dowelled to a central piece of pine, which can be veneered with a different kind of wood from the rest of the cabinet, or, alternatively, dressed leather could be glued on.

Desk.—In Fig. 456, E shows the moulded top, F desk bottom, and G desk sides. Fig. 456 is the plan of the top with the lid off, showing the method adopted for locking the desk. The front part is cut away to about ¾ in., and the moulding cut in and returned on itself. The lock is then fixed, and the yellow pine F fitted to the space which has been cut out. A division is fitted into the back part of the desk and bradded on the angle to the top and bottom. Light shelves are put in between, and either raggled to the sides and division or carried on a small fillet. The drawer front is moulded to the same shape as the gables, and dovetailed. The small projections on the front edge of the gables are rounded over and bradded on. The construction of the shelves is seen in Fig. 453. A piece of pine is screwed on to the back of them, and another piece immediately at the back of the hole for drawing them out is screwed up from the under side. Fillets to carry them are screwed to the gables.

Music Stand.—A music stand would be almost a necessary accompaniment to the cabinet, and details of it are shown in Fig. 457, in which A shows the brass rod, B pieces of flanged brass for the rod to slide in, and C the music easel. It consists of a brass tube A, bent at the top to receive the easel C, and is intended to be portable.

Fig. 459.

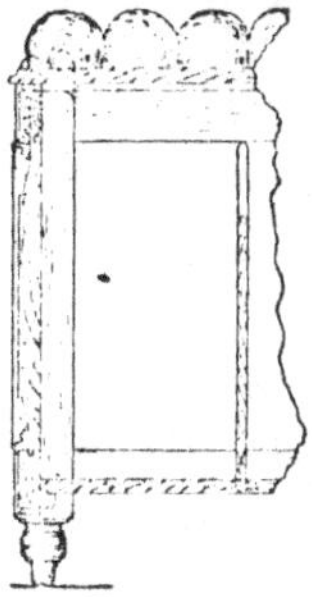

Fig. 460.

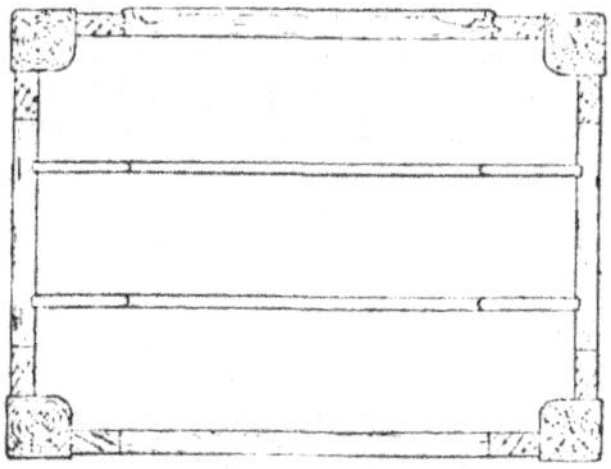

Fig. 461.

Figs. 459 to 461.—Front Elevation, Part Vertical Section, and Horizontal Section of Music Stool.

Pieces of brass B are bent to suit the size of the tube and flanged. Two small pieces are screwed to the back side of the easel, and two small pieces and a larger piece are screwed to the back of the cabinet, and near to one end. A hole is bored through the large piece and tapped to suit a butterfly tap. This prevents the stand from slipping down when in use. The bottom piece of brass will require to be plugged up to keep the rod from slipping to the floor. Fig. 458 is a design to suit this music stand.

Music Stool.

The music stool shown in elevation in Fig. 459 is intended to hold sheet music, the space inside being divided into three. The

Fig. 464.

Fig. 463.

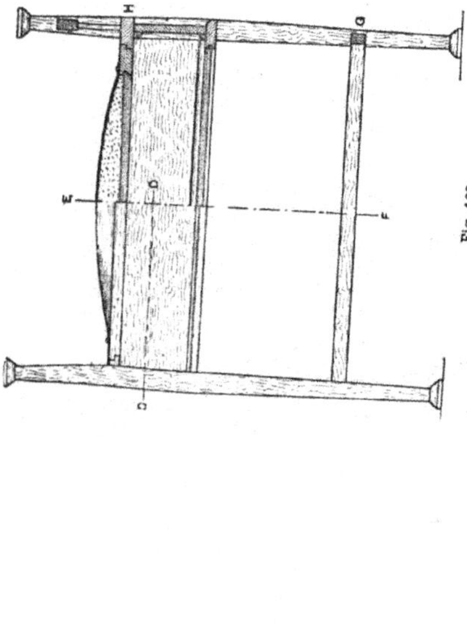

Fig. 462.

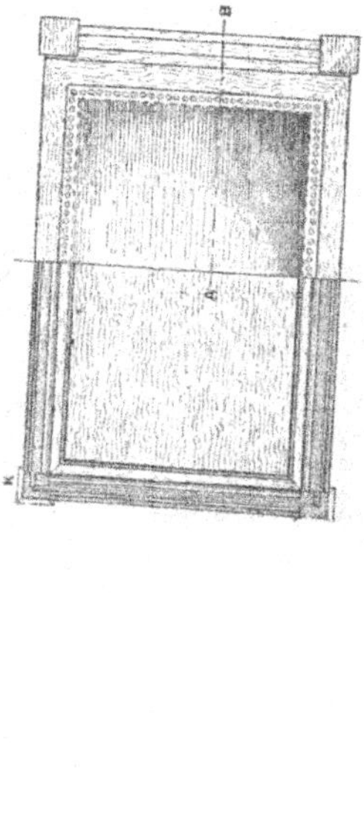

Fig. 465.

Fig. 462.—Half Front Elevation and Half Vertical Section (on Line A B) of Music Stool with Box Seat.

Figs. 463 and 464.—End Elevation and Cross Section (on Line E F) of Music Stool.

Fig. 465.—Half Horizontal Section (on Line C D) and Half Plan of Music Stool.

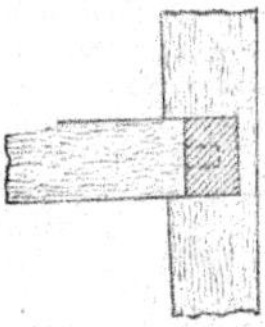

Fig. 468.—Joint of Music Stool Rails at G (Fig. 462).

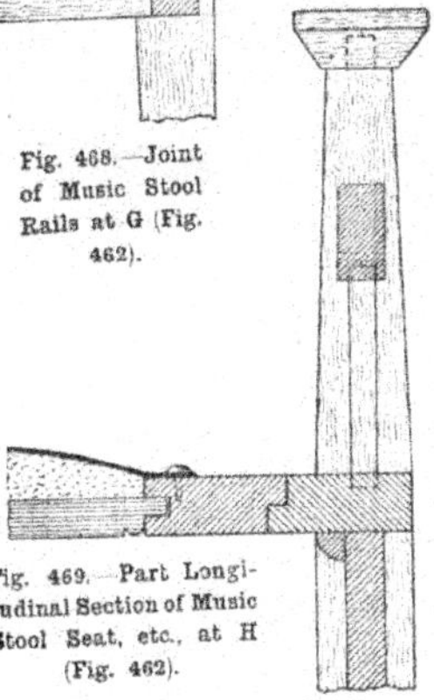

Fig. 469.—Part Longitudinal Section of Music Stool Seat, etc., at H (Fig. 462).

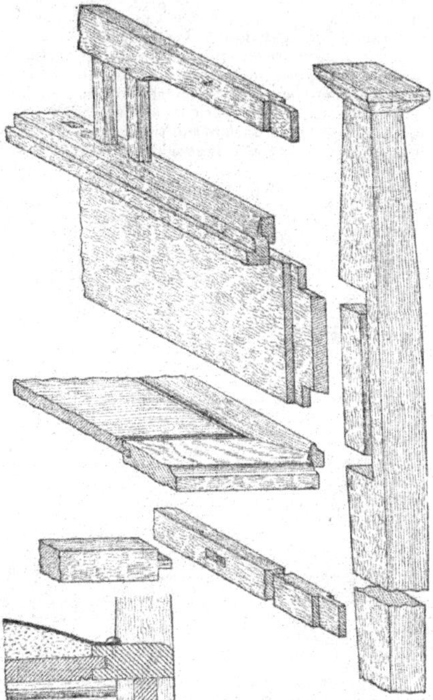

Fig. 466.—Enlarged Detail View, showing Method of Framing Up Music Stool.

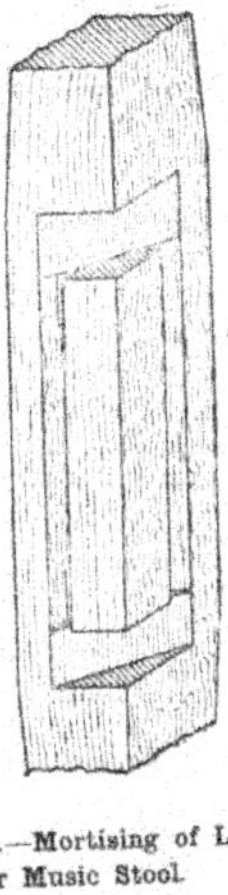

Fig. 467.—Mortising of Legs for Music Stool.

Fig. 470.—Part Cross Section of Music Stool Seat, etc., at J (Fig. 464).

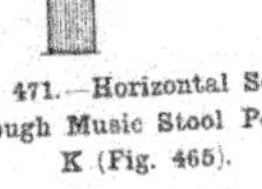

Fig. 471.—Horizontal Section through Music Stool Post at K (Fig. 465).

stool is intended to be 20 in. long by 15 in. broad by 20 in. high, and the method of construction is as follows: Two side and two end frames are mortised and tenoned together, and grooved for the panel to suit the sinkage of the moulding intended to be used. The panel is then fielded, and afterwards rebated on the back to suit the groove. When the frame is ready, it should be glued and cramped up, and allowed to set, after which it may be cleaned on both sides and stripped to the width and length. The corner posts, which have been previously turned at the foot to some design, have a line drawn upon them to represent the bottom edge of the frames. The frames are then taken and laid upon the posts (the bottom edge of the frame coinciding with the line previously drawn upon the posts), and lines are drawn for the dowels across the edge. These lines are squared across the inside face of the post and across the edge of the frame. A marking gauge is then set to half the thickness of the frame, and the centre of the dowel hole is marked with it. Another gauge is then set $\frac{1}{8}$ in. farther on, and the posts are gauged with it. After the holes are bored and the dowels inserted, the carcase may be cramped up, the frames showing themselves recessed back $\frac{1}{8}$ in. The bottom is supported on a small fillet screwed to the inside of the frames, the bottom in turn being screwed to it. Previous to the fixing of the fillet and the bottom, the divisions must be inserted and glued to the raggles which have been prepared for them in the end frames. After the bottom has been fixed in, the small bead may be planted on the under side of the frames, between the posts. The carcase may now be flushed off on the top edge, the posts rounded towards the inside, and the bead planted on. It will be seen, on examining the vertical section at Fig. 460, that the bead on the top edge is in two, one part being fixed to the top of the carcase, and

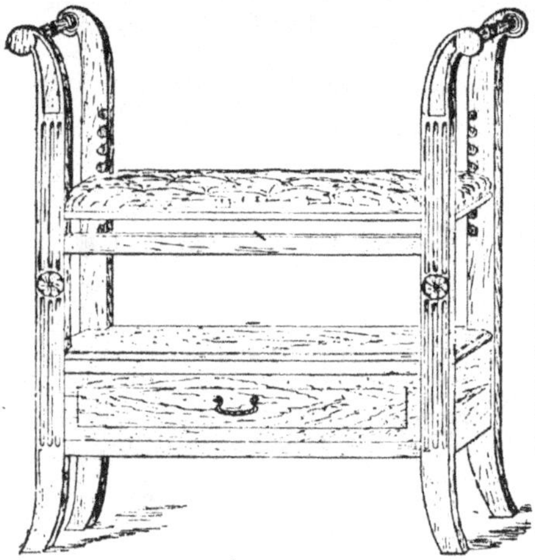

Fig. 472.—General View of Music Stool with Shaped Posts and Rising Seat.

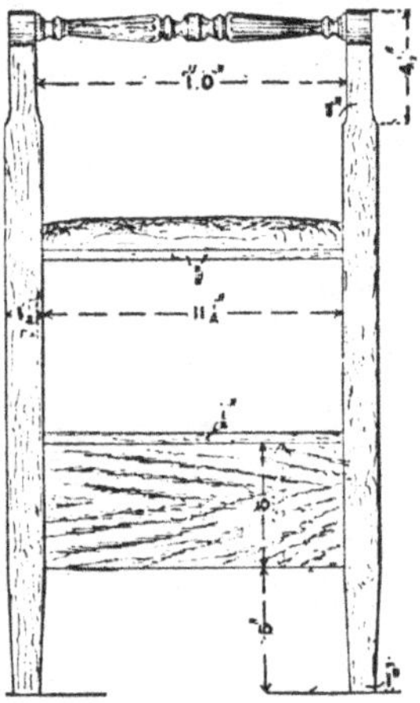

Fig. 473.—End Elevation of Music Stool with Shaped Posts and Rising Seat.

the other to the under side of the top. This serves to keep the top rigid. The top itself is formed of a hardwood board with cross ends mitered to it, so as to show side wood all round the moulding. The top is then hinged, and the lock fixed on, after which it may be taken off, and the whole stool polished. The moulding should be polished at the same time, but stained a slightly

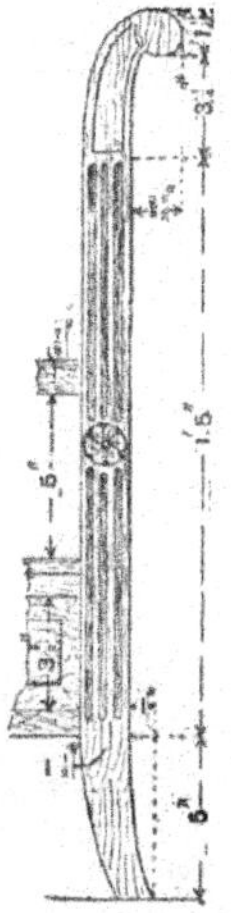

Fig. 474.—Shaped Post for Music Stool.

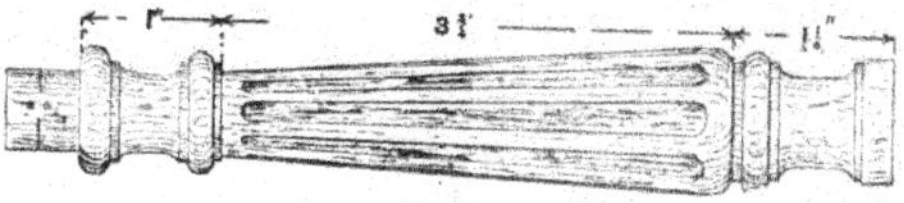

Fig. 475.—Half of Music Stool Stretcher.

Fig. 476.—Music Stool Rails Tenoned to Posts.

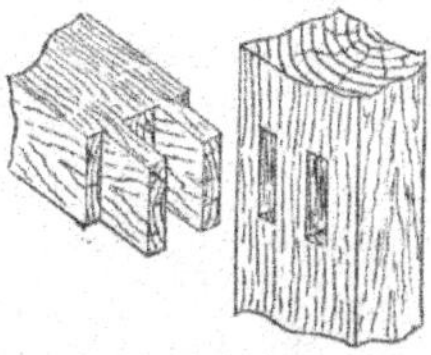

Fig. 477.—Double-tenon between Drawer Framing and Post of Music Stool.

darker colour. After the top is polished, the seat should be upholstered, and the mouldings planted. The inside of the stool should be stained and slightly treated with polish, so that it may harmonise with the outside when the lid is opened. Fig. 461 is a horizontal section through the stool, showing the general arrangement of the inside. One of the sides in Fig. 461 is shown with the fielded panel, but the maker can vary it with a small surface moulding if the field is likely to entail too much labour.

Music Stool with Box Seat.

A music stool with a box seat is illustrated in half elevation and half vertical section by Fig. 462, end elevation by Fig. 463, vertical cross section by Fig. 464, and in half horizontal section (on line C D, Fig. 462) and half plan by Fig. 465 (see p. 136). Fig. 466 is an enlarged detail showing the method of framing up. The method of mortising, etc., for the legs is shown in Fig. 467, and enlarged details at G, H, J, and K respectively by Figs. 468 to 471.

Music Stool with Shaped Posts and Rising Seat.

Figs. 472 and 473 illustrate a music stool with a seat that can be adjusted to any suitable height; a drawer is also arranged below the seat for holding music, etc. Mahogany or walnut would be a suitable wood. First prepare a mould of thin wood to Fig. 474. Line this out on 1½-in. wood, and cut and dress it to the mould. The turned stretchers, half of one being shown by Fig. 475, are fixed to the heads of the scrolls, holes being bored to fit the end pins. They may be enriched by fluting the plain parts. The side and back rails, $\frac{3}{4}$ in. thick, are tenoned into the posts as shown in Fig. 476, and are kept back $\frac{1}{4}$ in. from the flush. The two fore edges for the drawer are $1\frac{1}{4}$ in. deep by $\frac{3}{4}$ in. thick, and are double-tenoned into the posts (see Fig. 477), and kept $\frac{1}{4}$ in. from the front. Two top rails, $1\frac{1}{4}$ in. by $\frac{7}{8}$ in., are tenoned and screwed from the inside. The drawer is dovetailed together in the usual way. The stuffed seat consists of a frame of birch or other suitable hardwood, the stiles and rails of which are 2 in. by 1 in. A bead D (Fig. 480), $\frac{3}{8}$ in. deep, is glued on the edge to form the check for the stuffing. The seat is kept in place by being checked

Fig. 478.

Fig. 479.

Figs. 478 and 479.—Sections of Music Stool Drawer.

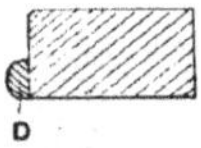

Fig. 480.—Bead on Music Stool Seat.

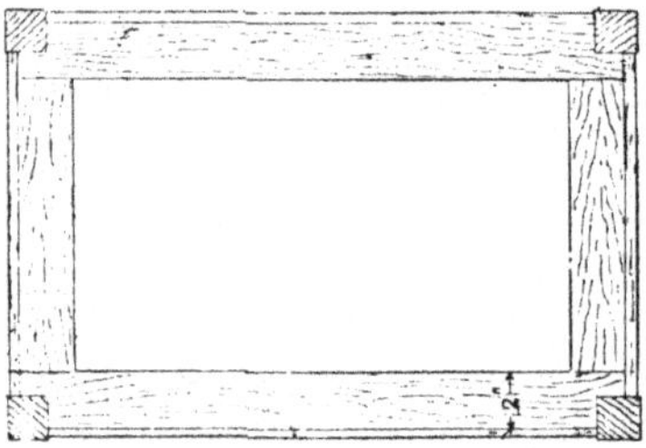

Fig. 481.—Plan of Music Stool Seat Frame.

into the posts at the front and back in the same manner as the drawer fore edges. Figs. 478 and 479 show sections of drawer. Two bearers A (Fig. 479) are glued and sprigged to the side rails to carry the drawer sides, and guides B make up the thickness of the posts. Two guides C (Fig. 479), of the thickness of the top fore edge, are glued to the inside of the rails. A top, $\frac{1}{2}$ in. thick, with a thumb moulding on all four edges, is carefully fitted round the posts at the corners at the corners (Fig. 481), and two brass rods are placed between the end posts; the last have racks cut in them to the depth of $\frac{1}{4}$ in. (see Fig. 472). To raise or lower the seat, shift the rods which form a support for the seat to the required position. The fronts of the posts are fluted, and the scroll parts are channelled. The stool would look well if french polished, the seat being covered with velvet; but many other materials would also be suitable.

Umbrella Stand with Turned Posts.

A CONVENTIONAL view of an umbrella stand to be made in oak is presented by Fig. 482. The following are the general dimensions: 2 ft. 6 in. long over the posts by 10 in. deep, and height from floor to top rail 2 ft. 4 in. It is divided into three equal spaces by the two cross rails as shown in Fig. 483. First prepare the four posts (Fig. 484). Square them up to 1⅝ in. by 1⅝ in., and mark off the squares and mortice for the top and bottom rails before turning. The top rails are 1¼ in. deep by ⅞ in. thick. Two flutes are worked on the face. The flat round (similar to that shown in Fig. 485 at the top) is added after the stand is glued up, for simplicity in working. Tenon

Fig. 482.—Umbrella Stand with Turned Posts.

the rails to the posts, keeping the latter ⅛ in. over the rails. Fig. 485 gives a section of the bottom rails, which are 2 in. deep by 1½ in. thick. They are rebated to receive the ½-in. pine bottom B, which is fitted and fixed with screws to the bottom rails. The rails are tenoned to the posts and set back ⅛ in. The flat round on the top edges is left off until the job is glued together. The two cross rails, 1¼ in. deep by ⅞ in. thick, are fixed to the two long top rails with short rebated tenons. When the stand is cramped up, the top rails can be finished off with the flat round moulding, mitering them into the long rails (see Fig. 486). The ends are butted between the posts. The zinc pan fits between the bottom rails, and is divided into three spaces. Part of the zinc pan is shown in section in Fig. 485, above. When finished, the stand would look well stained lightly and dull polished.

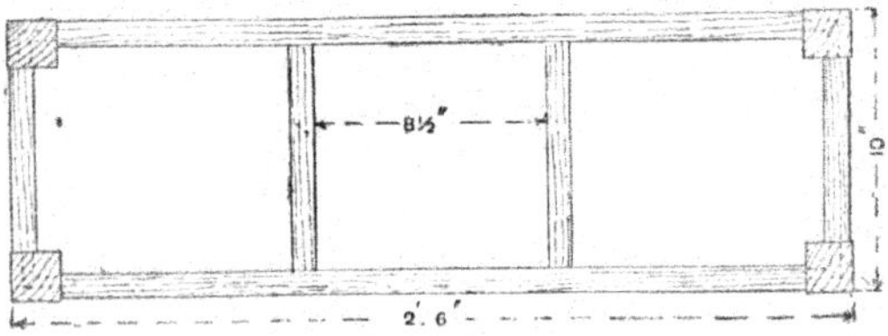

Fig. 483.—Plan of Umbrella Stand.

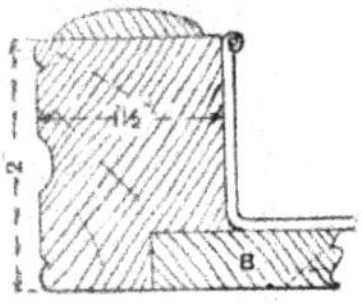

Fig. 485.—Section of Umbrella Stand's Bottom Rail and Pan.

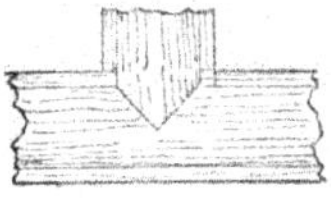
Fig. 486.—Plan of Umbrella Stand's Top-rail Joint.

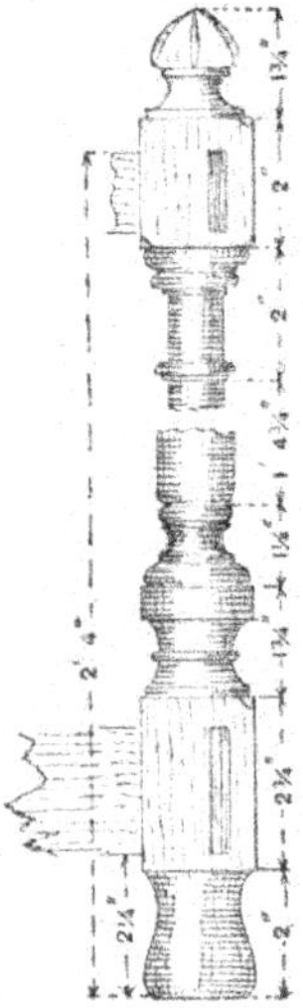
Fig. 484.—Turned Post of Umbrella Stand.

Oak Umbrella Stand.

The strong and serviceable umbrella stand of which Fig. 487 is a general view is 3 ft. high and 2 ft. 3 in. long, and is best made of oak. The ends are 10 in. wide at the bottom, tapering to 6 in. at the top, and are curved and fretted as shown in the illustration. The lines in the diagram (Fig. 488) will enable a full-size outline of the fretted portions to be made, each square representing ½ in. The capping pieces A (Fig. 487) are 8 in. by 2⅝ in. by ⅝ in., and are chamfered on the under side as shown at Fig. 489, being secured to the ends with two small dowels. The long rails B (Fig. 487), 1¼ in. by ⅝ in., are stump-tenoned to the ends, the middle dividing rail being stump-tenoned into the long rails. The lower framework, having rails 1½ in. by ⅝ in., is divided into three portions, to take a zinc pan in each end and a tile or panel between, an enlarged section of the rail, tile, and pan being given at Fig. 490. The method of constructing the framing is set out at Fig. 491. The end rail is dovetailed into the long rail about ½ in., the middle rails being housed in. Both the long rails and the middle rails are grooved to receive the tile-frame. A rebate is made in the latter to receive the tile, and a small fillet A nailed round the under side to hold it in position, the corners of the frame being mitered and skew-nailed. The framework is secured to the ends with screws driven through the outside. A few reeds along the face of the rails would improve the appearance of the stand, or any other kind of ornamentation may be introduced.

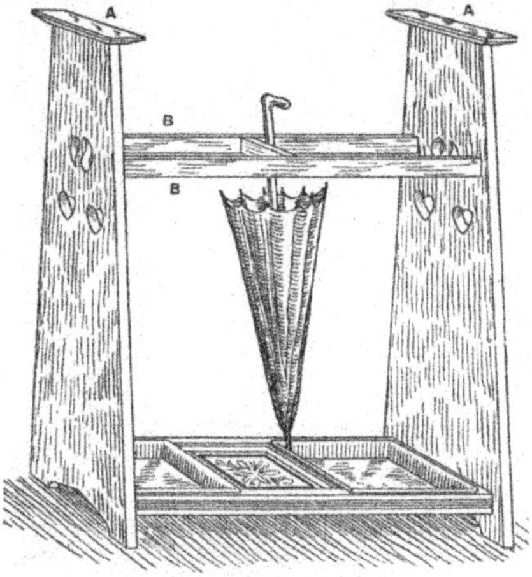

Fig. 487.—Oak Umbrella Stand.

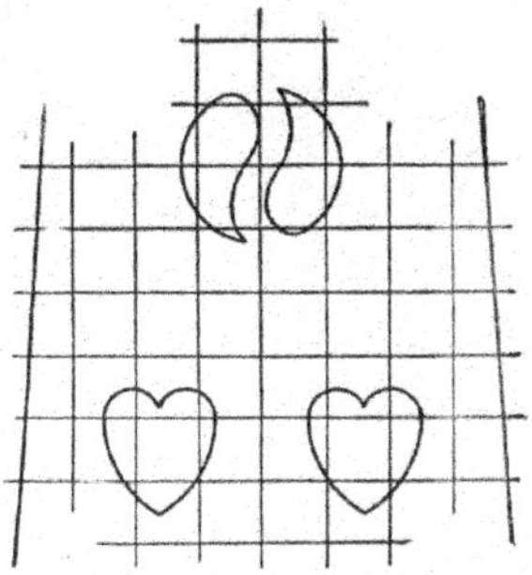
Fig. 488.—Fretted Part of Umbrella Stand.

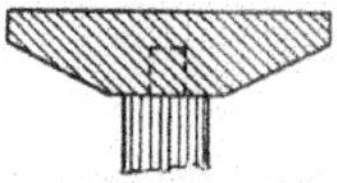
Fig. 489.—Section of Umbrella Stand's Cap.

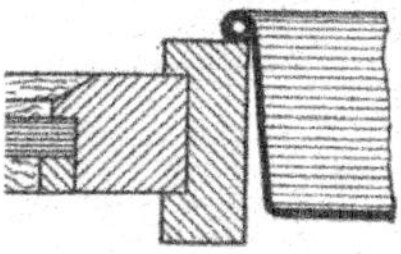
Fig. 490.—Rail, Tile, and Pan of Umbrella Stand.

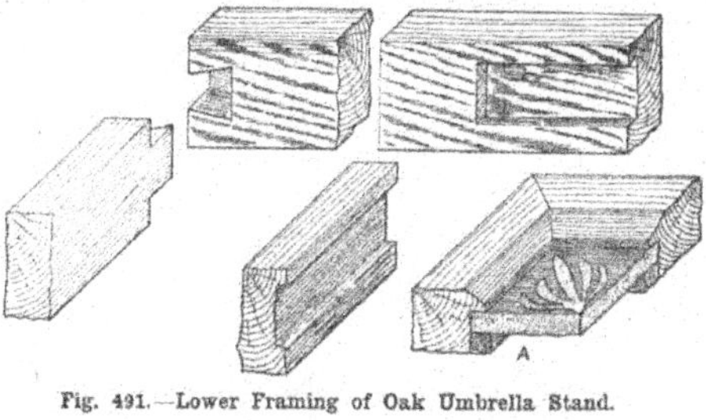

Fig. 491.—Lower Framing of Oak Umbrella Stand.

More Elaborate Oak Stand.

The stand shown by Figs. 492 and 493 would look well if made in figured oak, fumed or stained a dark colour, and oiled.

The two end boards (Fig. 494) may be cut from stuff $11\frac{1}{2}$ in. wide by $\frac{3}{4}$ in. thick finished; mark off the dimensions, cut the mortices $\frac{1}{2}$ in. wide by $1\frac{1}{4}$ in. long, the top in being driven home before tightening the end pieces to the shoulders of the rails. If pine is used instead of oak, the length of the tenons should be increased. Fig. 496

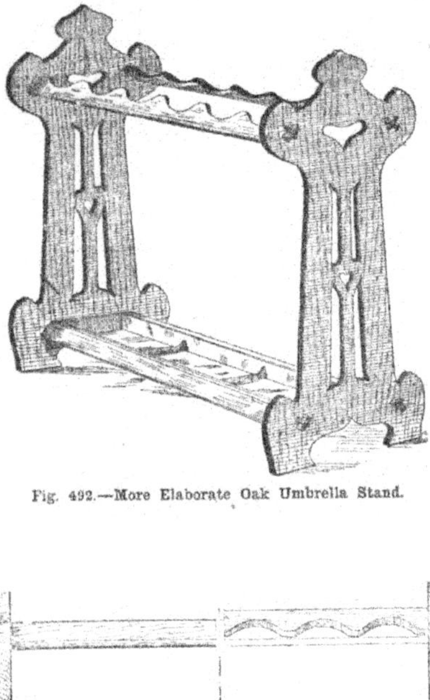

Fig. 492.—More Elaborate Oak Umbrella Stand.

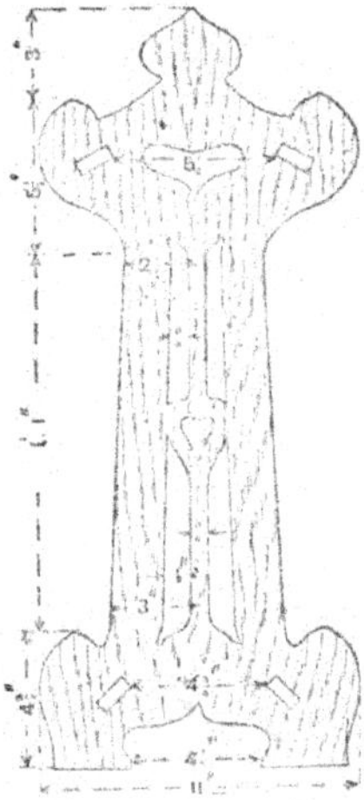

Fig. 494.—End Board of Umbrella Stand.

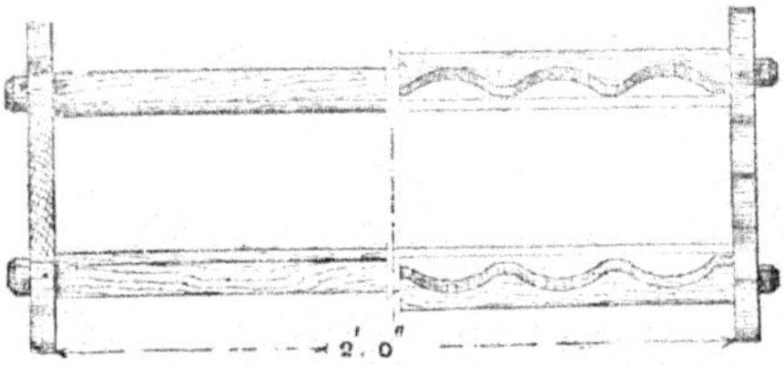

Fig. 493.—Part Plan and Part Section of Umbrella Stand.

Fig. 496.—Rail and Pan of Umbrella Stand.

Fig. 495.—Tenon and Keyway on Umbrella Stand Rail.

pair being inclined at 15°, the lower pair at 45°. Next square all the rails to one length; they shoulder to the end pieces, and the tenons (see Fig. 495) project about $\frac{7}{8}$ in. The keyway should be slightly under the face line to allow the key sufficient "draw," otherwise it would probably choke shows a section of the lower rail and pan, only part of the latter being illustrated.

Corner Umbrella Stand.

The corner umbrella stand illustrated by Fig. 497 may be constructed from pitchpine and varnished. The finished thickness of

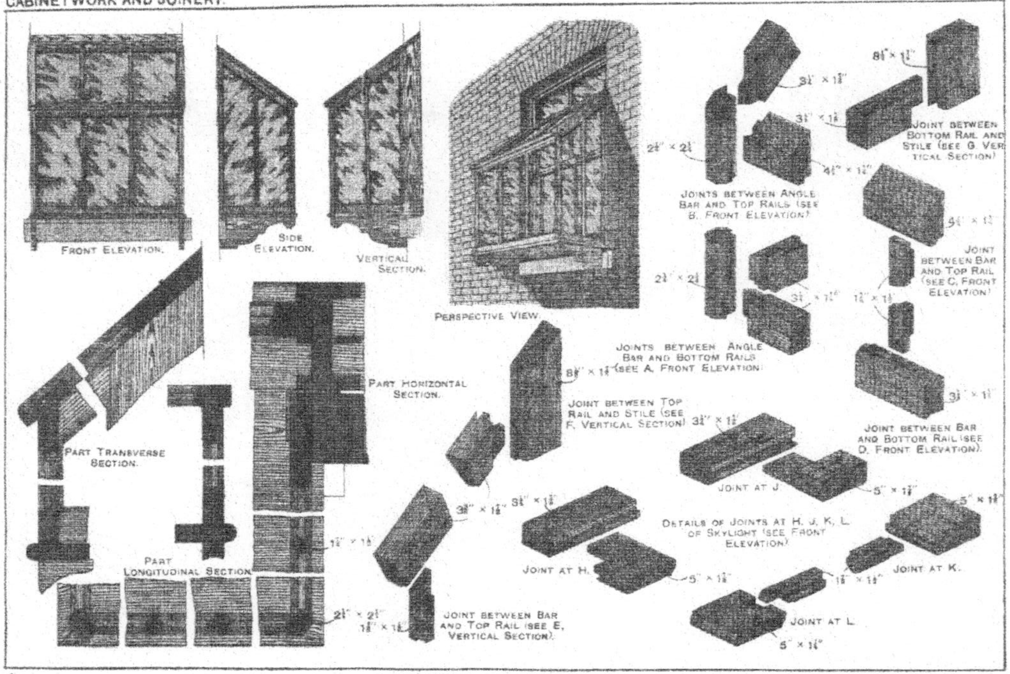

A WINDOW CONSERVATORY

the rails (Figs. 498 and 501) should not be less than $\frac{7}{16}$ in., and the front legs may be cut from board 3 in. wide and $\frac{7}{8}$ in. thick of jointing when square front legs are adopted. The front faces may be reeded, or the part between the top and bottom

Fig. 497.—Corner Umbrella Stand.

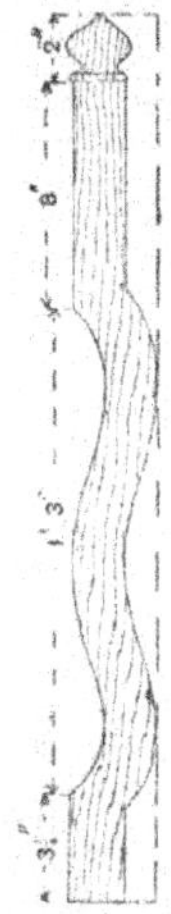

Fig. 499.—Leg of Corner Stand.

Fig. 502.—Joints of Front Legs of Corner Stand (Alternative Method)

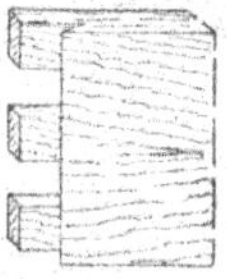

Fig. 503.—End of Rail of Corner Stand.

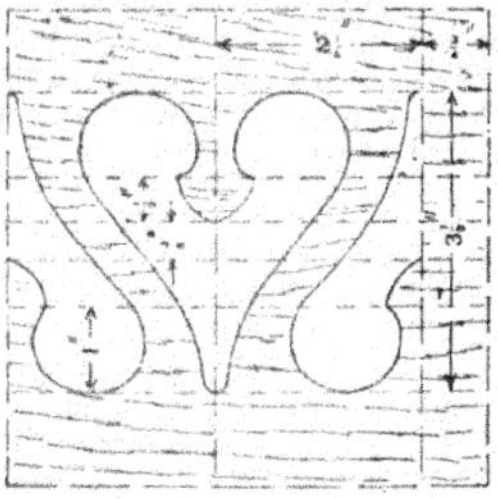

Fig. 498.—Fret Design for Corner Stand Rail.

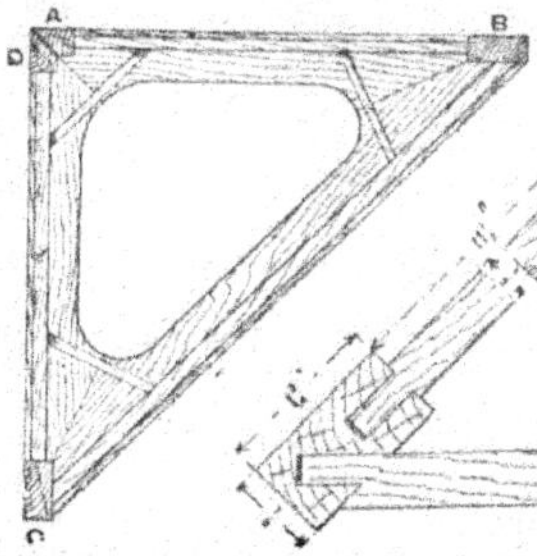

Fig. 500.—Horizontal Section of Corner Stand.

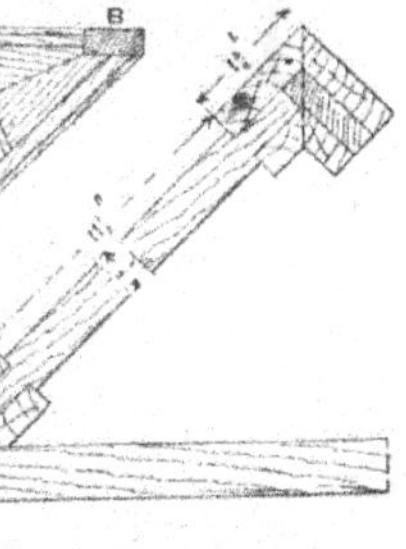

Fig. 501.—Part Enlarged Section of Corner Stand.

finished. The finials are made separately, dowelled and glued on. Figs. 500 and 501 are sections, Fig. 501 being taken through the joints. Fig. 502 is an alternative plan rails could be turned. If the method shown in Fig. 501 is adopted, some care must be exercised in cutting the oblique mortices. A full-size plan should be made, and the dimensions marked off from it to the material; then square the ends and edges of the legs, set the bevel to an angle of 45°, mark two lines on each end representing the width of the mortices, set the marking gauge to the face ends of the lines, and scribe from them; this will give the diagonal or face widths of the mortices. A small gauge, similar to a bevel, could be made from a thin piece of hard wood; if small enough to enter the mortice it would be useful for correcting purposes. There are

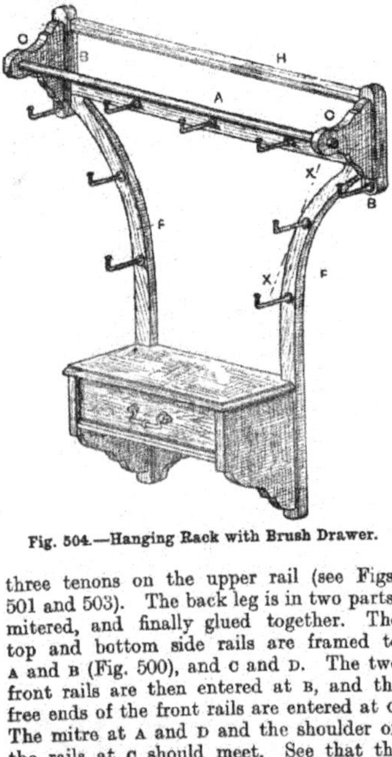

Fig. 504.—Hanging Rack with Brush Drawer.

three tenons on the upper rail (see Figs. 501 and 503). The back leg is in two parts, mitered, and finally glued together. The top and bottom side rails are framed to A and B (Fig. 500), and C and D. The two front rails are then entered at B, and the free ends of the front rails are entered at C. The mitre at A and D and the shoulder on the rails at C should meet. See that the shoulders fit well to the legs; then remove the top front rail, and space it out for the fretwork. Starting at the centre, make a template of cartridge paper similar to Fig. 498, marking the curves with a lead pencil; then after the fretting is finished fix the parts with fresh hot glue. Next fit the lower shelf as in Fig. 500, which shows a plan of the under side of the base. The shelf is formed of three boards ⅜ in. thick, glued and bradded to the rails and mitered at the angles; three fillets are shown, and to stiffen the mitre extra blocks may be glued under the joint. When the glue has set, trim off the top and the space for the drip tray; the latter should lift out easily and be given a coat of enamel paint. The cap moulding is ploughed to fit on the fretted rail, and mitered to the front legs, a side elevation or profile of one of which is given by Fig. 499.

Hanging Rack with Brush Drawer.

Fig. 504 is a general view of a hall rack with brush drawer, which can be made of deal, stained and varnished, oak, mahogany, walnut, or any other similar wood. Fig. 505 is a side elevation, and Fig. 506 a transverse vertical section. The rack is 3 ft. wide at the top and 1 ft. 6 in. at the bottom, and has a total height of 4 ft. The round rail A is 1¼ in. in diameter; the pieces B (Fig. 507) are 1⅛ in. thick, 1⅞ in. wide, and 1 ft. long; the shaped pieces C (Fig. 507), the shelf D under the drawer, are ¾ in. thick, and the top is ⅞ in. The sides and back of the drawer, and the ornamental back piece, are of ½-in. stuff, the drawer bottom being ⅜ in. thick. All the other wood may be ⅞ in. thick. The curved stiles F are 1⅝ in. by ⅞ in., and are cut out so that the grain runs tangential to the curve, or parallel to the line X X. The top end is stub-tenoned to fit the piece B, which is mortised to receive it and also the stub-tenon on the rail G. This joint is shown in Fig. 507. The upper rail H is 1⅝ in. by ⅞ in., and is dovetailed into the uprights as shown in Fig. 507. The ornamental supports for the round rail having been shaped and bored are screwed to the uprights; the positions of the screws are as shown on the piece B in Fig. 507. The brackets supporting the drawer are stop-housed to receive the lower shelf, whilst the upper ends are tongued to fit the grooves in the under side of the top shelf, which has a moulded edge and ends. These joints are shown at J, K, and L; where it will also be seen that the straight ends of the curved stiles are notched out to receive the back corners of both shelves. The back of these straight ends should be rebated for the ornamental piece, and provision should also be made in the same way for the backing of the drawer division. The drawer is of the usual construction. Suitable hooks and drawer handle are fixed in position as shown.

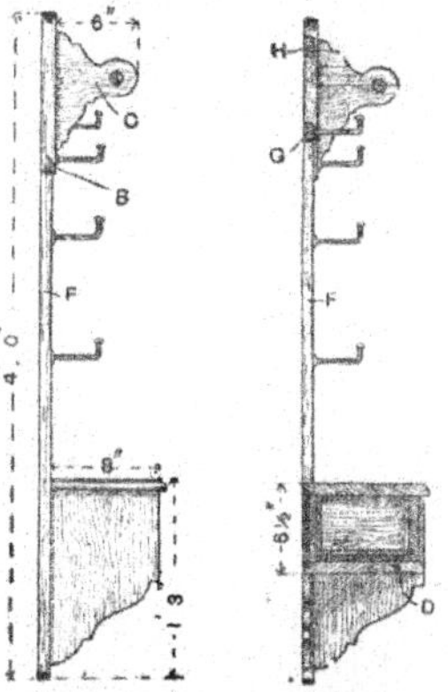

Fig. 505. Fig. 506.

Figs. 505 and 506.—Side Elevation and Cross Section of Hanging Rack.

6-ft. 6-in. Hall Stand with Cupboard, Drawers, Shelf, and Mirror.

Figs. 508 and 509 give elevations of a novel hall stand which is of easy construction, and quite rigid and substantial. Fig. 510 shows a vertical section. Fig. 511 is a half-horizontal section at A (Fig. 508), and Fig. 512 a half plan of the top. Fig. 513 shows the umbrella rack in half plan, and Fig. 514 is a half section showing drawers. The stand combines a hat-, coat-, and umbrella-stand, a cupboard for small articles, glove and brush drawers, a dressing mirror, and a small table or shelf to hold a flower vase, etc. The top may also be similarly utilised. The design is severe, most of the mouldings having straight members, the chamfer predominating. Almost any hard wood would be suitable; fumigated oak or pine, stained a rich brown red, being perhaps the best. The back is upright, the sides and front tapering to the top, to ensure steadiness.

Sides, Back, etc.—The two sides are prepared from 1-in. board, tongued into a solid plinth at the base (see Fig. 515), and housed

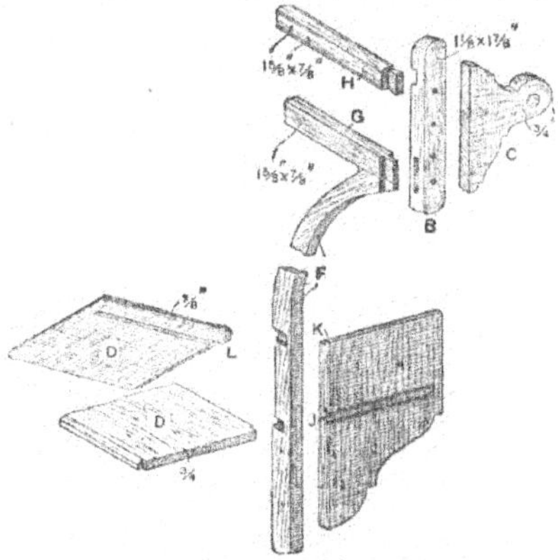

Fig. 507.—Joints for Hanging Rack.

solid into the top, which overhangs about 7 in. at the ends to carry revolving coat hooks. The sides should be checked or rebated out ½ in. to receive the framed back, as shown in Fig. 513, the checking to receive the mirror frame being made 1⅛ in. deep between the shelves at F and H (Fig. 510). For cheapness, the back might be filled in with matchlining, but the appearance would not be so good. All the horizontal members, shelves, rack, tray, brackets, etc., should be housed in the sides ⅛ in. full deep, and well glued; if they are all fitted tightly, no nails or other extraneous fastenings will be required. The frame shown at O (Fig. 510), which carries the metal umbrella tray, is mortised and tenoned together, and secured to the sides by angle-blocks; but a more

Fig. 508.

Fig. 509.

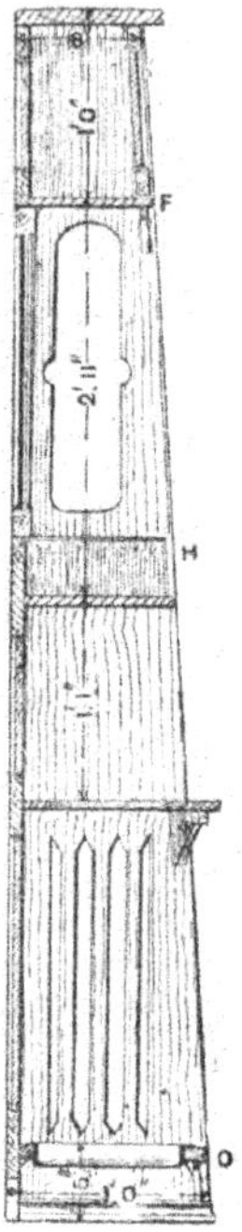

Fig. 510.

Figs. 508 to 510.—Front Elevation, Side Elevation, and Vertical Section of 6-ft. 6-in. Hall Stand with Cupboard, Drawers, etc.

Fig. 511. Fig. 512.

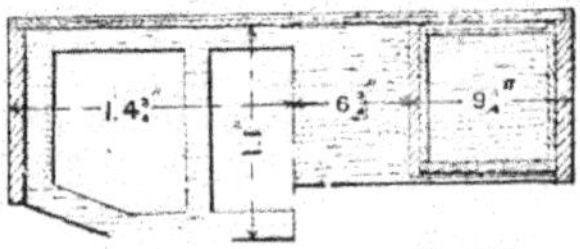

Fig. 513. Fig. 514.

Figs. 511 to 514.—Half Section at Bottom, Half Plan at Top, Half Section at Centre, and Half Section through Drawers of Hall Stand.

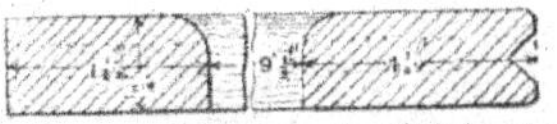

Fig. 516.—Umbrella Rack of Hall Stand.

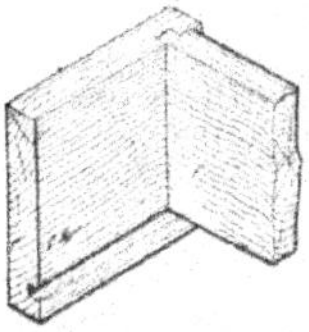

Fig. 519.—Joint in Drawer of Hall Stand.

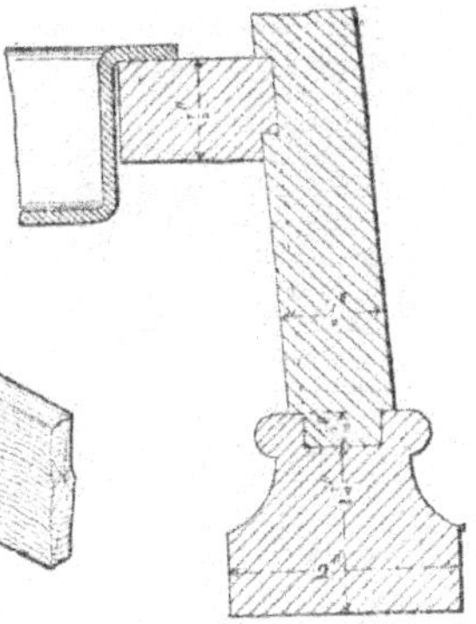

Fig. 515.—Section through Foot of Hall Stand.

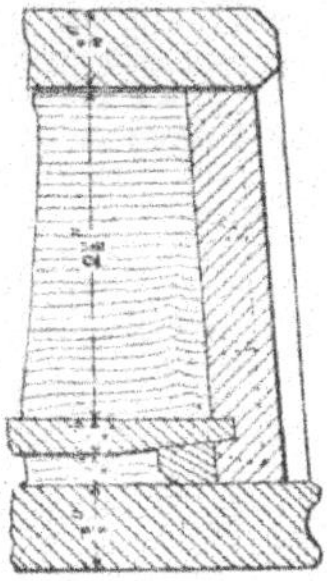

Fig. 517.—Details of Hall Stand Frame.

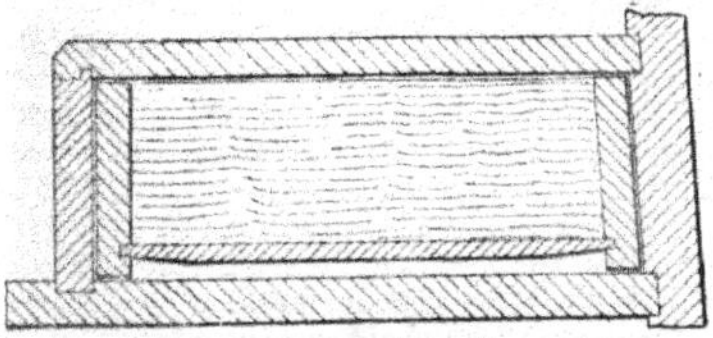

Fig. 518.—Section through Hall Stand Drawer.

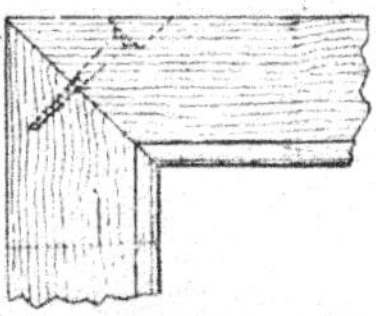

Fig. 524.—Joint of Hall Stand Mirror Frame.

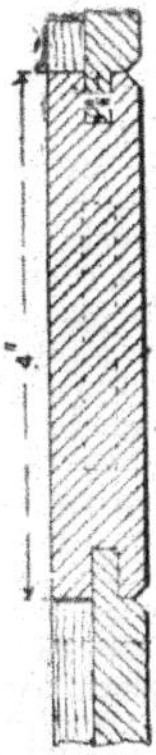

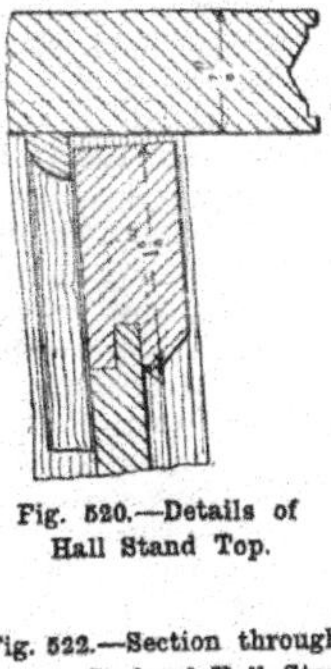

Fig. 520.—Details of Hall Stand Top.

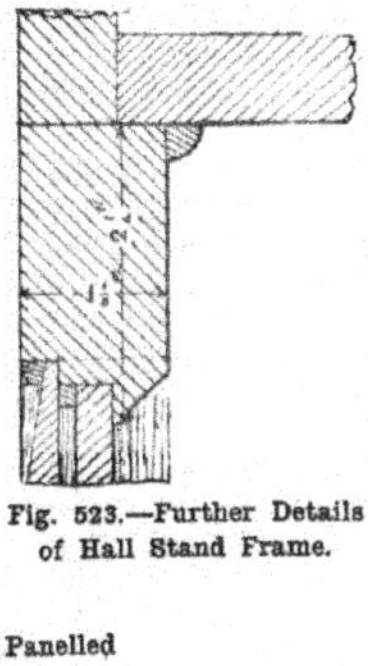

Fig. 523.—Further Details of Hall Stand Frame.

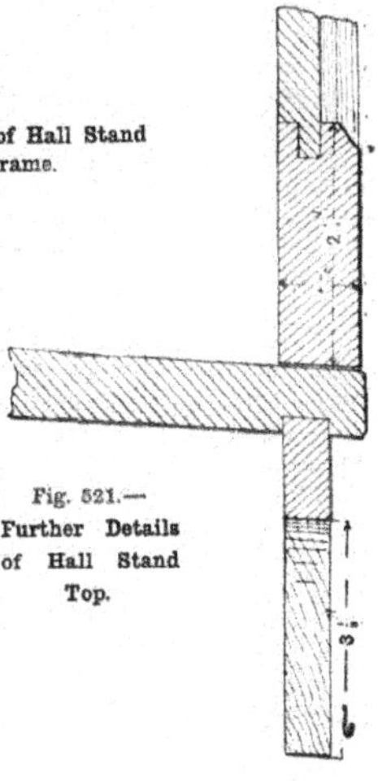

Fig. 521.—Further Details of Hall Stand Top.

Fig. 522.—Section through Panelled Back of Hall Stand.

substantial joint may be made by cutting the housing to a dovetail section and driving in the frame from the back as at Fig. 515.

Umbrella Rack, Shelf, etc.—The umbrella rack shown in half plan by Fig. 513 projects in the centre, and is divided into three or more bays by division rails. This rack is made of ¾ in. stuff, mortised and tenoned together as shown by the dotted lines in Fig. 516. The back tenons may be taken through the rails and wedged, but the front ones should be stopped; the inner edge of the frame is rounded and the front rail V moulded as at Fig. 516. The two brackets under this frame should be grooved in the sides, dowelled to the frame, and inserted in the sides with the former. The shelf H (Fig. 510) is of ¾-in. stuff, its edge having a wave mould and chamfer (see Fig. 517).

Drawer Cases.—The drawer cases, of ⅝ in. stuff, are put together as shown in Fig. 518, which is an enlarged section. These cases should be glued together and into the shelf first, and when dry can be inserted in the sides of the stand as part of the shelf. A block should be used to keep the top of the case at its proper distance from the shelf at the ends.

Drawers.—The drawers (see Figs. 514, 518, and 519) are lap dovetailed at the front, as indicated in Fig. 517, and the back may be also dovetailed; but a simpler and an equally effective joint for small drawers is shown in Fig. 519. The outside of each drawer, as well as the front, must be inclined so as to fit the slope of the case.

Cupboard Doors.—The cupboard door is made of ¾-in. stuff, mortised and tenoned together, a ¼-in. chisel being used. The size of the tenon is indicated in dotted lines at Fig. 520, a continuation of which is shown by Fig. 521. The top edge of the door must be square from the face, and should be kept down, as shown in Fig. 520, to clear the front edge of the top; the open joint will not be discernible at the height, but if desired the joint can be made close by chamfering off the under edge of the top until it is square with the pitch of the front. A quadrant stop should be glued round the two sides and the top of the case. The panelled back, which should be screwed in dry, is made in three pieces, framed and flush-panelled inside as shown in the enlarged section (Fig. 522), the dotted line indicating the tenon.

Mirror Frame.—The mirror frame (Figs. 510, 523, and 524) is mitered and screwed at the angles, and should be twice checked—once for the glass and once for the back panel; a small slip is bradded between the glass and the back.

Construction of Hall Stand.—To set the case out, make full-sized drawings of Figs. 510 to 514. Plane all the stuff to size, true and gauged, taper the sides to the pattern, and place one of them on the rod, and square up the width of all the housings on the front edge, and the shoulders at the top and bottom as shown in the details. Pair the other side with it, and repeat the marking on the edge; then, from the back or upright edge, square all the lines over on the inside with a knife. Stop all the housings ¼ in. back from the face edge, gauge the rebate for the back, and work the housings. To do this, bore a few centre-bit holes in the front end of the housings, and square them out with a 1-in. chisel; then run in the tenon saw, cutting from the back edge to the notches, down the lines, and cut out the core with chisels, finishing to a regular depth with an "old woman's tooth"; or use a grooving plane. Then work the tongues on the ends with a rebate plane, and also plough out the back rebate. Next mark on the open and lattice panels, and cut them out with a bow saw. Their edges are best cleaned off by shooting straight and square a piece of 1-in. stuff of convenient size; hold it firmly by hand or handscrews to the edge to be cleaned off, and draw a 1-in. chisel firmly along its edge, keeping the face of the chisel close to the piece; repeat the process on the opposite side till the cuts meet, which will produce a clean, square surface. Next prepare the shelves and frames, their lengths being taken off the plans. Mark one side first, allowing ⅛ in. for housing, and mark the centre line; then turn the piece over, adjusting it again on the centre line, and on the other side repeat the marking. The open frames should have the sight lines of the rails squared up to give the size of the mortices. The front rail of the umbrella rack should be shaped out of the solid, the edges being cleaned off before the mortices are set out. The fronts of the cupboard may be grooved in, or simply cut in tight between the stops and nailed through top and bottom. If the first course is decided upon, the fronts must be set out and made before the carcase, and in this case an elevation of the top part must be set out to obtain the size. The back frames may be left until the carcase is together, when they can be set out from it. The shelves, etc., being prepared, commence fitting in; all should be fitted tightly, making any easing required underneath. Square them with a bevel set to the elevation slope; the bevel should be used with the blade on one side of the stock to prevent it being applied wrongly, and should be tried on the face sides of the work. When the shelves have been fitted individually, insert the shaped rails in their grooves, and mark their ends on the sides and groove them in. Next try all the divisions in place and cramp up the sides to see that all is right. If the sides are not straight, shorten a shelf where required, then place the top in position and mark and sink the housings and clean up all the parts; when these are finished, the case is ready for knocking together. The bottom frame, if dovetailed, should be inserted first, the ends being glued and knocked in from the back. Then the other divisions are glued and inserted in the grooves on one side, which should be laid on the bench for that purpose, the other side being lifted slightly to allow them to enter. Then the side is brought down on the ends, which are glued and inserted. Cramp the shelves, then fix the top by nailing, square the carcase, and leave the cramps on until the glue is dry. The back can then be set out by laying the stiles with their outside edges on the rebates, placing the rails across them in position as shown in Fig. 508, and marking the edges of each on the other. The inside lines of the mirror frame can be obtained in like manner; mark the mitres on the face, work the rebates and moulding, cut and shoot the mitres, and screw them together as shown in Fig. 524.

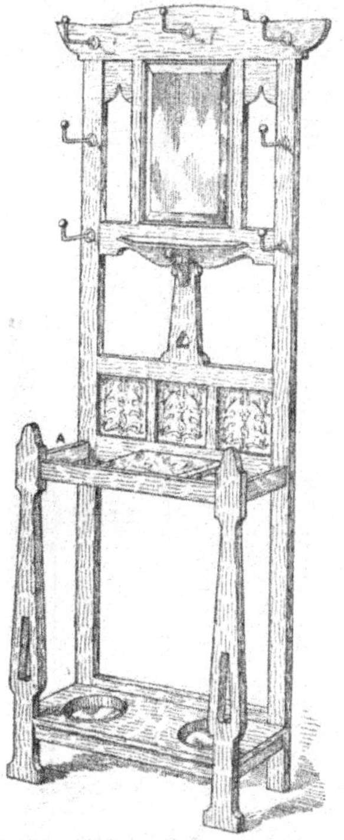

Fig. 525.—Tiled Hall Stand with Mirror and Marble Slab.

Tiled Hall Stand with Mirror and Marble Slab.

A perspective view of a tiled hall stand complete is given by Fig. 525, and elevation and plan by Figs. 526 to 528. All the necessary details of construction are illustrated on a larger scale by Figs. 529 to 536. The mirror, tiles, and marble slab, it will be seen, greatly improve the appearance of the stand, which can be made by anyone of average ability. A suitable hardwood, oak for preference, should be used. The rails and stiles for the back frame should be squared up to the required size, and tenons 1½ in. long cut on both ends of the four bottom rails. The top rail is mortised on the under side to receive the 2-in. tenons on the stiles. The vertical bars on each side of the mirror and between the tiles form the main part of the framework, the

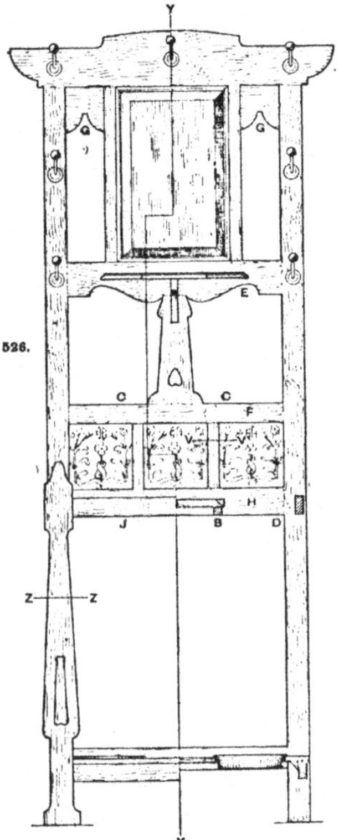

Fig. 526.

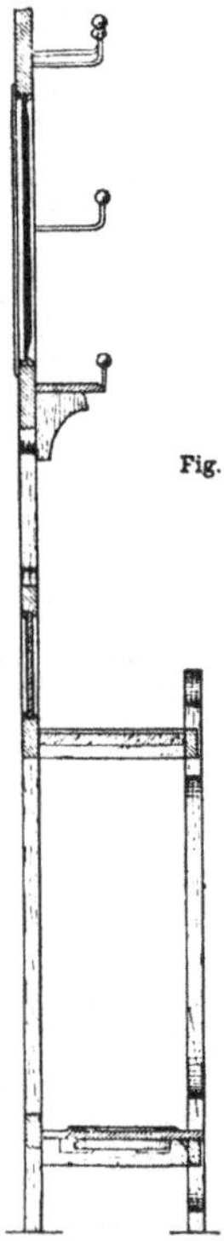

Fig. 527.

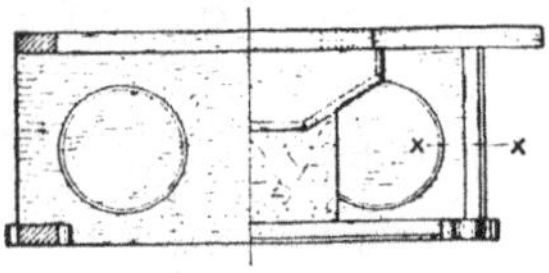

Fig. 528.

Figs. 526 to 528.—Half Front Elevation and Half Longitudinal Section, Cross Section (on Line Y Y), and Half Horizontal Section and Half Plan of Tiled Hall Stand.

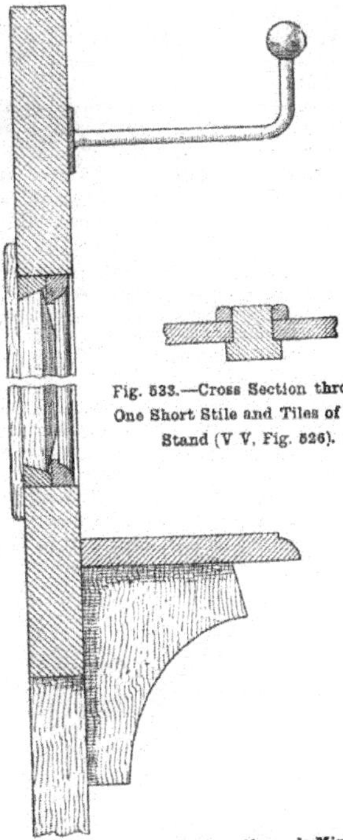

Fig. 533.—Cross Section through One Short Stile and Tiles of Hall Stand (V V, Fig. 526).

Fig. 529.—Part Vertical Section through Mirror of Tiled Hall Stand.

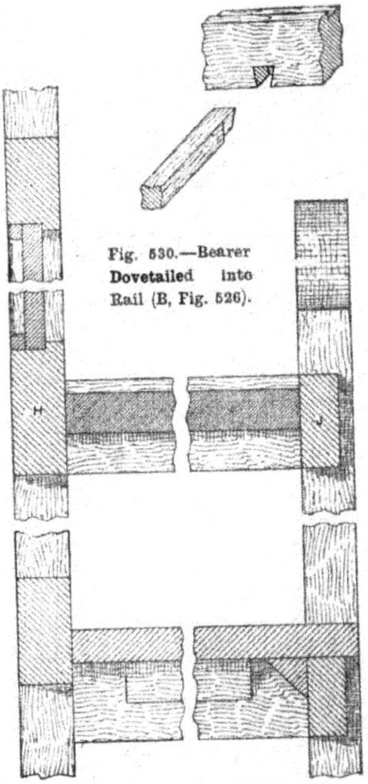

Fig. 530.—Bearer Dovetailed into Rail (B, Fig. 526).

Fig. 534.—Part Vertical Section through Front and Back Frames of Hall Stand.

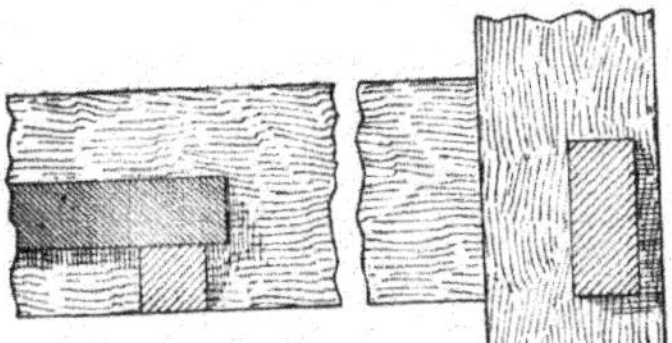

Fig 531.—Detail of Tiled Hall Stand Rails, Bearer and Shelf (B and D, Fig. 526).

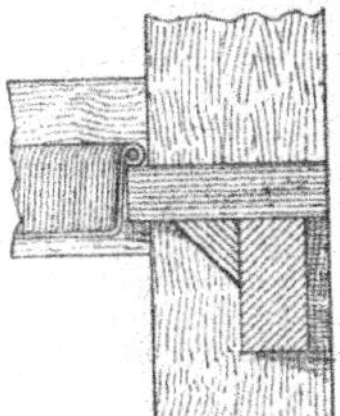

Fig. 532.—Detail of Drip Pan, etc., of Hall Stand (X X, Fig. 528).

former being secured by ⅜-in. dowels and the latter by stump tenons. The shaped piece between the rails E and F is halved on, and screwed from the back. The pieces G, which are ⅜ in. thick, are placed ⅛ in.

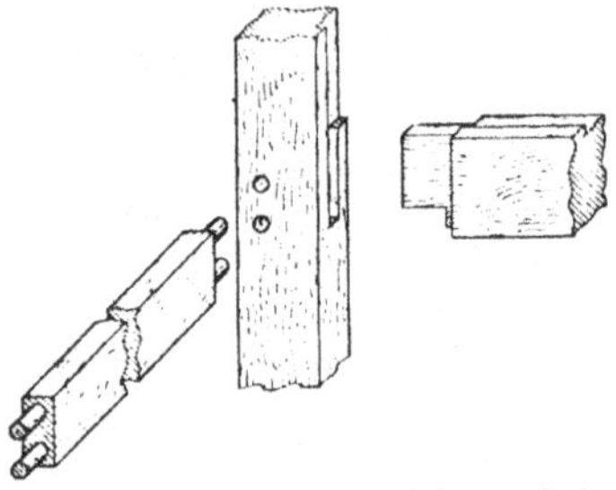

Fig. 535.—Joints of Side and Back Rails to Back Upright of Tiled Hall Stand.

from the face of the stiles, etc., and are glued and blocked behind. A shallow rebate for the tiles is worked on the back edges of the rails F and H, as shown in the enlarged part section (Fig. 534). The two vertical bars between the tiles are rebated along both the back edges. A section through the tiles and one of the bars shows this detail. The stiles also are stop-rebated for 6 in. to receive the end tiles. The front uprights and connecting rails are mortised and tenoned together. The rails are ⅞ in. thick, and are set back ⅛ in. from the face, the tenons being 1½ in. long. Either ordinary or bare-faced tenons may be used. Four rails, 8½ in. long, connect the front uprights to the back. Two ⅜-in. dowels should be used in the end of each rail, these being let into the frame as far as possible. The method of connecting the front and back frames is shown in the enlarged detail (Fig. 536). The ends of the bearers B, which support the marble slab, are dove-tailed about ⅝ in. into the cross rails J and H, as shown separately in the detail (Fig. 530). The slab should be cemented to the bearers with plaster-of-Paris mixed with thin glue. The bottom board, which has two holes, 6 in. in diameter, cut in it for the trays, should rest on the front and side rails and butt against the back rail, the corners being cut to fit the two front uprights. It should be screwed to the rails, and glued and blocked underneath. The shelf below the mirror is supported by a bracket, both being secured to the frame with screws inserted from the back. A thumb moulding is worked round the edge of the shelf. The tiles are held in position at the back by ¼-in. beading, which is mitered at the corners and secured with ½-in. panel pins. The beading used to secure the mirror is shown in the enlarged part vertical section through the mirror. The circular trays are made of stout zinc, with a strong wire flange, and should be enamelled all over. The hooks should be fixed in the

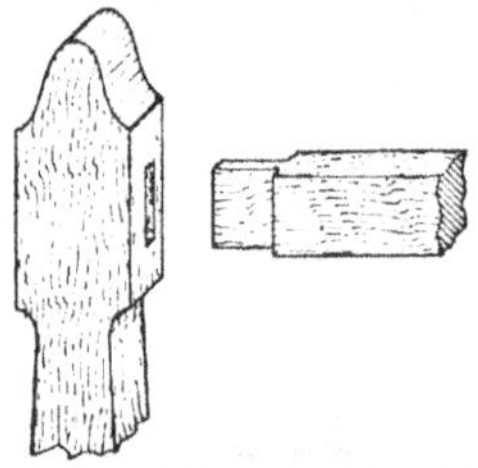

Fig. 536.—Joint between Front Rail and Upright of Hall Stand (A, Fig. 525).

positions shown after the work of construction is finished. The woodwork may be left in its natural state; but if it is to undergo any finishing process, this should be done before the mirror, tiles, slab, and hooks are fitted.

7-ft. Wardrobe in Stained Wood.

THE conventional view presented by Fig. 537 shows a wardrobe forming part of a complete suite which will be illustrated and described in this chapter. Figs. 538 to 541 are the working drawings. The suite may be made either in ash or American pine, and is to be stained green. As a preliminary to the construction, a series of full-size working drawings must be set out accurately, but not necessarily with completeness of detail. It may be noted that Fig. 538 need not be drawn to its full depth from back to front, but broken, as shown in Fig. 541, which is a section at D D (Fig. 542), but the full depth must be given in another section. This method may be applied to all sections which are too wide to go on the board, but all must be full and unbroken in one direction at least. In copying the drawings, take the main dimensions from the complete sections (Figs. 538, 539, and 540), and the minor dimensions and sizes of the components from the enlarged sections. Fig. 538 is a section at A A (Fig. 540), Fig. 539 a section at B B (Fig. 540), part being shown enlarged at Fig. 543, and Fig. 540 a section at C C (Fig. 539).

Carcase of Wardrobe.—The carcase is made in two pieces, the cupboards being framed separate from the drawer case for convenience in handling; the cornice is also framed separately, and lifts on and off. The method of fitting together the cupboard carcase is shown in Fig. 542, F being a side of the division, and E and F the top and sub-top respectively. The bottom is fitted similarly, but runs through the two compartments, and the division is grooved into it as shown in the transverse section (Fig. 541). Lap dovetails, about 2½ in. wide,

Fig. 537.—7-ft. Wardrobe.

and stopped back ¼ in., secure the angles, and are also used in the lower case (Figs. 544 and 545). The sub-top F (Figs. 541 and 542) is housed solid into the division at one

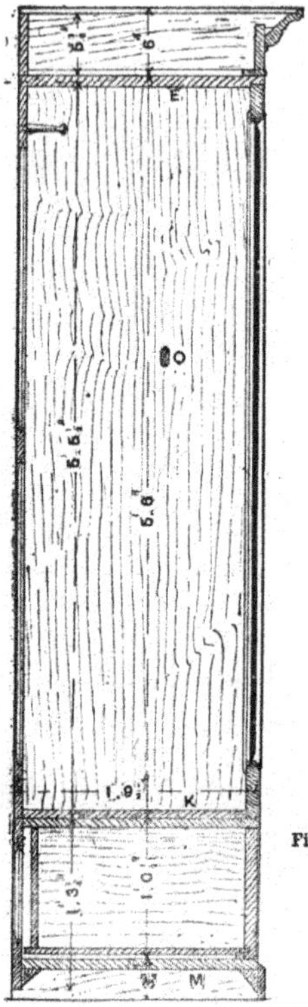

Fig. 538.

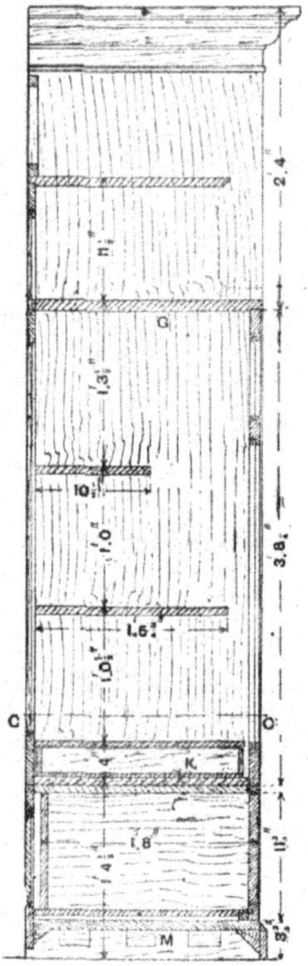

Fig. 539.

Figs. 538 and 539.—Vertical Sections of 7-ft. Wardrobe.

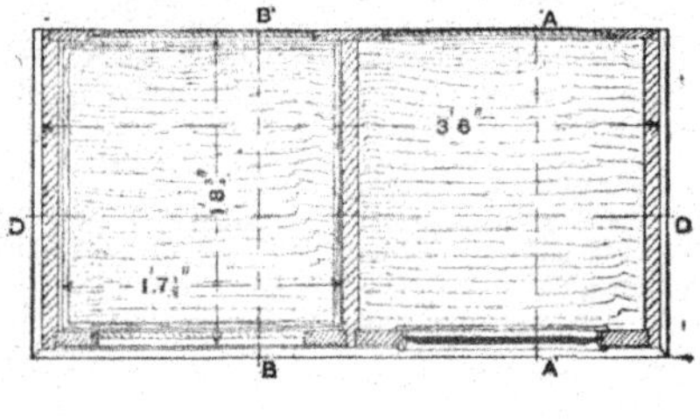

Fig. 540.—Horizontal Section through 7-ft. Wardrobe.

end, and stop-grooved to receive the side P at the other, this arrangement being necessitated by the overhanging moulded edge shown enlarged in Fig. 546. The hanging cupboard is fitted with a 1-in. square and sunk framed door, with a $\frac{5}{16}$-in. silvered and bevelled-edge glass panel, and a $1\frac{1}{2}$-in. by $\frac{3}{4}$-in. rounded rail for hanging garments. The dwarf cupboard is fitted with a 1-in. square and sunk panelled door, three fixed shelves, and a sliding tray for shirts, etc. The back (Fig. 547) is a $\frac{3}{4}$-in. square and sunk panelled frame. The lower case is fitted with a large drawer, a $\frac{3}{4}$-in. panelled back, and a $\frac{1}{2}$-in. chamfered plinth. The cornice consists of a 4-in. by $\frac{3}{4}$-in. frieze, a 5-in. by 1-in. cavetto-ovolo cornice mould, and a 2-in. by $\frac{5}{8}$-in. astragal necking, with cover and back boards complete. Prepare all the stuff, gauging it to size, and place

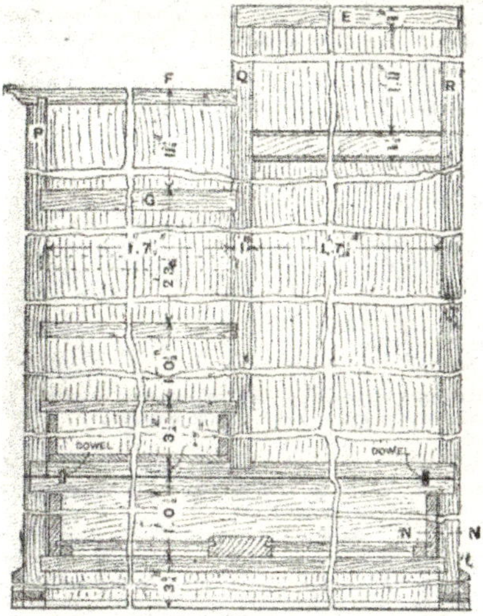

Fig. 541.—Longitudinal Vertical Section of Wardrobe.

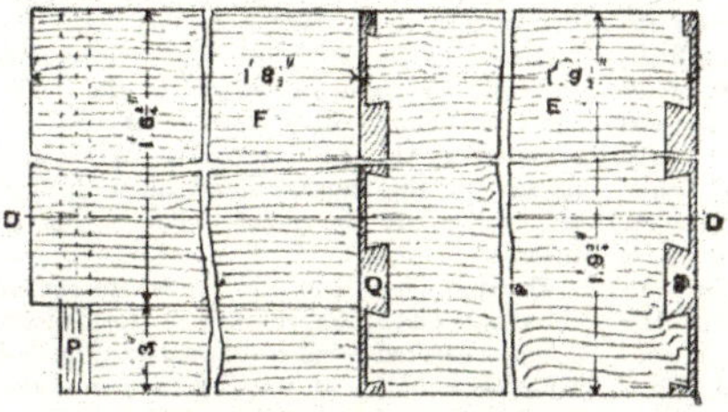

Fig. 542.—Plan of Top of Wardrobe.

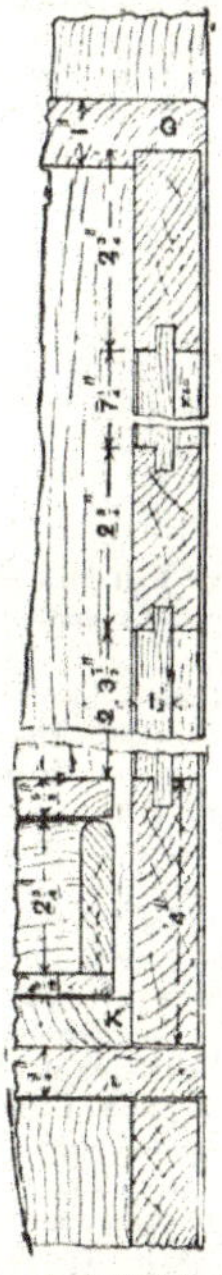

Fig. 543.—Part Vertical Section of Wardrobe.

the face marks of the carcase stuff on the inferior sides, as the working is from the insides; and, if the sides require jointing, endeavour to make the grain run in the same direction in both pieces. Set out the sides, marking all the dimensions that occur on both pieces on the front edge in pencil, pair the two sides, place the division between them, and square the lines across the three edges; next square these across the sides with a knife where required. The insides of the top and bottom should be cut in, but the line representing the outsides should be marked $\frac{1}{16}$ in. long to allow for cleaning off. In marking the sizes of the grooves, keep the upper side exact, and make the lower rather tight to the thickness of the shelves, so that these may be bevelled slightly underneath to fit tightly. Stop all grooves $\frac{1}{4}$ in. behind the front edges of the shelves, and gauge the rebate for the doors $1\frac{1}{8}$ in. from the face edge, the extra $\frac{1}{8}$ in. being for the sinking; the back rebate should be gauged also from the front edge. The division Q (Figs. 541 and 542) is cut flush with the rebate to the top F on its back; above this it oversails and lies flush with the outside of the back, this part being rebated on the right-hand side to receive the back as in Fig. 547. The door rebate of the left of the division stops at G (Figs. 539 and 541). The top E (Fig. 538) is rebated at both edges, and the bottom K is cut in the clear between the rebates. In setting out the top, allow $\frac{1}{4}$ in. extra on the shoulders at each edge to fill the spaces formed by the rebates in the sides as shown in Fig. 542. In putting this part together, after having fitted each portion individually into place, smooth up but do not glasspaper all the insides; place the side P on the bench and glue in the shelves, carefully squaring them upright; let the top end overhang the bench, and fix on the top F, which may be bradded on, as it is too high to be seen. Then drive the division Q on the ends of the shelves, and brad these through from the top side. Next stand the case on the floor on its top end, packing the shorter side level, and drive on the bottom, nailing the division through; then turn the carcase over and drive on the top, first inserting the hanging rail O (Fig. 538) in its sinkings. The carcase, after being squared diagonally with a rod, should be stood aside to dry; then the back may be fitted in and the outside cleaned off.

Fig. 544.—Half Plan of Wardrobe's Lower Case.

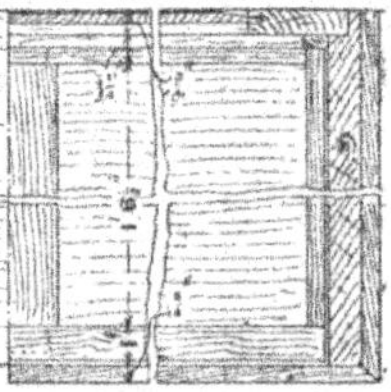

Fig. 545.—Horizontal Section of Wardrobe's Drawer, etc.

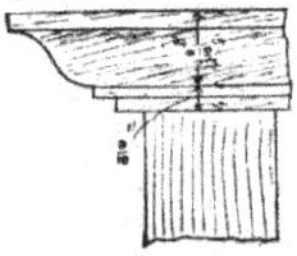

Fig. 546.—Edge of Sub-top of Wardrobe.

Back of Wardrobe.—The back may be prepared as in Fig. 547; this inside back elevation is not necessary for the setting out, but is included to make the disposition of the rails clearer. The lengths of the stiles may be taken from Figs. 538 and 539, and the lengths of the rails from Fig. 540. The top rail on the left side is shown tenoned through the muntin, which is the better way, although, if preferred, it may be stubbed, as are the other rails. When done as shown, the muntin must be wedged to the rail before the panels on that side are inserted.

Doors.—The doors are framed up square, the rebate on the longer one being formed with a small cocked bead as shown in Fig. 548, which is a section at A A (Fig. 540) enlarged; this is glued and sprigged round after the

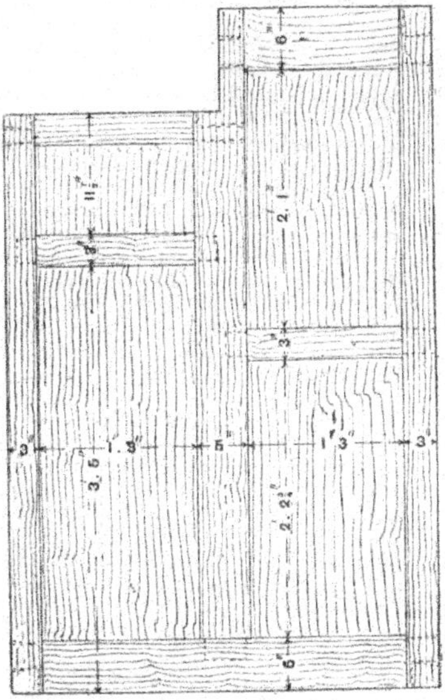

Fig. 547.—Back Framing of Wardrobe.

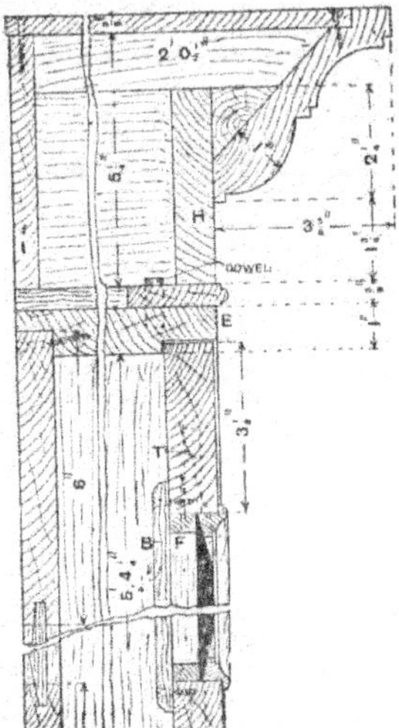

Fig. 548.—Broken Enlarged Cross Vertical Section of Wardrobe.

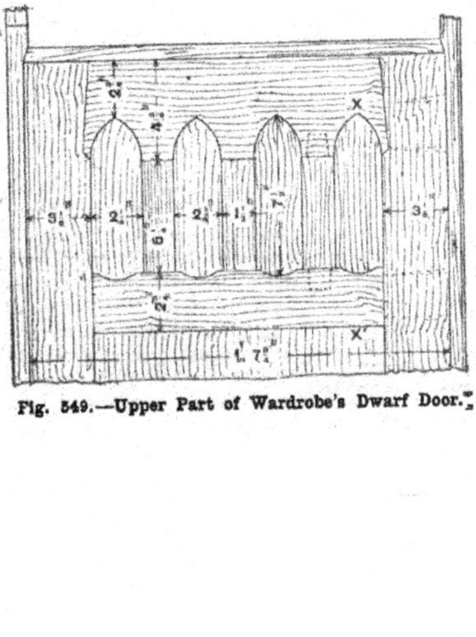

Fig. 549.—Upper Part of Wardrobe's Dwarf Door.

door is cleaned off; a glazing fillet F being sprigged round inside the glass, and all covered with the 1¼-in. backboard B, which is screwed on. Tenons one-third of the thickness and one-half the width are used, as shown by the dotted lines at T (Fig. 548). The shorter door is sunk-panelled on the

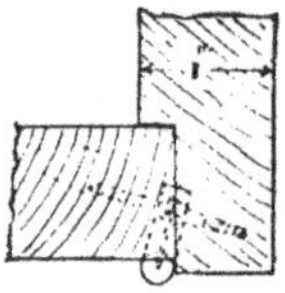

Fig. 550.—Hanging Stile of Wardrobe Door.

face, and bead-butt panelled on the back, the upper part being divided into four narrow upright panels with lancet heads. The construction is shown in Fig. 549, where the muntins go through, and are grooved to receive the panels; but in the cheaper class of furniture all the rail and muntins between X and X′ (Fig. 549) would be planted on the face of the panel, the latter then running from the top to the bottom rail. The top rail must be cut with a taper shoulder to prevent the edge breaking away in cutting the arch. The doors should be fitted in the case upright, with a joint all round equal to the thickness of a piece of brown paper, and should hang on opposite edges as shown in Fig. 550, the centre of the knuckle of the hinge being in line with the edge of the case; the doors will then open flat back. Three hinges should be used for the mirror door, on account of the weight; and in glazing the door, pack a piece of washleather, or blotting paper folded, under the bottom edge of the glass close to the hanging stile, so that the weight may be taken at the inside lower end of the door, thus preventing racking.

Lower Case of Wardrobe.—The lower case is shown in Figs. 544 and 545, the first-named being a half plan of the top, and the latter a horizontal section at N N (Fig. 541). The sides should be rebated for the back, but not at the front, and the outside dimensions must be exactly the same as for the upper case. The bottom is grooved solid into the sides, and blocked in the angles underneath; it should be cleaned off outside before the plinth is fixed, and the two lower ends shaped afterwards. The case should be squared before the back is fitted, and the latter then nailed in tightly. The drawer front can then be fitted hand-tight into the opening, and the back set out from it. The construction of the drawer is shown in Fig. 551, the sides being dovetailed in the usual manner. A strengthening batten is dovetailed to the front, and rebated over the back, because the drawer is rather long and the bottom would be liable to sag without it; a section of this rail will be seen in Fig. 541. The plough grooves for the bottom should be made with a ¼-in. iron, about ⅜ in. up from the edge; the bottom itself is rebated on three sides as shown in Fig. 541 to receive a hardwood blocking slip, which is glued to the sides and front of the drawer (not the bottom), and cleaned off flush to form a runner. The ends of the drawer sides should be finished square, and made to butt against a small slip glued to the back of the case. The drawer, like the doors, sets back ⅛ in. from the face of the case.

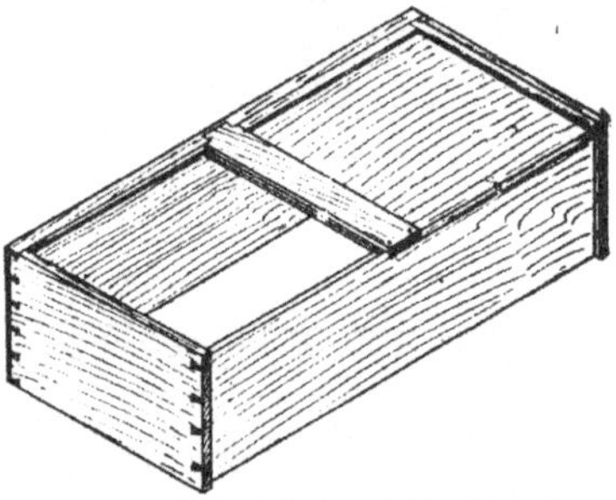

Fig. 551.—Wardrobe Drawer, Bottom Upwards.

Cornice, etc.—The cornice is begun by making a frame with the frieze rail H (Fig. 548) and the back rail I, and these may be mitered and blocked in the angles at the front side, and the backboard rebated into the frieze at the back. The cornice mould, worked from 1-in stuff, is mitered round the frieze and blocked in the angle as shown. The cover board should be screwed tight to the front cornice, and slot-screwed to the end pieces and back, so that when it shrinks it will not split. The fretted back rail to the sub-top is a conventional representation of the rays of the rising sun; it is cut out

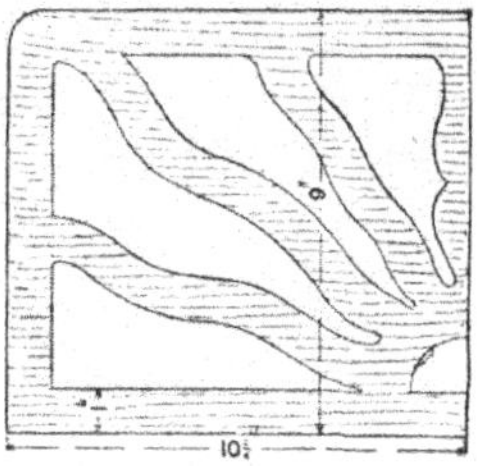

Fig. 552.—Half Design of Fretted Back to Wardrobe.

of a board ¾ in. thick, the margins and the rays being ⅞ in. wide. The half-design given in Fig. 552 may be enlarged to scale, or by the usual geometrical methods, and the rail is fixed by screwing to the shelf and side of the case.

Cutting List for Wardrobe.—The rough cutting list for the wardrobe is as follows:—Main carcase: Two sides, 5 ft. 6½ in. by 1 ft. 10 in. by 1 in. one side, 4 ft. 8½ in. by 1 ft. 10 in. by 1 in.; one top, 1 ft. 9½ in. by 1 ft. 10 in. by 1 in.; one top, 1 ft. 9 in. by 1 ft. 7 in. by ¾ in.; one bottom, 3 ft. 5 in. by 1 ft. 8 in. by ¾ in.; one shelf, 1 ft. 8½ in. by 1 ft. 9½ in. by 1 in.; one shelf, 1 ft. 8½ in. by 11 in. by ¾ in.; one shelf, 1 ft. 8½ in. by 1 ft. 5¾ in. by ¾ in.; one shelf, 1 ft. 8½ in. by 1 ft. 5½ in. by ⅝ in. Back: One stile, 5 ft. 7 in. by 3⅛ in. by ¾ in.; one stile, 4 ft. 9 in. by 3⅛ in. by ¾ in.; one muntin, 5 ft. 7 in. by 5⅛ in. by ¾ in.; one top rail, 1 ft. 9 in. by 6⅛ in. by ¾ in.; one top rail, 1 ft. 11 in. by 3⅛ in. by ¾ in.; two mid rails, 1 ft. 8 in. by 3 in. by ¾ in.; one bottom rail, 3 ft. 5½ in. by 5⅛ in. by ¾ in.; two panels, 2 ft. 3 in. by 1 ft. 4 in. by ½ in.; one panel, 3 ft. 1 in. by 1 ft. 4 in. by ½ in.; one panel, 9½ in. by 1 ft. 4 in. by ½ in.; two tray sides, 7 ft. by 2⅜ in. by ½ in.; one bottom, 1 ft. 8 in. by 1 ft. 7 in. by ½ in. Doors: Two stiles, 5 ft. 6½ in. by 3⅛ in. by 1 in.; two stiles, 3 ft. 9½ in. by 3¼ in. by 1 in.; one rail, 1 ft. 8½ in. by 3⅜ in. by 1 in.; one rail, 1 ft. 8½ in. by 5⅛ in. by 1 in.; one rail, 1 ft. 8½ in. by 4⅛ in. by 1 in.; two rails, 1 ft. 8⅛ in. by 2⅞ in. by 1 in. Panels: One, 2 ft. 5 in. by 1 ft. 2 in. by ½ in.; four, 1 ft. 4 in. by 2⅜ in. by ½ in.; one rail, 1 ft. 8 in. by 1⅜ in. by ¾ in.; and one glass back, 4 ft. 10½ in. by 1 ft. 2½ in. by 5/16 in. Drawer case: Two sides, 1 ft. 5 in. by 1 ft. 10 in. by 1 in.; one top, 3 ft. 6 in. by 1 ft. 10 in. by ¾ in.; one bottom, 3 ft. 5 in. by 1 ft. 9 in. by 1 in.; two back rails, 3 ft. 5½ in. long; two stiles, 1 ft. 4 in. by 3 in. by ¾ in.; and one plinth, 7 ft. 4 in. by 3½ in. by ½ in. Drawer: One front, 3 ft. 4 in. by 1 ft. by 1 in.; one back, 3 ft. 4 in. by 11⅜ in. by ⅝ in.; two sides, 1 ft. 6 in. by 1 ft. by ½ in.; two

Fig. 553.—Dressing Table.

bottoms, 1 ft. 8 in. by 1 ft. 8 in. by ½ in.; one rail, 1 ft. 9 in. by 3 in. by 1 in.; and blocking oak, 6 ft. 6 in. by ¾ in. by ½ in. Cornice: Moulding, 8 ft. 11 in. by 5½ in. by 1 in.; frieze, 7 ft. by 4 in. by ¾ in.; astragal, 7 ft. 1 in.

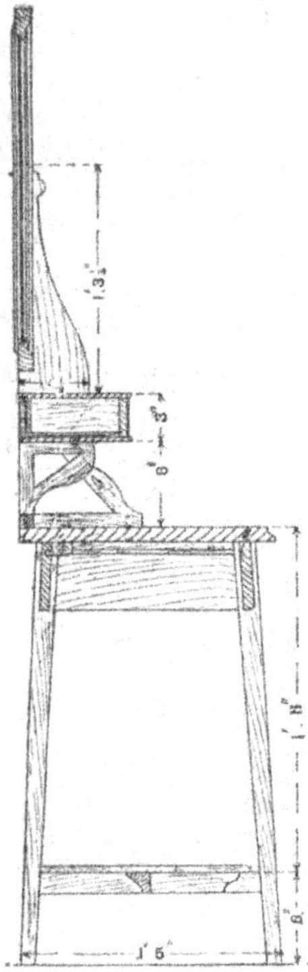

Fig. 555.

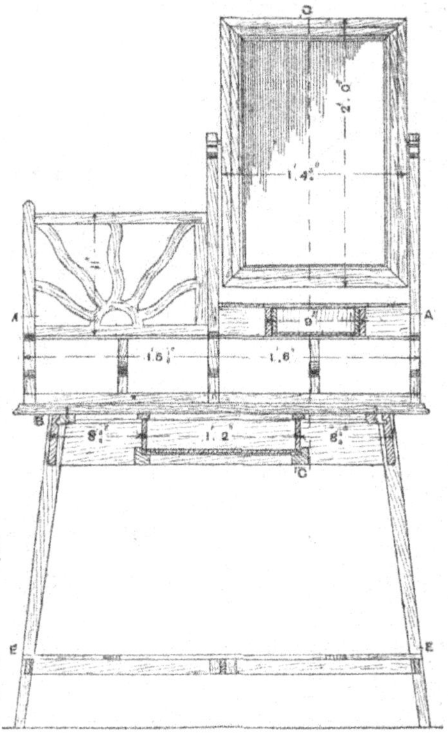

Fig. 554.

Figs. 554 to 556.—Sectional Elevation, Cross Section, and Plan of Dressing Table.

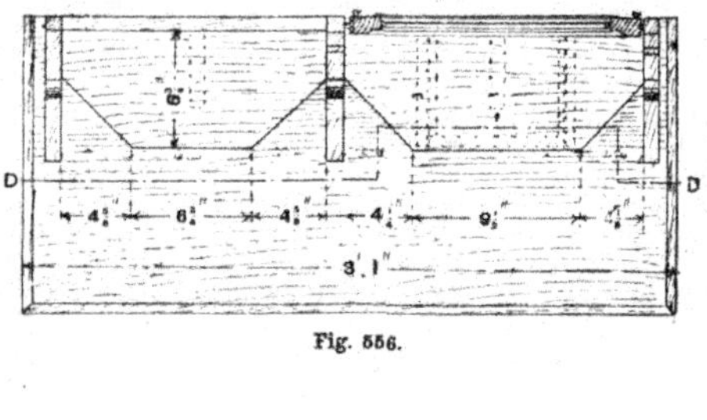

Fig. 556.

by 2 in. by $\frac{3}{8}$ in.; back, 3 ft. 6 in. by $5\frac{3}{4}$ in. by $\frac{1}{2}$ in.; and cover, 3 ft. 11 in. by 1 ft. $0\frac{1}{2}$ in. by $\frac{5}{8}$ in. Sundries: Glass, 4 ft. $9\frac{1}{2}$ in. by 1 ft. $1\frac{1}{2}$ in. by $\frac{3}{16}$ in.; glazing

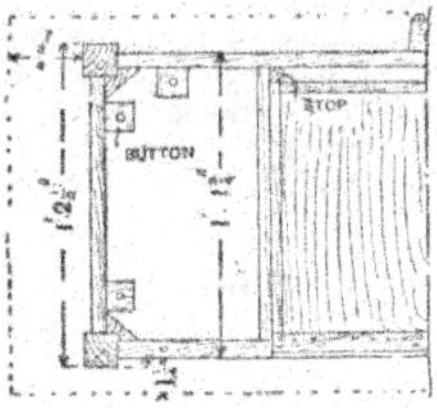

Fig. 557.—Half Horizontal Section of Dressing Table.

mould, 7 ft. 6 in. by $\frac{3}{4}$ in. by $\frac{3}{8}$ in.; two and a half pairs of $2\frac{1}{2}$-in. rolled brass butts and screws, two 2-in. door locks, one drawer lock, two antique drop door handles, one pair of similar drawer handles, two wardrobe hooks, and three hanging pegs.

Dressing Table in Stained Wood.

The dressing table illustrated by Fig. 553 forms part of the bedroom suite of which

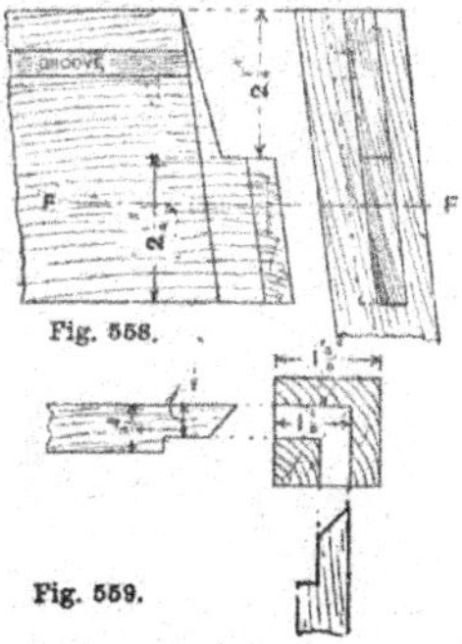

Figs. 558 and 559.—Joint of Rail and Leg of Dressing Table.

the wardrobe has already been described. The method of preparing the working drawings need not be again described. Drawings similar to Figs. 554, 555, 556, and 557, which are sections respectively at D D (Fig. 556), C C (Fig. 554), A A, and B, will be required; set them out carefully to full size before preparing the stuff. All the chief dimensions are here shown, and no difficulty will be found in filling in the minor ones. The table measures over all 3 ft. 1 in. by 1 ft. 5 in., and is 2 ft. 5 in. high. The top is of 1-in. stuff, ogee moulded, and fixed to the frame by buttons. The legs are thrown in 3 in. from the ends at the top, and $1\frac{1}{2}$ in. from the front and back; they are cut out of $1\frac{3}{4}$-in. stuff, and tapered off

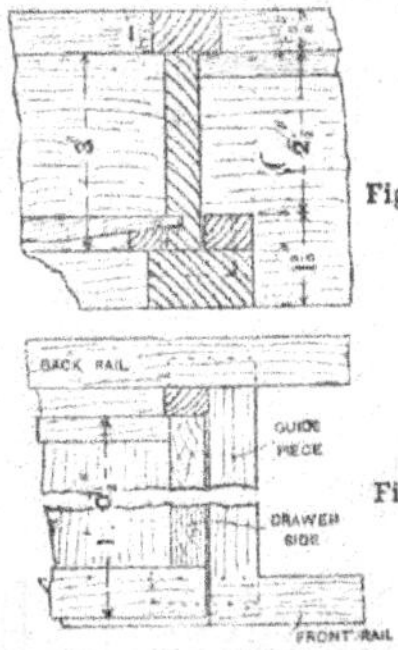

Figs. 560 and 561.—Part Section and Plan of Dressing Table's Bottom Drawer.

to 1 in. at the bottom, and spread at the bottom to the size of the table top. The top rails are of $\frac{7}{8}$-in. by $4\frac{1}{2}$-in. stuff, and are framed to the legs with bare-face tenons, and mitered together as shown in Figs. 558 and 559. The front rail has an opening cut in for a drawer 1 ft. 2 in. by 3 in., and two runners are framed between the rails to carry the drawer, as shown in Figs. 554 and 560. These runners are stub-tenoned into the rails, and may be rebated out of the solid, or formed by nailing a guide piece to the runner, as shown in Figs. 560 and 561. A tilting piece I (Fig. 560) should be fitted across the rails, flush with the under edge of the opening, to prevent the drawer tipping. A $\frac{3}{8}$-in. by $\frac{1}{4}$-in. groove should be run round the inside top edges of the back and end rails to receive the buttons, and the front rail is screwed direct to the top from the inside, so that when shrinking it shall not alter its position in front. The cross stretchers are kept up 6 in., and stub-tenoned into the legs, and the long stretcher is housed into them and fixed with angle blocks at the back. The bottom shelf, being wide in the middle, requires a rail to keep it from twisting, and this rail should be first, and then polish all sunk parts or re-entrant angles, as these cannot be finished properly afterwards ; of course, they can be botched over with a brush, as cheap furniture is prepared. At the same time it will be wise not to do more polishing than necessary before fitting together, as it is liable to get damaged in the working. When the frame is made, prepare the top, if possible, in one piece; but if there must be joints,

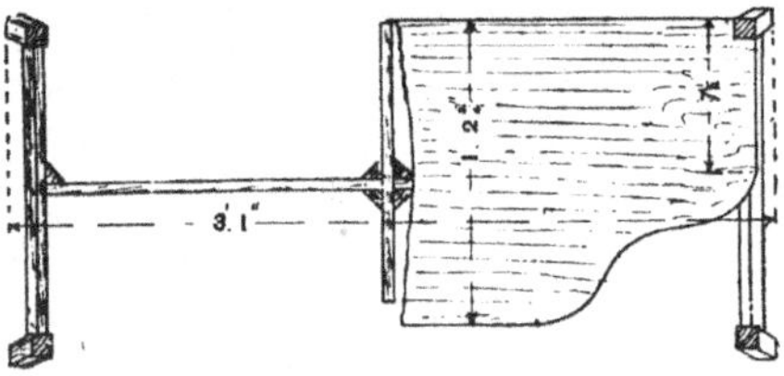

Fig. 562.—Part Plan of Dressing Table Shelf.

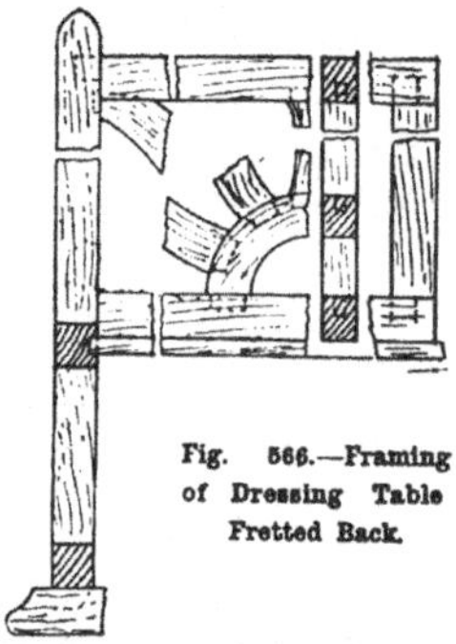

Fig. 566.—Framing of Dressing Table Fretted Back.

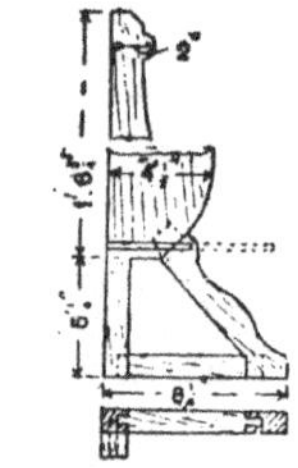

Fig. 563.

Fig. 564.

Figs. 563 and 564.—Right End of Top Fitting of Dressing Table.

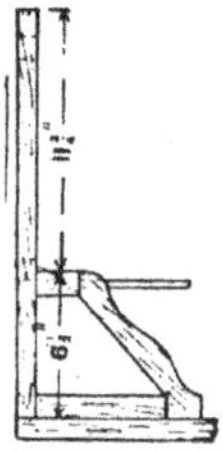

Fig. 565.—Left End of Top Fitting of Dressing Table.

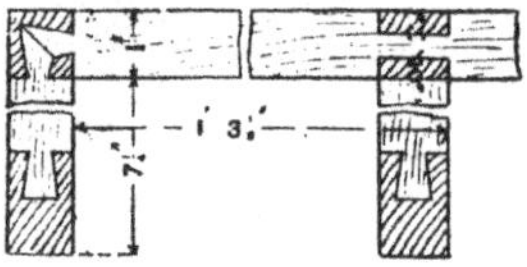

Fig. 567.—Joints of Back and Bottom Rails of Dressing Table.

notched over the stretcher. If the notching is done the other way, the rails will probably collapse when loaded. Angle blocks can be glued all round the joint as shown in Fig. 562, which is a section at E E (Fig. 554), and blocks should also be glued to the under side of the shelf.

Method of Construction.—In constructing the table, make and glue up the frame first, and put the end legs together and let them dry before gluing the sides together. If the work is to be french-polished, it will be better, before gluing together, to fit it up dowel them at 6-in. intervals, starting at 2 in. from the ends. The drawer should be fitted before the top is fixed on. Next cut the lower shelf to shape, as shown broken in Fig. 562, and fit it in place. The top fitting should then be proceeded with, and the two shaped standards that support the mirror may either be cut out of the solid, or be partly solid and partly framed (see Figs. 563 and 564), the latter method being more difficult, but stronger. The curving of the front edge of the standard is continued down to within ⅝ in. of the shelf groove, where it finishes abruptly in a straight line at right angles to the back edge of the shaped arm, to which it forms a shoulder; it is mortised to receive a tenon, as shown in Fig. 563, and the back edge of the standard continues down to the bottom rail, to which it is secured by a dovetail as shown in Fig. 564. The front edge of the rail is also fastened by a dovetail to the front arm, and the back bottom rail is connected to the standard by a $\frac{3}{8}$-in. dowel. The bracket at the opposite end (Fig. 565) is framed together in a similar manner, and the short straight rail carrying the shelf is dowelled at each end. The fretted back is framed as shown in Fig. 566, and is secured to the shelf by brads or screws inserted from below. The joints in the bottom rail are shown in Fig. 567, the lower end of the central standard lipping over a notch tenon on the back rail. The two back rails are shown mitre-dovetailed into the end standard, but may be dowelled if preferred. The intermediate brackets supporting the shelves are cut out of the solid and dowelled at each end, and the shelves are housed $\frac{1}{8}$ in. deep at the ends and run over flush with the back, the middle portion of the two under the mirror being rebated $\frac{1}{8}$ in. deep to receive the $\frac{1}{4}$-in. back of the drawer case, as shown in Fig. 568. The sides of this case are housed $\frac{1}{8}$ in. into the shelves, but are stopped $\frac{1}{4}$ in. from the front. The construction of the drawer is illustrated in Figs. 568 to 570. The top fitting is secured to the table, as shown in Fig. 555, by screws from the under side.

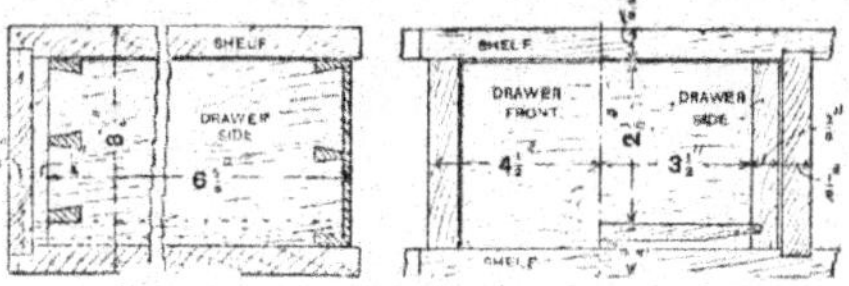

Fig. 568. Fig. 569. Fig. 570.

Figs. 568 to 570.—Side Elevation and Sections of Dressing Table's Small Drawer.

Mirror Frame.—The mirror frame (shown in section by Fig. 571) is mitered together and secured by a screw in the ordinary manner, the holes for the screw heads being sunk $\frac{1}{4}$ in. and pelleted. The frame is rebated $\frac{1}{8}$ in., and slipped at the back of the glass, the whole being covered in with a $\frac{3}{16}$-in. rounded edged backboard screwed on. The mirror is hung 1 in. above its centre with a pair of patent brass catches, shown in Figs. 572 and 573. The pivots are fixed to the standards, and the locking sockets to the mirror are sunk in its edges until the hole stands in the middle of the thickness.

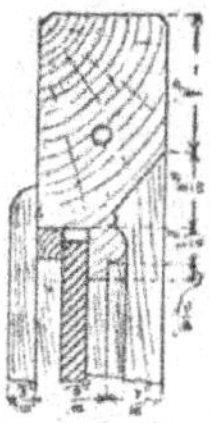

Fig. 571.—Part Section of Dressing Table's Mirror Frame.

Washstand in Stained Wood.

The washstand shown by Figs. 574 to 578 forms part of the suite of which two articles have been described above. It has a Sienna marble top, 2 ft. 11 in. long by 18 in. wide and $\frac{3}{4}$ in. thick, with an ogee moulded edge; this is surmounted by a pair of square bracketed standards, carrying a round curtain-rod, and having a $\frac{1}{2}$-in. by 6-in. shaped splash-board framed between them. The table frame has splayed legs, and is fitted with a pot-board, an enclosed cupboard, and a towel drawer. Fig. 575 is a sectional elevation on C C (Fig. 577), showing the

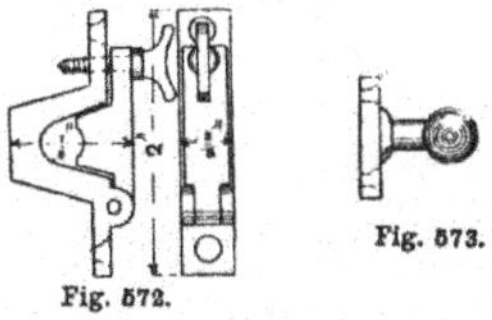

Fig. 572. Fig. 573.

Figs. 572 and 573.—Mirror Pivot for Dressing Table.

back and door of the cupboard. Fig. 576 is a cross section, Fig. 577 is the half plan at the top of the table frame, Fig. 578 is a section at B B (Fig. 575), and Fig. 579 is a half plan above the top. Fig. 580 illustrates one of the back legs, and the ends of the back and side rails, showing barefaced tenons mitered at the ends. Fig. 581 gives the edge and side views of a door stile, showing the setting out required for the rails and panel, and Fig. 582 a side and edge view of the bottom rail, showing the tenons full length.

Construction.—The construction of the frame and brackets is fully explained above, but it should be noted that the back legs are upright in the side elevation, and will require square shoulders on the back ends of the side rails. The sides of the cupboard are housed $\frac{1}{8}$ in. into the shelf or pot-board, and are carried up flush with the top edges of the frame, being notched out and sunk into the latter $\frac{1}{4}$ in., as shown in Fig. 583. At the front edge the notched part should be reduced to a $\frac{1}{4}$-in. tongue: if housed in the full thickness, one side of the housing will be cut away when the opening is made for the drawers. The runners for the drawer (Fig. 584) are screwed to the cupboard sides, and are grooved to receive a dust panel forming the top of the cupboard. A similar rail to the runners is glued and nailed to the inside of the front rail as shown in Fig. 585, and carries the front edge of the dust panel, forming also a fixing for the door stop underneath. A rail is not needed at the back, the hinder edge of the panel being grooved into the frame rail. Two rails are tenoned into the sides of the cupboard as shown in Fig. 575, to carry the back panel, which is also grooved into the sides (see Fig. 578). As the top is of marble, buttons are not required to fix it, its own weight being sufficient to keep it in place; therefore grooves are not required at the top edges of the frame. Two end spreaders and one stretcher are required to carry the shelf as shown in Fig. 578, but a central cross rail is not necessary, as the shelf is fixed to the cupboard sides. The top fitting is mortised and stub-tenoned together and kept in position by dowels, as shown in Fig. 576.

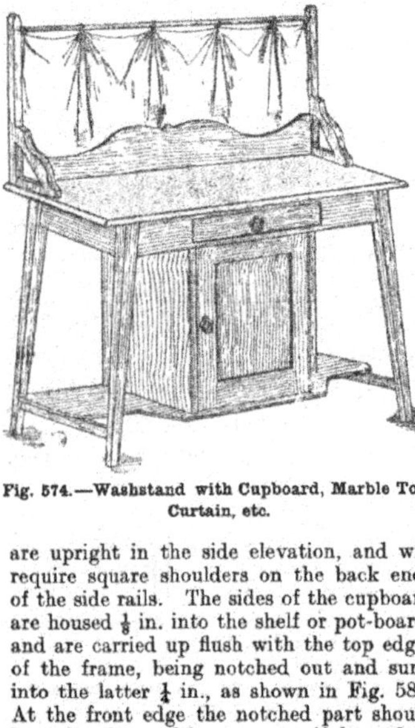

Fig. 574.—Washstand with Cupboard, Marble Top, Curtain, etc.

Drawer of Washstand.—In making the drawer, cut the aperture in the rail first to the required size, as shown in Figs. 575 and 585, then fit in the drawer front tightly, and cut the back to the same length as the front, but $\frac{5}{8}$ in. narrower; then prepare the sides to the same width as the front, and to the length shown in Fig. 585, cutting the front ends accurately to the bevel of the legs. Next plough a $\frac{1}{4}$-in. groove in the drawer front and sides, $\frac{3}{8}$ in. from the bottom edges, and $\frac{1}{4}$ in. deep in the front and $\frac{3}{16}$ in. deep in the sides. Proceed to lap dovetail in the usual way. Figs. 586 to 588 should be consulted. The parts should fit fairly tight, but not sufficiently to cause splitting. Having ascertained that the parts come together accurately, separate them, clean off the insides, and then glue them up, trying the drawer for squareness with a rod; afterwards fit in the bottom, which should be cut exactly to the length between the backs of the grooves, its grain running parallel with the drawer front. Chamfer off its ends and front edge to fit the groove as shown in Figs. 584 and 585, and drive it in; bore a small hole in the middle of the back edge, cut a slot, and screw the bottom in place. Fit in strips of oak or deal about $\frac{1}{2}$ in. square on the front and sides (see Figs. 584 and 585), and run a series of saw cuts in these, nearly through from the bottom surfaces, to enable them to bend freely over any irregularity in the bottom, and glue them to the drawer sides. When these side play. Then, with a knife, the front should be marked all round on its edges by running the tool round the margins of the opening, cleaned off to the knife marks, and

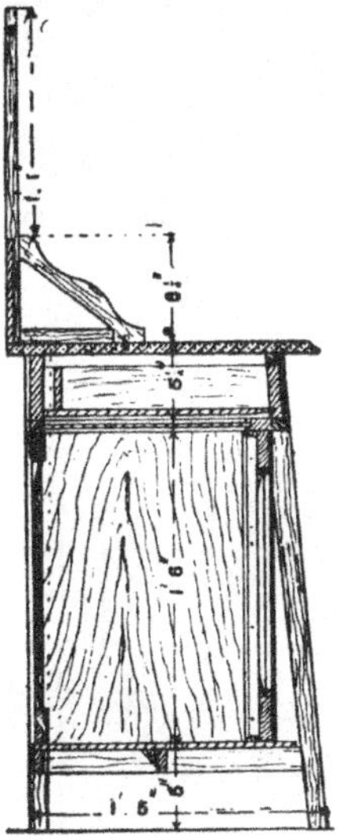

Fig. 576.

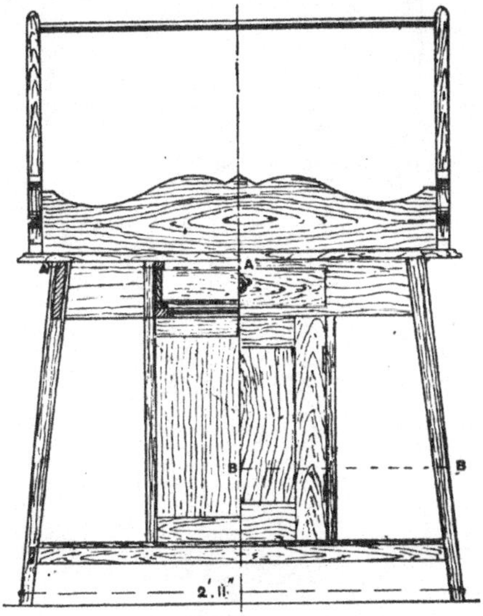

Fig. 575.

Figs. 575 to 578.—Front Elevation and Longitudinal Section, Cross Section, and Two Half Horizontal Sections of Washstand.

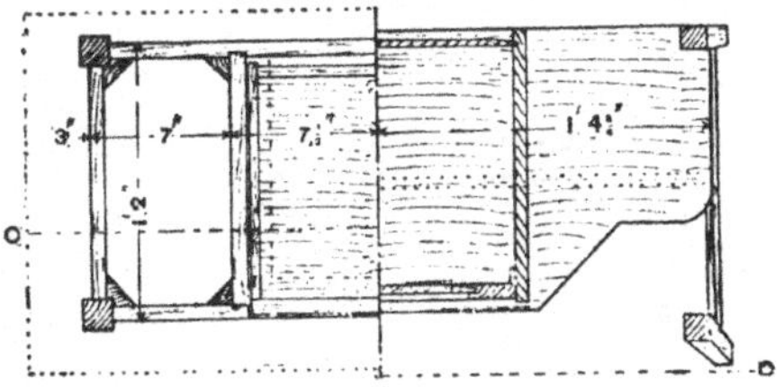

Fig. 577. Fig. 578.

blocks are dry, the drawer may be fitted in place. It should be planed down in the bench screw with the trying plane until it will pass freely into the opening, but without then be stopped ⅛ in. from the face by thin hardwood stops glued on the face of the front rail as shown at s (Fig. 585). The drawer front, the door framing, and the

Fig. 579.—Half Plan of Washstand Top.

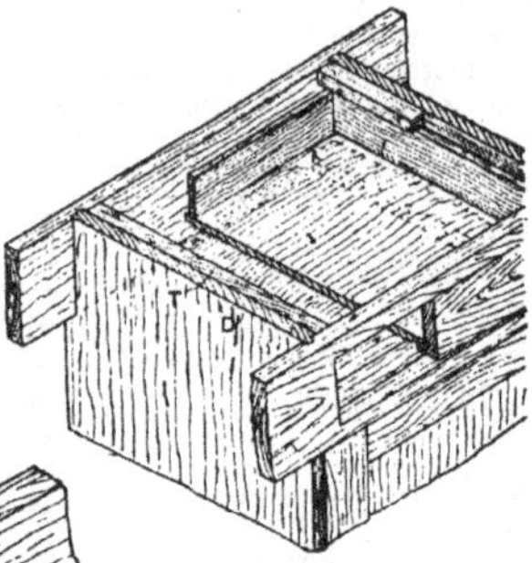

Fig. 583.—Washstand Drawer and Fittings.

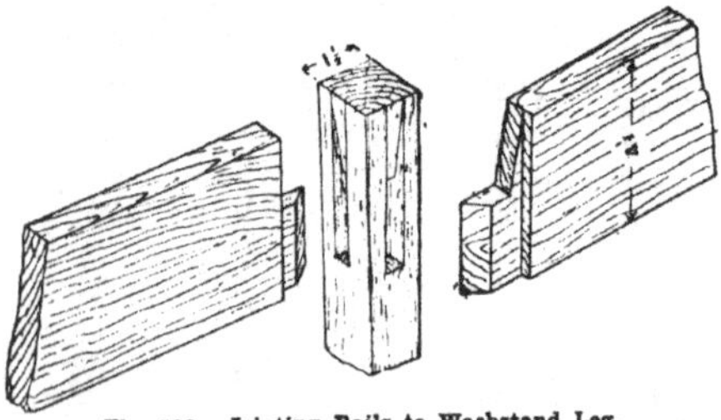

Fig. 580.—Jointing Rails to Washstand Leg.

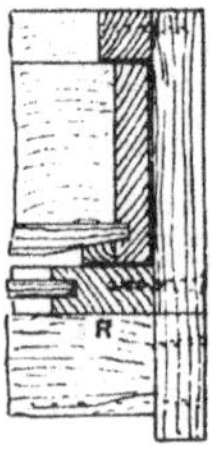

Fig. 584.—Part Section of Washstand Drawer, etc.

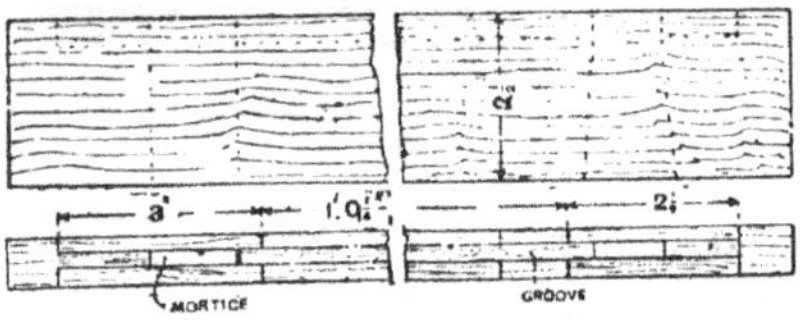

Fig. 581.—Edge View and Side View of Washstand Door Stile.

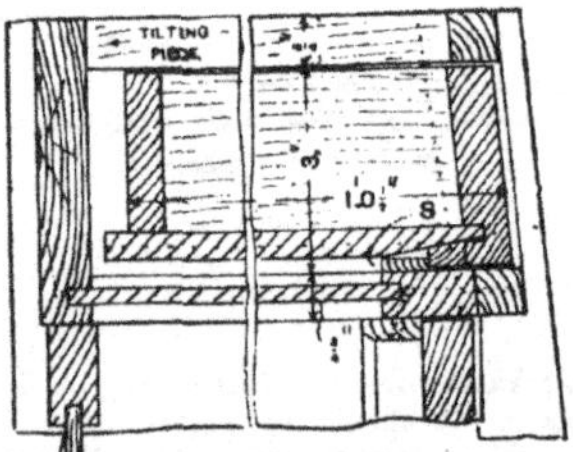

Fig. 585.—Part Section of Washstand Drawer, etc.

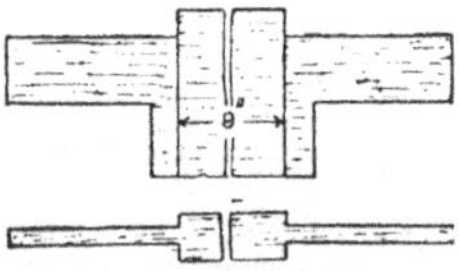

Fig. 582.—Edge View and Side View of Washstand Bottom Rail.

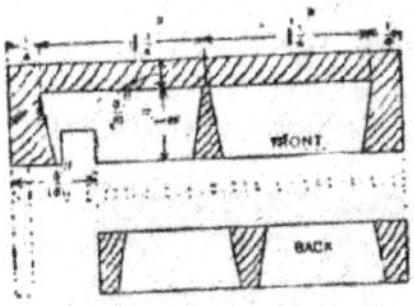

Fig. 586.—End Views of Washstand Drawer.

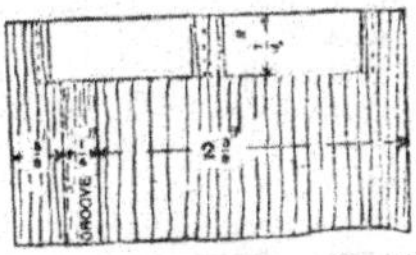

Fig. 587.—Inside Elevation of Drawer Front.

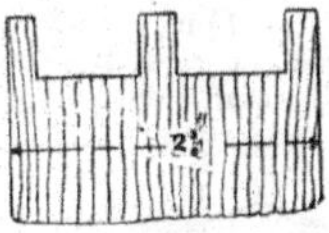

Fig. 588.—End of Drawer Back.

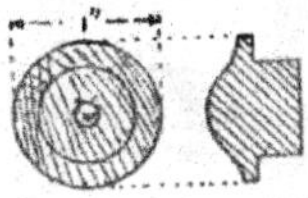

Fig. 590.—Patera Covering for Bed-screws.

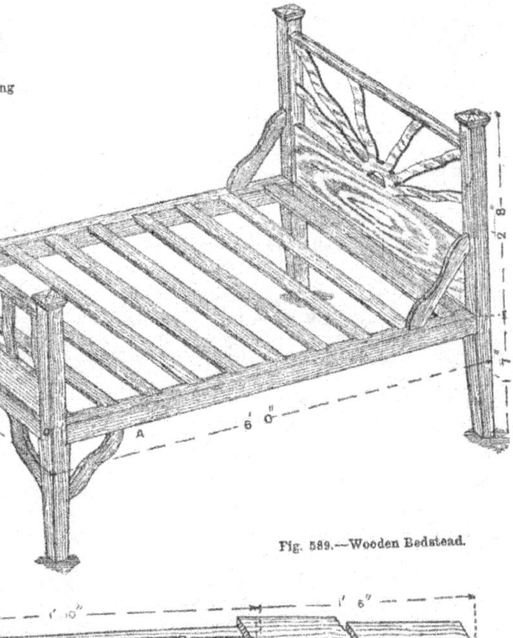

Fig. 589.—Wooden Bedstead.

Fig. 591.—Leg at Foot of Bedstead.

frame rails are of ¾-in. stuff; table legs 1½-in. by 1½-in.; shelf, drawer sides and bottom, door, back panels, and dustboard of ½-in. stuff; splash-board is ⅝ in. thick, and standards are 1 in. by ¾ in. in section.

Bedstead, Chair, and Towel Horse in Stained Wood.

To complete the stained wood bedroom suite, of which the wardrobe, dressing-table

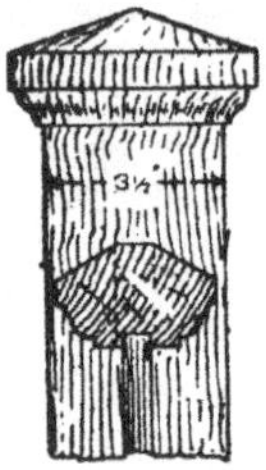

Fig. 593.—Bedstead Post (B, Fig. 589).

Fig. 597.—Plan of Chair Back.

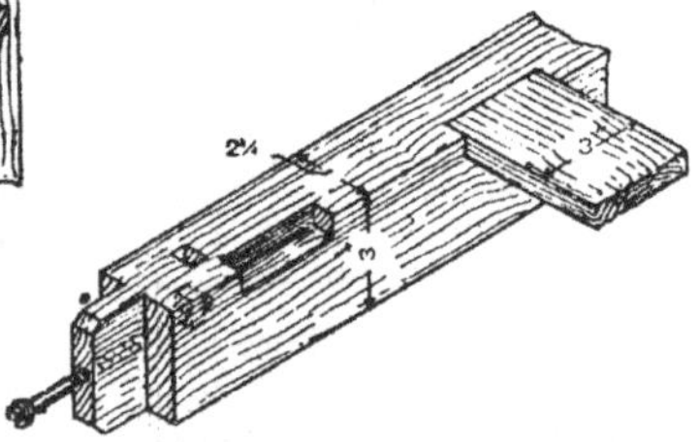

Fig 592.—Detail of Bedstead Construction (A, Fig. 589).

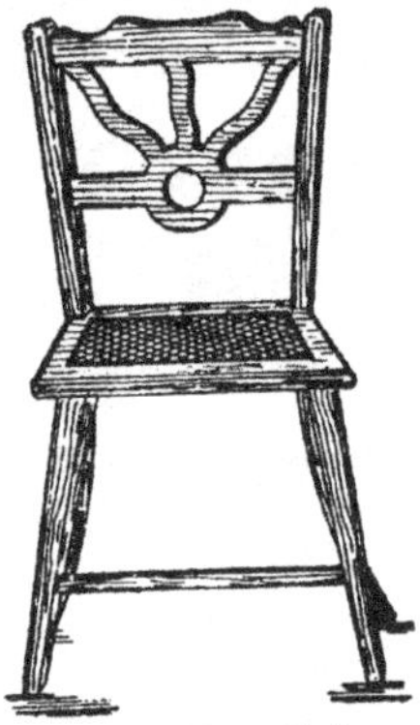

Fig. 594.—Bedroom Chair.

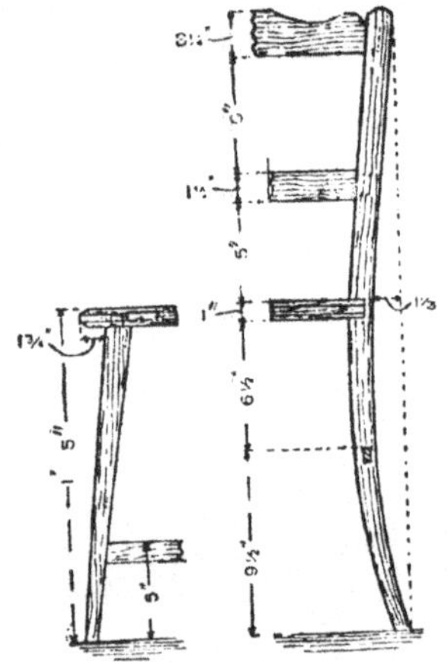

Fig. 595.—Chair Framing.

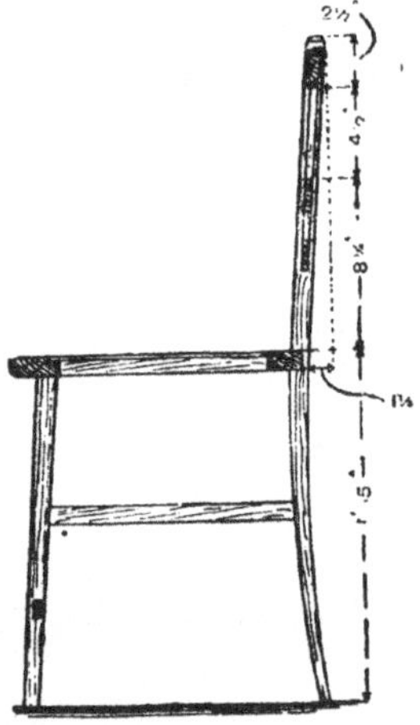

Fig. 596.—Cross Vertical Section of Chair.

and washstand have been described in detail, a bedstead, chair and towel horse may be required. Of the bedstead, a general view is presented by Fig. 589. Fig. 590 shows the patera covering for the bedscrews; Fig. 591, a detail of the leg at the foot of the bedstead; Fig. 592, a detail at A (Fig. 589); and Fig. 593, a detail of the post at B (Fig. 589). Views of the chair are presented by Figs. 594 to 596. A plan of the back is shown by Fig. 597, a plan of the seat frame by Fig. 598, and a detail of the joint C (Fig. 598) by Fig. 599. The towel horse is illustrated by Figs. 600 and 601.

Wardrobe of Modern Design.

In designing the wardrobe shown in elevation by Figs. 602 and 603, the purpose has been to guard against sacrificing beauty to the prevailing fashion of severe plainness. The wardrobe is part of a complete suite—

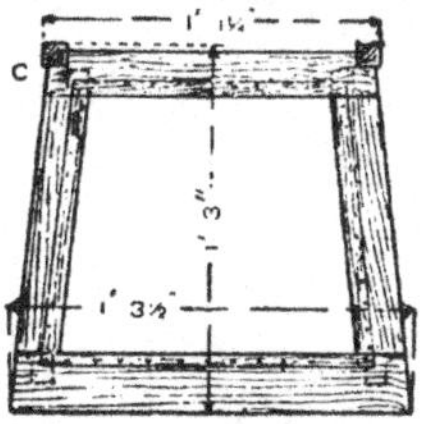

Fig. 598.—Plan of Chair Seat Frame.

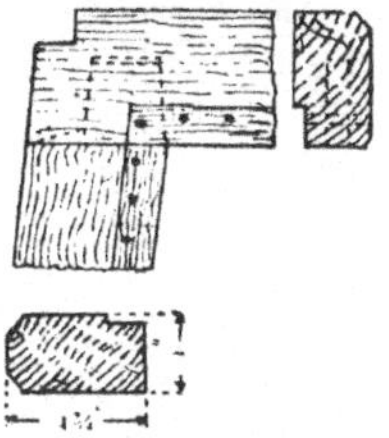

Fig. 599.—Joint in Chair Seat Frame C, Fig. 598).

Fig. 600.—Towel Horse

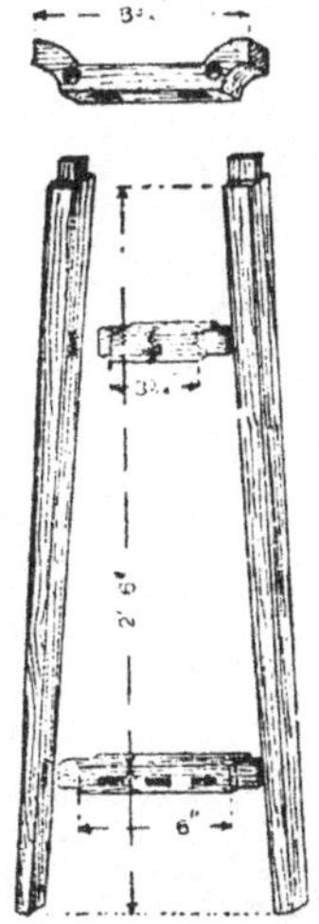

Fig. 601.—Framing of Towel Horse.

the second described in this chapter. It stands 6 ft. high, and the two sides, which should be got out of the solid hardwood, will be 5 ft. 11 in. long by 1 ft. 6 in. by ⅞ in., and are shaped at the lower ends, as shown by Fig. 603, not deeper than 5 in. Fig. 604 shows a cross-section of the wardrobe. The back edges must then be rebated on the inside for the back (see Fig. 605), and two shelves of 1-in. deal, trued up to 3 ft. 5 in. by 1 ft. 5 in., are let into half-dovetail grooves, ¼ in. deep, made in the sides and stopped ½ in. from the front edge; one is 6 in. from the floor, the other 11 in. above it. A rail of solid stuff, 3 ft. 4½ in. by 3 in. by 1 in., is fitted at the top, ⅛ in. back from the front, and is fixed with dowels, and a rail of deal is dovetailed as Fig. 605 shows. The rails may now be taken out, and the shelves faced on the front edge with ½-in. stuff, then finally put together with glue. The full size of the top is 4 ft. by 1 ft.

Fig. 602. Fig. 603

Figs. 602 and 603.—Front and Side Elevations of Modern Wardrobe.

9 in. by 1 in.; it may be made from the solid, or a piece of deal 3 ft. 6 in. by 1 ft. 6 in. may be clamped 3 in. on each end with hardwood, then 3 in. on the front edge. It is moulded as shown by Fig. 606, and fixed with screws through the rails and sides, which are thumb-notched on the inside for the purpose.

Front Frames.—For the front frames, four stiles 4ft. by 2 in., and four rails 8½ in. by 2 in., must be made from 1-in. solid stuff, the rails being tenoned to 6 in. and let into mortices in the stiles at the extreme lower ends, and 1 ft. 4 in. from the top ends; these are then dowelled to the wide rail, the frames being secured in place with screws through the shelf and the side stiles, which must be behind. The ornaments of the top panels (Fig. 608) are of ⅛-in. stuff, fixed with glue and needle points. It may be remarked that the bottom shelf is dovetailed to the side in the ordinary way.

Brackets, Drawer, etc.—The brackets are of ½-in. material, put on in pairs 1½ in. apart,

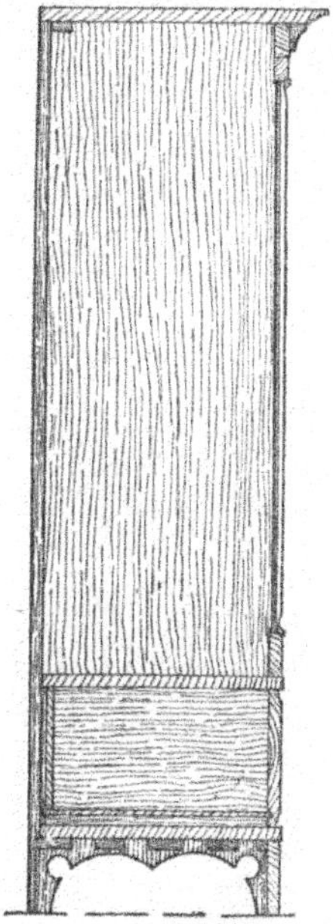

Fig. 604.—Cross Section of Modern Wardrobe.

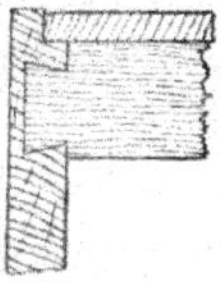

Fig. 605.—Back Corner of Ward robe (Top Removed).

Fig. 606.—Section of Wardrobe Top.

Fig. 607.—Section of Lower Panels of Wardrobe.

Fig. 608.—Section of Top Panels of Wardrobe.

Fig. 609.—Section of Wardrobe Drawer Front.

Fig. 610.—Section of Wardrobe Door.

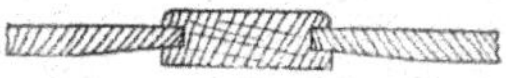

Fig. 611.—Section of Part of Wardrobe Back.

notched. Use also glue between the joints. An ovolo beading is mitered round the panel spaces ⅛ in. from the front, being fixed with glue and needle points. Panels of ⅜-in. material are then fitted, the long ones (see Fig. 607) being ornamented with pieces 3¼ in. wide by ½ in. thick, which are bevelled from the centre to ⅛ in. at the edges, and are put on with glue and small screws from as shown in Figs. 602 and 603; the long bracket at the foot is cut from ¾-in. board, being fixed with glue and blocks placed behind. For the drawer (see Fig. 604), choose a piece of figured 1-in. board for the front; the sides and back may be of ½-in. deal, and the bottom of ⅜-in. stuff. The front is bevelled (Fig. 609) after it has been dovetailed and before being put together.

Door.—To make the door, the two stiles should be 2 in. by 1 in., and the rails 4 in. by 1 in. They are mortised and tenoned in the usual way, and a slight sweep is cut out of the top rail. In putting on the beading, it will be best to rebate it to fit the corner, so as not to cover the bevel of the mirror, which should fit loosely, and be regulated with small triangular blocks. It is then protected with a panel of ⅜-in. board kept in by a beading as in Fig. 610. A strip of wood, 1 in. wide by ¼ in. thick, should be put behind the stile of the left frame to form a rebate for the door, which may now be hinged, and the handle added.

Back, etc.—For the back, three muntins 2½ in. wide by ¾ in., to be grooved ¼ in. deep on each edge, and four 9-in. boards ½ in. thick, will be required; one muntin should be put quite central, and a board should be fitted each side (see Fig. 611), then the other muntins, and finally the other boards; they are nailed to the backs of the shelves, the top rail, and the sides. When the handles are put on, the drawer may be adjusted, and the wardrobe is then ready for lining. A piece of the material is cut ½ in. larger all round than the inside measurement of the top; this ½ in. is doubled in, and it is tacked in place with fine tacks. The bottom, back, and sides are treated in the same way, and then the hooks may be put in. The wardrobe is then ready for polishing.

Fig. 612.

Fig. 613.

Figs. 612 and 613.—Front and Side Elevations of Modern Dressing-chest.

Dressing Chest of Modern Design.

The dressing-chest illustrated by Figs. 612 to 614 is intended to match the wardrobe just described. The extreme measurements are: Height, 5 ft. 4 in.; width, 3 ft. 6 in.; depth, 1 ft. 7½ in. The two sides may be got out first in the solid hardwood 2 ft. 7 in. long by 1 ft. 6 in. wide by 1 in. When cleaned up and squared, the lower ends are cut to the shape shown by Fig. 613, by two rails of deal 3 ft. 3½ in. by 2 in. by 1 in., the front one being faced like the bottom; they are then dovetailed as shown by Fig. 616. The drawers are to be 9 in., 7 in., and 5 in. deep, so the sides must be grooved accordingly for the rails and bearers (see Fig. 617, which also shows how they are

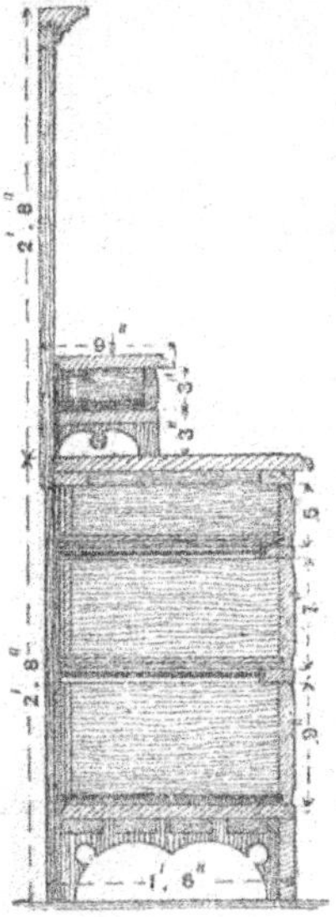

Fig. 614.—Vertical Cross Section of Modern Dressing-chest.

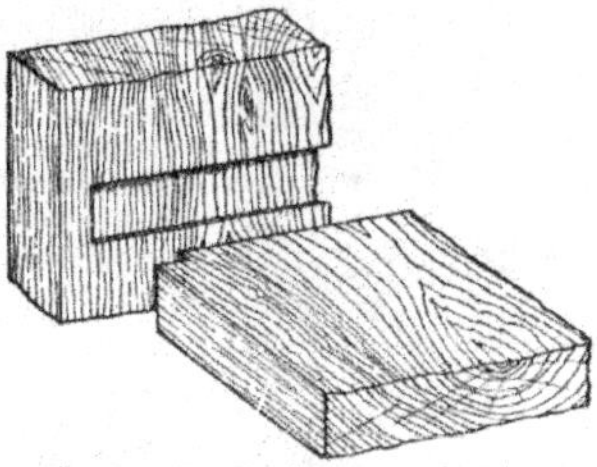

Fig. 615.—Joint of Bottom to Side of Dressing-chest.

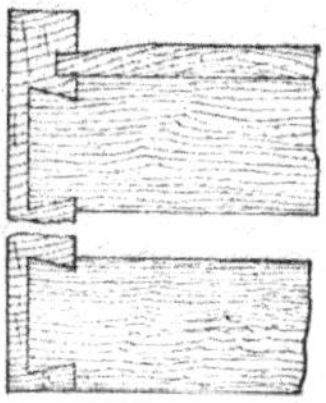

Fig. 616.—Joint of Top Rails of Dressing-chest.

Fig. 617.—Fixing Framing of Dressing-chest Rails and Bearers.

not working into the wood more than 4½ in.; they are then grooved on the inside at the back edges about ¼ in. deep, with the inner side of the groove 1 in. from the edge. A bottom of 1-in. deal is now required, 3 ft. 2¾ in. by 1 ft. 4¾ in., and this is joined to the sides as shown by Fig. 615, by being let into half-dovetail grooves ⅜ in. deep, 6 in. from the floor, and stopped ¼ in. from the front, the bottom faced with a slip of ¼-in. hardwood to bring it flush with the sides. The top ends of the sides are connected grooved to receive the dust-boards). The division for the top drawers should be mortised and tenoned, and faced with the hardwood like all the rails. The whole may then be taken apart and afterwards glued together.

Top.—The top is made in the solid stuff, and measures 3 ft. 6 in. by 1 ft. 7½ in. by 1 in. An ovolo mould is worked on the top side of both ends and front, working about ⅜ in. on the top side and ½ in. on the edge. It is fixed to overhang 1 in. at the front and sides, by being screwed through the rails; long blocks should be glued under at the top ends of the sides. A centre bearer for the top drawers, grooved on both edges, and the dust-board of ¼-in. deal, are now made and put in. For the back, two deal boards 2 ft. 7 in. and one 2 ft. 2 in. long, all bottom ends as shown by Fig. 619, and when in place they may be either screwed or sprigged. The shaped plinth piece at the front should be of 1-in. stuff, well fitted, and glued in place ⅛ in. in from the edge of the bottom and sides; a sprig may be driven in at the ends, and blocks glued behind.

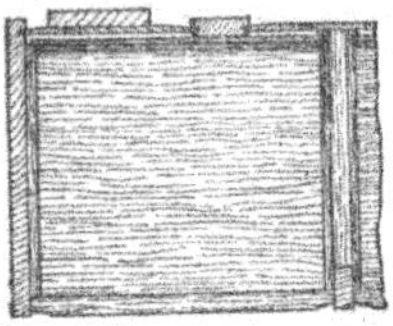

Fig. 618.—Part Horizontal Section through Dressing-chest's Top Drawers.

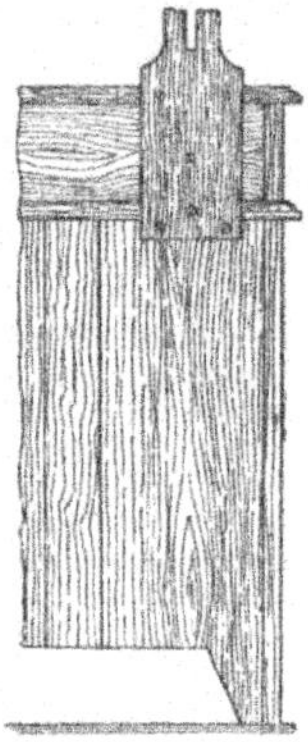

Fig. 619.—Part Back View of Dressing-chest.

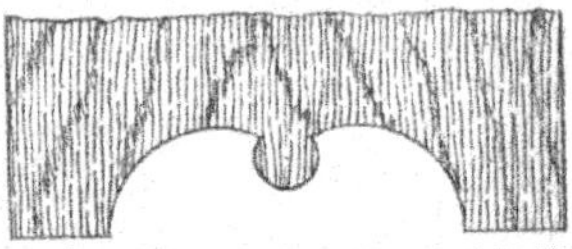

Fig. 620.—Shaped End of Dressing-chest's Jewel Drawer Casing.

Fig. 622.—Part Plan of Top of Dressing-chest.

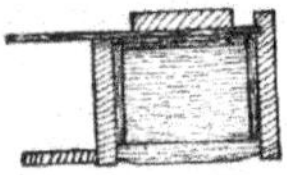

Fig. 623.—Part Horizontal Section through Dressing-chest's Jewel Drawers.

Fig. 621.—Framing of Dressing-chest's Jewel Drawer Casing.

Figs. 624 and 625.—Dowel Joints in Dressing-chest's Mirror Frame.

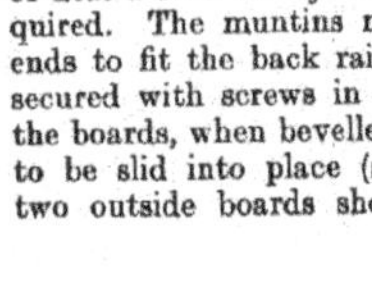

Fig. 624.

Fig. 625.

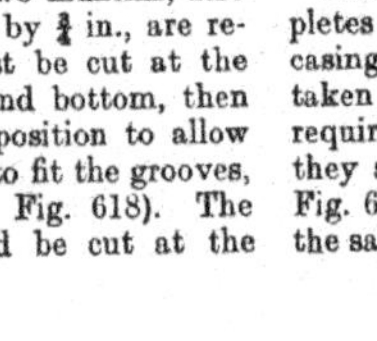

Fig. 626.—Part Cross Section of Dressing-chest's Mirror.

12 in. wide by ½ in., and two muntins, also of deal 2 ft. 2 in. by 3 in. by ¾ in., are required. The muntins must be cut at the ends to fit the back rail and bottom, then secured with screws in a position to allow the boards, when bevelled to fit the grooves, to be slid into place (see Fig. 618). The two outside boards should be cut at the

Casing for Jewel Drawers.—This completes the carcase except the drawers, so the casing for the jewel drawers can now be taken in hand. Two sides in the solid are required 6⅝ in. long by 8½ in. wide by 1 in.; they should be cut to the shape shown by Fig. 620, not deeper than 2 in., and grooved the same as the carcase sides; the two inner sides are made the same, but 1 in. narrower and not grooved, the 1 in. to be off the back edges only. The top will be 3 ft. 6 in. long by 9½ in. by 1 in., moulded like the carcase top and grooved to match the sides, to receive the top edge of the back. The sides are let into grooves made in it, to allow for drawers 9 in. long, the shelves for these being made of deal faced with hardwood and joined to the sides exactly the same as the carcase bottom (Fig. 621), allowing 3 in. for the depth of the drawers. A piece of ½-in. board, 3 ft. 2½ in. by 6¼ in., must be got out for the back, which should be of hardwood, and made to slide in place, being fixed to the back edges of the inner sides with screws; then the brackets may be made of ½-in. board glued in place and sprigged.

Cornice.—The next thing will be to make the cornice, and for this a piece of solid stuff is required 3 ft. by 2 in. by 1 in., and another 3 ft. 4½ in. by 3¼ in. by 1 in., the latter to be moulded on the under side and screwed on the top edge of the former, flush at the back and overhanging equally at the ends; then the small brackets can be made of ½-in. material and glued under. For the standards, two pieces, 2 ft. 8 in. by 6 in. by 1 in., are cut to the shape and well finished up, then dowel-jointed to the cornice and screwed at the back to both tops, which must be cut out as shown by Fig. 622, also to the shelves of the casing and the top back rail (see Figs. 618, 619, and 623). Two screws should also be driven through the top from underneath into the ends of the inner sides. These screws, and those at the back which go into the carcase only, are to be withdrawn when removing the upper part.

Mirror Frame.—The extreme measurements of the mirror frame will be about 1 ft. 8 in. by 1 ft. 10½ in.; but to be accurate in the width, the space between the standards should be measured and the pivot joints allowed for. The stiles and lower rail are of 1¼-in. by ⅞-in. stuff, but the top rail will require to be 2 in. wide, so that when the curve is cut out it will not be less than 1 in. in the centre. It is dowel-jointed together as shown by Figs. 624 and 625, then a beading of 1-in. by ⅜-in. material rounded on the front edge is mitered round, and a small ovolo moulding rebated at the back is planted to make a rebate for the mirror. This should fit the frame loosely, to be regulated by small triangular blocks so as not to take from the width of the bevel; then the back of the thin board is put in and beaded (see Fig. 626), but it is not put in permanently till the frame is polished. The pivot joints may be put on, and the drawers made in the usual way, already fully shown, using 1-in. material for the fronts, ½-in. material for the sides and backs, and ⅜-in. material for the bottoms. The centre bearer can be kept in place by nailing through the back, a guide piece being fixed on the top side. The jewel drawers, of course, must be made lighter in proportion to size, say ⅝ in. for the fronts and ¼ in. for the sides, backs, and bottoms. The job is now ready for polishing, and it is advisable that the rebate of the mirror frame should be stained black, so as not to show any reflection. This done, the mirror may be put in and swung, and the fitting of the copper drawer handles completes the work.

Washstand of Modern Design.

The washstand shown in front and end elevations by Figs. 627 and 628 is designed to match the wardrobe and dressing-chest above described and illustrated. The overall measurements are: Height, 4 ft. 2 in.; width, 3 ft. 6 in.; depth, 1 ft. 7½ in. The two sides, 2 ft. 7 in. by 1 ft. 6 in. by 1 in., are made first; they are of solid hardwood, cut to the shape shown, and grooved for the back. The two shelves, which may be of deal slipped on the face edge with hardwood, are 3 ft. 2⅝ in. by 1 ft. 5 in. by 1 in., ⅜ in. of each end being let into the sides in the same manner as in the dressing-chest, the top side of the bottom one being 6½ in. from the floor; a space of 1 ft. 5½ in. is allowed for the cupboards. The two top rails and the division piece for the drawers, as well as the front plinth piece and the carcase back, are done in the same way.

Top of Washstand.—For the top, a piece of deal is got out 3 ft. 2 in. by 1 ft. 6 in. by ⅝ in. It is made 4 in. longer by clamping the ends with pieces of the hardwood 2 in. wide and the same thickness, which may be dowelled on or tongue-and-groove jointed, but not nailed; a $1\frac{1}{2}$-in. piece is plain jointed on the front edge to make it the full width. The top, as shown by Fig. 629, is tiled; will be necessary to glue hardwood the thickness of the tiles, $\frac{3}{8}$ in., the front corners being mitered as shown. An ovolo moulding is then worked round (see Fig. 630), and the

Fig. 627.

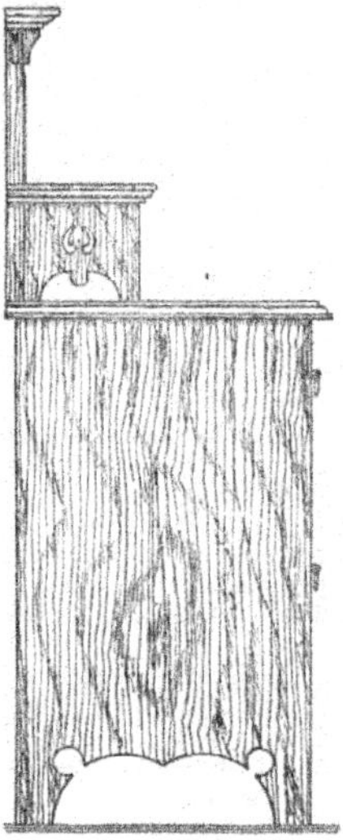

Fig. 628.

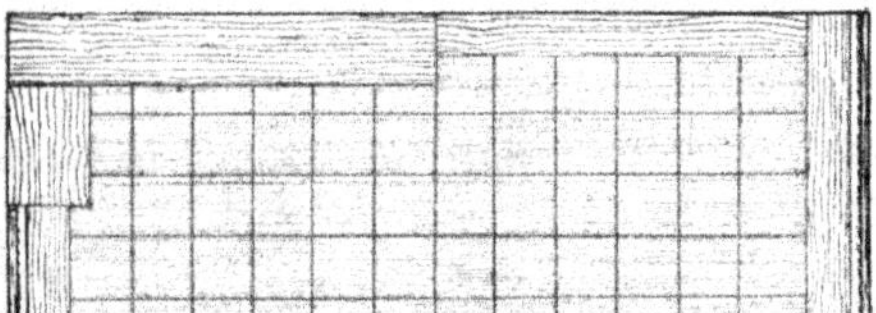

Fig. 629.

Figs. 627 to 629.—Front and Side Elevations and Half Plans of Modern Washstand.

3-in. square tiles of a medium green shade will contrast well with either oak or mahogany. They are cemented on with a mixture of plaster-of-paris and glue made to the consistency of thick cream, leaving a margin of 3 in. at the front and ends, over which it top fixed to overhang 1 in. at the front and sides.

Door Frames, Shelves, etc.—The door frames are made of 2-in. by 1-in. section, dowel-jointed, with panels of $\frac{1}{4}$-in. material fitted to them. Pieces of $\frac{1}{2}$-in. stuff, bevelled off from the centre to $\frac{1}{8}$ in. at the edges, are glued on the front, leaving an equal margin of $1\frac{1}{2}$ in. When the moulding has been not be fixed permanently, but by screwing cleats of $1\frac{1}{2}$-in. by $\frac{1}{2}$-in. section to the sides, they may be placed on and removed at

Fig. 630.—Section of Modern Washstand Top.

Fig. 632.—Shaped Front of Middle Shelf of Modern Washstand.

Fig. 635.—Part Section of Modern Washstand Back.

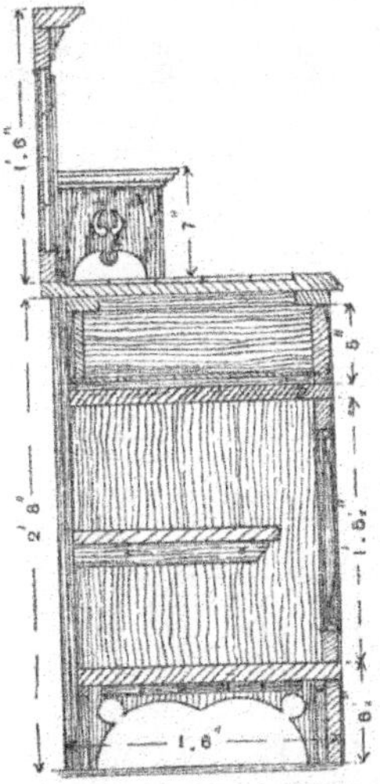

Fig. 633.—Cross Section of Modern Washstand.

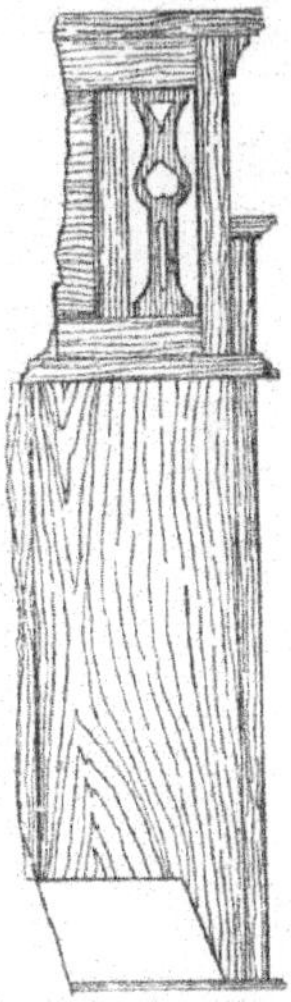

Fig. 634.—Part Back Elevation of Modern Washstand.

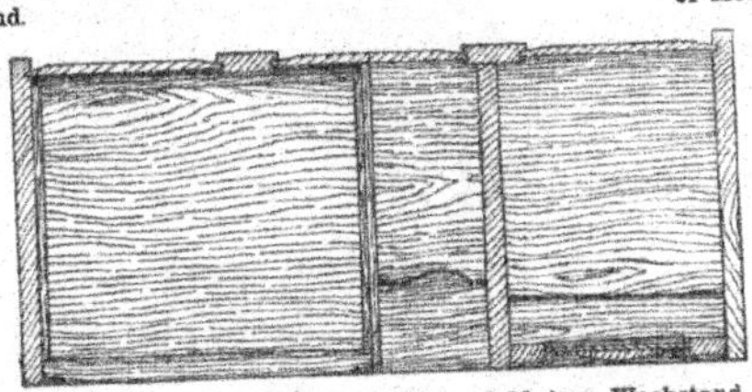

Fig. 631.—Half Horizontal Section of Modern Washstand.

secured to the frames to form the rebate, the panels are put in and beaded behind. A shelf is fitted to each of the cupboards, and also one in the centre (see Fig. 631), the latter being shaped as shown by Fig. 632, and all three rounded off. They need will (see Fig. 633). The cupboard shelves are 1 ft. 2 in. wide, and the middle one is 1 ft. 1 in. Before fixing the two brackets, which are made of $\frac{1}{2}$-in. stuff, a $\frac{1}{2}$-in. brass rod is fitted for the curtain.

Curtain Rod.—One way of fixing the brass

rod is to cut a length of brass tube just to fit between the sides, then to drive a screw into the left side, leaving it standing up about ½ in.; a hole should then be made in the right side to take a thumbscrew. The curtain, which should be of a pale green material with a hem at the top to take the rod, is then put on, one end of the rod being put over the screw, and the thumbscrew screwed into the other from the inside of the right-hand cupboard. Or small rings could be sewn on the curtain, and the rod put on two brass cup hooks screwed above.

Back of Washstand.—The drawers are next made and fitted; then the back is taken in hand. For this, a frame made of the following pieces is prepared: two uprights 1 ft. 5 in. by 1½ in. by 1 in., and two 1 ft. by 1½ in. by 1 in.; one bottom rail, 2 ft. 11 in. by 2 in. by 1 in.; and one top rail, 2 ft. 11 in. by 3 in. by 1 in. The two short uprights are dowel-jointed to the rails, allowing a space of 2 ft. for the tiles; the outside uprights are then jointed on likewise. The ornamental pieces are of ⅜-in. material; they should be tightly fitted, glued in place, and secured with fine springs.

Fig. 636.—Modern Wood Bedstead.

A piece of ¼-in. stuff, 2½ in. wide, is cut to a curve and fitted under the top rail, flush with the front, to fit the tile space, and when the moulding is fixed round, a ¼-in. backboard is prepared, on which the tiles are cemented; this is placed in and beaded behind (see Figs. 634 and 635).

Cornice and other Details.—The cornice piece is 3 ft. 6 in. by 3½ in. by 1 in., fixed on, as already explained, with brackets beneath. Two side pieces must be made 6 in. long by

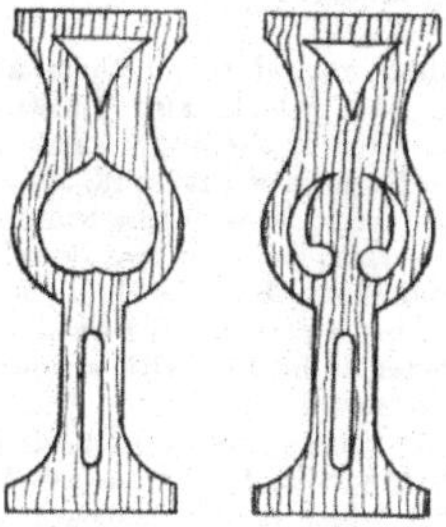

Fig. 637. Fig. 638.

Figs. 637 and 638.—Stiles for Bedstead.

Fig. 639.—Section of Bedstead's Top Rail.

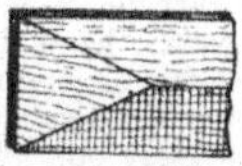

Fig. 643.—Panel Ornament for Bedstead.

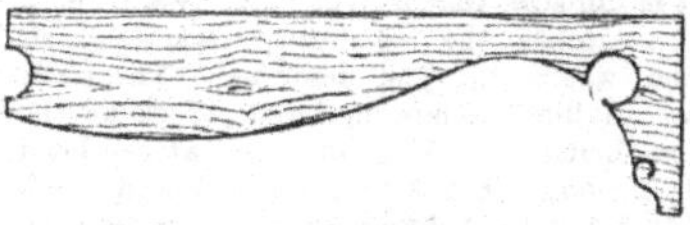

Fig. 640.—Half of Bedstead Bracket Piece.

Fig. 641.—Post of Bedstead.

Fig. 642.—Section of Bedstead Panel.

8 in. wide by 1 in. thick, and when these are cut to the shape they should be screwed to the uprights, the screw-heads being sunk below the surface and the holes filled with wood. Shelves are made for the side pieces 9 in. by 3½ in. by 1 in., moulded on the under side and fixed with dowels. The back is placed on the carcase flush with the back edge of the top, through which screws must be driven from underneath into the bottom rail, a screw on the slant also being driven through each end of the top to catch the foot of the side pieces. The doors may be hinged on, but should be taken off for polishing; the panels and tiled back must also be removed. When the polishing has been finished, they are replaced, and the drawer and cupboard handles affixed.

Wood Bedstead of Modern Design.

The wooden bedsteads now in use are very different in design and construction from the old-fashioned forms which became so unpopular on account of their tendency

Fig. 645.—Section of Bedstead's Pediment Cornice.

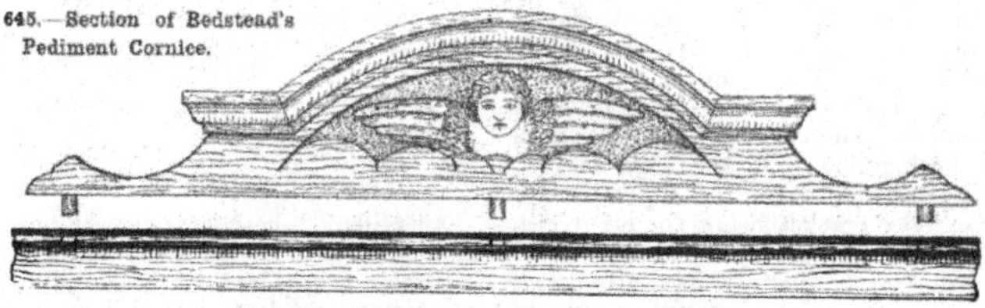

Fig. 644.—Pediment and Top Rail of Bedstead.

to harbour vermin. The modern bedstead, however, has a bottom resembling that of an iron or brass bedstead instead of the heavy wood side rails and laths which form the great objection to the old form. Fig. 636 shows a very effective design, which looks particularly well in mahogany or oak; it is full-size, that is, 6 ft. 6 in. by 4 ft. 6 in. It matches the wardrobe, dressing-chest, and washstand just described. The posts are of 2-in. square material, those at the foot being 3 ft. 6 in. long, and at the head 4 ft. long. Eight rails are required, each 4 ft. 5 in. by 2 in. by 1¼ in., which are tenoned 1½ in. at each end, making them 4 ft. 2 in. long in the clear. The foot-posts are then mortised to receive the tenons, one rail being 6 in. from the floor, another above it allowing a space of 5 in. for the panel, and the others are 4 in. from the top ends, with a 4½-in. space allowed for the spindles. These are plain turned, ½ in. thick in the centre, tapering to ⅜ in. at the ends; thirteen will be sufficient, and they are simply let into ⅜-in. holes bored in the rails at equal distances apart. The ornamental pieces (Figs. 637 and 638) are made of ¾-in. material, 5 in. wide, and should be well finished; five are required for the foot, and should be tightly fitted between the rails. Whilst in their exact positions, they, with the rails, should be marked for dowels, two in each end. An additional mark should be put on each to ensure finding their right places as fitted. The frame may now be taken apart, but the ornaments are first dowelled between the rails, glued, and cramped up close. Two saw kerfs should be cut in each tenon of the rails, as they should be blind-wedged. When quite ready for being finally put together, the spindles should be placed between the two top rails, using a touch of glue. The cramps may be taken off the rails, and used on the posts to bring up the mortised joints. A coping, shaped as shown by Fig. 639, is made of 2-in. by ¾-in. material, and fixed on the top rail with dowels. The bracket below the lower rail, of which Fig. 640 gives a half view, is cut from ¾-in. board; it should be neatly fitted, and fixed with glue, with a sprig or two at the ends, and small blocks should be glued behind. The lower ends of the posts have a moulding worked round them as shown in Fig. 641, which also illustrates the ornamentation of the upper end. This top consists of a piece of ¾-in. stuff, 3½ in. square, the edges being shaped to a thumb mould; it is then fixed on with two dowels, and a scotia moulding, 1 in. by ½ in., is mitered round below it; then the turned ball, 2 in. in diameter, is held on with a dowel screw. An ovolo moulding, ¾ in. by ½ in., is fixed round the panel space, ⅛ in. back from the front of the rails, to form a rebate; then the panel, which is of ½-in. board, is placed in and beaded behind; see Fig. 642. It is ornamented on the front by a piece of ¾-in. stuff, 2½ in. wide, bevelled off to ⅛ in. at the edges and ends, and is fixed with screws through the panel (Fig. 643).

Head of Bedstead.—The head is made the same as the foot, except that the space for the ornaments will be 6 in. higher; of course, the ornaments themselves are 6 in. longer, but they need not be slotted, and the panel may be quite plain, or even omitted altogether; also the bracket may be left out. Fig. 644 shows the shape of the pediment with the carved design of a cherub. It illustrates also the method of fixing on the top rail with dowels glued in the pediment to project ¾ in. The pediment may be taken off for convenience in removing, etc. A cornice of ogee moulding is worked on as shown in Fig. 645.

Completing Bedstead.—A set of good castors should be added, when the bedstead will be ready for polishing. Bed bottoms are manufactured specially for wood bedsteads, and consist of head and foot and two side angle-irons, laths, and stretcher. The head and foot angles are secured to the posts with four strong screws at each end, the upper side of the angle-iron being 1 ft. 5 in. from the floor. They are then connected by the side angles, as in an ordinary iron bedstead.

"Tallboy" Chest of Drawers.

A "tallboy" chest of drawers, as illustrated at Fig. 646, is useful where ample accommodation is required for holding bed linen, blankets, etc. It is intended to be made of solid mahogany, inlaid with satinwood stringing and finished with french polish.

The out-of-sight parts may be made of pine or American whitewood, but for a good job baywood is more suitable. To obtain the dimensions of the various pieces of wood required, a full-sized drawing should be made of half the front elevation and the end elevation (Fig. 647). The main dimensions are as follows: Extreme height, 5 ft. 6 in.; width from end to end, 4 ft.; depth from front to back, 1 ft. 9 in.; height of plinth, including moulding, 6 in.; height of drawer fronts, 11 in., 10 in., 9 in., and 8 in.; thickness of bearers between drawers, ⅞ in. (bare); top, about 1½ in. thick, and projecting 1¼ in. over the front and ends. The front corners, with the quarter-circle fluted columns (see enlargement of left-hand corner, Fig. 648), and including the fillet A, are 2½ in. wide. A sectional plan of the corner is shown at Fig. 649. The fillet A and ends B are got out of 1-in. stuff, as thick as the working will allow, the centre filling C making up the 2½ in. required. The columns are a quarter of a 3-in. circle; this should leave a bare ¼-in. fillet down each edge. The tablets D (Fig. 648) are 4 in. long; and the turned capitals and bases of the columns are each 1 in. in height. The stiles and rails of the doors are 1⅞ in. wide, including the ovolo moulding (see section, Fig. 650). The inlaid lines of stringing form a 1¼-in. margin round the door panels, the corners breaking inwards 1¼ in. The margins of stringing on the drawer fronts are 1⅜ in., and the corners 1⅜ in. The diamond-shaped stringing in the door panels measures about 9 in. by 6 in.; this is shown enlarged at Fig. 651. The margins of stringing on the tablets above and below the columns are ½ in.; those on the carcase ends are 2½ in., the corner squares breaking inwards 2½ in., and the diamond stringing is about 2 ft. by 9 in.

Carcase of Chest of Drawers.—First prepare the carcase ends out of 1-in. stuff, and joint on the pieces top and bottom to form the face of the front tablets. Next get out the upright fillets A (see section, Fig. 649), 3¼ in. wide, and the same length as the ends; then the packing pieces C, 1½ in. wide, with pieces jointed on the top and bottom to form the centre part of the face tablets. Glue the three parts A, B, and C together. Rebate the back edges of the ends to receive the back, as shown in Fig. 649. The top front bearer above the cupboards and the one below the bottom drawer are 4½ in. wide, and are dovetailed to the ends and side fillets as shown. The other bearers are 3¼ in. wide, and are tenoned through the side fillets; the quarter-columns hide the ends of the tenons. The back top and bottom bearers E are dovetailed into the ends. The back is made up of three muntins about 3½ in. wide, grooved on the edges to receive the ½-in. backs F. The runners and guides for the drawers are made in the usual way. The carcase should be put together temporarily, and the sizes obtained for the doors, the top, and the plinth.

Plinth.—In making the plinth, a foundation framing (shown in part plan at Fig. 652) is required ½ in. shorter than the carcase, and 1 ft. 8¾ in. from front to back, the 1-in. rails G being 5 in. wide. The shaped front and ends of the plinth, mitered at the corners, are glued to this framing, as shown at H. The contour of the face is first worked with suitable hollow and round planes; then the pieces are marked and cut to shape with a bow saw. If desired, a straight-faced plinth may be substituted. The moulding on the top edge of the plinth is formed on strips of 1-in. stuff, 2½ in. wide, mitered at the front corners, and screwed and glued to the top edge of the plinth. The plinth is secured to the carcase with screws driven from the under side of the strips. The top is in two parts, the upper J (see section, Fig. 653) being of ¾-in. stuff, and the moulded lining strips K of 1-in. stuff, about 3 in. wide, mitered at the corners. The moulded strips may be fixed to the carcase, and the top J secured with screws from underneath, inside the cupboards.

Fluted Columns, etc.—Before fitting the carcase together, the corners, with the fluted columns, must be finished off, and the stringing inlaid in the carcase ends and on the tablets. In cases where a large number of columns are required, it is usual to have them turned, for which purpose the four quarters are jointed together, with paper between the joints; after they are turned, a thin knife is inserted in the joints, and the four quarters separated. But for only two columns, the simplest way is to work them with a hollow plane. The flutes are carved with a gouge, and finished with glasspaper. Above the turned capitals and below the grain of the wood are channelled with a steel cutter fixed in a cutting gauge; but when the stringing crosses the grain or is at any angle to it, the sides of the channels are cut with the aid of a sharp penknife

Fig. 646.—"Tallboy" Chest of Drawers.

bases are square pieces L (Fig. 648), $\frac{3}{16}$ in. thick, rounded on the two outside edges. These having been placed in position, the capitals and bases are butted against them, and the fluted columns fixed between, the whole being secured with glue.

Stringing.—In putting in the stringing, those lines which run the same way as the and a straightedge, and the channel routed out with a $\frac{1}{16}$-in. chisel. The latter method is used for the diamond pattern in the doors and carcase ends. The small circular dots are cut out of solid satinwood, about $\frac{3}{16}$ in. thick. The straight lines are first inlaid and then $\frac{1}{4}$-in. diameter holes are bored with a centre-bit (see section, Fig. 654).

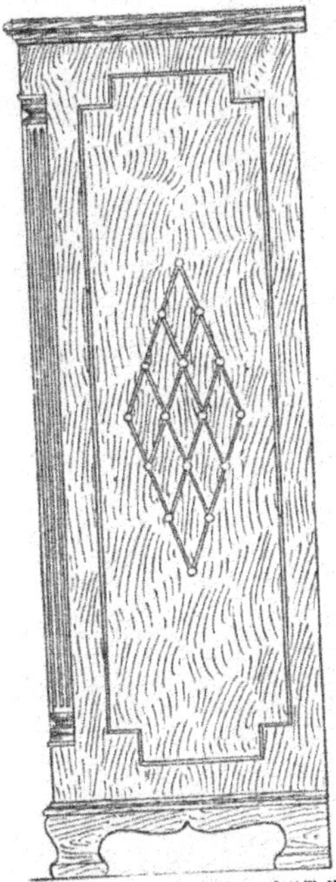
Fig. 647.—End Elevation of "Tall-boy" Chest of Drawers.

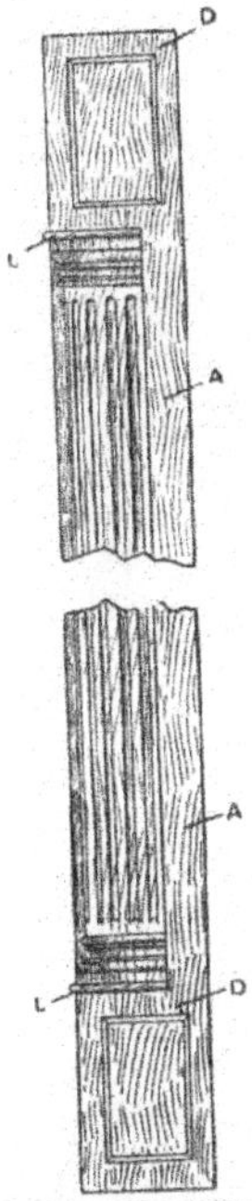

Fig. 648.—Fluted Column, etc., on Front of Chest of Drawers.

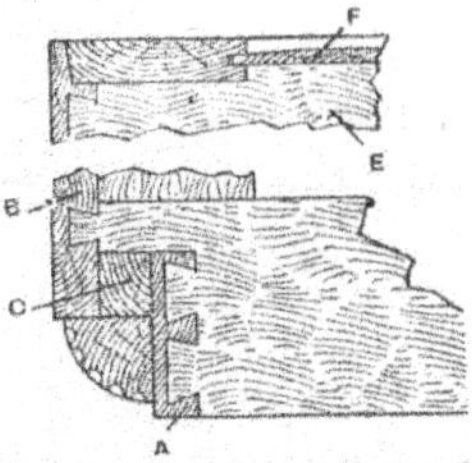

Fig. 649.—Part Plan of Left-hand Front Corner and Carcase Back of Chest of Drawers.

Fig. 650.—Section of Door for Chest of Drawers.

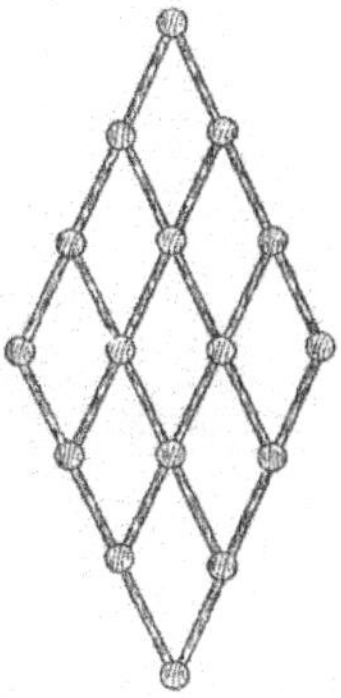
Fig. 651.—Diamond Stringing in Doors of Chest of Drawers.

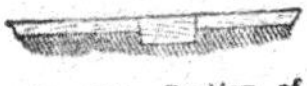
Fig. 654.—Section of Circular Dots in Stringing.

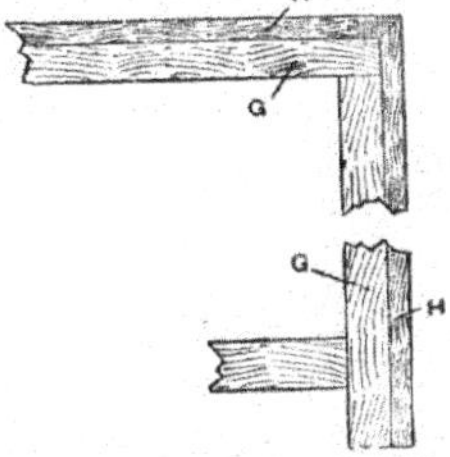

Fig. 652.—Part Plan of Plinth of Chest of Drawers.

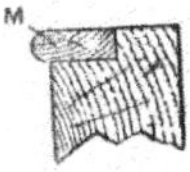

Fig. 655.—Section of Cocked Bead on Drawer Front.

Fig. 653.—Section of Top of Chest of Drawers.

Completing Chest of Drawers.—When the carcase is glued together, the doors may be fitted and hinged, the lock fixed, and the drawers made. When the drawers are fitted, the fronts must be level with the bearers, and then rebated to receive the cocked beads, as shown in section at M (Fig. 655), these being mitered at the corners.

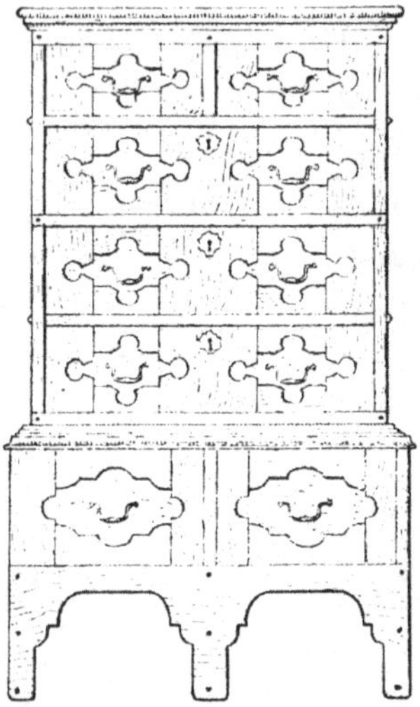

Fig. 656.

Fig. 657.

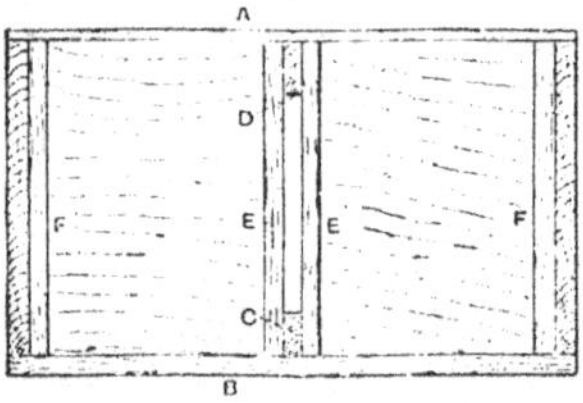

Fig. 658.

Figs. 656 to 658.—Front and End Elevations, and Horizontal Sections of Queen Anne Chest of Drawers.

Queen Anne Chest of Drawers.

Fig. 656 shows the front elevation of a Queen Anne chest of drawers to be made in two separate parts. The lower part is 2 ft. 3 in. wide, 1 ft. 4½ in. deep, and 1 ft. 6 in. high; and the upper part is 2 ft. wide, 1 ft. 3 in. deep, and 2 ft. 5½ in. high. A suitable material is pine, which can be ebonised. The lower part must be strongly built, the end pieces being of 1-in. stuff. They are 1 ft. 5¼ in. high and 1 ft. 4½ in. wide; two widths dowelled together will probably be required for each. In the back edge of the end piece (see Fig. 657) an opening, ½ in. wide, and running downwards from the top 1 ft., is cut at A to receive the backboard; its lower part is shaped to form legs, and in the front edge is cut an opening B $\frac{3}{4}$ in. wide and 9 in. high

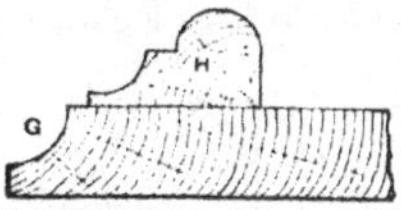

Fig. 659.—Mouldings of Base of Chest of Drawers.

for the front piece. This front piece is of $\frac{3}{4}$-in. stuff 2 ft. 3 in. long by 9 in. high. It is shaped to form three legs, and is fixed with screws into the openings in the end pieces. Behind the middle leg is a support, shown in section at C (Fig. 658), which is of $\frac{3}{4}$-in. stuff 3 in. wide and 1 ft. $5\frac{1}{4}$ in. long. In the lower part of its front edge is an opening, $\frac{3}{4}$ in. by 9 in., for the front piece, with which it comes flush above and forms the division between the lower drawers. The front piece is screwed to it, and at the back there is a corresponding support D (Fig. 658), which is screwed to the backboard. To C and D are screwed the middle runners E, $\frac{3}{4}$ in. by 1 in., which carry the drawers, and similar runners F are screwed to the inner side of the end pieces. The backboard A (Fig. 658) is of $\frac{1}{2}$-in. stuff 2 ft. 3 in. long by 1 ft. wide, and is screwed to the end pieces and back support D.

Top and Upper Part of Queen Anne Chest of Drawers.—The top is of $\frac{3}{4}$-in. stuff 2 ft.

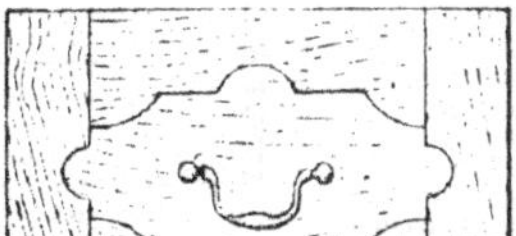

Fig. 660.—Front of Lower Drawer in Queen Anne Chest.

$3\frac{1}{2}$ in. by 1 ft. $4\frac{3}{4}$ in., as it overhangs $\frac{1}{4}$ in. at the front and the ends. As shown enlarged at G (Fig. 659), a hollow is run along its upper edge at the front and ends; it is screwed to the end pieces, the backboard, and the supports C and D, on which it rests. Within the hollow is screwed a $\frac{3}{4}$-in. moulding H (Fig. 659), which keeps the upper chest in position. The upper part has end pieces of $\frac{3}{4}$-in. stuff 1 ft. $2\frac{1}{2}$ in. wide and 2 ft. 5 in. long. Fig. 657 shows how one of these end pieces is pierced at J with mortices for the tenons of the horizontal partitions. The cuts at K are made for horizontal strips. There are two horizontal partitions, those on which the two upper small drawers and the middle long drawer slide. The partitions are of $\frac{3}{4}$-in. stuff, 1 ft. $2\frac{1}{2}$ in. wide and 1 ft. $10\frac{1}{2}$ in. long from shoulder to shoulder, beyond which they have tenons 1 in. long at each end; for the tenons project, as shown, $\frac{1}{4}$ in. beyond the end pieces. In addition to the tenons, these partitions are fixed with strong round-headed screws driven

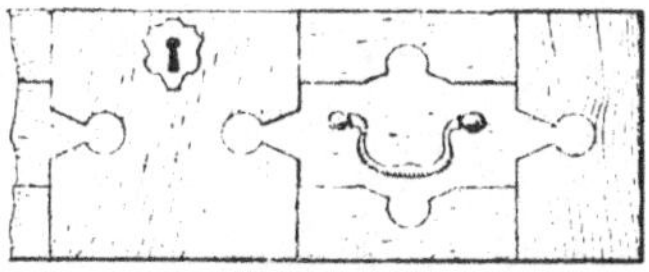

Fig. 661.—Front of Upper Drawer in Queen Anne Chest.

into their ends through the end pieces. The lower and third drawers slide on $\frac{3}{4}$-in. square runners screwed on the inner sides of the end pieces, flush with which come the horizontal front strips K, which are also $\frac{3}{4}$ in. square, and which are screwed into the openings in the end pieces. Above the top drawers is another strip L (Fig. 657), which is 2 in. wide and fixed in the same manner. The backboard is $\frac{1}{2}$ in. thick, 2 ft. long, and 2 ft. 5 in. high. The upright partition between the two upper drawers is $\frac{3}{4}$ in. thick, and is fixed with screws driven into it through the upper horizontal partition, the backboard, and upper front strip L, an opening being cut in the front of the partition for the latter. The top of the chest is of $\frac{1}{2}$-in. stuff 2 ft. by 1 ft. 3 in., and is screwed to the end pieces, backboard, and upright partition. Its edges are hidden by the cornice moulding, which is fixed over them and mitered at the corners.

Drawers.—The size of the drawers will allow them to be lightly made, say with $\frac{3}{8}$-in. stuff for the backs and fronts, to which $\frac{1}{2}$-in. sides and bottoms are screwed. From front to back the outside measurement of those of the upper part will be 1 ft. 2 in. only, as they are to be ornamented with an outer layer of $\frac{1}{4}$-in. stuff, as shown in Figs. 660 and 661. This will bring them flush with the end pieces, etc., and also cover anything unsightly in their construction. A similar allowance must be made in the drawers of the lower part. The ornamental layer is more safely fixed with small round-headed screws. Figs. 656, 657, and 658 are reproduced 1 in. to the foot; Fig. 659 is half size; and Figs. 660 and 661 are to a scale of 2 in. to the foot. These scales are only approximate.

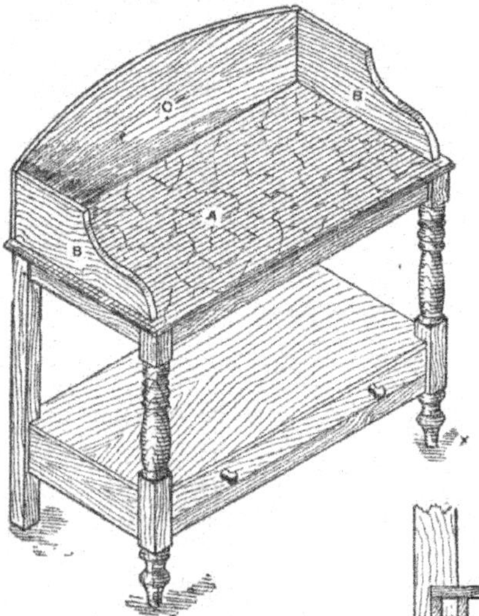

Fig. 662.—Simple Washstand.

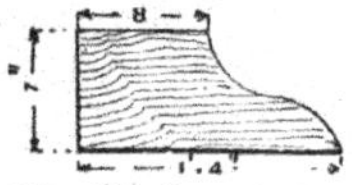
Fig. 663.—Side Support of Simple Washstand.

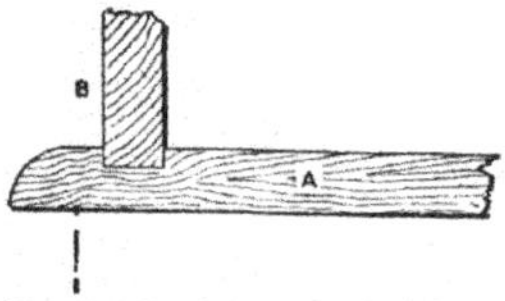

Fig. 664.—Section showing Washstand Sidepiece let into Grooved Top.

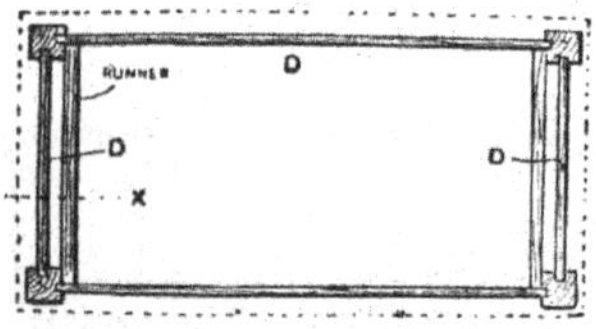

Fig. 665.—Horizontal Section through Washstand's Drawer Runners.

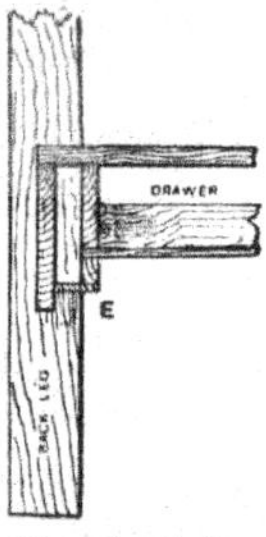

Fig. 666.—Section through Washstand's Drawer Framing.

Fig. 667.

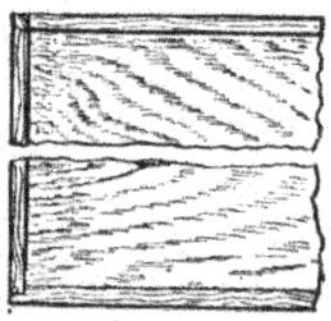

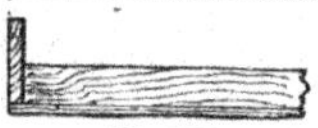
Fig. 668.

Figs. 667 and 668.—Part of Drawer for Washstand.

Simple Washstand.

The washstand shown at Fig. 662 has the advantage of being very easily made.

It is 3 ft. long, 1 ft. 6 in. wide, and 3 ft. 2½ in. high (at the back) over all. The wood used is whitewood, ½ in. thick. The top A is the full width and length, and should be screwed to the legs, the screws being well countersunk, and the holes filled in. The side pieces B may be fitted flush, but it is better to run shallow grooves for

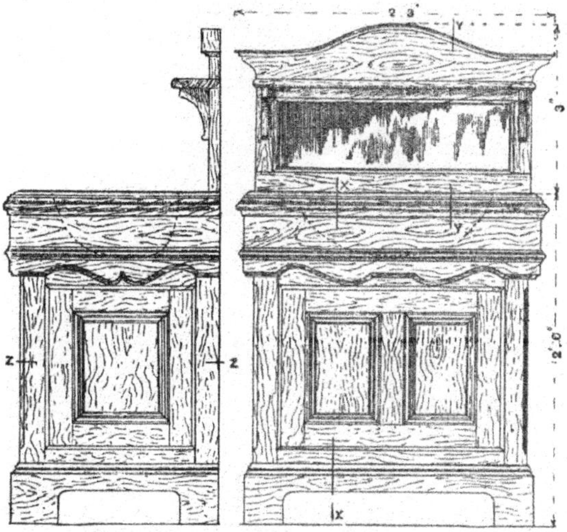

Fig. 670. Fig. 669.

Figs. 669 and 670.—Front and Side Elevations of Lavatory Washstand.

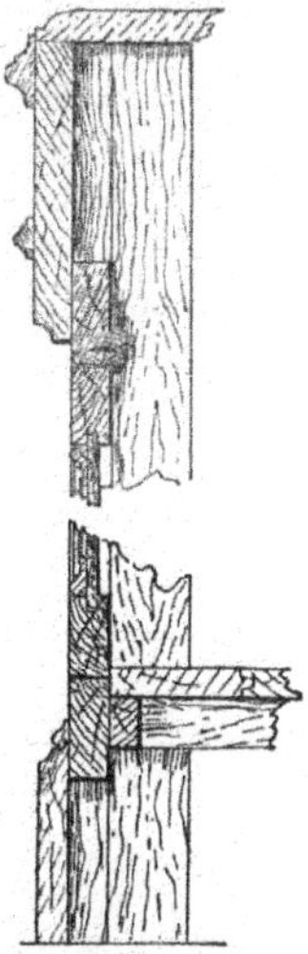

Fig. 671.—Vertical Section of Lavatory Washstand on Line X X (Fig. 669).

Fig. 672.—Vertical Section of Lavatory Washstand on Line Y Y (Fig. 669).

Fig. 673.—Horizontal Section of Lavatory Washstand on Line Z Z (Fig. 670).

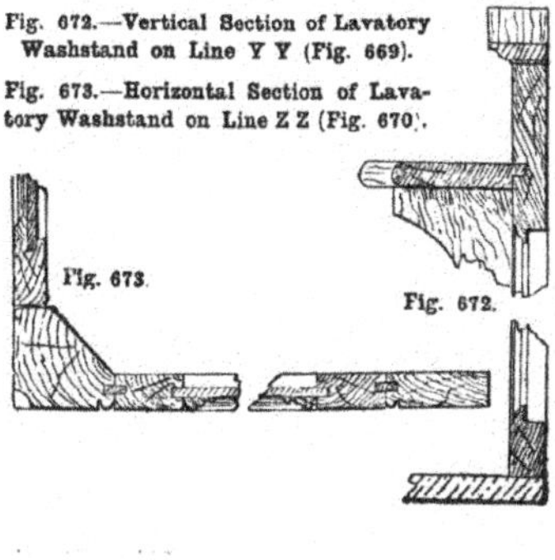

Fig. 673. Fig. 672.

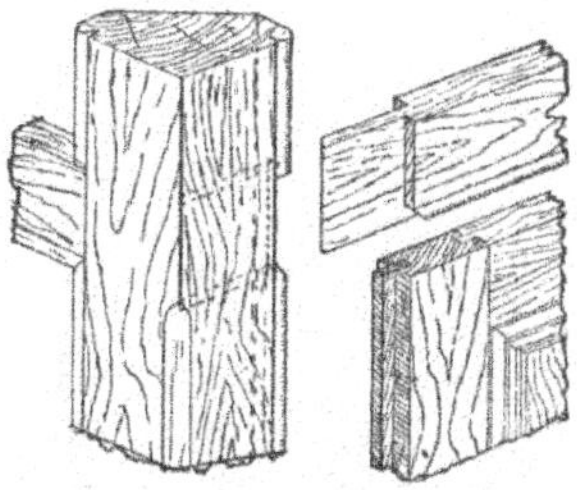

Fig. 674.—Fixing Sides of Lavatory Washstand to Posts

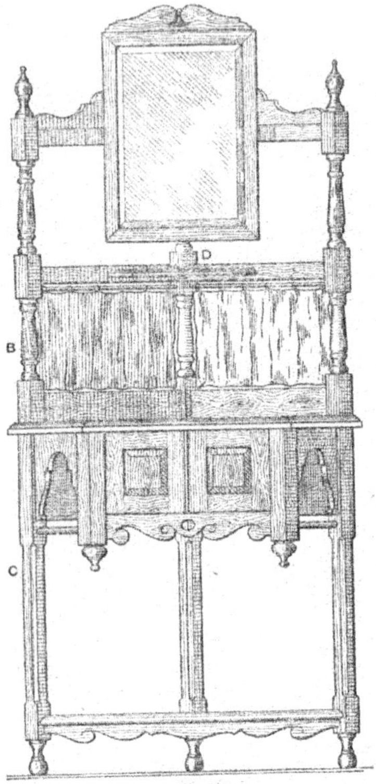

Fig. 675.

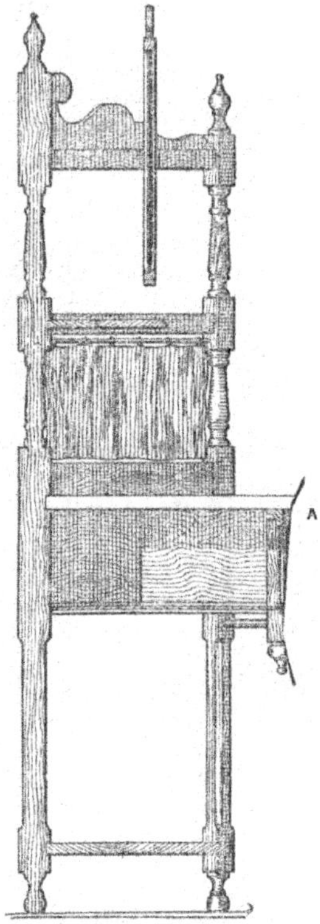

Fig. 676.

Figs. 675 to 677.—Front Elevation, Vertical Cross Section, and Horizontal Section of Corner Washstand.

Fig. 677.

Fig. 678.—General View of Corner Washstand.

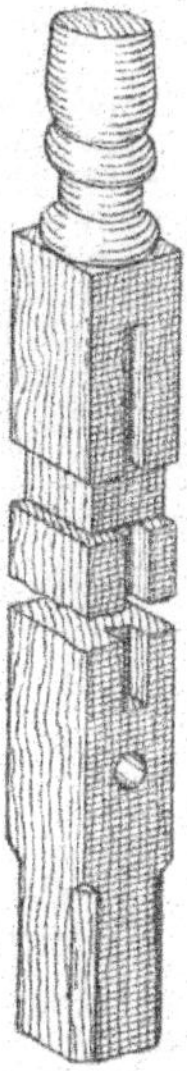

Fig. 681.—Corner Washstand Post cut to receive Rail (see B and C, Fig. 675).

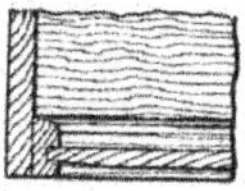

Fig. 680.—Section through Cupboard Bottom of Corner Washstand.

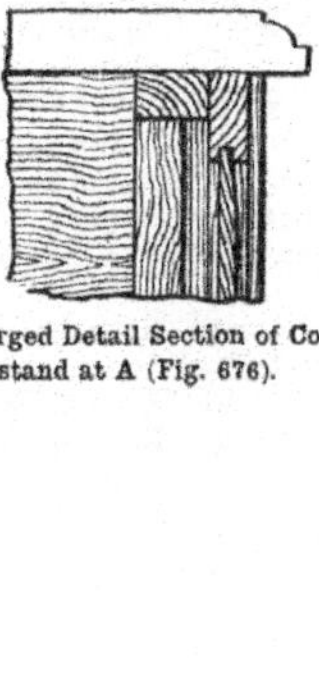

Fig. 679.—Enlarged Detail Section of Corner Washstand at A (Fig. 676).

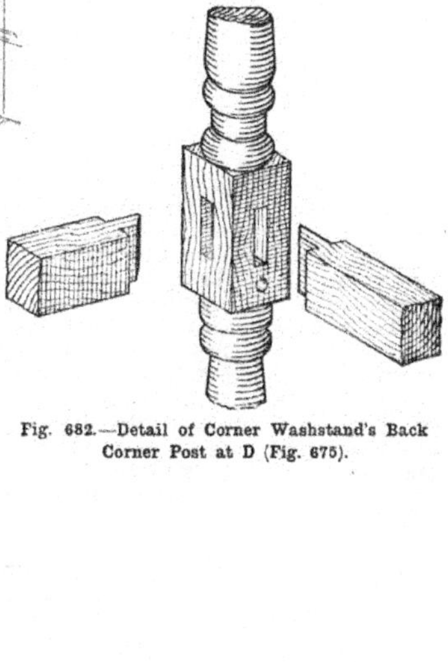

Fig. 682.—Detail of Corner Washstand's Back Corner Post at D (Fig. 675).

them in the top. These side pieces are shown separately at Fig. 663, and in section at Fig. 664; they are 1 ft. 4 in. long at the bottom, 8 in. at the top, and 7 in. deep, the curves being cut out with a bow saw. The back C (Fig. 662) is simply fixed on flush, and need not be let into the top. The front and two side edges of the top are rounded as shown in Fig. 664, which also shows how the side pieces are let into the top, these being fixed with screws driven in from underneath the top. The back legs are rectangular in section, 1¾ in. by 1½ in., while the front legs, if not purchased ready made, may be turned from a piece 1⅞ in. square to the pattern shown. The extreme length of all the legs is 2 ft. 4 in., the square part in the two front ones being 5½ in. from the bottom and 7½ in. long. The legs are connected below by stiffening rails D (see sectional plan, Fig. 665), ½ in. thick, and let into ¼-in. grooves. As indicated by the dotted outline in Fig. 665, the top projects ¾ in. beyond the legs all round. By marking out the position of the legs on the under side of the top, the length of the stiffening rails can be obtained, allowance being made, of course, for the ¼-in. grooves. The width of the side and back rails is 3½ in., and that of the front rail under the drawer 2 in. Two runners, ⅝ in. square, are fixed inside the legs for the drawer. To strengthen the legs at the top, stiffening rails 2 in. wide are fixed between them just under the top, similar to the lower rails. The back of the washstand may be curved to any outline desired, and a hole for the basin may be cut in the top if preferred. The method of constructing the drawer is shown at Figs. 666, 667, and 668. The sides, front, and back are ½ in. thick, and the bottom is ¼ in. thick. The depth of the drawer is 2¾ in. outside, making the sides 2½ in. deep, but the back is only 1¼ in. deep. The pieces are framed together as shown in Figs. 667 and 668, after screwing a couple of china knobs in the front. Over the drawer is a shelf or top 2 ft. 8⅞ in. long by 1 ft. 3 in. wide by ½ in. thick, and notched out at the corners to fit the angle of the legs. To prevent dust getting into the drawer, a dust strip E (Fig. 666) should be nailed on the bottom of the ⅝-in. runners, as shown; and blocks are put in under the shelf as required. Paint the washstand, with the exception of the top, a light oak colour which should be suitably grained, or paint it a light green; the top should be painted white, in imitation of marble. The top could be primed with pure whitelead paint, say two coats, and finished with best white enamel paint.

Lavatory Washstand.

Figs. 669 and 670 are the front and side elevations of a lavatory washstand, Figs. 671, 672, and 673 being sections on X X (Fig. 669), Y Y (Fig. 669), and Z Z (Fig. 670) respectively. The two front posts in the lower part are 3 in. square, and the two back posts are 3 in. by 1 in.; they should be cut to shape, chamfered, moulded, and beaded as shown in Fig. 673. There are two front rails, and two in each side, 2½ in. by 1 in., the posts being mortised to receive them. Frames and panels are prepared for the door and the two sides, the latter being fixed with cross tongues let into the stiles and posts (see Figs. 673 and 674). The boarded bottom of the cupboard and a fillet screwed to the upper rails of the frame (see Fig. 671) form a rebate for the bottom and top of the sides, and act as a stop for the hinged door at the front. A shaped piece of 8-in. by 1-in. material is mitered round the top, on which the basin rests, moulds also being mitered round; and a shaped plinth is secured at the bottom (see Figs. 669, 670, and 671). The upper portion of the washstand is composed of a frame of the shapes shown in Figs. 669 and 670, rebated and moulded to receive a slab of marble or tiles, with a shelf supported by brackets (see also Fig. 672). The supply pipe and the waste pipe in the cupboard may be boxed up if this is thought to be desirable.

Corner Washstand.

Elevations and plan of a corner washstand are presented by Figs. 675 to 677, a general view being shown by Fig. 678. Fig. 679 is an enlarged detail at A (Fig. 676); Fig. 680, an enlarged detail showing the fixing of the cupboard bottom; Fig. 681, an enlarged detail of the post from B to C (Fig. 675); and Fig. 682 is an enlarged detail at D (Fig. 675), showing the joints.

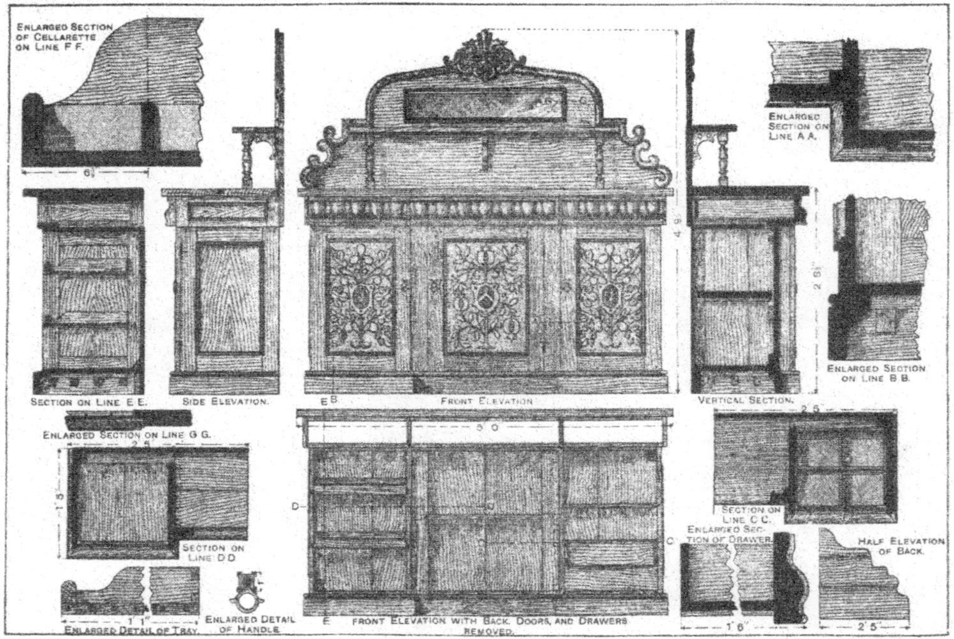

Enlarged Section of Cellarette on Line F F.
6½
Section on Line E E.
Side Elevation.
Front Elevation
4 9
Vertical Section.
2 8½
Enlarged Section on Line A A.
Enlarged Section on Line B B.
Enlarged Section on Line G G.
2 5
1 3
Section on Line D D
1 1"
Enlarged Detail of Tray.
Enlarged Detail of Handle
5 0
Front Elevation with Back, Doors, and Drawers removed.
Section on Line C C.
Enlarged Section of Drawer.
1 6"
Half Elevation of Back.
2 5"

CHIFFONIER.

Bedroom Lavatory with Reservoir.

The bedroom lavatory about to be described consists of a framed pedestal with a pair of doors (see Figs. 683 to 687), a sloped top (see Figs. 684 and 686) with two flaps, and a raised shelf with a shaped plinth; it contains a stoneware lavatory basin and splash tray fitted with waste pipe and metal receiver, and with an iron reservoir to contain about 6½ gal. of water. The lower part of the pedestal has divisions and shelves as indicated in Fig. 683, and one door is fitted with brackets and a towel roller, both doors being hung to the central division to fold in on each other (see Fig. 687), and be fastened with cupboard locks. The upper part, on the side not occupied by the wash basin, may be fitted with toilet accessories, a mirror being usually fastened under the lavatory flap, or a stationery case may be formed and the slope utilised as a writing desk. If the reservoir were reduced to half the length, the remaining portion of the top enclosure might also be fitted as a stationery case, with a falling flap hinged at the bottom; and with slight alteration and addition to the lower internal arrangement, the water in the reservoir might be heated with a paraffin lamp. The tank is filled through an aperture by the aid of a spout water-can, either by lifting off the shelf E (Fig. 686) or by providing in the shelf a hole which may be covered by a wood cap. Suitable woods for the design would be as follows: To finish in polish, American birch or black walnut; to finish in paint or enamel, yellow deal or American pine. All the interior parts, such as shelves, bottom, back, etc., may be made of white deal for economy. The sides or ends consist of framing of 1-in. stuff, mortised and tenoned together, the tenons on the back edges coming through and being wedged, while those on the front edges are stopped ½ in. back from the front sinking; if not fitted tightly, they should be secured with short, stout screws on the inside. The stiles are shaped as shown in Fig. 684, after they are cramped up and cleaned off, and the top rails are shaped to fit the slope, the dimensions being given on Fig. 683. A moulding (Fig. 688) is worked on the rail either with hollow and round planes, or with a special scratch tool. The raised and chamfered panel is of ⅝-in. stuff, and is flush inside. The framing is stop-grooved to receive the shelves at the bottom and top, and is rebated at the inside to receive the back. A ⅛-in. sinking is made across the level portion of the rail to receive the return ogee moulding of the top, and the panels may be either ploughed in or inserted in rebates and fixed with beaded slips as shown in Fig. 689.

Reservoir Case, etc.—The reservoir case is formed separately, and inserted after the pedestal is glued up. The piece forming the front may, if of hardwood, be jointed just below the rail G (Fig. 686) to a piece of deal, this piece (see Fig. 685) being grooved ⅛ in. into the sides, and running down and resting on the lavatory top, to which it is glued and screwed. The return ends of the enclosure are made lying (that is, with the fibres of the wood disposed horizontally), and may be jointed to the front piece either by groove and tongue or by mitre dovetailing. The flaps are shown made in the solid, with mitre-clamped ends. A pair of 2-in. butts should be used for each flap, and these must be fixed in the side grain, not end grain, of the clamps. The back is framed up with ⅝-in. stuff with ½-in. flush panels, the end stiles running from top to bottom, and the three rails shown in Fig. 686 tenoning into them. The muntin, shown in Fig. 687, is framed between the two lower rails, but there is no muntin in the top panel, which is a lying panel. The back legs should be strengthened by solid blocks endways of the grain, as shown in Fig. 690.

Completing the Lavatory.—In putting the carcase together, after each part has been properly fitted and brought to size, nail together the interior division and shelving, keeping them flush at the front; then fix the top and bottom to the upright division, inserting at the same time the standard F (Fig. 687). It may be noted that the hollow for the basin in the division J will be cut only approximately at first, the final cutting being made with a keyhole or pad saw after the basin has been scribed in position. Next turn the interior on one end and drive on the end framing, putting a little glue in the grooves; then reverse the position and treat the other end similarly; then stand the pedestal on its feet, cramp up, and square the carcase. For painted work, nail through into the top and bottom, but in polished work do not nail, but use angle blocks wherever possible without showing. Leave the cramp on until the glue is dry. Having fitted and fixed the reservoir case together, drive it down into position and glue the joints. Drive in the rail H (Figs. 685 and 686), gluing its lower edge, and brad the edge O (Fig. 686) into it. Fix the rail R and the back, which may be bradded on but not glued. Flush off the top edges and fit in the shelf E. Hang the flaps; mitre moulding I (Fig. 683) up to them. Fix the shaped

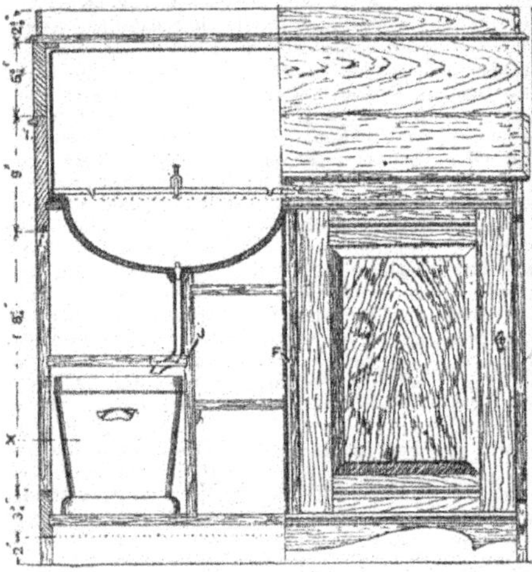

Fig. 683.

Fig. 684.

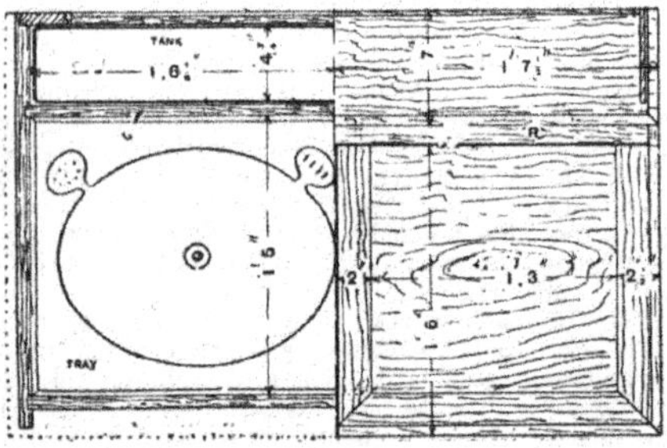

Fig. 685.

Figs. 683 to 685.—Part Front Elevation and Part Longitudinal Section, Side Elevation, and Part Horizontal Section, and Part Plan of Bedroom Lavatory with Reservoir.

bottom rail, fit and hang the doors, bore holes for the tap and waste pipe, and the fitting up will be complete. In the case of polished work, all parts below surrounded parts should be bodied in with polish before being put together.

Night Commode with Folding Arms.

Details of a night commode with folding arms are given in Figs. 691 to 695, the first three figures being produced to a scale of 1 in. = 1 ft. The further explanatory

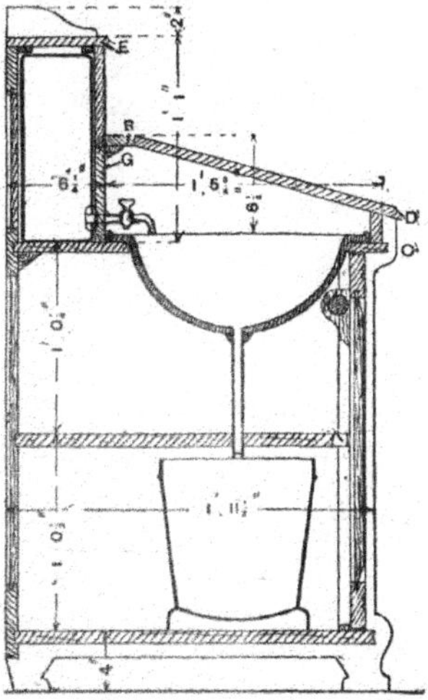

Fig. 686.—Cross Section of Bedroom Lavatory with Reservoir.

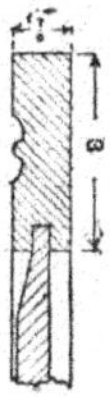

Fig. 688.—Part Vertical Section of Bedroom Lavatory Door.

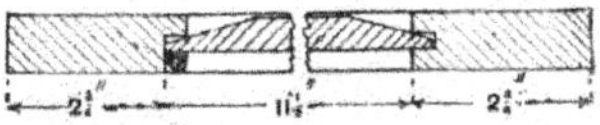

Fig. 689.—Part Horizontal Section of Bedroom Lavatory Door

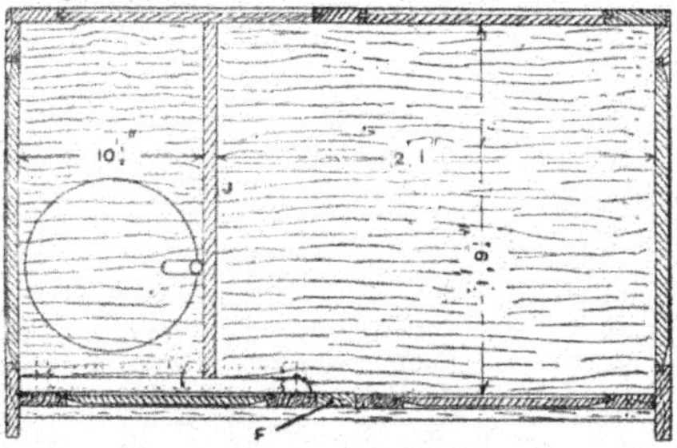

Fig. 687.—Horizontal Section through Bedroom Lavatory.

Fig. 690.—Detail of Back Leg etc., of Bedroom Lavatory.

figures (Figs. 696 to 701) give all necessary information. Figs. 696 to 699 are drawn to a scale of 4 in. = 1 ft., and Figs. 700 and 701 to a scale of 2 in. = 1 ft.

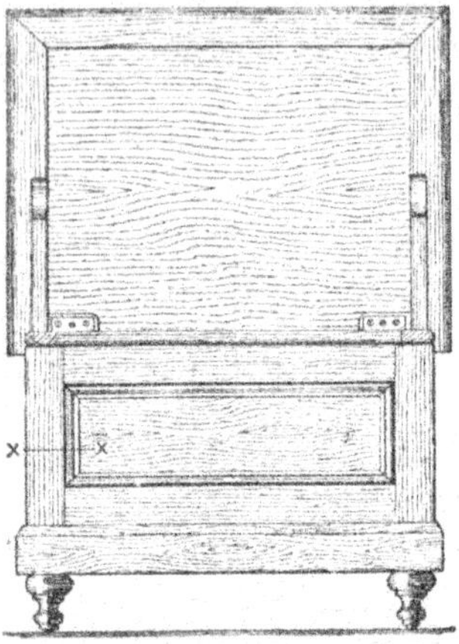

Fig. 691.

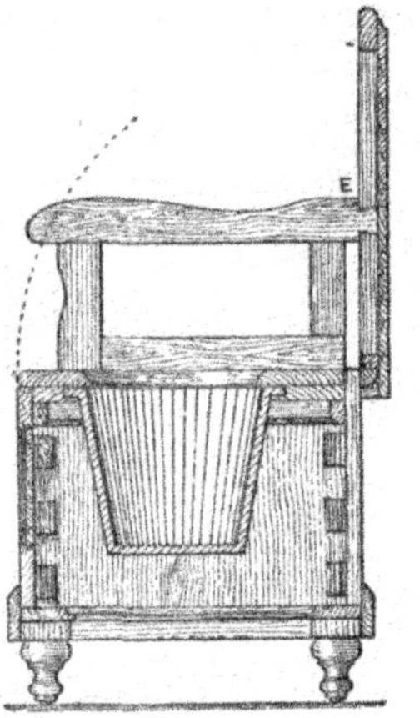

Fig. 692.

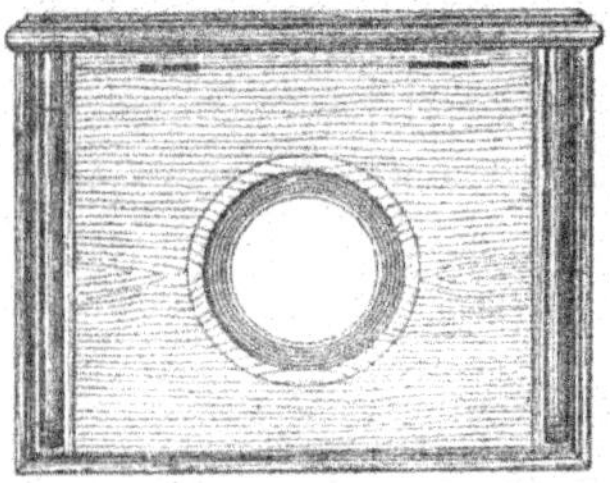

Fig. 693.

Figs. 691 to 693.—Front Elevation, Vertical Section, and Plan of Night Commode with Folding Arms (open).

Wardrobe and Mantel Fixtures for Bedroom.

Where a bedroom is large and spacious, and the fireplace is in the centre of one side of the room, ample wardrobe accommodation can be obtained, and a highly effective decoration be added, by such an arrangement as illustrated in Fig. 702. The recesses, 1 ft. 6 in. deep, on each side of the chimney-breast, are arranged of equal width, and are fitted with wardrobes as shown, the fireplace having a wood mantelpiece surmounted by an overmantel. In this case the wardrobe frames and the mantel framing have their principal surfaces in the same plane, the joints between them being covered by a pilaster, around which the plinth necking, and cornice moulds are broken. Fig. 703 is a vertical section through the wardrobe, and Fig. 704 a vertical section through the mantel and overmantel. The various

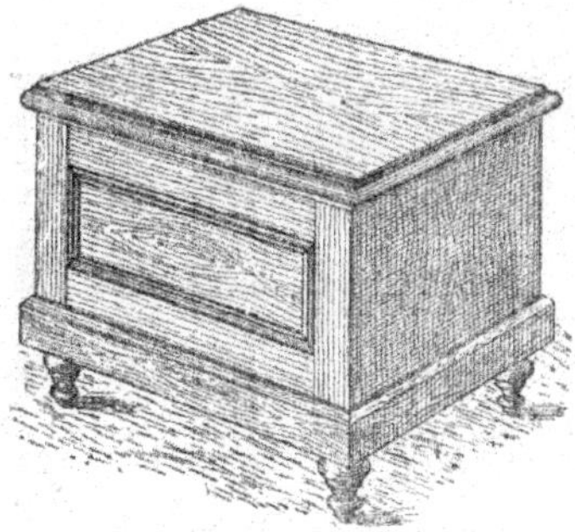

Fig. 694.—Night Commode Closed.

Fig. 695.—Night Commode Open.

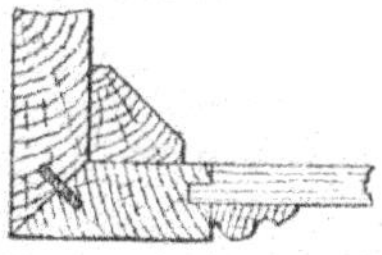

Fig. 696.—Horizontal Section through Corner of Commode (X X, Fig. 691).

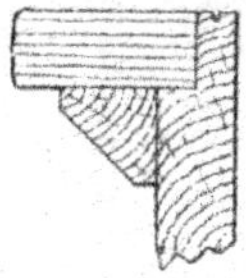

Fig. 697.—Fitting Side of Commode to Back.

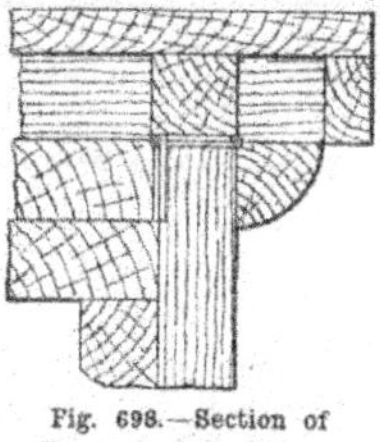

Fig. 698.—Section of Commode Lid, Closed.

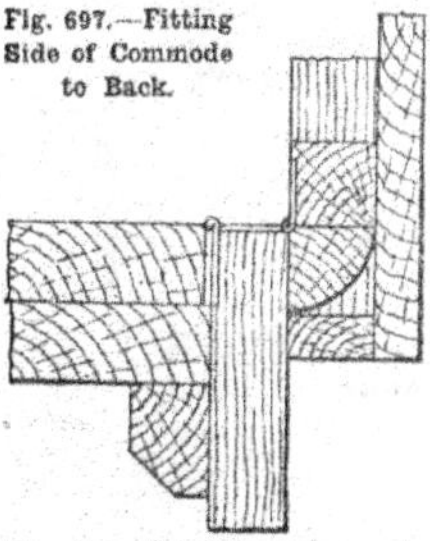

Fig. 699.—Section of Commode Lid, Open.

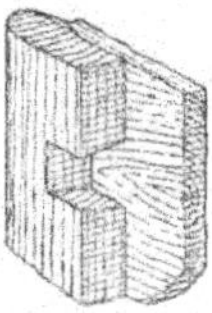

Fig. 701.—Socket (E, Fig. 692) for Dovetail of Arm of Commode.

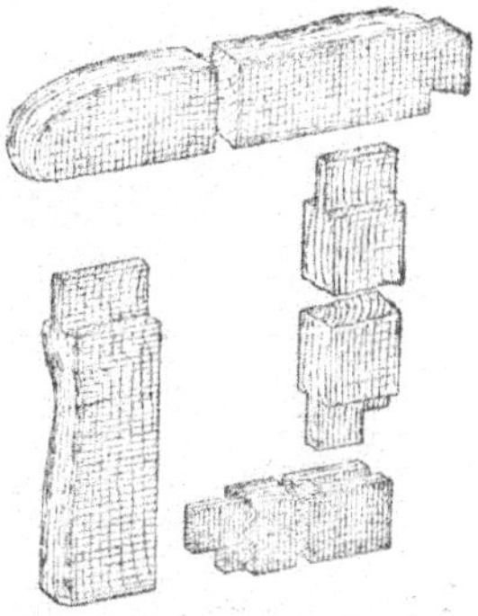

Fig. 700.—Framing of Commode Arms.

Fig. 702.

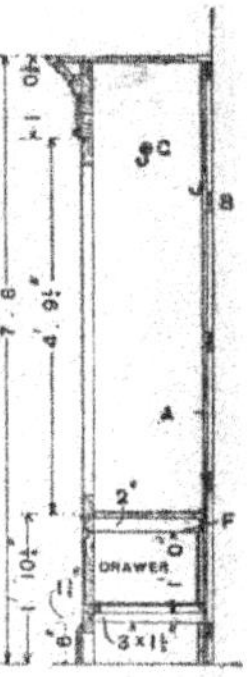

Fig. 703.

Fig. 702.—Front Elevation of Wardrobe and Mantel Fixtures.

Fig. 703.—Vertical Section through Wardrobe.

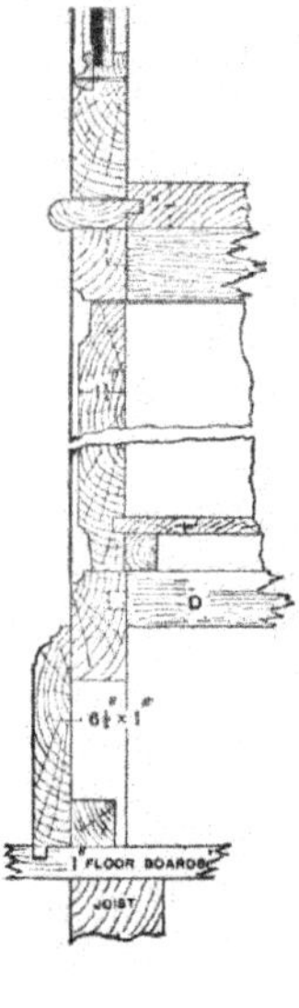

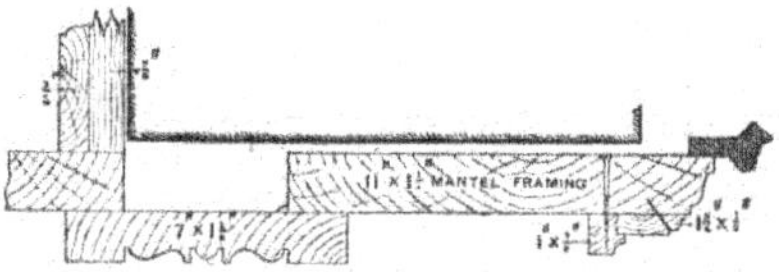

Fig. 705.—Horizontal Section through Mantel and Pilaster.

Fig. 706.—Vertical Section through Drawer, Plinth, etc., of Wardrobe.

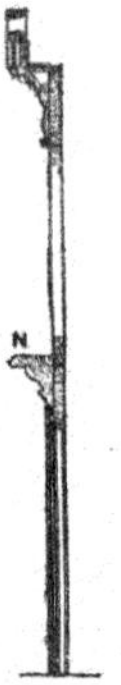

Fig. 704.—Vertical Section through Mantel and Overmantel.

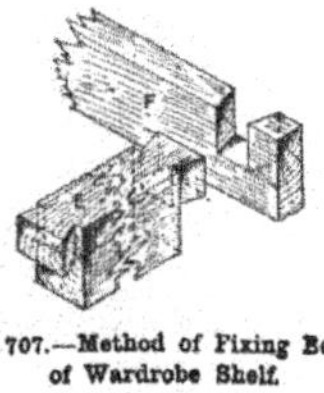

Fig. 707.—Method of Fixing Bearer of Wardrobe Shelf.

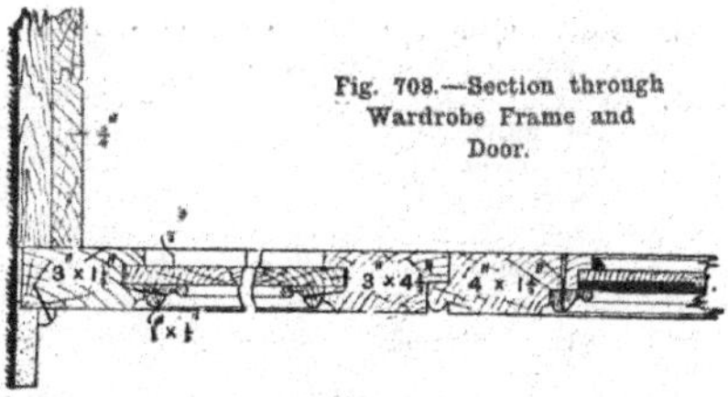

Fig. 708.—Section through Wardrobe Frame and Door.

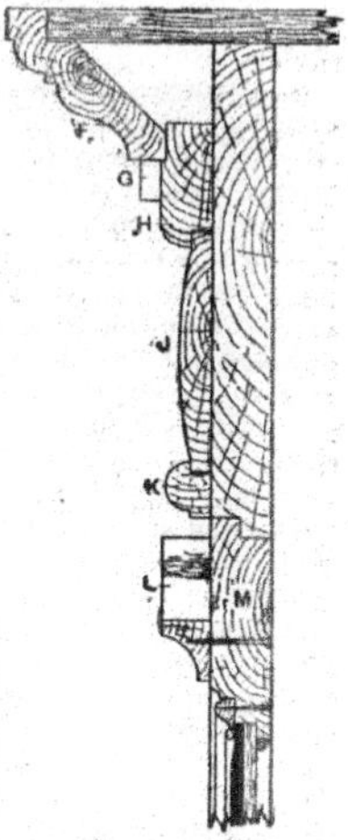

Fig. 709.—Section through Cornice, etc., of Wardrobe and Mantel Fitment.

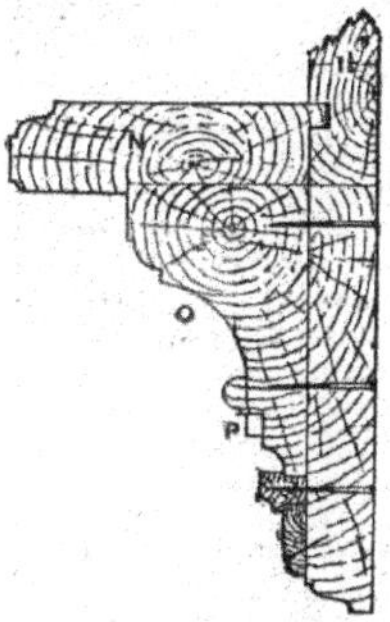

Fig. 710.—Section through Mantelshelf, etc.

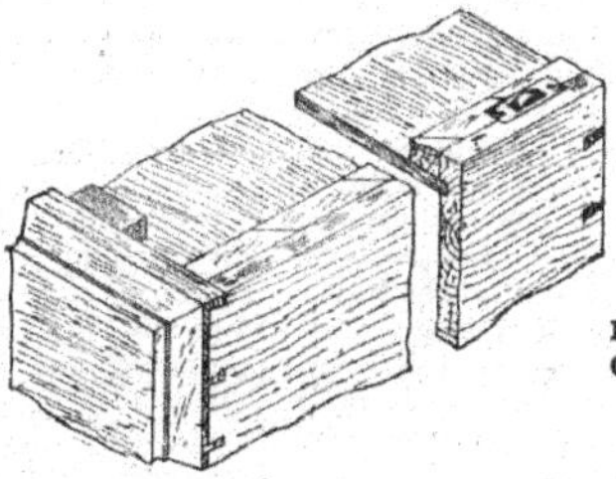

Fig. 713.—Part of Wardrobe Drawer with Friction Roller.

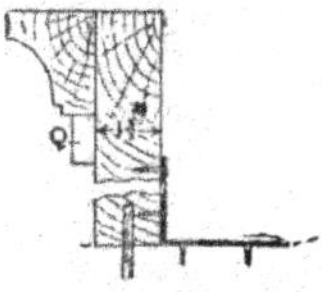

Fig. 711.—Section through Pediment of Wardrobe and Mantel Fitment.

Fig. 712.—Sash Friction Roller.

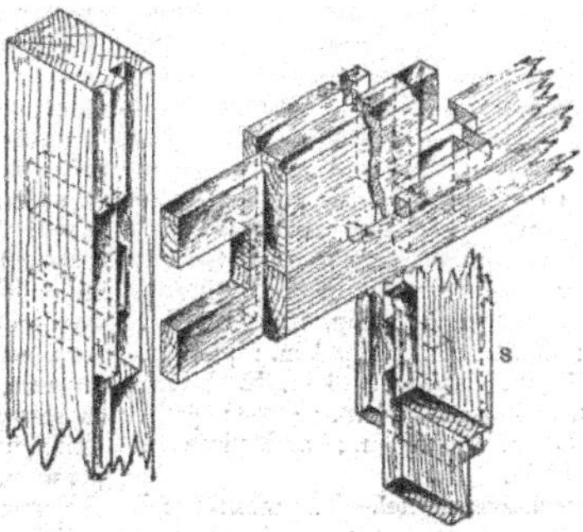

Fig. 714.—Joints in Wardrobe (see Fig. 702).

details of construction are also fully illustrated.

Construction of Wardrobes.—The wardrobes are formed by first lining the recess at the back and sides with ¾-in. jointed or beaded boards A (Fig. 703) fixed to 3-in. by 1-in. grounds B, one of the latter being placed at a convenient height to receive the wardrobe hooks, so as to allow of longer screws than could be employed if screwed through the thickness of the boards only. A good system of obtaining hanging accommodation is to fix a brass rod or tube across the centre, with sliding hooks, as indicated at C (Fig. 703). The rod may be supported by pateræ screwed to the boarding at each end. To the ends of the side grounds the wardrobe frame is secured (see Fig. 705), the drawer runners D (Fig. 706), 3 in. by 1½ in., being inserted at the same time; the back end of the latter rests on packing pieces of the necessary height, and is nailed on or dovetailed. The bearers E (Fig. 707) for the shelf are 2 in. by 1¼ in., and are dovetailed at the front into the rail of the frame, and at the back into a piece of stuff F, 2 in. by 1 in., screwed to the lining of the cupboard (see also Fig. 703). A thickness of 1½ in. is necessary for the frame and door, the detail of rebates and mouldings being such as to allow of the ¼-in. bevelled plate mirror being inserted from the face, and the $\frac{5}{16}$-in. backboard and mirror fixed by screwing the mouldings from behind (see Figs. 706 and 708). To prevent dust entering the wardrobe, the top should be boarded with 1-in. tongued and grooved boards and covered with strong paper. It will be found best to have the length of the boards at right angles to the face of the frame, sufficient overhang being given to provide for the proper securing of the top edge of the cornice (see Fig. 709). The wardrobe doors are rebated all round their edges to assist in the exclusion of dust. The cornice F (Fig. 709) is 5½ in. by 1¼ in.; dentils G, 1 in. by ½ in.; piece H, 3 in. by 1¼ in.; frieze J, 6 in. by ⅞ in.; necking K, out of stuff 1¼ in. square; pediment L, 1½ in. by 1¼ in.; and piece M, 5½ in. by 1½ in.

Mantel and Overmantel.—The mantel and overmantel are prepared to similar detail to the wardrobe frame, and fixed in the same plane, and a sunk-moulded pilaster covers the edges of the frames. The shelf N (Fig. 710), 7 in. by 2 in., is tongued to the bottom rail of the overmantel, and is fixed to the bed mould O, 6½ in. by 4 in., which is provided with dentils P, ½ in. by ⅜ in., the mould having been previously secured to the mantel frame with screws from behind. Where the bed mould is of large size, it may be built up with advantage, or may take the form of a sprung mould and be left hollow behind; this would necessitate the returned ends of the mould being mitered on instead of being worked in the solid.

Cornice, Frieze, Pediment, etc.—The cornice, frieze, and necking are built up of medium-size sections, and rebated or housed together to avoid open joints through warping or shrinkage (see Fig. 709), and are fixed by nailing through the rebates or screwing from behind, a deep top rail being provided in the framing for that purpose. The pediment is built up separately, as indicated by Fig. 711, and is dowelled to the cornice and further secured with 4-in. by 4-in. by 1-in. angle irons screwed to the pediment and cornice mould. Dentils Q, 1 in. by ½ in., are glued on as shown.

Drawer Rollers and Runners.—The wardrobe drawers, being large, and liable to become heavy, should be provided with friction rollers (see Figs. 712 and 713) to ensure easy motion, and it is imperative that hardwood runners be used, so as to prevent the rollers working a groove in the material. Fig. 714 is a detail of the joints as seen at R in Fig. 702, S representing the muntin rebated for the door.

Special Points.—If the work is to be painted (which is the usual mode of decoration for fixtures), the whole of the face-work should be executed in thoroughly well-seasoned American yellow pine. If it is to be polished, any of the ornamental hardwoods of good figure might be chosen. In that case the whole of the moulds and built-up portions must be fixed from behind, as shown in the details. This method of fixing, indeed, is advisable in all cases where possible, as, if this is carefully done, and if the screws are inserted in the most effective position, warping will be prevented and a better face will be produced. Where there is considered to be any danger from fire, owing to the close proximity of the woodwork to flues, and the thinness of the walls, good breeze bricks should be inserted for nailing to, and sheets of asbestos placed between the brickwork and the woodwork. Where the walls are 9 in. thick, under average conditions no special precaution is necessary. The design illustrated is somewhat elaborate, and suited to a good style of house, but the arrangement allows of many modifications without detracting from its usefulness.

Combination Wardrobe Bedstead.

The combination wardrobe and bedstead illustrated by Figs. 715 to 729 is a convenient piece of furniture where space is limited. Its outward appearance is a well-made wardrobe of early English design, perhaps somewhat taller than usual, with a pair of panelled doors and a large drawer beneath. On opening the door, however, it will be seen that at the back of the hanging cupboard a full-size bedstead lies folded up, occupying about 3 in. of the depth, as shown in Fig. 716, the rest of the space being available for hanging clothes. The bed-frame is hinged at the bottom, and when in use the foot is supported by hinged legs, as shown in Fig. 719, thus providing a strong and easily adjustable bedstead, 6 ft. long and 3 ft. wide, with head- and foot-boards complete. Fig. 715 is a front sectional elevation, the left half showing the outside; the right half is a vertical section through the case, just in front of the folded-up bedstead. Fig. 716 is a vertical cross section, and Fig. 717 a horizontal section through the upper part of the cupboard. Fig. 718 is a sectional plan of the drawer pedestal, the left half being a plan of the top, and the right half a section below. Fig. 719 is a side view of the lower part of the wardrobe with the bedstead down. The remaining illustrations are enlarged details:—Fig. 720 is part plan of the top of the cupboard; Fig. 721 a broken section through the doors; Fig. 722, elevation of the end of the middle rail of the doors; Fig. 723, a section at A (Fig. 722); Fig. 724, elevation of the end of a bottom rail; Fig. 725, section at B (Fig. 724); Fig. 726, longitudinal section of the bed-frame, showing method of fixing the webbing; Fig. 727, plan of corner and middle of the bed-frame, showing method of framing the corners and fixing the web fillets; and Fig. 728 a section of the same.

Pedestal.—The pedestal is made separate from the cupboard for convenience of handling; the sides and bottom are of 1-in. deal, the top and back of ¾-in., the bottom being housed in as shown in Fig. 715. The top is lap-dovetailed, as shown in Fig. 718. The plinth, ¾ in. by 3 in., is planted on, the front piece being fixed by angle blocks glued to the bottom (see Fig. 716). The back is a plain piece of ¾-in. board placed in rebates on the sides and on the edges of the top and bottom; a ½-in. cocked bead is planted round the front and each end of the pedestal top to break the joint, and also to form a well for the cupboard to fit in; this is shown at C (Fig. 718). The drawer which is intended to receive the bedding is made with a 1-in. front, ¾-in. back, ⅝-in. sides, and ½-in. panelled bottom, grooved, blocked, and dovetailed in the usual manner.

Cupboard.—The cupboard consists of two solid sides of 1-in. deal, with top and bottom of the same material, dovetailed together as shown in Fig. 720. The back edges of the sides are rebated out to receive the panelled back. The bottom is kept back 1 in. to form a rebate for the doors, the remainder of the rebate being formed by planted slips. The doors are framed up from 1-in. pine, as shown in the details, with ⅜-in. circular stop chamfers and ½-in. V-jointed panels, and hung with 2½-in. brass butts, and fastened with two edge-flush thumb-bolts and a 2-in. brass cupboard lock. The back is a square sunk panelled frame, ¾ in. thick, with ⅜-in. panels, 6-in. rails, and 3-in. stiles, mortised and tenoned together. The cornice, made as shown in detail by Fig. 729, is mitered and screwed to the top of the cupboard, rebated for a ⅜-in. dust cover board, and finished with a ½-in. shaped and moulded cresting. At B (Figs. 716 and 717) are shown two brass hanging rails screwed to the doors; H (Figs. 715 and 716) is a movable wood bar over which clothes may hang.

Bedstead.—The bedstead is composed of

two side rails 2 in. by 2 in., and three cross rails of 3-in. by 2-in. pitchpine, mortised and tenoned together as shown by Figs. 727 and 728. A rebate, $\frac{3}{4}$ in. by $\frac{1}{2}$ in., is worked round the inner top edge to receive the fillet that fixes the webbing w, which is made up into a series of loops or endless bands, passing over and under each other,

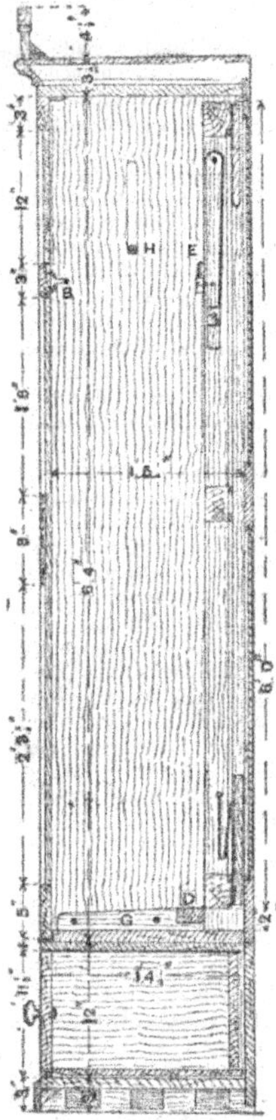

Fig. 716

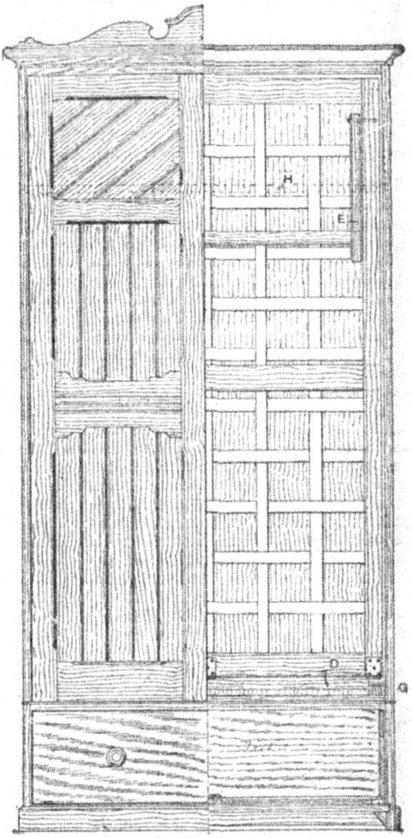

Fig. 715

Figs. 715 and 716.—Half Front Elevation and Half Longitudinal Section and Vertical Section of Combination Wardrobe Bedstead.

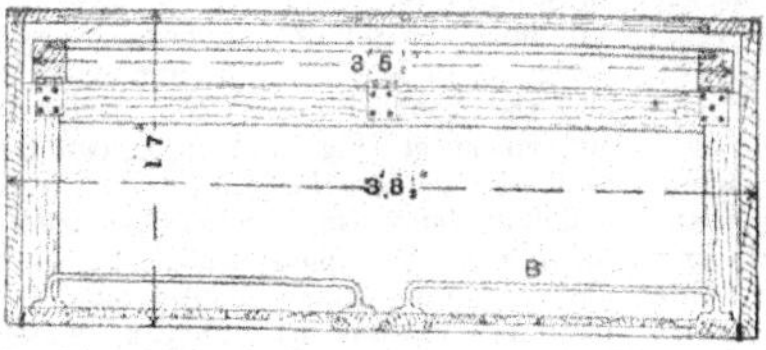

Fig 717.—Horizontal Section of Combination Wardrobe Bedstead.

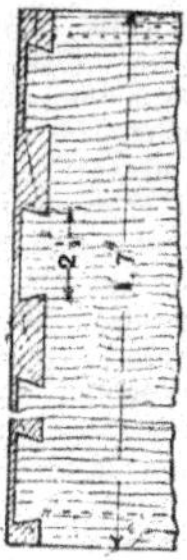

Fig. 720.—Detail of Wardrobe Top.

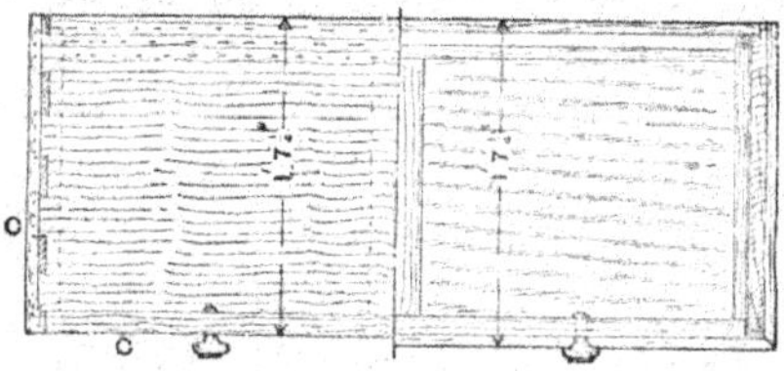

Fig. 718.—Half Plan and Section of Drawer Pedestal of Combination Wardrobe Bedstead

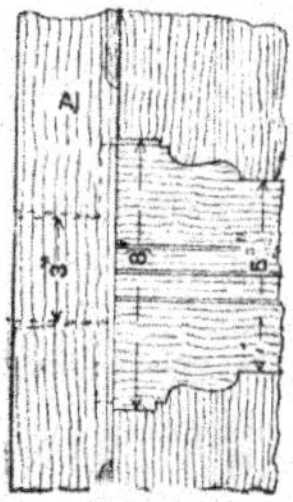

Fig. 722.

Fig. 723

Figs. 722 and 723.—Details of Middle Rail of Wardrobe Door.

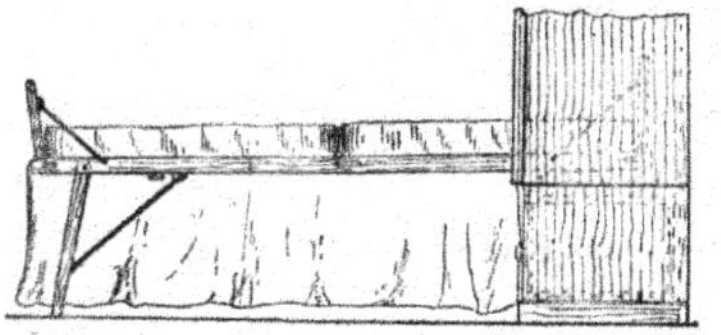

Fig. 719.—Side Elevation of Bedstead, Down.

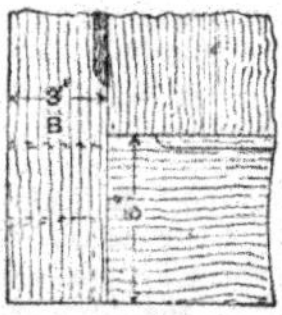

Fig. 724.

Fig. 725.

Figs. 724 and 725.—Part Elevation and Section at Bottom of Wardrobe Door.

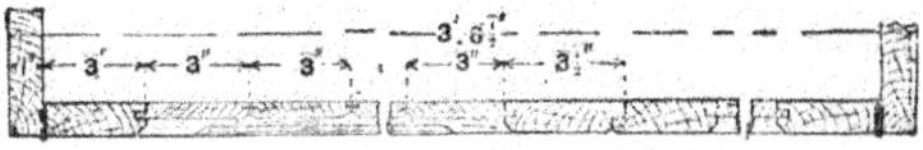

Fig. 721.—Horizontal Section through Wardrobe Doors.

their ends being rove on the fillets, which are then screwed into the frame (see Fig. 726). The bands should be the length of the opening when they lie flat together; the spreading necessary to introduce the side fillets will then shorten them somewhat. One side fillet is first fixed, then the opposite fillet is pinched up with hand screws and fixed in position with wood screws. One end fillet is next fixed, the webbing interlaced, and the other end served the same as the side, pulling all up taut. The join in the webbing should be lapped and brought under the fillet, and it is an improvement to make a slight notch in the bottom of the rebate under each band of webbing, so that the fillet may sit close down in the rebate.

Bedstead Frame.—The bedstead frame is hinged to a rail 3 in. by 1½ in. (D, Figs. 715 and 716) fixed inside the cupboard to two elbow-pieces G, which are sloped off slightly to prevent the frame riding on them when it is down. The two legs E are framed together by a chamfered cross rail, of such a

Fig. 726.—Section of Bed-frame.

size as just to pass easily within the frame, to which they are pivoted with 3-in. by ⅜-in. screw-bolts. The leg frame is prevented from slipping when in use by two iron hooks fixed under the bed and fitting into two eyes screwed in the legs; these hooks should be hung so that they enter the eyes from outside, otherwise they may be accidentally knocked out by anyone who is standing at the bed-side. The head- and foot-boards, 12 in. and 9 in. by ¾ in. respectively, are hinged to the frame by back flaps, and are held in position by cords, as shown in Fig. 719. The bedstead is held securely when in the wardrobe by two turn-buttons, screwed under the frame as shown near E (Fig. 716). Figs. 715 to 718 are drawn to a scale of ¾ in. to 1 ft., Figs. 720 to 726 are 1½ in. to 1 ft., and Figs. 727 to 729 at 3 in. to 1 ft.

Combination Cabinet-bookcase and Bedstead.

In the cabinet-bedstead about to be described, the bedstead and mattresses are enclosed behind doors, and the drawer below will hold the pillow and bedclothes. The design is of a plain and simple character, and the front elevation (Fig. 730) and the end elevation (Fig. 731) are drawn to the scale of 1 in. to a foot.

Cabinet-bookcase.—The cabinet-bookcase, being entirely separate from the interior fittings, may be first considered. It will look well if made of mahogany stained a dark red in imitation of Chippendale mahogany, or, in fact, any of the usual hardwoods, to match the rest of the furniture. The flat top or the cornice, the outer and inside ends, the bottoms, the shelves, the drawer front, and the shaped pediment and span-rails are all of 1-in. stuff. The door

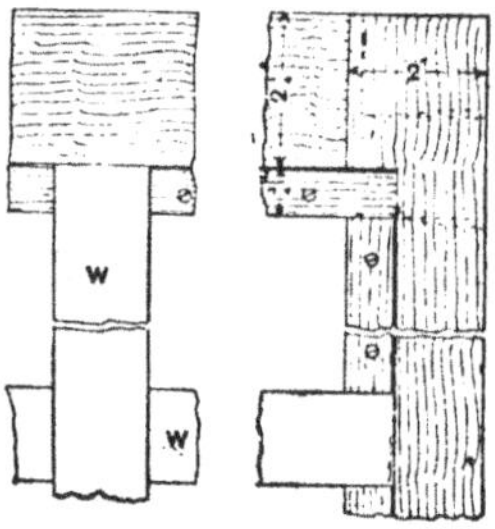

Fig. 727.—Plan of Corner of Bed-frame.

stiles and rails are of 1¼-in. stuff. The tracery pattern in the doors may be made with the usual bead and fillets to the section shown in Fig. 732, or an effective way is to cut out the pattern in very thick brown paper, and gild the face with Ardenbrite gold paint. The gilded paper is placed against the glass, and to keep it in position a frame covered with some suitable fabric, such as serge, cloth, or silk, is placed behind. In getting out the outer and inside ends to width, the construction of the carcase back must be considered. The side brackets A (Fig. 730), the framing of the back panel over the doors, and the upright muntins B (Fig. 733), are all of 1-in. stuff. A (Fig. 734) is a section of the side brackets resting against the top above the bookshelves, and C is the back framing, at the same height.

The outer ends are rebated to receive the muntins, and the latter are grooved to accommodate the ½-in. backs D (Fig. 733). The bracket A and back panel C, and the muntins B and back D (Figs. 733 and 734), rest equally against the edge of the top above the doors, and are secured with screws. Fig. 733 also shows a sectional plan of the carcase back fitting against the bottoms above and below the drawer. The outside ends stand back 3 in. from the front edges of the inside ends. The latter are cut through at E (Fig. 731). The bottoms and top over the door are dovetailed across the inside ends as at F (Fig. 735), which view is looking from the back. The groove must

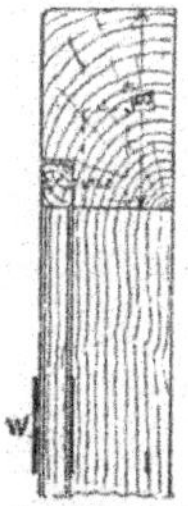

Fig. 728.—Section of Bed-frame showing Webbing.

finish ⅜ in. short from the front so that the dovetail does not show. The bottoms, tops, and shelves are fixed in the same way to the outer ends, but the ends of the bottoms, and the shelves which intersect with the inside ends, must be mortised and tenoned as shown in Fig. 736. The shaped span-rails stand back ⅛ in. from the face of the ends, and are housed ⅛ in. deep; also the frieze G (Fig. 730). The cornice is fixed to the ends with screws driven through from the top side; the small brackets H (Figs. 730 and 731) are also secured with screws. If desired, the bookshelves may be movable, and supported on brass pins as shown in Fig. 737, which represents a portion of the end with holes bored for the pins which support the shelf. Fig. 738 shows a peg or pin which can be obtained of any ironmonger. The drawer is made in the usual way, and the lines with square corners are hollow in section as shown in Fig. 739. The doors and the back panel above may have ovolo mouldings worked on their inner edges as shown in Fig. 740. The shaped valances under the shelves are of leather, and are fixed by first gluing the top edges to a strip of pine about ¾ in. square. A groove in the under side of the shelves is made to receive the strip (see the section Fig. 741). The valance should stand back $\frac{3}{16}$ in. from the edge of the shelf. The ornamental part in the centre of the shaped pediment is carved, each petal being half-circular in section, and the ends rounded as shown. The usual french-polishing will complete the job.

Folding Bedstead.—Fig. 742 is a side elevation, and Fig. 743 a part plan drawn to 1-in. scale, of the folding bedstead. The

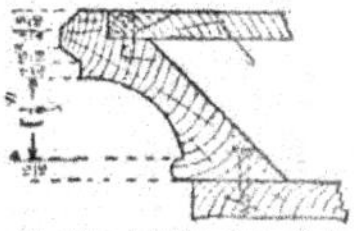

Fig. 729.—Section of Cornice of Combination Wardrobe Bedstead.

best material for the framing is well-seasoned birch 2 in. wide and 1⅝ in. thick. The side rails J (Fig. 742) are connected to blocks K (Fig. 743), which are 2 in. wide and 1½ in. thick. These blocks must be firmly fixed with screws driven into the ends and the bottom above the drawer. It will be seen that the whole of the strain of the bedstead framing on the cabinet lies at the bolts L (Fig. 743). The side rails M (Fig. 742) are connected to the rails J by ⅝-in. diameter bolts, and the end rail N (Fig. 743) is dovetailed into the side rails M. Pine laths O (Figs. 742 and 743), 1½ in. wide by ⅜ in. thick, are screwed to the side rails, with a space of 2½ in. between the laths. Two legs P (Fig. 742) are bolted to the rails; they are 2 in. by 1⅝ in. at the top, and taper to 1¼ in. at the bottom. These legs are fixed to the other sides of the rails J, the other legs Q (Fig. 742) being fixed to the inside of the rails M. Stop pegs R (Figs. 742 and 743) are fixed

Fig. 730.—Front Elevation of Cabinet-bookcase and Bedstead.

Fig. 731.—End Elevation of Cabinet-bookcase and Bedstead.

Fig. 732.—Section of Door Moulding for Cabinet-bookcase and Bedstead.

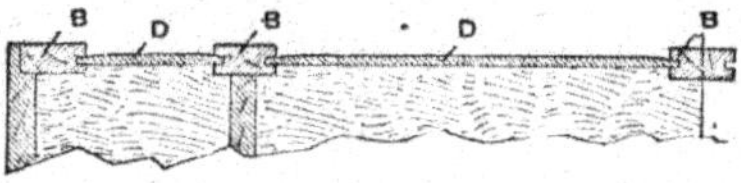

Fig. 733.—Section through Back of Cabinet-bookcase and Bedstead.

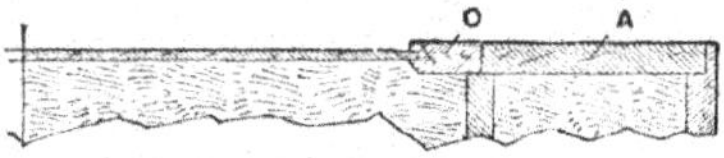

Fig. 734.—Section through Back of Cabinet-bookcase and Bedstead.

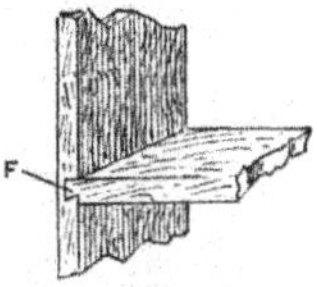

Fig. 735.

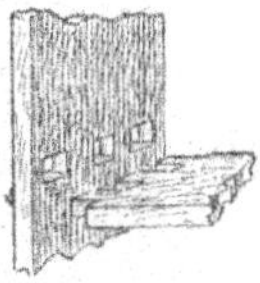

Fig. 736.

Figs. 735 and 736.—Methods of Fixing Carcase Tops, Bottoms, and Shelves of Cabinet-bookcase.

Fig. 737.—Holes in Side of Cabinet-bookcase to receive Movable Shelves.

Fig. 738.—Movable Shelf Fitting.

Fig. 739.—Section of Hollow Lines on Drawer Front.

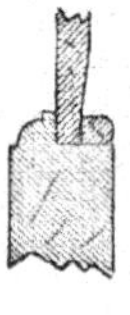

Fig. 740.—Section of Door and Panel Moulding.

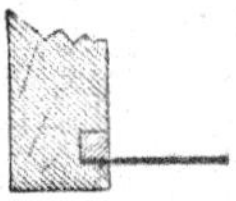

Fig. 741.—Section, showing Fitting of Valance to Cabinet-bookcase Shelves.

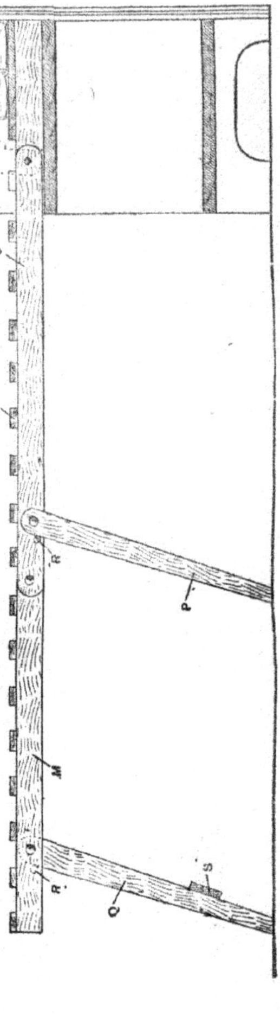

Fig. 742.—End Elevation of Folding Bedstead

to the rails to keep the legs at the angle as shown. To prevent the bedstead framing **collapsing sidewise, a cross rail s (Figs. 742** and 744), 2½ in. wide and ¾ in. thick, is screwed to the legs Q. Also, to keep the frames at right angles, a lath T (Fig. 743) is screwed anglewise to the under side of the laths O. When the bedstead is folded into the cabinet, the legs P fall alongside the rails J. The legs Q are lifted up alongside the rails M, and with the latter fold between the rails J. To make up the level of the

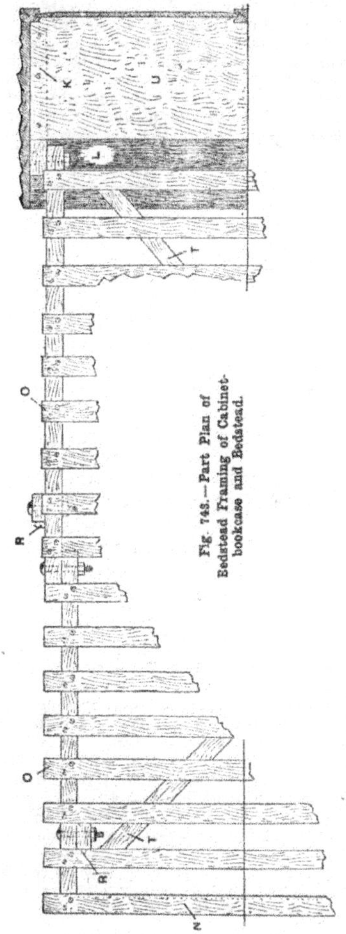

Fig. 743.—Part Plan of Bedstead Framing of Cabinet-bookcase and Bedstead.

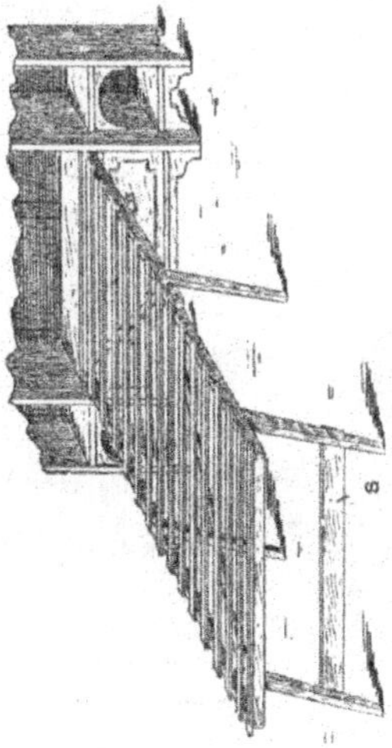

Fig. 744.—General View of Bedstead, Unfolded.

bedstead framing, a ⅜-in. board U (Fig. 743) is fixed across the rails K. Two flock or hair mattresses about 4½ in. thick are laid on the bedstead framing, and when not in use are placed on their ends inside the cabinet as shown at V (Fig. 742). The length of the bedstead from the foot to the head is 6 ft. 9 in.

Combined Wardrobe, Washstand, and Cupboard Fitment.

Figs. 745 and 746 illustrate a bedroom fitment which combines a wardrobe, a washstand, a mirror, two sets of drawers, a boot and shoe cupboard, and a towel rail.

It should be made of clean dry whitewood, and finished in enamel, the panels being lined out with curved corners. The back framework, with sections of the rails, is shown by Fig. 747, the several parts being mortised and tenoned together. The back for the wardrobe and cupboard is panelled with $\frac{1}{2}$-in. rough boards, glue-jointed together. To obtain the width, 1 ft. 6 in. for each end, three widths of 6-in. by $\frac{7}{8}$-in. stuff are tongued and glued; the face side is trued and cleaned, and the inner side cleaned and grooved for the bottom (see Fig. 748). The inside end of the wardrobe is $\frac{7}{8}$ in. thick, grooved and tongued to the floor, which is afterwards nailed to it (see Fig. 749). The dotted lines in Fig. 749 show the hanging stile of the small cupboard, and the plinth over the division. The wardrobe door is framed of $\frac{7}{8}$-in. stuff, and is grooved to receive a panel $\frac{3}{8}$ in. thick, which should be fitted in the grooving when the framework is knocked together. It is finished at the front with a bolection moulding, shown in section in Fig. 750. The door is finished inside with a beaded fillet nailed round the panel opening to the framing. Hang the door with $2\frac{1}{2}$-in. brass butts, and furnish with a knob and fastening. Battens of $2\frac{1}{2}$-in. by $\frac{3}{4}$-in. stuff are fixed inside the wardrobe to carry some brass hooks for clothes.

Mirror.—The swing mirror is of bevel-edge plate glass, mounted in a framework of stuff 2 in. by $\frac{7}{8}$ in.; the rails and stiles are secret tenoned and wedged together, and rebated at the back for glazing. A backing board $\frac{3}{16}$ in. thick protects the glass at the back, and fillets nailed in the rebate keep the glass in position. The inside front edge of the framework is moulded in the solid, and on the face another moulding is planted, being so arranged that the square edge of the solid moulding forms part of the planted moulding (see Figs. 751 and 752). The mirror is hung with brass pivoted plates let into the woodwork flush with the surface and screwed. Fig. 753 shows a pivoting plate; one is fitted at each side of the mirror, and Fig. 754 shows the drilled plate to receive the pivots. One plate must be drilled to the solid lines, and one plate must be slotted as indicated by the dotted lines in Fig. 754.

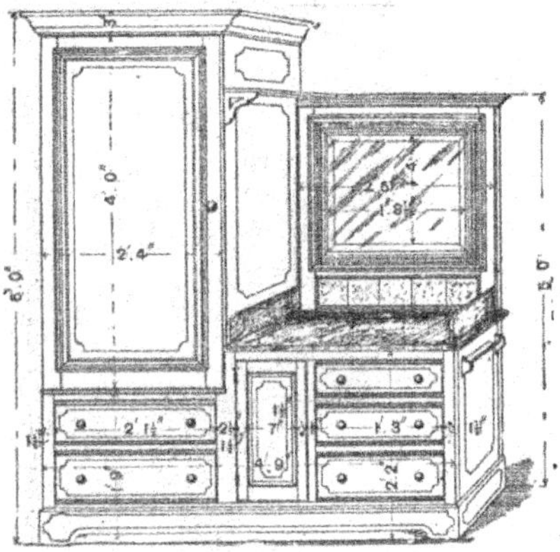

Fig. 745.

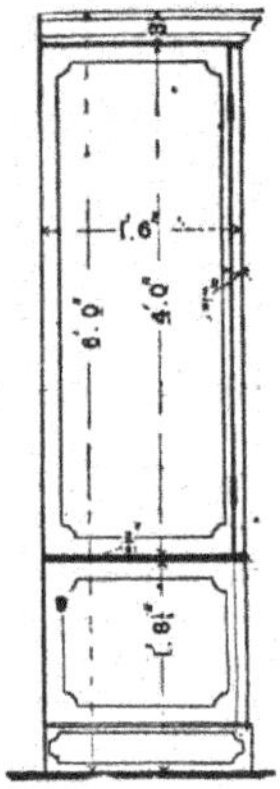

Fig. 746.

Figs. 745 and 746.—General View and End Elevation of Combined Wardrobe, Washstand, and Cupboard Fitment.

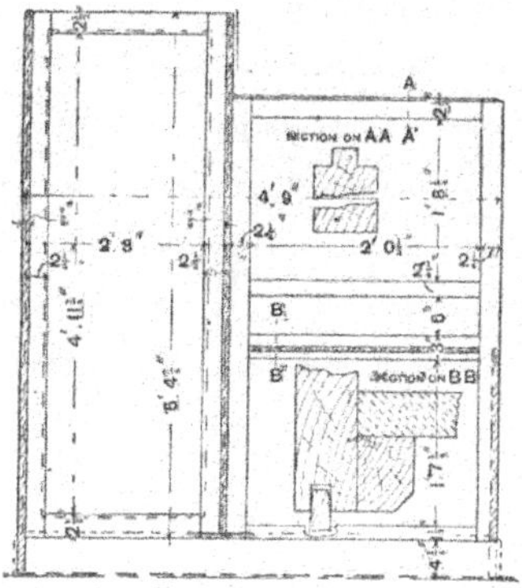

Fig. 747.—Back Framework of Combined Fitment.

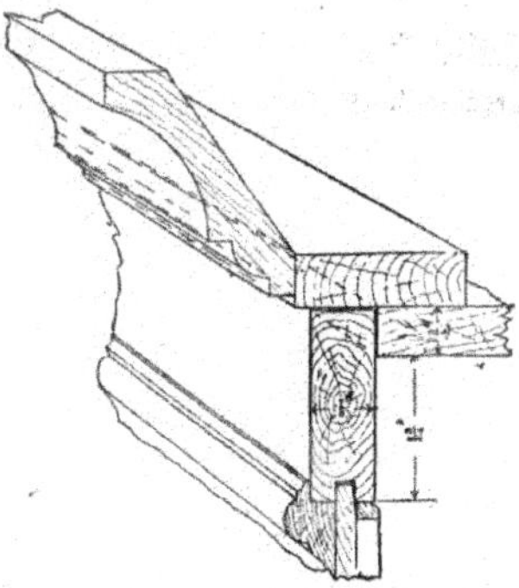
Fig. 750.—Section of Wardrobe Door and Cornice of Combined Fitment.

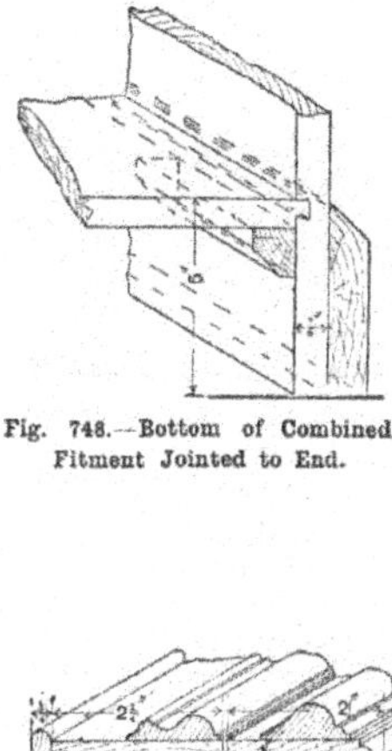
Fig. 748.—Bottom of Combined Fitment Jointed to End.

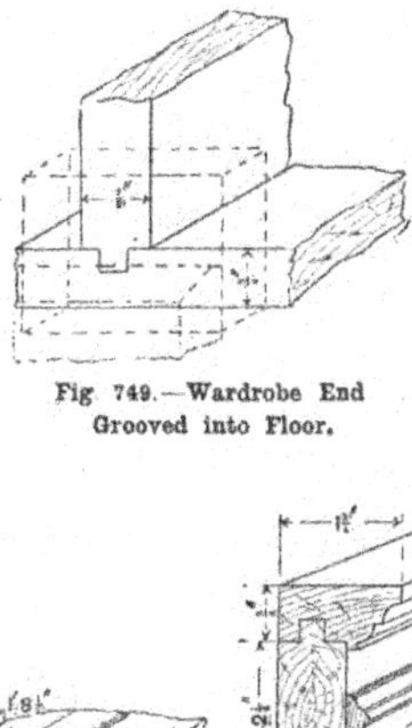
Fig. 749.—Wardrobe End Grooved into Floor.

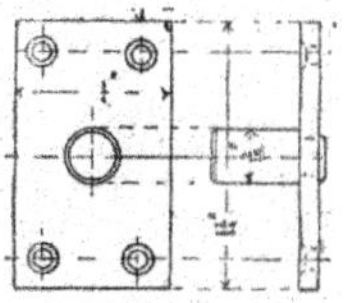
Fig. 753.—Pivoting Plate.

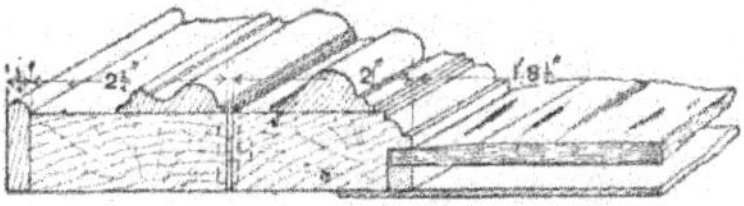
Fig. 751.—Section of Mirror Framing.

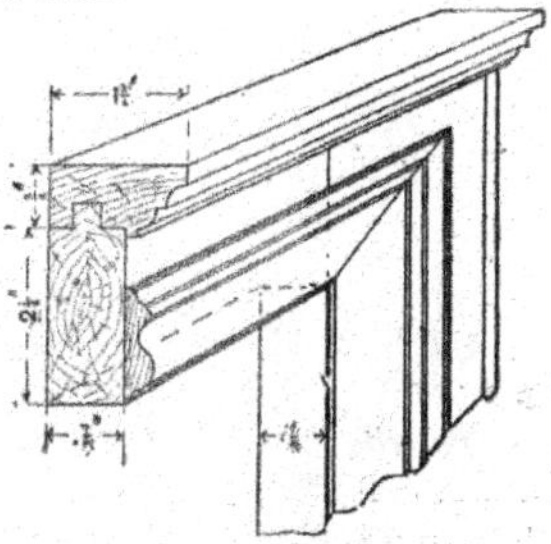
Fig. 752.—Part of Framing of Combined Fitment, showing Mouldings in Section.

These plates are let into the surrounding framework, and the wood is cut away to fit the slot, so that the mirror can be taken out if required. Cut the plates from sheet iron

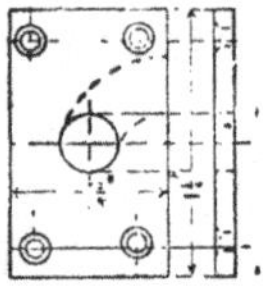

Fig. 754.—Plate to receive Pivot.

or brass, round up the pivots, and rivet them into the drilled piece.

Washstand Top, etc.—The tiles at the back of the washstand top fit between two rails in the carcase frame, and are fixed against a small angle moulding mitered in and pinned at the front, and by means of strips pinned behind the tiles. The washstand top and side fences are of marble (white or coloured) fixed together with screws and plaster-of-Paris. Fillets screwed to the wardrobe end and back rail support the marble top (see section on B B, Fig. 747). For economy a wooden top and fence could readily be adapted. A cornice moulding, 1¾ in. by ¾ in., surmounts the mirror portion; groove it to fit the top rail (see Fig. 752), and return the outer end, the inside being

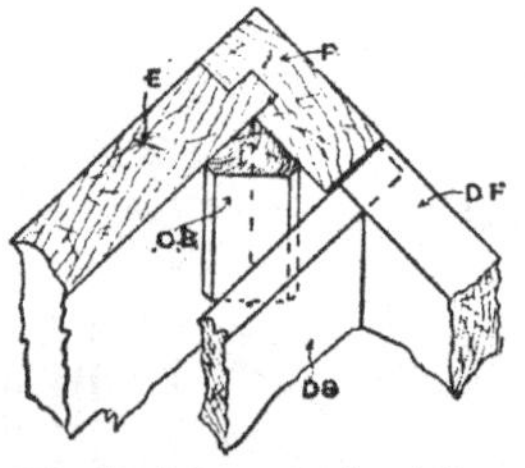

Fig. 755.—Front Corner Joint and Drawer of Combined Fitment.

cut square to the wardrobe. A nosing 1 in. by ¼ in., glued and pinned on the vertical edge of the framing above the marble top, gives it a finished appearance (see Figs. 751 and 752). The low carcase end is ⅞ in. solid, and to it the towel rail can be fixed with wooden blocks, shaped and glued on, or fancy brass brackets. If wooden blocks are used, they should be recessed to receive the ends of the towel rail. When they are glued to the carcase, drive a screw into them from the inner side. The joint with the end and front corner stile is tongued and grooved, and when it is glued and nailed together, a number of small blocks should be glued in the angle to support it; see Fig. 755, in which E is the carcase end, F the front corner stile, D F the drawer front, D S the drawer side, and C B the glued corner blocks. Putting together the various parts of the

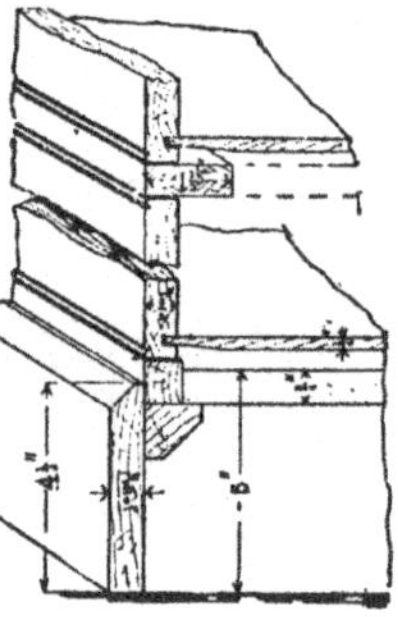

Fig. 756.—Drawers and Plinth of Combined Fitment.

carcase should leave a flush surface at the front (with the exception of the projecting tongue on the right-hand corner), on which the whole of the front framing for the drawers and cupboard door can be fixed. In framing this part together, the bottom rail projects ¼ in. above the flooring, and shows ½ in. above the top edge of the plinth or skirting (see Fig. 756). The hanging stile of the small cupboard door is cut away at the top left-hand edge to admit the wardrobe door in folding, consequently a corner block must be glued inside, in the angle, down to the point at which the stile again covers the division (see Fig. 757, and also the dotted line in Fig. 749). In Fig. 757, C B is the corner block, D the end division of the wardrobe, W D the wardrobe door, and S the hanging stile of the small cupboard door. Below, where the portion is cut away, the stile is fixed to the division by screws driven from the inner side, and it is advisable to leave out the final fixing of the back until the front is finished. The drawer fronts are prepared from stuff $\frac{5}{8}$ in. thick, the sides from $\frac{1}{2}$-in. stuff, and the bottoms from $\frac{3}{8}$-in. stuff. They are dovetailed and grooved together, and the front top and bottom edges are relieved with a small bead. Knobs or handles can be fitted, and locks added if desired. The drawers move on runners let into the framing and grooved into the adjoining sides.

Cupboard Door, Partitions, etc.—The cupboard door C D (Fig. 757) has $1\frac{1}{2}$-in. by $\frac{7}{8}$-in. stiles and top rail, and 2-in. by $\frac{7}{8}$-in. bottom rail; these are mortised and tenoned and grooved for a panel $\frac{1}{4}$ in. thick. A moulding is planted on the front as a finish to the panel (see Fig. 757); inside, the door should be left square. Hang it with $1\frac{1}{2}$-in. brass butts, and also fix a knob and catch. The cupboard partitions are formed on one side by the wardrobe end, and on the other by boards nailed to a fillet screwed to the door-closing stile or jamb, and another batten is screwed to the carcase backing. If this batten is recessed into the floor and framing rails, a stronger job will result. This batten is also used to take the ends of the drawer runners, as the cupboard division is not strong enough to carry them, and if fixing it to the back only, mark it in place. Try it in position, and screw it up before raising the back, the drawer runners meanwhile being held by a waste strip. If a wooden washstand top is used, it could be grooved along the bottom to receive the division boards fixed in vertically. In fixing the various parts, the front, bottom, and divisions should be put together, and the part of the back framing carrying the mirror screwed to the wardrobe division as shown in Fig. 758. The back of the wardrobe can then be screwed to the rebate and ends. The framing of this part being $\frac{3}{4}$ in. thick, a $\frac{1}{8}$-in. strip must be nailed to make up the difference at the bottom, compared with the rest of the framing, which is $\frac{7}{8}$ in.; or the rebate can be cut $\frac{7}{8}$ in. deep, and may show a $\frac{1}{8}$-in. projection above the framing.

Plinth.—The plinth is of $4\frac{1}{2}$-in. by $\frac{3}{4}$-in. stuff, with chamfered top edge, mitered at the corners to return the ends and nailed to the carcase. Along the front a number of small blocks are glued in the angle to support it, as shown in Fig. 756.

Cornice Moulding.—The cornice moulding is 3 in. deep; a suitable section can be obtained ready made. This is built on a framework of $2\frac{1}{2}$-in. by $\frac{5}{8}$-in. stuff (see Fig. 750), and is kept in position on the wardrobe by four small blocks glued at the angles and ends. It must not be fixed to the wardrobe, but should be left detachable for moving purposes. On a level with the lower cornice a shelf 6 in. wide is placed, and the near end is supported on a bracket, which may be of wood or brass. At the opposite end it is lapped into the top part of the cornice moulding. A small angle moulding is fixed in the joint at the wardrobe end and back framing, as shown in Fig. 758.

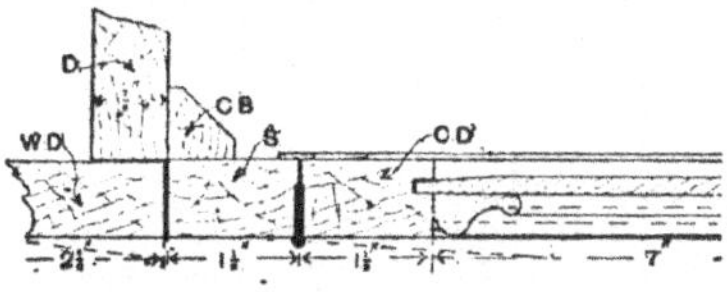

Fig. 757.—Section of Cupboard Hanging Stile and Door of Combined Fitment.

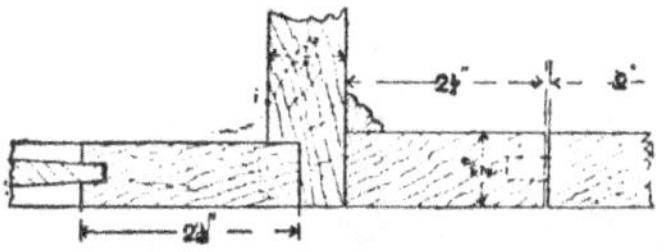

Fig. 758.—Section of Carcase Back and Jointing of Framing with Wardrobe Division.

Shaving Pedestal.

The shaving pedestal illustrated by Figs. 759 to 773 looks effective if constructed of light wood, such as Hungarian ash or light walnut; dry white pine, stained and polished

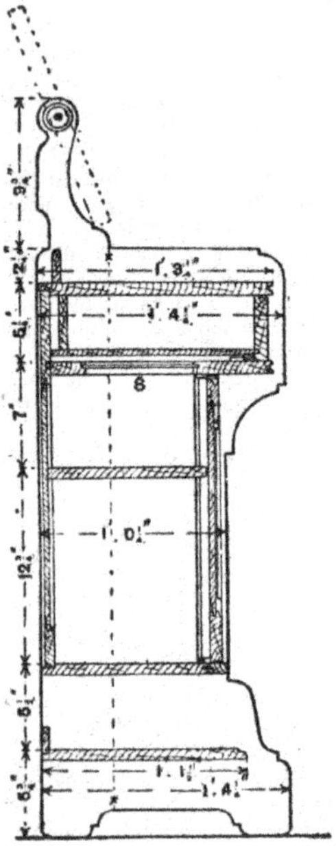

Fig. 759.

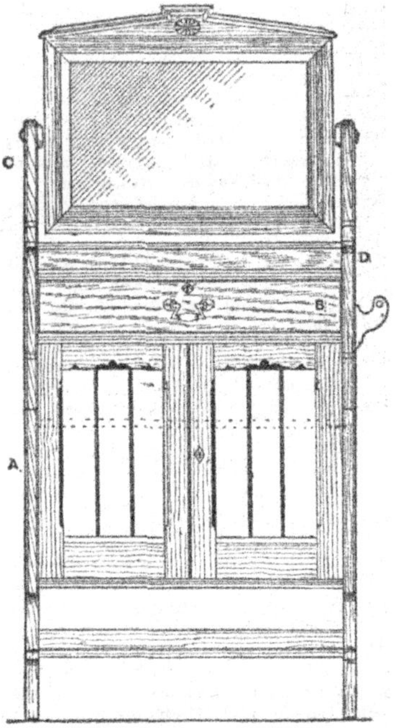

Fig. 760.

Figs. 759 to 761.—Vertical Section, Front Elevation, and Horizontal Part Sections of Shaving Pedestal.

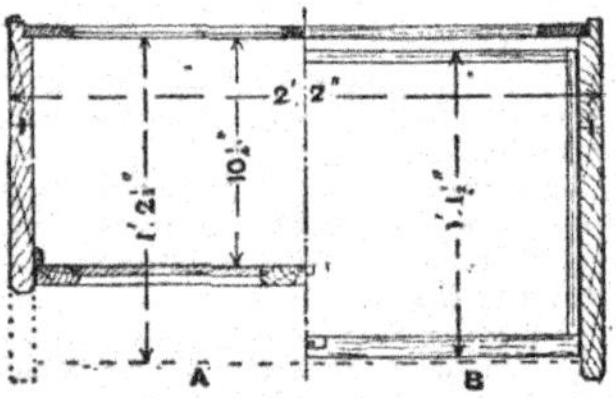

Fig. 761.

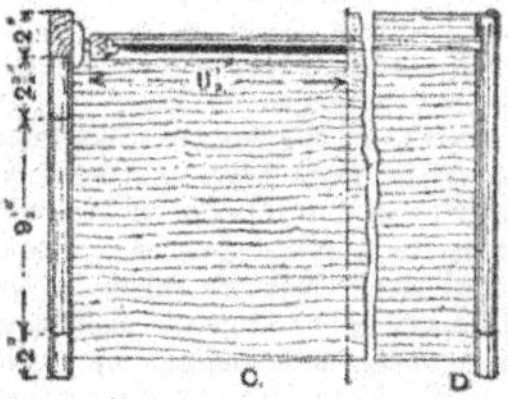

Fig. 762.—Part Horizontal Section of Shaving Pedestal.

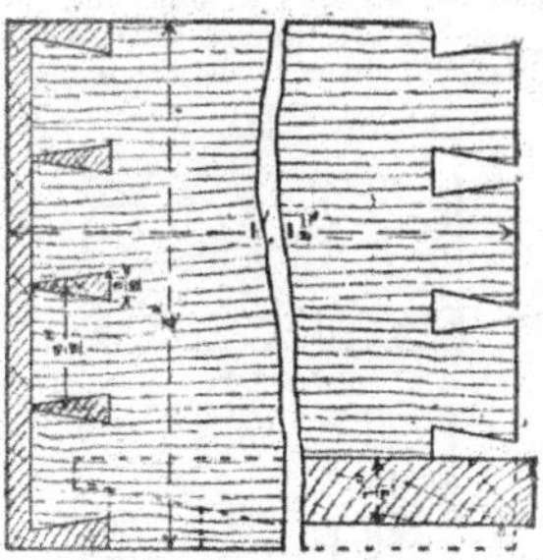

Fig. 763.—Details of Shaving Pedestal Drawer.

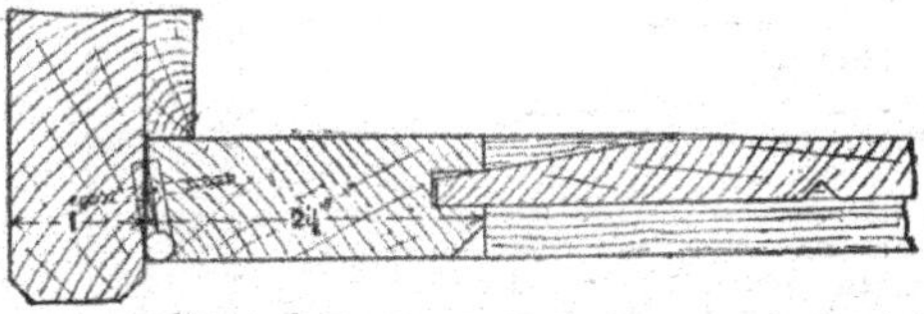

Fig. 764.—Section of Hanging Stile of Shaving Pedestal's Door.

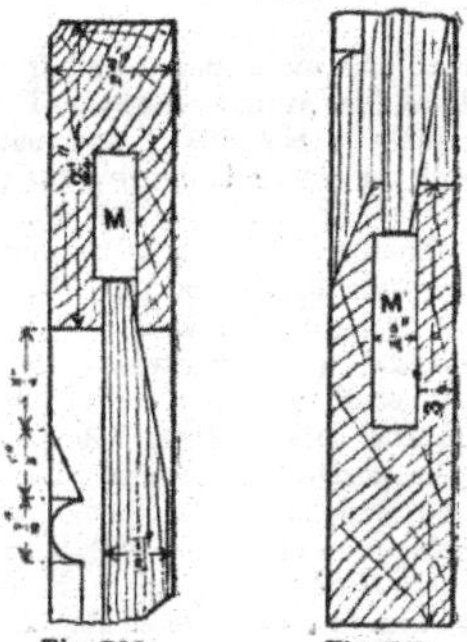

Fig. 766. Fig. 767.

Figs. 766 and 767.—Sections through Top and Bottom Rails of Shaving Pedestal Door.

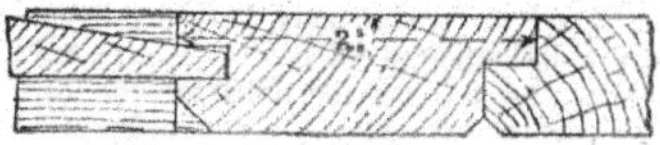

Fig. 765.—Section of Meeting Stiles of Doors of Shaving Pedestal.

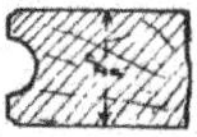

Fig. 768.—Section of Top of Shaving Pedestal.

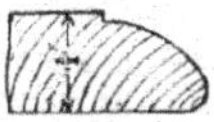

Fig. 769.—Section of Lower Shelf of Shaving Pedestal.

or enamelled, is less expensive. The pedestal consists of two solid shaped ends, a solid and moulded top, framed soffit, a drawer, two panelled doors, enclosing a cupboard

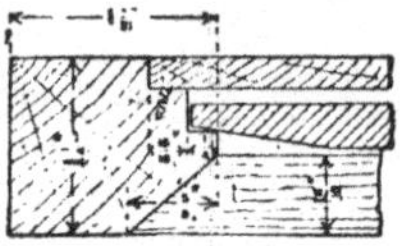

Fig. 770.

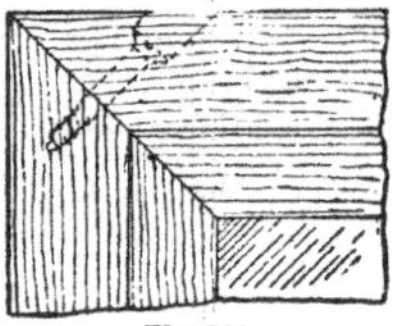

Fig. 771.

Figs. 770 and 771.—Section and Elevation of Corner of Shaving Pedestal's Mirror

containing a deal shelf, a framed and panelled back, a swing mirror, and guard rail. Prepare the two ends in the solid to the shape shown in Fig. 759. The mirror elbows may be economically produced by jointing on the dotted lines x x (Fig. 759), the plough groove then necessary being stopped at the upper end. If this method is adopted the elbow may be worked and the top end squared off before jointing up; all the lines for the housings for the top, soffit, and shelves should be struck across on the worse side of the stuff in pairs, keeping the best edges to the front. Mark

Fig. 772.—Section of Capping of Shaving Pedestal's Mirror.

all the housings to the width of the thinnest part of the particular piece going in, and stop these ½ in. back from the front edge. Gauge the checks for the back and the housing for the upper and lower guard rails from the front edge. The housings should be sunk $\frac{3}{16}$ in. deep. Prepare the top, soffit, and shelves of equal lengths, and to the widths shown in Fig. 759. Prepare their ends to fit the grooves, then mould and shoulder the front edges and work a small chamfer on each front edge of the ends as far down as the moulded foot at the bottom of the cupboard. Next fit together, clean up the face sides of the top and shelves and the outsides of the ends, and glue up, leaving the clamps or cleats on until dry. Then turn the carcase upside down and skew-brad

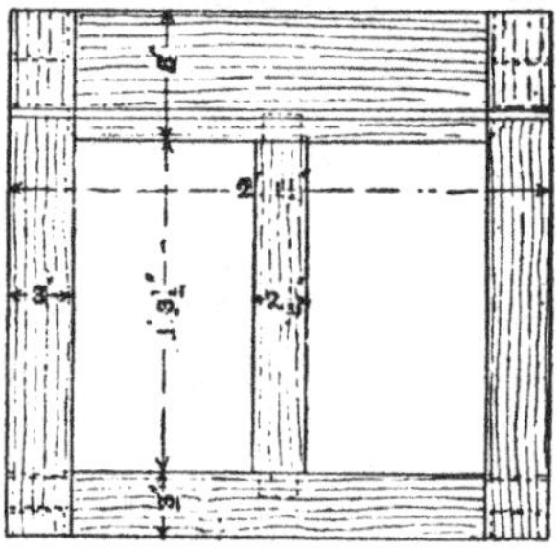

Fig. 773.—Framed Back of Shaving Pedestal.

through all the under sides, and frame up the back, of ¾-in. stuff, as shown in Fig. 773. Fit it tightly in the checks, and mark the position of the groove to receive the tongue on the soffit s (Fig. 759), which is fixed by screws. Then fix the door stops and clean off the outside of the carcase. The back should fit in tightly, as the rigidity of the case depends on this. Prepare the doors from 1-in. stuff, stopping the chamfers as shown in the details (Figs. 766 and 767), M being the mortices in the rails. The tenons should be wedged in the usual manner. The panels are in one piece, with false joints made with a V or rebate plane. Rebate the meeting stiles in the centre as shown in Fig. 765. Fit the doors in with a joint that will just take a piece of stout notepaper all round and hang ⅜ in. below the face, and with all the knuckle of the butt in the door, as shown in Fig. 764. Use a pair of 2-in. brass butts and a $2\frac{1}{2}$-in. neck bolt.

Drawer.—Fig. 763 is a side view of the drawer, showing size and spacing of the dovetails. The front ends are spaced $\frac{7}{8}$ in. apart, made small and lapped in front; the back ones are larger and cut through; they also finish flush with the bottom. The left half of Fig. 763 is an outside view, and in the right half the side and back are supposed to be removed. Glue a fillet round the two sides and the front (see Fig. 763). This should be of hardwood and be glued to the drawer sides only, not to the bottom. The grain of the bottom must run parallel with the drawer front. The extra depths of grooves in the front and the screw slot at the back are to allow for the shrinkage. Fit the drawer in the case $\frac{5}{16}$ in. below the flush of the top, and glue stops on the soffit with the end grain as shown in Fig. 759.

Fig. 774.—Toilet Glass.

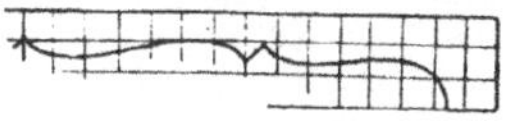

Fig. 775.—Outline of Underframe of Toilet Glass.

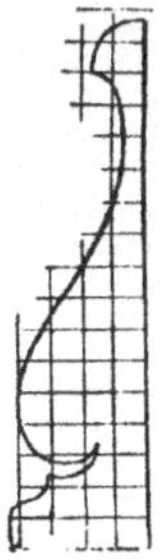

Fig. 776.—Outline of Side Standard of Toilet Glass.

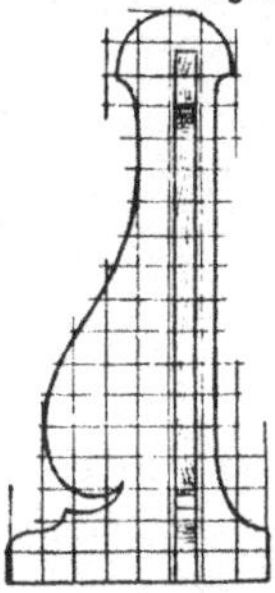

Fig. 777.—Outline of Front Standard of Toilet Glass.

Fig. 778.—Side View of Pediment and Moulding of Toilet Glass.

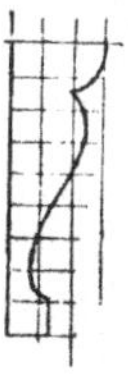

Fig. 779.—Outline of Pediment of Toilet Glass.

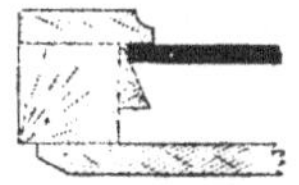

Fig. 780.—Part Section of Toilet Glass Frame.

Mirror.—A corner of the mirror frame is shown in section and elevation by Figs. 770 and 771. It is double checked $\frac{1}{4}$ in. for the glass and $\frac{5}{16}$ in. for the back; the checking is so arranged that there is a space between the glass and wood. A deep chamfer is worked round the inside of the frame; a plain mitre joint is used at the corners, secured with a $1\frac{1}{4}$-in. screw as shown in Fig. 771. The screw holes should be in the top and bottom pieces and pelleted. When the frame is made, the shaped panel can be glued in the centre, and the mirror capping, shown in section by Fig. 772, mitered round and fixed by means of dowels, shown by dotted lines. The mirror may be hung in the old-fashioned pivots or pegs of hardwood working in holes in the disc shown in Fig. 759. A bracket towel-rail of hexagonal section, screwed on the right-hand side, will complete the fitment. Figs. 759, 760, 761, 762, and 773 are drawn to a scale of 1 in. to 1 ft., and Figs. 763 to 772 half full size. A (Fig. 761) shows the section at A (Fig. 760), and B (Fig. 761) the section at B (Fig. 760). Similarly C (Fig. 762) is part horizontal section at C (Fig. 760), and D (Fig. 762) is part section at D (Fig. 760).

Light Toilet Glass.

In many toilet glasses the standards are made of extra stout wood, but in the glass shown by Fig. 774 the thickest stuff measures 1 in., and the rest is only ⅝ in. The wood suggested is pine; but, of course, hardwood, such as mahogany, ash, or walnut, will look well. The shaped under-framing A, of ⅝-in. stuff, is 2 ft. 6 in. long, and measures 9 in. from front to back; it is dovetailed at the corners. Half an elevation is given in Fig. 775. The top B is ½ in. bigger all round than the under-framing, and is of ⅝-in. stuff, with an ovolo moulding worked on its edge. As an alternative, a bevelled edge would look well. The top, if painted, is fastened to the under-frame with glue and nails. The standards (Figs. 776 and 777) are of ⅝-in. stuff nailed and glued at the sides, and then screwed from the under side of the top. Their shapes may be drawn by copying the illustrations, but using 1-in. squares. Fig. 778 illustrates the pediment, which should similarly be set out to Fig. 779. The mirror frame is 1 ft. 7⅝ in. by 1 ft. 3⅝ in., sight inside measure, this allowing for a 1-ft. 8-in. by 1-ft. 4-in. silvered plate glass. The facing on the frame (see Fig. 780) is 1¼ in. wide, and the frame underneath is 1 in. wide and deep. The 1-in. pediment moulding (Fig. 778) is mitered round the frame, and the pediment is of ⅝-in. stuff. A thin backboard (Figs. 778 and 780) is screwed on the back of the glass frame, which is supported by brass screw centres fixed to the frame and standards. These should be fitted and fixed before the standards are screwed to the top, there being then no fear of the standards being too wide apart or too close.

WRITING TABLES AND OFFICE FURNITURE.

Pedestal Writing-desk.

FIG. 781 shows an ordinary pedestal writing-desk with drawers. Figs. 782 and 783 are elevations, while Fig. 784 is a cross section. For the carcase, each of the four sides should be made by jointing together three 1-in. boards, as indicated planed to their breadths; afterwards the sides and backs should be ploughed and tongued together, these joints being plainly illustrated by Figs. 785 and 786. The four sides should then be set out for the housings, which extend from the back edge to within ½ in. of the front edge (see Fig. 787). It will be an advantage to place all the sides together

Fig. 781.—Pedestal Writing-desk.

in Fig. 783; a much stronger job results if the joints are ploughed and cross-tongued and then glued. The two backs to the pedestals should be similarly formed, as shown by Figs. 785 and 786. The different rails forming the horizontal divisions between the drawers should be prepared from 1-in. stuff finishing about ⅞ in. thick. The four sides and the backs of the pedestals may be trued to a thickness of about ⅞ in. and then and square down the front edge with a square and pencil line to ensure all the corresponding divisions being alike. To make the housings, the simplest way is to cut with the mallet and chisel a portion, A (Fig. 788), taking care to pare exactly to the lines; this makes an entrance for the tenon saw, which can then be used to cut along the lines B C and D E. The waste may be removed with a chisel, and the housings made to an even depth by means of a router, as in Fig. 788.

Front Rails, etc.—The front rails must be accurately cut off to length just between the housings, and then pieces notched out to form the shoulders, as illustrated at Fig. 789; these front rails and also those fitting into the housings will require tenoning together and ploughing to receive the divisional panels as indicated. To prevent shrinkage, cut small notches in the back, as shown at Fig. 787, so that the ends of the rails may enter; or the rails may be kept a little short. The rail along the front immediately under the top must be dovetailed at each end into the two inner sides

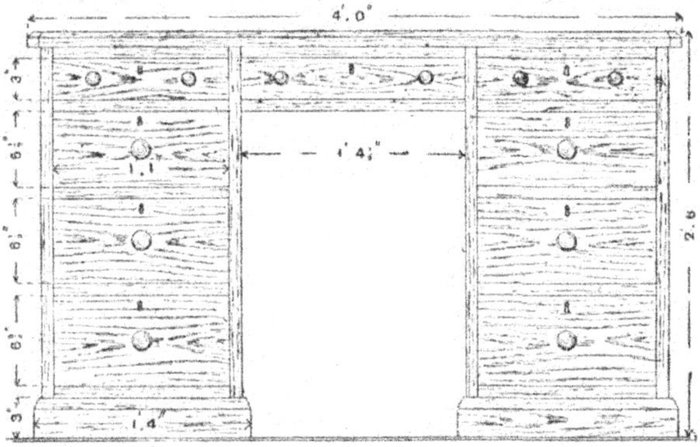

Fig. 782.—Front Elevation of Pedestal Writing-desk.

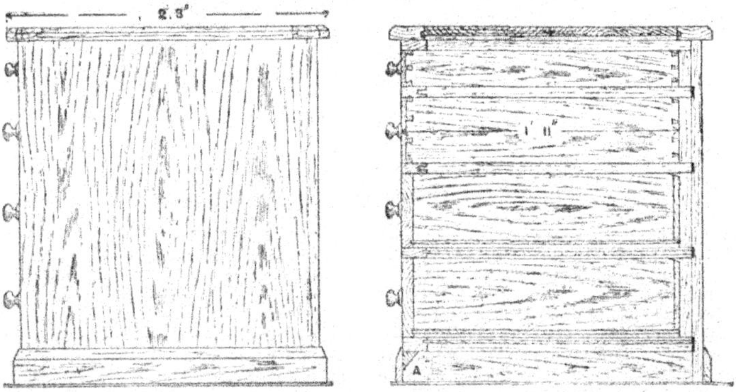

Fig. 783. Fig. 784.

Figs. 783 and 784.—End Elevation and Cross Section (through Drawers) of Pedestal Writing-desk.

as shown at Fig. 790, and the divisional panels should next be jointed, the best way being to shoot the edges on a shooting board, after which they should be glued together. The grain of these panels runs lengthwise of the table so that the end grain fits into the plough grooves of the rails, as illustrated at Fig. 786. When the glue is dry, these panels should be smoothed off on each side and cut to length and breadth and then mulleted; that is, the edges are bevelled to fit in the plough grooves. The whole of the carcase may next be fitted together and glued up, this of course being done in sections. The pieces for the plinth can be sawn out and planed to thick-

Fig. 785.

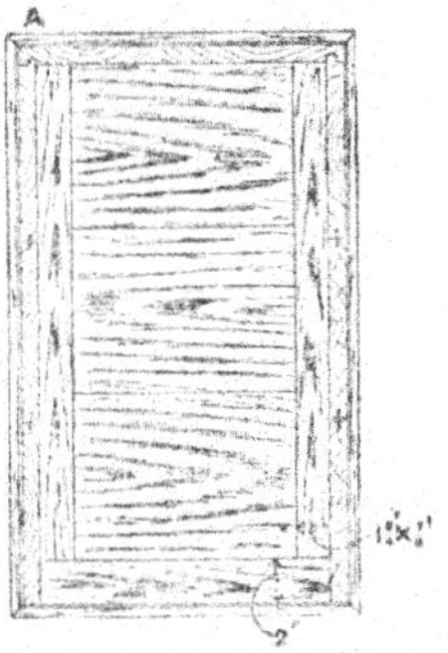

Fig. 786.

Figs. 785 and 786.—Horizontal Sections of Pedestal Desk, with and without Drawer.

Fig. 787.—View of End of Pedestal Desk.

ness and breadth, and then chamfered and mitered, each to its proper length, and fixed by gluing to the sides and back; they may be further secured by a few nails or screws driven from the inside of the backs and plough them for cross tongues, and glue them up. The outer edges of the top are formed by two stiles and two rails, these being stub-tenoned, haunched, and mortised together, the inner edges of the

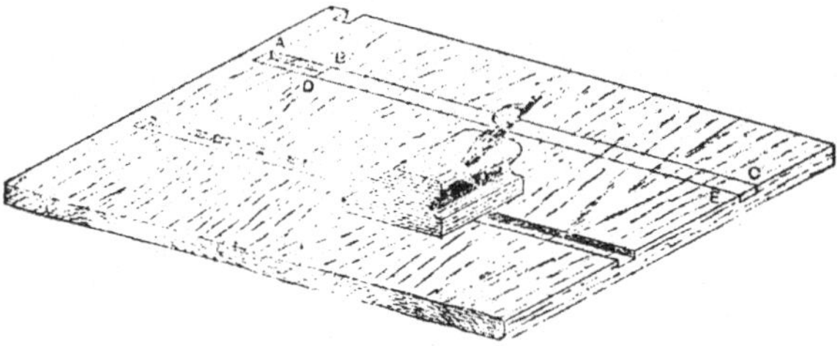

Fig. 788.—Method of making Housings in Side of Desk.

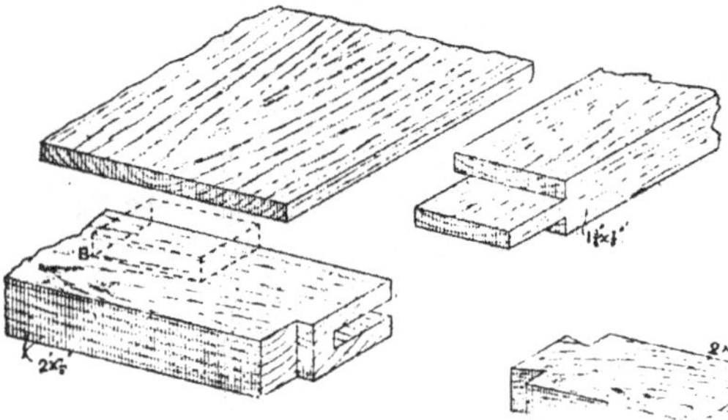

Fig. 789.—Jointing Horizontal Divisions of Pedestal Desk.

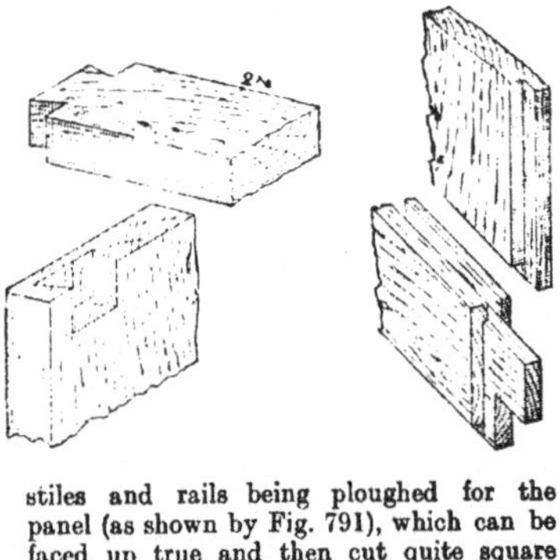

Fig. 790.—Joints of Top Rail with Ends of Pedestal Desk.

sides. To the front lower rails the plinth can be fixed by a few triangular blocks A (Fig. 784) glued on.

Top of Desk.—For the panels of the top, obtain two perfectly seasoned pieces of pine 11 in. wide and $\frac{7}{8}$ in. thick; joint and stiles and rails being ploughed for the panel (as shown by Fig. 791), which can be faced up true and then cut quite square to length and breadth; then the rebate should be set out accurately to the distances between the stiles and rails. The top surface of the panel should be about $\frac{1}{16}$ in. below the stiles and rails when finished, but at this stage a little more should be allowed. The rebate may be made with a side fillister or a rebate plane. The under side of the top towards the ends and edges should be placed so that the tongued part may just fit into the plough grooves. The framing and panel must be fitted together, and the panel should be narrowed nearly $\frac{1}{8}$ in., so that the shoulders of the frame can be cramped up tight. When this is found to be correct, the joints of the frame should be well glued (without gluing any part of the tongues of the panel or the grooves), and held together with a couple of cramps until the glue is dry. If iron cramps are not to hand, two strips of wood with blocks nailed on and pairs of wedges (Fig. 792) will answer the purpose. When the glue is dry, the top of the frame should be planed off true, to project $\frac{1}{16}$ in. or a little less above the panel. Of course, the exact distance will depend on the thickness of the leather or American cloth with which the top is to be covered. The edges can next be trued up, and then chamfered. The top is secured to the sides, ends, back, etc., by gluing on blocks, which if carefully done will make a strong job.

Drawers.—The various pieces for the fronts, sides, backs, and bottoms of the drawers can next be sawn and planed to thickness and breadth. The bottoms should be made from two pieces of 11-in. stuff jointed and glued as indicated at Fig. 785, from which it will be seen that the grain of the bottoms runs crosswise. The sides and backs may be cut off and planed to length; then the fronts must be carefully fitted in by planing the edges and ends, all the sides and fronts being ploughed for the edges of the bottoms. The pins for the lap-dovetailing at each end of the fronts should be set out, and made by cutting with a dovetail saw just by the side of the line in the waste, and removing the waste with a mallet and chisel to leave the pins and sockets finished as shown at A (Fig. 793). The pins B to the backs must be set out and made, the pins of the fronts and backs being next marked on to the sides to obtain the shape of the sockets; these should be carefully sawn in the waste just inside the

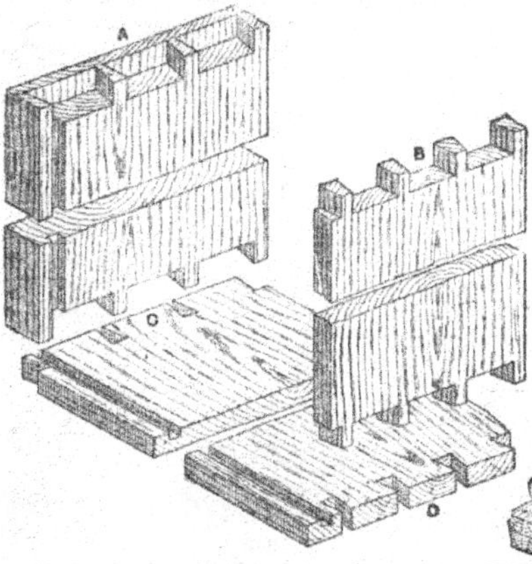

Fig. 793.—Joints of Drawers of Pedestal Desk.

Fig. 791.—Method of Jointing Top Rail to Panel of Desk.

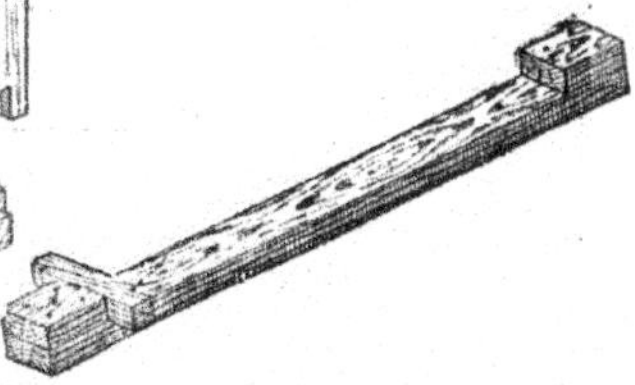

Fig. 792.—Improvised Cramp for Top of Desk.

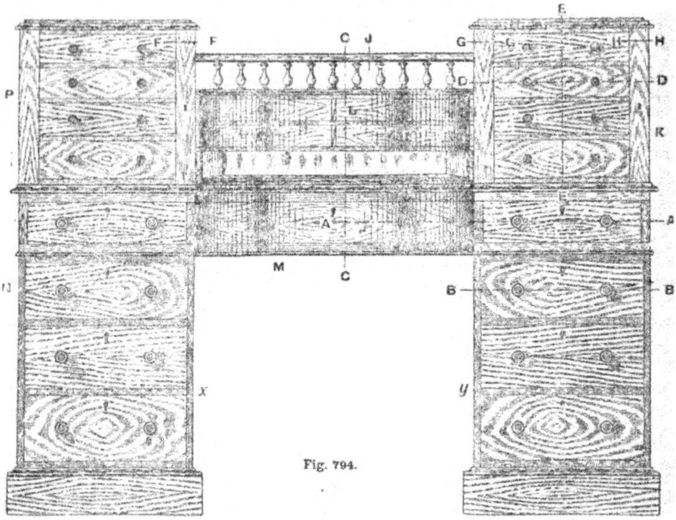

Fig. 794.

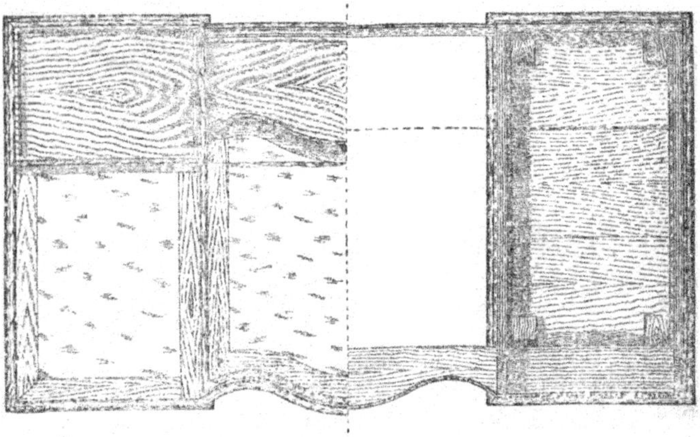

Fig. 795.

Figs. 794 and 795.—Front Elevation and Half Plan and Horizontal Section (on Line A A) of Registered Pedestal Desk.

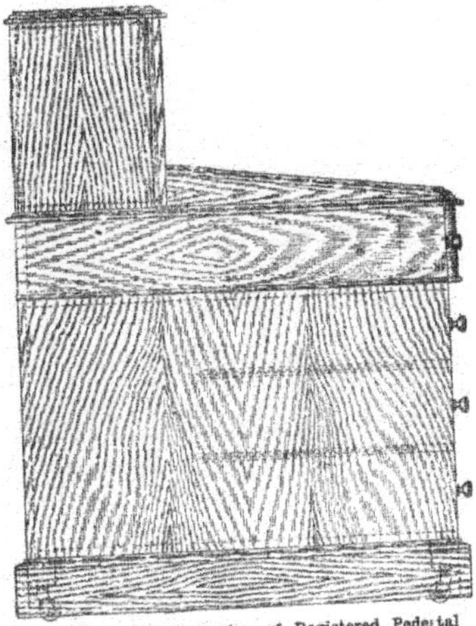

Fig. 796.—Side Elevation of Registered Pedestal Desk.

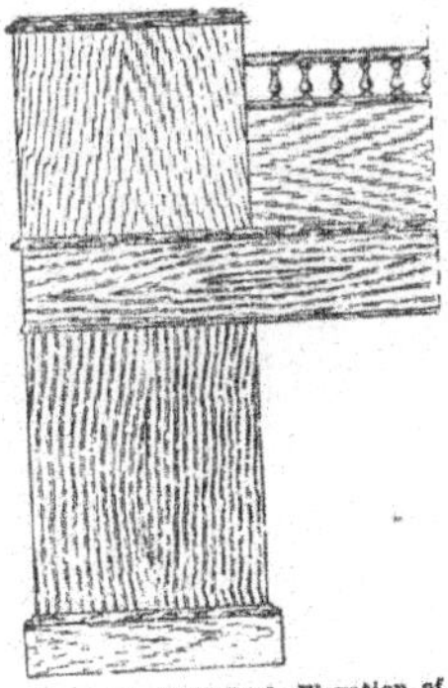

Fig. 797.—Half Back Elevation of Registered Pedestal Desk.

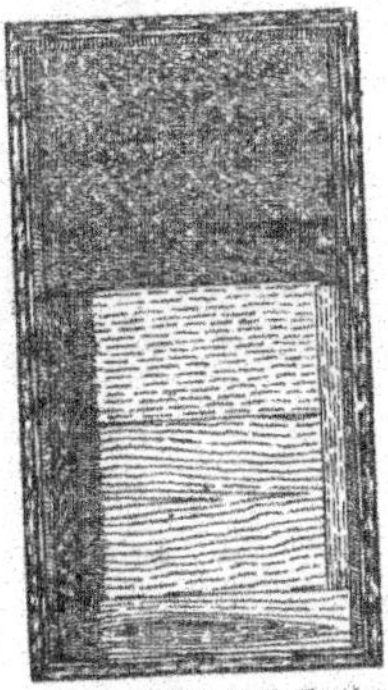

Fig. 799.—Horizontal Section of Registered Pedestal Desk on Line B B (Fig. 794).

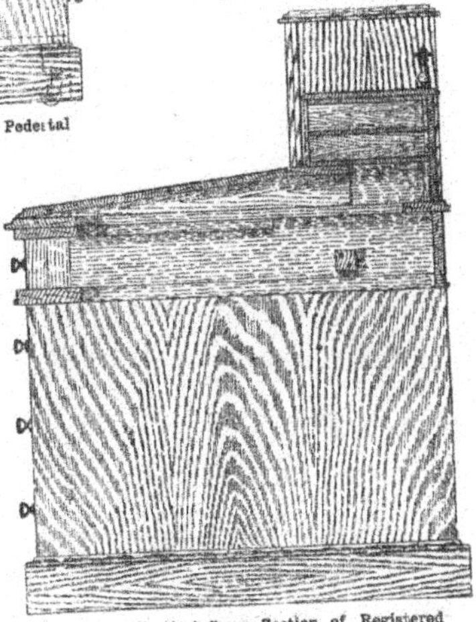

Fig. 798.—Vertical Cross Section of Registered Pedestal Desk, with Drawers removed, on Line C C (Fig. 794).

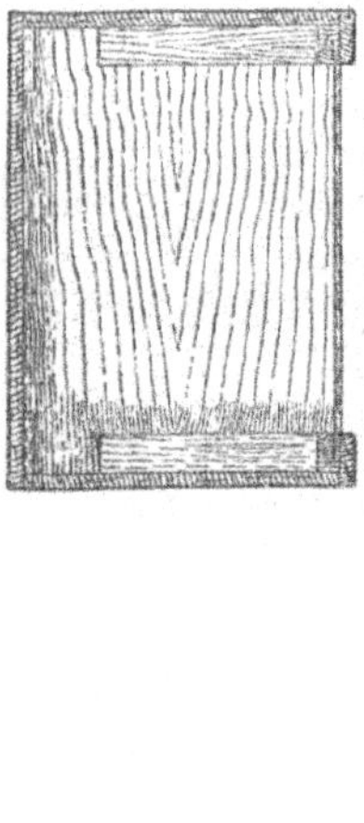

Fig. 800.—Horizontal Section of Right-hand Pedestal of Desk on Line D D (Fig. 794).

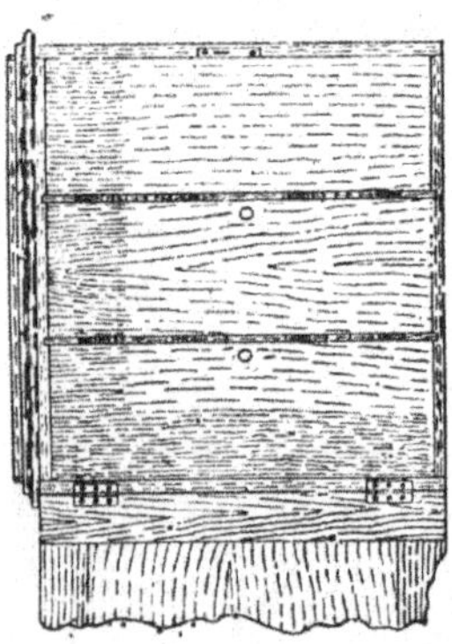

Fig. 804.—Front Elevation of Left-hand Pedestal of Desk, with Door open.

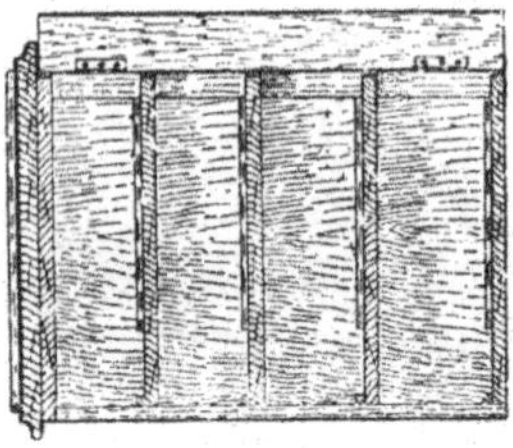

Fig. 801.—Vertical Cross Section of Right-hand Pedestal of Desk on Line E E (Fig. 794).

Fig. 802.—Horizontal Section through fixed Pilaster of Right-hand Pedestal on Line G G (Fig. 794).

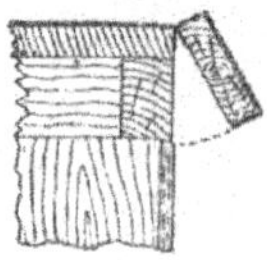

Fig. 803.—Horizontal Section through Locking Stile or Hinged Pilaster of Right-hand Pedestal on Line H H (Fig. 794).

Fig. 805.—Horizontal Section through Pilaster fixed to Door of Left-hand Pedestal on Line F F (Fig. 794).

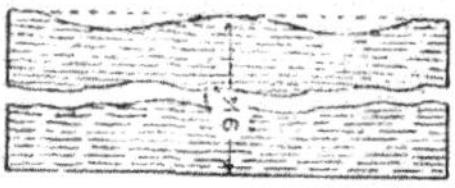

Fig. 806.—Partition for Left-hand Pedestal Cupboard (see O, Fig. 804).

Fig. 807.—Section of Rail over Spindles of Desk on Line J (Fig. 794).

Fig. 813.

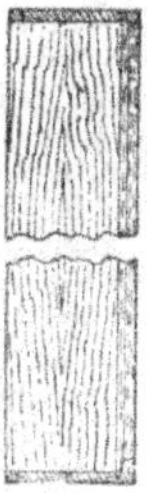

Fig. 814.

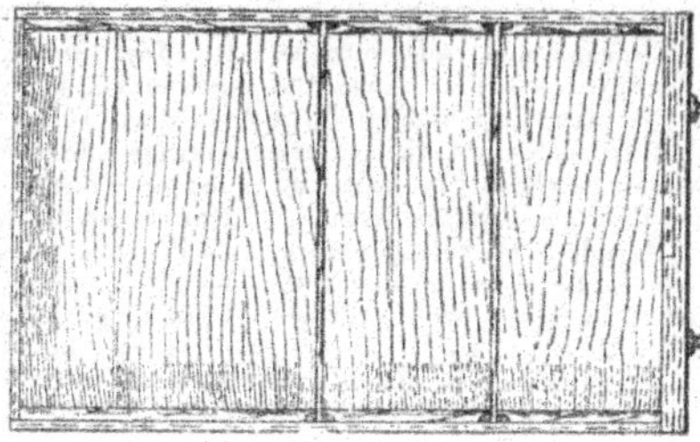

Fig. 812.

Figs. 812 to 814.—Plan, Side Elevation, and Back Elevation of Desk's Side Drawer (see N, Fig. 794).

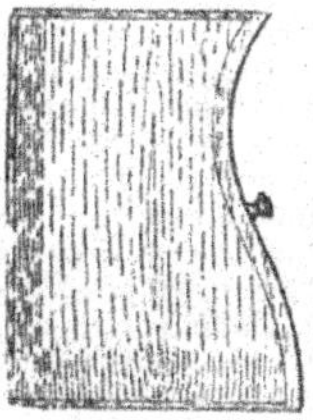

Fig. 811.—Plan of Small Drawer in Middle Pedestal of Desk (see L, Fig. 794).

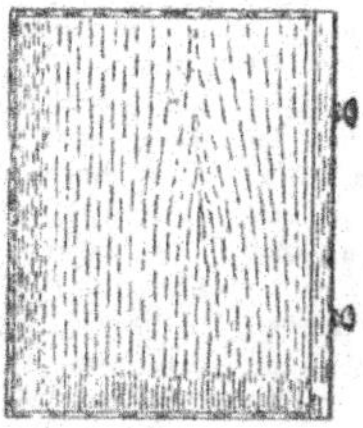

Fig. 815.

Fig. 816.

Fig. 817.

Figs. 815 to 817.—Plan, Side Elevation, and Section of Drawer in Right-hand Pedestal of Desk (see K, Fig. 794).

Fig. 809.

Fig. 810.

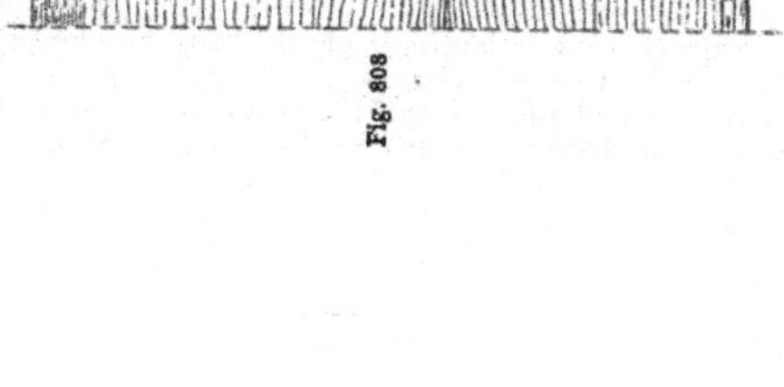

Fig. 808

Figs. 808 to 810.—Half Plan, Side Elevation, and Section of Desk's Middle Drawer (see M, Fig. 794).

lines, and cut out with a chisel, as at C and D. The dovetail joints will require gluing together, but the sides and ends must be quite square with each other. Then the bottoms must be planed up and trued to size so as just to slide into the plough grooves. Some strips about $\frac{3}{4}$ in. wide, and a little thicker than the distance from the bottom edge of the side to the plough groove, should be prepared, these having saw kerfs in them about 3 in. apart, so that when they are glued on to the sides and bottom they will bed better. When the glue is dry, the sides of the drawers should be smoothed off and the drawers then carefully fitted into place. To prevent their being pushed in too far, stops about $\frac{1}{4}$ in. thick and 2 in. by $1\frac{1}{4}$ in. should be glued and further secured with a couple of small sprigs to the front rails, one (B) being indicated in Fig. 789 by dotted lines. These stops catch against the bottom inner edge of the front of the drawers, and two of them will be required for each drawer, each being fixed about 2 in. from the ends of the rails. To make the drawer fronts and rails quite flush, all the drawers should be pushed into place, all the stops then butting against the fronts.

Completing Desk.—Then the carcase, with drawers in, should be placed back downwards, and all the fronts carefully planed off flush with the rails, the plane being set fine so as not to split off any of the edges. Finally, the leather for the top should be cut accurately to size, and then the wooden panel quickly covered with an even layer of glue. This should be done in a warm place, and quickly, the glue being made rather thin; the leather must be laid in position, and well rubbed down, working from the middle to the edges so as to rub out any superfluous glue. Any glue on the frame can be removed with a cloth dipped in hot water. After this process it may be necessary to take off a few shavings from the top so as to bring it flush with the cloth or leather; this should be done with a smoothing plane set very fine.

Material Required.—The quantities of material are as follows:—Vertical casing, 9 in. by $\frac{7}{8}$ in. by 40 ft.; plinth, 3 in. by $\frac{3}{4}$ in. by 15 ft. Horizontal divisions: front rails, 2 in. by $\frac{7}{8}$ in. by 10 ft.; stiles, $1\frac{3}{4}$ in. by $\frac{7}{8}$ in. by 40 ft.; and panels, $7\frac{1}{2}$ in. by $\frac{1}{4}$ in. by 27 ft. Top: frame, 3 in. by $1\frac{1}{16}$ in. by 13 ft.; and panel, 11 in. by $\frac{3}{4}$ in. by $7\frac{1}{2}$ ft. Drawers: bottoms, 11 in. by $\frac{3}{8}$ in. by 19 ft. Top drawers: fronts, 3 in. by $\frac{7}{8}$ in. by $3\frac{3}{4}$ ft.; and sides and backs, 3 in. by $\frac{1}{2}$ in. by 16 ft. Next to top drawers: front, $5\frac{1}{2}$ in. by $\frac{7}{8}$ in. by $2\frac{1}{3}$ ft.; and sides and backs, $5\frac{1}{2}$ in. by $\frac{1}{2}$ in. by $10\frac{1}{3}$ ft. Next to bottom drawers: fronts, $6\frac{1}{4}$ in. by $\frac{7}{8}$ in. by $2\frac{1}{3}$ ft.; and sides and backs, $6\frac{1}{4}$ in. by $\frac{1}{2}$ in. by $10\frac{1}{3}$ ft. Bottom drawers: front, $6\frac{3}{4}$ in. by $\frac{7}{8}$ in. by $2\frac{1}{3}$ ft.; and sides and backs, $6\frac{3}{4}$ in. by $\frac{1}{2}$ in. by $10\frac{1}{3}$ ft.

Registered Pedestal Desk.

Full working drawings for a registered pedestal desk with curved front will now be presented. Fig. 794 is a front elevation, Fig. 795 a half plan and section (on line A A, Fig. 794, with drawer removed), and Fig. 796 a side elevation. Fig. 797 is a half back elevation (to a smaller scale), and Fig. 798 is a vertical cross section on line C C (Fig. 794), with drawers removed. Fig. 799 is a horizontal section on the line B B (Fig. 794), also with drawer removed. In explanation of the pedestal of small drawers above the desk top to the right, attention is directed to the section on lines D D, E E, G G, and H H (Fig. 794), shown by Figs. 800, 801, 802, and 803 respectively. The locking stile is clearly shown in the last-mentioned illustration. To the left of the desk, the pedestal takes the form of a three-compartment cupboard, details of which are given in Figs. 804, 805, and 806. The rail over the spindles at the back of the desk over the four small drawers with curved fronts is of the section shown by Fig. 807. Of the drawers, Figs. 808 to 810 show that at M (Fig. 794); Fig. 811 is a plan of that at L (Fig. 794); Figs. 812 to 814 show that at N (Fig. 794); and Figs. 815 to 817 show those at K (Fig. 794). Figs. 794 to 796 and 798 are produced to a scale of approximately 1 in. = 1 ft. Fig. 799 and all the detail views of drawers are to a scale of approximately $1\frac{1}{2}$ in. = 1 ft. An accurate scale for the chief views can be constructed by noting that the distance in the clear from x to y (Fig. 794)—between the two lower pedestals—is exactly 24 in., and for all the other views by noting that the drawer M (Fig. 794), shown separately by Figs. 808 to 810, is exactly 24 in. wide over all. In description of the desk it may be said that the left-hand portion of the upper part of the desk has a dummy drawer front to correspond with the other side, which contains a nest of four drawers. The difference between the two sides will be readily seen if Figs. 801 and 804 are compared. The arrangement of the right-hand upper part of the desk is as follows :—The left-hand pilaster is fixed, while the right-hand pilaster is locked, overlapping the drawer fronts sufficiently to prevent them from being withdrawn. The locking is accomplished by means of a lock in the right-hand pilaster. The lock fastens into a projecting staple in the stile behind, the staple entering a recess in the pilaster containing the lock. The fronts of the central drawers are built to the shape illustrated in the various figures, and veneered, no bent wood being required. The leather can be fixed to the table top with thin glue or shoemaker's paste. To prevent buckling, first brush over the back of the covering a thin solution of the adhesive, the thicker preparation being spread over the wood, and the leather while still damp being well pressed into position. A more pleasing finish may be given by the aid of gilt-edge chain banding; this consists of leather, ½ in. wide, gilt on the outside edges. The major portion of the leather is laid as usual, but, whilst still damp, a strip of ½-in. wood, corresponding with the width of the gilt bandings, is laid along the outer edge; a sharp knife is drawn along this, the surplus leather withdrawn, and the gilt banding inserted in its place; the corners of the banding are mitered.

Registered Pedestal Desk with Side Cupboard.

A conventional view of another registered pedestal desk is given by Fig. 818. There is a cupboard in the lower pedestal on the right, it will be noted, the cupboard door containing a carved panel. A front elevation is shown by Fig. 819 (scale, 1 in. = 1 ft.). The left-hand pedestal is clearly shown in the vertical section (Fig. 820) taken on the line G G (Fig. 819)—the flap H being also shown—and in the horizontal section (Fig. 821) on the line A (Fig. 819). A half plan showing the framing of the top of the desk is presented by Fig. 822 (scale, ½ in. = 1 ft.). Enlarged details (scale, 5 in. = 1 ft., approximately) of B, C, D, E E, and F (Figs. 819 and 820) are shown by Figs. 823, 824, 825, 826, and 827 respectively.

Knee-hole Writing Table with Turned Legs.

Figs. 828 and 829 are front and end elevations respectively of a knee-hole writing table intended to be made in walnut, and containing five drawers for holding writing materials, stationery, etc. The top is covered with leather, showing a wood margin 2 in. wide. First plane up four posts, the finished sizes of which are 2 ft. 5¼ in. by 2 in. by 2 in. The turning extends from the rail of the bottom drawer to the floor. The two end rails (Fig. 830) are 9 in. deep, and may be made up of ½-in. walnut clamped with pine on the inside, and stub-tenoned into the posts, keeping the rail ⅛ in. in from the outside to break joint, and flush inside to act as a guide for the drawer. The back rail is of the same depth as the end rails, ⅞ in. thick, tenoned into the posts, and kept back ⅛ in. from the outside. It need not be walnut; pine stained to match the rest may be substituted. The front rails are all 2¼ in. wide by ⅞ in. thick. B and C (Fig. 831) are each in one length, B being lap-dovetailed into the posts at the ends, while C is tenoned to the posts. The two short fore-edges D are fixed to the post in the same way, and to the upright pieces E by lap-dovetailing. These uprights are ⅞ in. thick, fitted to pieces of pine of the same thickness, and joined by grooving the two edges and gluing in a feather, and cramping. They are housed in the back rail to the depth of ¼ in. C is screwed to the top edge of E, and the division F is tenoned at the bottom and screwed through B at the top. The front rails are all flush with the posts at the front. When the table is cramped up, the runners G may be carefully glued and sprigged in their places. The drawer spaces between the runners should be slightly wider at the back to

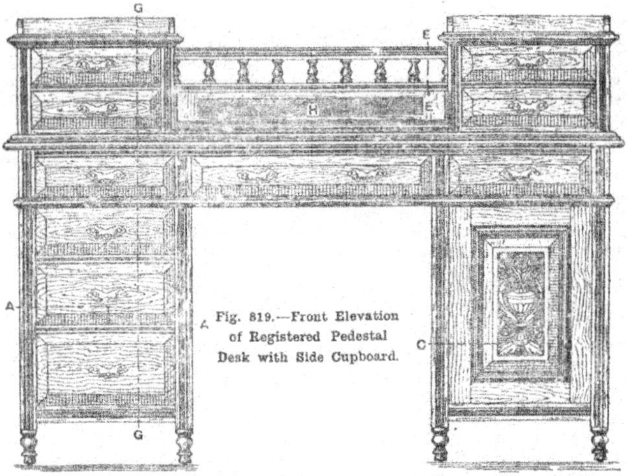

Fig. 819.—Front Elevation of Registered Pedestal Desk with Side Cupboard.

Fig. 818.—Registered Pedestal Desk with Side Cupboard.

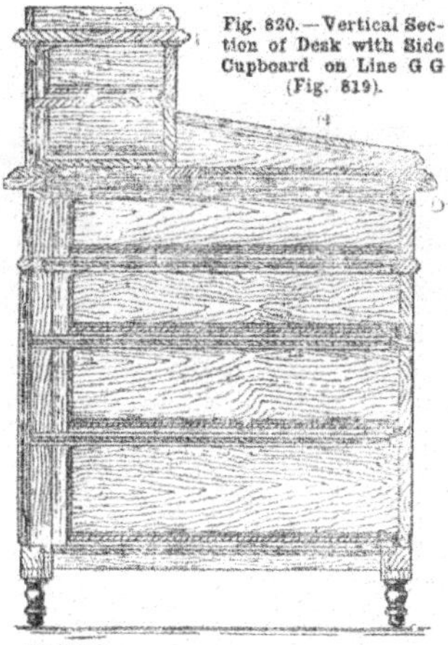

Fig. 820.—Vertical Section of Desk with Side Cupboard on Line G G (Fig. 819).

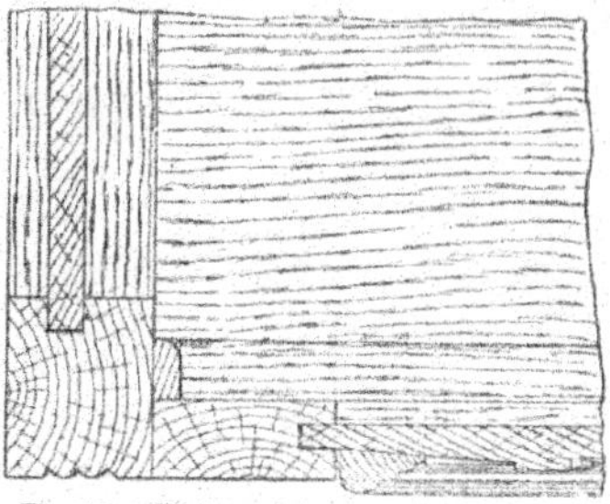

Fig. 824.—Horizontal Section of Part of Desk Cupboard on Line C (Fig. 819).

Fig. 826.—Cross Section of Desk Rails, and Spindle on Line E E (Fig. 819).

Fig. 822.—Half Plan of Framing of Desk Top.

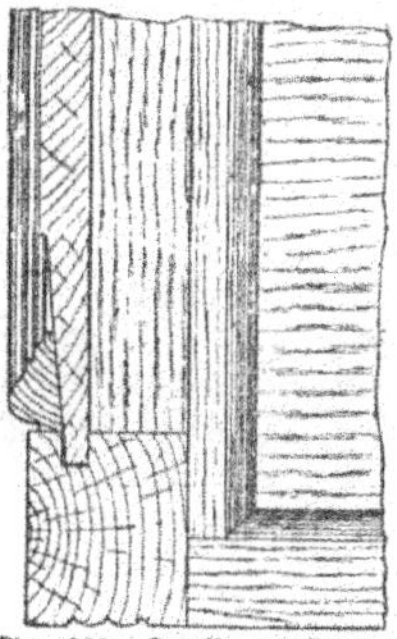

Fig. 823.—Details of Joint in Desk at B (Fig. 821).

Fig. 827.—Cross Section through Top of Desk Pedestal at E (Fig. 819).

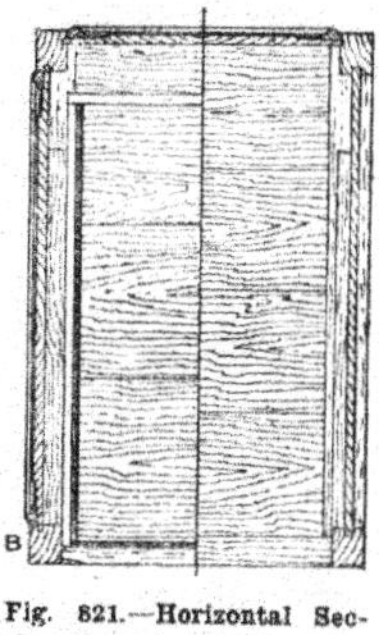

Fig. 821.—Horizontal Section through Desk on Line A (Fig. 819).

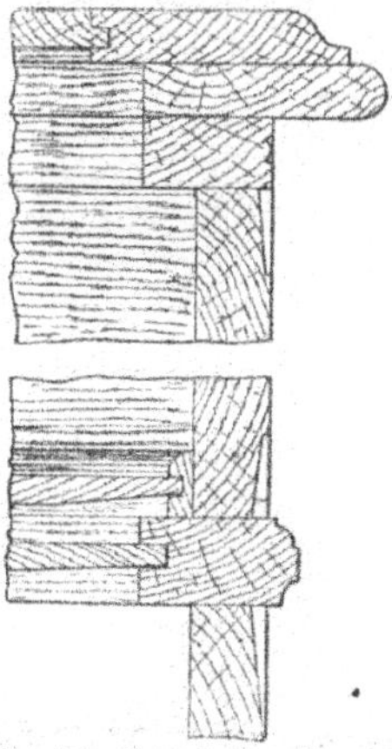

Fig. 825.—Cross Section through Drawer Fronts, etc. (see D, Fig. 820).

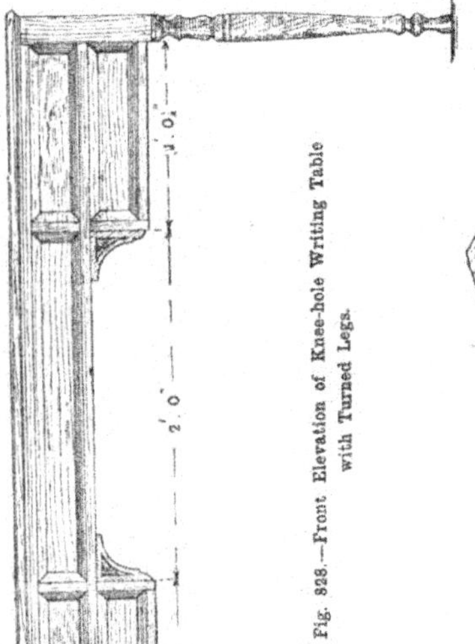

Fig. 828.—Front Elevation of Knee-hole Writing Table with Turned Legs.

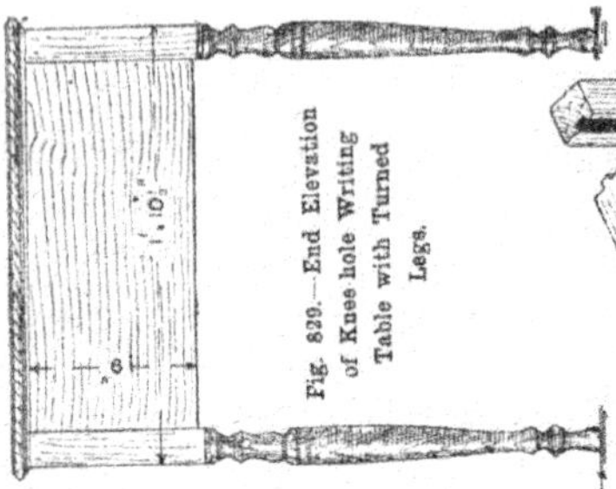

Fig. 829.—End Elevation of Knee-hole Writing Table with Turned Legs.

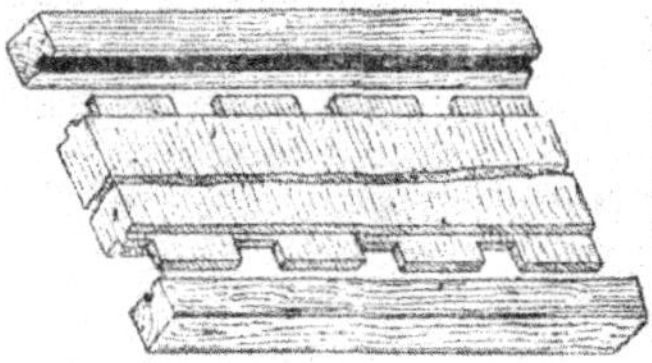

Fig. 830.—End Rail of Knee-hole Writing Table.

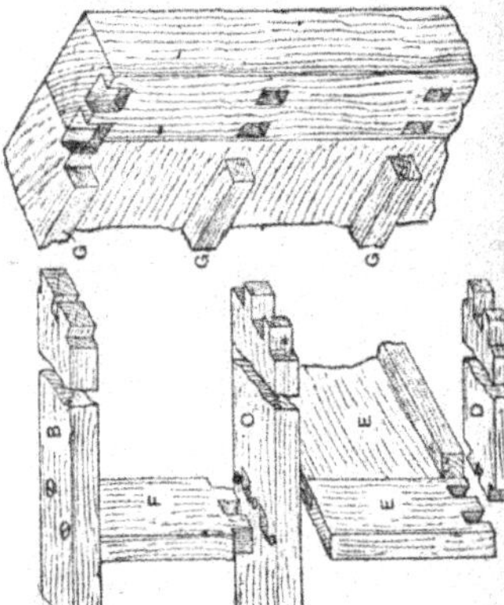

Fig. 831.—Framing of Knee-hole Writing Table.

prevent the drawer sticking as it reaches the back. The runner H (Fig. 832) should be wide enough to carry the drawers on each side, and is glued and sprigged to the top

Fig. 832.—Drawer Runner of Knee-hole Table.

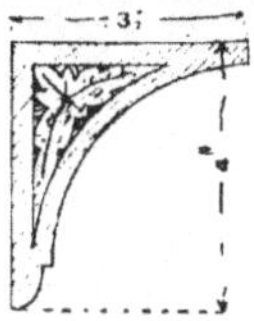

Fig. 833.—Bracket for Knee-hole Table.

edge, the guide I being glued and sprigged to the top edge of the runner H. The two brackets (Fig. 833) are ¾ in. thick, and fixed by two screws, one through the top edge and another from the inside of the upright. The drawer fronts can now be fitted. The middle drawer is 2 ft. long by 3¼ in. deep, the top drawer on each side is 11½ in. long by 3¼ in. deep, and the two bottom drawers are 3½ in. deep. The fronts are ¾ in. thick, and bevelled round the edges as shown in Fig. 834. The drawers are dovetailed together in the usual manner, the backs being kept ¼ in. below the tops of the sides. The sides and back should be ⅜ in. thick, and the bottom may be of the same thickness, and is grooved into the sides and front,

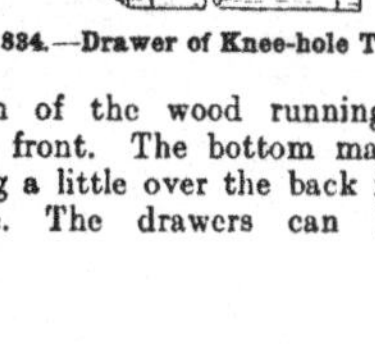

Fig. 834.—Drawer of Knee-hole Table.

the grain of the wood running parallel with the front. The bottom may be left projecting a little over the back in case of shrinkage. The drawers can be made 1 ft. 8 in. in depth, and are stopped by gluing blocks at the back; they are set back ⅛ in. from the fore-edges. The top, of dry yellow pine, ⅞ in. thick, is jointed round the edges with walnut, and, when finished, measures 4 ft. 6 in. by 2 ft. wide. It is made up in the following manner:—Plane up the pine to 4 ft. 2 in. long by 1 ft. 8 in. wide, any jointing being dowelled. Pieces of walnut for the ends are next jointed by grooving the edges of both the pine and walnut, and fitting in a feather and cramping up. The grain of the walnut for the ends should run in the same direction as the pine. The two pieces for the front and back edges should next be put on by plain

Fig. 835.

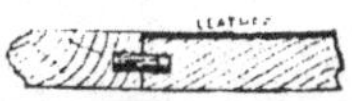

Fig. 836.

Figs. 835 and 836.—Part Plan and Section of Top of Knee-hole Table.

jointing. The walnut edging round the top shows a margin of 2 in., including the moulding, and stands above the pine the thickness of the leather which covers the top (see Figs. 835 and 836). Use morocco leather, or, if this is considered too expensive, American leather-cloth of a green colour may be substituted. The moulding is run on all four edges to enable the table to be placed in the middle of a room if this should be desired. The bracket (Fig. 833) is shown with a little carving, but as an alternative it may be panelled by sinking the ground and leaving a band all round the edges.

Davenport with Rising Top.

The davenport writing desk shown in elevation by Figs. 837 and 838, and in section by Figs. 839 to 843, may be constructed in any fancy hardwood, or in

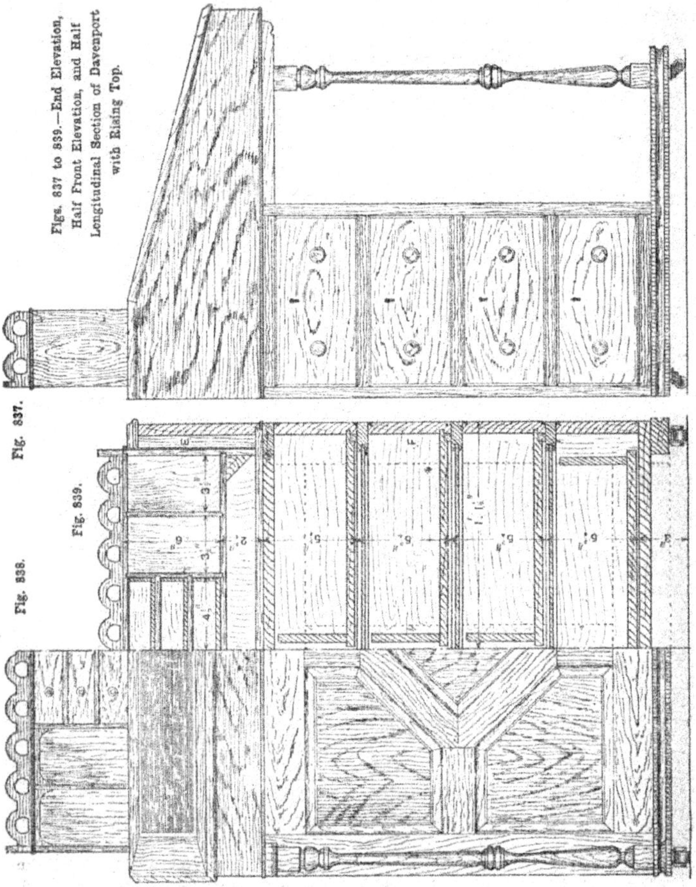

Fig. 837. Fig. 838. Fig. 839.

Figs. 837 to 839.—End Elevation, Half Front Elevation, and Half Longitudinal Section of Davenport with Rising Top.

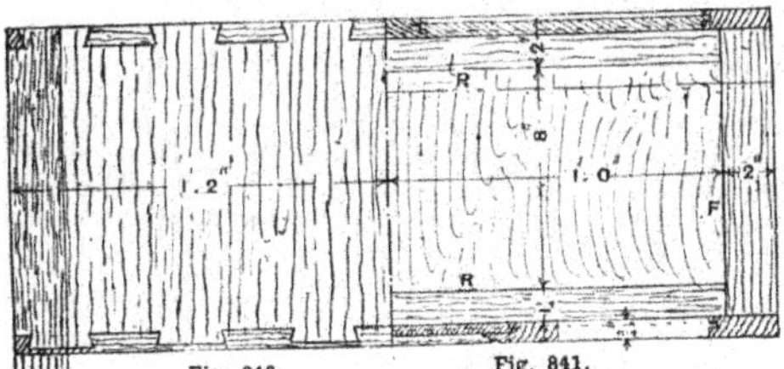

Fig. 840. Fig. 841.

Figs. 840 and 841.—Half Plan and Half Horizontal Section of Pedestal of Davenport.

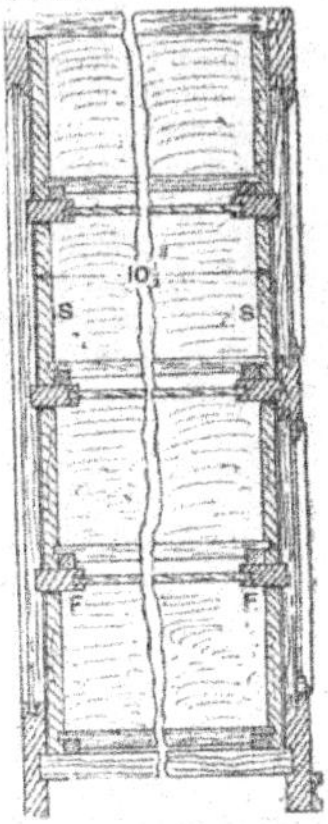

Fig. 842.—Vertical Cross Section of Pedestal of Davenport.

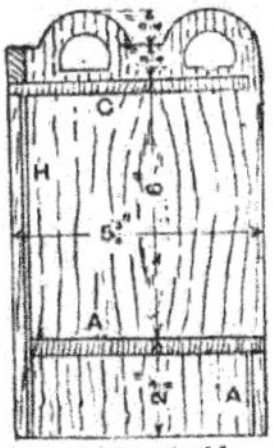

Fig. 844.—Inside Elevation of End Stationery Case of Davenport.

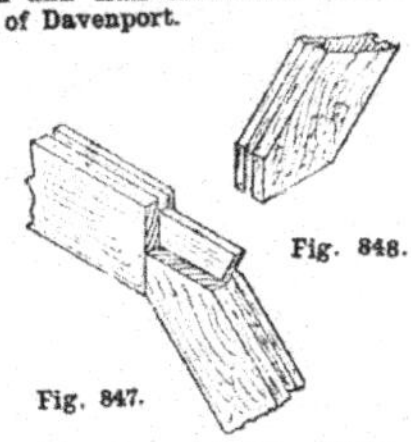

Fig. 848.

Fig. 847.

Figs. 847 and 848.—Joints in Panel Frame of Davenport.

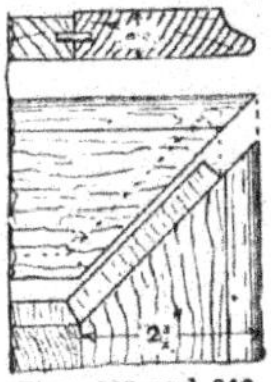

Figs. 845 and 846. —Corner of Davenport's Flap.

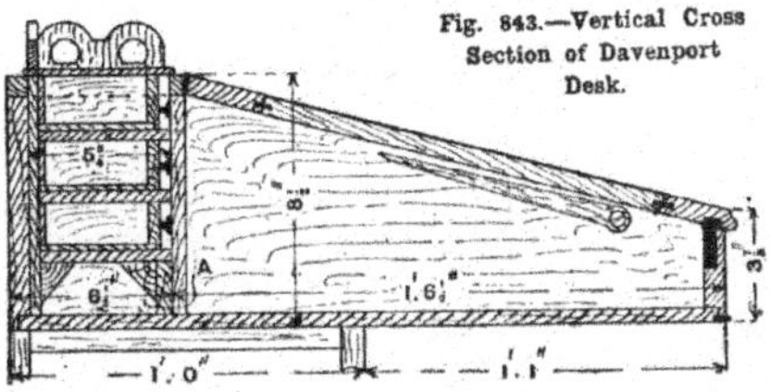

Fig. 843.—Vertical Cross Section of Davenport Desk.

Fig. 849.— Bottom Cross-rail of Davenport.

Fig 850. Fig. 851.

Figs. 850 and 851.—Top Cross-rail of Davenport.

American pine veneered. Fig. 844 is an interior elevation, enlarged, of one end of the stationery case, and Figs. 845 and 846 give enlarged details of one corner of the desk flap.

Cutting List.—The following is an approximate list of quantities: Pedestal.—Front: Two stiles, 2 ft. 1 in. by $2\frac{1}{2}$ in. by $\frac{3}{4}$ in.; $\frac{3}{4}$-in. rails, one, 2 ft. 4 in. by $3\frac{1}{2}$ in.; one, 2 ft. 4 in. by $2\frac{1}{2}$ in.; two, 10 in. by 2 in.; two, 7 in. by 2 in.; four, 9 in. by 2 in.; four panels, $9\frac{1}{2}$ in. by 11 in. by $\frac{3}{8}$ in.; one panel, $9\frac{1}{2}$ in. by $9\frac{1}{2}$ in. by $\frac{3}{8}$ in.; moulding, 15 ft. by $\frac{3}{4}$ in. by $\frac{1}{2}$ in. Back: Two stiles, 2 ft. 1 in. by $2\frac{1}{2}$ in. by $\frac{3}{4}$ in.; one rail, 2 ft. 4 in. by $3\frac{1}{2}$ in. by $\frac{3}{4}$ in.; and one, 2 ft. 4 in. by $2\frac{1}{2}$ in. by $\frac{3}{4}$ in.; one muntin, 1 ft. 10 in. by 3 in. by $\frac{3}{4}$ in.; two panels, 1 ft. $7\frac{1}{2}$ in. by $10\frac{3}{4}$ in. by $\frac{1}{2}$ in.; one deal top, 1 ft. by 2 ft. by $\frac{3}{4}$ in.; and hardwood, two pieces, 1 ft.

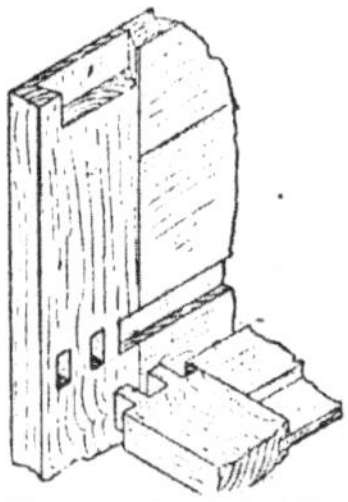

Fig. 853.—Joints in Side of Davenport Pedestal.

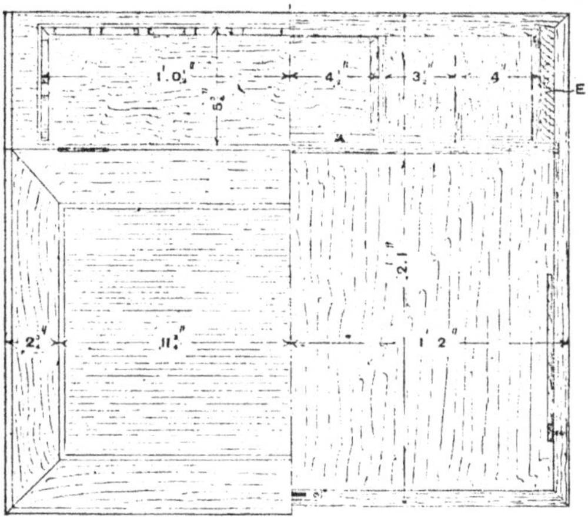

Fig. 854. Fig. 855.

Figs. 854 and 855.—Half Plans of Davenport Desk, with and without Top.

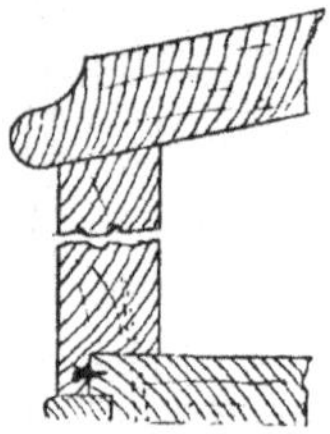

Fig. 857.—Front of Davenport Desk.

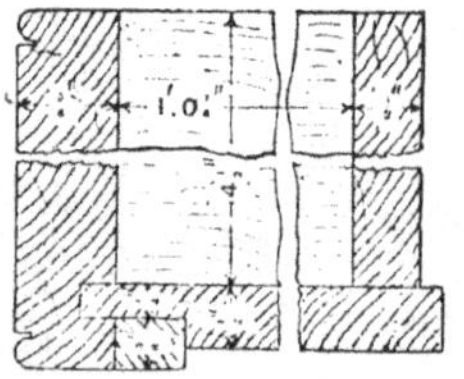

Fig. 852.—Section through Davenport Drawer.

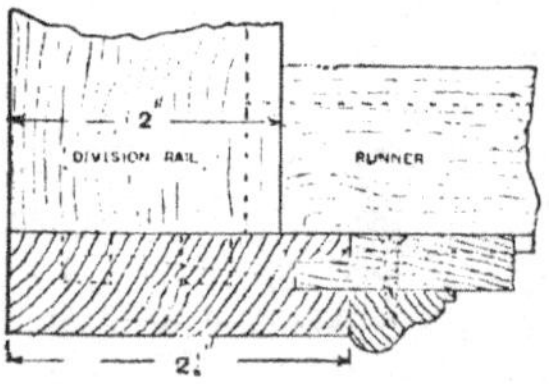

Fig. 856.—Section of Front of Davenport Pedestal.

by 2 in. by ¾ in. Six divisions, 11½ in. by 2 in. by ¾ in.; six deal runners, 2 ft. 1 in. by 1½ in. by ¾ in.; three dust-boards, 8¾ in. by 2 ft. 1 in. by ⅜ in.; eight drawer fronts, 10¾ in. by 5¼ in. by ¾ in.; seven drawer backs, 10¾ in. by 4½ in. by ½ in.; drawer sides, twelve, 1 ft. 1½ in. by 5¼ in. by ½ in.; two, 2 ft. 2 in. by 5¼ in. by ¾ in.; drawer bottoms, six, 10 in. by 1 ft. 1½ in. by ½ in.; one, 10 in. by 2 ft. 1½ in. by ½ in.; two bottom bearers, 1 ft. 9½ in. by 2¼ in. by 1¼ in.; two top bearers, 11½ in. by 2¼ in. by ¾ in.; two pillars, 2 ft. 1 in. by 1½ in. by 1½ in.; one plinth, 2 ft. 1 in. by 1¼ in. by ½ in. Desk: Two ends, 2 ft. 1 in. by 7¾ in. by ¾ in.; one front, 2 ft. 4 in. by 3 in. by ¾ in.; one back, 2 ft. 4 in. by 8 in. by ¾ in.; one deal division, 2 ft. 3 in. by 7½ in. by ½ in.; one deal bottom, 2 ft. 3¾ in. by 1 ft. 1 in. by ½ in.; one top, 2 ft. 5 in. by 7¼ in. by ¾ in. Flap: Two, 2 ft. 5 in. by 2¾ in. by ¾ in.; two, 1 ft. 7 in. by 2¾ in. by ¾ in.; one deal, 1 ft. 11½ in. by 1 ft. 1¾ in. by ¾ in. Case: Two ends, 10 in. by 5¾ in. by ⅜ in.; one back, 2 ft. 1 in. by 8 in. by ⅜ in.; one top, 2 ft. 1½ in. by 6¼ in. by 5/16 in.; one bottom, 2 ft. 1 in. by 5½ in. by ⅜ in.; four 5/16-in. divisions, 6 in. by 5½ in.; two, 9 in. by 5½ in.; one plinth, 2 ft. 1 in. by 2 in. by ⅜ in.; three drawer fronts, 9 in. by 1¾ in. by ⅜ in.; three backs, 9 in. by 1½ in. by ¼ in.; six sides, 5 in. by 1½ in. by ¼ in.; three bottoms, 9 in. by 5 in. by ¼ in.; one fretrail, 2 ft. 1 in. by 1½ in. by ⅜ in. Also sixteen 1-in. turned knobs, six ½-in. knobs, seven 2-in. brass drawer locks, eight escutcheons, one 2½-in. desk lock, one pair of 2½-in. brass butts, and four 1¼-in. screw plate castors.

Pedestal, etc.—The pedestal and desk are made separate, and screwed together. The pedestal itself is constructed with ¾-in. panelled sides, and open framed ends to receive drawers, the top and bottom being solid. The top is dovetailed to the sides as shown in Fig. 840, and the bottom is grooved in as shown in Fig. 842. The division rails are fixed to the sides with double stub-tenons, and the drawer runners are housed in ⅛ in., the ends of the runners being tenoned into the division rails; also ¼-in. dust panels are inserted in grooves in the rails and runners. The uppermost drawers in the pedestal are in three pairs, drawing from each end, but the lowest drawer moves from the left-hand end, and is a through drawer, the front at the opposite end being a dummy. The top and bottom rails of the front and back frames are stub-tenoned to the stiles, and screwed from inside; the tenons cannot come through, because the edges of the stiles are seen. The interior rails in the front frame are framed together as shown in Figs. 847 and 848, tenons being formed on the ends of the muntins, and the inclined rails forked over the tenon. The bottom cross-rails (Fig. 849) are screwed underneath the pedestal, the front plinth piece being moulded similarly and mitered into them. The top cross-rail (Figs. 850 and 851) is tenoned into the side of the pedestal and bradded. The pillars are tenoned through the rails and wedged. The drawers are dovetailed, and finished flush with the case, and beaded all round; the bead across the end is glued into a rebate. Fig. 852 shows the method of blocking the bottom at the front. The drawers are prevented going too far in by means of two thin oak stops glued and bradded on the front division rails. Fig. 853 illustrates the method of connecting the top and drawer rails of the pedestal. Secret dovetailing is employed in the angles of the desk (Figs. 854 and 855). Fig. 856 is a section of the stile of the front framing, etc., the dotted lines showing the tenons and sinkings.

Desk.—The desk has a hinged framed and moulded flap, the pine panel being sunk 1/16 in. below the hardwood margin to receive a leather top, and is cross-tongued all round to the margin, the mitres of which are also tongued. The back part of the desk has an enclosed well, into which slides the stationery case.

Stationery Case.—This case is fitted with three shallow drawers and four pigeon-holes, and can be drawn to the top, where it will remain sustained by the friction of its ends until pushed down. Should the case work too stiffly, rub a little powdered French chalk on its sides. Some cases are made with balance weights in the enclosure E in Fig. 839 and Fig. 855. The ends of the case should have the grain running up and down, and the top, bottom, and back should be housed in. The two holes on the fretted top are for the fingers when lifting the case, and the necessity of having the grain running in this direction will be apparent. The back fret is a separate piece fixed on the top, the back of the case being tongued and grooved into the end under the top, as shown at H (Fig. 844). A narrow rail (Fig. 843), fitted under the bottom of the case in the groove A (Fig. 844) prevents the case tilting when drawn up to its full height. The top of the case is made ¼ in. wider than the ends, and its edges are rounded. Two slips are mitered to the edges of the top and glued across the ends to hide the joint in the top of the desk when the case is down.

Completing Desk.—The top of the desk is cut out flush with the sides and end of the well; see Fig. 855. The sides of the desk are dovetail-mitered at the angles, the bottoms being rebated in as shown in Fig. 857. The bottom is allowed to project ¼ in., and after it is fixed a ¼-in. cocked bead is planted round the ends and front, a portion of the bottom being rebated out to receive it. It is best to make the stationery case before preparing the well; then, placing the case in position on the desk top, mark all round it with a pencil, and cut the aperture tight to the mark. Next arrange the top so that the inside of the aperture is flush with the back of the desk, mark down the opposite side on the desk ends, house in the front of the well to this mark, fix the top by glue and screws to the sides of the desk, and fill in pieces at E (Figs. 839 and 855) flush with the ends of the aperture in the top. Next fix in the bottom, and then hang the flap. A light spur working on a screw is fixed on the right-hand side of the desk, as shown in Figs. 843 and 855, and a shallow rack is formed in the under side of the flap to receive it.

Bureau.

A bureau as in Fig. 858 consists of a solid carcase with framed divisions and back, two large drawers, a hinged flap or table which rests upon sliding brackets when down, two small drawers, a cupboard, and six pigeon-holes for stationery, etc. The dimensions are:—Height to top, 3 ft. 2 in.; to table, 2 ft. 3 in.; width, 3 ft.; and depth, 1 ft. 10 in. A bureau is made when in the solid of oak, walnut, or mahogany; when veneered, of pine, covered with rosewood, maple, or satinwood, with covered inlays. The present design is suitable for solid treatment. The sides are 1 in. thick; the division rails, sub-top, flap, bottom, and back, are ¾ in. thick; the moulded top measures 1¼ in.; the two sliders are ⅝ in.; the stationery fittings ¼ in.; while the door is ⅝ in. and the drawer fronts are ½ in. thick. Fig. 859 is a front elevation, Fig. 860 a section on B B (Fig. 859), and Fig. 861 a section on A A (Fig. 859). Fig. 862 is a vertical section of the side, Fig. 863 section through top fitting, Fig. 864 section through plinth. Fig. 865 shows the upper part of one of the pigeon-holes enlarged, Fig. 866 is a section through the hanging stile of the door, Figs. 867 and 868 show a portion of the flap and mitered clamp, Fig. 869 is part of the side showing the method of framing the divisions, Fig. 870 shows the taper dovetail at the end of a division rail, and Fig. 871 is a section of the moulded top. The top is fixed to the sides by a groove-and-tongue joint stopped in front (see Fig. 862). The table, divisions, and bottom are grooved into the sides $\frac{3}{16}$ in. deep, the groove being stopped back from the front by the width of the division rail, that part of the side being worked into a tapered undercut groove. This is also stopped ¼ in. back to hide the sinking (see Fig. 869). Each rail and the table must be cut to Fig. 870 and fitted separately; the rails are then fitted round the panel or dust-board. The runners have tenons at each end as shown by dotted lines in Fig. 869, and are glued up. Then each frame can be treated as a solid division and fitted into the carcase bodily. The bottom need not be dovetail-grooved, as it can be nailed through from outside, the nails being covered by the plinth. This should be sunk $\frac{3}{16}$ in. and glued and screwed on from inside. Plough grooves are made at each side of the top drawer openings, and ¼-in. by $\frac{3}{16}$ in. oak slips are inserted to form runners for the sliders. These must be stopped back ⅝ in. to prevent the grooves showing; and if the slides are not to come right out, a ¼-in. groove should be worked across the side, corresponding studs being glued on the tail end of the sliders, which must be inserted before the back is fixed.

Drawers.—The two top drawers are arranged to fit between the sliders and the vertical division; the other two drawers run in the clear of the sides. The drawers are composed of 1-in. fronts, ½-in. backs and sides, and ⅜-in. bottoms; they are dovetailed at the angles, and the bottoms are chamfered and slipped with oak as shown in Fig. 862. The drawers are sunk ⅛ in. below the frame, and are checked by oak steps glued and nailed to the divisions as shown at D (Fig. 869).

Flap.—The flap is made of ⅞-in. stuff, mitre-clamped and stub-tenoned as shown in Figs. 867 and 868, and hung to the table with a pair of 1½-in. brass flush flaps. The inside of the flap is covered with leather, the margins, elbows, and table being polished. Two ½-in. by ¼-in. guard slips are fixed at each end of the flap, and a bevelled and rebated lock-rail, screwed under the top, forms a seat for the flap when the latter is closed.

Top Fitting, etc.—The top fitting (Fig. 863), of sycamore, is made separate from the carcase and slipped in tight. The top and bottom are dovetailed to the sides, the shaped crowning is sunk $\frac{1}{16}$ in. in the divisions, and the divisions are housed at the ends, etc. The drawers are grooved and tongued at the corners, and the bottom is grooved in flush with the sides. The small cupboard door is mitered around the panel, the mitres being dowelled or veneer slipped. The back is a panelled frame with 3-in. stiles and rails and 2-in. muntins, with ⅜-in. flush panels between them; it is mortised and tenoned together, and sits in a rebate in the sides and top.

Bureau Writing Desk and Shelves.

The bureau writing desk illustrated by Figs. 872 to 874 will have a neat appearance if of mahogany, walnut, or similar hardwood, or even pine, stained and varnished, or ebonised and polished. 1-in. stuff, finishing to ⅞ in., will be most suitable for the sides, broad shelves, flap, main vertical divisions, and drawer fronts, material ⅝ in., finished, being suitable for the other parts. The breadths and lengths of the various pieces may be ascertained from the illustrations. Both sides can be got out of a board 1 ft. 4 in. wide and 8 ft. long, thus preventing waste and jointing. The broad boards shown in section at A, B, and C (Fig. 875) may be of material of the same width as the sides, but it would be cheaper and as serviceable to joint the two shelves A and B; the shelf A and table top C should be without a joint. Having cut out the various pieces, plane them true to dimensions. The sides

Fig. 858.—General View of Bureau.

should be set out to shape, a stout paper or canvas template being made for each half. The housings also should be set out, and the two main vertical divisions as D (Fig. 873). Next make the housings, their forms being shown at Fig. 876; it will be noticed that they are stopped about ⅝ in. from the front edge, and that a piece is notched out of the boards as indicated at E, F, and G (Fig. 876). A rebate should be made in the back edges of the sides for the back, as shown at H (Fig. 876). The three broad boards A, B, and C (Fig. 875), when prepared to length, may be fitted into their housings, and then the shelves B and C should be set out for the housings for the division between the drawers, and also for the two vertical divisions. The upper shelves can also be fitted in. The carcase should be fitted together, and,

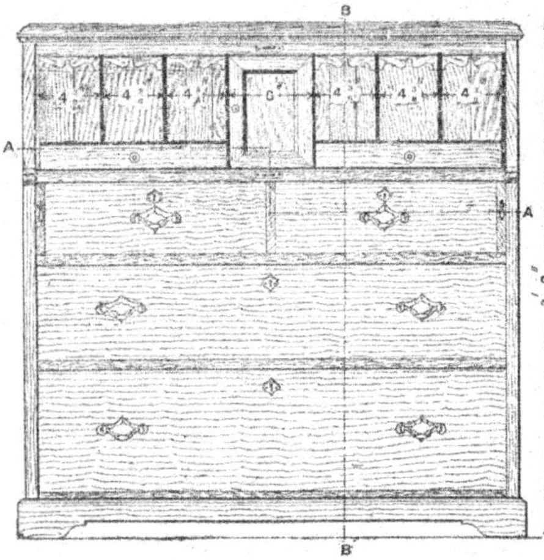

Fig. 859.

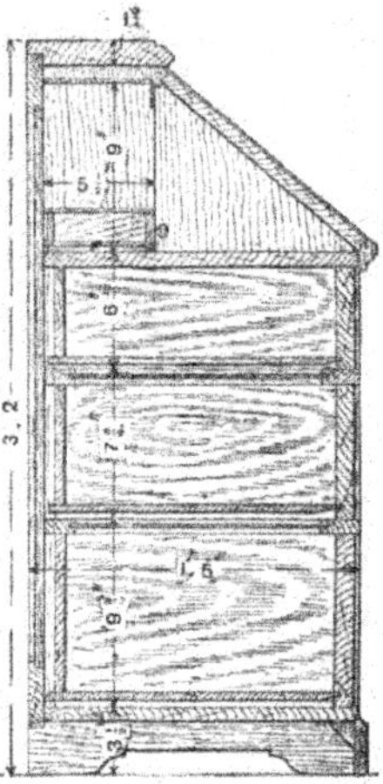

Fig. 860.

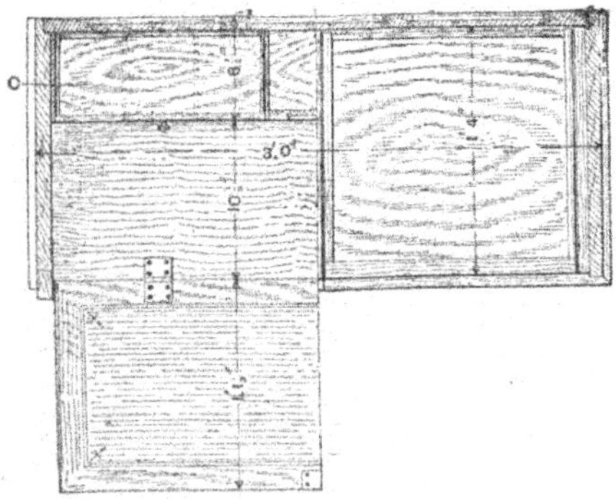

Fig. 861.

Figs. 859 to 861.—Front Elevation, Vertical Cross Section and Horizontal Section of Bureau.

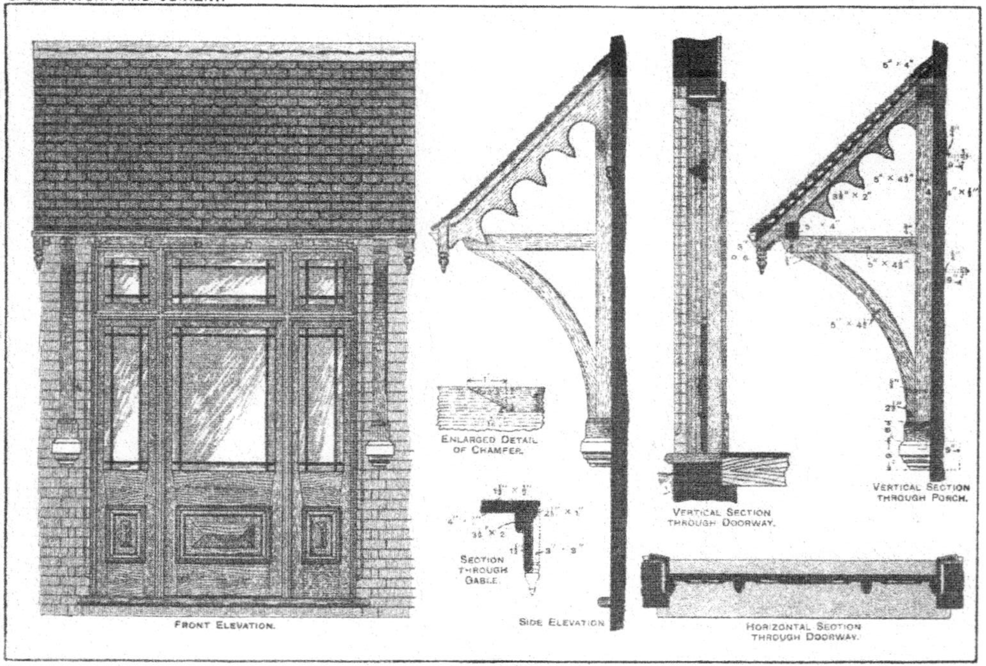

6.

PORCH AND DOORWAY.

if satisfactory, can be taken to pieces so that the curves of the sides may be sawn, a bow-saw being most useful for this purpose; finish off with a sharp spokeshave, a chisel, and fine glasspaper. The carcase may then be put together again, and, if nailing through the sides into the ends of the shelves is objectionable, nails may be driven obliquely from the under sides of the shelves into the sides. The shelf A and the board B (Fig. 875) may be further secured to the sides by a few neat triangular blocks glued underneath. The top middle shelf K (Fig. 873) can be nailed direct on the top

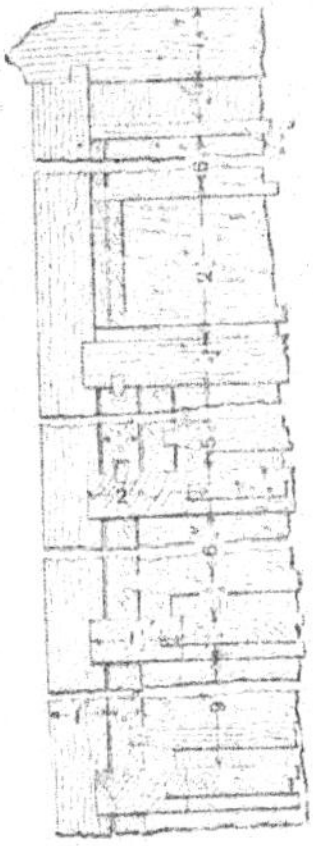

Fig. 862.—Vertical Section of Bureau Side, etc.

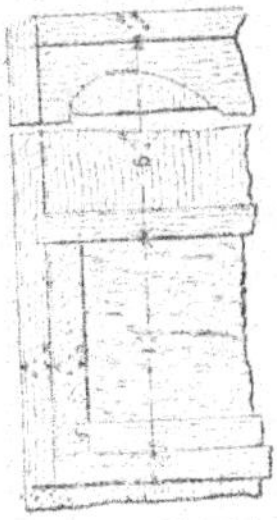

Fig. 863.—Section through Bureau Top Fitting.

Fig. 865.—Top of Bureau Fitting.

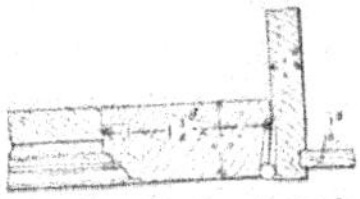

Fig. 866.—Section through Bureau Door Stile.

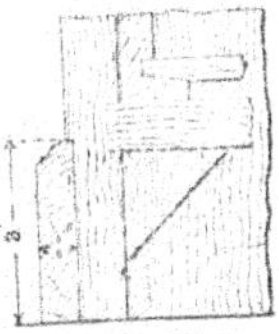

Fig. 864.—Section through Bureau Plinth.

Fig. 867.

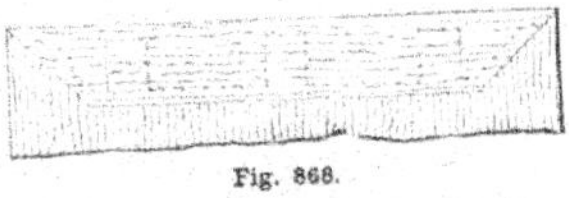

Fig. 868.

Figs. 867 and 868.—Section and Part Plan of Bureau Flap, showing Clamp.

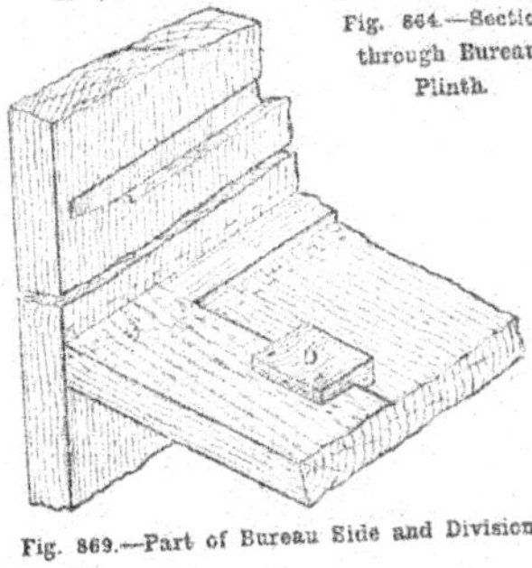

Fig. 869.—Part of Bureau Side and Division.

Fig. 870.—End of Bureau Division.

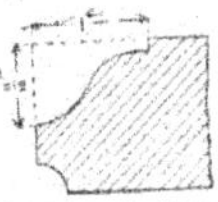

Fig. 871.—Section of Moulded Top of Bureau.

ends of the main vertical divisions. The two top side shelves (one of which is shown at L) have one end nailed on top of the side. These shelves are rebated as at M (Fig. 875) on the back edges to receive the back. For this, narrow matchboarding, smoothed on the face side before fixing, can be used, and should be nailed to the shelves. If the case is of hardwood, this back should be stained to match. Next the various curved pieces shown can be prepared and fixed in position by glue, fine sprigs being inserted from the back wherever possible. Round the edges of the top shelves a neat piece of moulding should be mitered and fixed. The fronts of the drawers having been fitted in, and their sides and backs prepared, should be dovetailed together, and the sides ploughed to receive the bottoms; handles can then be fitted. For the flap,

Fig. 872.—Bureau Writing Desk and Shelves.

a well-seasoned piece of board should be clamped at the ends and fitted between the two main vertical divisions and also under the upper shelf, as illustrated at Fig. 875. It can be hung by a couple of 3-in. narrow brass butts N, and a suitable lock may be fixed to the edge as shown in Fig. 872.

Flap-front Writing Desk with Divisions.

Fig. 877 shows a writing desk which can stand on a table, and when not in use the flap may be raised so as to close in the divisions. Fig. 878 shows a front elevation. It might be made of hard or soft wood

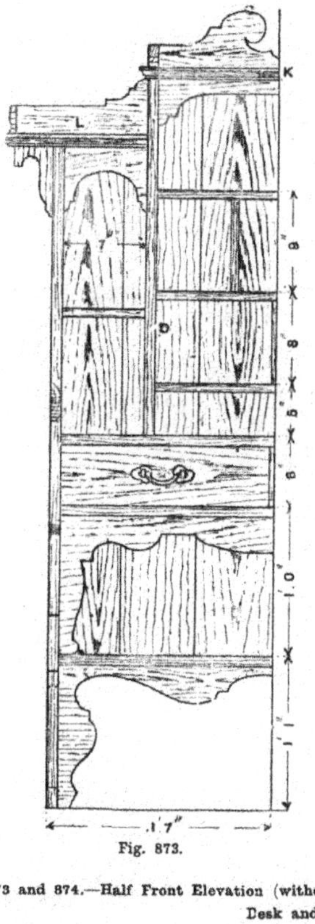

Fig. 873.

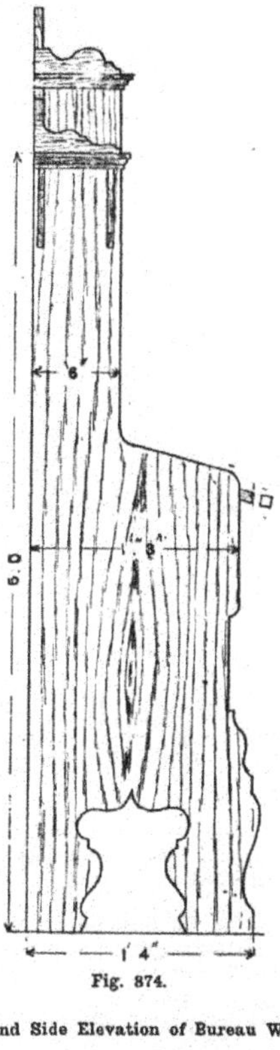

Fig. 874.

Figs. 873 and 874.—Half Front Elevation (without Flap) and Side Elevation of Bureau Writing Desk and Shelves.

according to requirements and taste; to save jointing, basswood would be suitable; but if pine is used, it will be necessary to joint up and glue the flap, bottom, back, and sides. As the divisions and shelves are to be housed together, the construction will not be difficult, and the leading points are as follows:—The shelves, divisions, and sides are house-jointed together, and the bottom, top, and sides are rebated at C (Fig. 879) to receive the back, which may be screwed or bradded in position. The flap should be made about ⅜ in. small all round, and, to keep it true, the inner surface should be grooved and two dovetailed keys should be inserted, as shown at Fig. 880. The lower edge of the flap requires bevelling as shown in Fig. 879. Leather, green baize, or other suitable material, may be glued on the inner surface of the flap, the cloth being fastened well round the edges. When the glue is thoroughly dry, the surplus material should be cut off flush with the edges and ends of the flap. Then four pieces of ⅜-in. stuff should be prepared, the width being equal to the thickness of the flap, and one edge

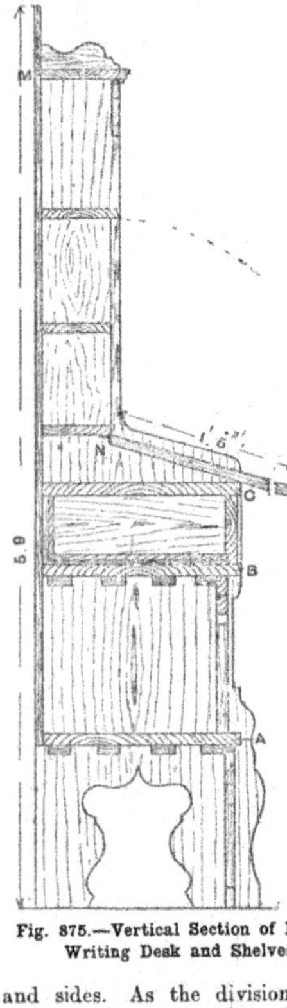

Fig. 875.—Vertical Section of Bureau Writing Desk and Shelves.

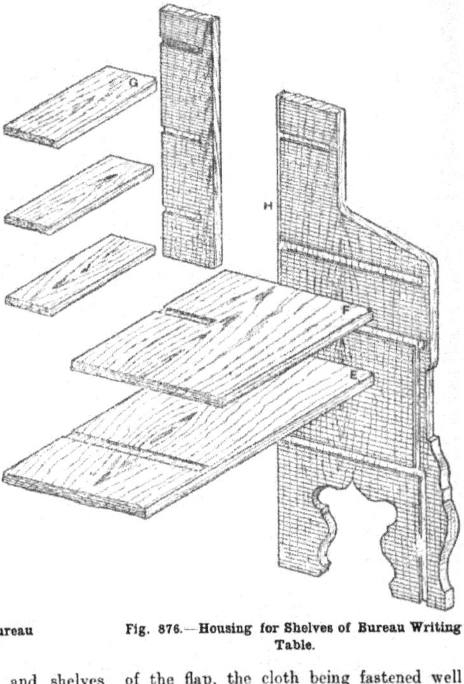

Fig. 876.—Housing for Shelves of Bureau Writing Table.

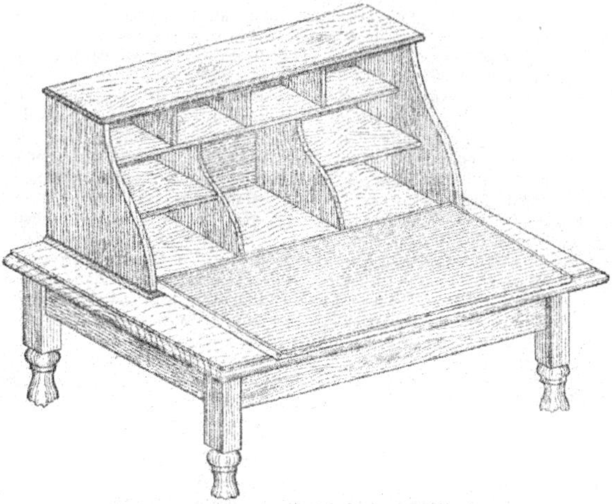

Fig. 877.—Flap-front Writing Desk with Divisions.

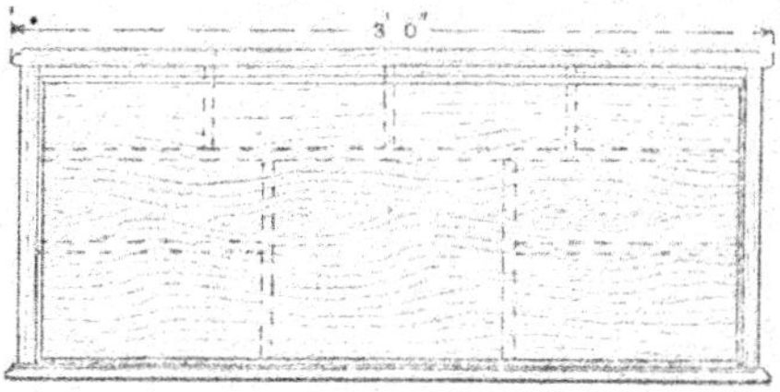

Fig. 878.

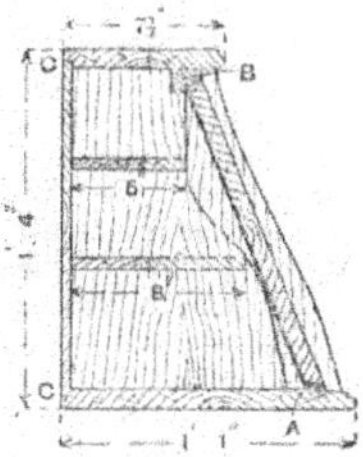

Fig. 879.

Figs. 878 and 879.—Front Elevation and Vertical Section of Flap-front Writing Desk.

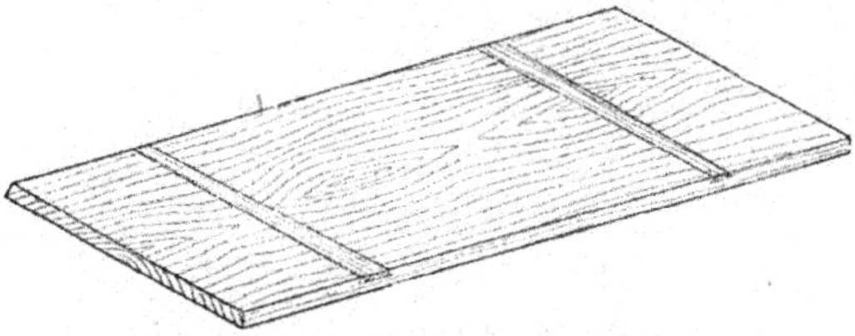

Fig. 880.—Writing Desk Flap with Dovetailed Keys.

of each piece should be rounded. The piece for the bottom should be rather wider than the others, and the inside edge will require bevelling as at A (Fig. 879). Then these pieces should be mitered, glued, and nailed to the inside edges and ends of the flap, the square inside edges fitting flush with the material glued on the flap. The top and bottom may be rounded, or may have thumb mouldings worked on them. To form a stop for the upper edge of the flap, a piece of wood should be rebated as shown at B, and this fillet will also improve the appearance of the front and assist in strengthening the top shelf. The flap should be hinged to the bottom with a pair of 2½-in. butt hinges, and a suitable lock should be obtained and fixed on the flap.

Pedestal Desk for Office Use.

Figs. 881 and 882 represent, respectively, a side view and a half end view of a pedestal desk intended to be executed in mahogany. The upper part of the desk consists of a double slope and flat, with solid returned ends and framed fronts and bottom; it is fitted with ten drawers, and provided with brass standard rails for books. The six fitted pedestals on which it is mounted have framed, moulded, and returned ends. It will be seen by reference to the plan (Fig. 883)—the right-hand half of which represents a section through the upper portion of the desk, and the left-hand half a section through the pedestals—that the pedestals are framed separately, and placed back to back with about ¾ in. between them; the face side of one pedestal runs over the back and fits close to the other, a ¼-in. bead being used at the junction. This is a much more convenient and economical method of construction than to make each pair of pedestals in one fitting, as, when made as above described, they are lighter to handle, there is less danger of the sides splitting through shrinkage, and the drawers are easier to fit, as the backs can be seen. The sides of the two end pairs of pedestals, which are covered by the framed ends, may be of deal, but both sides of the centre pair must be of mahogany. It is usual in narrow fittings like these to make the drawer divisions solid, as the value of the stuff is less than that of the labour in preparing framed divisions; they can be of white deal, edged with mahogany from 1½ in. to 2 in. wide, the joints being ploughed and tongued. There is no necessity to dovetail-groove for framed divisions, as they will be quite strong enough if fitted tight and glued. The ends in the deal sides can be secured by nails, and, if the other ends are secured by angle-screwing them to the side at the back edge, there will then be no danger of the drawers starting them. The tops are fitted into the mahogany sides with lap dovetails, and with common dovetails into the deal sides. They must be kept flush with the rebates at the back, as also must the deal sides; the backboards lie on the edges of these, and fit into the rebates in the mahogany sides (see A and B, Fig. 883). The divisions are housed ¼ in. into the sides, and the grooves stopped ⅜ in. from the front edge. The front plinth is wrought 1½ in. thick, and framed between the sides with a mortice-and-tenon joint, as shown in Fig. 884, the piece marked C being sunk and lipped over the front edges, and mitered to the return plinth, which is ⅝ in. thick, and also sunk ⅛ in. to hide the joint. When setting out the divisions, mark a small mortice on the under side of each for the bolt of the locks, as the mortices are awkward to make after the framework is fitted together. The position for the mortice can easily be found by laying the lock with its keyhole in the centre of the opening; keep the mortice in a trifle, so that there will be no play if a front happens to be thin. The beads are stuck on the drawers after they are fitted; in the case of the cupboard, they are stuck on the sides of the pedestal and stopped and mitered at the division, thus necessitating a bevelled shoulder. All the other divisions have square shoulders.

General Construction.—Before setting out the framed slope it will be well to run over the plan (Fig. 883), and the section (Fig. 885) which is taken on the line D D (Fig. 883), and to note how the various parts are put together. The top is formed of two wide pieces of 1-in. mahogany running lengthwise, and overhanging the frame ¾ in. all round; they are secured to the flat with a glued and tongued joint. The fronts are framed, by means of rails and cross-divisions, into a number of openings for the reception of drawers; the cross-divisions are formed with upright pieces of 1-in. mahogany G clamped on each end of shaped deal pieces E (Fig. 886) that are notched halfway through in the middle to receive longitudinal centre division F (Fig. 883). Top rails K are of the same width as the clamps G (Fig. 886), and 1¼ in. thick; they are double mortised for the divisions and dovetailed into the ends as shown at Fig. 887. The bottom rails L, ⅞ in. by 1½ in., are tongued and glued to the framed bottom M, flush on the top side, and, as the latter is 1 in. thick, a rebate 1½ in. by ¼ in. is formed to receive the cocked bead R, which breaks the joint of the desk and also hides the ends of the tenons on the divisions (see Fig. 886). The ends of the fronts must equal in width the combined thickness of the framed end and pedestal side, so as to line with the latter when in position. They are dovetailed to the rails and mitered to the solid ends, which are shaped to the outline of the top and rebated to receive the ends of the bottom.

The Case.—In setting out the case, great accuracy must be observed, as errors in double-faced work are very difficult to correct. Take one of the top rails, and lay it on the plan rod, face up. Mark the divisions a trifle small, the shoulders of the end divisions, and two other lines ⅞ in. and 1¾ in. respectively beyond for the dovetails shown in Fig. 887; gauge the line H at each end to the thickness of the end piece, and draw the dovetails. Two mortice gauges will be required for the division mortices; use ¼-in. chisel ⅜ in. from each edge; no wedging need be allowed, as the paring of the tenons for entry will be sufficient. Gauge the front edge ⅝ in. thick from the underside; the rail will be bevelled to this after mortising, etc. Pair the corresponding rail to this one, also pair the bottom rails; handscrew them all together, and square the lines over; transfer these to the faces, and gauge; it will be found that only one mortice will come in the bottom rail, the other one coming in the deal bottom when it is glued on. The division F may be set out from one of these rails, the mortices giving the lines for the notches to receive the cross-divisions (see Fig. 888); do not cut to the length till the case is together. The divisions next claim attention. The clamps will be set out from the section (Fig. 885), the length between the shoulders being equal to the width of the drawers, the tenons at each end being gauged with the rail gauges. A stump mortice must be set out on the back edge 1½ in. by ¼ in., and this should be in the middle of the thickness; if mortice gauges are scarce, one of the others may be used, but care must be taken to gauge both deal and mahogany from the same side. The deal portion of the divisions should be set out from the same section, with shoulder lines at E, the notches for the centre division in the middle, and tenons at the ends; one of these may be cut to shape at top, and used as a template for marking the others. The two ends must be paired, the total width of the case outside squared up, and a bevel set to 45° applied at the top edge, the inside bottom edge gauged for rebating ⅜ in. by ⅝ in., and the top edge marked with the template. The bottom will be set out from the plan; the two side rails should run through, and be mortised to receive the end-rails and muntins, by which arrangement end grain will not come in the joints. The mortices can come through and the tenons be wedged, but their ends must be cut back ⅛ in. The framing is to be ploughed from the face side with a ¼-in. groove ⅜ in. deep for the panels, the lengths of which will be taken from the muntins and the widths from the rails; gauge from the face side a full ¼-in. tongue. The top can be marked from the plan, cut to size, full in width, the joints shot to the bevel obtained from the section, and ploughed square with the edge from the under side, and the groove stopped 1 in. from the ends. If it has to be ploughed by hand, a piece of stuff should be tacked on the back and planed square with the joint; this will form a fence for the plough to work against. Fig. 889 is a part view of the under side of the desk, showing joints. Fig. 890 is a view of the bottom of the pedestal side, showing the sinking, etc., for the plinth.

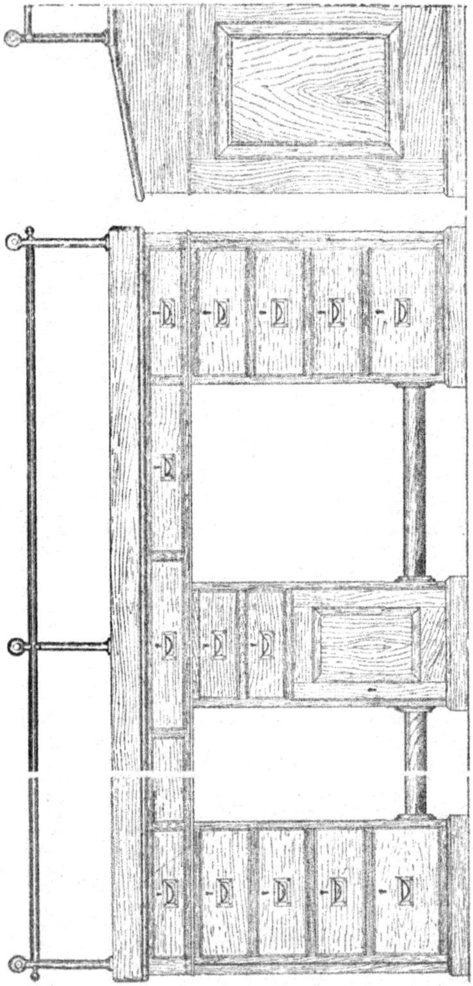

Fig. 882.

Fig. 881.

Figs. 881 and 882.—Side Elevation and Half End Elevation of Pedestal Desk for Office Use.

Fitting Up.—When all mortising, tenoning, and ploughing are done, fit and glue the clamps on the divisions, clean off flush when dry; glue the ends of fronts on the solid ends by means of angle blocks temporarily glued on the faces, and pinched together with handscrews. Fit the divisions into the front rails, marking each as done; cut and fit the dovetails in the ends. Glue up the bottom, and put a screw in the tenons not wedged. When the work is dry, level off, and shoot to the exact width,

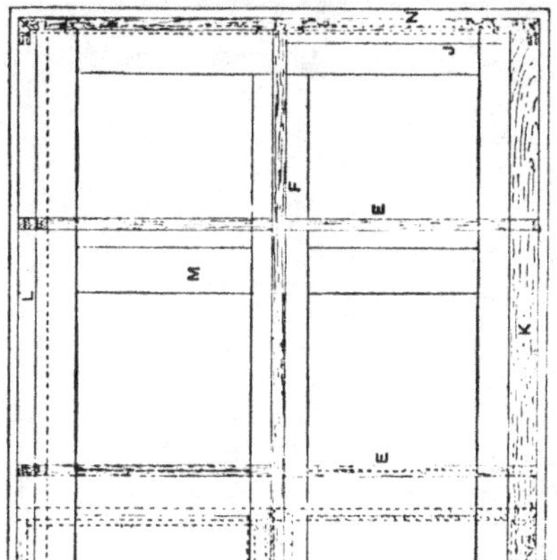

Fig. 883.—Half Horizontal Sections of Pedestal Desk for Office Use.

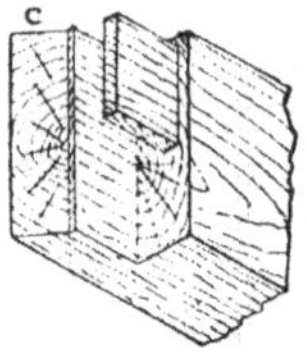

Fig. 884.—End of Plinth of Pedestal Desk.

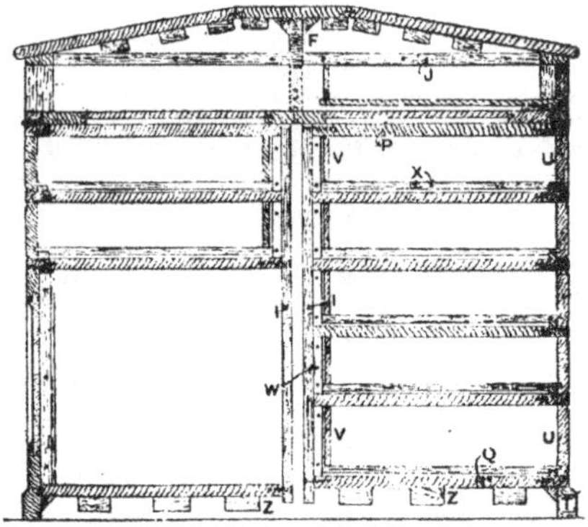

Fig. 885.—Vertical Section of Pedestal Desk on Line D D (Fig. 883).

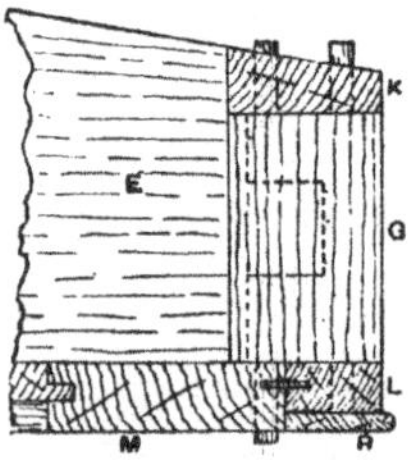

Fig. 886.—End of Cross Division of Pedestal Desk.

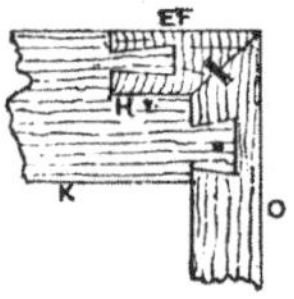

Fig. 887.—End of Top Rail of Pedestal Desk.

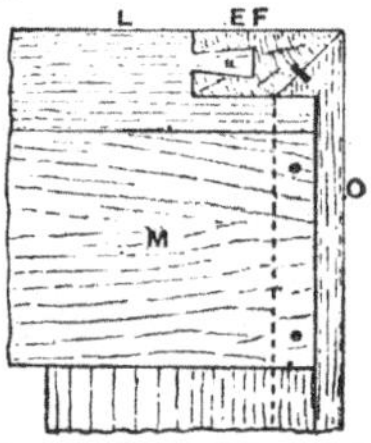

Fig. 889.—Under Side of Desk, showing Joints.

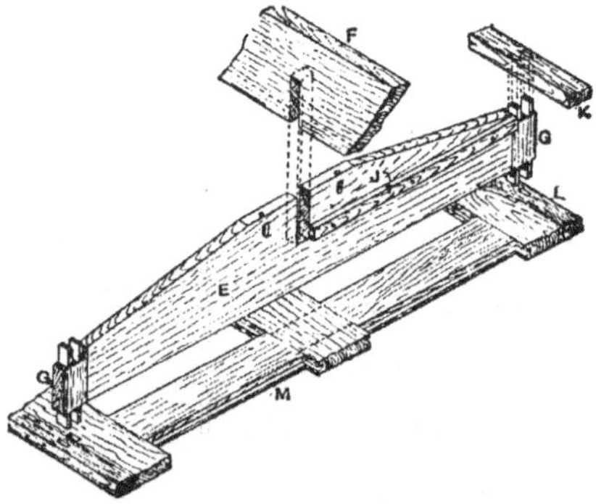

Fig. 888.—Cross Division, Framed Bottom, etc., of Pedestal Desk.

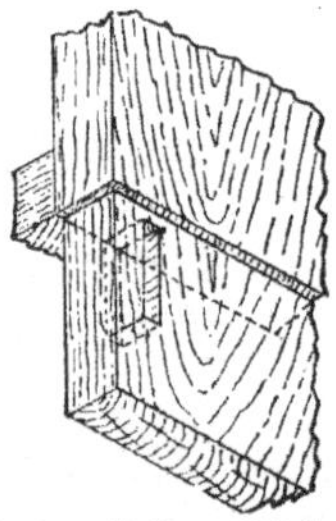

Fig. 890.—Bottom of Pedestal Side, showing Sinking, etc., for Plinth.

and glue on the bottom rails flush on the top side; when these are dry, the remaining mortices will have to be made for the divisions. Nail or screw the drawer guides or tilting pieces J (Fig. 888) upon the cross-divisions parallel with the bottom edge, and make the lock mortices in the top rails.

Gluing Up and Completing Desk.—Lay the bottom on the bench, face side up, glue the lower tenons of the divisions, insert in the bottom, drive the ends on, and screw; put the top rails on, glue, cramp, and wedge up carefully, trying for square in both directions. Next apply the centre division in place, mark to length, and cut the dovetails in the end (these need only go halfway down), glue, and drive home; clean the case off, turn it over, clean out the rebate, and mitre round the cocked bead, bradding it in. At this stage it will be convenient to fit the drawers if they are ready, as it will be easier to see where they bind before the top goes on. To fit on the top, lay it flat on the bench, face down, and turn the case up on it; put a handscrew on the end, to keep it steady, and turn in the screws; glue in plenty of angle blocks. Then turn the case right side up, and fit the slopes, correcting the nosing or joints where required. A cramp should be provided for each 2 ft. of length, and a stiff piece of quartering to run along the top for the cramps to pull against; hollow out several pieces of stuff for the nosing end of the cramps, and put some shavings inside under the joints to catch the glue that may fall. Rub a little chalk on the joints, glue in the tongues and joints, and cramp up both sides equally; leave the cramps on for at least six hours, then block the under side, and clean off. If plenty of help is not available, it will be advisable to glue only one side at a time, letting the first side dry before the second side is done. All that is now required to finish off the case is to bead the drawers and fit in the locks and handles; the brass fitting on the top is screwed in position after the desk is fixed. The pedestals can be brought to the exact width of the desk, placed back to back, and the plinth fitted, but not fixed, the sockets for the foot-rails screwed on, and the rails cut in; as soon as the pedestals are fixed in position, the plinths may be fixed also. The following are the letter references not mentioned in the text:—N, nosing of top; O, pedestal side; S, plinth; U, drawer front; V, drawer back; Y, panelled end framing; P, pedestal top; X, drawer bottoms; Q, pedestal bottom; Z, angle blocks; W, drawer stops; T, front plinth; E F, end of front; I, pedestal back.

Portable Desk.

The desk shown by Figs. 891 and 892 has been designed for use as an entering, invoicing, or checking desk. It may be made of pine, with ash sides, and painted and grained; or of hard foreign woods and polished. The choice must be governed by the style of furniture in the office in which it is to be placed. The pieces forming the bottom stands B S (Figs. 891 and 892) are first proceeded with. Fig. 893 shows a part sectional elevation. Each piece is 2 ft. 1½ in. by 6½ in. by 3 in., with an ovolo moulding and V groove worked on the face edges (see Fig. 894). A groove 1⅜ in. wide by 1½ in. deep is cut for the reception of the side-piece S P (Figs. 891, 892, and 894). The top framework T F (Figs. 891 and 892) is moulded and grooved (see Fig. 895) in a similar manner to the bottom stands, but the addition is made of the back bar shown by B B (Fig. 893). This is tenoned into the sides, the moulded edges being cut back to get square shoulders for the tenons. This bar is moulded on one side only, as seen in section in Fig. 893. The two side-pieces S P (Figs. 891 and 892) are now set out to the drawing, the ornamental sweeps being cut with a pad-saw. The front edges are heavily chamfered to relieve the thickness, and at the back edge immediately above the bottom stands a recess is cut, 2¾ in. wide by 1 in. on and ⅞ in. deep, to receive the dovetailed end of the bar B (Figs. 891 and 892); this ties the two side-pieces together. The lower part, or stand for the desk proper, may now be framed together, the side-pieces being fixed into the bottom stands and top frame by screws which are driven on the splay, and afterwards filled up.

Middle Part of Portable Desk.—The middle part, or desk proper, is now taken in hand. It consists of four corner pillars of the panels, the arrangement of the dovetailing of these being shown in Figs. 896 and 897. In letting in the back panel B P (Fig.

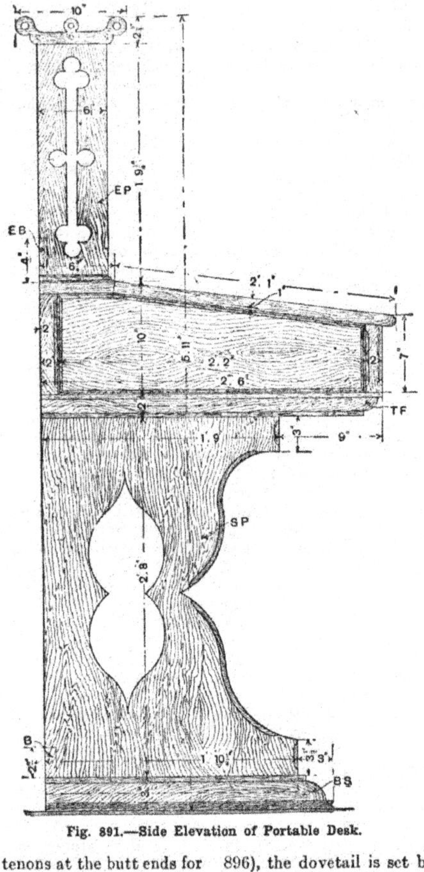

Fig. 891.—Side Elevation of Portable Desk.

2 in. square, with tenons at the butt ends for fixing them into the top frame moulding. Dovetail grooves are cut for the reception 896), the dovetail is set back from the face line of the panel ⅛ in., to allow greater thickness of wood in the pillar. The same

arrangement is carried out in the front panel. The panels are ¾ in. finished; the various widths can be obtained from the drawings. The front and back panels are left plain. The end panels have a small corner moulding fitted; a section of this is shown in Fig. 897. The bottom edge of the front panel is rebated to receive the floor (see Fig. 898); in the other panels rebating is not necessary, as the floor-boards rest upon the top framing and butt against the panels (see Fig. 893). The slope is formed by a board 5 in. by 1 in., fixed at each end, and a hinged flap working between them. Round the corners as shown in Fig. 893, and run a bead along the inside edge to break the joint. The hinged

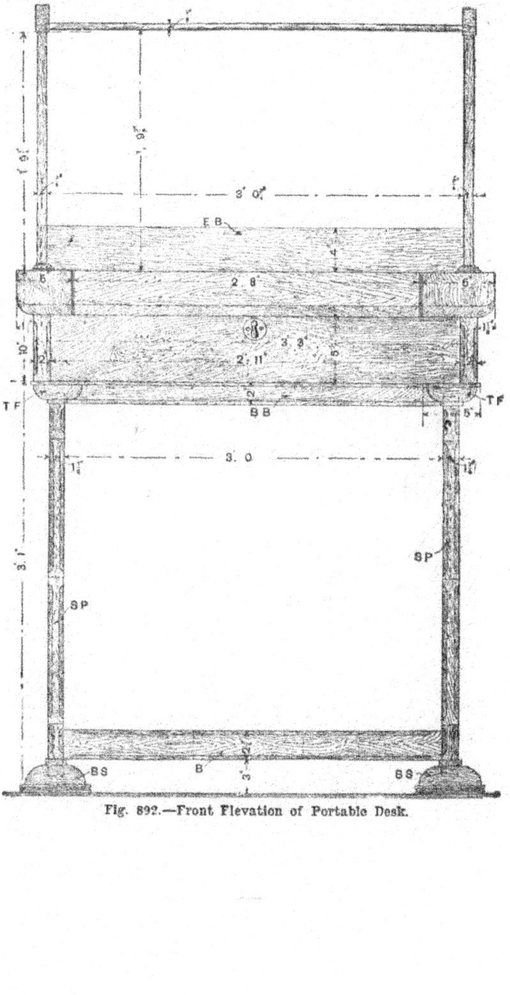

Fig. 892.—Front Elevation of Portable Desk.

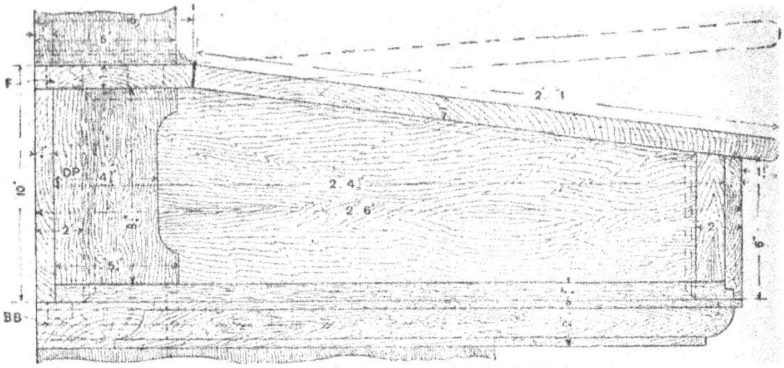

Fig. 893.—Part Vertical Section of Portable Desk.

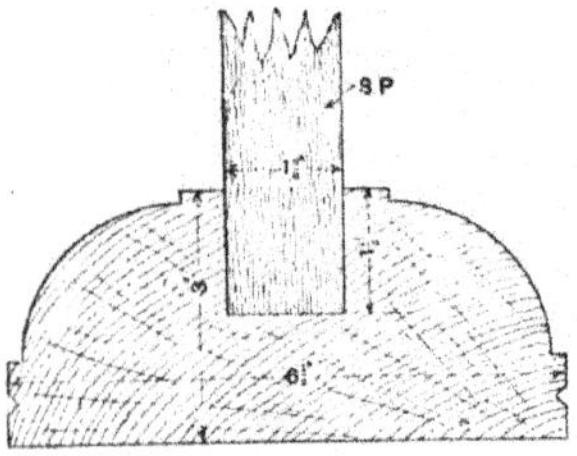

Fig. 894.—Section of Bottom Stand and Side-piece of Portable Desk.

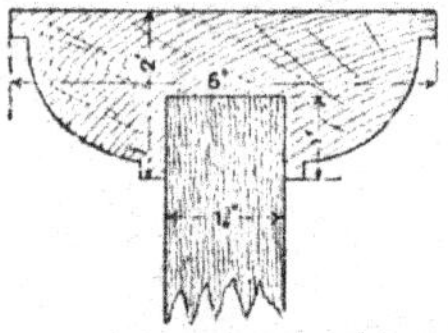

Fig. 895.—Section of Top Frame and Side-piece of Portable Desk.

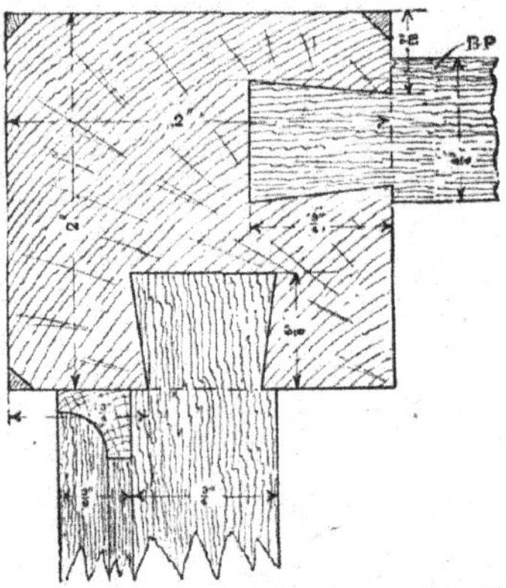

Fig. 896.—Section of Pillar and Dovetails of Side and Panels of Portable Desk.

flap is made of 1-in. stuff, jointed and glued to the required width—2 ft. 1 in. The ends are then clamped, and the hinges fitted. A flat piece F (Fig. 893), 6¾ in. by 1 in., provides accommodation for inkstands, etc., and acts as the hinge-piece for the flap or lid. Inside the desk under this flat piece the space may be conveniently partitioned off for different lots of invoices, with divisions ¼ in. thick cut as shown at D P (Fig. 893). Glue will be sufficient to keep them in place.

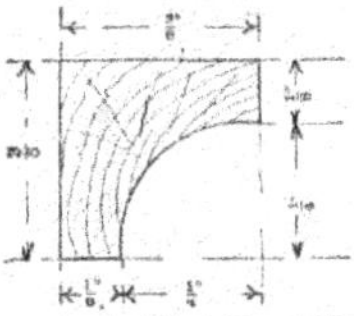

Fig. 897.—Section of End Panel Moulding of Portable Desk.

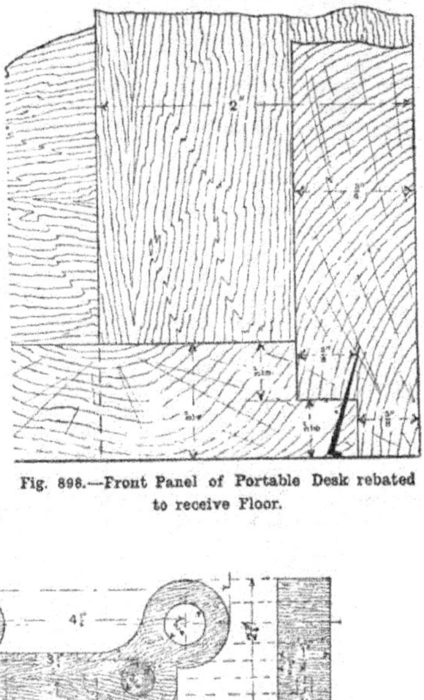

Fig. 898.—Front Panel of Portable Desk rebated to receive Floor.

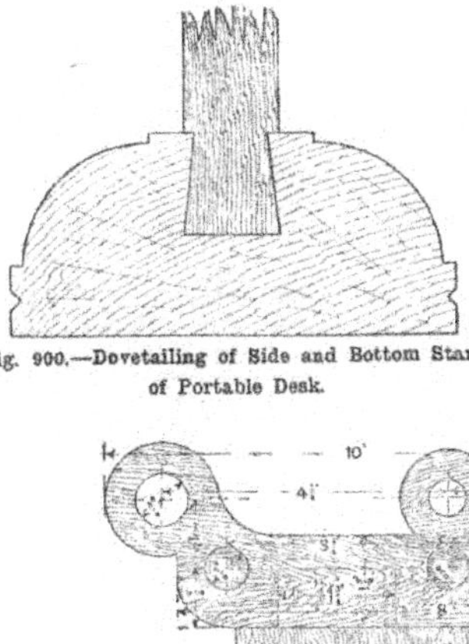

Fig. 900.—Dovetailing of Side and Bottom Stand of Portable Desk.

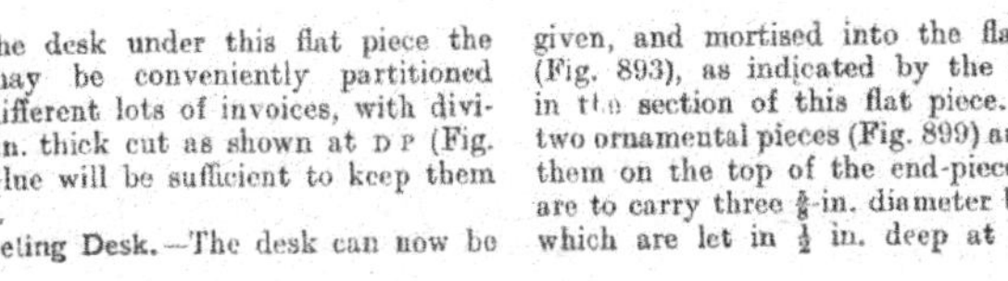

Fig. 899.—Ornamental Piece of Portable Desk, carrying Brass Rails.

Completing Desk.—The desk can now be placed upon the stand, the tenons at the butt ends of the corner pillars fitting in corresponding mortices in the top frame mouldings to keep it in position. These tenons are indicated by dotted lines (Fig. 893). Two end-pieces E P (Fig. 891) are cut out to the size given, and mortised into the flat piece F (Fig. 893), as indicated by the lines seen in the section of this flat piece. Cut out two ornamental pieces (Fig. 899) and mortise them on the top of the end-pieces. These are to carry three ⅜-in. diameter brass rods, which are let in ½ in. deep at each end.

These rods are useful for books and papers to lie upon when not in actual use. At the foot of each piece a small ovolo moulding is fixed, to give a finish, and also add to their support. An edge board E B (Figs. 891 and 892), 4 in. by $\frac{3}{4}$ in., is fixed along the back of the desk flat, to guard against things being pushed over. It fits flush on the top, and is dovetailed into the uprights. The making of the desk is finished with the fitting of a lock and escutcheon. An alternative and better method of putting the bottom stands and side-pieces together is shown in Fig. 900, the side being dovetailed instead of grooved in, as shown in Fig. 894.

Office Stool.

Fig. 901 is a general view of a common form of office stool which could be made in deal, mahogany, birch, pitchpine, and similar woods, to match the other fittings of the office. Figs. 902 and 903 are elevations. The dimensions here given are frequently adopted, although, of course, they can be varied to the requirements. The legs are made of stuff $1\frac{3}{4}$ in. thick, and it will be noticed that they are $1\frac{7}{8}$ in. wide for the greater part of their length, spreading out at the bottom to $2\frac{1}{2}$ in. To economise material they should be cut out in pairs as shown at Fig. 904. All the lower horizontal rails should be trued up to $1\frac{3}{8}$ in. by $1\frac{1}{8}$ in., the top rails being 3 in. by $1\frac{1}{8}$ in. When all the pieces are planed, the rails should be set out for stub-tenoning and the legs for

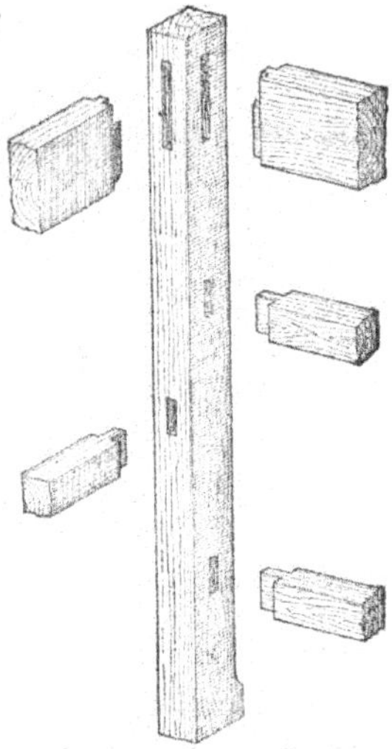

Fig. 905.—Joints in Office Stool.

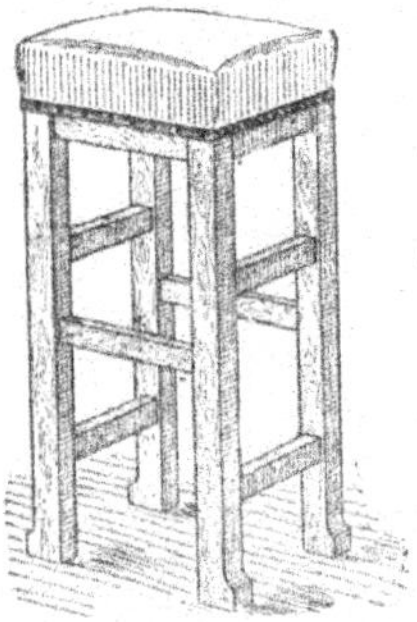

Fig. 901.—General View of Office Stool.

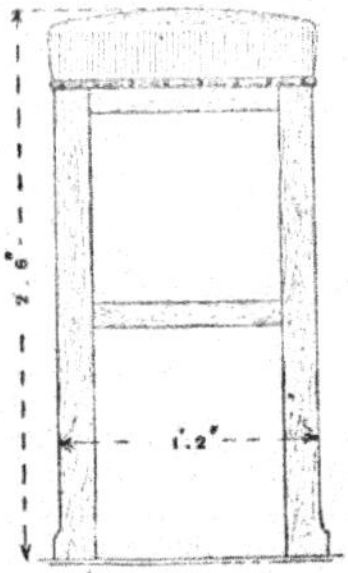

Fig. 902.

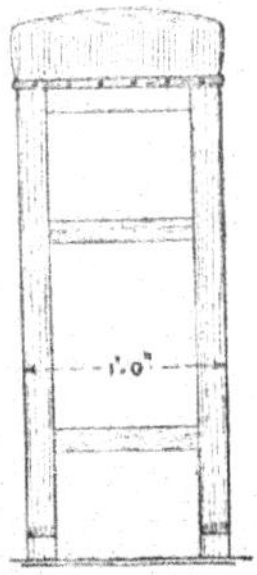

Fig. 903.

Figs. 902 and 903.—Front and Side Elevations of Office Stool.

Fig. 904.—Leg of Office Stool.

mortising. The best way to set these out so as to have them all the same is to place the legs together and mark out the position of the mortices. The same method applies to setting out the rails for tenons, care being taken to have the proper rails together. The joints are shown conventionally at Fig. 905, which illustrates one leg and the ends of three side rails and two back rails. When the joints are made, the whole of the framework should be fitted and knocked together, and the several joints numbered, so as to get them into their proper places again. The whole of the framework may now be glued together; or perhaps an easier way is to glue together the two front portions, leaving them in a light cramp until dry. This cramp may simply be a piece of wood with two cleats nailed on and tightened with a wedge. In fixing the parts together, see that they are square with each other. The joints of the side rails may be treated in the same way when the fronts are dry.

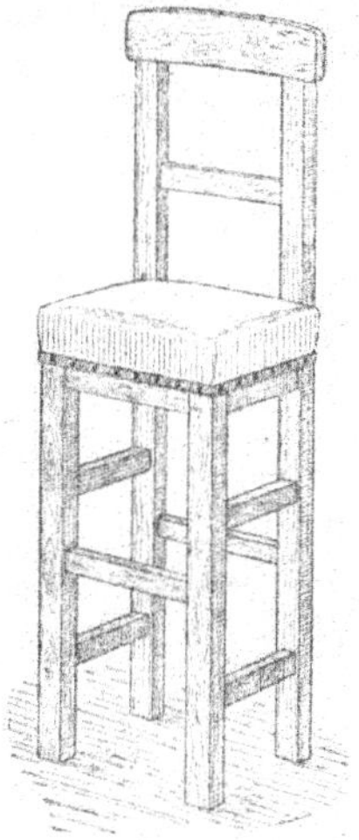

Fig. 906.—General View of Office Chair Stool.

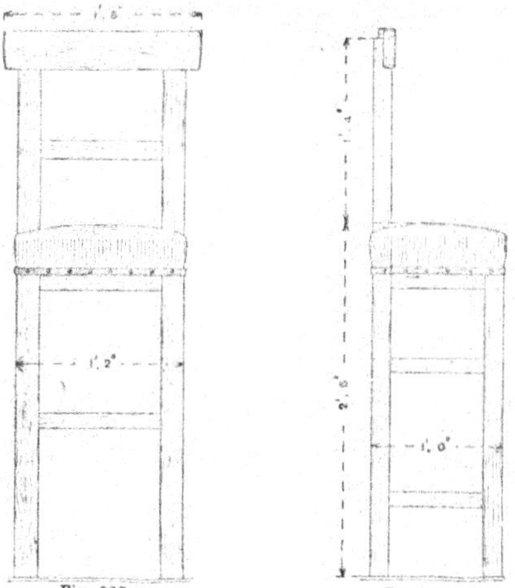

Fig. 907. Fig. 908.

Figs. 907 and 908.—Front and Side Elevations of Office Chair Stool.

Office Chair Stool.

The stool shown in Figs. 906 to 908 is so similar in construction to the foregoing that a detailed description is unnecessary. The sizes of the legs and rails are the same as those of the stool shown by Fig. 901. The back, which is 3 in. by 1 in., should be hollowed in plan to give greater comfort to the user. The top ends of the back legs are notched out to receive this rest, two screws being inserted through the legs to hold the rest.

Table-desk with Rounded Corner.

The desk shown by Figs. 909 to 911 is suitable for fitting in an internal angle formed by two walls or screens; the outer corner is rounded, and the central compartment of the desk top is fitted with a hinged flap; when a flap is not fitted, the upper portion is termed a slope. Desks of this class are usually mounted upon legs similar to those of a table, hence the name table-desk. Figs. 909 and 910 are respectively front and end elevations showing the desk resting upon the table, and surmounted at back and ends with a skirting; end rails are framed between the legs near the bottom, and between these a footrail is provided about 9 in. from the front. The portion of Fig. 911 to the left of the centre line C is a plan of the desk; that to the right of the centre line is sectional, showing the framing of the desk; the dotted lines represent constructional details, but the framed bottom has been omitted in order to avoid confusion.

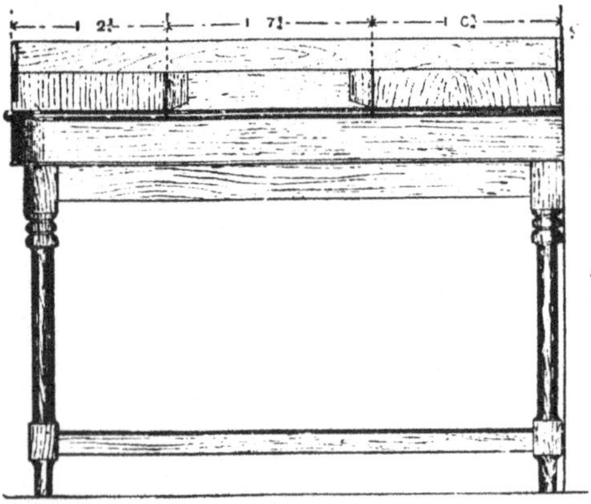

Fig. 909.

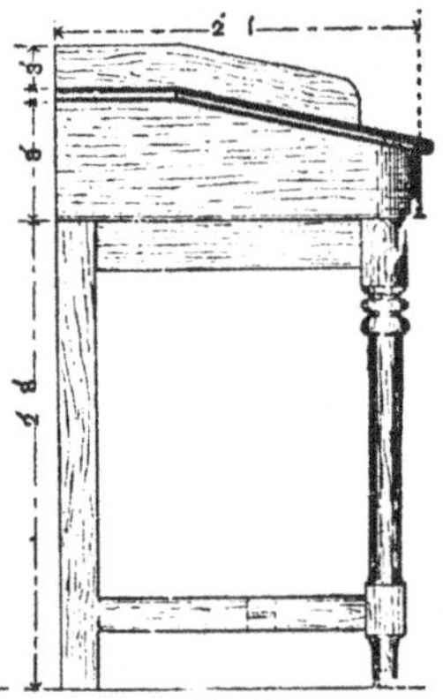

Fig. 910.

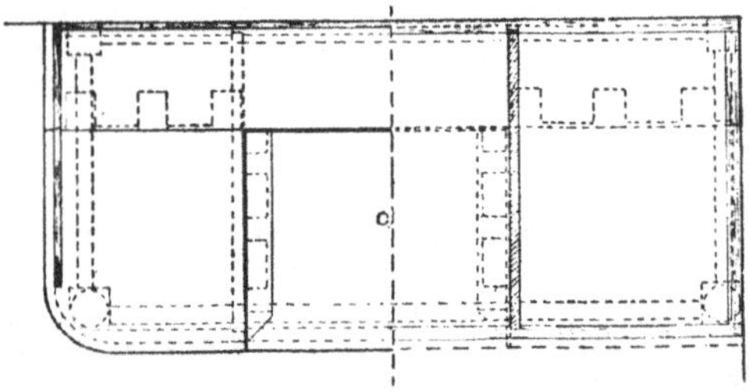

Fig. 911.

Figs. 909 to 911.—Front and Side Elevations and Half Plan and Half Horizontal Section of Table-desk with Rounded Corner.

Fig. 912 shows a cross section through the centre of the desk top; Fig. 913 is a plan, to an enlarged scale, of the block at the corner of the desk front: Fig. 914 is a development of the front and end of the desk ready for being bent round the block; Fig. 915 is a sketch of the top of the corner leg, showing the method of framing in the rails. The figures will provide almost all the necessary data for the full-sized setting-out from which to get the quantities and sizes of the stuff. The corner of the desk top should be set out to a 4-in. radius, and the framing should be of 1-in. stuff, and the skirting

½ in. thick. The flat portion at the back of the slope is 8 in. wide; the wings, it will be noticed, are of unequal width, and the flap is mitre-clamped with 2-in. clamps. The divisions, which are 1 in. thick, are kept flush with the edges of the wings, although in some instances these are kept over ¼ in. to form rebates for the flaps. The front legs are 3 in. square, the square of the corner leg being rounded to the same sweep as the shaft. The desk bottom is ¾ in. thick, with ½-in. panels; the back is 7¼ in. by 1 in., and the front, which is 4¼ in. by 1 in., is rebated to receive the bottom, and is kept up ¼ in. to form a rebate for the edging bead.

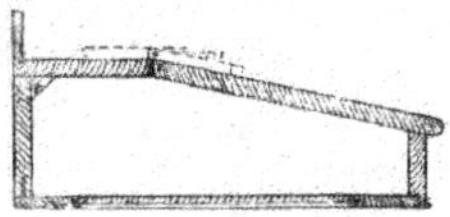

Fig. 912.—Cross Section through Desk.

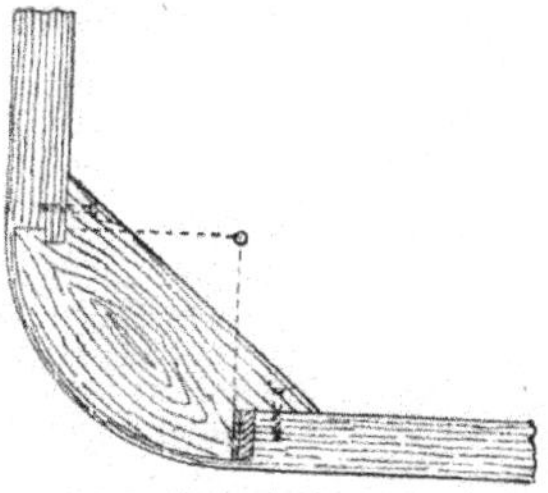

Fig. 913.—Block at Rounded Corner of Desk Front.

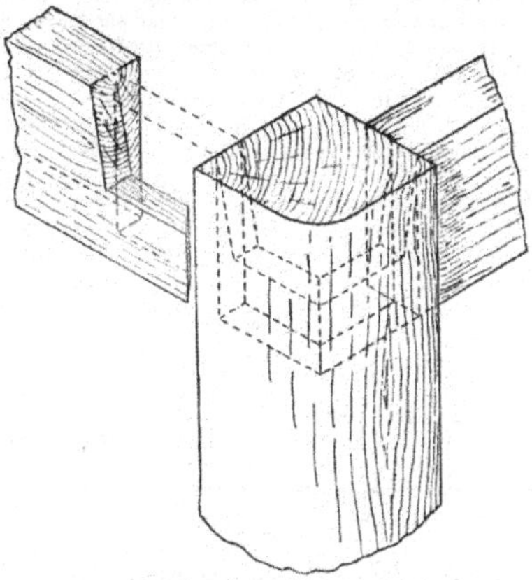

Fig. 915.—Framing of Table-desk Legs.

Fig. 914.—Front and End of Table-desk before bending.

Stand.—In setting out the stand, the front and back legs, being of different substances, should be marked in pairs, best sides out; lay one of the front ones on the rod, and square up the top and floor lines, and the lines of the rails, marking the line of the top rail on both faces, but the bottom rail on one face only; pair the other legs with this one, and square over all the lines, all but the floor line being in pencil. Set a mortice gauge to a ⅝-in. chisel, and gauge it ½ in. from the face; this will bring the mortices nearer the outside of the leg; and as the mortices are made on adjacent sides, in which the tenons meet, a longer tenon is ensured. The mortice should be half as wide as the rail, and the haunching should be sloped from nothing at the top to ⅜ in. at the bottom, as shown by Fig. 915. The mortice for the bottom rail will be kept down ¼ in. so as to clear the rounding, and the back legs can be mortised through. The profile of the turning may be marked on one leg, and the members knife-marked over. The rails will have the lines of the insides of the legs squared up, the shoulder lines being drawn on the insides of the rails, and a ⅝-in. barefaced tenon gauged on the outside; square over a 2-in. by $\frac{5}{8}$-in. mortice in the bottom rails for the foot rail, and in setting out this rail allow $\frac{1}{8}$ in. extra at each end for housing.

Legs.—After the legs are turned, they should be mortised, the rails tenoned and fitted, and the foot rail rounded, housed, and mitered into the end rails; and after being fitted together and marked, the frame should be knocked to pieces, and have the rails and mortised sides of the legs polished. The two end legs may then be glued up and be left cramped up till dry, when they should be cleaned off and have the front, back, and foot rails glued in.

Desk Top.—The desk top may now be proceeded with. The framed bottom is rebated $\frac{1}{2}$ in. into the front and shaped end of the desk, but runs over the back and the wall end as shown by Fig. 912. Two intermediate muntins should be arranged beneath the divisions; the tenons can all come through and be wedged, and the panels will be kept flush on the top side. The back of the desk is a plain board dovetailed into the ends, the pins being cut on the back so that the ends can be driven on. Mark the housings for the divisions, and sink them $\frac{1}{4}$ in. deep. The two divisions and the wall end-piece can be marked alike, between the sight lines of the front and back, allowing $\frac{1}{4}$ in. extra at each end of the divisions for housing, and 1 in. at back, and $\frac{3}{4}$ in. at front extra on the end for the dovetails, which are stopped at the front end as shown by Fig. 914. One of the divisions, after being marked to length, should be laid on the section (Fig. 912), with its lower edge in line with the face side of the bottom; the respective sight lines on the back and the front, and the lines of the under sides of the top and the slope, can then be accurately marked with square and straight-edge; shoot off to the lines, and use this as a template with which to mark the remainder. Next take the front and round end, which, as will be seen by Fig. 914, is all in one piece; shoot the bottom edge, gauge the rebate for the bottom, and square over a line $1\frac{1}{16}$ in. from the right-hand end for sight-line of back (all lines to be on the inside); then take the division and set its bottom edge to the rebate gauge line, and its back sight-line to the line just squared over, and mark the shape of the top. There is no need to mark the front end, but measure the exact distance from the inside face of the back to the centre from which the corner was struck, and set this off on the end, and square a line across; this is the springing line. Knife-cut another line $\frac{3}{16}$ in. farther along, which will be the end of the tongue seen in Fig. 913; then make a template out of thin stuff to the shape of the block (Fig. 913). To do this, lay it in position on the plan, and strike the sweep with the compasses from the same centre. Care should be taken to get the two edges forming the abutment of the template square with each other, and exactly at the springing of the curve, otherwise the desk will not be square when finished. When the template is ready, hold it upright on the front, with its tongue end to the springing line, and steadily roll it along until the other end is reached, at which point square over a line, and allow $\frac{1}{4}$ in. more for wedging room; the wood between these two lines is eventually to be cut away to a veneer for bending round the block. Gauge the front 4 in. wide up to the first springing line, and as a portion of the slope comes in the corner, more width will be wanted at the bend, and it will be found best to cut the stuff rather full, finishing off to the required line after bending.

Bending Veneer Round Block.—Mark and work the rebates, dovetails, and grooves; form the veneer for the corner by gauging $\frac{1}{16}$ in. from the face on each edge, sinking a series of grooves to the required depth with a router, and cleaning out the core with a rebate plane. Care must be taken not to make the veneer either hollow or round on the back, as any such fault will be reproduced on the face of the finished work. Next cut the tongue to fit the groove in the block; then prepare the block itself, for which a piece of clean yellow deal or pine 6 in. long, $2\frac{1}{2}$ in. wide, and 5 in. thick will be required; where possible, the grain should run the same way as that of the veneer, to minimise the danger of splitting through unequal shrinkage. Mark the shape of the block from the template, keeping its edge fair with the back of the block and square down the rebates, then place the templates on the other side, keeping the rebates exactly to the lines and the edge fair at back as before, and mark this also. After the face is worked, using the same precautions as advised for the veneer, lay the tongue end on the veneer, mark the thickness of the groove, and cut it to a rather tight fit; bore holes for two screws on each side, and make ready for gluing up. Well soak the outside surface of the veneer with boiling water for about five minutes, then turn the front over and secure it firmly to the bench with hand-screws. Score the face of block and inside of veneer with a bradawl, glue the tongue and abutment, and drive home the two screws at that end; then thoroughly glue the faces of both block and veneer, and, steadily but quickly, bend up the end, at the same time pressing hard on the bench to squeeze the glue out. When the veneer is bent quite round, and the abutment of the block is well down, put in a pair of folding wedges previously glued, and gently drive them with a hammer above and below. This is the critical part; for if the wedges are overdriven, the veneer will be torn off; and if they are insufficiently driven, blistering and buckling of the veneer will occur when dry. If, when the face is tapped lightly with the handle of the hammer, the sound is solid and everywhere alike, stop further wedging, and turn in the two screws; then turn the work on edge, brace it square, and leave it until the glue is thoroughly dry. The back may then be glued in, the end driven on, the divisions put in from the top, the rebates flushed off, the bottom dropped in and screwed, and the whole cleaned off and papered; then level off the framing with a straightedge ready for the top.

Preparing and Fixing Top.—To prepare the top, get out the back rail and plough a $\frac{1}{4}$-in. groove for the skirting, stopping it at the return end; work the nosing on the end; then set out about three mortices for each wing, $2\frac{1}{4}$ in. wide, 3 in. deep, and $\frac{5}{16}$ in. thick, make a table haunching at the seen end and an ordinary one at the wall end. Set out the wings, marking the shoulders with the requisite bevel, allowing $\frac{3}{16}$ in. extra on top side for bead. To gauge the tenons, set a bevel as shown by dotted lines in Fig. 912, and from that work off the top ends from the shoulder line; gauge the tenons from this face, and cut and fit them in position on the desk. Mark round the margin, also mark the shape of the nosing, then take the wings out and work the nosing, also a $\frac{3}{16}$-in. bead round the flap opening. To fix the top, lay it face downwards on the bench, unscrew the desk-bottom, turn the framing down on the top, and fix the back rail in position with screws and angle blocks. Again turn it face upward, glue the tenons of the wings and drive them in, and cramp the work up till it is dry; then screw each tenon from underside, fix the wings to the framing with blocks, and finally screw in the bottom. Plant in the edging bead on front and end, mitre in and fix the back head in the flap opening, and finish the skirting grooves, stopping them $\frac{3}{8}$ in. from the end of the skirting. Clean off the top, and fit up and hang the flap, which should be mitre-clamped at the front sides, the mitres starting from the finish of the nosing, and the tenons coming through. The grain of the flap will, of course, correspond with the desk. The skirting must be cut to fit the top, and be rebated at the back to fit the groove in the top, and should be dovetailed at the corners, the outer end being secret dovetailed. The desk may now be either screwed or dowelled to the frame, the latter method being best. Put one $\frac{1}{2}$-in. dowel projecting $\frac{1}{2}$ in. in each leg, black the ends with a little oil from the stone, carefully place the desk in position, and the places for the dowel holes will be accurately marked.

Secretary's Knee-hole Desk.

The desk described below contains cupboards accessible from the front; the two end cupboards are deep, but the centre cupboard is a shallow one, because of the knee-hole, the back of which is indicated by the dotted lines in the end elevation (Fig. 916); the shelves are not shown,

Fig. 917.

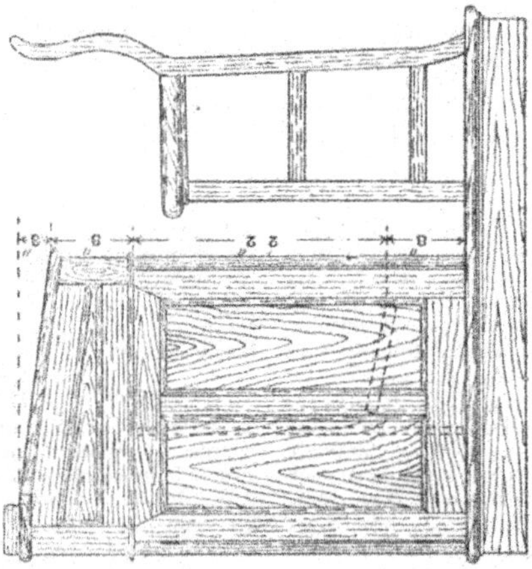

Fig. 916.

Figs. 916 and 917.—End and Back Elevation of Secretary's Desk.

but the end cupboards may each contain two shelves, which should be trenched into the ends and divisions, and divided to suit the owner's convenience. Along the top of the desk, and opening behind it, are three drawers. The length of the desk is 4 ft. 6 in., and it is divided into three equal portions; or the centre opening may be made larger than the others. The width of the desk is 2 ft. 6 in., and the height is 3 ft. 9 in. Fig. 917 shows a back elevation, Fig. 918 front elevation, and Fig. 919 section of the framing. The desk is and the bottom are dovetailed into the ends and mortised for the divisions; a tongue is worked up the back edges of these divisions, and the back framing is grooved to receive them, a $\frac{1}{8}$-in. bead being run up both edges of the ends, to break the joint of the doors and the back framing. In putting together, the backs are screwed, glued, and blocked on, and the holes cored up. An 11-in. by $1\frac{1}{4}$-in. footboard, with a 2-in. slope, should be inserted in the knee-hole, as shown. The top carcase is formed of framing similar to that already

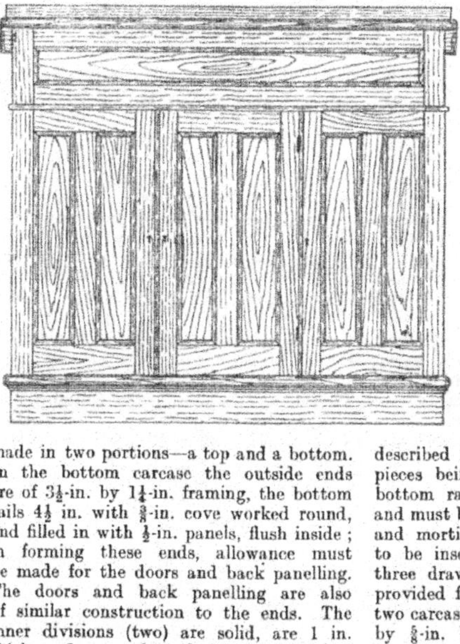

Fig. 918.—Front Elevation of Secretary's Desk.

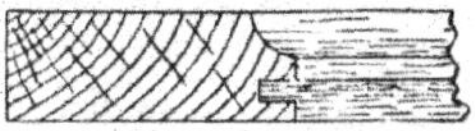

Fig. 919.—Section, showing Jointing of Desk Framing.

made in two portions—a top and a bottom. In the bottom carcase the outside ends are of $3\frac{1}{2}$-in. by $1\frac{1}{4}$-in. framing, the bottom rails $4\frac{1}{2}$ in. with $\frac{3}{8}$-in. cove worked round, and filled in with $\frac{1}{2}$-in. panels, flush inside; in forming these ends, allowance must be made for the doors and back panelling. The doors and back panelling are also of similar construction to the ends. The inner divisions (two) are solid, are 1 in. thick, and come through on the front face to receive the cupboard doors; the top and the bottom are also of 1-in. stuff, the bottom being the full width, but the top of this carcase may be formed of three pieces, 5 in. or 6 in. wide; both the top described for the bottom, the front angle-pieces being 3 in. square; the top and bottom rails are about 4 in. by $1\frac{1}{4}$ in., and must be dovetailed into the end framing and mortised for the divisions that have to be inserted to form openings for the three drawers; proper runners must be provided for these drawers. To keep the two carcases in their proper position, a 2-in. by $\frac{5}{8}$-in. bead should be mitered round the underside of the top carcase, and blocks corresponding to the size of the beading should be fastened to the top side of the bottom carcase; a few screws from the inside of the top carcase into the bottom will secure the blocks. The desk top (which should be buttoned on) is 1¼ in. thick, and may be as shown (Fig. 917), or it may be clamped; two stretchers should be dovetailed into the front framing and the back top rail, to support the top, and on the flat portion, which is 6 in. wide, a rim should be fixed. In this flat portion, receptacles should be made for the ink pots, with a hinged lid to cover them when not in use. A raised platform is required, and the desk can be secured to this platform by screws through the cupboard bottoms. The desk may be made of red deal, yellow pine, or pitchpine varnished—pitchpine for preference.

Draughtsman's Knee-hole Writing Table.

Fig. 920 shows a knee-hole writing-table which is specially suitable for a draughtsman, as the top can be raised so as to be used standing, and it can be made to slope for colouring purposes; moreover, the table is easily taken to pieces for travelling. Fig. 921 is a cross section showing the method of construction. The four drawers are shown in the block at the right-hand side, the spaces on the opposite block being closed with flaps hinged at the top and fastened at the bottom with spring pins, similar to the large flap on the top portion, which is seen in section at Fig. 921. The gables of the top part can be screwed from the under side of the bottom piece, and the long rail immediately below the top dovetailed to them. Fig. 921 shows the top lying level; and to slope it for use when sitting, it is lifted up and drawn forward, thus taking the pin which is fixed to the top out of the socket which is screwed to the back, and inserting it into the socket immediately above it. The bar B, shown in section bearing against the front rail, prevents it from slipping forward. To slope it for use when standing, the pin is placed in the upper hole, and the strut S inserted into one of the spaces in the saw teeth T. The space immediately below the top, with the flap F, can be used to stow away drawings. To render the table as portable as possible, the top part is dowelled to the carcases below, and can readily be taken apart. The back portion is fixed by means of brass plates, which can easily be taken off and the back laid on the top. Immediately below the top is another long rail dovetailed to the gables. Fig. 922 is a vertical section through the drawer block, showing the construction. The gables are dovetailed to the top and bottom pieces, and fillets are nailed to the under side to keep the carcases free from possible dampness. The carcases are covered by a small base, carried along the front and ends. A (Fig. 923) is an enlargement of the pin and socket shown in Fig. 921; B (Fig. 923) showing a modification.

Pedestal containing Two Drawers and a Cupboard.

The solid mahogany pedestal, fitted with two drawers and a cupboard for ledgers, etc., shown by Figs. 924 to 926, conforms to the following specification: Prepare and fix, in Cuba mahogany, a pedestal, 3 ft. by 1 ft. 6 in. by 2 ft., constructed in the strongest possible manner, one side and the front to be of 1-in. mahogany, one side of 1-in. yellow deal, top and sub-top of 1-in. yellow deal edged with 2-in. by 1-in. mahogany slips, and the bottom and back of 1-in. yellow deal. The drawer division to be framed and panelled. The front rail to be of mahogany; the back one of yellow deal; the sides of oak; the panel of basswood. The door to be framed, panelled, and moulded with 1-in. by ⅜-in. mahogany ogee moulding. The bottom to be finished off with 2½-in. by 1½-in. chamfered plinth, with ½-in. by 2-in. return plinth, sunk. The door to have 1-in. by ¾-in. chamfered mahogany stops. The drawers to have 1-in. mahogany fronts, ⅝-in. basswood backs, and ½-in. basswood sides and bottoms; the whole to be rebated and blocked with clean oak slips. The ironmongery will consist of two 2-in. lever locks, two 3-in. brass handles, one 2½-in. lever door lock, a 1-in. brass knob, and a pair of 2½-in. patent brass butts.

Rods.—Given the specification and the drawings (the latter would be similar to Figs. 924, 925, and 926), the first thing to do will be to set out the rod. Three sections will be required: one vertical, as Fig. 925, from which to obtain the heights of the door, drawers, etc.; one horizontal as Fig. 926, showing the width; and one horizontal as Fig. 927, showing the depth from back to front. These may be all be taken through the cupboard. Work that is required above or below the line of section should be shown in dotted lines.

Beginning the Construction.—The pedestal illustrated by Fig. 924 is supposed to stand

Fig. 920.—Draughtsman's Knee-hole Writing Table.

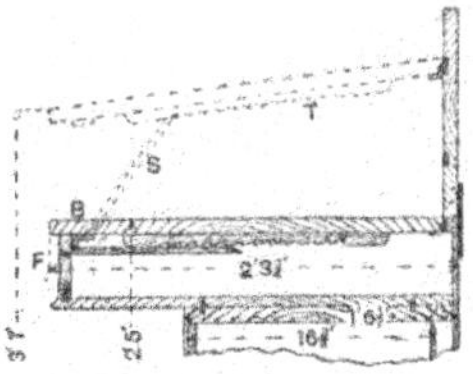

Fig. 921.—Part Cross Section through Draughtsman's Table.

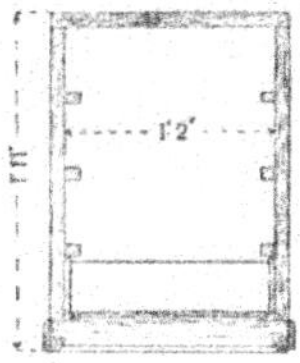

Fig. 922.—Vertical Section through Drawer Carcase of Draughtsman's Table.

Fig. 923.—Alternative Arrangements of Pin and Socket of Draughtsman's Table.

drawn with broken lines, so as to get them upon a 9-in. board, the correct dimensions being, however, marked where the sections are broken. In determining where to make a section, select a point from which most work can be done; in the present instance, a more useful section can be taken through the line X X than could under a desk with its back and left end against the walls. It would be unnecessary to go piece by piece through the table of materials required; the worker will be able to obtain the quantities from the rod, always remembering to allow ½ in. extra in the lengths for squaring. Assuming the stuff to be cut out and planed up out of

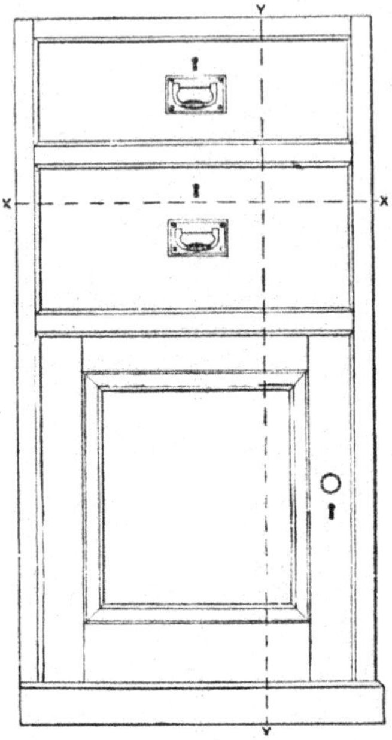

Fig. 924.

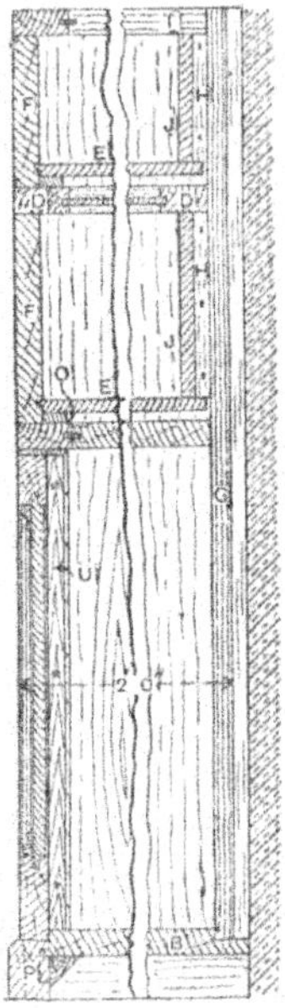

Fig. 925.

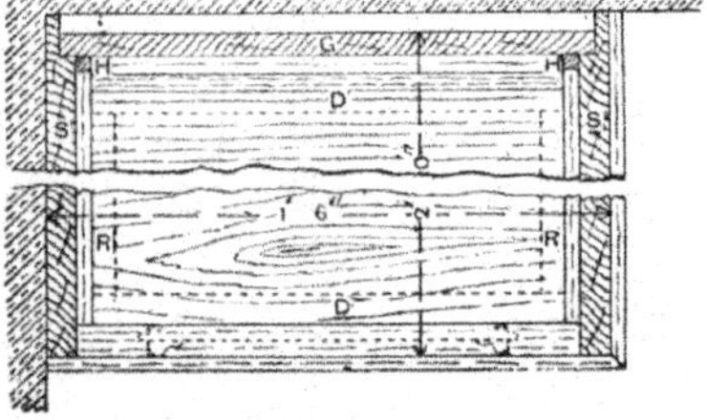

Fig. 926.

Figs. 924 to 926.—Front Elevation, Vertical Section (on Line Y Y), and Horizontal Section (on Line X X) of Pedestal containing Two Drawers and Cupboard.

winding, with the edges shot, take one of the sides, lay it face down on the section (Fig. 925), and mark on the edge the floor line, and lines to indicate both sides of the top, the bottom, and the two divisions. Mark, also, the top of the plinth; pair the other side with the one in hand, and transfer the marks. Next lay the two sides inside up on the bench, and, with the marking knife, square the lines across the face. The top line of the top, also that of the plinth, should be squared over on the outside. The under line of the top, and the top lines of all the rest, should be struck exactly to the mark on the edge, but the other lines should be marked rather bare, in order to ensure the grooves being small enough to allow of the divisions, etc., fitting tight after having been cleaned off. Gauge a stop for the grooves 2 in. from the front edge, and, in the two top grooves, an additional stop, ½ in. from the edge; the portion

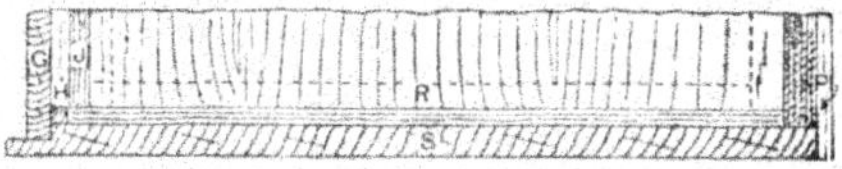

Fig. 927.—Part Horizontal Section of Pedestal, showing Depth.

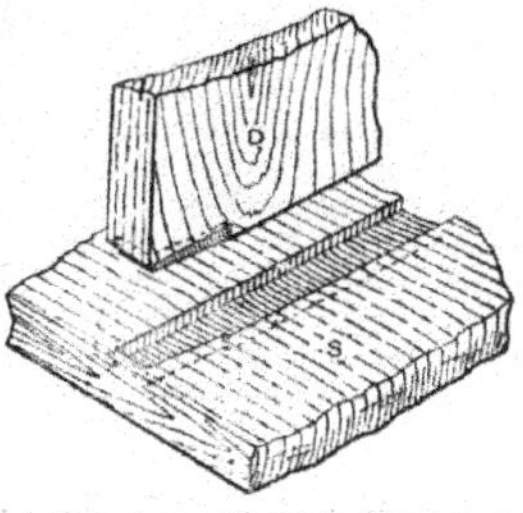

Fig. 928.—Taper Dovetail Housing of Solid Division to Side of Pedestal.

Fig. 929.

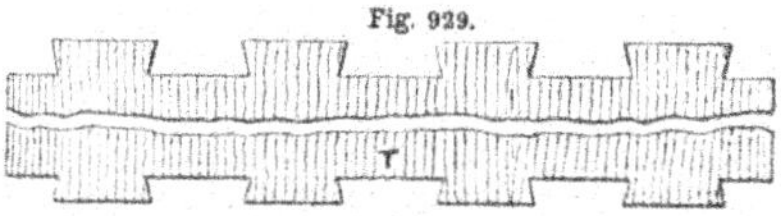

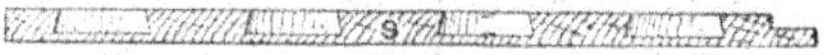

Fig. 930.

Figs. 929 and 930.—Top Dovetailed to Sides of Pedestal.

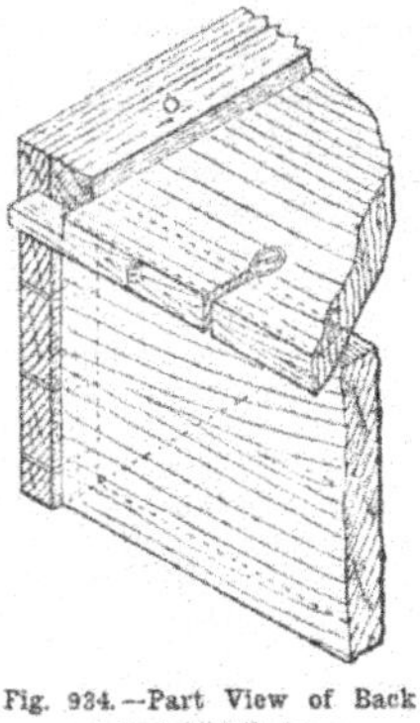

Fig. 934.—Part View of Back of Pedestal Drawer.

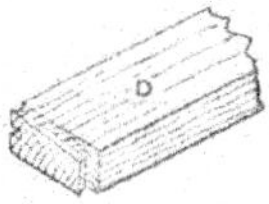

Fig. 932.—Secret Dovetail on Top Division Rail of Pedestal.

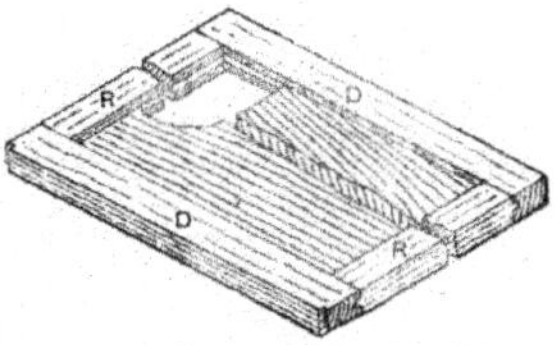

Fig. 933.—Framed Division of Pedestal.

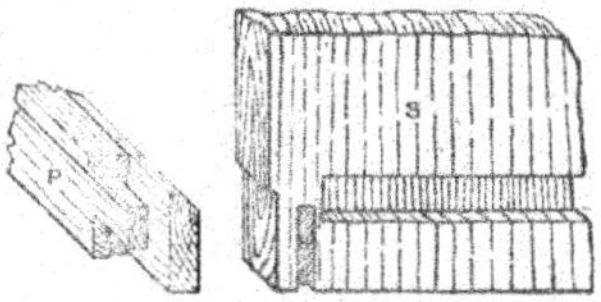

Fig. 931.—Method of Fitting Plinth to Pedestal.

between these two stops will be formed into a taper dovetail housing, as shown in Fig. 928, and, if well fitted, will form an immovable joint. The bottom, being held together by the plinth, does not need this dovetail, so the groove can be housed up square to a 1-in. gauge mark; the grooves, or housings, should be ¼ in. deep. The top ends will be dovetailed as shown in Figs. 929 and 930, but must be left until the pins are cut on the top. Gauge for the rebate to receive the back; this should be done from the front edge in order to ensure the depth being parallel. Obtain the distance from the section (Fig. 927); the rebate should be only deep enough to take a screw—½ in. will be ample. It will be seen by reference to Fig. 931 that the plinth, being solid, affords an opportunity of making a mortice-and-tenon joint at the bottom, the wedging being covered by the return plinth. The oversailing piece of the front plinth is sunk ⅛ in. into the face edge of the side, so as to provide against shrinkage. Set out the mortice ¼ in. below the line of the plinth; this will be sufficient to hide the wedge. Set the gauge to a 5/16-in. chisel, 1 in. from the face; also gauge the ⅛-in. sinking. Run this also across the face of the mahogany side, in order to provide for the reception of the return plinth.

Top.—The top may now be taken in hand. Lay it worse side down on the plan (Fig. 926), and mark over the inside lines of the sides. Next set off on the right-hand end a line ¾ in. farther out; this is for the lap dovetails (see Fig. 930). As the left side is against the wall, the dovetails may be allowed to run through, so set off 1 1/16 in. here, square the lines over the face, and proceed to set out the dovetail pins. As these will not be seen, they may, for the sake of both strength and economy, be made rather large. About 3 in. is the maximum width for safety; greater width might result in the failure of the dovetail through shrinkage. Square a line over in the middle of the length of the proposed dovetails, and set a pair of compasses to 3 in. or whatever width is determined. Commencing at one end of the line, with half the width in the compasses oversailing the stuff, mark off on the line a series of points; commence with half a socket and finish with half a socket. Next set a bevel to 72°, and mark off the pins through the points just made; gauge the width of top with the gauge used for the back rebate in the sides.

Divisions, etc.—The divisions and bottom may now be laid in order, the top placed upon them, and the shoulder lines squared over. It is assumed that the bottom and solid division are glued up, but it is not necessary that the framed division should be glued up before setting out. On all these divisions, allow an extra ¼ in. at each end for housing. The divisions will be gauged the same width as the top; the bottom will require different gauging. Gauge the front edge for a ¼-in. tongue and a ¼-in. rebate; alter the width gauge to ¾ in. less, and gauge the rebate at the back (see B, Fig. 925). Run on another line 1½ in. farther out, this being the cut-off line. The back may be left to be fitted in when the case is together. In actual work the setting out of the drawers would next be proceeded with, but these directions will probably be rendered more clear if the carcase is finished first. Cut the dovetails in the top with a dovetail saw, and, before removing the core, place the top on the sides and mark the sockets in the usual way, care being taken to keep the top to the lines, and flush at the front, otherwise the carcase will be out of square, and there will be trouble with the drawers. Next cut the sockets and remove the core with a bow-saw and chisel; keep the bottom of the lap sockets square. Work the housings with a grooving plane, or bore out about 2 in. with a ¾-in. centre-bit. Bite a straightedge on the line with handscrews or clips, or even a couple of fine brads, and run the tenon saw tight up to it to the required depth. Remove the core with a chisel and router, then cut the sockets for the secret dovetails (see Figs. 928 and 932). If many had to be done, it would pay to make a template, but in this case it is not important that they should be all alike, as each pin is fitted to its own socket. They should be tapered about ¼ in., and undercut ⅛ in.; and sometimes the front edge of the pin is also dovetailed. The body of the divisions should fit fairly tight, but not tight enough to prevent them being driven through from the back. Glue up the panelled division, and, for fitting, treat it as solid. The panel, or dust-board, can be ploughed in $\frac{3}{8}$ in., and stub tenons may be cut on the side rails to fit the groove. When glued up, the division will be found stiff enough to handle. One end of the front rail is illustrated by Fig. 932, the form of dovetail being clearly shown. (Fig. 933 is a view of a framed division.)

Plinth, etc.—Mortice the sides for the plinth, and cut the sinking; the shoulder lines of the plinth will be the same as at the top. Gauge a $\frac{5}{16}$-in. barefaced tenon on the back, and a line $\frac{7}{8}$ in. from the back for the front sinking. Let the right end run over an extra $\frac{1}{2}$ in. to form a mitre with the return plinth. Having fitted the plinth, drive the bottom in place and mark the tongue upon the back of the plinth. Run a $\frac{1}{4}$-in. ploughed groove to this mark, and the carcase will be ready for gluing up. The plinth should be glued on to the bottom, care being taken to keep the shoulder lines right. Enter the tenons in the mortices, and glue up quickly. Place the divisions in their grooves just up to the entrance of the dovetails, glue the latter, and drive on, the assistant holding a piece of stuff against the end of the groove to prevent it being burst. Cramp up the plinth. Glue the top dovetails, and drive on the top. Give all the shoulders a squeeze with the cramp, and a spare cramp may with advantage be left on at each shoulder. Then serve the back the same, turning in a $1\frac{1}{2}$-in. screw in the back edges of the divisions and the bottom. Turn the case down on its side, and hammer the sides well down on to the ends of the divisions; then try the inside with a straightedge. The wall side may be nailed if desired. Clean out the superfluous glue, and stand the case aside to dry. After it is dry, the top, face, side, and front should be cleaned off, the return plinth fitted and glued on, and the bottom blocked, after which the case will be ready for the drawers and the door to be fitted.

Drawers.—On referring to the illustrations, it will be noticed that the drawer backs are kept about $\frac{1}{2}$ in. away from the pedestal back; this is to allow room for the insertion of a $\frac{1}{2}$-in. square block on each side, against which the end of the drawer sides abut. These blocks are marked H in the sections. Cut the drawer fronts off square, $\frac{1}{16}$ in. longer than the opening, and, in gauging the width, make the same allowance. Always face up the worse side and edge of the drawer fronts, as these will go inside and to the bottom. The outside need not be planed till the drawers are fitted in. The backs will be cut off to the same length as the fronts, and to the widths shown in the section (Fig. 925) at J, the illustrations being to a scale of $1\frac{1}{2}$ in. to a foot. The sides will be squared off to the length shown in Fig. 927, namely, from the blocking at H to the front, less $\frac{1}{4}$ in. which was left on for lap; gauge them to the same width as the fronts. Plough all the lower inside edges, with the exception of the back, with a $\frac{1}{4}$-in. groove, $\frac{7}{8}$ in. up from the bottom, $\frac{1}{16}$ in. deep in the sides, and $\frac{1}{4}$ in. in the front. The sockets must next be cut in the sides; for this purpose, set a cutting edge to $\frac{3}{4}$ in., and lightly gauge both sides at the front end. Run the same gauge, somewhat deeper, upon the ends of the fronts from the inside. A second gauge should be set to the thickness of the sides, and run on the insides of the fronts and both sides of the backs, and a third gauge to the thickness of the backs, plus $\frac{1}{16}$ in. for clearance, as shown in Figs. 926, 927, and 934. Run this gauge on each side of the sides at the back ends. Fix the sides in the bench screw in pairs, and set out the dovetails. If for the front end, mark off $\frac{1}{4}$ in. at each side, and, with the same bevel used for the case top, draw a half-socket similar to Fig. 929. Divide the intervening space into equal parts not exceeding $1\frac{1}{2}$ in. each, square the lines over, and draw the sockets as at Fig. 934. Cut them in with a dovetail saw; on the back ends set off half a socket in line with the top side of the ploughed groove, and a whole one $\frac{1}{2}$ in. down from the top edge, and divide the others equally between. The back dovetails may be $\frac{1}{8}$ in. wide at the outside, the front ones $\frac{1}{32}$ in. Fix one of the drawer fronts in the bench-

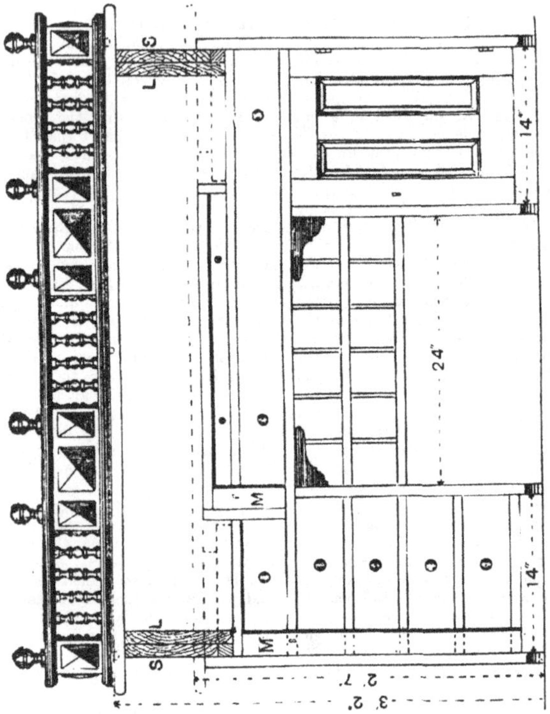

Fig. 935.

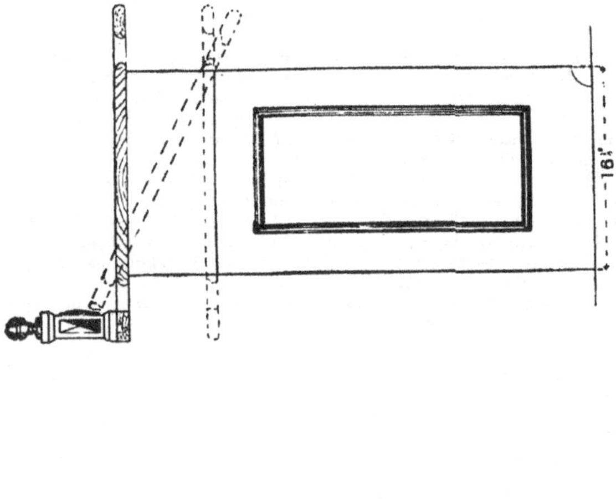

Fig. 936.

Figs. 935 and 936.—Front and End Elevations of Draughtsman's Adjustable Table.

screw, put a $\frac{1}{4}$-in. wood slip in the ploughed groove, and drop the side upon it. This slip will keep it in position laterally. Keep the end up to the gauge line, draw in the pins, and run in the cuts with a dovetail saw. Mark a corresponding number on the front and side, turn the work over, and In preparing the bottoms, clean up the best side of the stuff, and cut it off so that the bottoms fit tight between the grooves at the back of the drawers. Set one gauge $\frac{1}{4}$ in. full, and another $\frac{3}{8}$ in., and run along the back side for the rebate for the blocking (see Fig. 934). The front edge will be

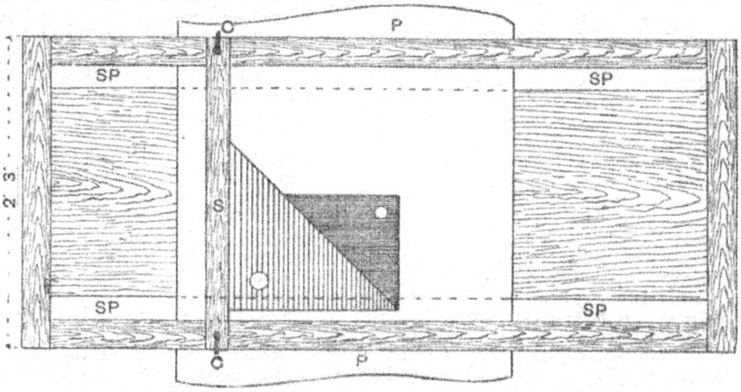

Fig. 937.—Plan of Top of Draughtsman's Adjustable Table, with Balustrade removed.

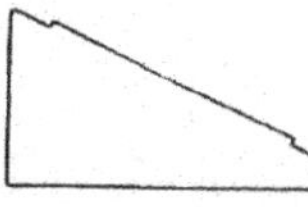

Fig. 938.—Diagram of Sloping Bracket for Draughtsman's Table.

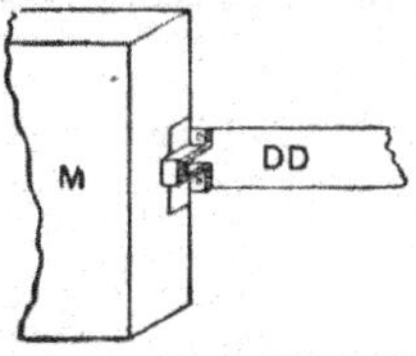

Fig. 939.—Locking Drawer of Draughtsman's Table.

cut the other end. Repeat the process upon the backs, keeping the bottom edge pressed tight up to the wood slip in the groove. Cut the pins down outside the marks, so that they will fit tight, and cut away the core with a bow-saw, finishing up square with chisels. Next clean up the insides; take a slight chamfer off the insides of the sockets, enter the pins, and glue up, set square, and stand aside to dry.

rebated $\frac{1}{16}$ in. deeper, as the front is ploughed deeper than the sides. The grain of the bottoms should run in the direction of the length of the drawers. The bottom must be driven in tight, and slot-screwed at the back, the oak blocks fitted tight in their rebates and glued in, but to the sides only, not the bottom; the front block may be glued to both. When thoroughly dry, the drawers may be fitted in, the fronts cleaned off flush with the carcase, a ¼-in. bead stuck on the edges, a return planted on the ends, and the locks and handles put on. Cut the stop blocks in tight at the back, glue and brad them in, clean off flush with the rebate, and screw in the back of the case. A ⅜-in. air-hole should be bored in the back, opposite each drawer, otherwise the drawers will take a long time to close.

Completing the Pedestal.—Have the inside edges of the door framing, the face of the panel, and the moulding polished before putting them together. It will be noticed that a ¼-in. bead is stuck on both stiles, and a similar one planted on the top rail; this must be done, as the bead cannot be stuck on the ends of the stiles; it may be rebated or taken right across, as shown in Fig. 925. The butts should be kept flush with the bead, and the whole of the knuckle sunk in the door. After everything has been fitted, the furniture should be taken off, and the case damped down and given a final rub with No. 0 glasspaper, when it will be ready for the polisher. In the illustrations, T represents the top of the case, B the bottom, D the divisions, G the back, F the drawer front, E the drawer bottoms, J the drawer backs, H the drawer stops, P the plinth, R the drawer runners, O the drawer blocks, S the sides of case, and U the door stops.

Draughtsman's Adjustable Table.

The drawing table illustrated by Figs. 935 and 936 is a very handy piece of furniture for a draughtsman's room, as regards its receptacles for storing drawings, note-books, etc. With the exception of the balustrade on the top, it is devoid of ornament. It is constructed in such a manner that the drawing-board, which forms the top, can be used level while sitting, or, by turning up a bevelled bracket, it can be used sloping; then, by turning up another bracket alongside the sloping one, it can be raised level for use when standing. Fig. 935 is the front elevation, Fig. 936 the end elevation, and Fig. 937 the plan of top with the balustrade removed, showing the method of using the board. Fig. 938 shows an elevation of the sloping bracket. The construction is complicated, and would require great care in drawing in the stuff. The gables are plain, with what is known as a surface moulding planted on the outside. The bottom shelves of the carcase, as well as the upper ones, would require to be solid to prevent vermin getting access to the interior. The rest of the drawer divisions would be 3 in. broad, and would be in pairs—one at the front and one at the back, though the one at the back could, in the majority of cases, be dispensed with. These drawer divisions are kept back so as to bring the margin stiles M M flush with the outside edge of the gables. The margin stiles are hinged, and have a cupboard lock, the bolt of which is turned into a socket fixed to one of the drawer divisions. This locks all the drawers at once, though the saving is questionable, as it entails extra labour in the construction. This method is shown at Fig. 939. The sloping bracket S and the level one L (Fig. 935) are hinged at the bottom; the sloping one to a small block, and the level one to the top shelf or runner. The dotted lines show the position of the top and brackets when down on the top of the carcase. The balustrade, though portable, is dowelled to the back rail of the top, and can be easily lifted off or on. Fig. 936 shows the different positions of the top, while Fig. 937 shows the manner of using. The paper P is passed through the spaces marked S P (for these spaces see also the end elevation at Fig. 936). A straight-edge S is fixed to the top by clamps C. Fig. 938 is an elevation of the sloping bracket, showing a raised part in the centre to correspond with a groove in the top. This prevents the board from slipping downwards. Fig. 939 is a sketch of a part of a margin stile, showing the method of locking the drawers.

Writing Table and Cabinet.

Fig. 940 shows in elevation a small writing table and cabinet which would look well if constructed of walnut and polished, the interior fittings being of white wood. The table part should be taken in hand first. In setting out the legs, see that the mortices are in their correct places and the legs in pairs with one another. Taking the right-hand end first (see Fig. 941), the rails and panel are made flush with the inside of the legs (see Fig. 942). Two mortices are cut in the back leg for the back rails (see Fig. 943), and one mortice in the front leg takes the bottom rail. Two grooves are also cut right across, and receive the partitions that form the cupboard; they should be stopped ½ in. from the front, and the partitions shouldered to fit them. The top front rail, of ⅞-in. stuff, is tenoned to each outside leg, and mortised to take the centre one (see Fig. 944). The centre legs are joined by two rails, the top rail being 5½ in. by 1⅜ in., and the bottom rail 3 in. by 1⅜ in., and are fitted with a bevelled panel, as at the right-hand end. In the back centre leg six mortices will be required, one each at the top and bottom to take the back rails, which carry the back panel (see Fig. 945), and one mortice at the bottom for the rail A (Fig. 940). The other mortices are for rails which carry the bevelled side panel, similar mortices being made in the front centre leg. Other mortices in this leg take the rail B and the rail under the left-hand drawer. The left-hand end is constructed like an ordinary table, and needs no detailed explanation, except that a rail similar to A and in line with it is mortised between the two legs. A section of the rails is shown at Fig. 946. In putting the framework together, the three pairs of legs should be glued up first, the right-hand and centre legs having their panels in place. They should be cleaned off, and the back panel and the part containing the cupboard and drawers glued up; then the remaining end should be glued on. The runners for the drawers should also be added now. The drawers are dovetailed together in the ordinary manner, and have sunk bevelled fronts fitted with turned wooden knobs. The door to the cupboard is also fitted with a sunk bevelled panel, held in place by ⅜-in. beads (see Fig. 947), and is hung with 1½-in. brass butts and fitted with a brass cupboard lock. The legs are turned to pattern, and the feet fitted with patent ball casters.

Table Top.—The top of the writing table is ⅞ in. thick, and has a thumb moulding worked on it (see Fig. 948), and should overhang the legs 1⅞ in.; it is also dovetail-grooved to receive the ends of the cabinet, and stop-rebated at the back edge C (Fig. 943) to take the back of the cabinet.

Cabinet.—For the cabinet the ends should be prepared first. The under top D (Fig 943) is face-dovetailed to the ends and cut back 1⅛ in. to take the doors (see Fig. 949). Mortices are cut in the ends to take the rails between the drawers (see Fig. 950), and grooves are also cut for the drawer runners. A groove is also made for the board E (Fig. 943), and a groove ¼ in. deep is worked to the sweep for the sliding cover. The partitions F (Fig. 940) are grooved into the board E (Fig. 943), and tenoned into the under top (see Fig. 949); they should be mortised and grooved to correspond with the ends. The back should be made in three widths, and the joints arranged so that they meet behind the two partitions.

Sliding Cover.—The sliding cover may be made in narrow strips, which should be perfectly straight, play being allowed in the groove so that the cover slides quite freely. The bevel for the strips can be obtained by setting out the sweep full size and drawing radial lines. The strips are fastened together by a sheet of sailcloth glued on the back, but the joints must all be free from glue. A ¼-in. bead (see Fig. 951) is put on the ends around the sweep, and a small chamfer is worked on the edge of each of the strips, as in Fig. 952. The bottom strip should be fitted with a brass lock having a hook bolt, and the plate should be let into the top of the table. The pigeon-holes, shown in elevation by Fig. 953, are constructed of ¼-in. stuff, which is grooved, glued, and bradded together. They may, of course, be altered so as to suit individual requirements.

Cupboard.—The cupboard is fitted with two shelves, and enclosed by a pair of glazed doors, hung with 1½-in. brass butts, a section through G H (Fig. 940) being shown at Fig. 954. The drawers are dovetailed together with plain fronts, and fitted with small wooden knobs. The top J and board E (Fig. 943) have a thumb moulding worked on the edge, the moulding on the top being returned in the solid; but

Fig. 940.

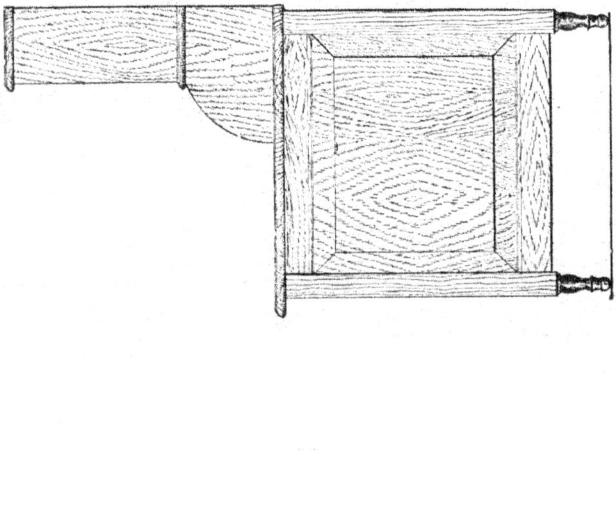

Fig. 941.

Figs. 940 and 941.—Front and Side Elevations of Writing Table and Cabinet.

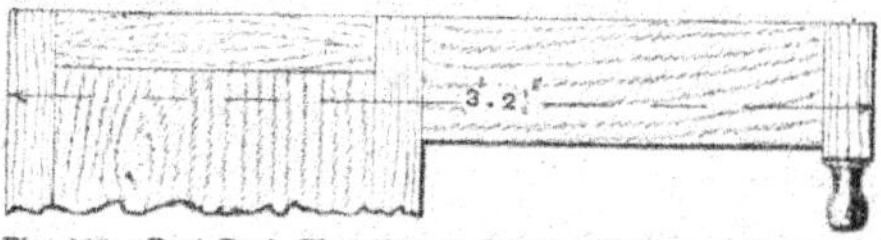

Fig. 945.—Part Back Elevation of Writing Table and Cabinet.

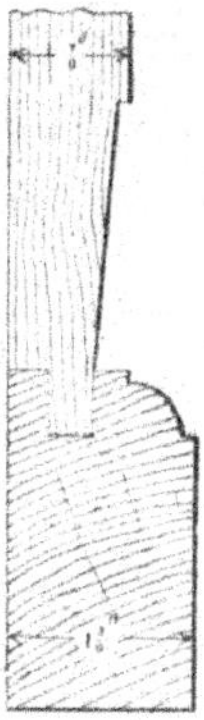

Fig. 942.—Section of End Panel Framing of Writing Table and Cabinet.

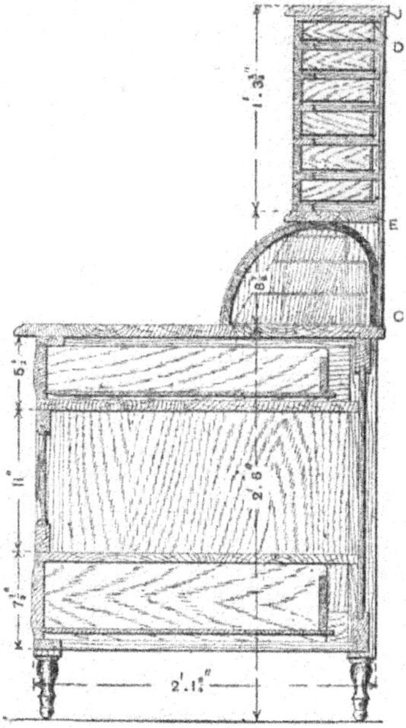

Fig. 943.—Vertical Cross Section of Writing Table and Cabinet.

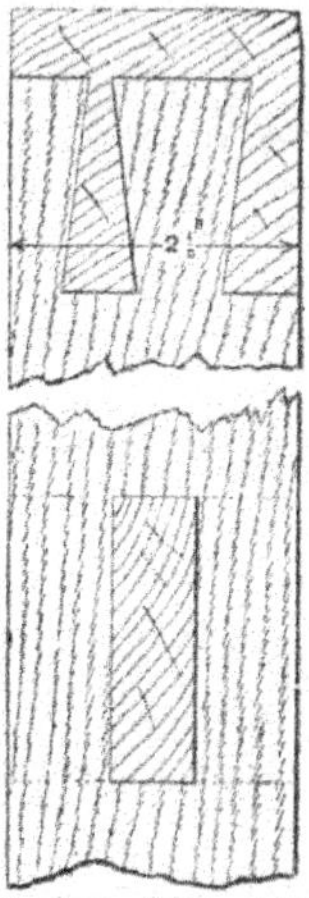

Fig. 944.—Joint of Top Front Rail and Leg of Writing Table and Cabinet.

Fig. 947.—Section of Door Stile of Writing Table Cupboard.

Fig. 946.—Section of Foot Rail of Writing Table and Cabinet.

the moulding on the shelf is mitered at the corner, and a separate piece returned to it (see Fig. 955) on the outside of the end.

Fig. 948.—Moulded Top of Writing Table and Cabinet.

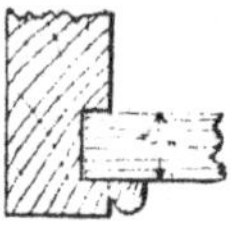

Fig. 951.—Joint of Cover and End of Writing Table.

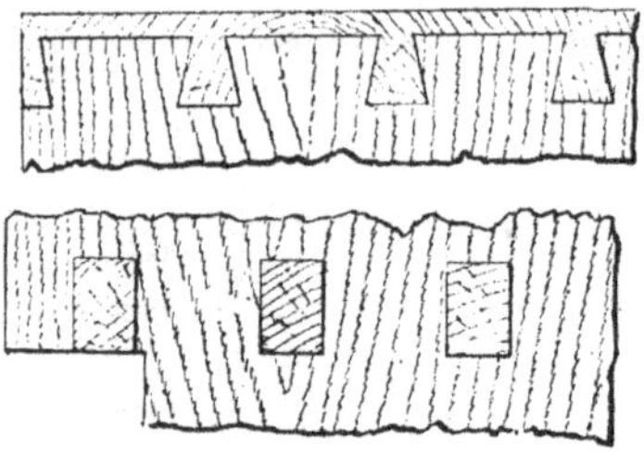

Fig. 949.—Joints of Partition and End of Writing Table and Cabinet.

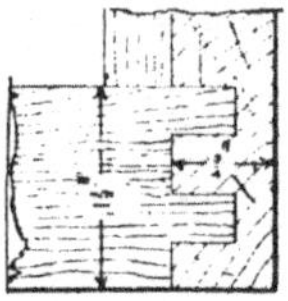

Fig. 950.—Joint of Rail to Cabinet End.

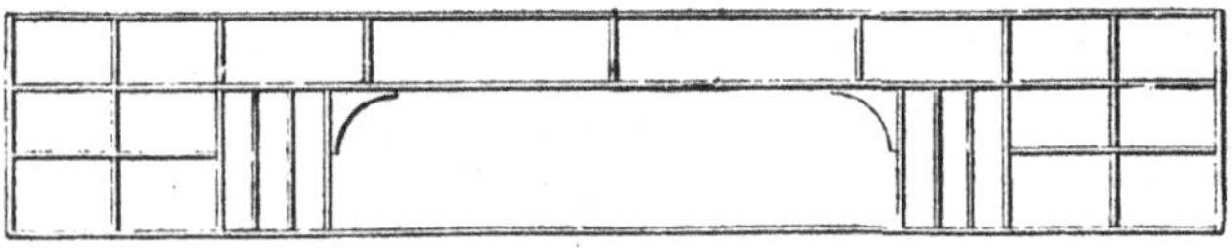

Fig. 953.—Elevation of Writing Table Pigeon-holes.

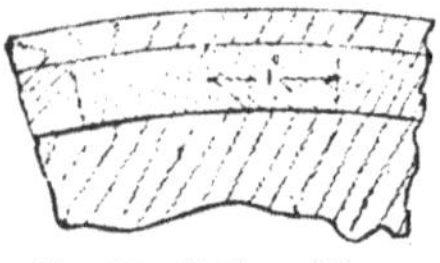

Fig. 952.—Section of Cover of Writing Table.

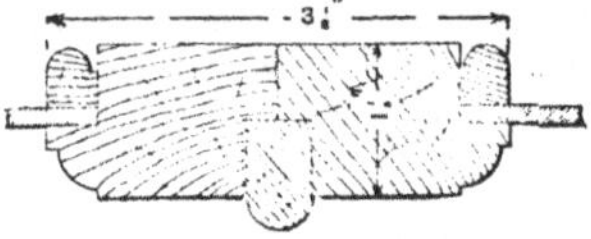

Fig. 954.—Stiles of Cabinet Doors.

Fig. 955.—Mitered Corner of Cupboard Shelf.

Bookcase Tables.

A bookcase table is an excellent substitute for a revolving bookcase, with the advantage of greater simplicity of form and consequent greater ease of construction. Fig. 956 gives a general view of a two-tiered bookcase table, and Fig. 957 an elevation of one with a single shelf. Fig. 958 is a horizontal section. Oak, walnut, or mahogany would be the most suitable material to use, but any cheaper kind of hardwood, stained and polished, will do. Both forms of table are 3 ft. square by 2 ft. 6 in. high; but these dimensions may, of course, be increased or diminished to suit special requirements.

Legs.—The legs, which are got out first, are 2 ft. 5¼ in. long by 2 in. square. They are shown square in the illustrations, but they may be turned if desired, taking care to leave two portions, in the case of the single-shelf table, untouched, the upper 5 in. and the lower 3 in. long at a distance of 1 ft. 1 in. from the lower extremity. Two 3-in. square parts are required if two shelves are fitted; they are 6½ in. distant from each other, the lower being 5½ in. from the bottom end.

Top Rails, etc.—The top rails of the table are 2 ft. by 4 in. by ¾ in., and are double-mortised into the legs, as indicated in Fig. 959. The mortices are 1½ in. long, ¾ in. wide, and 1½ in. deep, and are distant ½ in. from the outside faces. The haunched parts are ½ in. long and ½ in. deep. Figs. 959 and 960 show how the extremities of the rails are shaped. The rails supporting the shelves are 2 ft. by 1¾ in. by ¾ in., and are also mortised into the legs (see Fig. 961), the mortice being ¾ in. square by 1½ in. deep; the outer shoulder is ¼ in. wide. They are set back ½ in. from the outer edges of the leg. The position of the rails in the single-shelf table is 1 ft. 1¾ in. from the lower extremity; in the other, the bottom rail is 6¼ in. from the end, and the one above midway between that and the top rail.

Putting Table Together.—If the table has only one shelf, these various parts may now be glued together; but if there are two tiers, the shelves will have to be prepared before this is done. Each shelf is made up of a couple of boards 2 ft. 0½ in. by 1 ft. 0¼ in. by ⅝ in., glued together. Saw a rectangular piece 1¾ in. square from each corner to admit the table legs, and bore a ½-in. hole in the exact centre for the extremities of the pillars that support the laths against which the books rest. Then cut the notches on the four edges to take the vertical laths connecting the shelves and top rail, as in Fig. 962, which shows the shelf complete. These notches are ¼ in. deep by 1 in. wide, and ⅞ in. apart, the inner one being ½ in. from the middle point of the side. If there are two shelves, both will require to be notched. One side of the bookcase table should be built up permanently, and the remaining rails glued to the other legs. Then place the shelves in position, glue the two portions together, and secure the lower or single shelf.

Pillars.—Two pillars, 1 in. square, are required for the two-shelf bookcase, the lower being 9¾ in. long and the upper 1 ft. 0⅜ in. Both ends are fashioned into a pin, one ¼ in. long to fit the hole in the top shelf, and the other ⅝ in. long to fit holes in the lower shelf, and a block secured to the under surface of the table top (see Fig. 963). The length of the single pillar is 1 ft. 2¾ in. Pass each pillar through the centre of a couple of 5-in. squares of 1-in. stuff, and glue the latter to the pillars at equal distances from the extremities and each other (see Fig. 963). The book-supporting laths are to be attached to the edges of these squares, which obviate the necessity for making the pillars 5 in. thick. Glue the smaller pillar to the two shelves, and secure the top shelf with glue and screws passing through the rails from the under surface.

Table Top.—The table top is 3 ft. square by ¾ in. thick. It is made by gluing two or three lengths of material together, planing up, and working a suitable moulding on the edge. It is secured to the rails with screws driven into it obliquely through the rails, a cavity being first made to take the screw-head (see Fig. 964); or buttons (Fig. 965) may be employed at two or three points on each side. These have a tongue to fit a slot in the rail inner face, and are screwed to the under surface of the table top. Another method is to glue triangular blocks in the angle of the top and rail. Before the top is secured, the upper pillar must be glued to the shelf and block, and the latter glued and screwed to the top.

Vertical Laths.—The vertical laths are 1 in. wide by ¼ in. thick, and long enough to reach from the top to the upper edge of the lower or single-shelf rail. In the one case, therefore, they must be 1 ft. 10¼ in. long, and in the other 1 ft. 2¼ in. Sixteen are required. Similar laths, 1 ft. 0¾ in. long, are needed to form book supports, eight for the single-shelf table and sixteen for the double, a couple running outward from each side of the pillar squares to the innermost vertical lath, to which they are united with a lap dovetail (see Fig. 966). These laths should be at an equal distance from each other, and from the top and bottom of the tier. Glue and screw the vertical laths in place, using either round- or flat-headed brass screws. Let the nicks of the screw-heads be perpendicular, and make no attempt to conceal the flat-headed ones, which should be flush with the surface.

Writing Table with Four Drawers.

The writing table illustrated by Fig. 967 would look well in mahogany, with or without inlaid satinwood lines. An end view of the table, from the left side, is shown at Fig. 968. The extreme dimensions are: length, 3 ft. 2 in.; width, 1 ft. 8 in.; and height, 2 ft. 4 in. The six legs are 2 ft. 3 in. long, and 2 in. square in section at the widest part. The four legs on the right are left square for a distance of 1 ft. 5 in. from the top, and the other two for a distance of 7 in., from which points they taper to 1¼ in. at about 4 in. from the bottom; here the wood slopes outwards to the full thickness on all six alike. If casters are required, the taper should be carried to the end.

Legs.—Figs. 969 to 974 are views of the six legs, showing how they are prepared for the various joints. Fig. 969 is the front left leg. In the top end the mortices for the front and end rails are cut ½ in. from the outer edges. As shown in Fig. 975, the tenons of these rails meet each other at a mitre. The extreme length of the tenon is 1½ in., width ¾ in., and thickness ¾ in. The mortices for the second rails are of the same size, and are cut 4¾ in. from the top. Between the mortices for the end rails run a groove ¼ in. deep and ½ in. wide, ½ in. from the outer edge, for the end board (see Fig. 969); and in the rear face of the leg, and at a distance of 8 in. from the bottom, make a mortice for the bottom rail, ½ in. by ¾ in. by 1 in. deep. Finally bore two holes, ¼ in. in diameter and ¼ in. deep, for the dowels of the angle brackets, 1½ in. and 3 in. below the mortices of the second rails. The rear left leg (Fig. 970) is similar to the other, but it has an additional groove for the back boards and an additional mortice for the back bottom rail, while dowel holes for the bracket are needed

Fig. 956.—Two-tiered Bookcase Table.

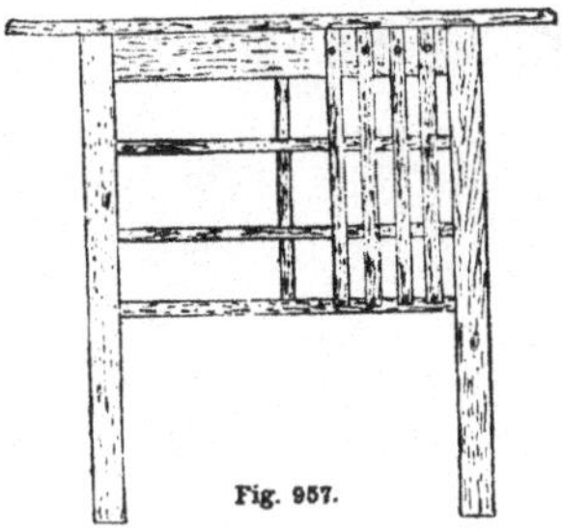

Fig. 957.

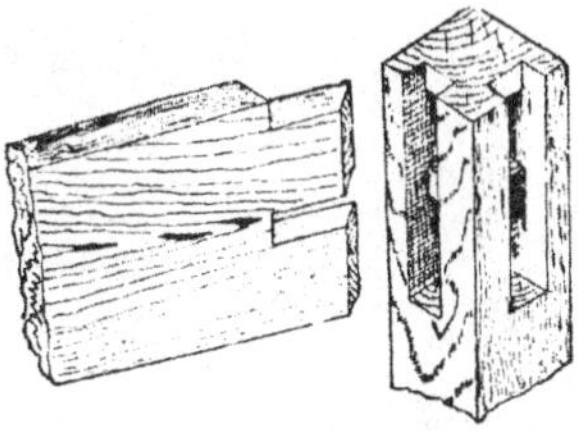

Fig. 959.—Joint of Leg and Top Rail of Bookcase Table.

Fig. 958.

Figs. 957 and 958.—Elevation and Horizontal Section of Single-shelf Bookcase Table.

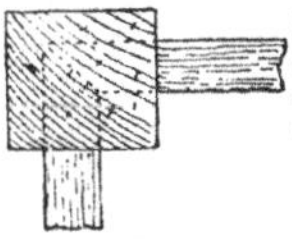

Fig. 960.—Plan of Leg and Rails of Bookcase Table.

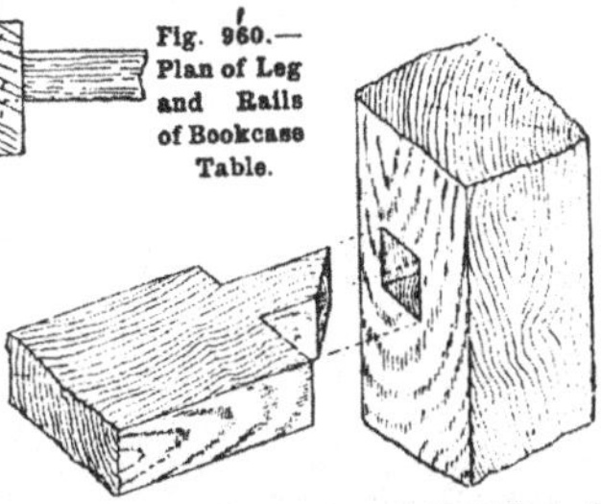

Fig. 961.—Joint of Leg and Shelf Rail of Bookcase Table.

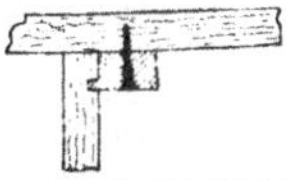

Fig. 965.—Button Securing Top of Bookcase Table.

Fig. 964.—Screwed Top of Bookcase Table.

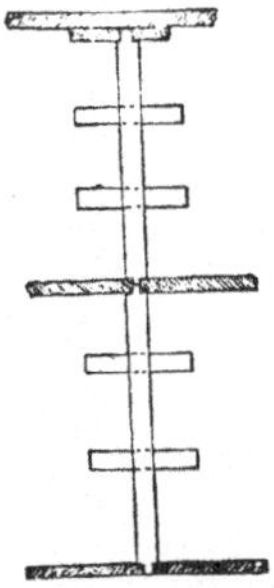

Fig. 963.—Pillars and Squares of Bookcase Table in Position.

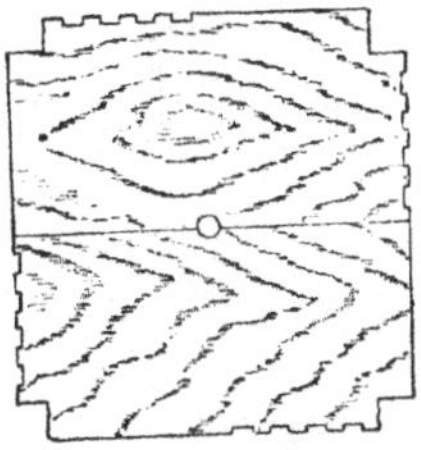

Fig. 962.—Shelf of Bookcase Table.

Fig. 966.—Inner Laths of Bookcase Table fixed to Squares and Uprights.

Fig. 967.—General View of Writing Table with Four Drawers.

Fig. 968.—End Elevation of Writing Table with Four Drawers.

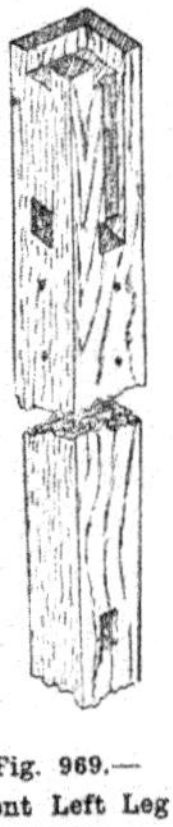

Fig. 969.—Front Left Leg of Writing Table.

Fig. 970.—Rear Left Leg of Writing Table.

Fig. 971.—Rear Right Leg of Writing Table.

Fig. 972.—Front Right Leg of Writing Table.

Fig. 973.—Front Middle Leg of Writing Table.

Fig. 974.—Rear Middle Leg of Writing Table.

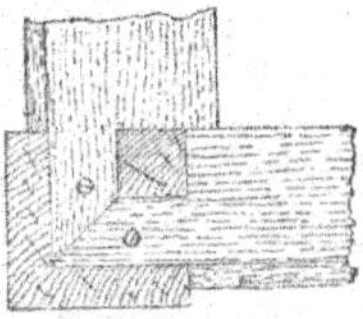

Fig. 975.—Joint of Top Rails and Leg of Writing Table.

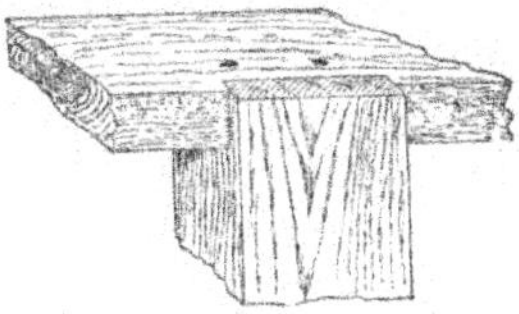

Fig. 976.—Joint of Front Rail and Middle Leg of Writing Table.

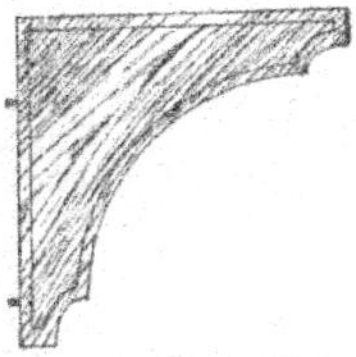

Fig. 980.—Inlaid Bracket of Writing Table.

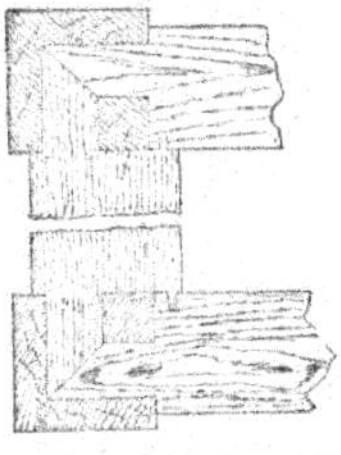

Fig. 977.—Section of Corner Legs and Rails of Writing Table.

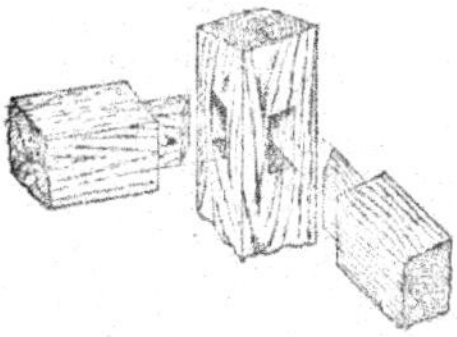

Fig. 978.—Joint of Bottom Rails and Leg of Writing Table.

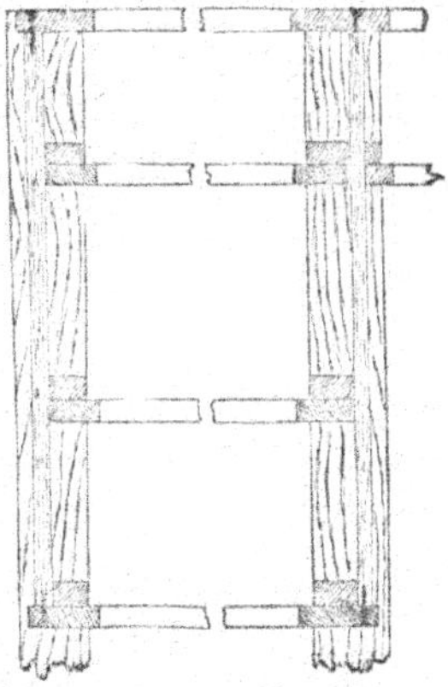

Fig. 979.—Part Vertical Section at Back of Writing Table.

on one face only. As the back and end boards on the right-hand side of the table are grooved to the rear right leg (Fig. 971), this leg must have a groove cut in the front and inner sides 1 ft. 3 in. long, reaching from the mortice of the top rail to that of the bottom rail. The sizes of the grooves and mortices are the same as those of the legs on the left. Fig. 972 shows the front right leg, which is prepared in the same way, with the difference that one of the grooves is replaced by mortices for the three rails running beneath the drawers. The bottom mortice is 1½ in. deep, and the others are 1 in. deep, all being ⅜ in. square. They are situated 4⅜ in., 10 in., and 1 ft. 3⅜ in. respectively from the top of the leg. Fig. 973 shows the front middle leg, with drawer rail mortices corresponding with the front right leg, a rear groove for the inner end board, and a mortice for the rail running beneath the inner end board (see Fig. 968). It has also a mortice for the tenon of the long drawer rail, this being a continuation of the mortice for the rail separating the drawers on the other side. The top of this leg is rebated, as shown, to ½ in., to take the front top rail, which extends from corner to corner. Dowel holes are made for the bracket to correspond with those of the front left leg. The rear middle leg (Fig. 974) is similarly cut at the top, but is grooved on three sides for the two back boards and the inner end board, and has mortices for the three rails.

Rails, etc.—The back of the right-hand portion of the table is 1 ft. 3 in. long by 1 ft. 0½ in. wide by ½ in. thick. The two ends have the same length and thickness, and a width of 1 ft. 2½ in. The left end and back are 1 ft. 2½ in. and 1 ft. 6½ in. long respectively, by 4 in. deep and ½ in. thick. The rails, with the exception of the bottom ones, may be cut from yellow pine, to which a facing of ¼-in. mahogany is glued. The front and back top rails are 2 ft. 11 in. by 1¾ in. by ¾ in. Fig. 975 shows how the ends are shaped to make union with the legs. A ¼-in. notch, 2 in. long, should be cut in the outer edge of each rail where the middle legs come, to bring the edge of the rails ¼ in. from the face of the legs (see Fig. 976). The end top rails are of the same thickness, and are 1 ft. 5 in. long by 2¼ in. wide, these being also the dimensions of the rails immediately below and the one connecting the two middle legs. These three rails have pegs at each end to fit into holes bored in the inner edge of the front and back rails (see Fig. 977). The rail under the long

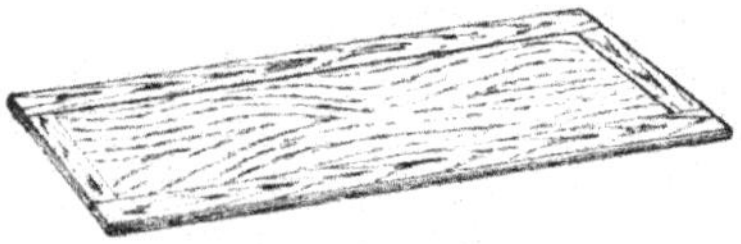

Fig. 981.—Top of Writing Table.

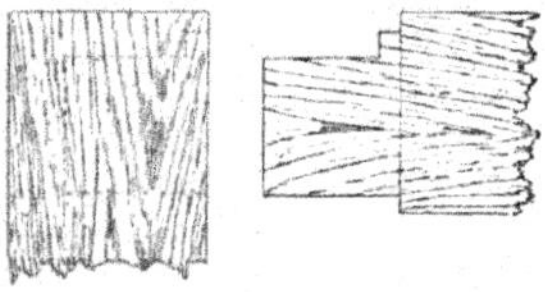

Fig. 982.—Joint of Top Framing of Writing Table.

drawer and the corresponding one at the back are 1 ft. 8½ in. long, and the rails beneath the first and second small drawers are 1 ft. 2 in., the one below and its companion at the back being 1 in. longer; all are 1⅜ in. wide by ⅜ in. thick. The outer edge of all these rails should be ¼ in. from the face of the legs. The back and left end bottom rails are respectively 1 ft. 8⅛ in. and 1 ft. 4½ in. long by 1¼ in. square. Bevelled tenons are worked on the ends to fit the mortices, as shown in Fig. 978. The tapering of the legs should be borne in mind when cutting the tenons of the two bottom rails, so as to make the shoulders a close fit. The width of the rail beneath the end of the long drawer, and that of the two beneath the bottom drawer, provides three drawer runners (see Fig. 979); the others consist of 1-ft. 2-in. pieces of ⅜-in. stuff, glued and screwed to the end boards, those

Fig. 983.—General View of Cross-legged Writing Table.

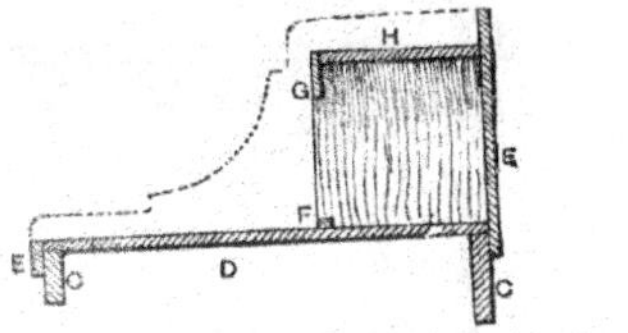

Fig. 987.—**Vertical Section through Top of Table (on Line X Y, Fig. 984).**

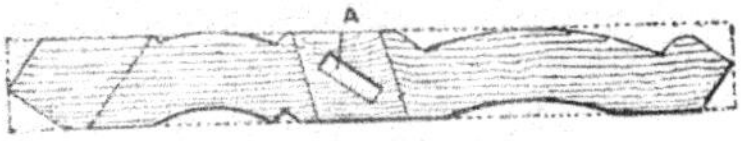

Fig. 986.—Leg of Cross-legged **Table.**

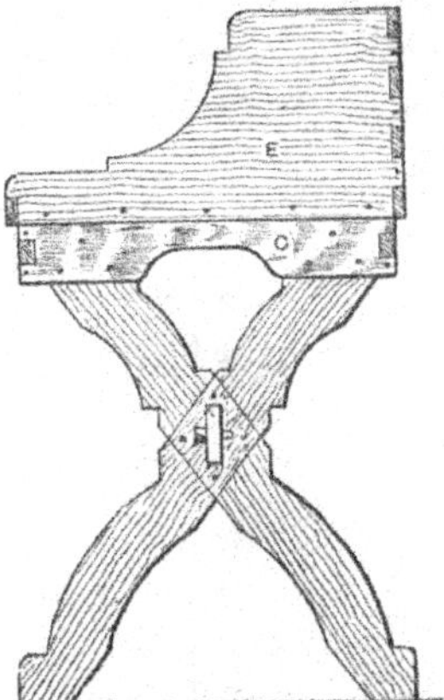

Fig. 985.

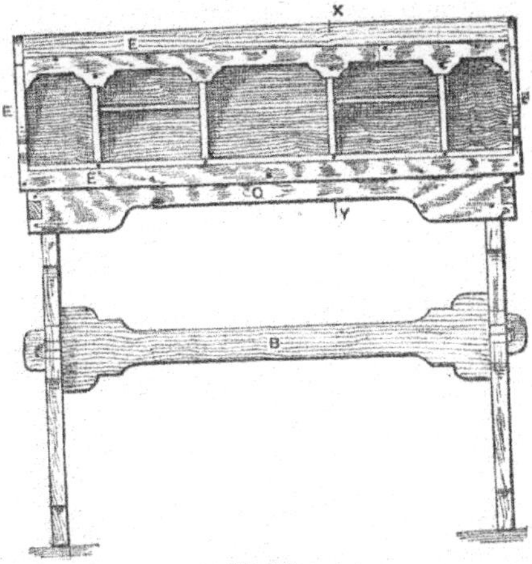

Fig. 984.

Figs. 984 and 985.—Front and End Elevations of **Cross-legged Writing Table.**

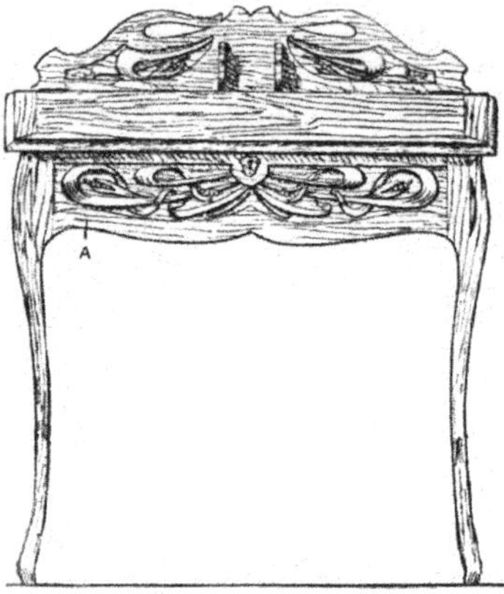

Fig 988.

Fig. 989.

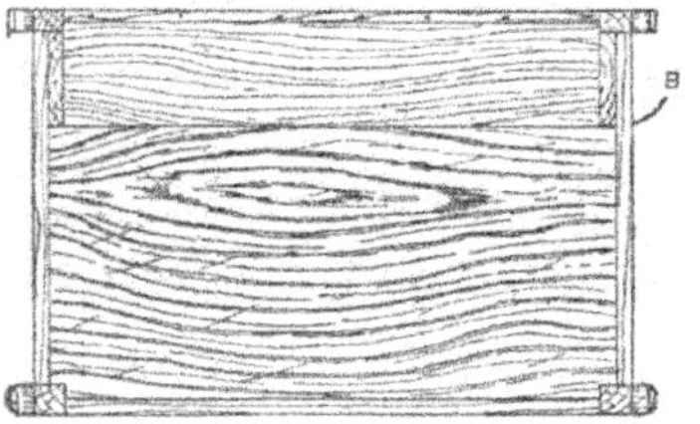

Fig. 990

Figs. 988 to 990.—Front and End Elevations and Inside Plan of Drawing-room Writing Table.

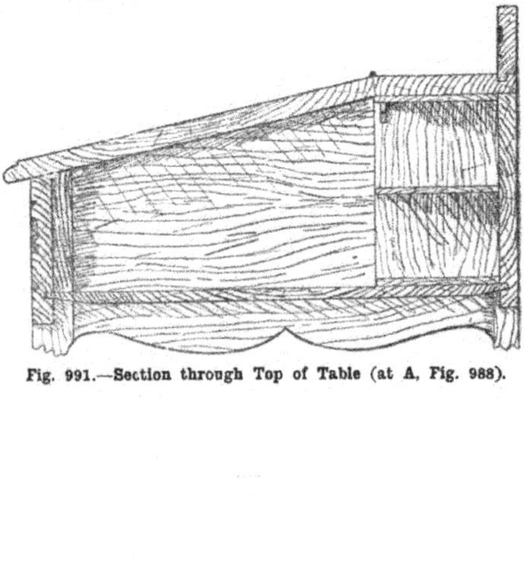

Fig. 991.—Section through Top of Table (at A, Fig. 988).

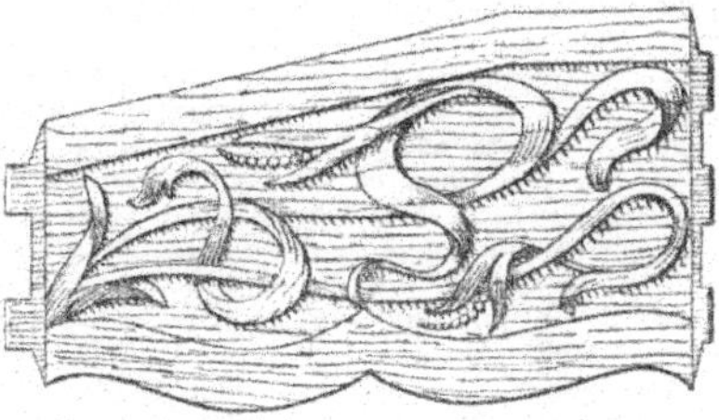

Fig. 992.—Carved End Rail of Writing Table.

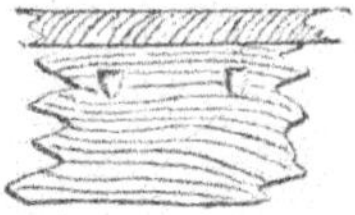

Fig. 996.—Securing Top of Writing Table.

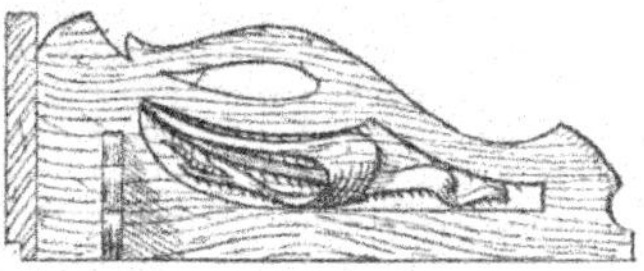

Fig. 994.—Half of Writing Table Pediment.

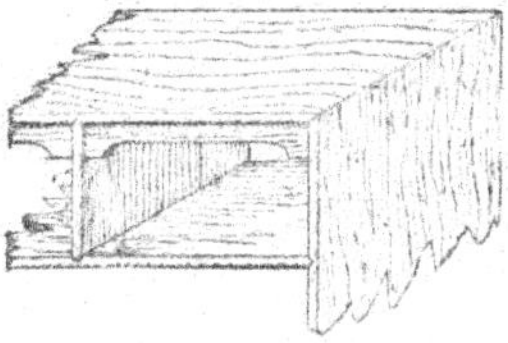

Fig. 999.—Part View of Pigeon-holes of Writing Table.

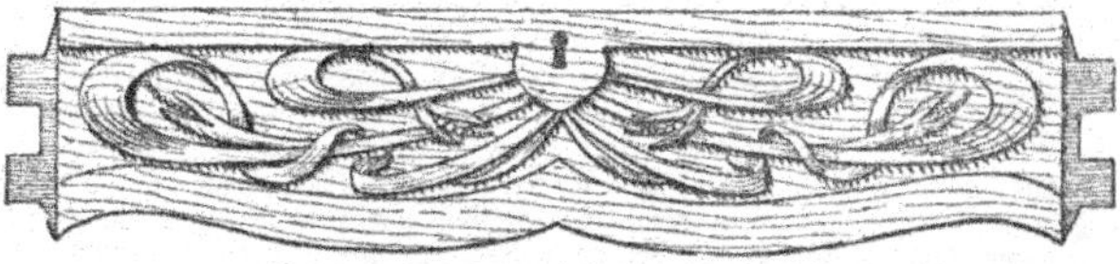

Fig. 993.—Carved Front Rail of Writing Table.

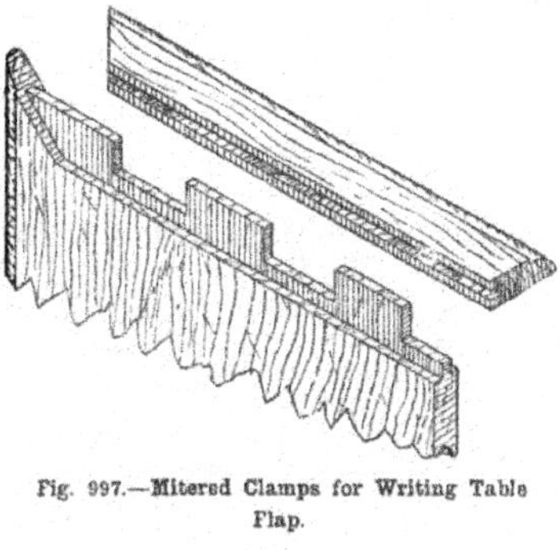

Fig. 997.—Mitered Clamps for Writing Table Flap.

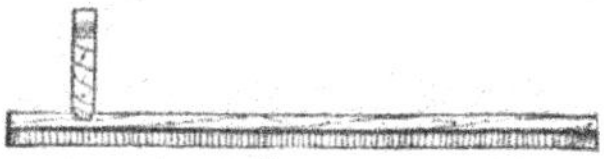

Fig. 995.—Fixing Bracket of Writing Table.

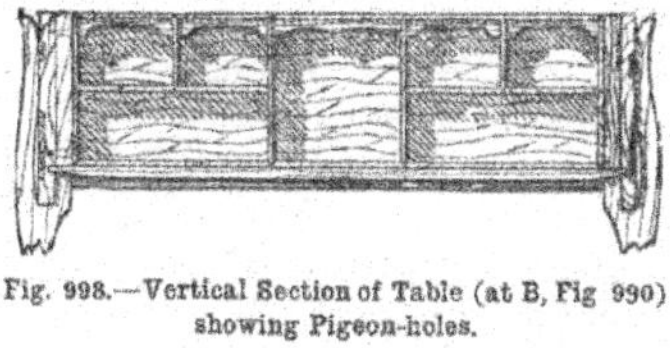

Fig. 998.—Vertical Section of Table (at B, Fig 990) showing Pigeon-holes.

for the short drawers being 1½ in. wide and that for the long drawer 1 in. The front end may be pegged to the front rail, or secured with a double-pointed nail. Drawer guides, 1 ft. long, and thick enough to be flush with the edge of the leg, are glued to the slide rails and end boards. To prevent the upper drawers rising at the inner end through the absence of a rail there, screw to the top edge of the inner end board a piece of ¾-in. stuff, 3 in. wide, long enough to reach from the front to the back rail. Fig. 979 is a sectional back view showing these details, the rear legs, back, and rails being omitted. Glue up these various parts, and secure the five boards to the rails with screws. Any inlaying must be done previously.

Brackets.—The brackets (Fig. 980) for the angles should have 6-in. sides; they are dowelled to the legs, and fixed with double-pointed nails to the rail above.

Table Top.—The table top (Fig. 981) is 3 ft. 2 in. long by 1 ft. 8 in. wide. It consists of a framed board 2 ft. 8¾ in. by 1 ft. 2¾ in. by 7/16 in. The framing is 3 in. wide by 1 in. thick; the front and back pieces are 3 ft. 2 in. long, and the ends 1 ft. 6 in. They are mortice-and-tenoned together, haunched tenons, 2 in. long by 2 in. wide, and ⅜ in. thick, being used. The haunch projects ½ in., is ⅜ in. wide, and starts ½ in. from the outer edge of the frame length (see Fig. 982). Run a groove, ⅜ in. deep and wide, down the middle of each inside face for the board, which is tongued so as to bring the underneath surfaces flush, thus leaving a depression above for the eather covering. The edges should be worked into a moulding. Secure the top to the table with screws passed into it from under the top rails. The drawers are made in the usual way, and are fitted with brass or copper handles.

Cross-legged Writing Table.

Fig. 983 is a general view, and Figs. 984 and 985 are front and end elevations of an easily constructed writing table, for which pine is the most suitable material. To ensure perfect firmness, so desirable in a writing table, the crossed legs should be cut from 1¼-in. stuff; if this is not easily obtainable, 1-in. stuff may be used. Fig. 986 shows how these legs are sawn. The board, indicated by dotted lines, is 2 ft. 9 in. by 4 in. Each leg is halved where the pair cross, and a mortice A, 3 in. by ¾ in., for the tenon of the middle bar, pierces both legs. The legs are further secured with four round-headed screws. The middle bar B (Fig. 984) is of ¾-in. board, 3 ft. by 6 in. Supposing the legs to be 1¼ in. thick, this bar will be 2 ft. 7 in. from shoulder to shoulder, and each tenon 2½ in. long. The tenons are pegged outside the legs. A frame C (Figs. 984 and 985), of ¾-in. board, encloses the top ends of the legs, to which it is screwed. Its front and back pieces are shown in section at C C (Fig. 987); these are 2 ft. 11 in. long, the end pieces being 1 ft. 8 in. long, and all are 4 in. wide. The front and end pieces are shaped as shown, but the back is left plain. The frame is dovetailed at the corners; and the upper edge, when fixed, comes level with the tops of the legs. The table top D (Figs. 983 and 987) rests on the frame, and is screwed down to the frame and legs. It is of ½-in. board, 2 ft. 11 in. by 1 ft. 8 in. Over this top, leather or American leather-cloth is stretched, and tacked down over its edges. This need not necessarily be carried farther back than the strip F (Fig. 987), as the latter can be made to hide the edge and the tacks which fasten it. Outside the frame comes the casing E (Figs. 984, 985, and 987). This is of ½-in. board; the back is 3 ft. by 11 in., and the ends are 1 ft. 9 in. by 11 in., the front being 3 ft. by 1¼ in. These pieces are dovetailed and screwed together at the corners; and the casing is screwed upon the frame, thus hiding the edges of the table top, and the tacks by which the covering is fastened. The upper edge of the front strip of the casing comes level, when fixed, with the top of the table. The upright partitions between the pigeon-holes rest on the table top as shown in Fig. 987. They are of ½-in. board, and are 7½ in. square. Two openings are cut in their front edges, one at the bottom, ½ in. square, for the strip F, and another at the top, 2 in. by ½ in., for the canopy strip G. The shelf H, which rests on the partitions, is ½ in. thick. The partitions are fixed with screws driven into them through the table top, back of the case, and the strips, and with dowels into the shelf H. The two horizontal partitions (see Fig. 984) are of ¼-in. board, and slide in V-shaped grooves cut for them in the uprights. All screws left showing should be round-headed. The illustrations, with the exception of Fig. 983, are reproduced to a scale of 1 in. to 1 ft.

Small Writing Table for Drawing-Room.

The writing table shown in front and end elevations by Figs. 988 and 989 should be of walnut or mahogany, and may be carved, as shown, or left plain if good figured wood is used. It is 3 ft. high, 2 ft. 6¾ in. long, and 1 ft. 7¾ in. from back to front. The front legs are 2 ft. 3¼ in. by 2½ in. by 1½ in., and are rounded as shown in Fig. 990, which is a plan with the flap and top removed. The back legs are 2 ft. 6 in. by 2½ in. by 1 in., and should be cut to the shape shown in Fig. 988, and left square and straight on the sides. The finished sizes of the various parts are as follows:—Front rail, 2 ft. 3 in. by 5⅞ in. of ¾-in. stuff; back rail, 2 ft. 3 in. by 8⅞ in. of ¾-in. stuff; the end rails are also of ¾-in. stuff, 1 ft. 5½ in. long by 9½ in. wide. The flap is made up of three pieces: one 2 ft. 6¾ in. by 1 ft. 2½ in., and two for clamps, 1 ft. 2½ in. by 2 in. by ¾ in. One piece, on which the flap is hinged, 2 ft. 6¾ in. by 5½ in. of ¾-in. stuff; one piece for carved pediment, 2 ft. 3½ in. by 5¾ in. of ¾-in. stuff; two pieces for brackets, 3 in. by 2½ in. of ½-in. stuff; one piece, 2 ft. 3¾ in. by 1 ft. 6¼ in. of ½-in. stuff, for the bottom, which may be of basswood, stained to match; and two pieces, 7 in. by 4¾ in. of ¾-in. stuff. The pigeon-holes also may be of basswood, ¼ in. thick; two pieces for top and bottom, 2 ft. 0⅝ in. by 4¾ in.; two end pieces, 7 in. by 4¾ in.; two middle uprights, 6¾ in. by 4¾ in.; two small vertical division pieces, 3¼ in. by 4¾ in.; two horizontal division pieces, 9⅛ in. by 4¾ in.; four small curtain pieces, 4⅜ in. by ⅞ in.; and one piece for the centre curtain, 5⅞ in. by ⅞ in. An enlarged section through A (Fig. 988) is given at Fig. 991, showing the pediment tongued into the top, and the basswood bottom fitted into the plough grooves of the rails. The rails are tenoned into the legs in the usual way; they should be fitted in dry, and then taken out and carved. Figs. 992 and 993 are enlargements of the carving on the end and front rails, together with the haunchings and tenons necessary; the tenons are shouldered on the front side only. Fig. 994 is an enlarged half of the pediment, giving a section, and a detail of the carving. These designs for the carving will be found easy to cut. The small brackets fixed on the top are shaped to form a support for pens; they are dovetailed into the pediment, as shown in Fig. 995, and are secured to the top by screws driven from inside. The top piece to which the flap is hinged is secured by screws driven through the ends inside, in the manner set out in Fig. 996. Fig. 997 illustrates the method of clamping the flap, the clamps being grooved, mortised, and mitered to receive the tenons and haunchings, and to fit the mitres of the flap; 2-in. brass butts should be used for hanging the flap, and a 2-in. brass box lock to secure it. Fig. 998 is a section taken at B (Fig. 990), showing the pigeon-holes in front elevation; these are mitered together as shown in Fig. 999, and fixed with small brads. The curtain pieces are cut in tight between the division pieces, and glued.

Small Pedestal Desk with Drawers and Pigeon-holes.

The pedestal writing desk illustrated by Fig. 1000 would look best in oak, with the internal fittings of the desk in mahogany. Fig. 1001 is a side elevation, while Fig. 1002 is a half plan of the top of the desk and a half plan of the bottom rails of the pedestal. The desk, which is 1 ft. 8 in. square, is fitted with five drawers and four pigeon-holes at the back, and has a loose tray at the front (see section Fig. 1003). The front, back, and sides of the desk are ¾ in. thick; the lid and top are ⅝ in. thick, both being clamped with stuff 1½ in. wide; and the bottom, which may be of pine, is ½ in. thick. The sides are dovetailed to the back and front, and the top A (Fig. 1003) is screwed to the back and sides, the screws

Fig. 1000.—General View of Small Pedestal Desk with Drawers and Pigeon-holes.

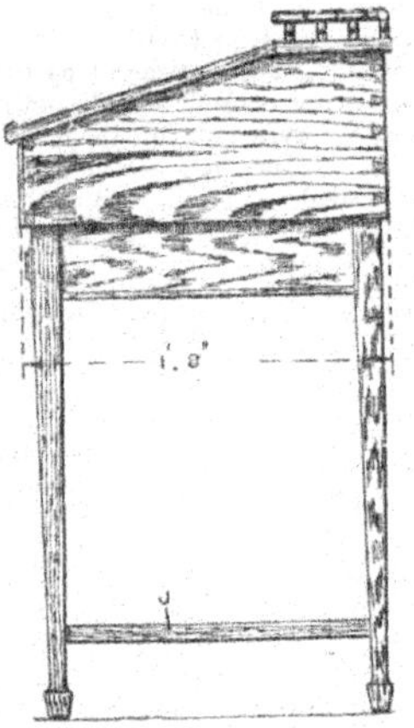

Fig. 1001.—Side Elevation of Small Pedestal Desk.

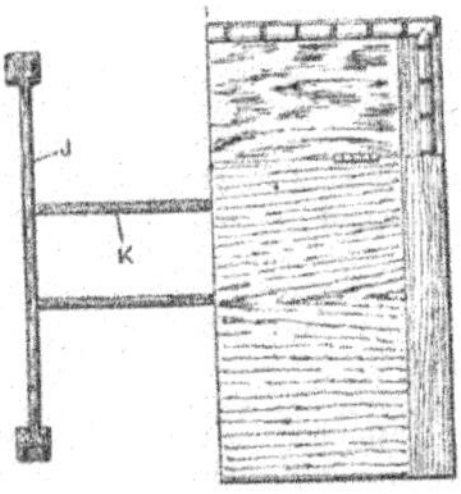

Fig. 1002.—Half Plans of Desk and Bottom Rails of Pedestal.

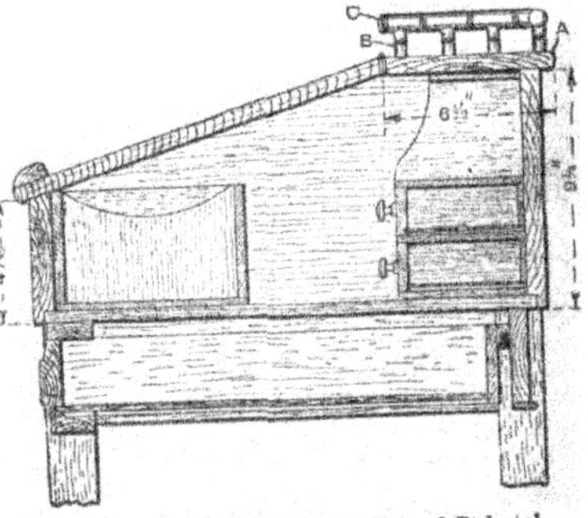

Fig. 1003.—Vertical Section of Top of Pedestal Desk.

7

SCREEN FOR BAY WINDOW

being hidden by the small turned pillars B. Three screws in the back and two in each side will be sufficient. The small pillars are ½ in. in diameter, and are relieved by having three turned grooves in the centre of each, as shown; they also have a ¼-in. diameter pin left on at each end for securing them to the desk top and to the top rails C. These rails are ⅝ in. in diameter, and are also relieved by small turned grooves. The lid and top should overhang ½ in. all round, and should be fitted with two 2-in. brass butt hinges and a desk lock. Fig. 1004 gives a front view of the internal fittings at the back of the desk. All the material is ¼ in. thick, with the exception of the drawer fronts, which are ⅜ in. thick. The joints D of the outside frame are either dovetailed or box pinned, and all the partitions are made a sliding fit in shallow grooves, as shown at F. A small bone knob should be fitted to each drawer. The joints for the loose tray at the front are the same as those shown at D in Fig. 1004, and there are five partitions in the tray. The edges of the partitions in both fittings should be nicely rounded. The pedestal is 2 ft. 3 in. high by 1 ft. 7 in. square over the legs, which are 1⅜ in. square for a length of 4¼ in. at the top, tapering to ¾ in. at the bottom. The small square feet, shown in section at Fig. 1005, are 1½ in. long, and are fitted separately, as shown, to facilitate working. A ⅜-in. square by ¾-in. long pin is left on the bottom of the leg for entering a corresponding hole in the foot. The front rails above and below the drawer are 1¼ in. wide by ⅝ in. thick, and are set back ⅛ in. from the front face of the legs. The upper rail is joined to the legs as shown in Fig. 1006, the joint of the lower rail being as shown in Fig. 1007, which is a view looking from the back. The groove is ¼ in. deep, and must be undercut on the lower edge, as shown, and extends to within ⅛ in. of the front face of the leg. This joint must be carefully done with the chisel. The back and side rails are 3⅝ in. wide by ⅝ in. thick, and are jointed as indicated at Fig. 1008. Each of the two side rails has a 1⅜-in. by ¼-in. fillet G screwed to the bottom edge, projecting ⅞ in. inside, to serve as the drawer runners, and projecting ¼ in. outside, the top edge being rounded (see Fig. 1009). A

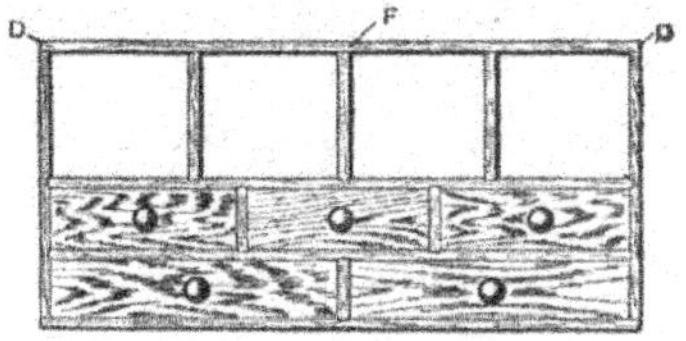

Fig. 1004.—Front Elevation of Pigeon-holes and Drawers of Pedestal Desk.

Fig. 1006.—Front Rails of Desk Jointed to Leg.

Fig. 1005.—Leg of Desk Jointed to Foot.

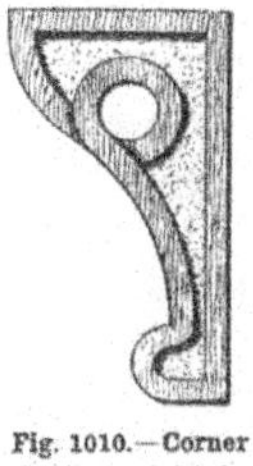
Fig. 1010.—Corner Bracket of Desk.

Fig. 1007.—Joint of Front Rail to Leg of Desk.

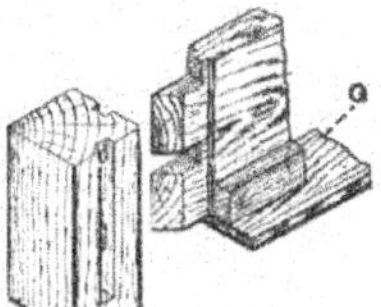

Fig. 1008.—Joint of Side Rail to Leg of Desk.

Fig. 1009.—Drawer Runner and Guide fitted to Side Rail of Desk.

13

guide H, of pine, must be glued and bradded to the inside of each side rail, and a hardwood drawer stop must also be glued to the inside of the back rail.

Rails.—The bottom side rails J (Figs. 1001 and 1002), which are ⅞ in. wide by ⅝ in. thick, are 4½ in. from the floor, and are stub-tenoned into the legs. The two inner rails K are also ⅞ in. by ⅝ in., and are 4 in. apart; they have round pins of ⅜ in. diameter left on each end for securing them to the side rails.

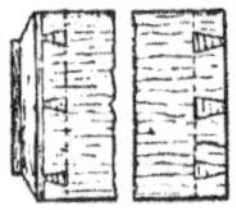

Fig. 1011.—Side View of Pedestal Drawer.

Fig. 1012.—Section of Side and Bottom of Pedestal Drawer.

Fig. 1013.—Securing Desk to Pedestal.

Brackets.—The two front brackets under the drawer, shown enlarged at Fig. 1010, are 3½ in. by 6 in.; these are not for strengthening purposes, but are intended merely to serve as a relief to the front. They are ⅝ in. thick, the hole being bored with a 1-in. centrebit. To get a clean hole, bore through till the centre can be seen; then turn the bracket over and bore right out. A margin of ⅜ in. should be left all round the ground being fetched out about ⅛ in. deep, and punched with a grounding tool. The brackets may be secured in place with screws driven through the leg and the rail from the back.

Drawer Front.—The drawer front, which is 3 in. wide by ⅞ in. thick, has a raised surface (see side view, Fig. 1011), with a brass drop handle fitted in the centre. The bottom of the drawer fits in a groove in the sides as shown at Fig. 1012.

Securing Desk to Pedestal.—The method of securing the desk to the pedestal is shown by Fig. 1013; the hardwood buttons L are screwed to the bottom of the desk, and when turned home engage in slots cut in the rails; two buttons in each side rail will be sufficient. The desk and pedestal should be stained or fumed, and finished by wax-polishing.

KITCHEN, LARDER, AND PANTRY FURNITURE.

Kitchen Dresser.

THE first example in this section is a dresser suitable for a house of medium size. Figs. 1014 and 1015 show front and end elevations; Fig. 1016 shows a section on line A B (Fig. 1014); Fig. 1017 a plan of the framing under drawers; and Fig. 1018 a plan of the cupboard in the dresser. The material used for the dresser should be good sound and dry Christiania white deal, free from knots, shakes, and resinous substances. The lower portion of the dresser is divided into three separate cupboards, with one shelf in each, as shown in Figs. 1016 and 1018. The pot-board and the divisions in the cupboard are $\frac{7}{8}$ in. thick, finished. The two ends of the lower portion of the dresser are framed and panelled as shown in Figs. 1015 and 1019. The top of the lower portion of the dresser is 1 ft. 6 in. wide by $1\frac{1}{8}$ in. thick, finished, as shown in Fig. 1020. The three drawer fronts are $8\frac{1}{2}$ in. wide by $\frac{7}{8}$ in. thick; the drawer sides, $8\frac{1}{2}$ in. wide by $\frac{1}{2}$ in. thick; the drawer backs, $7\frac{1}{2}$ in. wide by $\frac{1}{2}$ in. thick; the drawer bottom being $\frac{1}{2}$ in. thick (see Fig. 1020). The two end standards for the top portion of the dresser are $8\frac{3}{4}$ in. wide at the top, and 5 in. wide at the bottom, the standard at centre being $\frac{3}{4}$ in. less —the thickness of the matchboarded back. The two end standards are rebated to receive matching, as shown in Figs. 1015 and 1017. The top shelf is $8\frac{1}{2}$ in. wide; the middle shelf, $6\frac{3}{4}$ in. wide; the bottom shelf, $5\frac{1}{2}$ in. wide; all being $\frac{7}{8}$ in. thick, fini hed (see Figs. 1015 and 1016). The runners C D for drawers should be framed to the front and back rails, and panelled as shown in Figs. 1017 and 1020, the runner C being of 3-in. by $1\frac{1}{2}$-in. stuff, and the runner D 4-in. by $1\frac{1}{2}$-in. The adoption of this dust-proof method allows of the lower cupboard being kept clean. The drawer guides E are 2 in. by 1 in. The two ends of the lower cupboard are housed to receive the pot-board. The pot-board is also housed to receive the two standards, the housing being stopped $\frac{1}{2}$ in. from the front edge of the two ends and the pot-board. The front and back rails of the lower portion of the dresser are grooved so that the top of the dresser can be buttoned down (see F, Fig. 1020). The framing forming the end of the lower portion of the dresser is tongued as shown in Fig. 1019, and rebated at the back to receive the matchboarding, as shown in Fig. 1018. The small round at the angle of the dresser is stopped at the bottom as shown in the front elevation (Fig. 1014), and the corners of the dresser-top are rounded. Two bearers G of the same width and thickness as the drawer runners (4 in. by $1\frac{1}{2}$ in.) are fixed under the dresser top (see Fig. 1016). These bearers answer as tilting pieces for the three drawers, which, when pulled out, will not drop, but will keep in a level position. The drawers are $\frac{1}{2}$ in. shorter than the depth of the dresser, and are stopped at the front as shown at H in Fig. 1020—not at the back against the matching. In the latter case there is a tendency to push off the matching; the drawers then go back too far, and present an unsightly appearance. The drawers are fitted either with metal grip handles or with turn wood knobs. The doors are hung with two 3-in. steel hinges, and fitted with small

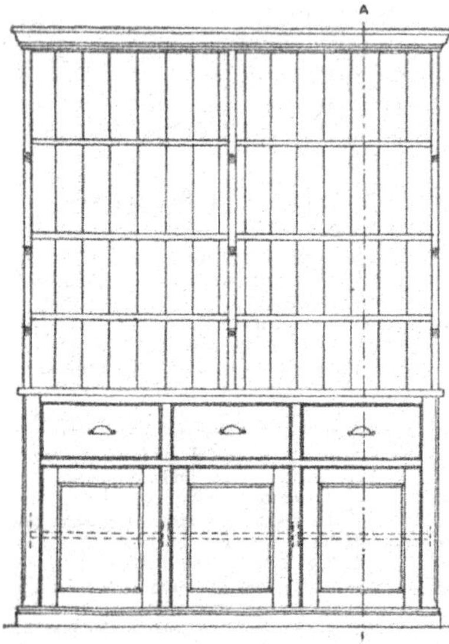

Fig. 1014.

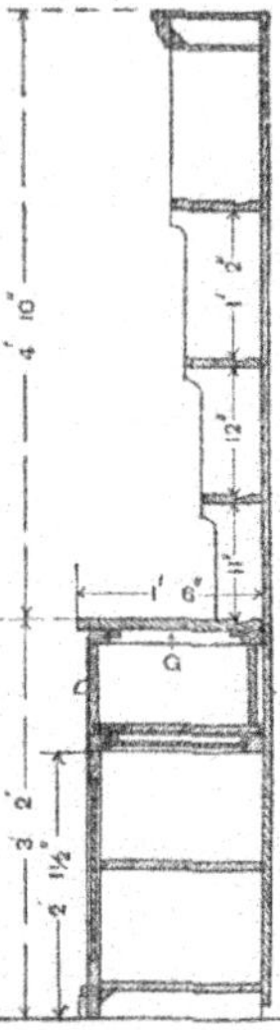

Fig. 1016.

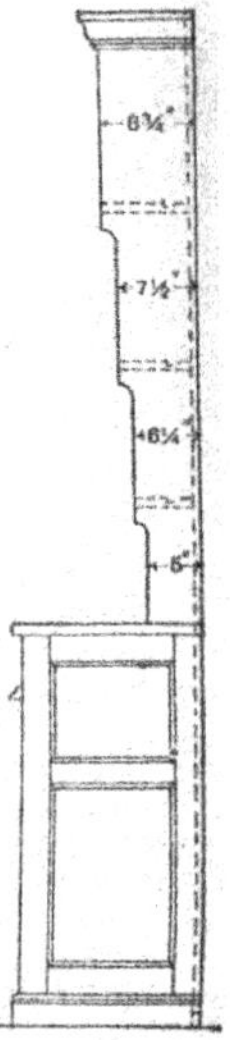

Fig. 1015.

Figs. 1014 to 1016.—Front and End Elevation and Vertical Section (on Line A B) of Kitchen Dresser.

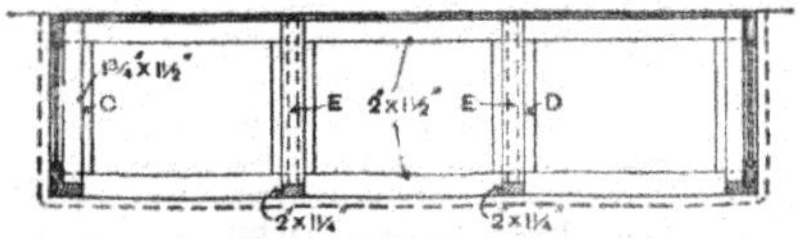

Fig 1017.

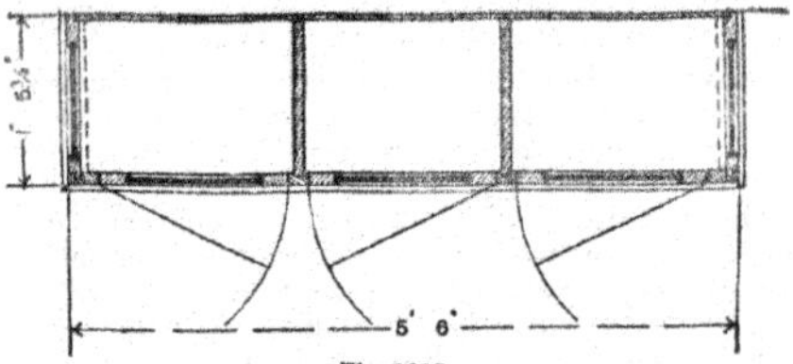

Fig. 1018.

Figs. 1017 and 1018.—Plan of **Kitchen Dresser Framing under Drawers and Plan of Cupboard.**

mortice locks or ordinary cupboard handles. The top of the lower portion of the dresser is sunk as shown in Fig. 1016, the sinking being stopped at each end in a line with the inside face of the standard. The three standards of the top part of the dresser should be set out and cut to shape as shown in Figs. 1015, 1016, and 1021. A sinking in the shelves is formed as shown in Fig. 1022. The standards are housed $\frac{3}{8}$ in. deep to receive the shelves, the housing being in each case stopped $\frac{1}{2}$ in. from the front edge of the standards. The housing is cut out to fit the sinking in the shelves. The soffit of the dresser is housed in the standard, and stopped as before described.

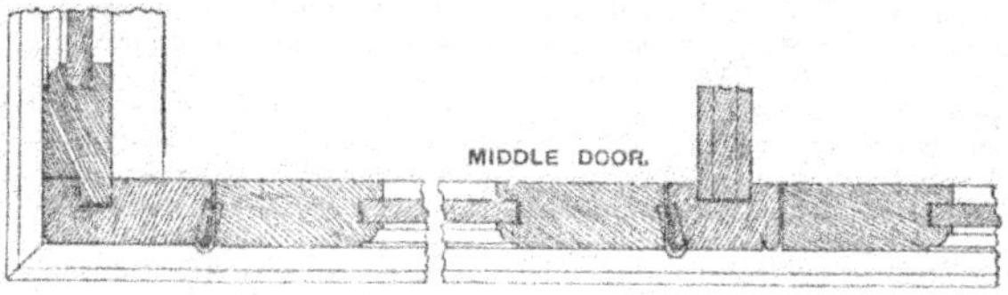

Fig. 1019.—Horizontal Section through two Cupboards in Front of Dresser Cupboard.

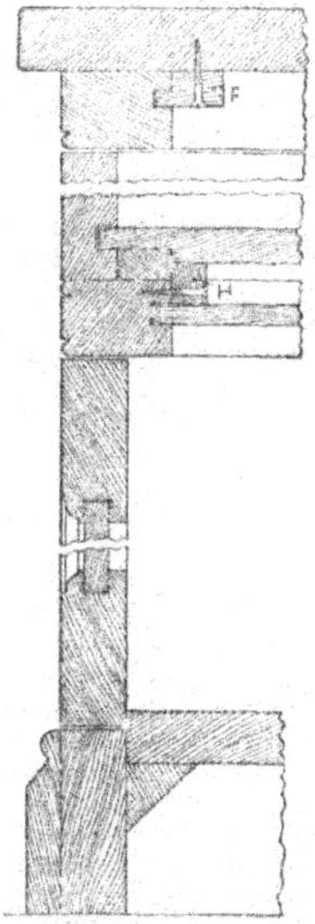

Fig. 1020.—Vertical Section through Dresser Drawers and Cupboards.

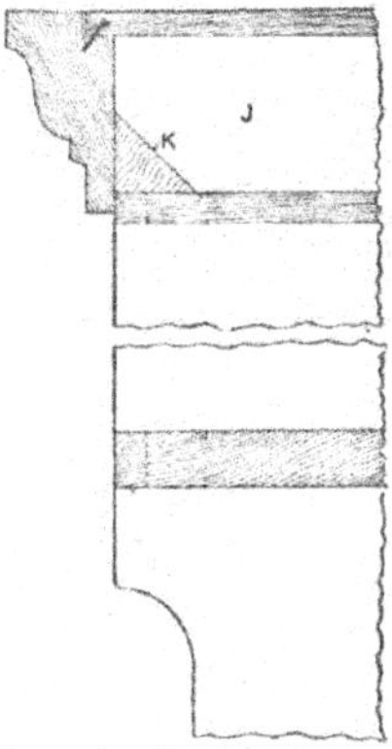

Fig. 1021.—Vertical Section through Dresser Cornice.

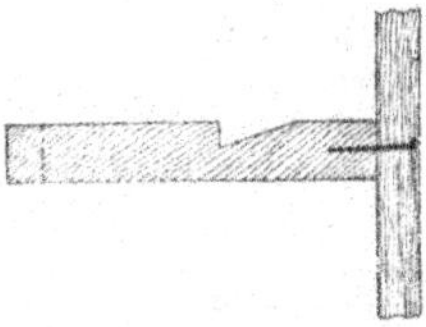

Fig. 1022.—Section of Dresser Shelves, showin Sinking.

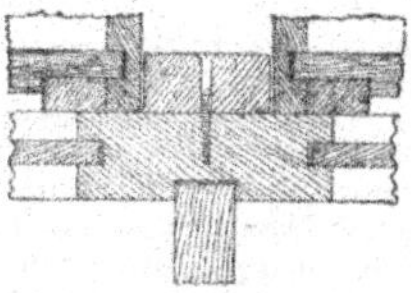

Fig. 1023.—Section of Dust-framing between Drawers and Cupboard of Dresser.

The top end of the standard is cut to length under the cover-board as shown in Fig. 1021. Two bearers J are fixed between the soffit and the cover-board. Glue blocks K are shown in Fig. 1021. The matchboarded back of the lower portion of the dresser is cut 1 in. short of the floor-line (see Fig. 1016), the top end of the matchboarding being in a line with the middle of the dresser top; the bottom end of the matchboarding for the top portion of the dresser being cut to suit (see Fig. 1016). The top of the lower portion of the dresser is left clean; the other parts are knotted, stopped, and painted in three oils, and finished to an approved tint; or they may be grained and varnished. The tongues of the matchboarding are painted in a colour that is as near as possible to the finished tint, because they are apt to shrink and show a white line. A sectional view showing the construction of the dust-framing between drawers and cupboard is presented by Fig. 1023.

Enclosed Dresser.

Figs. 1024 to 1026 illustrate the construction of a small kitchen dresser, built to fit in a recess, and having the upper part enclosed with a pair of glazed sliding doors, the lower part fitted with two drawers, and the cupboard enclosed with a pair of panelled doors rebated together. The chief dimensions are:—Height to dresser top 2 ft. 7¾ in., to cornice 6 ft. 11½ in., width of top enclosure 1 ft. 2 in., of lower enclosure 1 ft. 5 in., length 3 ft. 2 in. Minor dimensions are figured in the details. The glazed enclosure is designed for adaptation to an existing dresser, and is removable. Fig. 1024 shows a front elevation of the complete fitment; Fig. 1025 a vertical section on A A (Fig. 1026); Fig. 1026, a plan showing in one half a section through the cupboard, and in the other half a section through the drawer compartment; Fig. 1027, a horizontal section through the glazed enclosure; Fig. 1028, a half-plan of the top of the enclosure; Fig. 1029, an enlarged detail of Fig. 1027; and Fig. 1030, a similar detail of Fig. 1025. Fig. 1031 is a conventional view of one end of the dresser framing, showing the preparation for the drawer; Fig. 1032 is a similar view of the sliding rail of the upper case, with a portion of the case end. The dresser legs are 2½ in. square. The end rails are of 1-in. by 9-in. board, tenoned into the legs with two 2-in. by ½-in. tenons, mitered at the back end to meet the tenons of the back board, which is out of ¾-in. by 10-in. The drawer runners shown in Fig. 1031 at B are nailed on the bottom edge of the end rails, and project 1 in. in front of the legs.

Rails.—The rails are placed ½ in. from the outside of the legs, and the guide pieces C, 1 in. thick, are nailed on the ends to provide a flush surface for the drawers to run against. The top front rail, of ¾-in. by 2½-in. stuff, is dovetailed into the legs as shown in Fig. 1031, and the middle rail is fitted with double tenons, ⅜ in. thick, the front ends of the drawer runners being tenoned into this ¼ in. deep. The middle rail may be got out of 2½-in. by 1⅜-in. stuff and rebated as shown at D in Fig. 1031 for the doors, or out of 2½-in. by 1-in. stuff, with a ⅜-in. fillet nailed on as shown in Fig. 1030. The central division E (Figs. 1026 and 1029) is double-tenoned into the top and middle rails with ¼-in. tenons, and a guide piece of similar thickness is nailed on behind it to the runner as shown in the plan. The front bottom rail F (Fig. 1030) and the end rails, out of ¾-in. by 2¼-in. stuff, are dovetailed up from the bottom of the legs, and the pot-board is cut in tight between and around the legs when the carcase is put together, two cross bearers not shown in the illustrations supporting its middle portions.

Dresser Top, Sham Panel, and Doors.—The dresser top, which is formed of two 1¾-in. by 9-in. spruce boards, ploughed and tongued together, is fixed at the front by screws, and at the back and ends by buttons as shown at G (Fig. 1031). These buttons must not be thicker than the front rail, or they will interfere with the running of the drawers. The drawers are dovetailed together as usual, and fitted as shown in Fig. 1025, a sham panel being formed on the fronts by mitering round a ⅛-in. cocked bead, which may be either simply bradded and glued on, or sunk into a groove as shown in the section. These should be inserted after the drawer has been made and fitted. The panelled doors are framed together with mortice-and-tenon joints, out of 1¼-in. stuff, with ½-in. panels flush on the inside; a 1-in. by ½-in. bead and ovolo moulding being planted round on the outside. The doors are rebated together; therefore the meeting stiles will require to be ½ in. wider than the hanging stiles, and to have a ½-in. bead worked on the face side. The doors are hung with pairs of 2½-in. iron butts, $\frac{3}{16}$-in. below flush, the left-hand door being fitted with two 3-in. thumb-neck bolts, and the right-hand door with a 1-in. brass knob and turnbuckle. The bottom stop should be splayed as shown in Fig. 1030, for facility of sweeping out. The enclosure consists of two ends out of 1-in. by 13½-in. deal, one top 1-in. by 13½-in., one slider rail 1-in. by 2½-in., and ½-in. by 5½-in. matchlined back planted on. The top is dovetailed into the ends as shown in Fig. 1028, and the slider rail is dovetailed in a similar manner to withstand the shocks of the doors.

Oak Runners.—As shown in Fig. 1030, 1¼-in. oak slips or runners are inserted tightly into grooves in the slider rail and top of the case. These grooves should not be ploughed until the doors are cleaned off, because they should be arranged to fall in the middle of the thickness, and at such a distance apart that there is a clearance of ⅛ in., this opening being filled in, when the doors are closed, by the slip H (Fig. 1029) fixed to the inside of the outer door. Two other striking tongues, which may be of deal, are grooved into the sides of the case, and enter grooves in the stiles of the doors. These may be fixed, as there will be no necessity to remove them. The oak runners should be cut in two pieces, with the joint about ¼ in. under the door when right home. One piece may then be fixed, the door slid on to it, and the other piece placed in position by raking it in. The doors are prepared out of 1-in. stuff, with stiles 3 in. wide, rebated upon the inside ⅝ in. by ¼ in. for glass, and beaded on the face with a ¼-in. bead. The two middle stiles overlap, and only one shows, the size of the doors being alike. A similar bead may be worked round the front of the case to break the joint, if desired.

Completing Dresser.—The doors are framed together, the size of the tenons being shown in Fig. 1030. The shelves are housed in the sides of the case ⅛ in. each end, or may rest on fillets. None of the outside of the case except the front need be planed, as it will be hidden in the recess. The cornice is out of 4½-in. by 1-in. stuff cut in tight between the walls, and nailed to the top of the case. A cover-board may be nailed on to the top. The case is fixed to the dresser by screws through the runner rail, as shown in Fig. 1030, and in the quarter-round fillet nailed to the matchlining. Two 3-in. brass scoop handles are fixed in the door stiles, to open and close them. Of the accompanying illustrations, Figs. 1024 to 1028 are reproduced to the scale of ¾ in. to 1 ft., Figs. 1029 and 1030 to 2 in. to 1 ft., Figs. 1031 and 1032 to 1 in. to 1 ft. The letter references not explained in the text are as follows:—I, shelf; J, plate rack; K, runner for glazed sliding doors.

Dresser with Sliding Doors.

In small rooms, a dresser with sliding doors is generally to be preferred to one having the doors hung in the usual way and opening into the room. Fig. 1033 is a front elevation of such a dresser, and on it are marked dimensions that will be suitable for most purposes. Fig. 1034 is a side elevation, Fig. 1035 a vertical section, and Fig. 1036 a sectional plan. First set out the dresser to full size. The top A (Figs. 1033 and 1034) is 1 ft. 9 in. wide, and should be selected from 11-in. by 1½-in. seasoned red deal, the pieces being cross-tongue jointed with good glue and tongues. The framing B (Fig. 1035) should be of 1-in. red deal; the shelf C and bottom T (Fig. 1035) may be of 1-in. white deal, and all three should be glued and tongued. The angle posts D D (Fig. 1036), when finished, should be about 2¾ in. square; into these the top and bottom rails E F (see Fig. 1035) should be dovetailed, glued, and screwed. The middle rail should be tenoned into the angle posts, but should not go through. The top rail E (Fig.

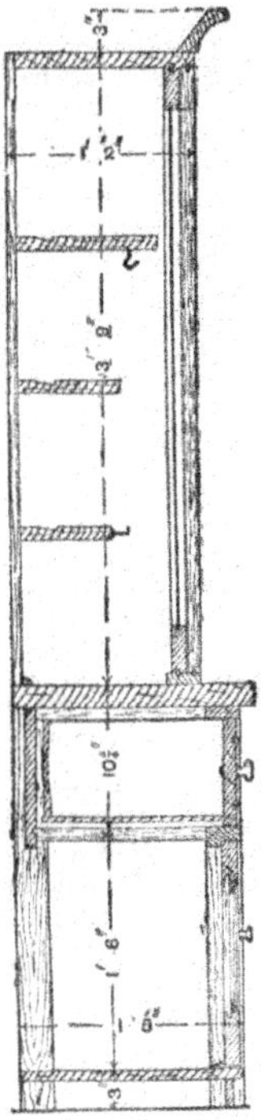

Fig. 1025.

Fig. 1024.

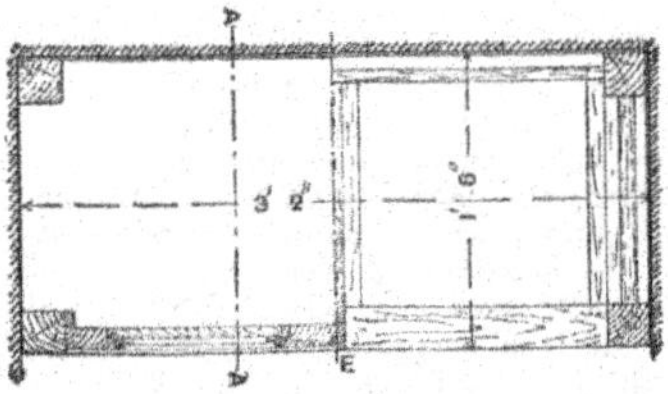

Fig. 1026.

Figs. 1024 to 1026.—Front Elevation, Vertical Section, and Half Horizontal Sections of Enclosed Dresser.

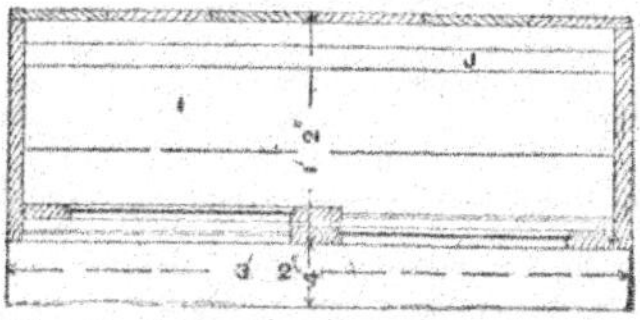

Fig. 1027.—Horizontal Section of Upper Part of Enclosed Dresser.

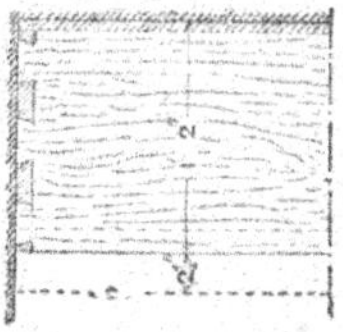

Fig. 1028.—Half Plan of Top of Enclosed Dresser.

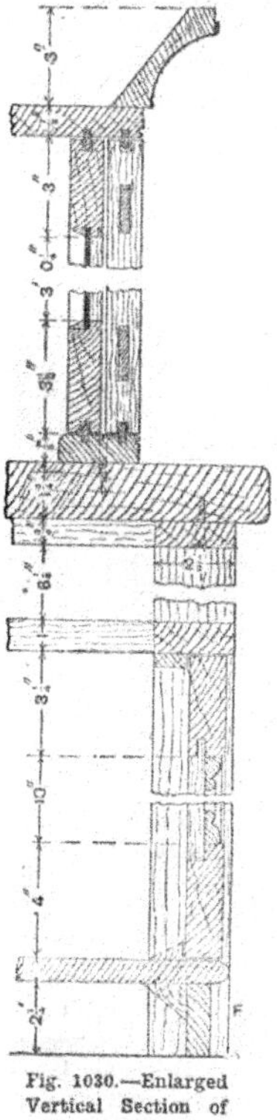

Fig. 1030.—Enlarged Vertical Section of Enclosed Dresser.

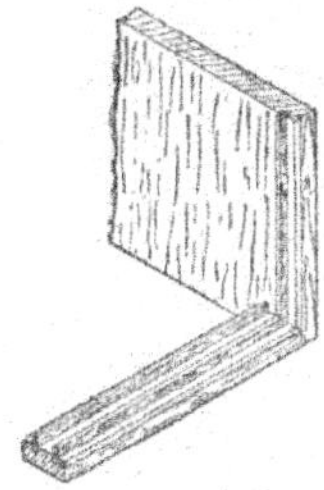

Fig. 1032.—Dresser's Door Carcase.

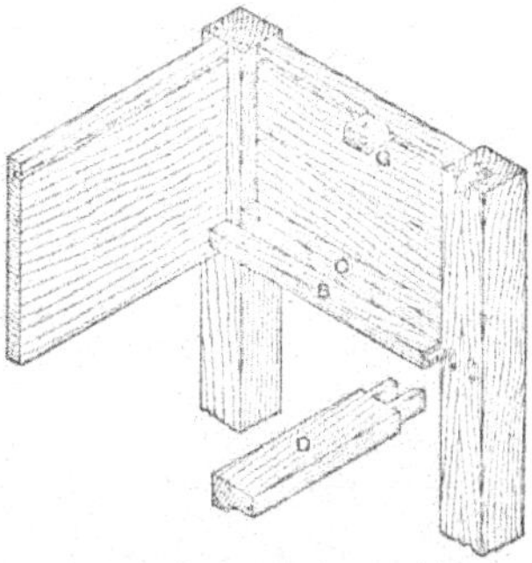

Fig. 1031.—Dresser's Drawer Framing.

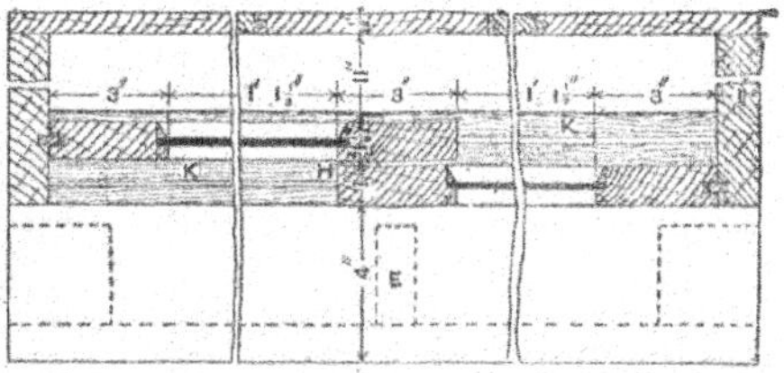

Fig. 1029.—Enlarged Half Horizontal Sections of Enclosed Dresser.

1035) should be about 3 in. wide by 1 in. thick, and in the centre and between the top rail and the framing B (Fig. 1035), frame in a vertical piece 3 in. by $1\frac{1}{4}$ in. and about 7 in. deep, to form a division for the drawers (see Fig. 1033). To construct the ends (Fig. 1034), prepare the top rails $3\frac{1}{2}$ in. wide by $1\frac{1}{4}$ in. thick, and bottom rails $4\frac{1}{4}$ in. by $1\frac{1}{4}$ in., and frame width of the top rail E would be $1\frac{3}{4}$ in., and of the bottom rail $2\frac{1}{4}$ in. Both rails are grooved $\frac{1}{2}$ in. deep, and into them is inserted a $\frac{7}{8}$-in. by $\frac{5}{16}$-in. bead P P, over which the rails of the door are rebated. The width of the outer linings G H should be the thickness of the rails E F, plus $\frac{1}{2}$ in. The linings should be glued and bradded on to the rails, which are set back the

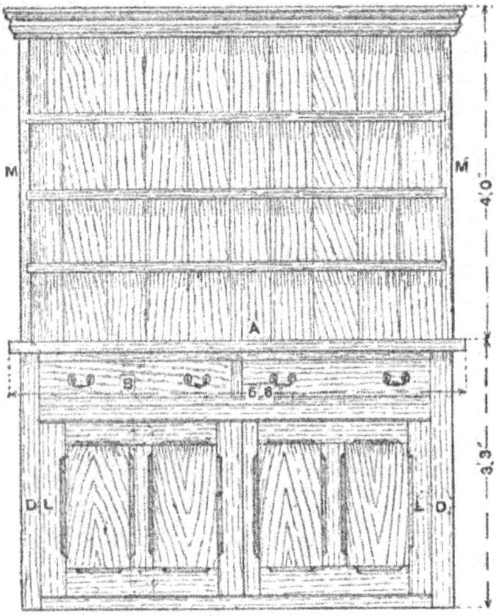

Fig. 1033.

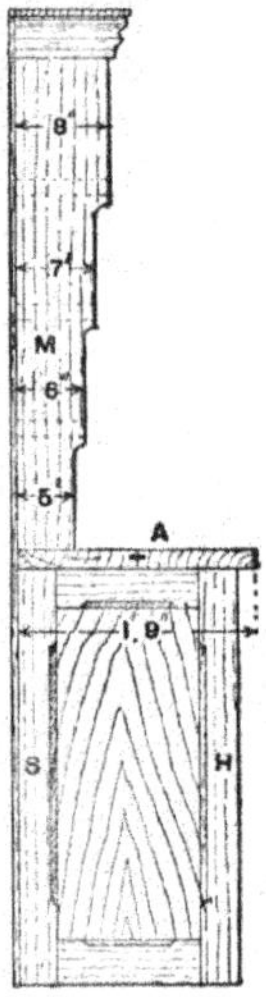

Fig. 1034.

Figs. 1033 and 1034.—Front and End Elevations of Dresser with Sliding Doors.

one of each into the stile S and into the angle post H, ploughing them for the panel. A B C D (Fig. 1037) are enlarged sections of the top and bottom rails of the doors, and E and F are enlarged sections of the rails in which the doors are made to run. Their thickness should not be less than $1\frac{1}{2}$ in., and their widths must be gauged according to the thickness of the doors. Thus, assuming that the doors are $1\frac{1}{8}$ in. thick when finished, and the outside linings G H (Fig. 1037) $\frac{1}{2}$ in. thick, the approximate thickness of the linings, in order that they may finish flush with the angle posts D D (Fig. 1036), and the outer door also will be flush when rebated over the linings, as shown. The stiles of the doors may be $3\frac{1}{2}$ in. wide, plus $\frac{1}{2}$ in., if tongued into the angle posts as shown at L L (Fig. 1036). Top rails and muntins are $3\frac{1}{2}$ in., and the bottom rails $4\frac{1}{4}$ in. wide, mortised, tenoned, and ploughed for panels, and stop-chamfered as shown in the lower part of Fig. 1033. Having fitted and secured the rails and linings to their respective places, clean off flush with the angle posts, and fit in the framed ends. Get the shelves to their proper lengths and widths, and secure the same to fillets which are screwed to the inside of the ends, as shown under the shelves (Fig. 1035). To fix the top, put screws through the top rail about 12 in. apart. To secure the top to the ends, plough a groove (see Fig. 1038), and make some wooden buttons of hardwood, preferably mahogany, on which cut a tongue. Insert the tongue into the groove, and screw through the buttons into the top, A being the top, B the top rail of framing, and C the mahogany button. The fronts of the drawers should be 1¼ in. thick, planed true and well fitted into the opening; the sides should be ½ in. thick and dovetailed together.

Dresser Standards.—The dresser standards, shown at M M (Figs. 1033 and 1034), are 1¼ in. thick. The shelves should be grooved for plates similarly to the top at N (Fig. 1035), the space between the shelves being arranged to requirements. They should be housed into the standards about ⅜ in. deep. Stopping the groove about ⅝ in. from the front edge, and providing the shelf with a shoulder, makes a better-looking job. The standards also should be grooved into the top, and pushed in from the back; and to make a good job, both shelves and dresser should be matchboarded at the back. The cornice is 3½ in. by ¾ in., fitted with a piece of 2½-in. architrave moulding mitered at the ends, and kept flush with the top edge of the ¾-in. piece. Fix these together with screws from the back, and on the top of all brad a board ½ in. thick, and of sufficient length and width to project ½ in. beyond the

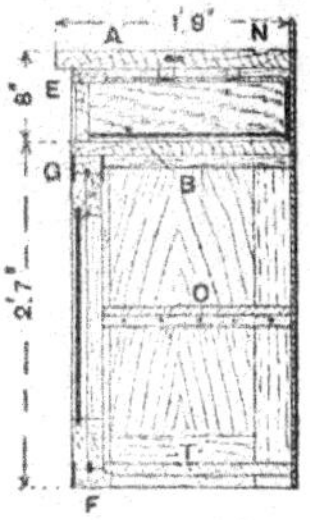

Fig. 1035.—Part Vertical Section of Dresser with Sliding Doors.

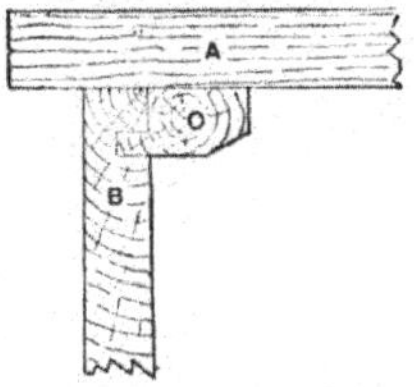

Fig. 1038.—Section showing how Top of Dresser is Fixed.

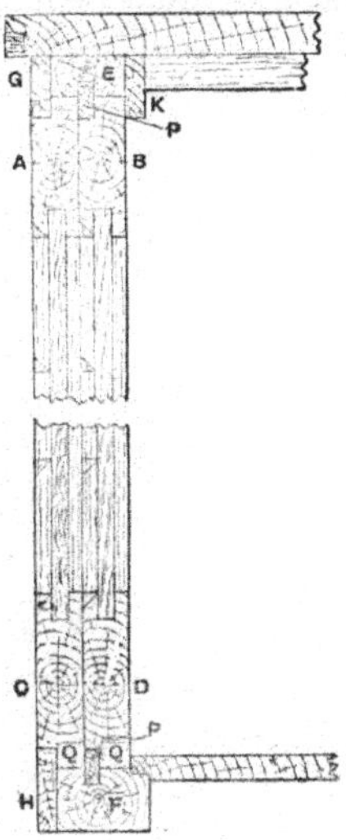

Fig. 1037.—Vertical Section through Dresser's Sliding Doors.

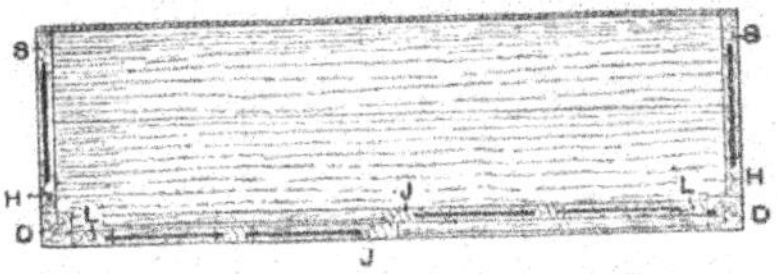

Fig. 1036.—Horizontal Section through Dresser with Sliding Doors.

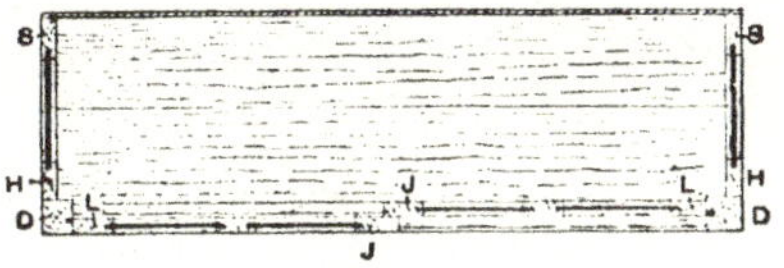

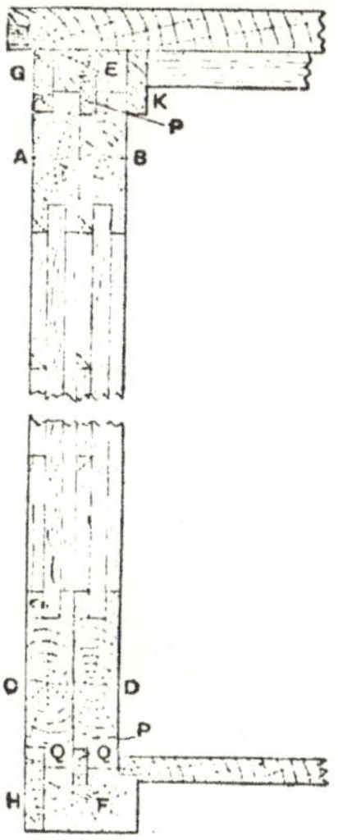

Fig. 1037.—Vertical Section through Dresser's Sliding Doors.

moulding both at the front and ends, as shown in Fig. 1033.

Completing Dresser.—Having cleaned off the doors and shot them to their respective widths and heights, remove the parting beads P P (Fig. 1037) and rebate the outer door over the linings G H. Replace the beads, and in a similar manner rebate the inner door, and brad on the fillet K (Fig. 1037). When fitting these doors, allow sufficient space between the beads and rebates for paint or varnish. The doors should be finished and quite dry before being put into place. The dresser, when it is completed, may be painted, or, if stained and varnished, should be well cleaned up, twice sized, and twice varnished; if painted, four coats should be given. The drawers should have a lock and two 3½-in. brass drop handles, and into the edges of the shelves should be screwed some brass dresser hooks, on which cups may be hung. To make the doors run easily, two small sash rollers to each door should be let into the under side of the bottom rails of the sliding doors at Q Q (Fig. 1037).

Fig. 1039

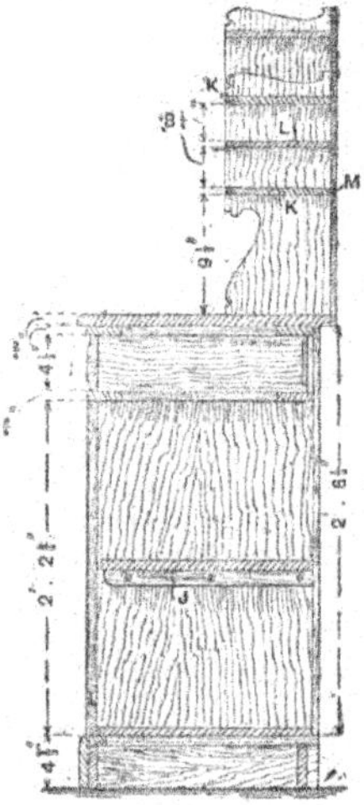

Fig. 1040.

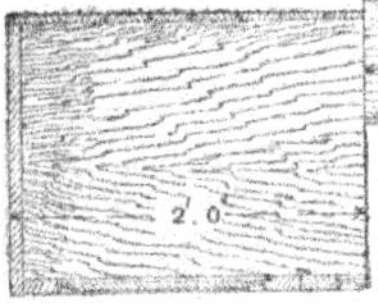

Fig. 1041.

Fig. 1042.

Figs. 1039 to 1042.—Front Elevation, Vertical Section, and Half Horizontal Sections of Dresser with Mirror.

Dresser with Mirror.

Figs. 1039 and 1040 illustrate the construction of a kitchen dresser which may be of pine or canary wood. The lower part (see Fig. 1041) consists of a cupboard, enclosed by two doors, which are divided by a fixed panel, and contains a shelf supported by fillets screwed to the gables. Three drawers of a handy size are arranged above the cupboard. A plinth, shaped at the front as shown, supports the lower carcase, and the upper part consists of four ends. Cut the dovetail grooves B in the ends of the plinth to receive the back, keeping it in 1 in. Then mitre the ends to the front, and glue them to the corner blocks. Glue a piece, 1½ in. wide by ⅞ in. thick, on the inside of the front at the top edge, and fix pieces to the ends in the same way. The shaping of the front and ends should be done before gluing the whole together, and a sash moulding should be run on the edge at the front and ends. The gables of the lower carcase should be squared up to 2 ft. 8½ in. by 1 ft. 7 in.

Fig. 1043.—Plan of Dresser Plinth.

Fig. 1047.—Section showing Position of Dresser Drawer Stop.

Fig. 1046.—Part Section of Dresser's Lower Doors.

Fig. 1044.—Corner Block of Dresser Plinth.

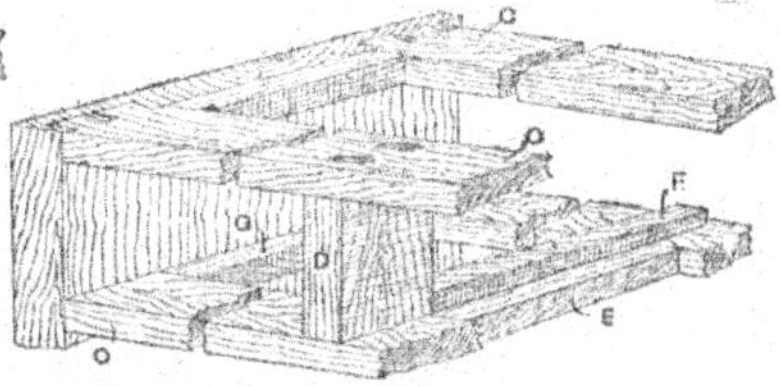

Fig. 1045.—Rails, Divisions, etc., of Dresser.

shaped gables, shown in Fig. 1040, the two inside being 5 in. longer than the outside gables, and forming a cupboard, which is enclosed by a door with a glass panel. Four small drawers (see Figs. 1039, 1040, and 1042) on each side of the cupboard complete the arrangement.

Construction of Dresser.—Beginning with the plinth, plane up sufficient wood for the front and back, and two ends 4½ in. wide by ⅞ in. thick. Cut the front to the sizes given in Fig. 1043, and make a mitre at each end. Dovetail together two pieces 4 in. long by 3½ in. wide by ⅞ in. thick (see Fig. 1044), and glue two of these corner blocks A (Fig. 1043) inside the front at the wide by ⅞ in. thick. They are rebated on the edges for the ½-in. back, which is made of tongued and grooved stuff 3 in. wide. The bottom is lap-dovetailed to the gables, and should be set back 1⅛ in. from the front to allow the doors to overlap. At the back, the bottom, which is ¾ in. thick, is narrowed to allow for nailing the back to the edge. The four long rails C (Fig. 1045) are 3 in. wide by ¾ in. thick, the two at the top, front, and back being dovetailed to the gables. The rail at the front under the drawers is fixed to the gables, with two short tenons on each end, and the back rail is housed in the gable with a dovetail groove. The two upright divisions D are 3 in. wide by $\frac{3}{4}$ in. thick, and are fixed to the rails by cutting square pins on the ends and mortising the rails to receive them. The two drawer runners E are $2\frac{1}{4}$ in. wide by $\frac{3}{4}$ in. thick, and are fixed to the front and back rails, with short tenons

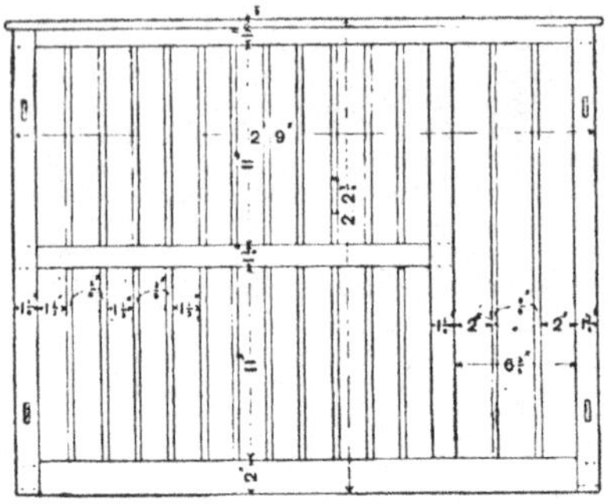

Fig. 1048.

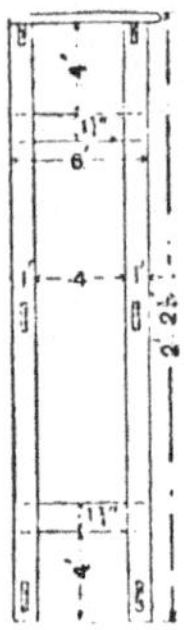

Fig. 1049.

Figs. 1048 and 1049.—Front and End Elevations of Plate Rack.

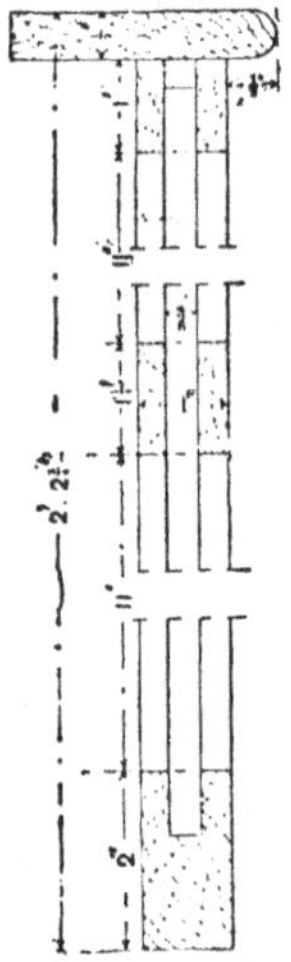

Fig. 1050.—Vertical Section of Plate Rack.

on the ends. The guide F is glued and nailed to the runner, while the drawer runners G are $1\frac{1}{4}$ in. wide by $\frac{3}{4}$ in. thick, and screwed to the gables.

Doors of Dresser.—The doors are mortised and tenoned together, the stiles and rails being $2\frac{1}{2}$ in. wide by 1 in. thick, with a sash moulding run on the front edges. Rebate the inside edges at the back for the panels, which are $\frac{1}{2}$ in. thick and bevelled. Fig. 1046 shows an enlarged section of the doors. The panels are fixed with beads nailed to the edges of the stiles and rails, and the panel between the doors is framed up in the same way as the doors, and screwed to the top and bottom rails. Work a $\frac{1}{4}$-in. bead on the edges of the stiles next the doors to break the joint, and glue and nail a fillet H on each edge of the dividing panel to form steps for the doors.

Drawers.—The three drawers above the doors are dovetailed together in the usual way, the fronts being bevelled all round the edges. The drawers are stopped by gluing and nailing thin pieces of wood—two for each drawer—about $1\frac{1}{2}$ in. square, to the rail. Fig. 1047 shows the position of drawer stop.

Dresser Top, Shelves, etc.—The top pro-

Fig. 1051.—Portable Larder or Safe.

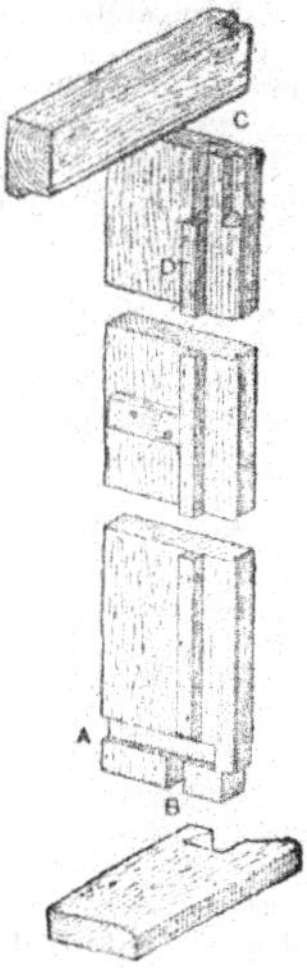

Fig. 1054.—Detail View of Side of Portable Larder.

Fig. 1052.

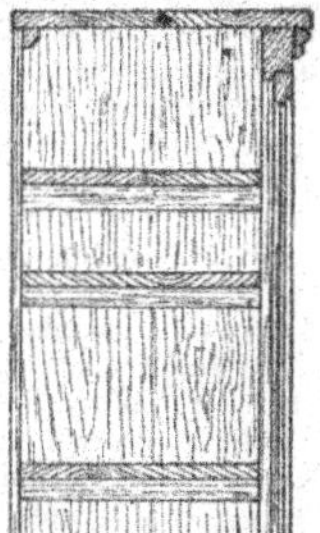

Fig. 1053.

Figs. 1052 and 1053.—Front Elevation and Vertical Section of Portable Larder or Safe.

jects 2 in. over the gables at the ends, 1¼ in. at the back, and ¾ in. at the front. It is 4 ft. 4 in. long by 1 ft. 9 in. wide by 1 in. thick, and is screwed to the rails at the front and back. Screw two fillets J (Fig. 1040) 1¼ in. wide by ⅝ in. thick to the gables, to support the shelf, which is ⅞ in. thick. The shaped gables for the upper part are shown in Fig. 1040, the two on the outside being 1 ft. 7½ in. long by 9 in. wide by ⅞ in. thick. Rebate the back edges for the ½-in. back. The top and bottom shelves K are ⅝ in. thick, and are housed into dovetail grooves cut in the sides of the gables, the grooves being stopped ½ in. from the front, and the shelves rebated. The shelf L between the drawers is ½ in. thick, and grooved in the gables, the short division between the drawers being 2 in. wide by ½ in. thick, and fixed to the shelves with short tenons. A fillet ½ in. thick is glued to the shelf at the back of the division to guide the drawer. The shelf above the door is fixed in the gables in the same way as the others. The back for the upper part is ½ in. thick, and is in two pieces half checked together at M, and screwed to the edge of the bottom shelf. It is fixed in the rebate in the outside gables, and placed over the edges of the inner gables, which are made ½ in. narrower for this purpose. Leave the back projecting over the back edge of the top of the lower part, and fix with screws. At each end glue a piece to the back edge of the lower carcase top, where it projects over the gables. The upper half of the back is fitted in the rebates in the gables, and shaped on the top edge as shown in Fig. 1039, and screwed in place. Turned knobs should be fitted to the drawers and doors, and the latter are hung with brass butt hinges. The dresser would look well if stained walnut colour and polished.

Plate Rack.

A rack for draining plates and dishes is illustrated by Figs. 1048 and 1049. It should be made of sound red deal or yellow pine, of which the following quantities are required: Two bottom rails, 2 ft. 10 in. by 2 in. by 1 in.; two top rails, 2 ft. 10 in. by 1 in. by 1 in.; two middle rails, 2 ft. 2 in. by 1¼ in. by 1 in.; four end rails, 7 in. by 1½ in. by 1 in.; four stiles, 2 ft. 3½ in. by 1¼ in. by 1 in.; two stiles, 2 ft. 1¼ in. by 1¼ in. by 1 in.; and one top board, 2 ft. 10 in. by 6½ in. by ½ in. The rails and stiles are mortised and tenoned together, and wedged from the outer edges. The upright bars are ⅜ in. in diameter, and are fixed in position after the framework is wedged together, the holes for them being previously bored in the rails. The rods are of birch, and those sold by house furnishers for curtains, etc., will be suitable. Fig. 1050 is an enlarged view, showing how the rods are placed in the rails. A handy shelf is formed by screwing a board 6½ in. by ½ in. to the top (see Fig. 1050). The plate rack is fixed on wall dogs, usually over the sink.

Portable Larder or Safe.

A convenient form of portable larder or safe is shown by Figs. 1051 to 1053, and should be made of good white deal. The sides, top, bottom, and back are each formed of boards ploughed and tongued together, the sides of the boards being beaded on their front edges to give a better appearance. Reference to A (Fig. 1054) shows the stopped housings to receive the bottom, which is cut to fit them, and also to continue over the edges of the sides, where it is mitered to fit chamfered fillets nailed to the lower ends of the sides. The top is screwed to the sides, but to give further support a fillet shown in section in Fig. 1053, and by Fig. 1054, is dovetailed to the sides. This fillet is rebated for the door and also beaded, this bead and the beads of the sides being mitered as shown at C. In Fig. 1053 it will be seen that the boards forming the back are continued to the floor, and are nailed to the back edge of the bottom of the larder. A chamfered fillet is screwed to the under side of the top, and to this the upper ends of the boards are nailed. Rebates should be made in the back edges of the sides to receive the edges of the back boards. To give a finish to the bottom of the larder, a flat plinth should be dovetailed into the lower ends of the sides as shown at B (Fig. 1054), the ends being mitered to strips nailed on the sides; whilst a moulding mitered round the under side of the top gives a better appearance to that part. The door is of a simple character, consisting of stiles, rails, and muntin rebated and chamfered, the joints at the corners being haunched mortice and tenon, and those connecting the muntin and rails being stub mortice and tenon. The netting, which is of a very fine mesh, is held in position by means of beads, mitered and screwed into the rebates. Door stops, one of which is shown at D (Fig. 1054), are screwed to the sides and bottom, the rebate in the top rail acting as stop for the upper part of the door. The door should be hung with a pair of 2-in. butts, and a suitable fastening should also be provided and fixed. The shelves are supported on fillets screwed to the sides at suitable heights, and should not be fixed to the fillets, but left free, so that they can be easily removed in order that they may receive thorough cleansing.

Fig. 1055.—Provision Safe with Semicircular Top.

Provision Safe.

A provision safe with a semicircular top is shown by Fig. 1055. Front and side elevations are shown by Figs. 1056 and 1057, details of the mouldings, etc., at A A and B B being presented by Figs. 1058 and 1059, whilst Fig. 1060 is an enlarged detail showing the construction of the semi-circular end. Fig. 1061 is an enlarged detail of the angle post, showing the joints with the rails.

Fig. 1056. Fig. 1057.

Figs. 1056 and 1057.—Front and Side Elevations of Provision Safe.

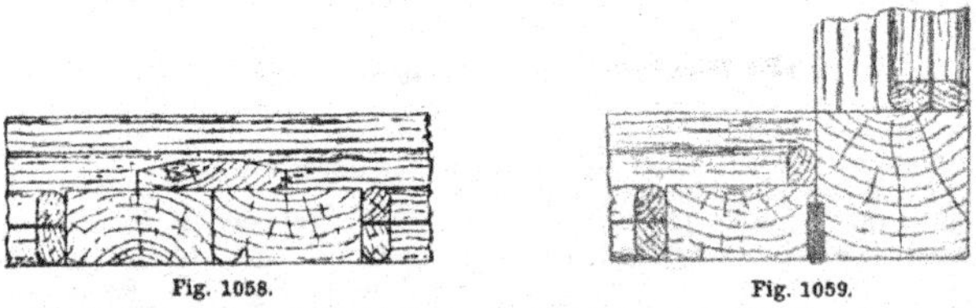

Fig. 1058. Fig. 1059.

Figs. 1058 and 1059.—Horizontal Sections through Door Stiles, etc., of Safe, on Lines A A and B B (Fig. 1056).

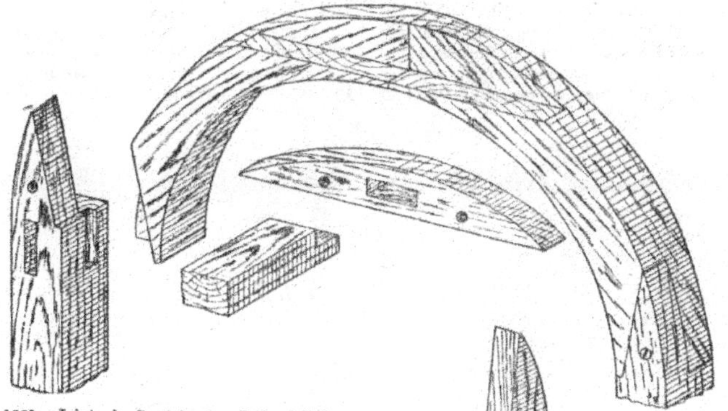

Fig. 1060.—Joints in Semicircular End of Safe.

Ship's Pantry-safe.

The pantry-safe illustrated by Figs. 1062, 1063, and 1064 is fitted with blinds and perforated zinc, and a ring plate for hooking up. This form of safe is used on ships and yachts, and the wood employed is usually teak or mahogany, but for domestic purposes pine is suitable, while the back, if fixed near a wall, could be matchboarded. Two methods of making the frame are shown in Figs. 1065 and 1066, which are alternative sections at A B (Fig. 1063). In Fig. 1065 there are four stiles, in which the eight rails are stump-tenoned and pinned, while in Fig. 1066 there are eight stiles and eight rails mortised, tenoned, and wedged, making four pieces, which are ploughed and tongued, and fixed with blocks screwed from the inside, or by nailing through from the outside and stopping the holes with putty. Prepare timber to the following dimensions, allowing ½ in. extra at each end on stiles, and also on the tenons of any rails that come through; the excess lengths are trimmed off after the work has set:—Four stiles, 2 ft. 1 in. by 2⅜ in. by 2⅜ in.; eight rails, 1 ft. 9 in. by 2⅜ in. by 1⅛ in.; two door stiles, 1 ft. 7½ in. by 2⅜ in. by 1⅛ in.; two rails, 1 ft. 5⅜ in. by 2⅜ in. by 1⅛ in.; and the shelf, 3 ft. 4 in. by 11 in. by ⅝ in. The top,

Fig. 1061.—Joints of Angle Post of Safe with Rails.

usually got out of one width, is 2 ft. by 2 ft. by 1¼ in. or 1½ in. thick in the centre, and worked down to ¾ in. on the edges, and the bottom can be made from two pieces, ploughed and tongued together, 1 ft. 11 in. by 1 ft. 11 in. by ¾ in. The blinds will take about 96 ft. of stuff, and ordinary venetian laths cut in two will answer the purpose. Plane the material and cut the mortices in the stiles and the tenons on the rails, fit temporarily, form rebates ⅜ in. wide by ¼ in. deep to receive the zinc and the wood slips for retaining it, and mark the divisions, raked at an angle of 45°, for the blind laths; eighteen are required for each of the three sides, and fourteen in the door. The grooves are $\frac{3}{16}$ in. deep, and are stopped $\frac{3}{16}$ in. from the front edges (see Figs. 1067 and 1068). Cut a notch 8 in. from the base in each stile to receive the shelf fillets, and bore and counter-bore three holes in each top rail (Fig. 1069) for fixing the roof (see Fig. 1068). The counter-bore is afterwards filled with wood

Fig. 1062.

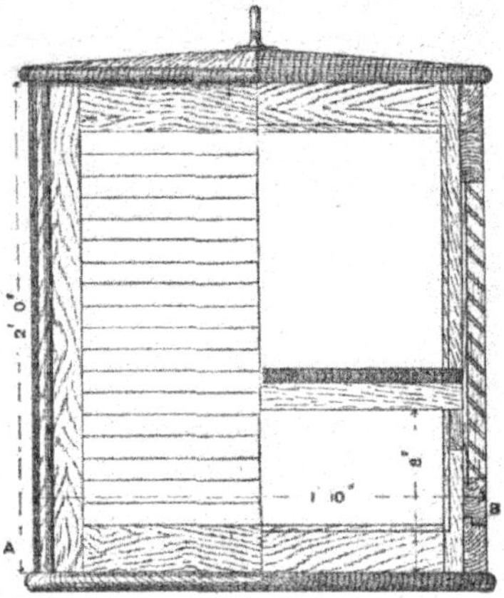

Fig. 1063.

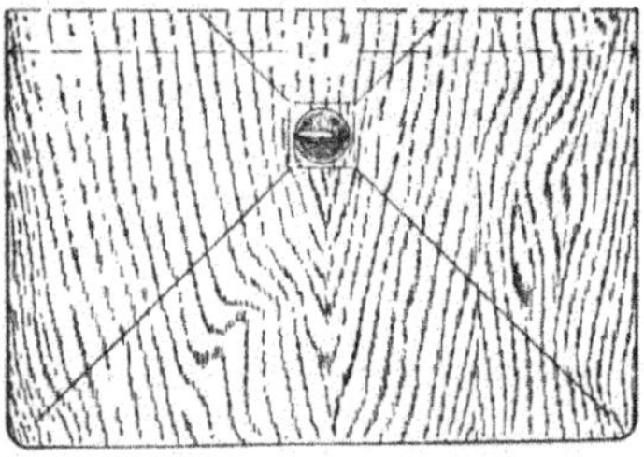

Fig. 1064.

Figs. 1062 to 1064.—Front Elevation, Half Side Elevation and Vertical Section, and Part Roof Plan of Ship's Pantry-safe.

plugs cut across the grain. Then form a ⅜-in. bead on the rails under and above the door, and frame the whole together, trim the ends of the stiles, shoot the top and bottom of the framework level, and are on the outside of the frame, and this method is adopted when the blinds are omitted. The fillets for the shelf are screwed on, the shelf being in two pieces (see Fig. 1070) for taking out when large

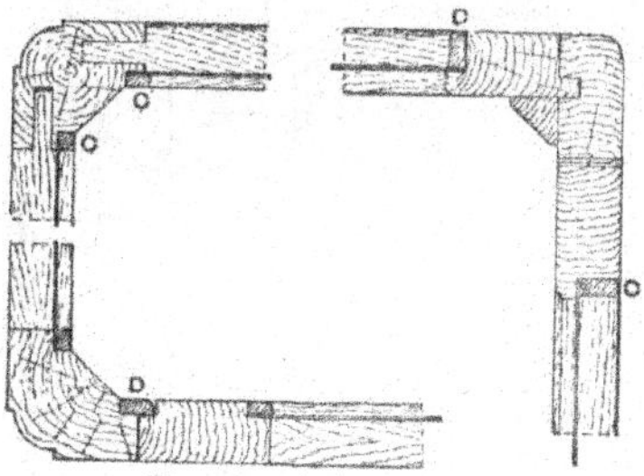

Fig. 1065. Fig. 1066.

Figs. 1065 and 1066.—Horizontal Sections of Ship's Pantry-safe at A and B (Fig. 1063)

Fig. 1067.—Blind Laths of Pantry-safe.

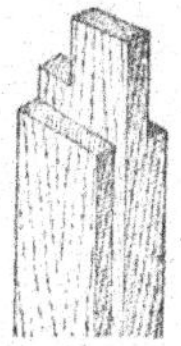

Fig. 1069.—End of Top Rail of Pantry-safe.

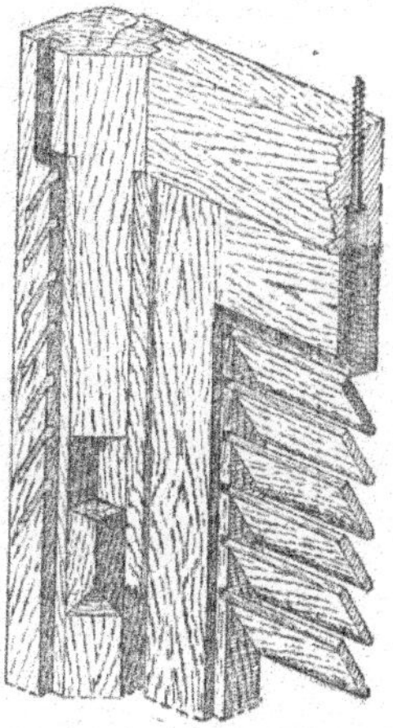

Fig. 1068.—Inside of Pantry-safe Frame and Hanging Stile of Door.

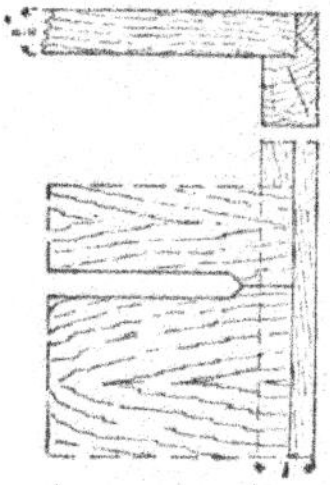

Fig. 1070.—Shelf and Batten of Pantry-safe.

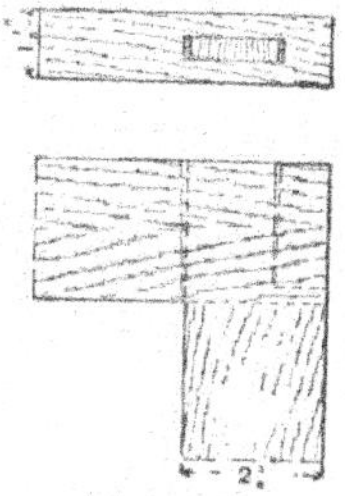

Fig. 1071.—Door Joints in Pantry-safe.

fix the top in position. The blinds may next be inserted (having been previously fitted, and painted or varnished), some allowance being made in the grooves for the thickness of the paint. Then the zinc is secured by wood slips C (Figs. 1065 and 1066) bradded on. In Fig. 1066 the rebate and the beaded slip for securing the zinc joints of meat require hanging. The bottom is screwed on from the under side, and the door (see Fig. 1071) is next fitted, a ⅜-in. bead being worked on the stiles only; the hinges are let in their full thickness on the hanging stile, the lock stile being rebated to meet the slip at D (Fig. 1065). It can be carried on at the lower rail and hanging

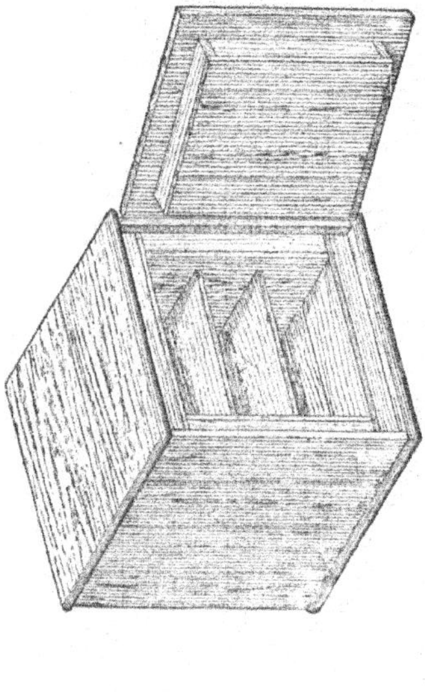

Fig. 1072.—General View of Cold Safe.

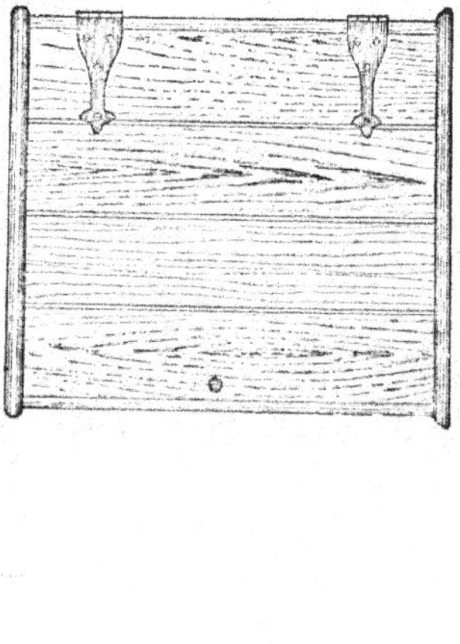

Fig. 1073.

Fig. 1074.

Fig. 1075.

Figs. 1073 to 1075.—Front and End Elevations and Cross Section of Cold Safe.

stile, the hinges being ⅞ in. wide at this case. The door should have a cupboard turn and lock, the eye bolt or ring plate, whichever is fitted, being galvanised. The screws, hinges, lock, etc., should be of brass if the safe is intended for sea-going purposes; and, finally, the dimensions given are extra for sea usage, and may be reduced for home purposes.

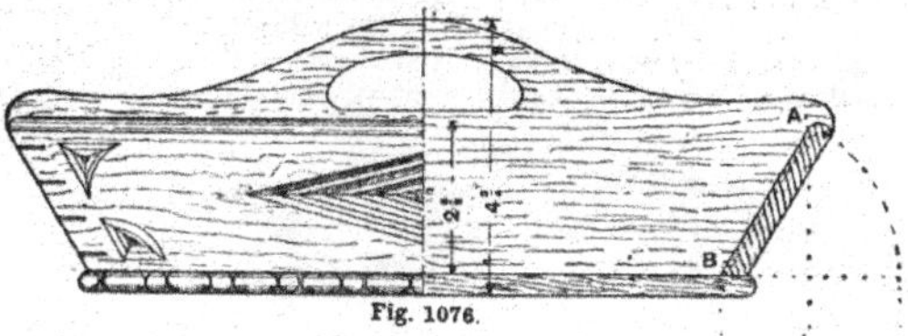

Fig. 1076.

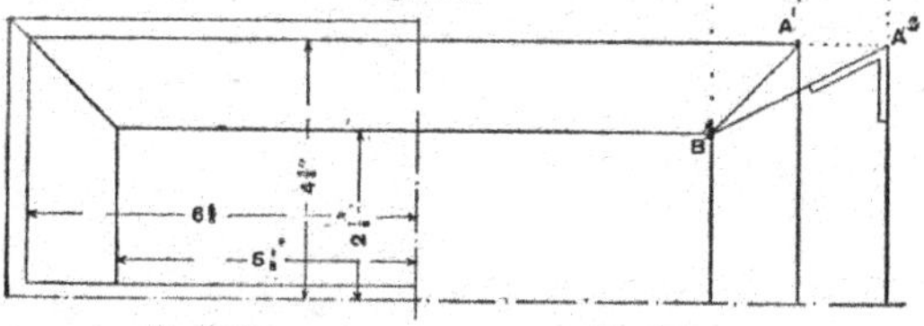

Fig. 1077. Fig. 1078.

Figs. 1076 to 1078.—Half Front Elevation and Vertical Section, Quarter Plan, and Method of Obtaining Bevels of Knife-box.

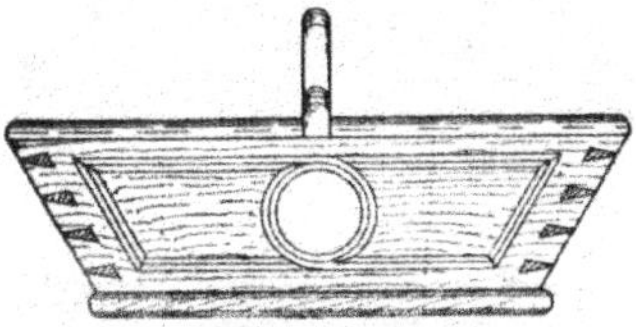

Fig. 1079.—**Alternative Design for End of Knife-box.**

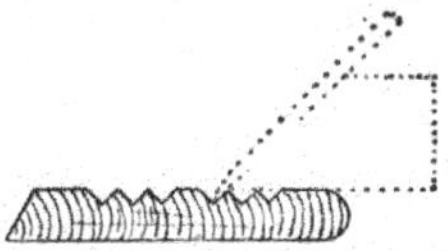

Fig. 1080.—Section through Side of Knife-box, showing Method of Incising.

Cold Safe.

The cold safe illustrated by Figs. 1072 to 1075 is suitable for keeping butter, meat, etc., cool during hot weather. A handy size is about 2 ft. 6 in. high, 2 ft. 3 in. wide, and 2 ft. deep, outside measurements; but of course the dimensions can be altered to suit requirements. All the outer boarding, except the top and bottom, should be ⅝ in. thick, ploughed and tongued, or rebated together. The boards for the inner casing may be about ⅜ in. thick, the joints being tongued or rebated as for the outer boards. Good matchboarding will be suitable. The bottom and top should be of 1-in. boards, jointed and ploughed, and tongued together, and the appearance will be improved if the outer edges are rounded as shown. Fillets of wood about 2 in. by 1 in. should be used, to which the inner and outer boarding can be nailed. The ends of some of these fillets are shown in section in Fig. 1075. In making the safe, the best plan will be to nail the inner casing and fillets together, and then to nail on the outer boards, filling in the spaces tightly with dry sawdust; but care must be taken not to force in so much as to bulge the boarding. The door will require careful work, and should be made to fit fairly close. Its construction is clearly shown in Figs. 1072 and 1075; it should be hinged with double garnet hinges. The fastening may be any of the ordinary forms as desired. Bearers and shelves can be fixed as required. A metal or earthenware dish to contain ice should rest in the bottom of the safe.

Fig. 1081.

Fig. 1082.

Figs. 1081 and 1082.—General Views of Ornamental Knife-boxes without and with Drawer.

Knife-box.

For making a knife-box such as is illustrated by Figs. 1076 to 1078, the best wood is sound, dry Honduras mahogany or white beech; and, relief carving being unsuitable for an article subjected to the rough usage of the kitchen, incised ornamentation can alone be suggested. Two designs are given, and each of the four sides may be finished either as Fig. 1076 or as Fig. 1079. Fig. 1080 is an enlarged section of the side in Fig. 1076 at the centre, and shows clearly the nature of the ornamentation. To make the incisions after they are drawn on the surface, prepare a block a few inches long and with one edge planed to an angle of 45°, as shown dotted in Fig. 1080. Holding this carefully to the lines, pare down its edge with a thin sharp chisel. Rub a little chalk on the under side of the block to prevent its slipping. The curved grooves are better cut with a V-shaped bent chisel or "veining tool." The edges of the bottom may be ornamented with beads, as in Fig. 1076, by drawing the circles and semicircles on a strip of thin paper and pasting this on the edge previously rounded; then little nicks are cut between the circles, and the ends rounded down with a small chisel. The dovetails should be made in the direction of the grain as shown, and the top edge mitered. The division, which also forms the handle, should be housed into the ends slightly, as shown to the right of Fig. 1076, and the bottom is screwed to it. The quarter plan (Fig. 1077) shows the half internal dimensions of the box, while Fig. 1078 illustrates the method of obtaining the bevel for cutting the shoulder lines on the end pieces. To make the drawing, turn down the edge A B (Fig. 1076), as shown by the dotted line, and project it into the plan to intersect a projector from the corresponding edge of the side in plan, the lettering indicating the same point in its various positions. Join the intersection A^2 to B^1, the intersection of the lower edges of the sides, and A^2 B^1 is a side of the true angle of the end, and also of the side, as the inclinations are equal. To set out, prepare the sides to the section shown in Fig. 1080, with the exception of the beading of the top edge, which is better done after dovetailing. Mark the lengths on the bottom edges, and set a bevel to A^2 B^1 (Fig. 1078), and apply this to the marks on the inside face; knife-cut these in, square this line over the edges, and pencil it in on the outside. Next allow the thickness of the stuff, $\frac{3}{8}$ in., and mark down a second bevel line, cutting the pieces off to this line square through. Next plane off the top edges to the same bevel as the bottom, being careful, however, to bevel the two edges parallel with each other. Then apply an ordinary mitre template to the top edges, and to the inside sight

Fig. 1083.

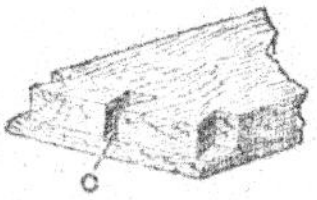

Fig. 1084.

Figs. 1083 and 1084.—Secret Dovetail for Knife-box.

Fig. 1085.

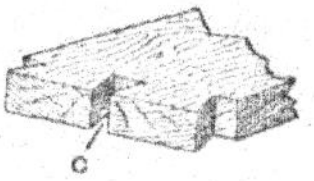

Fig. 1086.

Figs. 1085 and 1086.—Ordinary Dovetail for Knife-box.

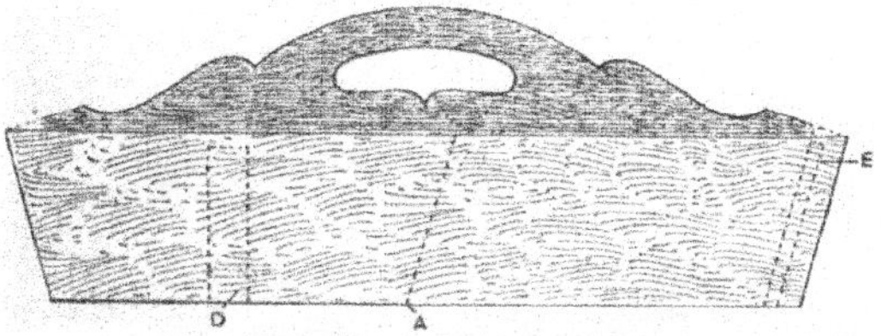

Fig. 1087.—Side Elevation of Knife-box.

Fig. 1089.—Moulded Edge of Knife-box Bottom.

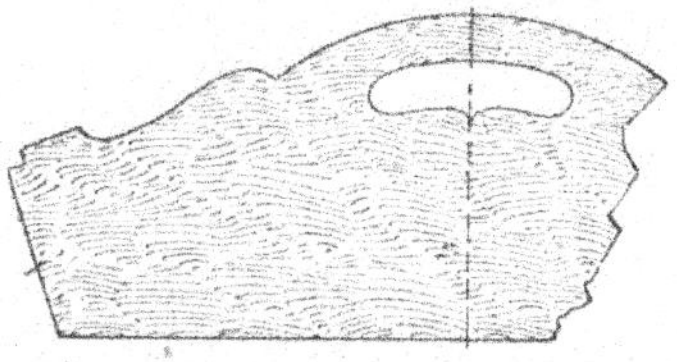

Fig. 1088.—Centre Division for Knife-box without Drawer.

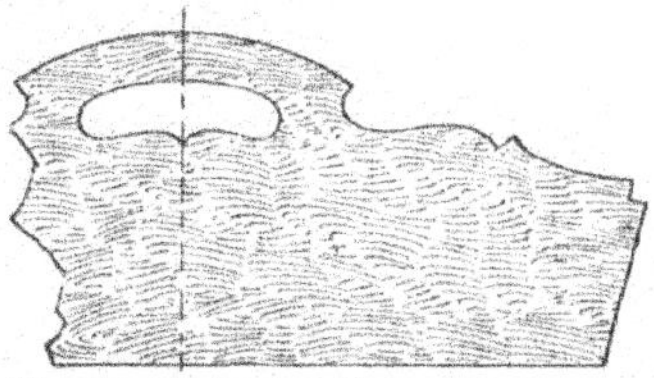

Fig. 1090.—Centre Division for Knife-box with Drawer.

lines, and mark the mitre on the edge; later, and before the edge is rounded, run in a fine-cut saw, not quite down to the dovetail, and set out the dovetail sockets on the ends. Make each outside space half the width of the interior ones, and from these points draw lines with the bevel used for the bottom edge, and on the inner line or back of the sockets mark the width equally on each side. The angle of the corner. Then bead the top edges, and cut out the dovetails with chisels; finish cutting the mitres down to the dovetails, and fit together. All being correct, mark the groove for the division in the two end pieces square from the bottom edge, knock to pieces, and sink the groove. Next glue up, fit the division, and clean off and fit on the bottom, rounding its edges previously, and fixing with glue and screws.

Fig. 1091.—Incised End of Knife-box.

Fig. 1093.—End View of Knife-box with Drawer.

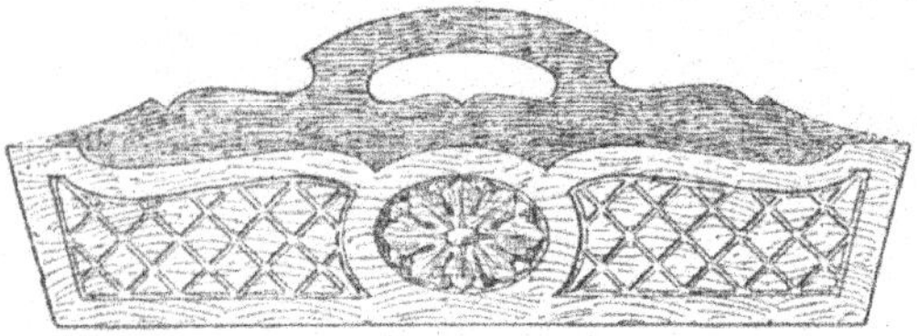

Fig. 1094.—Alternative Side Elevation of Knife-box.

Fig. 1092.—Moulded Edge of Bottom of Knife-box shown by Fig. 1082.

Fig. 1095.—Knob of Knife-box Drawer.

sides of the socket should be about 80°, and this method of marking will ensure the dovetails being all alike. Having marked both ends of one piece, place the two ends face to face in the bench screw, and line across their ends with the bevel; run in the dovetail saw within the lines, then fasten one side piece in the screw with its end level, and rest an end piece on it, keeping the inside sight lines in line with the inside face of the side, and the bottom edges flush. Hold the end down firmly, and draw the dovetail saw through the cuts, number the end, and repeat the process at each

Other Knife-boxes.

Figs. 1081 and 1082 show knife-boxes in perspective, the latter having the addition of a drawer underneath. Hardwood such as mahogany, oak, or walnut, finished with french polish, should be used; the corners are connected by mitre, or secret dovetailing, as shown in Figs. 1083 and 1084. If a more simple method of construction is desired, the boxes can be made of pine, nailed together or dovetailed in the ordinary manner as in Figs. 1085 and 1086. The principal sizes are the same for both designs: Extreme length, 1 ft. 3 in., and extreme width 8 in. In getting out the wood, a little extra in length must be allowed for working. The sides and ends are $\frac{5}{8}$ in. thick, and are gauged to 3 in. wide, with the top and bottom edges square. The correct angle for the ends can be taken from Fig. 1087, which is drawn one-quarter full size, or to the scale of 3 in. to 1 ft. The dotted lines at A (Fig. 1087) indicate the length of the ends. When the sides and ends have been taken to length and width, the next thing is to dovetail them together; the pin B (Figs. 1083 or 1085) is made on the ends of the box, and the sockets C (Figs. 1084 or 1086) to receive them are on the sides of the box. Before the box is glued together the ends must be grooved $\frac{3}{16}$ in. deep to receive the centre division; see the dotted lines D (Fig. 1087), and also at E. Note that the groove does not extend to the top edge of the sides. The half of the centre division (Fig. 1088), shown one-quarter full size, is got out $\frac{3}{8}$ in. longer than the inside measurement of the box, in order that each end F (Fig. 1088) will fit in the groove in the ends. The centre division is put in before the sides and ends are glued together. When the latter is done, the lower edges of the box must be planed level so as to fit against the bottom, which is $\frac{1}{2}$ in. thick, and provided with a half-round edge; see Figs. 1081 and 1089. The bottom is fixed to the sides and ends with screws driven from the under side. The turned feet G (Fig. 1081) are $1\frac{3}{4}$ in. in diameter and $\frac{1}{2}$ in. thick, and are fixed with glue and screws. It is usual to glue cloth or baize on the under side of the feet, so that the box may be laid on a polished sideboard or table without scratching the surface. An alternative pattern for the centre division is shown in Fig. 1090. The hand hole in the division piece should be slightly rounded on the inside edges. To relieve the sides and ends of the design (Fig. 1081) they can be panelled out in incised lines (V or hollow in section) as in the end view (Fig. 1091).

The construction of the knife-box with drawer (shown by Fig. 1082, p. 312) is nearly the same as described for Fig. 1081. The ends H (Fig. 1082) and the back are $\frac{5}{8}$ in. thick and $1\frac{3}{4}$ in. wide, the back corners being mitre-dovetailed as in Figs. 1083 and 1084. The drawer front is $\frac{3}{4}$ in. thick, and the sides, back, and bottom are $\frac{3}{8}$ in. thick. The drawer can be made in the usual way by grooving the sides to receive the bottom; or, to give more inside space, the latter can be rebated to the sides and front, the bottom standing up $\frac{1}{16}$ in. just to clear and avoid friction. The top portion of the box is fixed to the lower ends H (Fig. 1082), and the back with dowels and glue. The bottom J (Fig. 1082)—see enlarged section Fig. 1092—is fixed as in Fig. 1081. The lines K in the end view (Fig. 1093) are incised, but they can be further elaborated by the cross lines and the carved centre as in the alternative side view (Fig. 1094). It will be seen that the shaped outlines of the sides and the ends (Figs. 1093 and 1094) can be applied to the design Fig. 1081; also, the diamond centre of Fig. 1091 could be carved as a centre in Fig. 1094. Likewise the upper part of Fig. 1082 could be incised as in Fig. 1091. The turned drawer knobs of Fig. 1082 are shown in side view by Fig. 1095. Brass handles on the drawer instead of knobs would also look well.

CUPBOARDS.

Kitchen Cupboards.

FIG. 1096 shows an elevation of a kitchen cupboard over a sink; Fig. 1097 showing an end view and Fig. 1098 a section. Figs. 1099 and 1100 show respectively elevation and section of a kitchen cupboard for a recess. It has been assumed that only a front and doors are necessary in the latter case. All the framework should be made of wood 1¼ in. thick, working up to about 1⅛ in.; ½-in. boards will be required for panels, 1-in. boards for the bottom of the cupboard shown at Fig. 1096, and ¾-in. boards for the top of the same. The dimensions of the parts are given in the illustrations. The frames should first be made. The doors can be hinged with 2½-in. butts. To fix the first cupboard shown, prepare three pieces of wood about 3 in. by 2 in., and fix them in the wall as shown. The top should be fixed by driving two or three holdfasts into the wall, and screwing these to the top. The frame of the cupboard illustrated at Figs. 1099 and 1100 will have to be fixed to the wall or sides of the recess. Shelves are not shown, but can be added as desired. Some moulding fixed to the cupboards as shown will greatly improve their appearance. The following quantities of wood are required for the cupboard shown at Figs. 1096, 1097, and 1098: 50-ft. run of 2¼ in. by 1¼ in. for front and sides of frame, stiles, and top rails of doors; 5-ft. run of 3 in. by 1¼ in. for bottom rails of doors; 12 ft. of pine, 11 in. by ½ in., for panels of doors; 12 ft. of 7 in. by ¾ in. for top; 12 ft. of 9 in. by 1 in. for bottom; about 9 ft. length of 2¼-in. by 1-in. ogee moulding to fix round the top. The designs might also be adapted for cupboards in dining-rooms or bedrooms, in which case oak or other hardwood might be used.

Damp-proof Harness Cupboard.

Figs. 1101, 1102, and 1103 are front, end, and back elevations respectively of a cupboard for holding harness, Fig. 1104 being a horizontal section. The height from the floor to the top of the cornice is 7 ft. 6 in., and the width 3 ft. out to out of the ends; the depth in the clear is 1 ft. Owing to the frequent occurrence of damp in harness rooms that adjoin stables it is necessary to adopt some means by which the damaging effects may be obviated. The cupboard in which the harness is hung, therefore, should be air-tight and damp-proof; and the one illustrated has been designed to meet these requirements. Good yellow deal should be used. The front is composed of two 1¼-in. panelled folding doors. The lower panels of wood are bolection moulded, the upper panels of glass being divided into three squares each, with moulded bars, a section of which is shown at x (Fig. 1105). The meeting stiles are prepared with hook joints, and a moulded cover fillet is tongued to the face of the right-hand door, as shown in section at Fig. 1105. The outer or hanging stiles (see Fig. 1106) are prepared with a tongued heel, which fits into a shallow groove prepared in the rebate of the front frame of the ends, the lower edge of the doors shutting into a rebate prepared along the front edge of the pot-board. The upper end on the inside of the top rail of the door shuts on to a bead prepared on the fillet fixed on the inside of the cornice rail (see Fig. 1107). The ends are framed together with flush panels on the inside, the outside being sunk and bolection moulded. These ends are made the full height of the cupboard to the under side of the cover-board or top, the latter being fixed down to the top rails with screws (see Fig. 1107). The bottom ends are grooved to receive the pot-board (see Fig. 1108), which is held in position by being nailed through the ends, and additionally secured with glued angle-blocks on the under side, as shown. The plinth, when fixed, will cover the holes made by the nails in fixing the pot-board. The back is framed together in six panels, flush framed on the inside. The back stiles of the ends are prepared with rebates

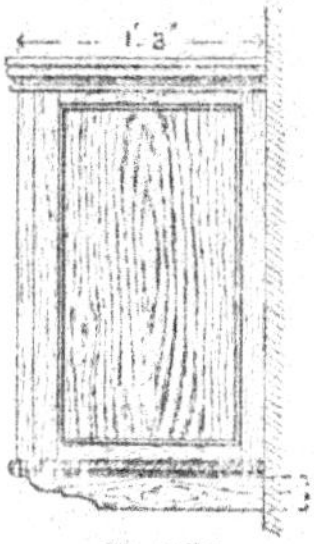
Fig. 1097.

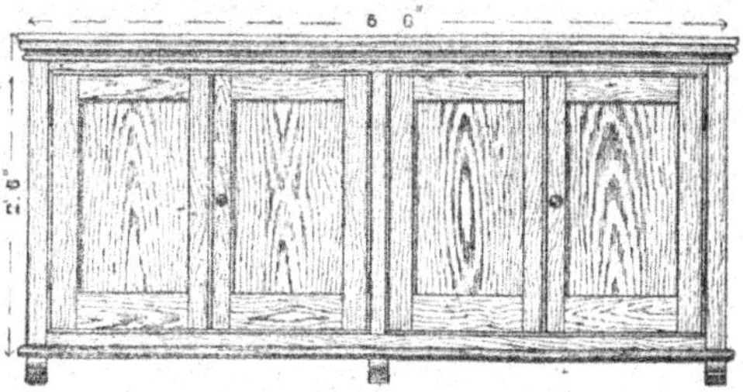
Fig. 1096.

Figs. 1096 and 1097.—Front and End Elevation of Cupboard to fit over Sink.

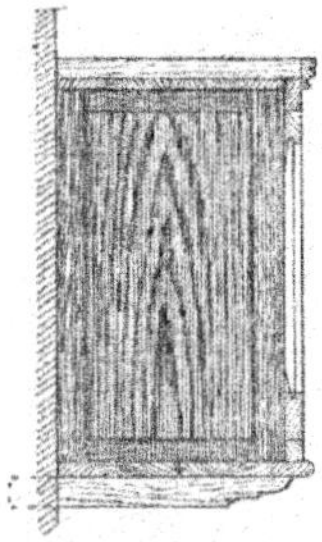
Fig. 1098.—Vertical Section of Cupboard to fit over Sink.

Fig. 1099.

Fig. 1100.

Figs. 1099 and 1100.—Elevation and Cross Section of Cupboard for Recess.

1 in. by ½ in. to receive the panelled back, the exact width of which is obtained after the ends have been fixed to the pot-board. It is essential that the back should fit in tight between the rebates, and for permanently fixing the edges should be glued and secured by inserting screws at close intervals, as shown in Fig. 1103. The plinth and cornice are mitered at the front angles, and screwed in position from the inside of the front and end framings. After the carcase has been put together, the inside must be treated with a damp-proof preparation, such as Palma cream, this being

Fig. 1101.

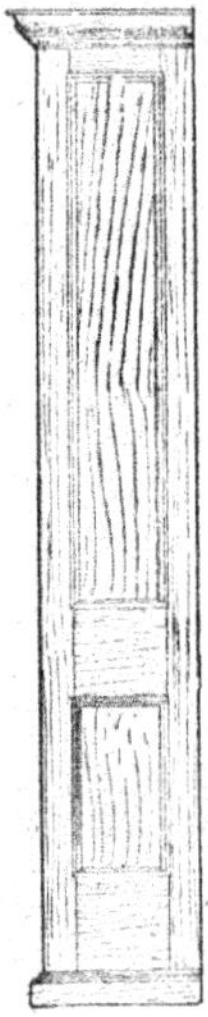

Fig. 1102

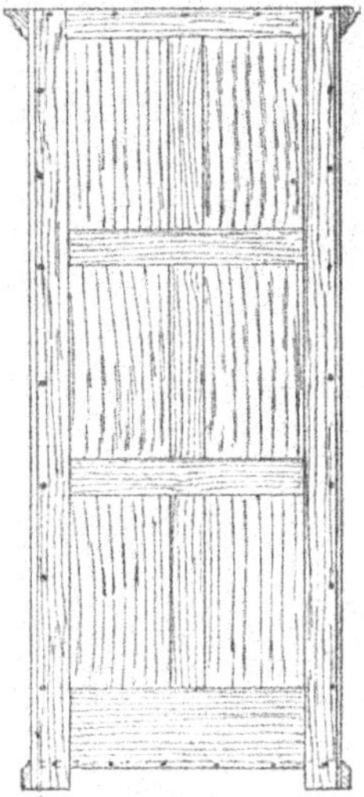

Fig. 1103.

Fig. 1104.

Figs. 1101 to 1104.—Front, End, and Back Elevations, and Horizontal Section of Damp-proof Harness Cupboard.

Fig. 1105.—Horizontal Section through Meeting Stiles of Cupboard Doors.

Fig. 1106.—Horizontal Section through Hanging Stiles of Cupboard Doors.

applied with an ordinary paint brush. The mode of procedure is as follows :—Having carefully rubbed down the face of the woodwork with glasspaper, and dusted over, give the work one coat of ordinary white-lead paint. When this is dry, apply two or three coats of the cream, allowing each coat to get perfectly dry and hard before applying the next. When the last coat is quite dry and hard, take some Willesden three-ply paper and cover the whole of the interior with it, hanging the paper in the usual way with strong paste. The edges of the paper are prevented from coming away by fixing rounded fillets along the edges and in the angles, as shown in the detail illustrations. The following is the table of quantities :—

		No.	ft.	in.	ft.	in.	in.	Remarks.
Doors.	Stiles	4	7	0		4	1¼	All the material to be yellow deal, specially selected for staining and varnishing. The whole of the deal to be wrought, and to hold the full nominal thickness and width. The inside of all framing, except doors, will be covered with paper, and must be flush framed and smoothed with the jack plane.
	Rails	4	1	6		8	1¼	
	,,	2	1	6		2½	1¼	
	Bars	4	1	6		1¼	1	
	Panels	2	1	6		10	⅝	
	Mouldings	1	10	0		1½	1	
	Beads	1	33	0		⅝	¾	
	Stop	1	7	0		1	¾	
Ends.	Stiles	4	7	6		3	1	
	Rails	4	1	2		8	1	
	,,	4	1	2		8	1	
	Panels	2	1	6		9	½	
	,,	2	4	2		9	½	
	Mouldings	1	30	0		1½	1	
Back.	Stiles	2	7	6		4	1	
	Rails	1	3	0		9	1	
	,,	3	3	0		4	1	
	Muntins	3	2	3		4	1	
	Panels	6	2	0	1	1	⅝	
	Pot-board	1	3	0	1	2	1	
	Top	1	3	0	1	2	¾	
	Plinth	1	5	9		4	1	
	Cornice	1	6	6		3½	2¼	
	Fillet	1	2	10		4	[illegible]	
	Top Rail	1	3	0		4	1¼	
	Fillet	2	7	0		[illegible]	[illegible]	
	,,	2	7	0		[illegible]	[illegible]	
	,,	2	3	0		[illegible]	[illegible]	
	,,	4	1	1		[illegible]	[illegible]	
		1		4				Mortice latch, left hand.
		2		6				Brass-necked barrel bolts.
			9	0				Super 21oz. clear sheet glass.
			42	0				Super Willesden 3-ply paper.
		1						Eccentric catch and handle.

Figs. 1101 to 1104 are reproduced to a scale of ½ in. to the foot, and Figs. 1105 to 1108 to 4 in. to the foot.

Corner Pedestal Cupboard.

As the cupboard illustrated in Fig. 1109 is intended to be painted, deal will serve

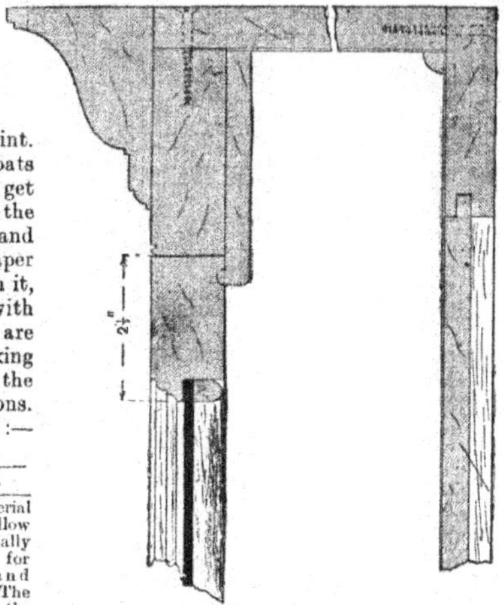

Fig. 1107.—Vertical Section through Upper Part of Harness Cupboard.

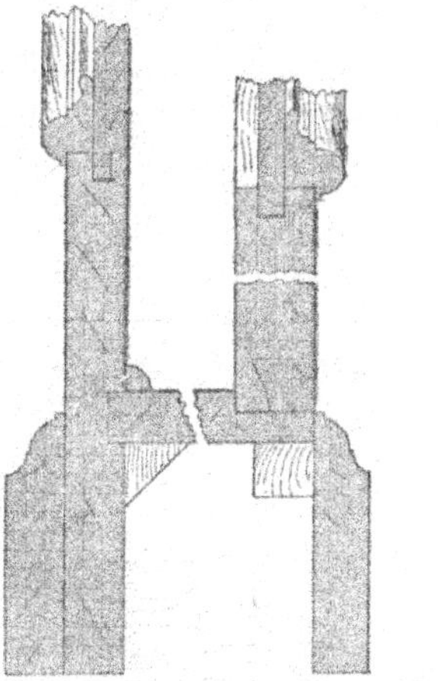

Fig. 1108.—Vertical Section through Lower Part of Cupboard Doors and Plinth.

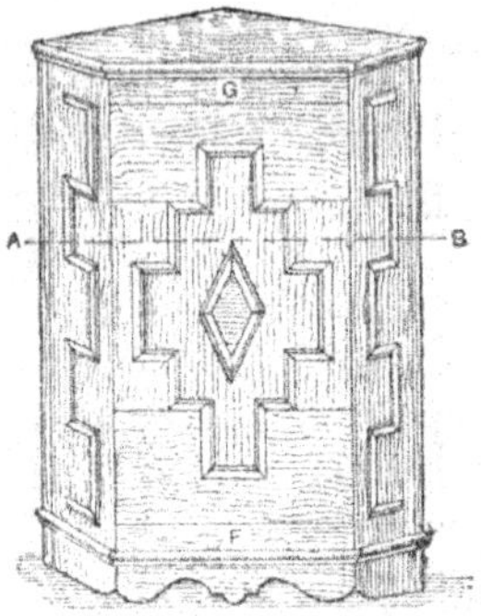

Fig. 1109.—Corner Pedestal Cupboard.

Fig. 1111.—Door of Corner Pedestal Cupboard.

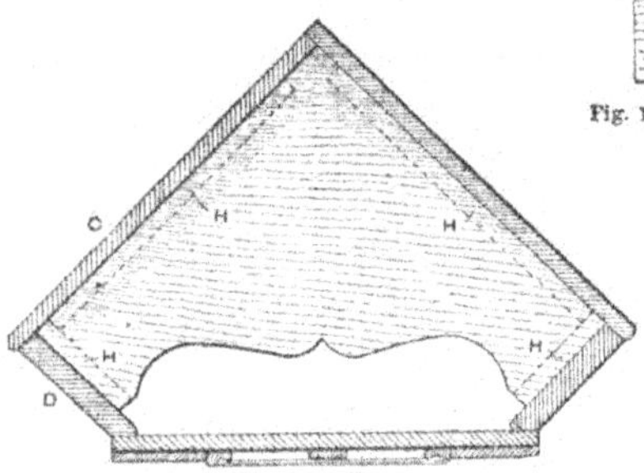

Fig. 1112.—Horizontal Section of Corner Pedestal Cupboard.

Fig. 1113.—Section of Moulding for Pedestal Cupboard Door.

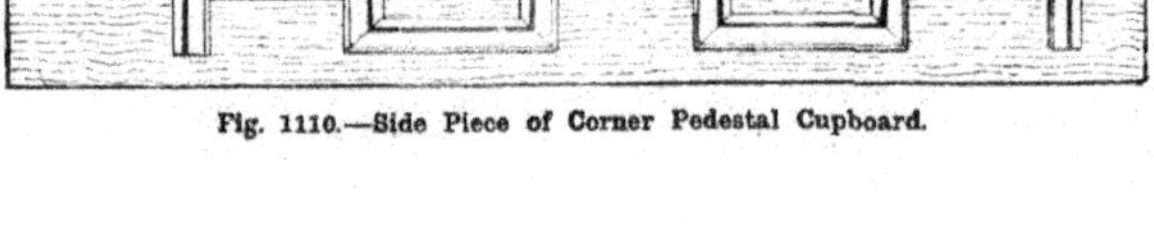

Fig. 1110.—Side Piece of Corner Pedestal Cupboard.

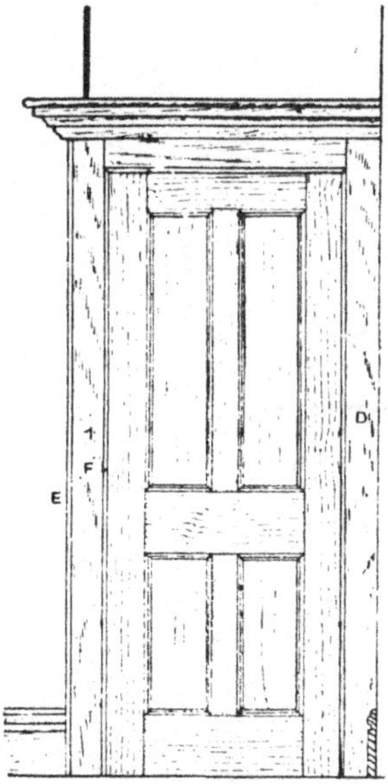

Fig. 1114.—Front Elevation of Dress Cupboard for Bedroom Recess.

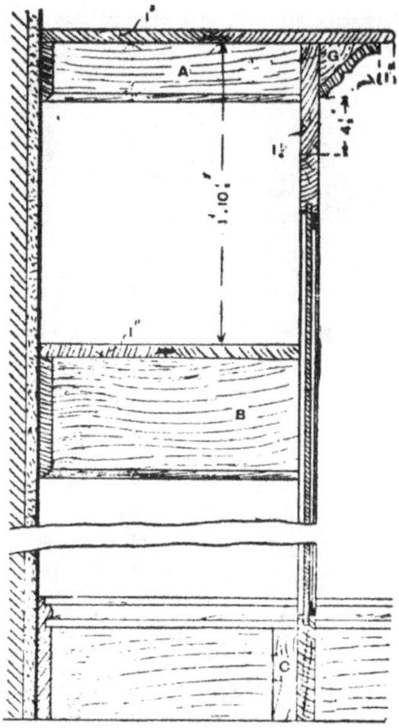

Fig. 1116.—Vertical Section of Dress Cupboard for Bedroom Recess.

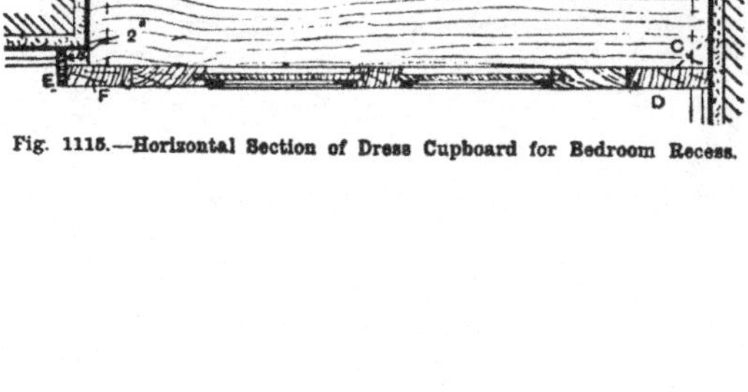

Fig. 1115.—Horizontal Section of Dress Cupboard for Bedroom Recess.

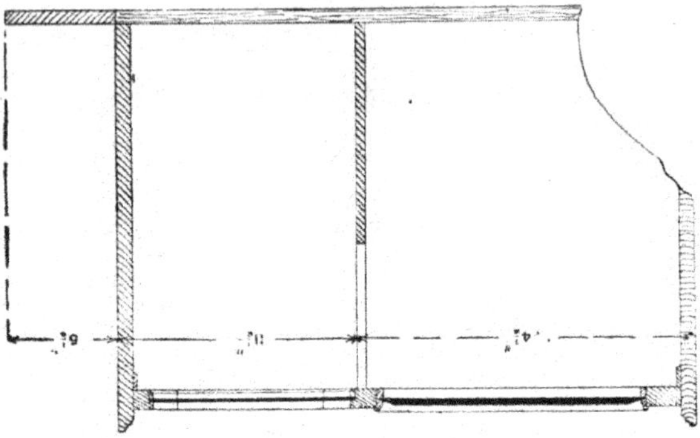

Fig. 1118.

Fig. 1117.

Figs. 1117 and 1118.—Front Elevation and Vertical Section of Hanging Corner Cupboard.

as material. The back is made of two pieces of $\frac{1}{2}$-in. board, 2 ft. $2\frac{1}{2}$ in. long, one 1 ft. wide, and the other wider by $\frac{1}{2}$ in. to allow for the overlap at the corner. These are screwed together and to the side pieces, one of which is shown separately at Fig. 1110. These side pieces are of $\frac{3}{4}$-in. stuff, of the same length as the back boards, and 4 in. wide. The front edge E of each is splayed off so that the door (Fig. 1111) and front strips may lie flat. Fig. 1110 also shows the arrangement of the moulded ornament on these side pieces. The back and sides are braced together in their lower parts by being screwed to the cupboard bottom, which is of $\frac{3}{4}$-in. board, and which is placed so that the door closes against one-third of its thickness. The front strip F (Fig. 1109), which continues the lines of the door, is also of $\frac{3}{4}$-in. board, and is 1 ft. long by 4 in. wide. It is fastened on the side pieces, and its lower edge is shaped. The upper part is held together by the cupboard top, which is made of $\frac{1}{2}$-in. stuff. This is screwed down on the tops of the back and side pieces, and on the upper front strip G which runs above the door, and is of $\frac{3}{4}$-in. board, 1 ft. long and $1\frac{1}{2}$ in. wide. It is fixed to the side

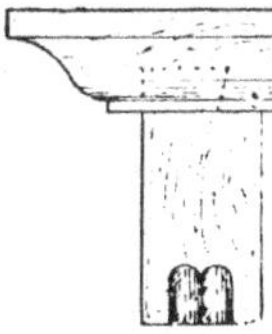

Fig. 1120.—Top End of Cupboard Standard.

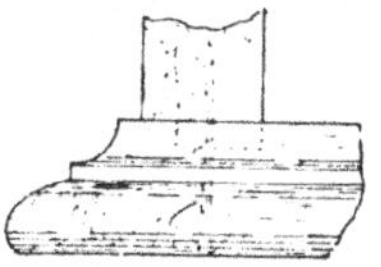

Fig. 1121.—Bottom End of Cupboard Standard.

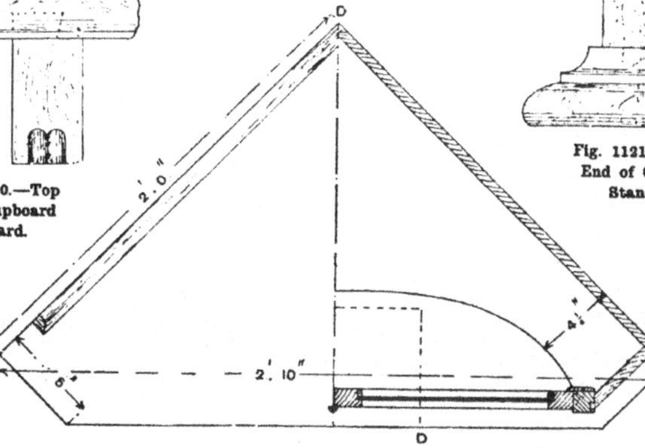

Fig. 1119.—Horizontal Section through Hanging Corner Cupboard.

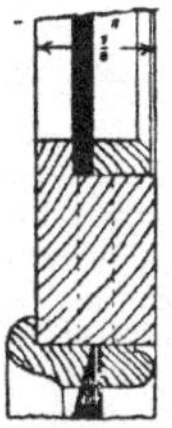

Fig. 1122.—Section through Glass of Cupboard Door.

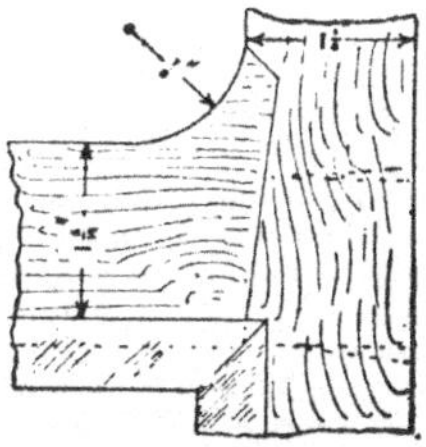

Fig. 1123.—Part of Middle Rail of Cupboard Door.

pieces, and along its under surface is fastened a slip of ¼-in. wood, which projects ¼ in. below its under edge, and against which the door closes. A line of moulding, as shown, runs along the front of the upper edge of the cupboard, and hides the fastening-on of the top piece. Another line of are fixed to the backs C and sides D. The door is a piece of ½-in. board, 1 ft. 9 in. high and 1 ft. wide, and on this are bradded the slips of moulding (Fig. 1113) which form the pattern. These are cut from ordinary steam-struck moulding, such as may cost about 3s. a 100-ft. run. Out-

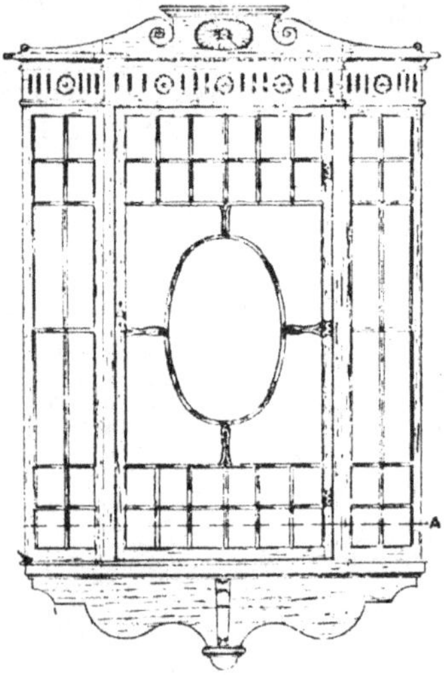

Fig. 1124

Fig. 1126.

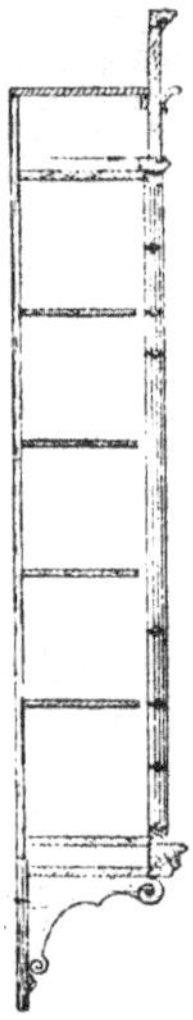

Fig. 1127.

Fig. 1125.

Figs. 1124 to 1127.—Front Elevation, Horizontal Section, Side Elevation, and Vertical Section of Cupboard for Antique China.

moulding runs along the lower part of the front, 1½ in. below the door, and shallow openings have to be cut through the slightly rounded front edges of the back pieces to accommodate these mouldings. The middle shelf, shown in plan in Fig. 1112, which is a section on A B (Fig. 1109), is of ¼-in. board, and is fixed 1 ft. above the cupboard bottom. For its support, strips H of ½-in. wood

side the rectangular figure, pieces of ¼-in. board, cut to shape, are fixed on the ½-in. base, the upper and lower ones crossing its grain at right angles. The space within the central diamond is filled in the same manner. In painting this cupboard, the mouldings, etc., may be kept of a darker shade than the other parts, but this is much a matter for private taste. Fig. 1109 is to no exact scale, but Figs. 1110, 1111, and 1112 are 2 in. to the foot, and Fig. 1113 is half size.

Dress Cupboard for Bedroom Recess.

A dress cupboard is shown in front elevation by Fig. 1114. Fig. 1115 is a horizontal section (enlarged) of the cupboard, and shows the shape of the recess. First prepare the front of clean yellow deal $1\frac{1}{4}$ in. thick and about 2 in. wider than the recess. If a ready-made door is used, the opening should be of standard size, showing as nearly as possible $4\frac{1}{2}$ in. margin all round. With a very wide recess, folding doors would be used. Brace the front securely and offer it up. Adjust the head level and scribe to the floor, and then scribe the stile D to the wall and skirting. Prepare and fix the ledges A (Fig. 1116) $4\frac{1}{2}$ in. wide and 1 in. thick round the recess, level with the top, allowing the ends to project the thickness of the skirting. Use 3-in. floor-brads for nailing to brickwork, and, in the case of a lath-and-plaster partition, feel for the studs with a bradawl, and screw to them with 3-in. screws. For the shelf, fix the ledges B, 9 in. wide and 1 in. thick, in a similar manner, and fit in the shelf so that the outside edge is flush with the ends. Fix the front in position, screwing to the ledges, the floor and skirting, and to the angle block C (Figs. 1115 and 1116) behind the stile D (Fig. 1115). Ascertain by how much the cornice overhangs, and prepare and fix the top so that the rounded outside edge projects 1 in. all round. The wide bead E is then scribed to the wall and skirting, and bradded to the stile F to hide the joint and the end of the rail tenon and to make a neat finish. The cornice is then mitered round and fixed with $1\frac{1}{2}$-in. oval wire brads to angle blocks G (Fig. 1116), glued round under the top. Then hang the door to the stile D, using 3-in. butts. The shelf acts as a stop, but, if thought necessary, a common stop may be nailed to the floor. Treat the cupboard to harmonise with the surroundings, and then provide and fix the necessary door furniture and also the dress hooks, which are screwed to the ledges B as required. If the cupboard is intended only as a temporary fixture, the use of screws as described

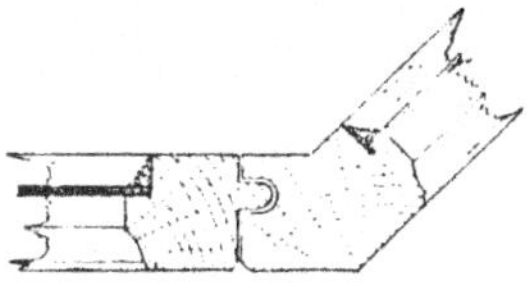

Fig. 1128.—Hanging Stile and Angle Post of Cupboard for Antique China.

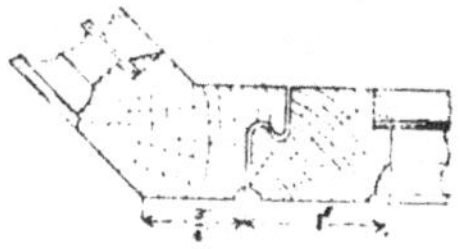

Fig. 1129.—Falling Stile and Angle Post of China Cupboard.

Fig. 1131.—Detail of Bar for China Cupboard.

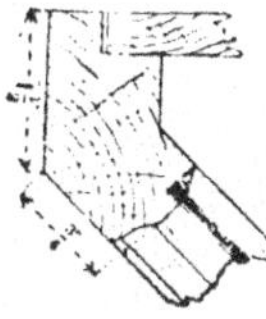

Fig. 1130.—Back Angle of China Cupboard.

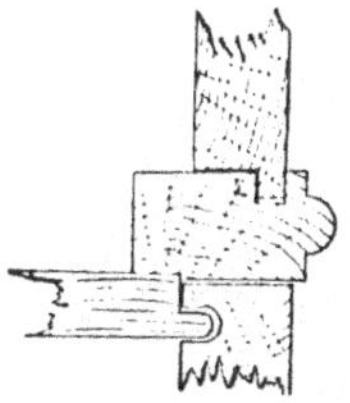

Fig. 1132.—Section of Necking to Frieze of China Cupboard.

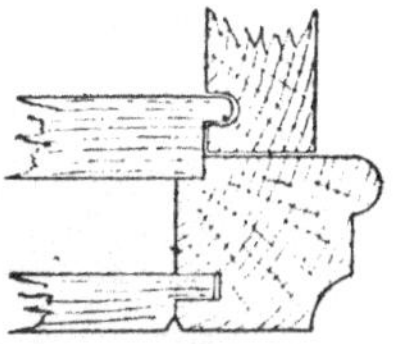

Fig. 1133.—Section through Base of China Cupboard.

is advised, but for a permanent fixture nailing would make a better job.

Hanging Corner Cupboard.

The corner cupboard shown in elevation by Fig. 1117, in vertical section by Fig. 1118, and in horizontal section by Fig. 1119, should preferably be executed in a dark, rich-coloured wood, such as mahogany or Italian walnut. The chief dimensions are: Width across front, 2 ft. 10 in.; width of sides, 2 ft.; central depth, 1 ft. 8½ in.; returns, 5 in.; and height over all, 3 ft. 1 in. The top shelf is of ¾-in. stuff, the bottom shelf of 1-in. stuff, and the middle shelf of ⅝-in. stuff. The sides are ½-in. boards, glued up to the width required, and may be of pine stained to match the remainder. The door standards are 1¼ in. square, and the doors themselves are of 1-in. stuff. The return ends are of ¾-in., and the shaped top plinth and the fretted brackets are of ½-in. wood. These two members are dowelled or screwed to the carcase, and the return ends are housed into the top and bottom shelves ¼ in. and bradded, as shown in Figs. 1120 and 1121, the top ends being preferably dovetail-housed from the back, as shown by the dotted lines. The sides of the carcase can run over the edges of the top and bottom, and be nailed directly thereto, as shown in Fig. 1118. The standards should be tenoned, fox-wedged at the top, and through-wedged at the

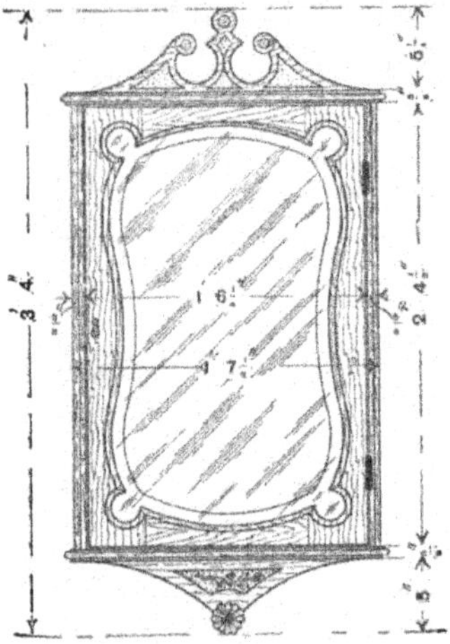

Fig. 1134.

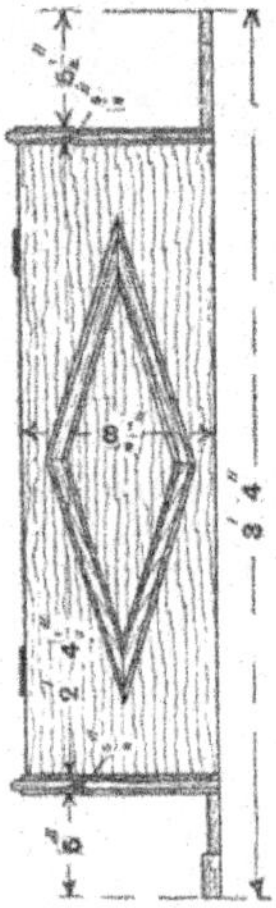

Fig. 1135.

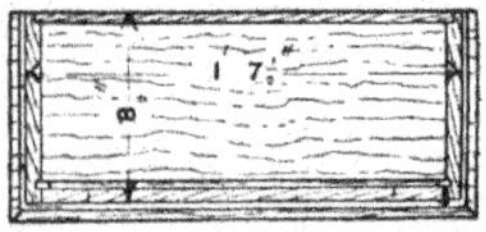

Fig. 1136.

Figs. 1134 to 1136.—Front and Side Elevations and Horizontal Section of Ornamental Wall Cupboard.

bottom, and the inner shelf may rest on bearers (not shown). The doors, made with shaped rails in the upper panels, are rebated for glass, and left square in front, as shown in Fig. 1122. The lower panels are left square. The rebate for the bevelled edge glass panels is formed with a $\frac{1}{2}$-in. bolection ogee moulding and a $\frac{3}{8}$-in. glazing bead. Fig. 1123 shows how the joints in the middle rails are made, a $\frac{5}{16}$-in. mortice and tenon being used to secure them. The meeting stiles of the doors are square, a $\frac{3}{8}$-in. planted bead covering the joint. Fig. 1119 is a section on A B (Fig. 1117), Fig. 1118 being a section on D D (Fig. 1119). Figs. 1117, 1118, and 1119 are to a scale of $1\frac{1}{2}$ in. to 1 ft., and the remainder are half full size.

Cupboard for Antique China.

The cupboard shown in front elevation by Fig. 1124 is intended to be made of mahogany, with plate-glass panels in the front and side lights. The door is hung with ornamental brass hinges, and is fitted with a brass lock and ornamental drop handle. The frieze is decorated with short flutings and sunk pateræ. The elliptical panel in the pediment is worked in low relief, and the scrolls are formed with the V-tool, the cornice

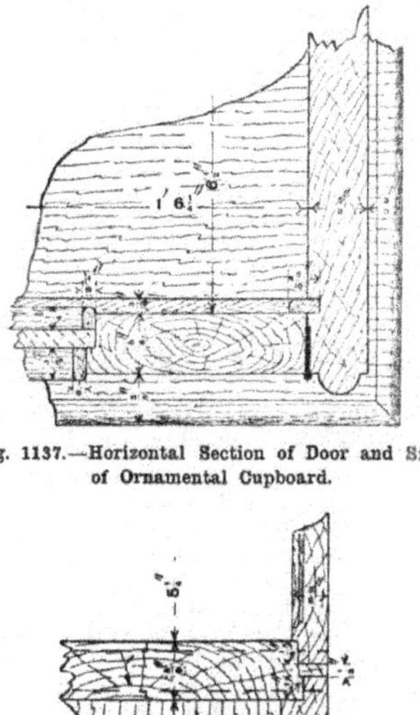

Fig. 1137.—Horizontal Section of Door and Side of Ornamental Cupboard.

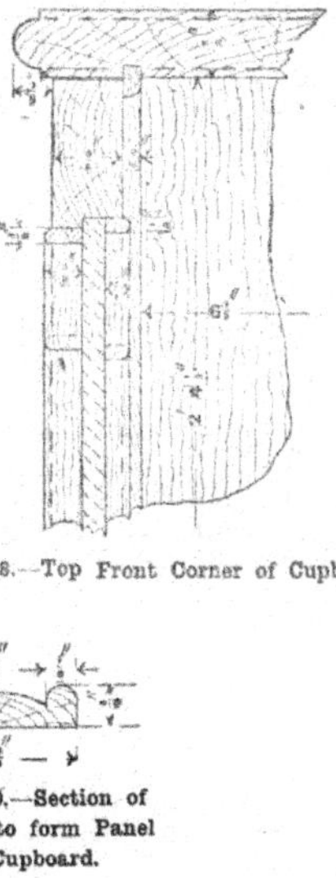

Fig. 1138.—Top Front Corner of Cupboard.

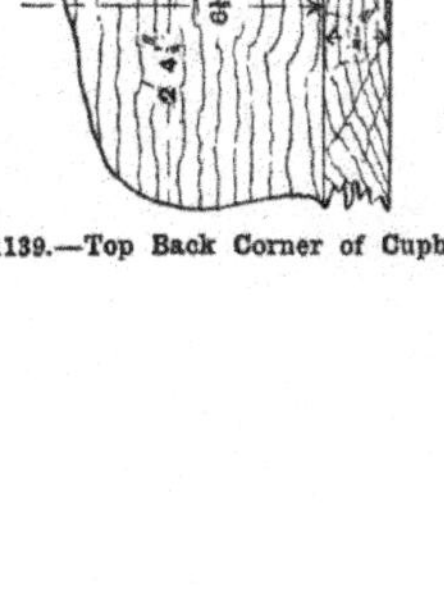

Fig. 1139.—Top Back Corner of Cupboard.

Fig. 1140.—Section of Moulding to form Panel on Cupboard.

Fig. 1141.—Cupboard Shelf with Plush Edging.

over the pediment being dentilled. Fig. 1125 represents a horizontal section above the bottom rail of the door on line A A (Fig. 1124). The angle pieces near the door are provided with a groove and hook joint to prevent dust entering the cupboard, the back angles being rebated to receive the backboard. The backboard may be framed, or if preferred it can be one plain board only. Fig. 1126 is a side view of the cupboard, and Fig. 1127 a vertical section. The inside of the cupboard is fitted with four shelves, ½ in. thick, and moulded on the front edge. The details of the cupboard are illustrated as follows:—Fig. 1128 is the angle post and hanging stile of the door; Fig. 1129 the angle post and falling stile of the door, with the hook joint; Fig. 1130 the back angle showing the glass panels, which are secured with coloured hard stopping on the inside; Fig. 1131 the section of the bars for the door and side lights; Fig. 1132 an enlarged detail of the necking to the frieze upon which the pediment is built, and Fig. 1133 shows the base enlarged. The plate-glass panels should be ¼ in. thick, and the woodwork, if light, should be toned down to represent Spanish mahogany, and french-polished.

Ornamental Wall Cupboard.

Fig. 1134 illustrates the front elevation of a small cupboard, made in polished walnut or mahogany, fixed to the wall with brass plates, or screwed through the back to wall plugs. Fig. 1135 shows a side elevation, and Fig. 1136 a sectional plan. The following pieces of timber are required:—For the top, 1 ft. 9 in. by 9¼ in. by ⅝ in.; bottom, 1 ft. 9 in. by 9¼ in. by ⅝ in.; two sides, 2 ft. 5 in. by 8½ in. by ⅝ in.; back, 8 ft. by 6 in. by ½ in.; two door stiles, 2 ft. 5 in. by 3¾ in. by ⅝ in.; two door rails, 1 ft. 2½ in. by 2 in. by ⅝ in.; door nosing, 10 ft. by ⅜ in. by ¼ in.; folding rebate slip, 8 ft. by $\frac{3}{16}$ in. by $\frac{3}{16}$ in.; glass fixing slip, 7 ft. by ¼ in. by ⅛ in.; top scroll, 1 ft. 6 in. by 5½ in. by ⅜ in.; bottom scroll, 1 ft. 6 in. by 5¼ in. by ⅜ in.; and moulding for diamond panel, 4 ft. 3 in. by ⅜ in. by $\frac{3}{16}$ in. The sides of the cupboard are rebated at the back edges to receive the back, which is of ½-in. stuff, placed crossways of the carcase, glued and pinned in the rebate, the joints of the boards being grooved and tongued. The face edges of the sides are ornamented with a moulding, shown in section by Fig. 1137. The top and bottom of the cupboard are similarly moulded, being fixed to the sides by screws either from the top or from the under side. A fillet $\frac{3}{16}$ in. by $\frac{3}{16}$ in. is grooved into the framework ⅝ in. on, to form the folding rebate for the door, as shown in Figs. 1138 and 1139. The scroll at the top and bottom is cut from ⅜-in. stuff, and the panelling is (with a carver's punch) cut ⅛ in. deep and decorated, and is glued and screwed into a rebate in the top and bottom, as shown in section in Fig. 1139. The diamond panel on each end of the cupboard is formed with a moulding ⅜ in. by $\frac{3}{16}$ in., shown in section by Fig. 1140; this can be worked with a router or moulding planes.

Cupboard Door.—The door is of ⅝-in. stuff, cut to the shape shown in Fig. 1134 with a band-saw. The stiles are 3¾ in. wide at the joint, and 2 in. at the widest part of the sweep. The rails are 2 in. wide at the shoulders, and 1¼ in. wide at the narrow part. The rebate for the glass can be cut with a router, after which the pieces should be framed together with secret mortices and tenons, wedged and glued. The nosing, ⅜ in. by ¼ in., is pinned in the opening as shown in Figs. 1137 and 1138. It can be worked in a straight length, and then soaked in hot water for an hour or so, when it can easily be bent into the shapes required. In the small corner circles it will be necessary to cut one or two small saw-kerfs on the inner side of the nosing to help in bending. The nosing also forms part of the glass rebate. The glass is $\frac{3}{16}$ in. or ¼ in. thick, and fixed with a slip ¼ in. by ⅛ in., shown in section in Figs. 1137 and 1138. The door is hung with 2½-in. brass butt hinges, and is closed with a brass lever lock.

Completing the Cupboard.—The interior of the cupboard is lined with plush, green and dark red being suitable colours. This is fixed with thin glue, brushed lightly and evenly over the woodwork, and the plush pressed against it. In pressing out the creases, place a sheet of tissue paper over the plush. The corners of the cupboard will look well if finished with ½-in. or ¾-in. plush roll. For the display of china and small curios it will be necessary to provide one or two shelves, which should be ⅜ in. thick, and fitted with clips, to enable them to be fixed at various distances. The shelves should be covered with plush, as shown in Fig. 1141, the fancy edging being glued to the slip which fits into the groove on the under side of the shelf.

Cupboard and Drawers for Recess.

Figs. 1142 to 1145 show a recess with drawers in the lower portion and a cupboard above them. The recess illustrated is 6 ft. 6 in. high, 2 ft. 10 in. wide. The illustrations show fully the principal parts of the construction, and will serve the purpose better than a lengthy description. In the following will be found the main features. The first part to construct will be the framework for the drawers, comprising the lower portion. The two stiles A and B (Fig. 1142), and the bottom and two intermediate rails, are housed together as illustrated at C, (Fig. 1146), whereas the joint at D (Fig. 1142), between the top rail and side, should be dovetailed together as shown at Fig. 1147. The runners E (Figs. 1146 and 1148) should tenon into the front rails F; and if it is desired to have panels to separate the spaces between the drawers the runners and rails should be ploughed. Next, the rails and stiles should be fixed together and then fastened temporarily to the sides of the recess. Fillets G (Fig. 1148), about ½ in. thick and 3 in. wide, should be prepared and cut so as to fit close against the back wall. The runners should next be cut off true to length, and to the back edges of these the fillets should be nailed. Now the front should be taken out, the runners attached to it, and the whole pushed into the recess. The runners should be carefully adjusted so that they are quite horizontal, and the fillet may be nailed to the wall. After this the front should be fixed to the sides of the recess. To make a good job, two brick joints on each side should be found, and small plugs driven in. To find the joints without damaging the plastering, drive in a moderately fine bradawl. The top of the drawers, which also forms the bottom of the cupboard, has a rounded edge; it should next be prepared and fixed. The material for the drawers should be got ready, the fronts being carefully fitted to the case. Then the sides and back are prepared and set out for dovetailing, which should be done in the usual manner. There is a lap-dovetailed joint between the front and side. After the drawers are fitted, guide strips should be prepared to go between the runner and fillet as shown in Fig. 1148, and a piece of skirting or plinth should be fixed to the bottom. The construction of the upper frame to receive the doors is quite simple. The joint between the head piece and stile is shown at Fig. 1149, the head piece serving as a top member to the cornice. The appearance of this framing will be much improved if a bead is worked on so as to break the joint between it and the doors. This framing should be fixed to the sides of the recess in a similar manner to the lower framing. A piece of suitable moulding should be fixed so as to form a cornice, and fillets K, L, and M (Fig. 1144) should be prepared and nailed to the sides of the recess to receive the shelves; for the doors, the stiles and rails require mortising and tenoning together. The stiles and rails should next be ploughed for the panels; then, when these are prepared, the parts should be fitted together. The joints should be glued, cramped, and wedged; and after the glue is dry the stiles and rails should be planed flush. The appearance of the doors will be improved if a small panel moulding is mitered and bradded in. Next, rebate the doors together as shown in Fig. 1142, and fit them in the opening, hanging them with 3-in. wrought-iron butt hinges. Finally, suitable drop handles and locks should be fixed on the drawers, and also a knob, lock, and bolt to the doors.

Collapsible Cupboard.

A collapsible cupboard, opened out for use as shown by Figs. 1150 and 1151, occupies a space of about 3 ft. 4 in. by 1 ft. 10½ in. by 1 ft. 3 in.; then, when closed up, as illustrated at Figs. 1152 and 1153, it occupies a space of 3 ft. 4 in. by 1 ft 10½ in. by 5 in. The thickness of the wood shown in the illustrations is 1 in. finished, except the panel, which would be about $\frac{5}{8}$ in. It will be seen that the front, sides, and back are hinged and fitted together with rebates and also beads to break joints and improve the appearance. The top and bottom are grooved, and rebated at the back edge, so that the front, sides, and back fit into them; and thus, when the several parts are opened out and fitted together and secured by a few hooks and eyes or similar fastenings (or even a few screws) fixed on the inside, the whole is

Fig. 1142.

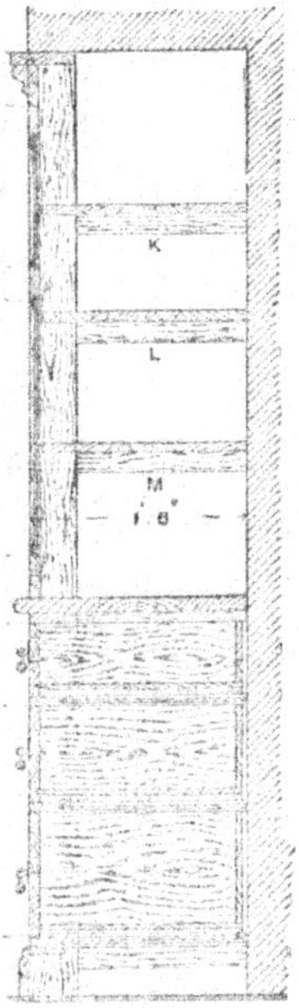

Fig. 1144.

Fig. 1143.

Figs. 1142 to 1144.—Front Elevation, Horizontal Section, and Vertical Section of Cupboard and Drawers for Recess.

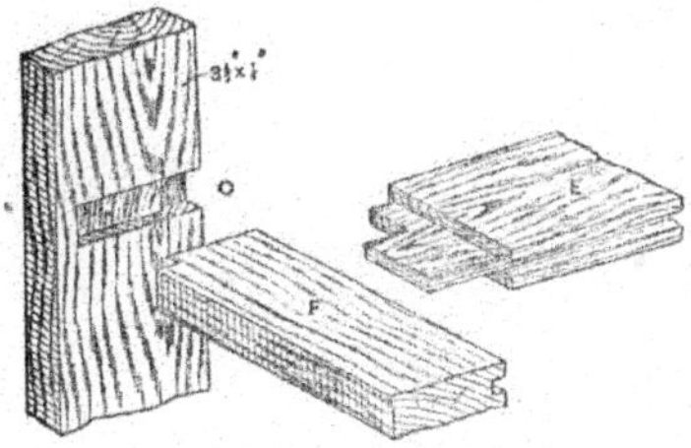

Fig. 1146.—Stile, Front Rail, and Runner of Cupboard Fitment.

Fig. 1145.—General View of Cupboard and Drawers for Recess.

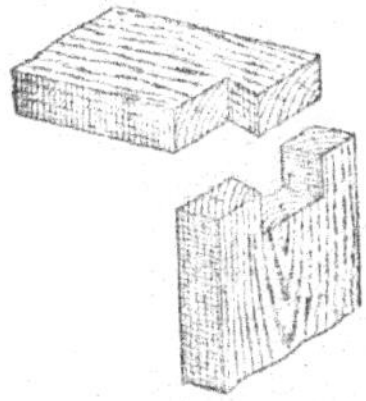

Fig. 1147.—Dovetailed Joint at D (Fig. 1142).

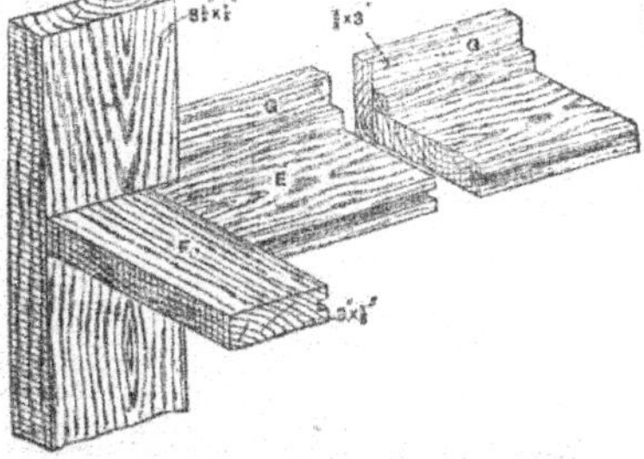

Fig. 1148.—Stile, Front Rail, Runner, and Fillet of Cupboard Fitment.

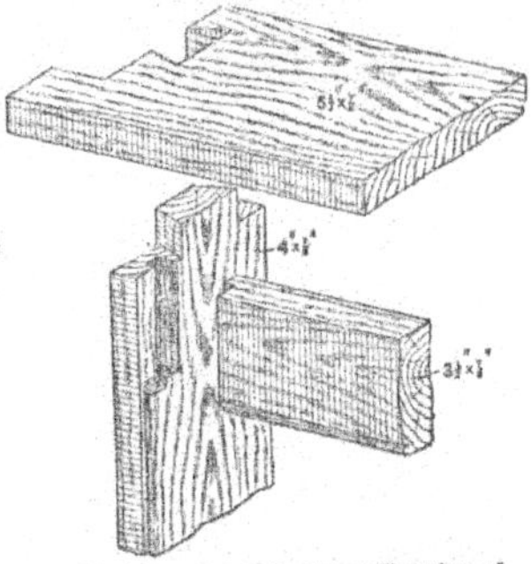

Fig. 1149.—Joints in Upper Framing of Cupboard Fitment round Doors.

Fig. 1150.—Collapsible Cupboard when in Use.

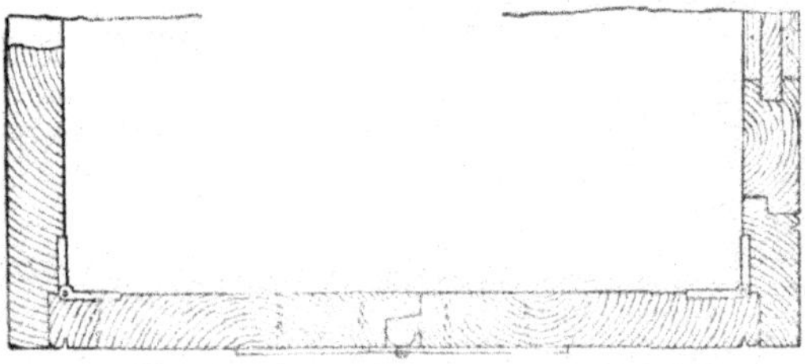

Fig. 1151.—Part Horizontal Section of Collapsible Cupboard.

firmly held together. The rebates and grooves also serve to make the cupboard more dust-proof. The object of the fillet marked A (Figs. 1153, 1154, and 1155), as will be seen, is to allow of the top and bottom folding back through an angle of 270 degrees. The sizes, of course, have to be varied to suit particular circumstances, and the particular kind of wood to be used is largely a matter of choice. Portable shelves could be made to fit into the cupboard in such a way that they could be taken out, or raised or lowered as required; the shelves would rest upon small

Fig. 1152.—Front View of Collapsible Cupboard (Folded).

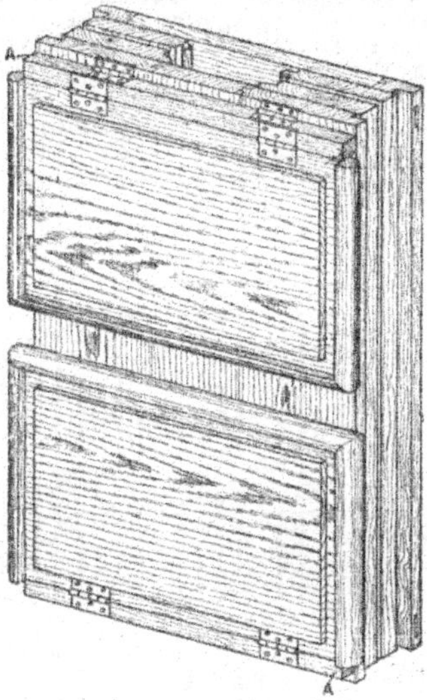

Fig. 1153.—Back View of Collapsible Cupboard (Folded).

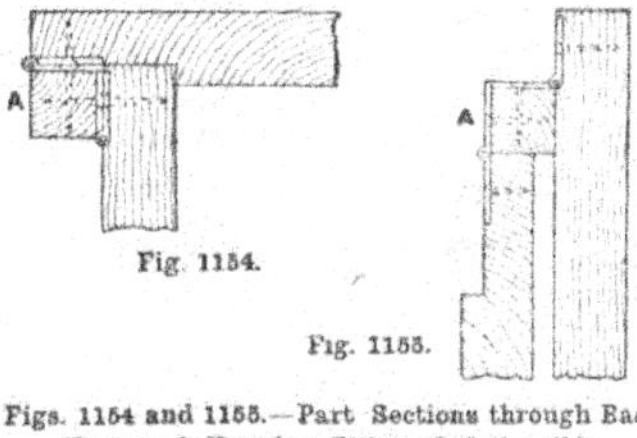

Fig. 1154.

Fig. 1155.

Figs. 1154 and 1155.—Part Sections through Back Top, and Hanging Fillet of Collapsible Cupboard.

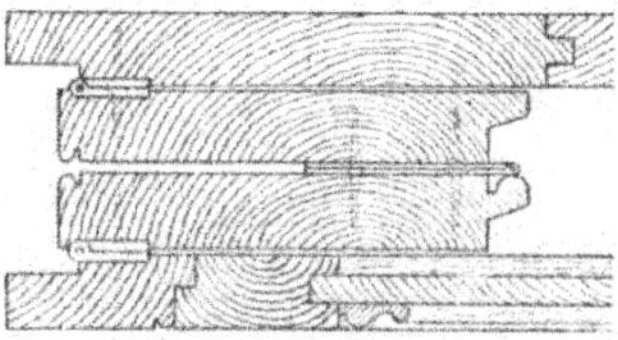

Fig. 1156.—Part Horizontal Section of Collapsible Cupboard (Folded).

turned buttons, in holes bored in the sides, as in bookshelf work. Fig. 1151 shows an enlarged part of horizontal section (opened), and Fig. 1156 an enlarged part of horizontal section (folded).

Linen Cupboard.

Linen cupboards should be made of good yellow deal, sound, dry, and free from large knots and shakes and other defects. The shelves are of Christiania white deal, of good quality. Figs. 1157 to 1161 are plans and elevations of a linen cupboard. Set out on a rod, full size, the width of the front (see Fig. 1158), and the height as shown in Fig. 1160; and set out on the same side of the rod, parallel with the front, the width of the division to the cupboard. The rod having been carefully set out, take off the quantity of material required, namely: Stile frame, two 9 ft. 2 in. by 4½ in. by 2 in.; one 9 ft. 2 in. by 5 in

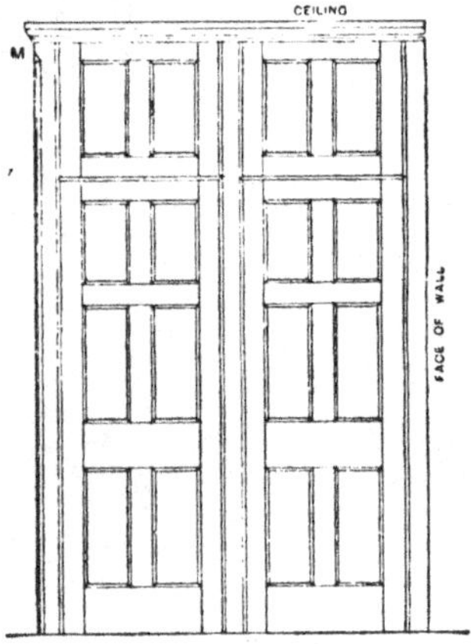

Fig. 1157.

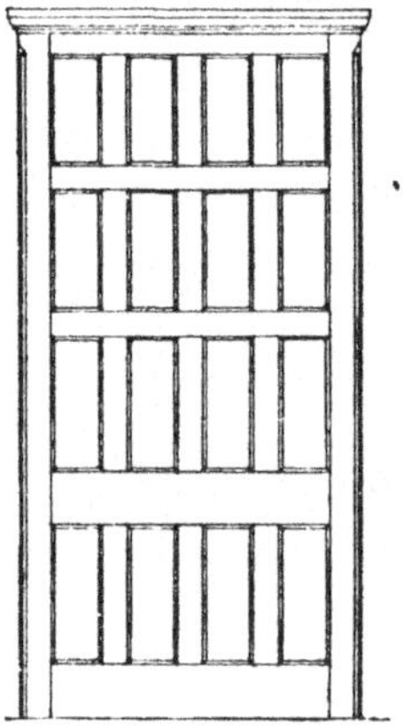

Fig 1159.

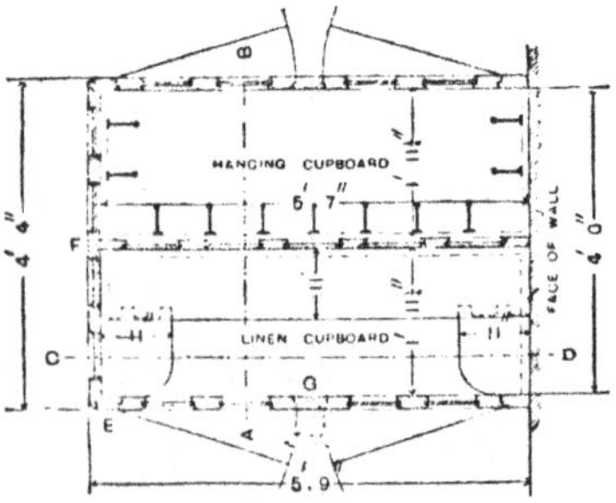

Fig. 1158.

Figs. 1157 to 1159.—Front Elevation, Horizontal Section, and Side Elevation of Linen Cupboard.

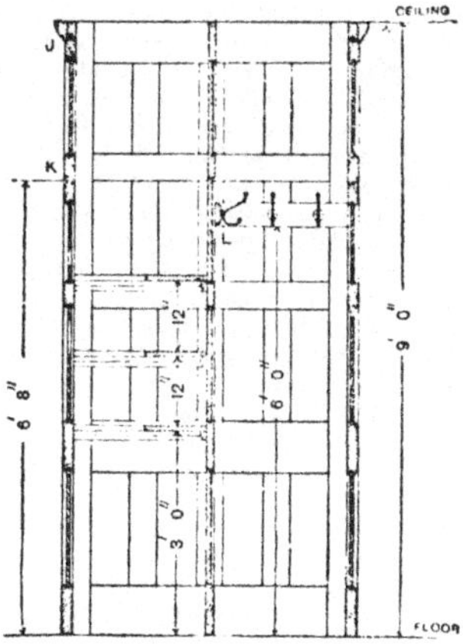

Fig. 1160.—Longitudinal Vertical Section of Linen Cupboard on Line A B (Fig. 1158).

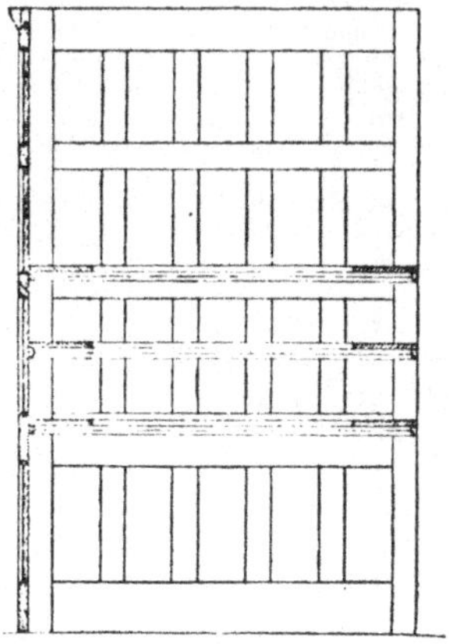

Fig. 1161.—Cross Vertical Section of Linen Cupboard on Line C D (Fig. 1158).

Fig. 1164.—Enlarge Part Horizontal Section of Cupboard (see F, Fig. 1158).

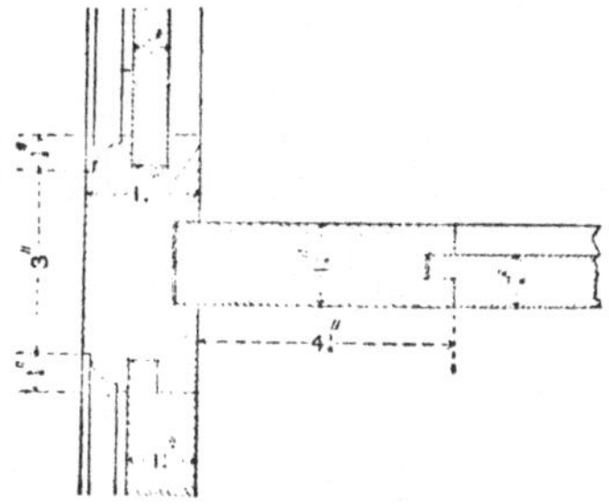

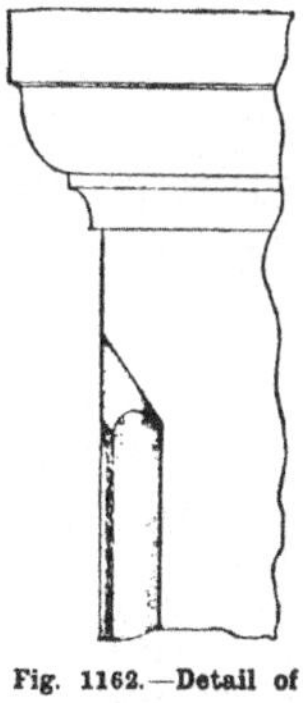

Fig. 1162.—Detail of Linen Cupboard Cornice (see M, Fig. 1157).

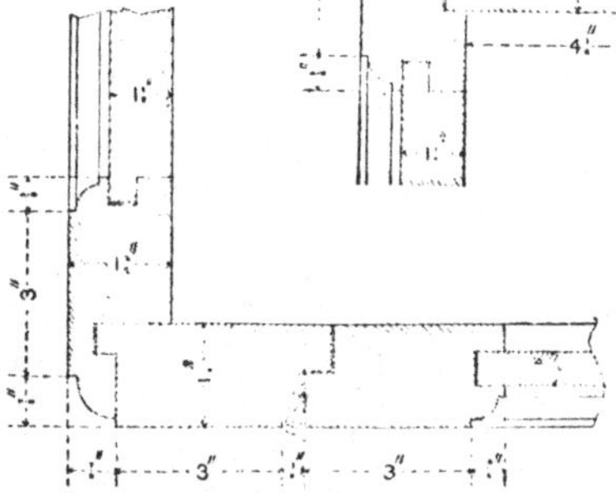

Fig. 1163.—Angle Joint in Linen Cupboard (see E, Fig. 1158).

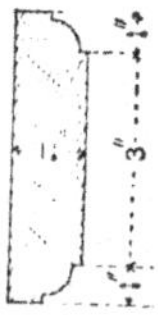

Fig. 1165.—Section of Hook Rail (see L, Fig. 1160).

by 2 in. Door, four 6 ft. 10 in. by 3⅞ in. by 2 in. Framing for door, two 2 ft. 6 in. by 3⅞ in. by 2 in. Top rails of frame, one 5 ft. 9 in. by 3¼ in. by 1⅝ in. Top rails of doors, two 2 ft. 5 in. by 3⅞ in. by 2 in. Frieze rails, two 2 ft. 5 in. by 4½ in. by 2 in. Rails

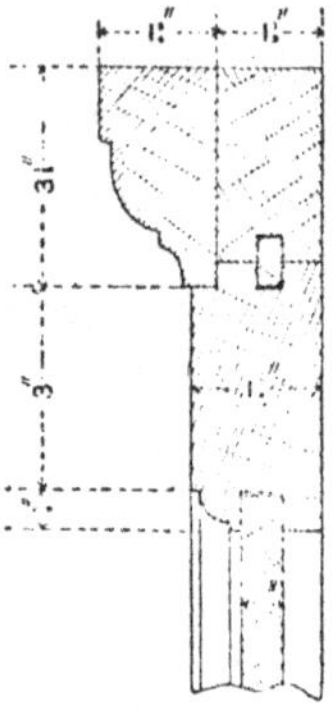

Fig. 1166.—Cross Section of Linen Cupboard Cornice (see J, Fig. 1160).

Fig. 1167.—Vertical Section of Linen Cupboard Framing (see K, Fig. 1160).

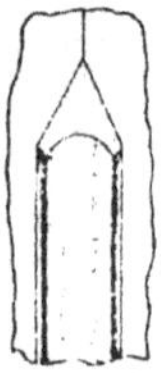

Fig. 1169. — Front of Angle of Linen Cupboard.

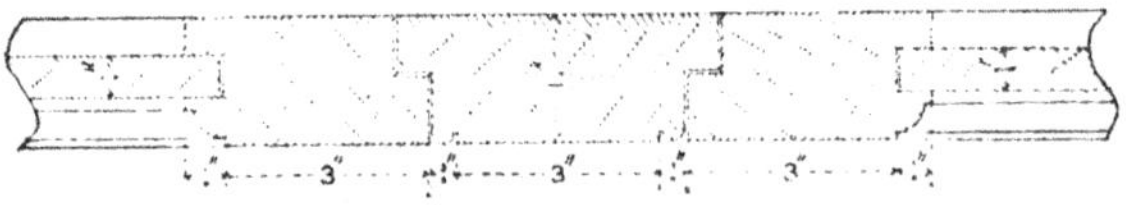

Fig. 1168.—Part Horizontal Section of Linen Cupboard Doors, etc. (see G, Fig. 1158).

of doors, four 2 ft. 5 in. by 9 in. by 2 in. Top rail framing, two 2 ft. 5 in. by 4¼ in. by 2in. Bottom rail framing, two 2 ft. 5 in. by 4¾ in. by 2 in. Muntins, four 2 ft. by 4¼ in. by 2 in.; two 1 ft. 8 in. by 4½ in. by 2 in.; two 1 ft. 5 in. by 4½ in. by 2 in. Panels, eight 1 ft. 9 in. by 9 in. by ¾ in.; four 1 ft. 3 in. by 9 in. by ¾ in.; four 1 ft. 5 in. by 9 in. by ¾ in. Cornice, one 17 ft. by 3½ in. by 2 in. These quantities are set out on a sheet, which is taken to the cutter-out, who gets out the material required, which is then marked, brought to the saw to be cut to the required size, then taken on to the planing machine to be faced and edged, then on to the three-cutter machine to be thicknessed and taken to width. The material, with the quantity sheet, should then be placed in a convenient position for the setter-out. It is only necessary to set out one rail or muntin of each length. In large and busy mills it is usual to face-mark each piece of wood, adding in each case the number of the job, otherwise the stuff might get mixed. The rails are then taken on to the tenon machine. The mortices for the muntins are marked after the rails have been shouldered and tenoned. One rail of each kind should then be marked for haunching. They are usually cut with the bandsaw, on the table of which a fence is fixed, to obviate the necessity of marking each rail separately. Set out the stiles for mortising, and gauge each stile, marking on each the number of the job, and the sight-line of the rails. The job is then taken to the spindle for moulding and grooving. The position of the stop on the stile should be marked as shown in Fig. 1162. Then trace off the various details of the work from which the machinist may prepare the cutters for the mouldings and beads, and mark also the position of the grooves for panels. The work, having been milled, is ready for the joiner to put together. The doors, end framing, and division, should be put together and allowed to stand for a time to dry. Each panel should be left ⅛ in. wider than the full dimension required. The panels of the framing in the linen cupboard are flush inside, as shown in Figs. 1163 and 1164. The shelves in the linen cupboard should

CHIMNEY-PIECE AND OVERMANTEL.

be framed to the size of the opening, and fixed on 2-in. by 1¼-in. chamfered bearers, as shown in Fig. 1158 by dotted lines, and in Figs. 1160 and 1161. The rail for the hooks in the hanging cupboard is shown in Fig. 1165. It is screwed on to the division and end, the screws being arranged so as to be hidden under the hooks. After the work has stood for a time, the framing for the front ends and division is glued up. The ¾-in. panels in doors and framing should be ⅛ in. narrower than the required width, while the 1⅛-in. panels in the hanging cupboard are got to the exact length and width, allowing $\frac{7}{16}$ in. all round for the tongue, as shown in Fig. 1164. Glue up the frame of the cupboard, and fix a stretcher at the bottom. Screw on the inside of the cupboard, the exact width of the opening at the top, and glue in the frame tongue as shown in Fig. 1166. Level off the doors, framing, and division. The doors and framing should be rebated as shown in Figs. 1163 and 1167. Fix in the two top framings, and fix the bead on the bottom edge of the rail, and mitre it into the frame stile. Glue and brad on as shown in Fig. 1167. The two doors should then be fitted and hung with 3½-in. wrought-iron butt hinges, as shown in Fig. 1163. In gluing up the end framing it is advisable to glue a block about 6 in. long on each stile, opposite each rail, to take the shoe of the cramp or cleat, and to screw each tenon from inside the framing. After the doors are hung and the end is fitted on, the edge of the division being shot straight and fitted into the groove as shown in Fig. 1164, the work is primed, knotted, and stopped. Only the edges of the shelves should be painted. It may be said that Fig. 1168 is an enlarged detail at G (Fig. 1158), and that Fig. 1169 is a detail of the moulding on the angle.

BRACKETS.

Oak Wall Bracket.

Figs. 1170 and 1171 are front and side views of a wall bracket made in fumed oak. The mouldings A and B (see section, Fig. 1172) are lengths of ordinary gold spoon picture moulding with the rebate cut away, while the strips C and D are

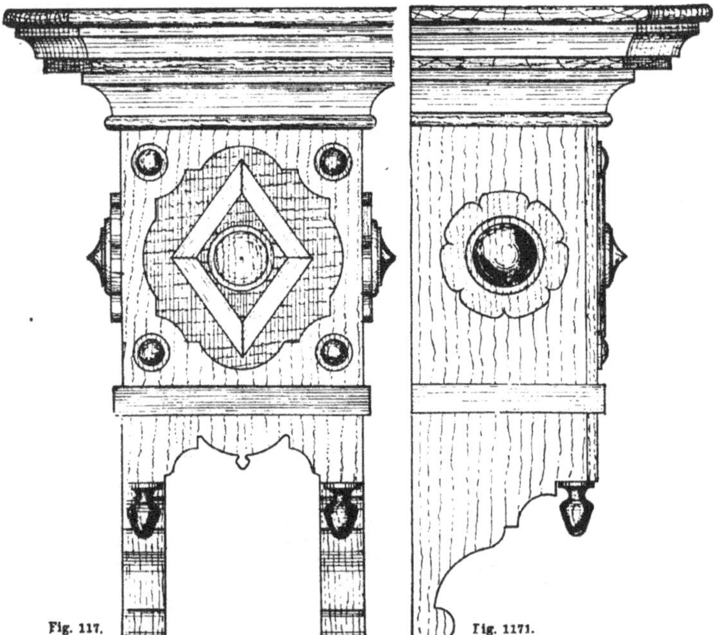

Fig. 117. Fig. 1171.

Figs. 1170 and 1171.—Front and Side Elevations of Oak Wall Bracket.

ordinary flat gold slips. The bracket sides are $8\frac{1}{2}$ in. by $2\frac{3}{4}$ in. by $\frac{5}{8}$ in. When finished, they are secured to two pieces of oak at the front and back, as shown at Figs. 1172 and 1173, the latter being an underneath plan; these pieces are $2\frac{1}{4}$ in. wide, and of any suitable thickness. The fretted front, which is $\frac{3}{16}$ in. thick, should be cut to the pattern shown in Fig. 1170. After this has been bradded in position, the $\frac{3}{8}$-in. wide gold slip should be cut and fitted to form a diamond, inside of which is a turned centre ornament (see also Figs. 1171 and 1172). The top of the bracket is 7 in. by $4\frac{1}{2}$ in. by 1 in., and is cut away at the corners as shown in Fig. 1172, a rebate $\frac{3}{4}$ in. wide and deep being then cut around the bottom edges. In this rebate the $\frac{1}{2}$-in. moulding is secured, after being correctly jointed. Screw the top in position, letting the screw-heads down $\frac{1}{4}$ in. and plugging the holes. Then mitre and fix the $\frac{5}{8}$-in. moulding around under the top, as shown. Next fix a mitered strip of oak $\frac{1}{4}$ in. thick by $\frac{3}{8}$ in. wide close up under the moulding so as to protect it. Cut the flower-shaped side ornaments from $\frac{3}{16}$-in. stuff, and brad in position; then turn the $1\frac{1}{4}$-in. bosses to go on them, and the drops under the front, $\frac{1}{2}$ in. in diameter, the front centre boss, 1 in. in diameter, and the four $\frac{9}{16}$-in. front buttons. A small pin is turned underneath each boss and knob, thus enabling them to be secured by gluing in

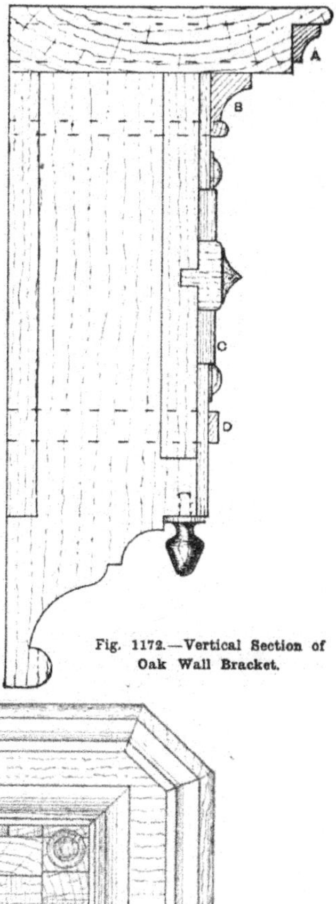

Fig. 1172.—Vertical Section of Oak Wall Bracket.

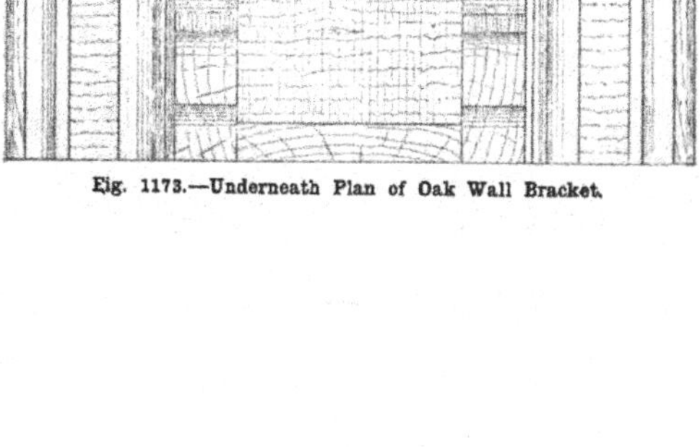

Fig. 1173.—Underneath Plan of Oak Wall Bracket.

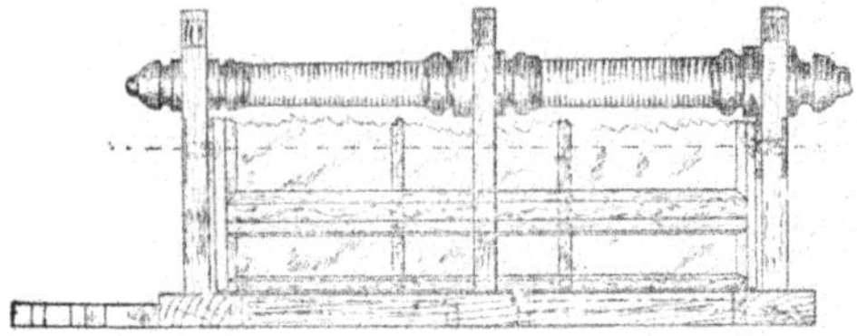

Fig. 1175.

Figs. 1174 to 1176.—Front and Side Elevations, and Horizontal Section of Wall Bracket with Mirrors and Revolving Glass Cabinets.

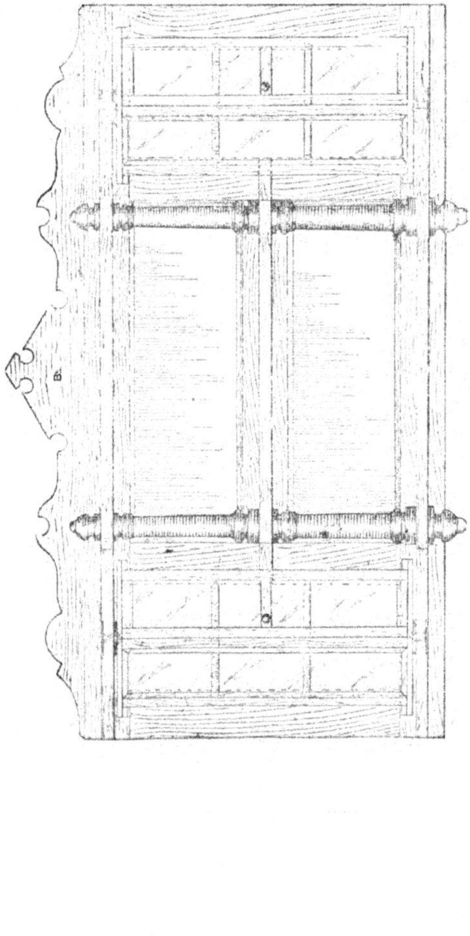

Fig. 1174.

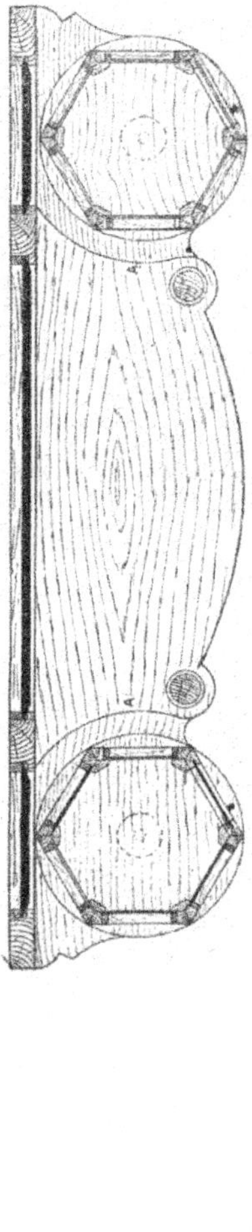

Fig. 1176.

holes bored to receive them. Now mitre and fix the ½-in. wide gold slip across the lower part of the front and sides. Fig. 1173 shows how the joints of the moulding are to be cut. A thin coat of shellac spirit varnish brushed over the gold mouldings and slips, after fixing them, will protect them from wear in cleaning the bracket. It should be said, though, that varnish tends to make gilding look brassy.

Wall Bracket with Mirrors and Revolving Glass Cabinets.

The wall bracket shown in front elevation by Fig. 1174 can be made of any suitable dark hardwood; if made of American pine it would be best stained a dark colour. Fig. 1175 is a side elevation with the revolving cabinet partly broken away to show the pillar, and Fig. 1176 a horizontal section. The body is made of 2-in. by ⅞-in. stuff, and is 2 ft. 9 in. wide and 1 ft. 4 in. deep, with two pieces fixed 7½ in. from the outside ends. To these a centre rail is secured, forming divisions for four mirrors. All the inside parts of the body should be rebated ¼ in. back and ⅜ in. deep to take the mirrors (see the horizontal section, Fig. 1176). The outside edges can be beaded with 3/16-in. bead if desired. The top and bottom shelves are 2 ft. 9 in. by 7½ in. by ½ in., cut to the shape shown in Fig. 1176. The centre shelf is shorter than the others, namely, 1 ft. 6 in. by 7½ in. by ½ in., being cut off to the lines A A (Fig. 1176). Holes ½ in. in diameter must be made for the pillar ends, and the top and bottom shelves are recessed ⅛ in. deep, 1½ in. in diameter, with a ½-in. hole for the dowel plate. The turned pillars are 6½ in. long, tapered from 1¼ in. to 1 in. in diameter. The end of one pillar is turned ½ in. in diameter, 1 in. long, to pass through the centre shelf into the top of the other pillar. The other ends are ½ in. in diameter for a distance of ½ in. The four turned pillar tops or knobs are then secured with fine cabinet screws. The shelves can be fastened to the body with screws from the back. A suitable top B (Fig. 1174) can be fastened to the body if desired.

Revolving Cabinets.—The dowel and socket joints for the revolving cabinets are of brass. The flange is 1¼ in. in diameter and ⅛ in. thick, with holes for the screws to secure them in position; the dowel is ½ in. in diameter and ½ in. long. The

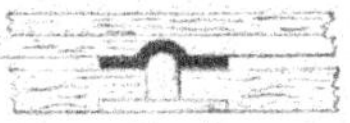

Fig. 1177.

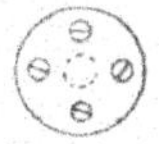

Fig. 1178.

Figs. 1177 and 1178.—Dowel and Socket Joint for Revolving Cabinets of Bracket.

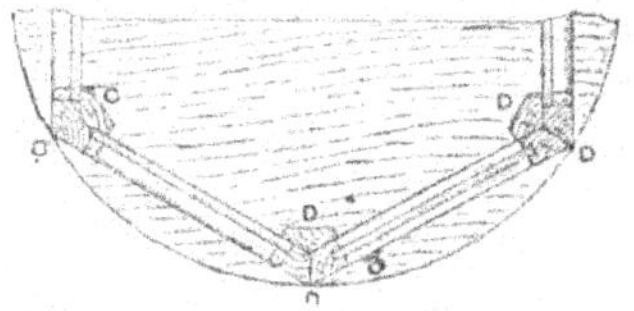

Fig. 1179.—Part Horizontal Section of Bracket Cabinet.

Fig. 1180.

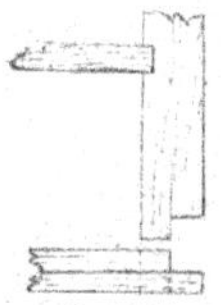

Fig. 1181.

Figs. 1180 and 1181.—Details of Upright and Shelves for Bracket Cabinet.

Fig. 1182.—Spring Catch for Door of Bracket Cabinet.

socket is recessed to take the dowel (see Figs. 1177 and 1178). The two cabinets are ¼ in. shorter than the distance between the top and the bottom shelves. The tops and the bottoms of the cabinets are 7 in. in diameter, ½ in. thick, and rebated back ¼ in. deep to form a hexagon, and 5¼ in. across the slips (see Fig. 1179). The four uprights are 1 ft. 1¼ in. long, made to section C (Fig. 1179); two uprights for the door are 1 ft. 1¼ in., made to the section D, the ends being secured by a mortice-and-tenon joint. The two shelves are cabinet, and are held in position by beading ¼ in. by $\frac{3}{16}$ in. The cabinet door is 1 ft. 0¾ in. by 3½ in.; the rails being of ½-in. by ⅜-in. stuff. The glass for each door is $5\frac{13}{16}$ in. by 2¾ in. The door rails should be

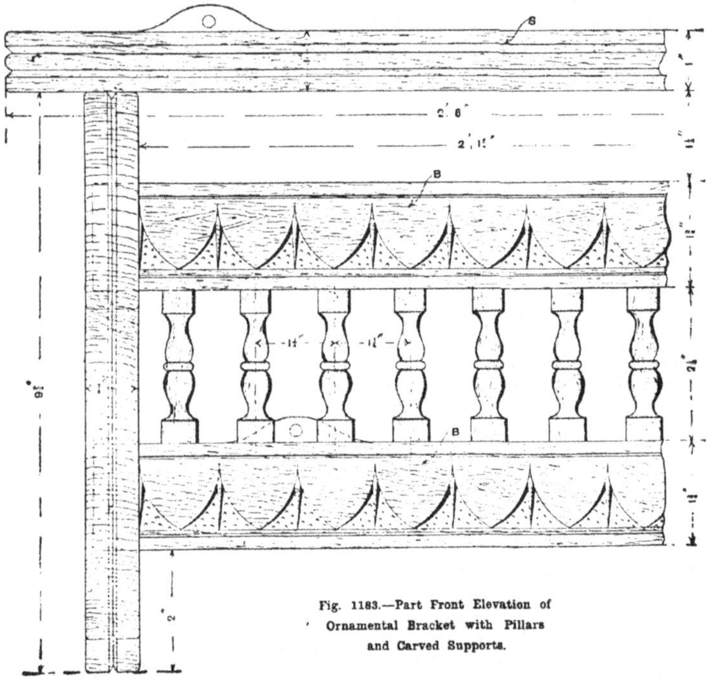

Fig. 1183.—Part Front Elevation of Ornamental Bracket with Pillars and Carved Supports.

equal distances apart, and 5¼ in. across the slips, hexagonal in shape, and ½ in. thick, a suitable rebate being made in the uprights (see Figs. 1180 and 1181). Secure the shelves to the uprights before fixing the cabinets together. Five pieces of glass, 1 ft. 0¾ in. by 2⅝ in., are required for each bevelled to fit against the uprights. Two small hinges, a knob ⅜ in. in diameter and ⅜ in. long from centre rail, and a small brass catch (Fig. 1182) let in the rail, will complete the door. The brass sockets for revolving the cabinets are secured to the tops and bottoms of the cabinets.

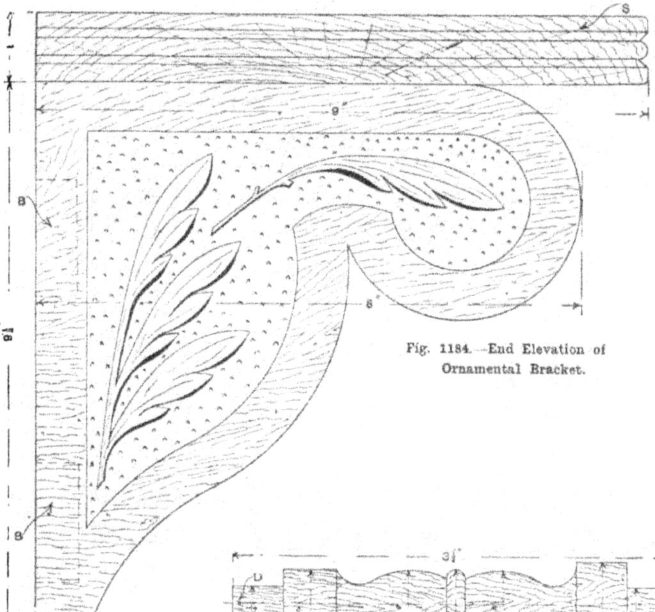

Fig. 1184.—End Elevation of Ornamental Bracket.

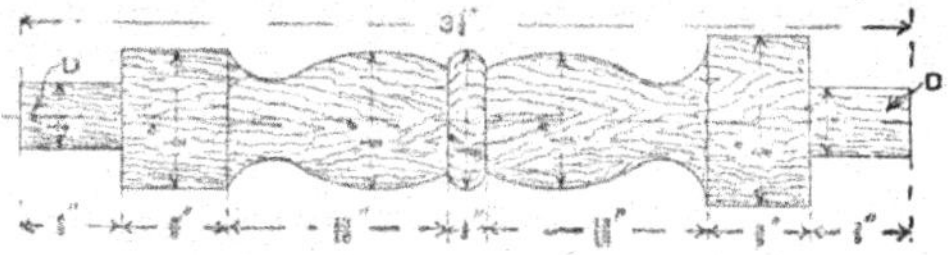

Fig. 1187.—Pillar of Ornamental Bracket.

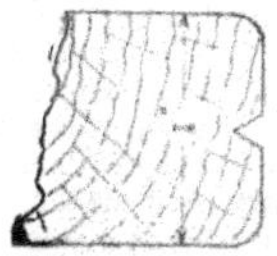

Fig. 1185.—Section of Front Edge of Bracket Support.

Fig. 1186.—Section of Bracket Bar.

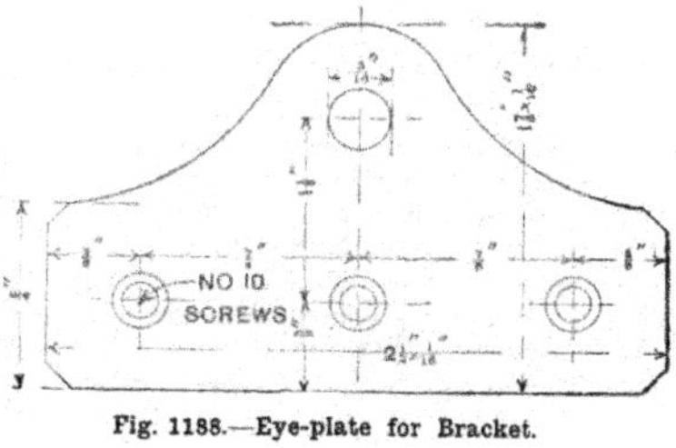

Fig. 1188.—Eye-plate for Bracket.

Mirrors.—The two centre mirrors are 1 ft. 2½ in. by 5½ in., and the two end mirrors behind the cabinets are 1 ft. 0½ in. by 6 in. These can be held in position by picture backing or with strips ⅜ in. by ¼ in., as shown in Fig. 1176.

smooth the edges, then along the front edge work a V groove, as shown in Fig. 1185, and slightly round the outer corners. These supports are carved on each side, a simple pattern being traced on and carved out ⅛ in. or $\frac{3}{16}$ in. deep, the wood between the pat-

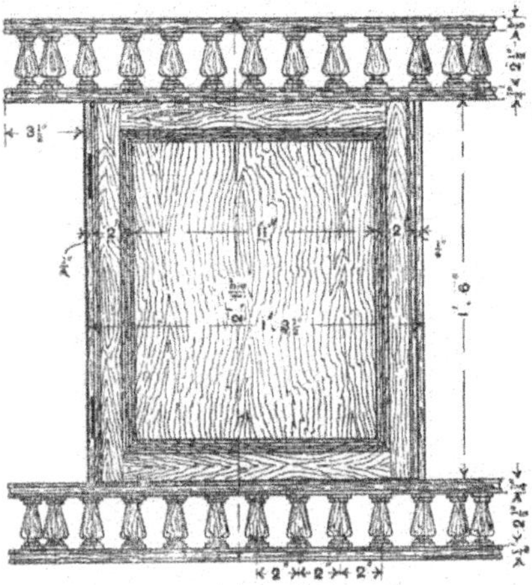

Fig. 1189.

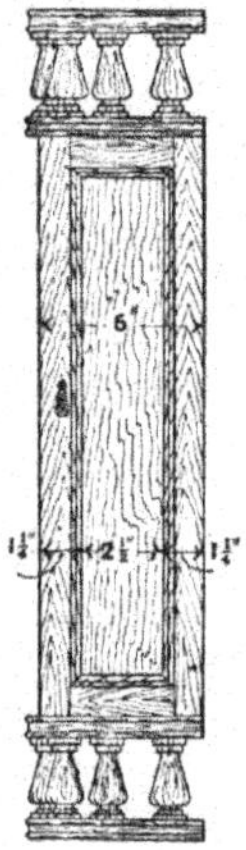

Fig. 1190.

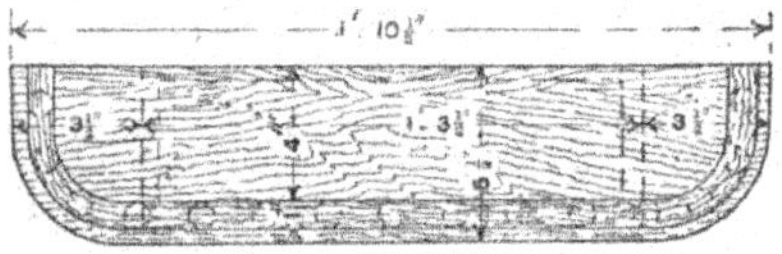

Fig. 1191.

Figs. 1189 to 1191.—Front and Side Elevations, and Plan of Fancy Wall-cupboard Bracket.

Ornamental Bracket with Pillars and Carved Supports.

Fig. 1183 is a part front elevation of a bracket from which it will be seen that s, the upper part or shelf, is 2 ft. 6 in. by 9 in. by 1 in., the edges being beaded to relieve the thickness; this also gives an ornamental finish. Having prepared this piece, cut out two supports (Fig. 1184), which will be best done with a band-saw; clean up and tern being cut away and left to form a moulding following the shape of the support (see Fig. 1184). The shelf is screwed to the supports with 1¾-in. No. 10 screws, the screw-heads being well countersunk and the holes afterwards filled in. Two bars B B (Fig. 1183), 2 ft. 3½ in. by 1¾ in. by ¾ in., pass along the back, and are let in flush with the back of the supports (see B, Fig. 1184). Fig. 1186 shows a section of the bar. These bars have a ¼-in. bead

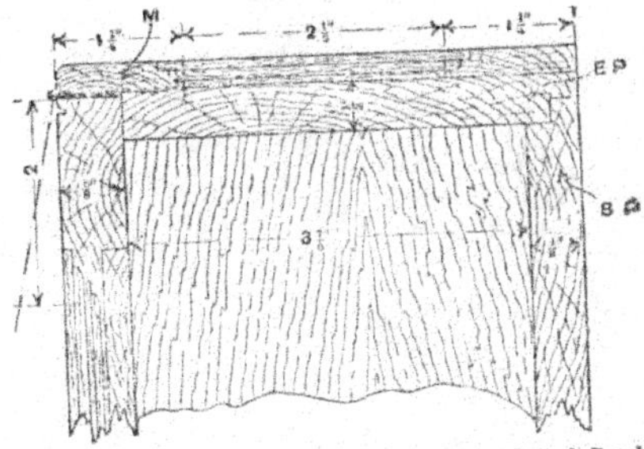

Fig. 1192.—Part Horizontal Section of Wall-cupboard Bracket.

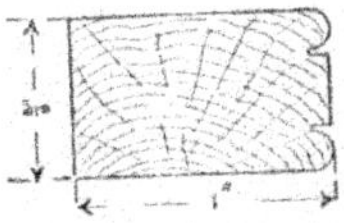

Fig. 1193.—Section of Bracket Fence Rail.

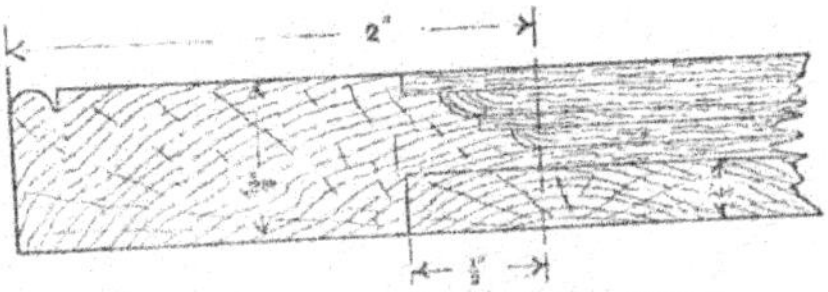

Fig. 1194.—Section of Bracket Door Stile

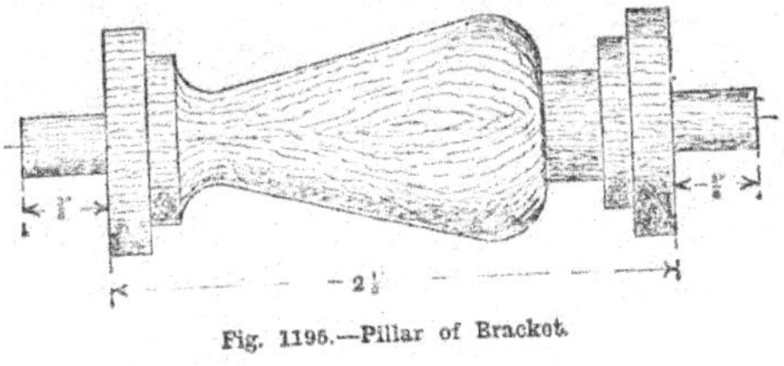

Fig. 1195.—Pillar of Bracket.

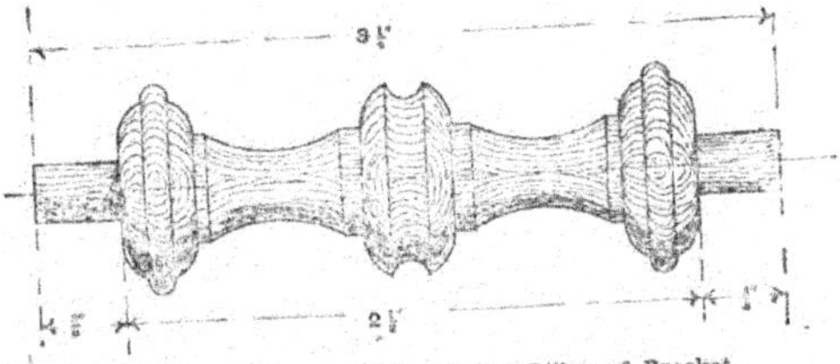

Fig. 1196.—Alternative Design for Pillar of Bracket.

run along each edge, and a band of carving 1¼ in. wide run along the face; the pattern may be the same as illustrated, or any pattern could be introduced which might be more in accordance with the maker's taste. A very good substitute would be a bar of carved moulding, which could be obtained from dealers in carved woodwork, the price of the width here shown being about 5d. per foot. Along the

Fig. 1197.—Bracket Frame for Three Photographs.

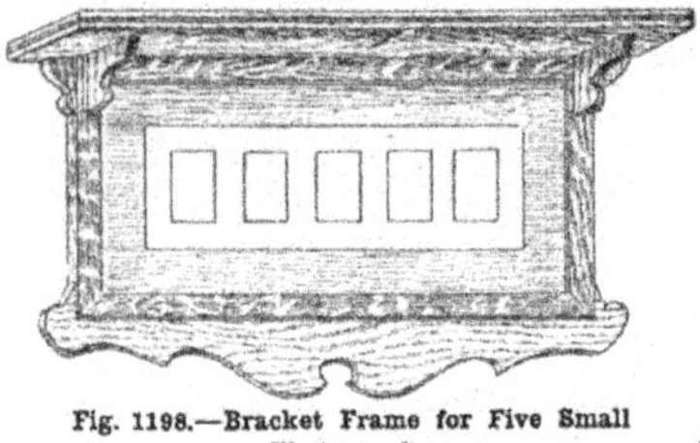

Fig. 1198.—Bracket Frame for Five Small Photographs.

inner edge of the bars, bore a row of ¼-in. holes at 1¼-in. centres, to receive the dowel ends of the pillars. The bracket pillars are turned with dowel ends D (see Fig. 1187), and should be glued in the two bars of carving before being fixed to the supports. The bracket may be made from good dry mahogany or walnut, either of which will have a handsome appearance when polished, but mahogany would show a better contrast against the black cases in which stuffed birds, etc., are usually placed. Four strong brass eye-plates (Fig. 1188) are used to fix the bracket upon the wall; these are screwed to the shelf and bottom bar with ¾-in. No. 10 screws, using brass screws throughout.

Fancy Wall-cupboard Bracket.

Figs. 1189, 1190, and 1191 illustrate, respectively, front elevation, end eleva-

Fig. 1199.—Corner Bracket for Photographs.

tion, and plan of a small cupboard bracket, made in white wood, stained or enamelled, or painted and varnished. It is suitable for a drawing-room, and the front panel should be embellished with a large floral scroll of a running design. The back piece B P (Fig. 1192) should be made first. It consists of ½-in. stuff running from top to bottom, the two side edges being ploughed to fit corresponding grooves in the end pieces E P (Fig. 1192). The ends are next fitted, using ½-in. stuff, with ploughed back edge. The top and bottom are screwed on, the screw heads being stopped up and cleaned with glass-paper. As shown at Fig. 1191, the top and bottom pieces project 3½ in. beyond the case, the corners being rounded and the edges beaded, similar to the fence rail, seen in section (Fig. 1193). The door is now put together by framing up some 2-in. by ⅝-in. stuff, using tenoned joints. A panel ¼-in. thick and one narrow flap, the wide flap reaching on to the case, the narrow one taking the door. This is necessitated by the moulding not being sufficiently strong for the screws. If any difficulty is experienced in purchasing this type of hinge, get a wide hinge, and cut down one flap to the required width. A row of holes is now bored along the top and bottom, ¼ in. diameter, ⅜ in.

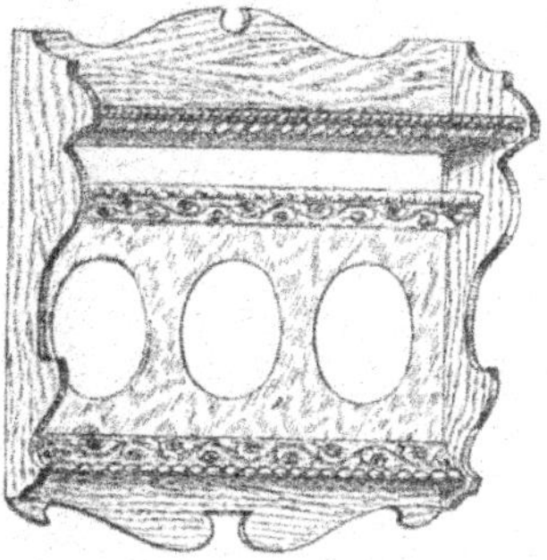

Fig. 1200.—Bracket Frame for Three Photographs.

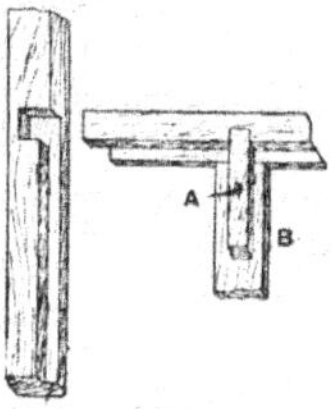

Fig. 1201.—Middle Bars of Bracket Frame (see Fig. 1197).

Fig. 1204.—Joints, etc., in Bracket Frame (see Fig. 1200).

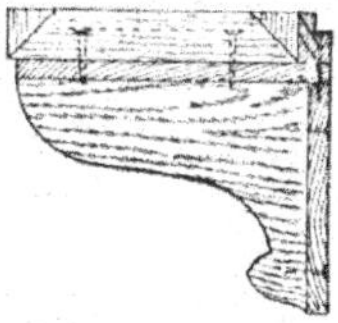

Fig. 1202.—Section of Lower Part of Corner Bracket (see Fig. 1199).

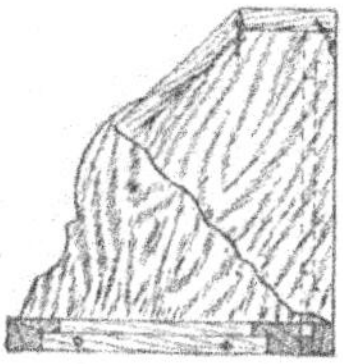

Fig. 1203.—Part Horizontal Sections of Corner Bracket, showing Top and Lower Shelves.

is ploughed in, the surrounding moulding being worked on the edges of the stiles and rails (see Fig. 1194). The two stiles are also beaded, so as to give a finish to the end moulding when the door is closed. When the door has been hung, the ends are panelled with a 1¼-in. by ¼-in. moulding M (Fig. 1192). This moulding projects beyond the front edge the thickness of the door, namely, ⅝ in. In hanging the door it is necessary to use hinges having one wide deep, and 1 in. on from the face edges, at 2-in. centres. These are for the dowel ends of the pillars forming the fences. These pillars may be similar to the designs given in Figs. 1195 and 1196, and can be purchased ready turned. Glue them in position, and, when dry, touch the reverse dowel end with some paint, and, while this is wet, gently lay on the fence rail. By this method the exact position for the dowel holes in the fences will be obtained. As the fence rails are only 1 in. by ⅝ in., it is advisable to have them bent to shape, which could be done at a trifling cost by a bent-timber merchant. Should it be decided, however, to saw them out, the best plan will be to get a board, and first mark out and bore the dowel holes; next cut the outside sweep and bead the edge; after it is cleaned up and practically finished, the inside sweep may be cut. After it is cut, the fence must be very gently handled; the grain running short across the corners renders it liable to be easily broken. One or two shelves are fitted inside, the distance between them being arranged to suit requirements. This completes the making of the cupboard, which is now ready for decorating. A few vases and plaques will stand upon the top, and a small bowl with a fern growing looks pretty on the bottom corner brackets. Use brass eye-plates, or screws driven through the back into wooden wall plugs, to fix it. By inserting a glass panel in the cupboard door, a cabinet for the exhibition of a few pieces of antique china is obtained.

Fig. 1205.—Wall Bracket with Copper Panel.

Photograph Brackets.

Figs. 1197 to 1200 show designs for photograph frames with shelves attached. That shown by Fig. 1197 is made from flat moulding, and in figured oak looks well when left in the natural colour. If preferred, stained or enamelled pine may be used. The frame is 1 ft. 8 in. wide by $8\frac{1}{2}$ in. high, and the shelf is 3 in. above the frame and 2 ft. 1 in. by $3\frac{1}{2}$ in. wide by $\frac{5}{8}$ in. thick. The vertical bars are 1 ft. 1 in. long, with small terminals dowelled and glued to the lower ends. If any difficulty is found in rebating the middle bars, make them as shown in Fig. 1201, where A shows a strip of deal let in flush with the part of the rebate on which the glass bears, and B a thin strip 1 in. wide glued to A. The trusses for the support of the shelf are secured by screws inserted from the back of the frame. The frames of the corner bracket (Fig. 1199) are 1 ft. 10 in. long and $10\frac{1}{2}$ in. wide, and are kept in position by the top shelf and the bottom bracket. The screws are driven from the inside of the frames, as shown in Figs. 1202 and 1203. The wood for the shelves should not be less than $\frac{1}{2}$ in. thick and 1 ft. wide. The frame shown by Fig. 1198 is 1 ft. 8 in. long, the shelf being 2 ft. 1 in. long by 3 in. wide and $\frac{5}{8}$ in. thick, with a $\frac{1}{4}$-in. bead round the edges. The height of the frame

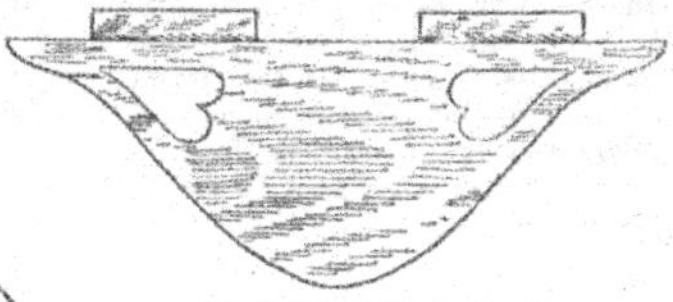

Fig. 1206.—Shelf of Bracket (see Fig. 1205).

Fig. 1211.—Method of Securing Copper Panel in Bracket.

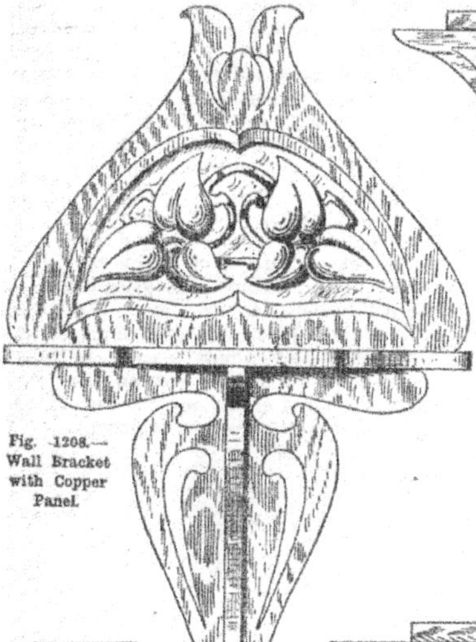

Fig. 1208.—Wall Bracket with Copper Panel.

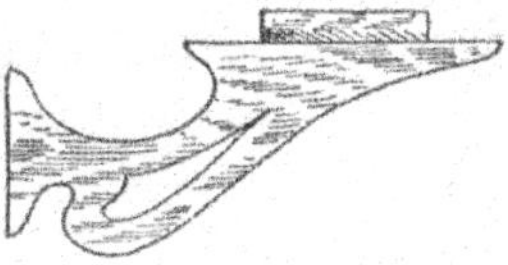

Fig. 1210.—Shelf Support for Bracket.

Fig. 1207.—Shelf Support for Bracket.

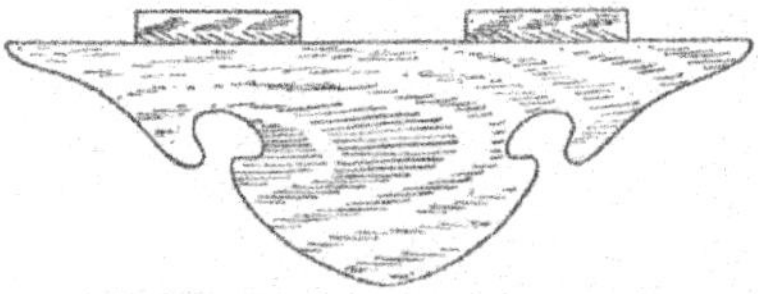

Fig. 1209.—Shelf of Bracket (see Fig. 1208).

is 11 in.; 1¼-in. moulding is used. The pediment is 1 ft. 10½ in. long and 3 in. deep in the centre curves. The frame at Fig. 1200 is designed to use short pieces of picture moulding. It is 1 ft. 3½ in. by 9½ in., and the shelf is 1½ in. above the frame and 14½ in. by 2½ in. by ½ in. thick. The moulding attached to the front of the shelf is ¾ in. thick. The trusses are 1 ft. 2 in. high and 3¼ in. wide where they join the shelf. No glass is required for this frame. The mount may be cut from $\frac{3}{16}$-in. or ¼-in. board, and covered with plush or Japanese wall-paper. The woodwork may be stained and varnished, or enamelled in colours to harmonise with the moulding used. The method of constructing the frame and fixing the parts is shown by the sections in Fig. 1204.

Wall Brackets with Copper Panels.

The brackets illustrated by Figs. 1205 to 1211 should be made from well-seasoned wood, oak being most suitable; it could be stained green, or oiled, as either result would harmonise with the copper panels. As dimensions for the brackets, the following would be suitable:—For Figs. 1205, 1206, and 1207, the back may be 10½ in. by 10½ in. by ⅜ in.; the shelf, 6 in. by 2¼ in. by ½ in.; and the stay, 1½ in. by 1½ in. by ½ in. For Figs. 1208, 1209, and 1210, the back may be 11¼ in. by 6¾ in. by ⅜ in.; the shelf. 6¾ in. by 2¼ in. by ½ in.; and the stay, 4½ in. by 1⅜ in. by ½ in. These dimensions are taken at the widest parts, but the tenons on the shelves and stays are not included. The outsides of the bevel shown in Figs. 1205 and 1208 will be the shape of the copper panels, but the holes in the backs will be to the inner lines. The panels can be made from thin copper in the following way:—Trace the inner lines on the copper, and sketch in the design. Then go round it with a tracing wheel, and put it on a cushion or something soft, and rub it up from behind with a modelling tool until it is like the design. The plate can then be polished up and fixed at the back (see Fig. 1211) and fastened by small pin points, after which plaster must be put in to stiffen it. The bevel can be put on the front with a carving chisel, and the rest of the bracket may then be put together.

WAITERS AND TRAYS.

Dumb Waiter.

The dumb waiter shown in elevation by Fig. 1212 should be made in oak or walnut, to be in keeping with the furniture of the room in which it is to be used. The dimensions are as follows: Main shelf, 1 ft. 10 in. in diameter; upper shelf, 7 in.; base, 11 in. Height to main shelf, 6½ in.; to upper shelf, 1 ft. 1½ in. The whole of the parts are turned in the lathe. The upper and lower pillars are separate, and are jointed as shown in the enlarged section (Fig. 1213). This joint must be perfectly true, and the shoulders turned equally true, with just sufficient play between, when pulled up tight with the bolt, to allow the main shelf to revolve easily. Recesses are hollowed in the pillars to receive the

Fig. 1212.—Front Elevation of Dumb Waiter.

Fig. 1215.—Plan of Top Pillar of Dumb Waiter.

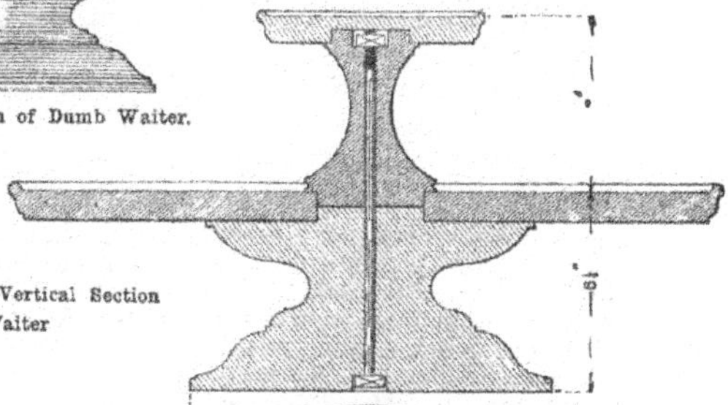

Fig. 1213.—Enlarged Vertical Section of Dumb Waiter

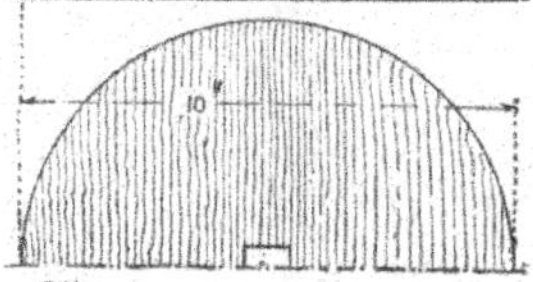

Fig. 1214.—Half Plan of Bottom of Dumb Waiter.

head of the bolt and the nut as shown. The nut of the bolt should be tightened up with a box-spanner. The upper shelf is fixed on the pillar with glue after the lower part has been put together. A half plan of the bottom is given at Fig. 1214, and a plan of the top pillar at Fig. 1215. The base of the lower pillar should be covered with baize to prevent scratching the table. The work should be french-polished in the lathe. The material required is as follows: One piece 11 in. by 11 in. by 6½ in.; one piece 7 in. by 4 in. by 4 in.; one piece 1 ft. 10 in. by 1 ft. 10 in. by 1¼ in.; one piece 7 in. by 7 in. by 1 in.; one ½-in. bolt, 10 in. between head and nut.

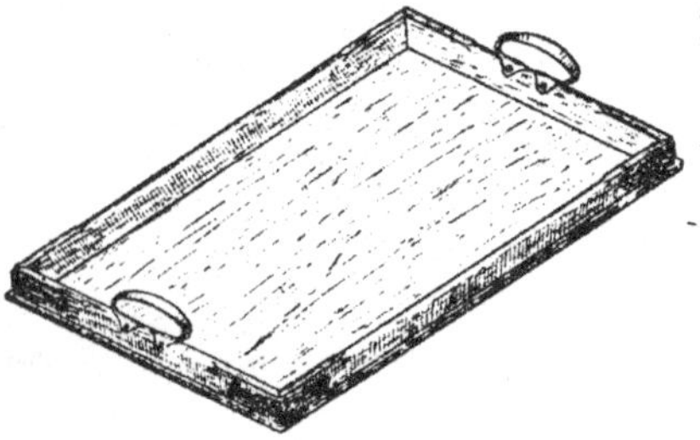

Fig. 1216.—Oak Tray with Copper Fittings.

Oak Tray with Copper Fittings.

The tray shown by Figs. 1216 to 1218 can be made of any hardwood, oak for preference. The size can be varied to meet requirements, the one illustrated being medium size. A section of the moulding for the sides is shown in Fig. 1219, and is easily made. It should be planed out of one length, if possible, and then mitered and glued. Each corner can be further secured with a nail or screw, which is hidden by the copper corner plates. At a distance of 2½ in. from each end, the top of the moulding should be rounded on each edge as shown. The base, which is a piece of ¼-in. thick figured oak fretwood, already planed, should be neatly rounded on the edges. It projects ¼ in. from the sides all round, and is screwed to the sides from underneath. It should be stained or fumed a rich brown, and then wax polished. The copper corner plates (see Fig. 1220, which is a half elevation) should be cut from sheet copper $\frac{1}{32}$ in. thick, and look well if hammered all over with the ball pene of a small hammer; or they can be left plain. A (Fig. 1221) shows the finished shape of the handles, and B and C (Fig. 1221) give sections, the half development being shown by D. These handles are $\frac{1}{32}$ in. thick, and they should be well annealed before being bent. This can easily be done by hand, after which they can be hollowed over a smooth ½-in. round rod, held in the vice, or on the handle of a flat-iron, the ears being bent to fit the moulding. The fittings should then be polished and lacquered, and secured to the tray with small brass snap-headed screws. Figs. 1217 and 1218 are reproduced to a scale of 3 in. to a foot, and Figs. 1219 to 1221 are half size.

Set of Waiters or Trays in Oak.

Figs. 1222 and 1223 show respectively part plan and elevation of a very useful waiter that will be found much more durable than japanned iron trays. The wood employed, oak, must be dry, straight-grained, well seasoned, and free from knots and shakes. The base-board B (Figs. 1222, 1223, and 1224) is first built up with 4-in. widths of ⅝-in. thick stuff. The boards are arranged as shown in Fig. 1225, with the grain of each board running in the reverse direction to its neighbour; the reason for this being that the shrinkage of a board is less at the butt end than at the branch end, and placing the boards as illustrated equalises the shrinkage and prevents splitting and twisting. The joints are shot square and true, and glued; they may be further strengthened by inserting small dowels before the gluing is done. The base-board now requires squaring to 2 ft. 6 in. long by 1 ft. 8 in. wide; a moulding is then worked on each edge, a section of which is seen at M (Fig. 1224). The fence mouldings are next prepared. Two designs for these are given in Figs. 1226 and 1227. Each one is worked out of stuff 1⅛ in. square. The timber for

Fig. 1218.

Fig. 1217.

Figs. 1217 and 1218.—Plan and Elevation of Oak Tray with Copper Fittings.

Fig. 1219.—Section of Moulding for Oak Tray.

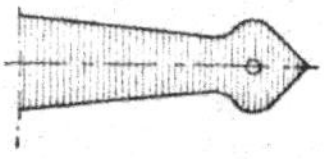

Fig. 1220.—Half of Corner Plate for Oak Tray.

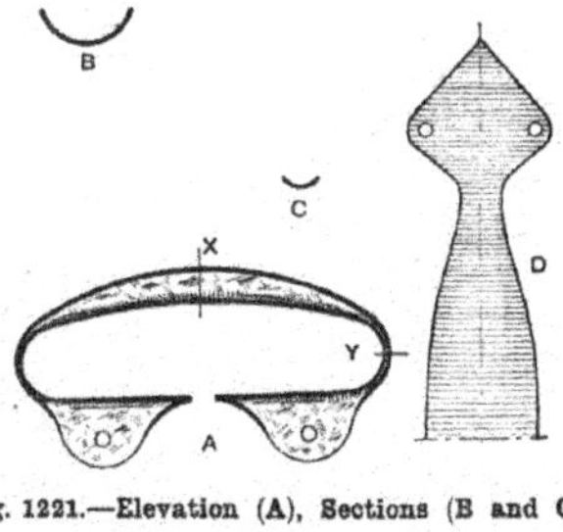

Fig. 1221.—Elevation (A), Sections (B and C), and Half Development (D) of Tray Handle.

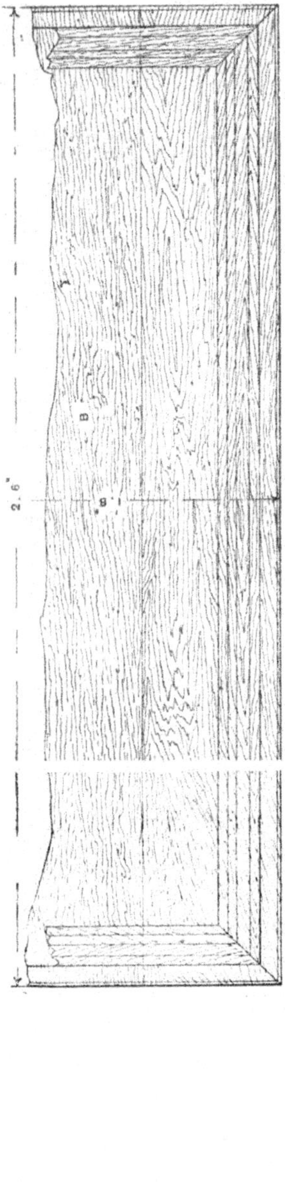

Fig. 1222.

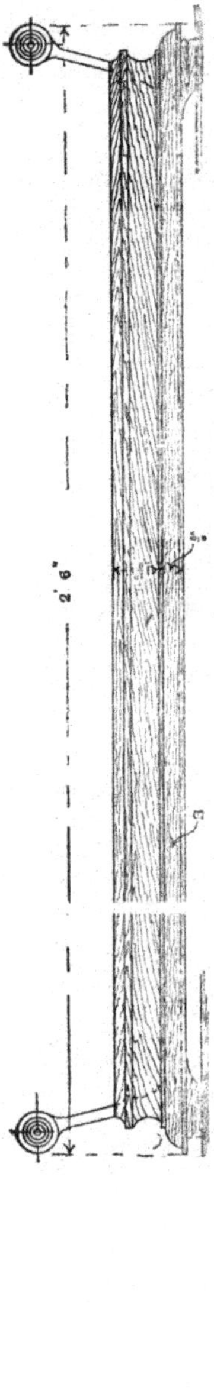

Fig. 1223.

Figs. 1222 and 1223.—Part Plan and Elevation of Oak Waiter or Tray.

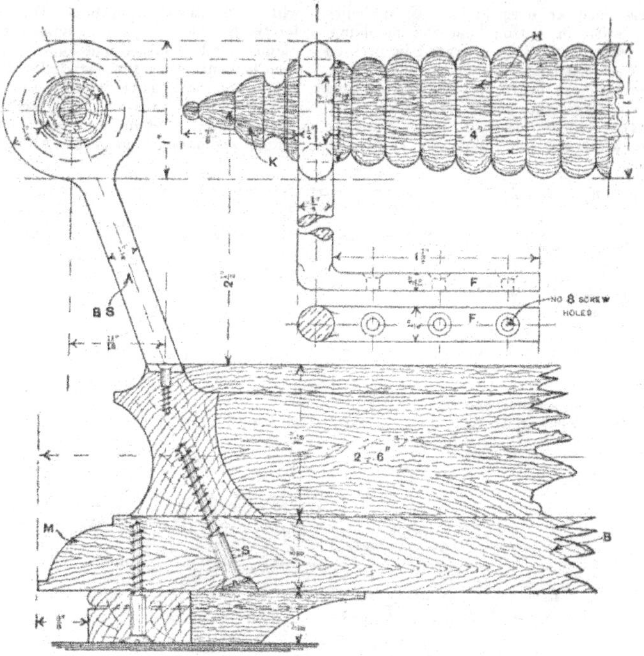

Fig. 1224.—Cross Section and Part Elevation of Tray Handle.

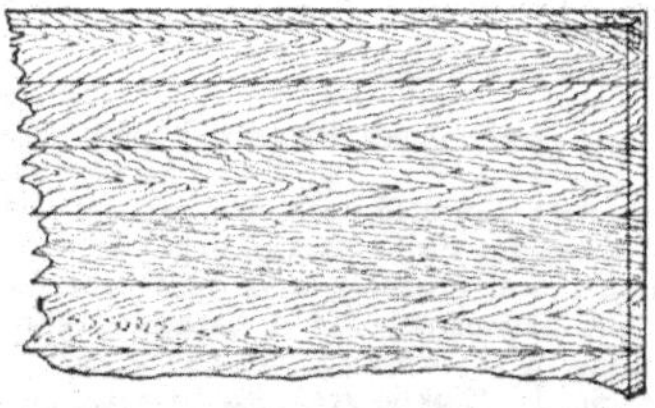

Fig. 1225.—Boards for Tray arranged in Narrow Widths and Reversed Grain.

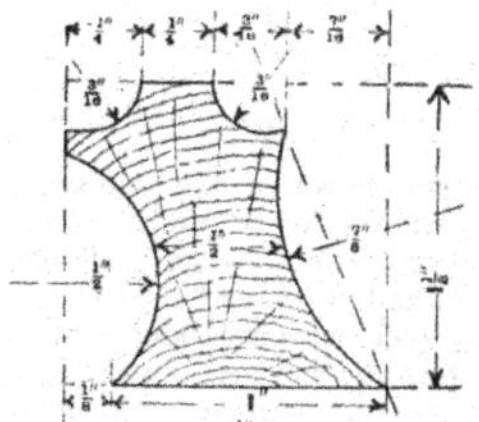

Fig. 1226.—Section of Fence Moulding for Tray.

these must be even grained, or difficulty will arise in getting out the moulding. The fence mouldings are fixed in position on the base-board by first gluing them, then they are further secured with screws driven from the under side (see S, Fig. 1224). Corner feet pieces are fixed to raise the board $\frac{3}{8}$ in. above the dead level. will be the brass supports B S (Fig. 1224) for the handles. A piece of brass rod $\frac{1}{4}$ in. diameter is first heated and cooled; it may then be bent over a $\frac{1}{2}$-in. mandrel with ease. When the end meets, it must be neatly fitted to form the eye or ring, and afterwards brazed. Set off the length for the upright and hammer out the flap F (Fig. 1224)

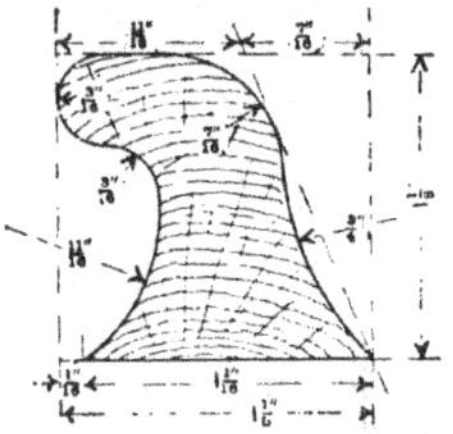

Fig. 1227.—Alternative Section for Fence Moulding for Tray.

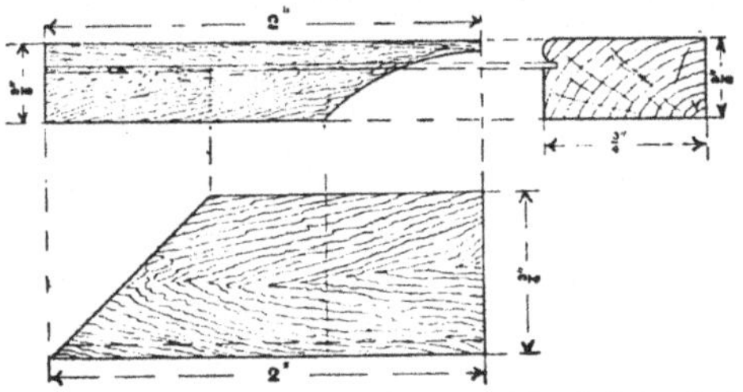

Fig. 1228.—Plan, Elevation, and Section of Tray's Corner Feet.

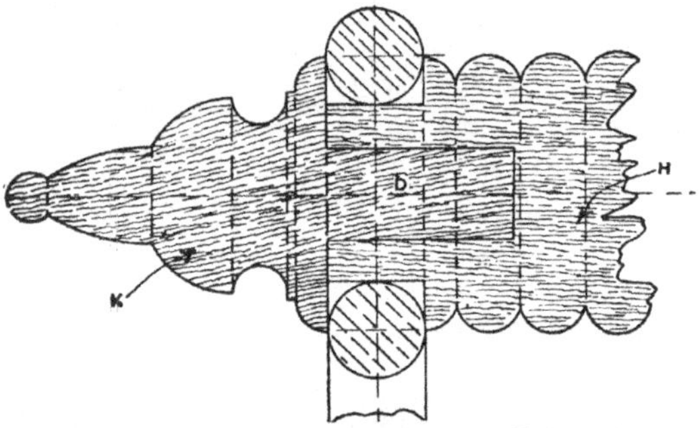

Fig. 1229.—Section of Tray Handle and Knob.

These are shown at Fig. 1228. Along the top outside edge of each a $\frac{1}{8}$-in. bead is worked to break the joint; the corners are mitered, and the ends finished with a sweep. These pieces are glued and screwed to the under side of the base-board $\frac{3}{8}$-in. from the edges; a piece of thick cloth or felt is afterwards glued to them to prevent scratches when moving the waiter on a table, etc. The next part to take in hand for the screw holes, which will finish about $\frac{1}{4}$ in. by $\frac{3}{16}$ in. The brass, having been properly annealed, will hammer out cold without fracturing, but during this operation it will probably work hard; the constant hammering has a hardening effect, owing to the consolidation of the particles of metal. This is remedied by again heating and cooling till finished. The screw holes in the flap must be accurately drilled and countersunk to fit the screw heads, a loose fit being dangerous to the security of articles that are carried upon the tray. The brass supports, before being fixed, must be well polished, or, if preferred, a coat of black lacquer may be applied. In fixing them use 1-in. No. 8 screws, and let the flaps into the moulding flush with the top surface. The handle H (Fig. 1224) is a piece of ebony or oak turned to give a good grip for the hand. It tapers from 1 in. diameter in the middle to ¾ in. diameter at each end, and is shouldered and reduced to ½ in. diameter to fit the hole in the brass support, the knob (K, Fig. 1224) being glued in after the handle is fixed in the support. The method of turning and fixing this knob is clearly seen in Fig. 1229, in which K is the knob, H the handle, and D the dowel turned on the knob. A convenient set of trays will be: one 2 ft. 6 in. long by 20 in. wide; one 2 ft. long by 15 in. wide, and one 1 ft. 6 in. long by 12 in. wide. In the smaller ones the base-boards may be about ½ in. thick, and the fence mouldings reduced to ⅞ in. and ¾ in. high, the thickness to correspond. The trays should now be cleaned off and rubbed over with two or three coats of linseed oil, and the polishing may then be proceeded with.

CHESTS.

Oak Linen Chest.

THE linen chest shown in conventional view by Fig. 1230 should be constructed in wainscot oak, fumed. A plan of the

Fig. 1230.—Oak Linen Chest.

chest is given at Fig. 1231, and various details are shown in Figs. 1232 to 1234. The dimensions are as follows:—Length outside angle posts, 5 ft.; width of ends, 1 ft. 10 in.; height from floor to top, 3 ft.; depth inside of chest, 2 ft. 0½ in. The angle posts are 3 in. square, tapering below the bottom of the chest to form the feet (see elevation, Fig. 1234). The inner edges are moulded and grooved to receive the panels, forming a stile for both the front and return framing. The back posts are the same size as the front posts, but are moulded, etc., on one edge only, to receive the return end framing; the back being solid, and flush with the face, the back posts will require grooving only on the inner edges (see Fig. 1231). The mouldings and grooves will require to be stopped at the lower ends, to enable a splayed shoulder to be formed on the tenon of the bottom rail; otherwise the leg would diminish the depth of the moulding, which is not desirable. The framing between the posts is 1¼ in. thick, and comprises bottom rails 3½ in. wide, top rails 3½ in. wide, and muntins 3 in. wide; the panels are 1 in. thick. The bottom rails are moulded and grooved for the panel on the top edge, and moulded and grooved for the bottom of the chest on the lower edge, the bottom being 1 in. thick. The top rails are moulded and grooved on one edge only, to receive the panel, the muntins being grooved on both edges. The panels are sunk, and a small moulding is worked on the solid round the raised parts. The whole of the framing should be properly mortised and tenoned together. The two outer faces of the front posts, and the return faces of the back posts, have a shallow sinking formed in them, as shown

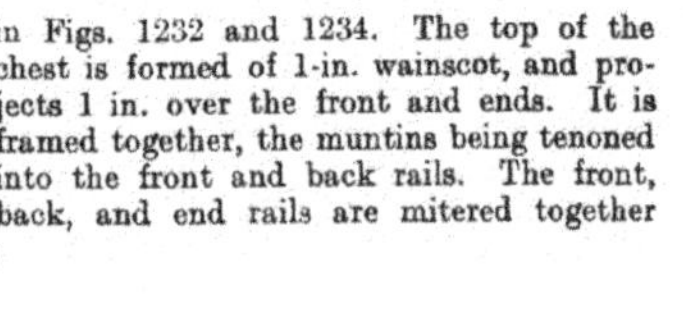

Fig. 1231.—Plan of Linen Chest.

in Figs. 1232 and 1234. The top of the chest is formed of 1-in. wainscot, and projects 1 in. over the front and ends. It is framed together, the muntins being tenoned into the front and back rails. The front, back, and end rails are mitered together at the angles. The joints are feather-tongued, and have in addition in each joint two $\frac{3}{16}$-in. double nut screws, the nuts being let in and turned from the under side. The inner edges of the framing are grooved to receive the tongue on the panels, which are flush-framed, with beaded joints. The front and return edges of the top are moulded as shown. The back of the chest is 1 in. thick. The ends are tenoned and haunched

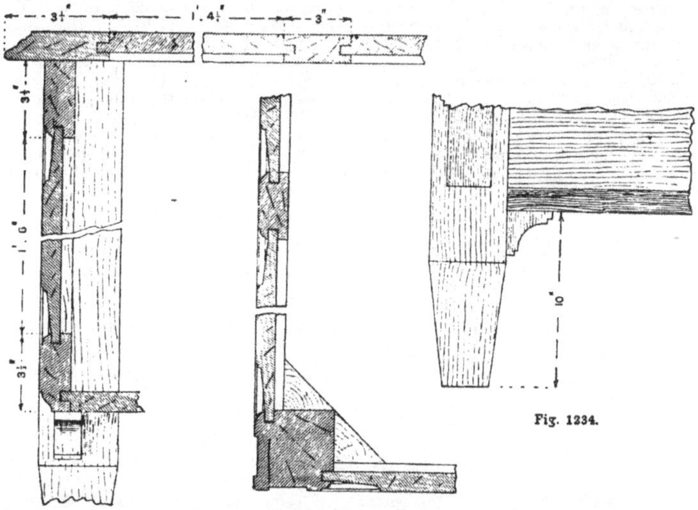

Fig. 1233.—Vertical Section through Front of Linen Chest.

Fig. 1232.—Horizontal Section through Post and Framing of Linen Chest.

Fig. 1234.—Elevation of Foot of Linen Chest.

	No.	Length.		Width		Thicknses.
		ft.	in.	ft.	in.	in.
Posts	4	3	0	0	3	3
Rails	2	5	0	0	3¼	1¼
Rails	4	1	9	0	3½	1¼
Rails	2	5	3	0	3½	1
Rails	2	2	0	0	3½	1
Muntins ...	2	1	11	0	3	1¼
Muntins ...	2	1	10	0	3	1
Panels	3	1	8	1	5	1
Panels	2	1	8	1	5½	1
Panels	3	1	5	1	5	0¾
Back	1	5	0	2	0	1
Bottom ...	1	4	11	1	9	0¾

6 small brackets.
3 wrought-iron chest hinges.
1 wrought-iron chest lock and key.

to fit into the mortice and groove in the angle posts. The bottom also is solid, and tongued all round. The brackets are of wainscot 3 in. by 3 in. by 1 in. thick (see Figs. 1233 and 1234). The materials required are given in the table opposite, all the wood being dry wainscot.

Clothes Chest.

Fig. 1235 is a perspective view of a clothes chest of a convenient size for use when travelling, Figs. 1236 and 1237 showing sections of the chest. The internal arrangement consists of a shelf A (Fig. 1237), 9½ in. wide by ⅜ in. thick, supported on two fillets

B screwed to the ends of the chest. The shelf should be made to slide easily from side to side. Another shelf C, 11 in. wide, placed in an opposite direction to the shelf A, is supported in the same manner. In the right-hand top corner is a small fixed tray suitable for small articles. This will be a convenient arrangement for the inside, preventing the crowding together of the contents of the chest. Good yellow pine should be used for constructing the chest. First plane and square up the two sides to 2 ft. 8 in. in length by 1 ft. 4½ in. wide by ¾ in. thick. Next prepare the two ends, 1 ft. 5 in. long, and the same width and thickness as the sides. The four pieces should deep, cut in the sides of the chest (see Fig. 1239). The tray lid is shown in section by Fig. 1240. A short pin is cut on each end of the lid and inserted in the hole H (Fig. 1239), thus forming a hinge for the lid. The tray must be inserted while the body of the chest is being glued up. The lid of the chest should be ¾ in. thick, with a projecting piece 1⅛ in. wide, and rounded on the outside edge, shaped as shown at J (Fig. 1238). At the back edge it is ¼ in. thicker to allow for the hinge. The inside would look well if stained and varnished, while the outside should be given a coat of red-lead. After puttying up the nail holes, the chest should have two coats

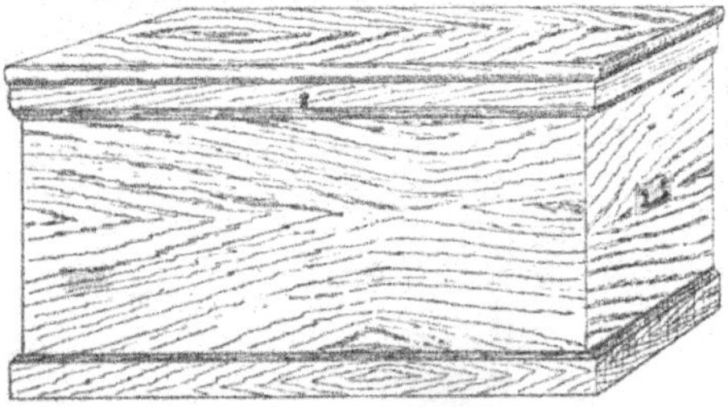

Fig. 1236.—Clothes Chest.

then be dovetailed together. The bottom should be of ½-in. lining, tongued and grooved and nailed to the edges, laying it across from back to front. Fillets 2 in. by ½ in. are nailed along the bottom and flush with the outside edges (see F, Fig. 1238). When the body of the chest is planed up and glass-papered, the plinth D (Fig. 1238), which is 2½ in. wide by ½ in. thick, can be nailed on. It should be mitered at the corners, and a sash moulding is run on the top edge. The moulding E (Fig. 1238) should now be nailed round the top edge, keeping it ⅜ in. below the edge of the chest. The fillets supporting the shelves are 1¼ in. wide by ½ in. thick, and are fixed to the sides with screws. The tray should be of ⅜-in. stuff, the inside measurements being 4 in. wide by 3 in. deep. The front and bottom are let into grooves ¼ in. of paint, smoothing between the coats with glass-paper. Japanned iron handles should be screwed to the ends. The lid is hinged with a pair of 3-in. brass butts, a suitable brass lock completing the fittings.

Another Clothes Chest.

Fig. 1241 is a view of the finished chest, whilst Fig. 1242 is a broken longitudinal section showing the details. The material should be good quality yellow pine, perfectly dry and free from knots or shakes. A suitable length for a chest is 2 ft. 10 in. outside measurement, and the width 1 ft. 5 in. The height given is 1 ft. 5½ in., but this may be varied slightly to suit the width of stuff obtainable, as it is advisable to have the sides and ends in whole pieces. The box, which should not be any thicker

than ¾ in., is dovetailed together and glued. The bottom is formed of feather-and-grooved lining, nailed to the under edge. Fillets F are carried all round the outside edges of the bottom to keep the box out of the damp. These fillets are nailed to the lining, but one or two screws should also pass right into the under edge of the box itself, so **are shown by Fig. 1243. A rim T is mitered** at the corners and nailed around the lid, the bottom edge projecting down ½ in. below the frame. The lower facing is fixed on to keep the lid a little clear of the top edge of the box. Brass butt hinges and a good lock, as well as good handles, should be used. If the frame of the lid inside is veneered

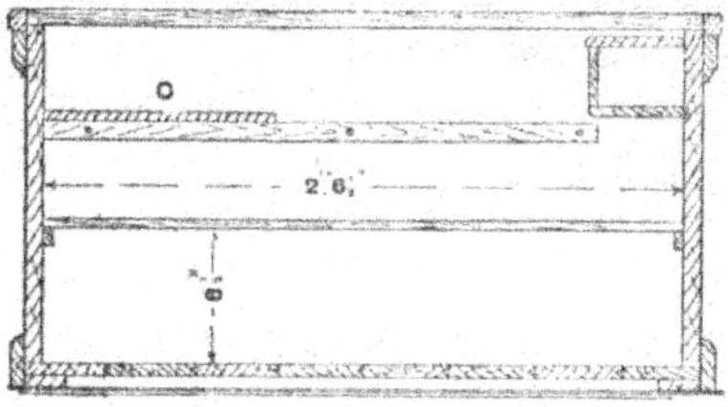

Fig. 1236.

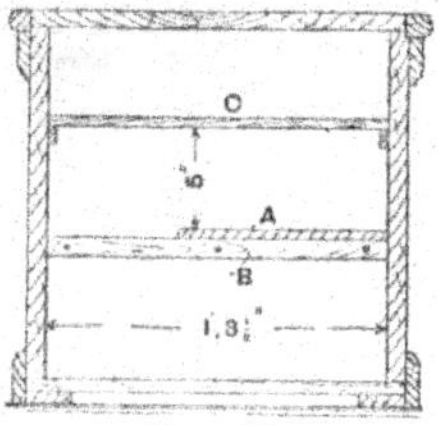

Fig. 1237.

Figs. 1236 and 1237.—Longitudinal and Cross Vertical Sections of Clothes Chest.

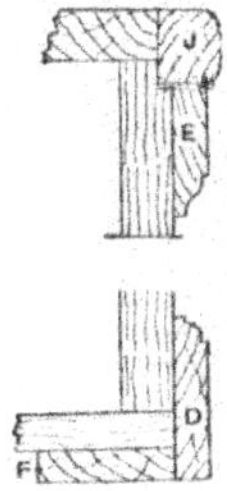

Fig. 1238.—Part Vertical Section through Back of Chest.

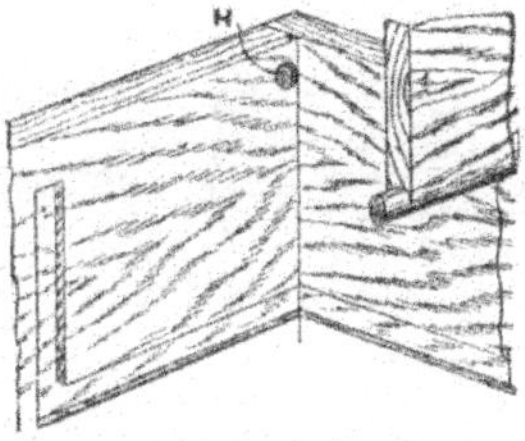

Fig. 1239.—Detail of Clothes Chest Tray.

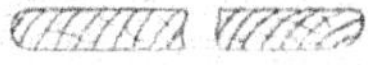

Fig. 1240.—Section of Tray Lid of Clothes Chest.

as to bind the bottom firmly on. The skirting S may be dovetailed at the corners, but mitre joints will do as well. It must not, however, be in contact with the floor, and for this reason is kept up ¼ in. The lid is formed of ¾-in. stuff, with 3-in. pieces clamped to each end. These pieces, being ½ in. thicker than the central portion, necessitate the planting-on of pieces of pine the same width as the cross ends, and flush with them, to form a mock frame, into which is planted a sunk moulding. These pieces with mahogany and the panel with bird's-eye maple, with a mahogany moulding planted in and the whole polished, it will give the interior of the box a very rich appearance. At one end of the box, as shown by Figs. 1242 and 1243, there is a till with secret recess, and having drawers below. Resting on a projection of the till at one end and on a fillet at the other end of the box is a tray, having a smaller tray inside. The bottom recess is intended for dress-shirts, collars, etc.; the top one for woollen articles. Suits are thus kept separate, which is an advantage. The till is formed by dovetailing the bottom raggled to within ½ in. of the top to receive the sliding part C, which has a square shoulder 1 in. in depth on the top edge to

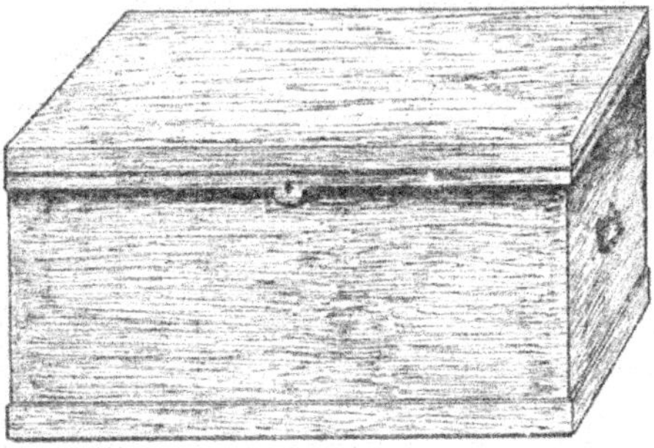

Fig 2141.—Another Clothes Chest.

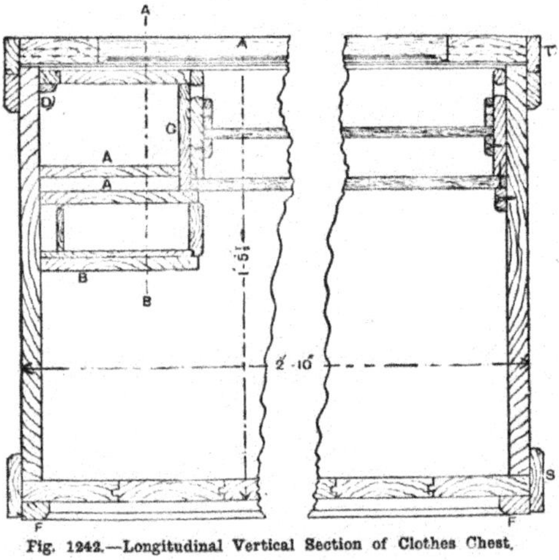

Fig. 1242.—Longitudinal Vertical Section of Clothes Chest.

B to the two gables G G. The two divisions A are raggled to the gables, the raggle of the bottom one being stopped about ½ in. from the face edge. A solid division separates the two drawers. The gables are allow it to slide up and reveal the recess below. A small piece of pine D (Fig. 1242) is checked down flush with the top edge of gables to carry a back stile, to which the flap is hung. The till carcase is screwed to the sides of the box. A small stiffening block is inserted into the recess immediately above the drawer division. The drawer bottoms are checked to facilitate lifting them out and in. Fig. 1244 is a broken plan of the interior of the box, having the till flap removed. If the inside of the lid is polished, it would be as

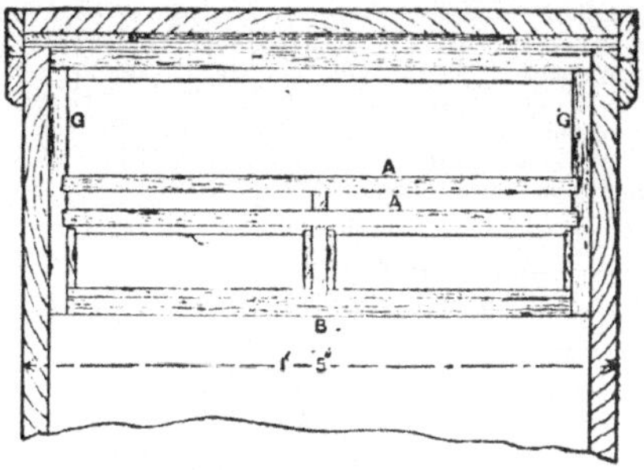

Fig. 1243.—Part Cross Section of Clothes Chest on Line A B (Fig. 1242).

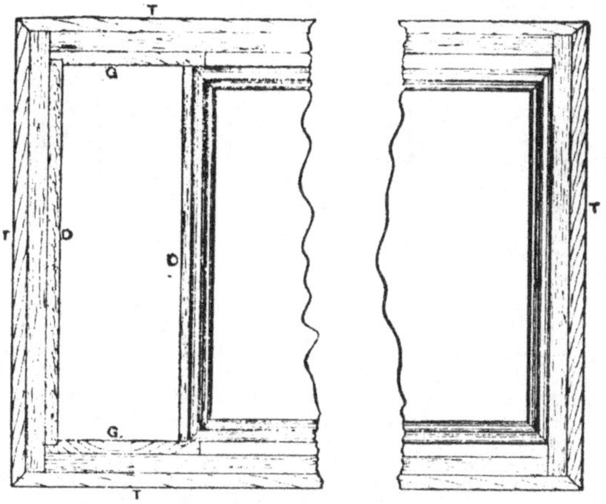

Fig. 1244.—Plan showing Interior of Clothes Chest.

into the front, and nailed to the front and under edges of the sides and back. The trays are dovetailed together and the top edges rounded. Holes are cut in both trays well to have a mahogany flap on the till. The outside of the box should be painted with good oil paint, and the interior should be left plain.

Sea Chest.

Most sea chests in a cabin or forecastle are usually found considerably larger at the base than above, with the idea, no doubt, that such a form prevents them from rolling over in bad weather. But this is quite unnecessary; chests slide across a cabin long before they capsize, and if lashed they do neither. It is necessary that the chest be watertight, as several inches of water, and sometimes enough to float the chest, continually wash to and fro over the cabin deck in bad weather; for which reason soft woods are better than hard, the latter being more liable to crack, while the former receive only a harmless dent from a blow. The wood should be 1 in. thick, free from knots, the corners being dovetailed and iron bound (see Fig. 1245). On the bottom two dunnage battens of teak or oak A A must be fixed with brass screws, and in their centres a groove cut to admit the chest lashing, thus preserving it from chafe and the chest from a wet deck. The top and bottom are fixed with brass screws, all joints and the insides of corner iron angles being painted before putting together. For the handles make two grommets of $1\frac{3}{4}$-in. or 2-in. rope; these are secured to the ends of the chest by a pair of oak or teak cleats (Fig. 1246), the edges of the groove being rounded off to prevent cutting the grommet. Brass hinges and lock are well worth the small extra cost. The till B is indispensable; let it be made of $\frac{1}{2}$-in. stuff, outside measurement $16\frac{3}{4}$ in. by 6 in. by 6 in. This will just fit across the chest at one end, resting on the strips D D (Figs. 1247 and 1248), which are nailed in such a position as will allow the chest lid to close over the projecting part of the till. The strips E and F are now nailed as shown in Fig. 1248; the partition G slides between them, and can be removed if desired when the till is lifted. G is another partition, which may be made in two parts, to slide between the strips H H in the same manner. The whole of the inside must now receive two coats of good copal varnish, and the outside three coats of paint. A canvas cover should also

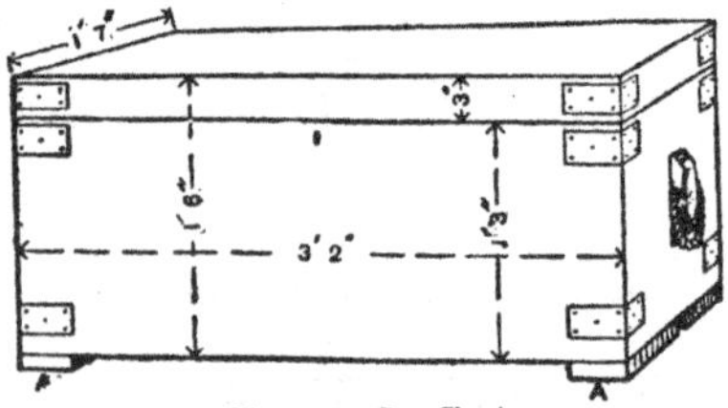

Fig. 1245.—Sea Chest.

Fig. 1246.—Cleat for securing Grommet Handle to Sea Chest.

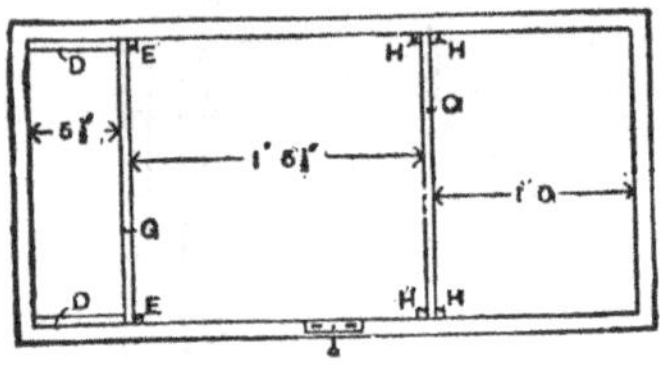

Fig. 1247.—Interior Arrangement of Sea Chest (Till Removed).

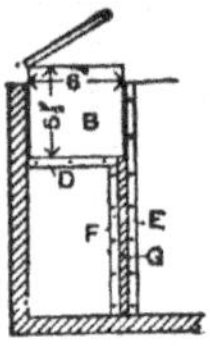

Fig. 1248.—Part Section of Sea Chest, showing Till, Partition Strips, etc.

be made, 3 ft. 2½ in. by 1 ft. 7½ in., with edges 4 in. deep to cover the lid joint; this may be painted black or white, or kept scrubbed; its use is to keep rain, or any other liquid, from running inside at the lid joint, and it is for many reasons better than a rebate. Dimensions not given in the text are shown on the illustrations.

Steward's Sea Chest.

Figs. 1249 and 1250 show one method of arranging a chest for holding a steward's outfit. There are two trays, which divide it into three spaces, so as to keep the various items separate. The bottom space is intended for trousers; the middle one (which is divided into three spaces by the raggling of two divisions into the sides) is for shirts and collars; and the top one for uniforms. The chest is made out of ¾-in. yellow pine, dovetailed together, with a ½-in. bottom nailed on, and 1½-in. by 1-in. pieces screwed to the bottom to keep it off the deck. The end and side wood at the bottom is covered by a base, which is mitered at the corners and nailed on. The lid is made of ¾-in. wood, with 3-in. cross-ends mortised and tenoned to it. There are two arrangements shown for the lid. A shows the ordinary arrangement for good chests, with one facing nailed to the lid and resting on the other, which is nailed to the box, and kept down ½ in. from the top edge. B shows a throating cut out of the top edge of the chest, and a small bead fastened to the lid to drip off any water which may find its way in. This throating would require to be extended to the outside of the chest to run the water off. The trays are dovetailed together, and a grip-hole is cut out at each end, and nicely rounded. The tray bottoms, which are feathered and grooved together, having the grain running the short way, are screwed to the under side of the trays. Small pieces are put in at the bottom of the box to carry the bottom tray, the top tray resting on the bottom one. The top edges of the trays are rounded, and should be mitered at the corners, the divisions being mitered to the middle of the round. The base is only shown on one part of the illustration, but it is intended to be carried all round.

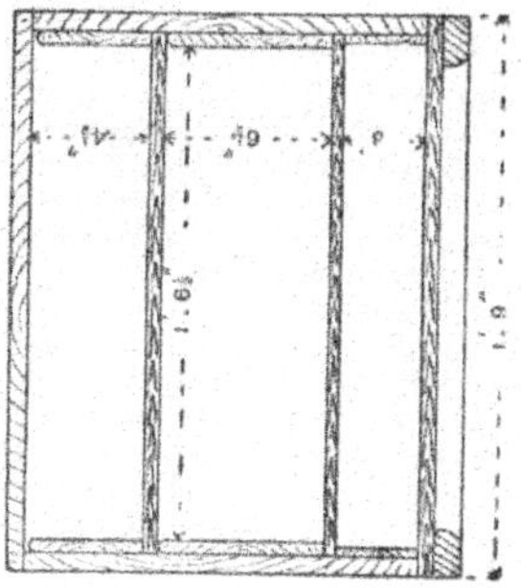

Fig. 1250.

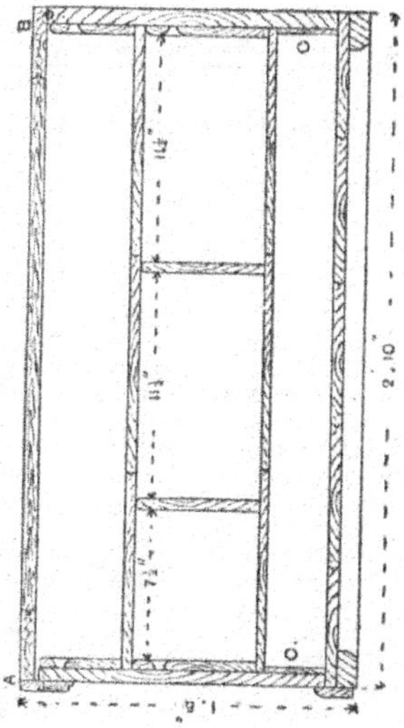

Fig. 1249.

Figs. 1249 and 1250.—Longitudinal and Cross Vertical Sections of Steward's Sea Chest.

Lead-lined Tea-chest.

The tea-chest about to be described is made preferably of mahogany, and is divided into two compartments. The insides of these are lined with lead, such as is used in large tea-chests. This lead can be bought at any grocer's for a small sum. Fig. 1251 is a longitudinal section of the box, showing the construction; Fig. 1252 being a section on A B, Fig. 1251. The body of the box is secret dovetailed, and the moulding on the top dowelled to the edge of it. Before fixing the moulding, a gauge line is run round the box where it has to be cut to form the lid. After the moulding is glued on, the top can be fixed with glue and small brads. The box can then be cut to the gauge line, and the division glued in a shallow groove which is cut for it before gluing the box up, after which the bottom can be screwed on. The fillets F, to support the lids, are fixed in after the joint of the lid and box has been faired up. The illustrations show them fixed with very small screws, but glue and small brads will make a strong enough job. The lids are then fitted, the holes bored in the centre, and the turned knobs glued in. A small moulding is mitered and glued to the inside of the lid. The lead is fixed in with small galvanised tacks. To make the job easier, a pencil line can be drawn all round the inside of the bottom, and the bottom unscrewed and the lead tacked on. After the lead has been fixed to the inside from the bottom edge of the box to the top edge of the fillets, the bottom can be replaced. The under sides of the small lids are also covered with the lead. The turned ball feet shown in the illustrations are screwed to the bottom with thin screws. Two beads are shown at the lid joint, one on each edge, but one of them can be dispensed with. The lid is hinged, but it is not intended to have a lock. If one is wanted, the division will require to be shifted clear of the centre, and one space made larger than the other; or the box must be constructed of sufficient width to justify two divisions.

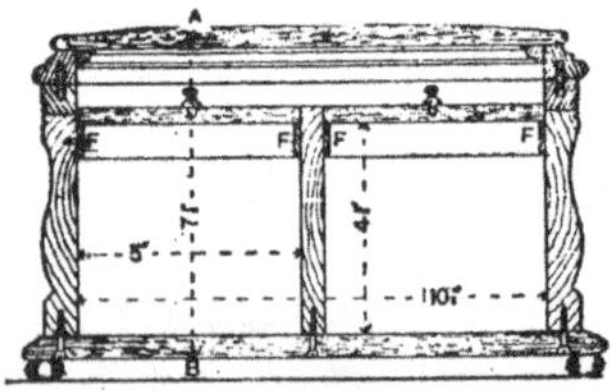

Fig. 1251.

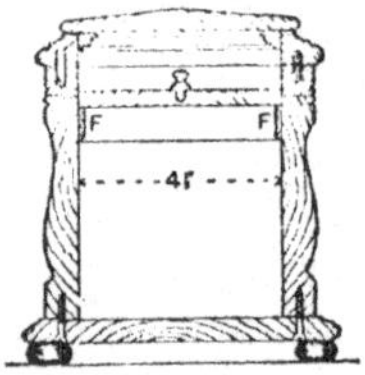

Fig. 1252.

Figs. 1251 and 1252.—Longitudinal and Cross Vertical Sections of Lead-lined Tea-chest.

Traveller's Sample Case.

Figs. 1253 to 1257 show the construction of a traveller's sample case, 36 in. by 18 in. by 24 in., to contain such things as leather goods. Good red-deal, birch, beech, or other similar hardwood, $\frac{3}{4}$ in. to 1 in. thick, may be used, according to strength and other requirements. The angles should be dovetailed together, and the boards jointed and cross-tongued, as shown at Fig. 1256. To prevent dust, etc., getting in, a fillet about $1\frac{1}{4}$ in. by $\frac{1}{2}$ in. should be nailed round so as to project into the lid when closed (see Fig. 1253). If the staples are made as shown at Fig. 1257, they can be screwed to the front of the rim of the lid, and the returned piece shown at A (Fig. 1257) can be let in and screwed to the under side of the lid; this will prevent it being broken off. The eye and plate can be made so that the eye passes through the front, the plate being screwed to the inside; it is thus not likely to be broken off or unscrewed from the outside. Two

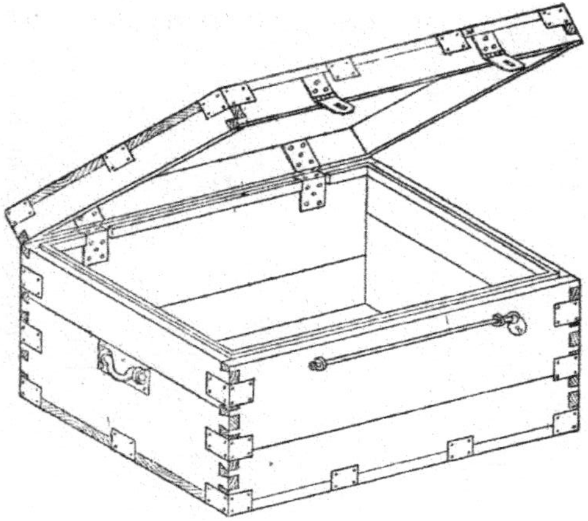

Fig. 1253.—Traveller's Sample Case.

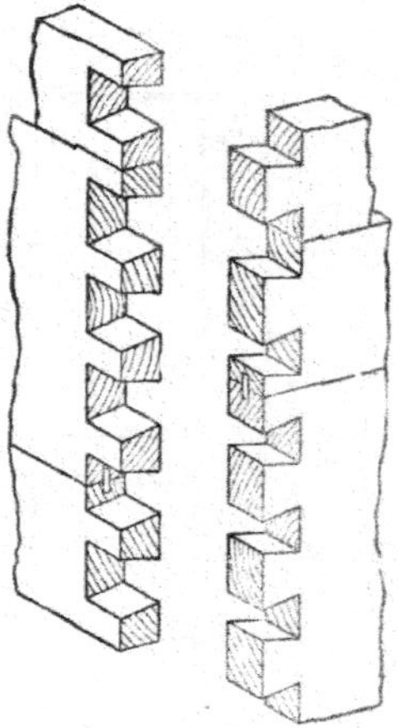

Fig. 1256.—Joints in Traveller's Sample Case.

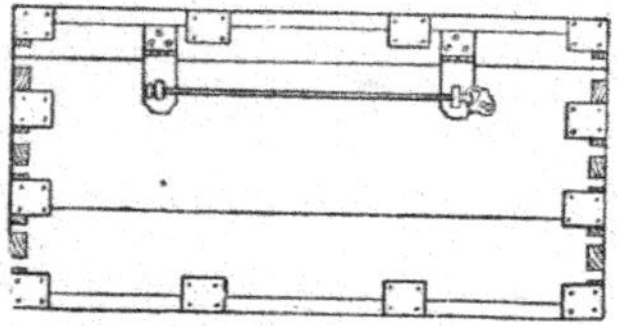

Fig. 1254.

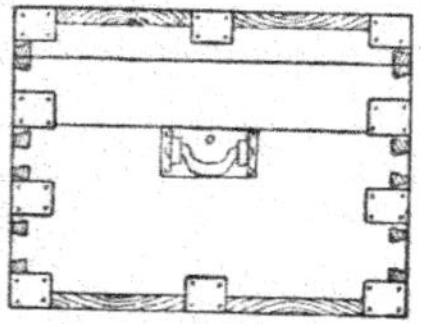

Fig. 1255.

Figs. 1254 and 1255.—Front and End Elevations of Traveller's Sample Case.

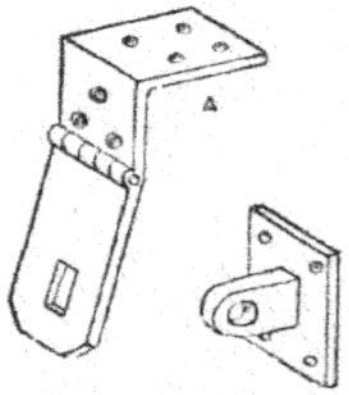

Fig. 1257.—Staple, Eye, and Plate.

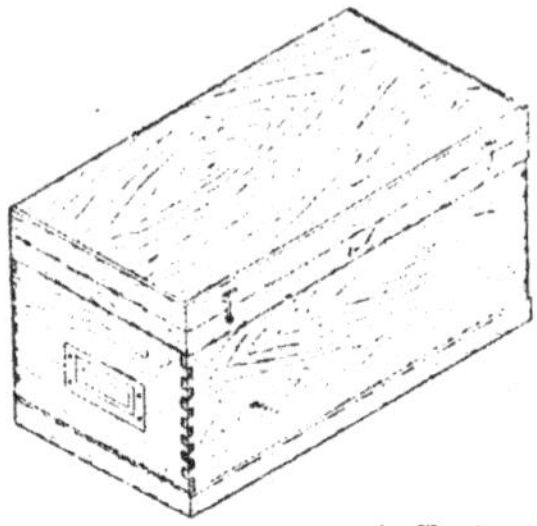

Fig. 1258.—Draughtsman's Chest.

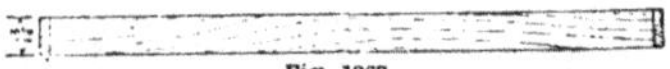

Fig. 1262.

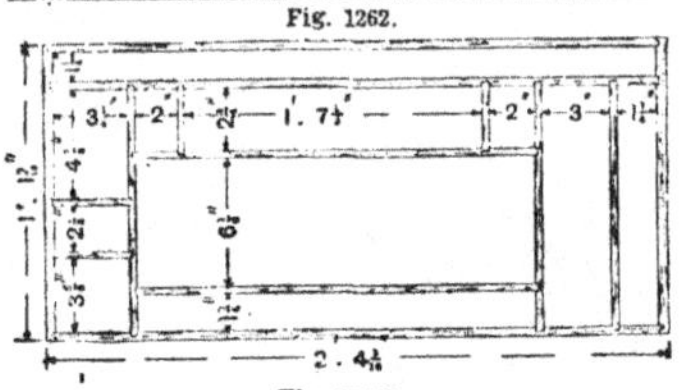

Fig. 1263.

Figs. 1262 and 1263.—Top Tray of Draughtsman's Chest.

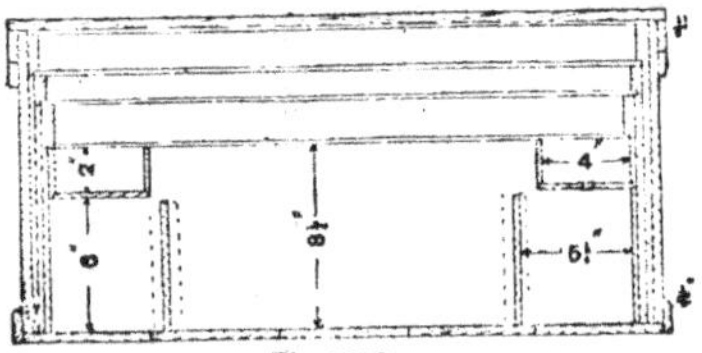

Fig. 1259.

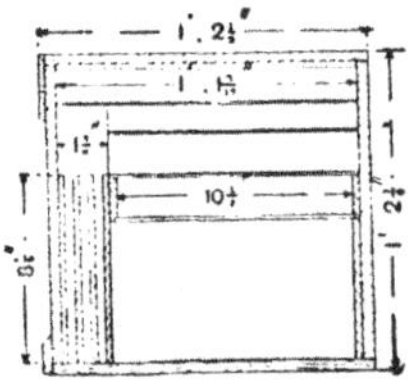

Fig. 1260

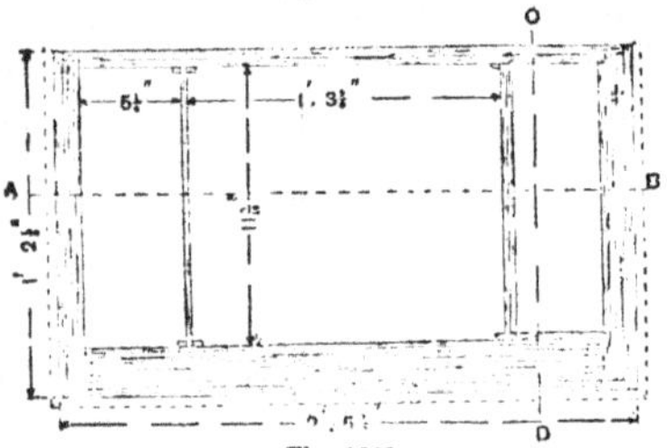

Fig. 1261.

Figs. 1259 to 1261.—Vertical Section, Horizontal Section, and Plan of Draughtsman's Chest.

Fig. 1266.

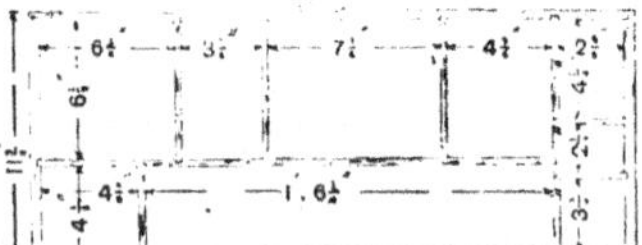

Fig. 1267.

Figs. 1266 and 1267.—Lower Tray of Draughtsman's Chest.

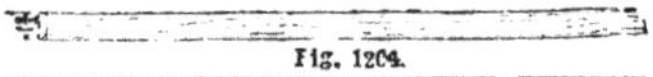

Fig. 1264.

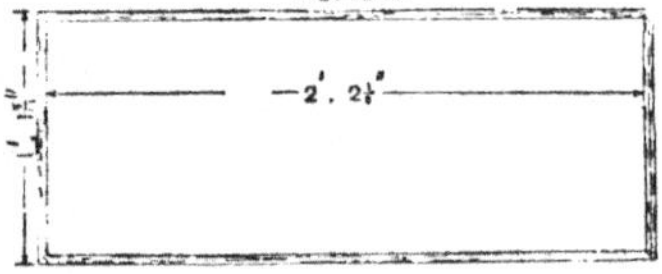

Fig. 1265.

Figs. 1264 and 1265.—Middle Tray of Draughtsman's Chest.

padlocks may be used, or a rod and one lock, as shown. For ordinary purposes, one staple, eye, and lock would be sufficient.

Draughtsman's Chest.

The chest shown in Fig. 1258 is designed to take various articles that could not be got into the draughtsman's instrument case. The wood shown is ½-in. pine (but ¾-in. would be better), dovetailed at the corners, and lightly stained and varnished. The ends of the trays are ½ in. thick, and the sides and bottoms are ¼ in. thick, finger-holes being bored in the ends for lifting. Figs. 1259 and 1260 are sections on A B and C D (Fig. 1261) respectively, Fig. 1261 being a plan of the box below the third tray, while Figs. 1262 to 1267 give plans and elevations of the three large trays. The racks shown in Figs. 1260 and 1261 are for large set-squares, the two end trays above are for ink, colour saucers, etc., and below are two divisions, with movable partitions, one for a large sponge and a water glass, and the other for a case of railway curves. The well between takes the instrument case. The top tray takes several beam-compass laths of different lengths, a 1-ft. 6-in. rolling parallel ruler, and a 1-ft. 6-in. hand level. The largest space to the right contains the pocket case for a 9-in. proportional compass, and the other divisions are devoted to special purposes, and the middle space to general use. The second tray is wholly devoted to curves, and the lower tray holds a speed indicator in case, a squeegee for use in mounting plans, a pair of clamps, and various odds and ends. If necessary, a lock and key can be added.

SCREENS.

Hall Screen with Fretwork and Leaded Lights.

THE hall screen illustrated by Fig. 1268 may be successfully introduced in a dwelling-house, private hotel, or similar building. A good position for the screen is about 4 ft. or 5 ft. back from the drawing-room and dining-room doors. Care should be taken to arrange the cornice, span rail, and dado

Fig. 1268.—Hall Screen with Fretwork and Leaded Lights.

Fig. 1269. Fig. 1270.

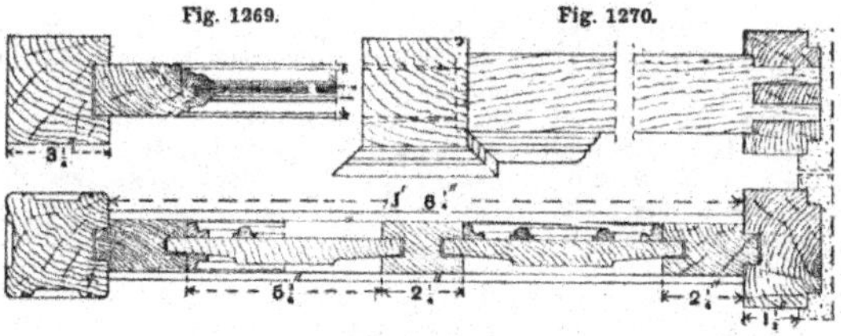

Fig. 1271.

Figs. 1269 to 1271.—Horizontal Sections of Hall Screen on Lines A B, C D, and E F (Fig. 1272).

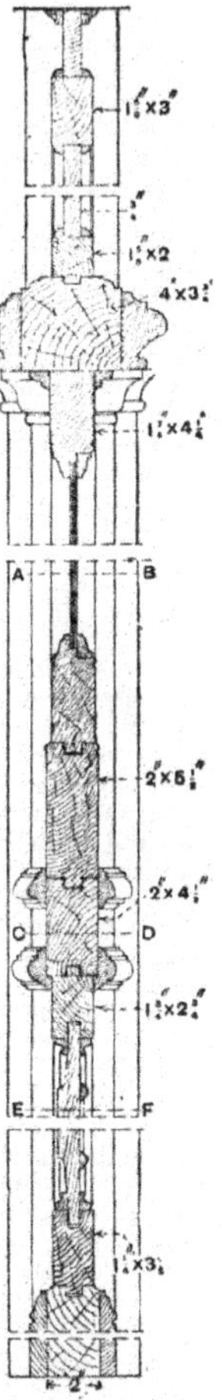

Fig. 1272.—Vertical Section of Hall Screen on Line G H (Fig. 1268).

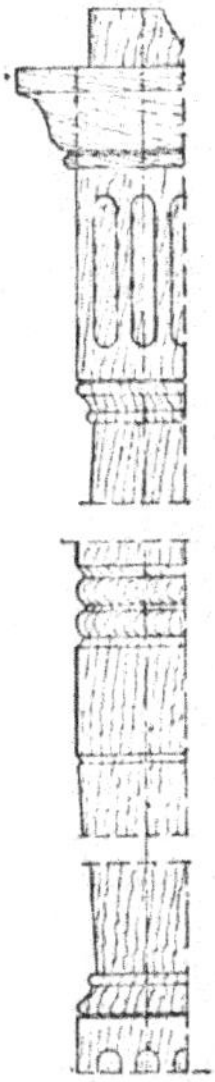

Fig. 1273.—Inner Stile or Pilaster of Hall Screen.

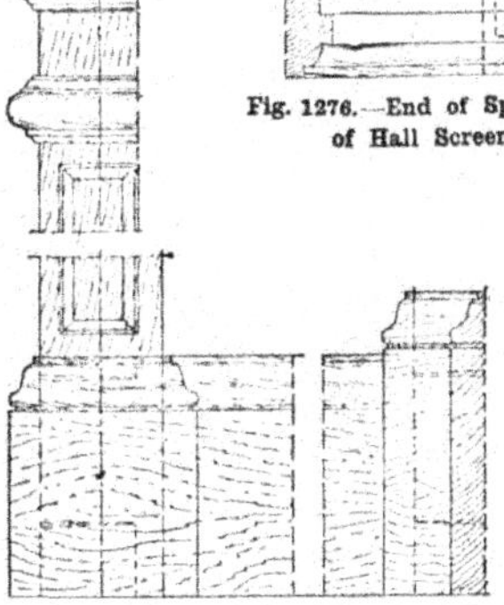

Fig. 1274.—Lower Part of Pilaster and Skirting of Hall Screen.

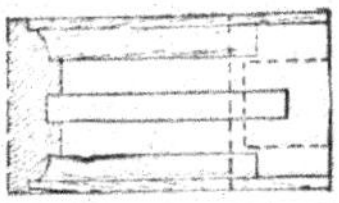

Fig. 1276.—End of Span Rail of Hall Screen.

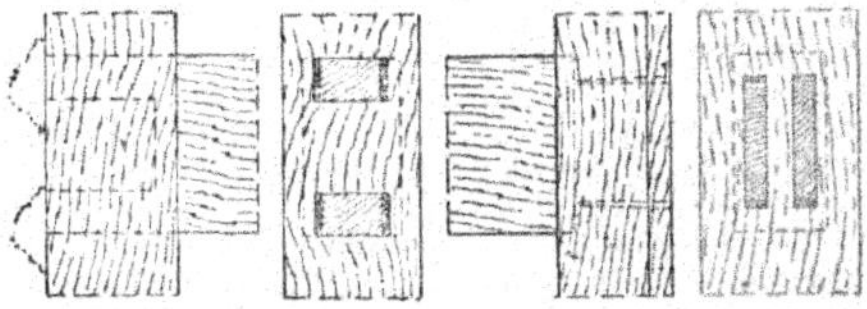

Fig. 1275.—Double Tenon Joints for Hall Screen (see Fig. 1270).

rail mouldings in line with and also of the same section as any existing details of the sort, otherwise the screen will appear to be disconnected from the original architecture of the house. Fig. 1269 is a section on A B ; Fig. 1270, section on C D ; Fig. 1271 section through E F, in Fig. 1272, which is a vertical section through G H. The dimensions are : Height, 8 ft. 10 in. ; width, 6 ft. 6 in. ; height from floor to under side of span rail, 6 ft. 2 in. The screen is intended to be constructed of hardwood with leaded light on each side ; two small panels are shown in the dado, but a single larger panel would look as well, and would lessen the labour. The upper framework may be of soft wood stained and varnished to match. The frame is divided by a centre span (the inner and outer radius are, respectively, 1 ft. 8½ in. and 1 ft. 11½ in.), surmounted by a small frieze band and cornice. Cut-through ornament is introduced in the divisions formed by the framing, but leaded lights of suitable design could be introduced with equally good effect if desired. The three parts of the screen may be framed up in the shop ready for erecting. It will be noticed from Fig. 1269 that the frame here is sunk into the pilaster its full width, owing to the irregular shape of the stile. It will be seen that the dado and bottom rails are tenoned to the inner pilaster (Fig. 1273), so that the moulding and skirting boards shall hide the joints (see Figs. 1274 and 1275). The two sides may be fixed first. The upper framework and the span rail (see Fig. 1276) (the latter being mortised to receive stump tenons from the wall and inner stiles or pilasters) having been removed, the wall should be plugged in positions where the screws for fixing will be covered by the attached mouldings ; while the floor may be mortised to receive tenons from the stiles (not shown in illustration), which are further secured by screws driven through the bottom rail before the skirting board is fixed. The span rail and the upper wall stiles are now placed in position. The frame (Fig. 1277) is tongued to the span rail and rebated to the upper wall stiles as shown by Fig. 1278. A rebate is worked round the frame to receive the fretwork, which is kept in position by a quarter-round slip, the top band of the fret being retained by slips attached on each side. The cornice moulding is attached to a batten fixed to the ceiling, the lower part abutting on the fretted band (see Fig. 1279).

Ornamental Dividing Screen.

Fig. 1280 illustrates a screen suitable for dividing a long room into two parts, or for fitting across a recessed window opening. Mahogany and walnut are suitable for use if the general tone of the room is dark ; but if it is light, whitewood should be chosen and enamelled in art colours. The pillars are about 4 in. square at the base, with turned shafts, finished by ornamental capitals, and above these the pillars are reduced to 2¾ in. or 3 in. square, and carried up through the frieze to the ceiling. The pillars against the wall are of half-round and half-square sections, and can be made from one whole pillar sawn down the centre. The several rails are mortised into the pillars, and the finishing mouldings mitered and carried round the pillars. The side bottom openings are fitted with a circular span rail ½ in. thick, cut to form two arches, and supported by a shaft 1⅛ in. or 1¼ in. square. This shaft is slotted at the upper portion, as seen in Fig. 1281, to fit the span rail. The bottom rails are moulded on each side (see Fig. 1282) and round the pillars to form a finish, the design corresponding with the room skirting. The pillar capitals are obtained by pinning and gluing on a piece of stuff 1¼ in. square, and then carving the design at the top of the capital. The centres from which the middle arch span rail is struck are shown in diagram form in Fig. 1283, together with the centres for the edging. Above the arch span rail a heavy moulding M (Fig. 1284) is fixed for mortising into the pillars. The under side is grooved to receive the arch span rail, and the top side grooved for the first frieze F. This frieze is ½ in. thick, and the floral wreath work is carved in wood and screwed on, or lincrusta decoration can with advantage be used. Above this a rail of square section is placed to carry the moulding x, and also the bottom edge of the second frieze A. This frieze is ⅝ in. thick, and is planted flush over the pillars on each side, and here again lincrusta decoration makes a very suitable finish. Overhanging this is a heavy moulding fitting close to the ceiling, and it is desirable to carry this round the room as a cornice. The pillars should, if possible, be let into the floor for support; but where this cannot be done, screw them to the floor before putting on the base mouldings. The tops of the pillars are tightly fixed with double wedges at the ceiling, the bearing being arranged under a joist and the wedges secured from movement. For this purpose the top frieze and moulding are fixed after the structure is in position; the side pillars are fixed to the walls, and where two vertical walls form the boundary, clearance must be left to allow for fitting. The pillars are then packed, and an angle moulding is used to cover the clearance. Brackets for flower bowls as shown in plan in Fig. 1285 are screwed on the rails R (Fig. 1280) above the small arches. In designing this screen, it has been borne in mind that draperies of plush curtains, etc., will be hung and looped in the archways in order to give the necessary softening finish. Loop them with heavy cords and tassels, and choose the colour to harmonise with the prevailing tone or colour scheme of the immediate surroundings.

Fig. 1278.—Section of Hall Screen's Frame and Upper Wall Stile.

Fig. 1279.—Cornice Moulding of Hall Screen.

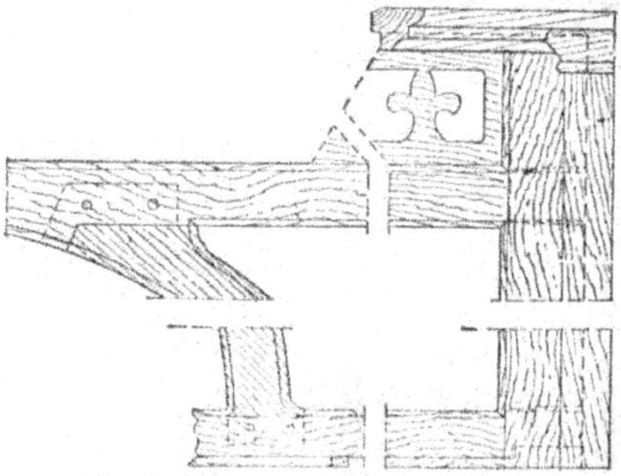

Fig. 1277.—Hall Screen Frame carrying Fretwork.

Vestibule Screen with Folding Doors.

The screen illustrated by Fig. 1286 is intended to be executed in Italian or American walnut, and french-polished. Fig. 1286 shows an interior elevation of the screen, Fig. 1287 a plan, while Figs. 1288 and 1289 show respectively vertical sections through doors, etc., and side framings. The screen is 10 ft. high from floor to ceiling, and there is a moulded and carved walnut cornice all round the vestibule and hall. The four jambs shown on the plan at Fig. 1287, and in the detail Figs. 1290 to 1295, are made out of 4-in. by 4-in. stuff rebated, grooved, beaded, etc. The transom is continuous and of the same thickness as the door and framings, and the jambs and transom are framed together in the ordinary manner. The jambs do not run up beyond the transom, and the framing above, which is 2¼ in. thick, forms an independent sash or fanlight. The head is cut to an elliptical shape, and marginal bars are cut to correspond. The head- or top-rail must be wide enough to take the moulded cornice on both sides. The spandril piece on each side of the exterior portion is sunk and carved with a fan or other pattern. The fanlight sash should be firmly fixed to the brickwork and the ceiling joists, and to the plaster at each end. The centre part below the transom is filled in with a pair of 2¼-in. folding doors, each having three panels; the lower panels are raised and mitered, and have bolection moulding on the outside and ordinary moulding on the inside, and the centre panels over the screws, and the upper panels are divided into squares, with marginal bar to follow the elliptical rail of the door, and are filled in with clear bevelled plate glass as before. The frieze rail is provided on

Fig. 1280.—Ornamental Dividing Screen.

are filled in with leaded lights. The glass is painted and burnt, and the centrepiece has a shield pattern in clear bevelled polished British plate-glass; the whole being held in position by brass saddle-bars, turned up at the ends, and fixed with screws passing into the rebate of the door. The glazing mouldings or beads are notched and fitted the outside with a moulding having a dentil course, and on the inside with a sunk moulding, as shown at Fig. 1295; moulding and apron lining are fixed on the middle rail. The framing on each side of the door is carried up to the transom, and is fixed in the centre of the jamb, the part below the middle rail being flush framed on the

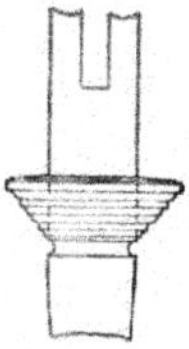

Fig. 1281.—Small Shaft of Screen.

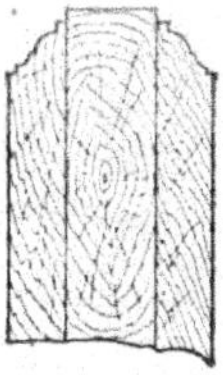

Fig. 1282.—Bottom Rail and Skirting of Screen.

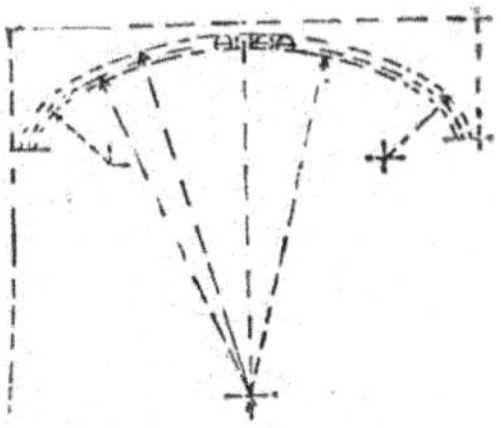

Fig. 1283.—Centres for Screen Arch.

Fig. 1284.—Section of Upper Mouldings and Frieze of Screen.

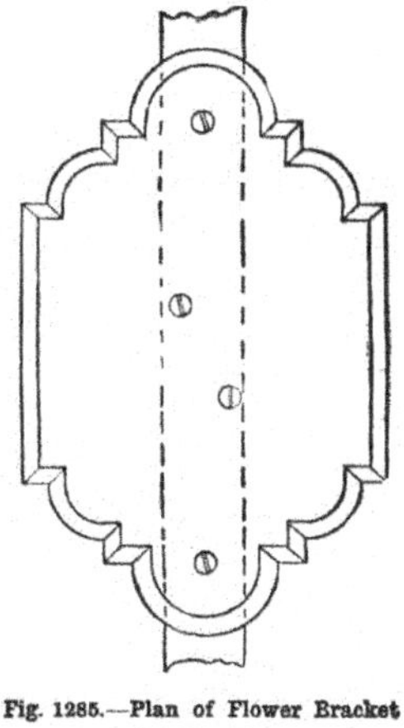

Fig. 1285.—Plan of Flower Bracket for Screen.

Fig. 1286.

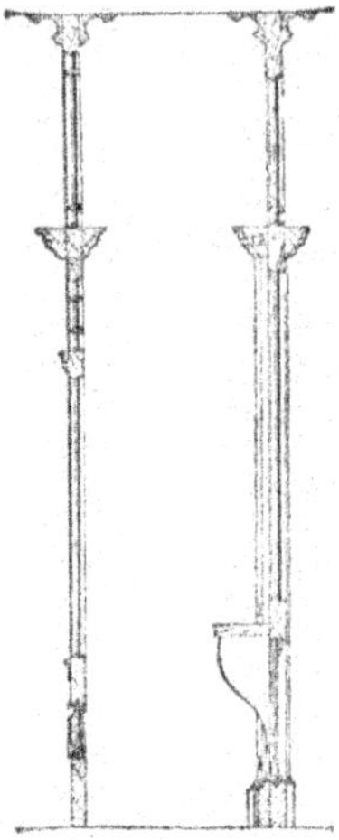

Fig. 1288. Fig. 1289

Fig. 1287.

Figs. 1286 to 1289.—Half Outside and Inside Elevations, Plan, and Vertical Sections of Vestibule Screen with Folding Doors.

Fig. 1290.—Horizontal Section through Lower Side Framing of Vestibule Screen.

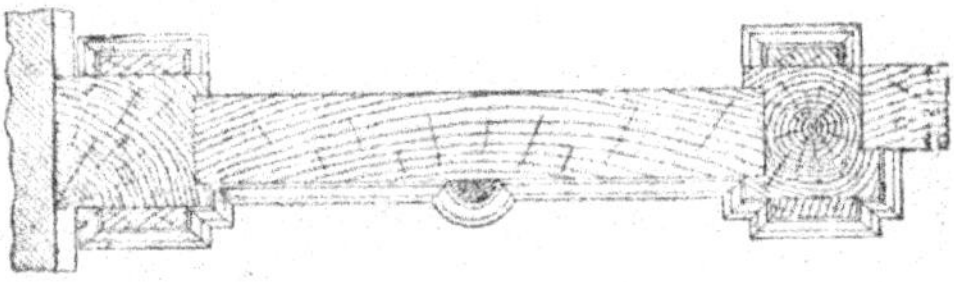

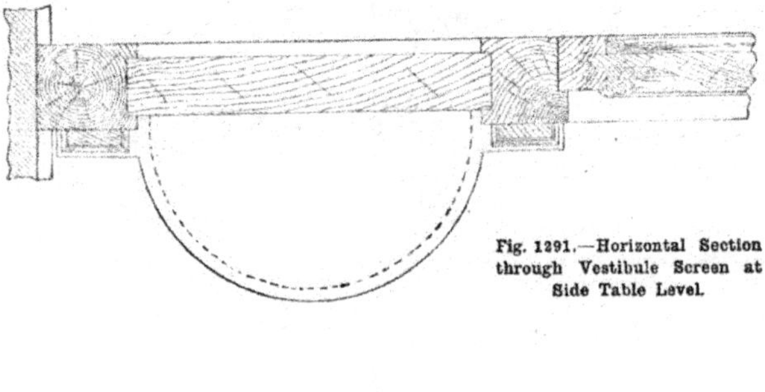

Fig. 1291.—Horizontal Section through Vestibule Screen at Side Table Level.

Fig. 1292.

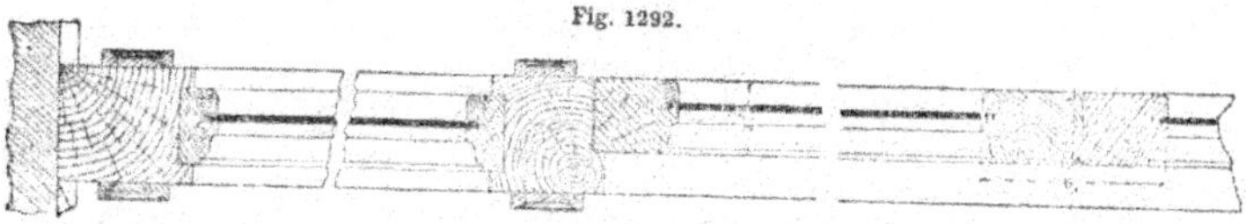

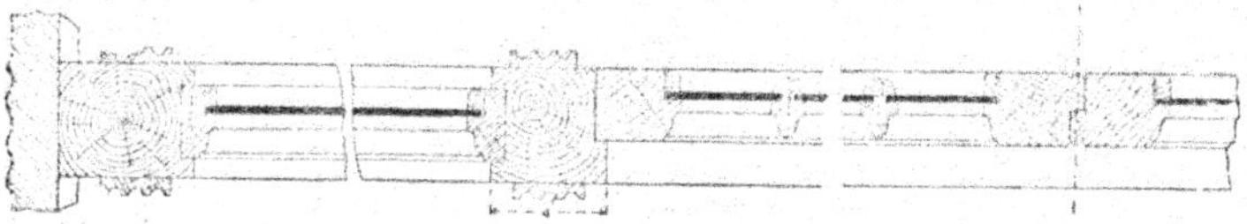

Fig. 1293.

Figs. 1292 and 1293.—Horizontal Sections of Vestibule Screen through Centre and Upper Parts of Door and Side Framings.

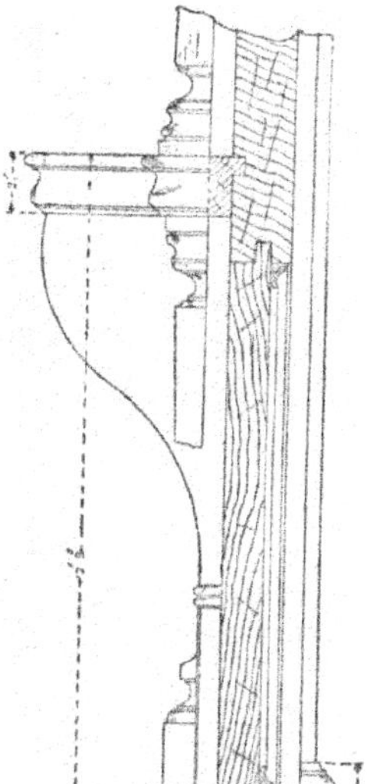

Fig. 1294.—Vertical Section through Lower Side Framing of Vestibule Screen.

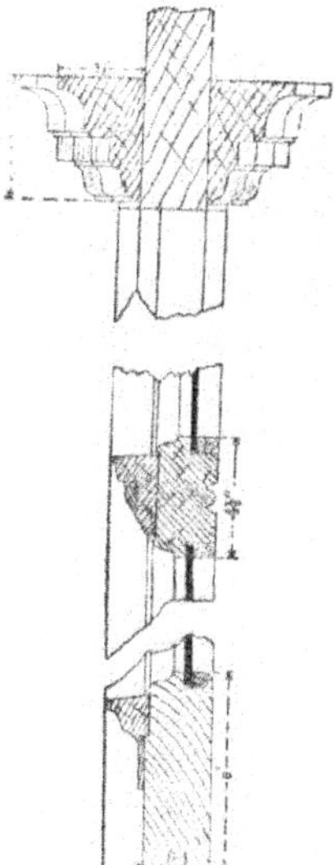

Fig. 1295.—Vertical Section through Door and Transom of Vestibule Screen.

outside, two moulded panels which carry the line of truss being sunk on the face; on the inside the panel stands back ⅜ in. from the face of the framing, and in the centre there is a false muntin framed at each end with a barefaced tenon. These panels are moulded as shown. The upper panel is filled in with leaded lights, the glass part painted and burnt, and the remainder fitted with cathedral glass of rich tints. On the face of the jambs are pilasters, reeded or fluted and panelled, with moulded bases and neckings. The skirting is broken round these pilasters and at the face of the framing. The centre part on each side is formed into a base for the shaft of the truss, the latter being turned as shown, and a carved lion's head planted and fixed on the face. Over the truss there is a shaped and moulded table or shelf, which is intended to carry a vase for plants or flowers. The side part of the table, which runs over the face of the jamb, also serves as a base for the pilaster above. A moulded cornice having a dentil course is fixed on the face of the transom, and is broken round the pilasters, for which

Fig. 1296.

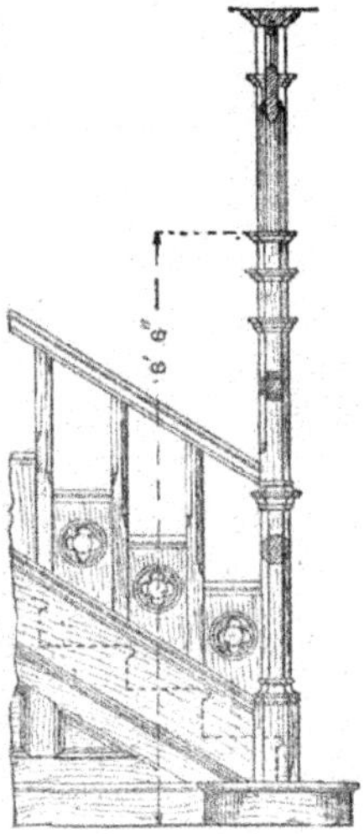

Fig. 1297.

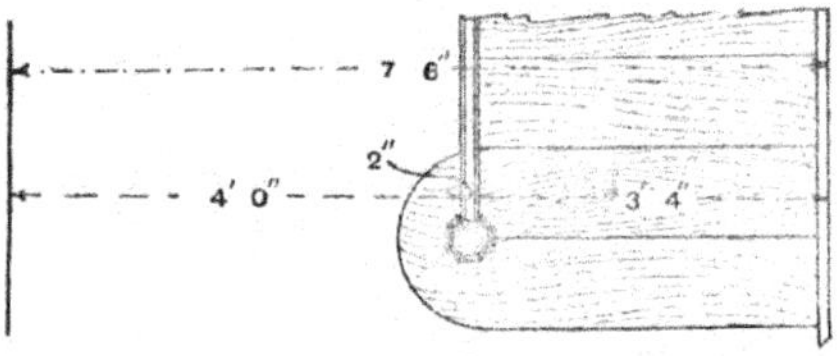

Fig. 1298.

Figs. 1296 to 1298.—Front and Side Elevations and Part Plan of Entrance Hall and Staircase Screen.

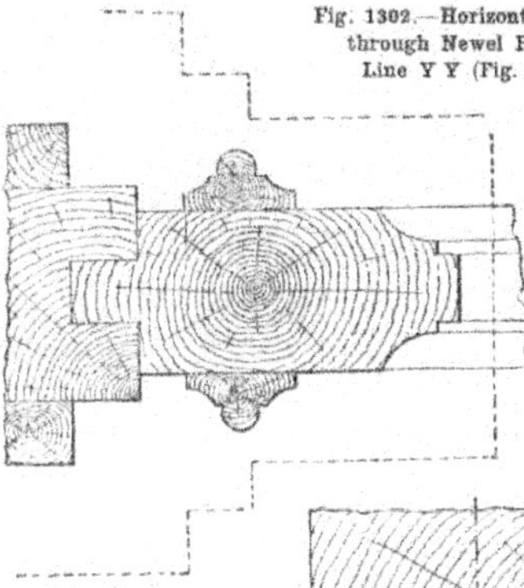

Fig. 1302.—Horizontal Section through Newel Post on Line Y Y (Fig. 1300).

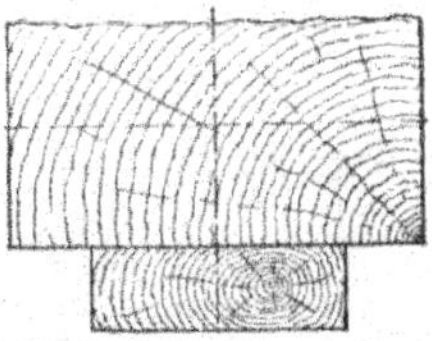

Fig. 1301.—Horizontal Section through Newel Post on Line X X (Fig. 1300).

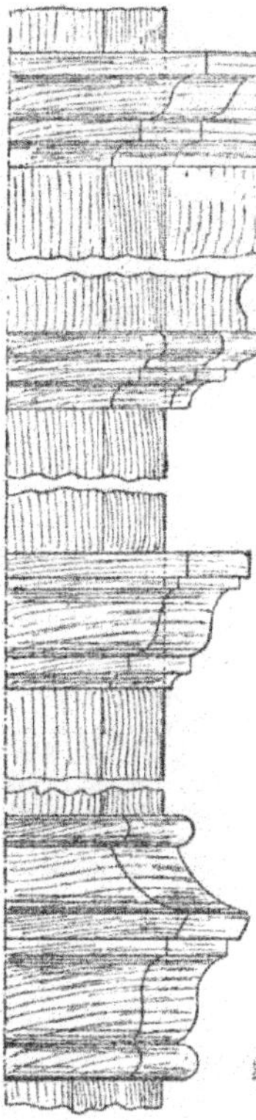

Fig. 1299.—Part Elevation of Newel Post.

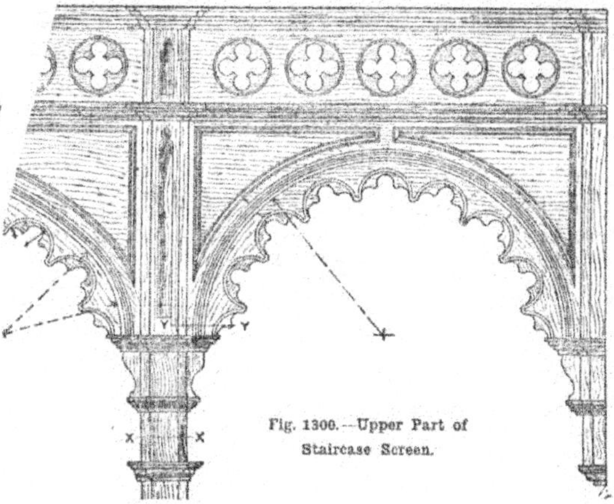

Fig. 1300.—Upper Part of Staircase Screen.

it forms a cap. The two centre jambs have turned terminals, between which a shaped and moulded pediment is fixed. The interior part of the screen is designed slightly differently from the exterior, the upper spandril carving and the dentil course in the moulding along the transom being omitted. The two lines showing below the floor level indicate the cement concrete and mosaic floor.

Fig. 1303.

Fig. 1304.

Figs. 1303 and 1304.—Front Elevation and Plan of Screen between Two Rooms.

Entrance Hall and Staircase Screen.

Front and side elevations of an entrance hall and staircase screen are presented by Figs. 1296 and 1297. The part plan of the hall and staircase given by Fig. 1298 shows the conditions which the screen has to fulfil. An enlarged part elevation of the newel post (A, Fig. 1296) is shown by Fig. 1299. This post helps to support the screen, as shown in the enlarged part elevation (Fig. 1300). Horizontal sections of the post at X X and of the screen and part of post at Y Y are presented by Figs. 1301 and 1302.

Screen between Two Rooms.

The screen about to be described is 10 ft. wide by 9 ft. high. Illustrations of it are presented by Figs. 1303 to 1313. It is ornamental in character, and of an unconventional type. It is divided into three

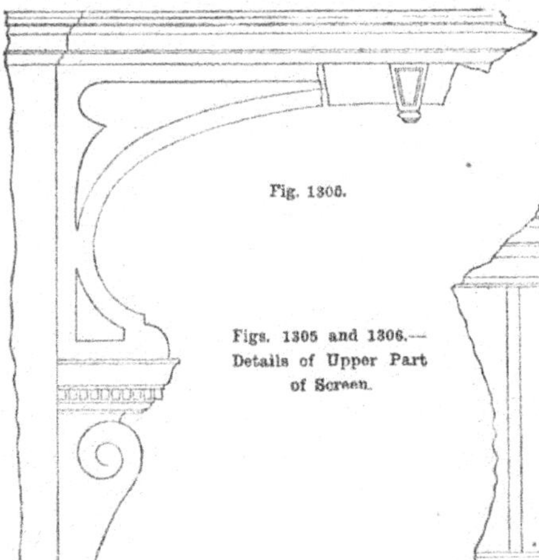

Fig. 1305.

Figs. 1305 and 1306.—Details of Upper Part of Screen.

Fig. 1306.

arches, the centre opening being 4 ft. 6 in. wide, the remainder divided equally. The panelling between the two side arched openings is arranged to carry the same line as the dado or chair rail round the room. If desired, the panelling may be entirely omitted without spoiling the design. The shafts of the column on each side of the central openings are 3 in. square, with a base to each 9 in. deep, boldly moulded at top, with mitered angles; the projection of the base beyond the shaft being 1 in., making the full size 9 in. by 5 in. by 5 in. (see Fig. 1307). In the same figure is shown the cap to the lower shaft, formed in the solid, sunk, and moulded and mitered

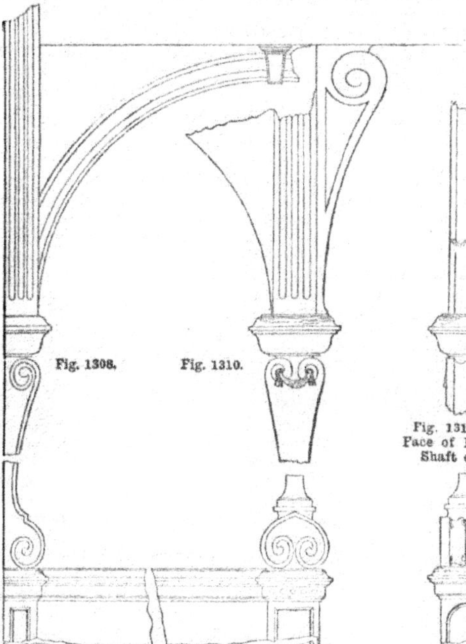

Fig. 1307. Fig. 1309.

Figs. 1307 to 1310.—Elevation of Parts of Screen on Enlarged Scale.

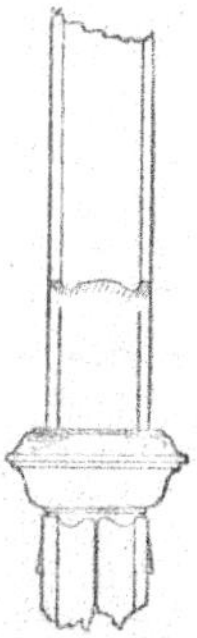

Fig. 1313.—Return Face of Intermediate Shaft of Screen.

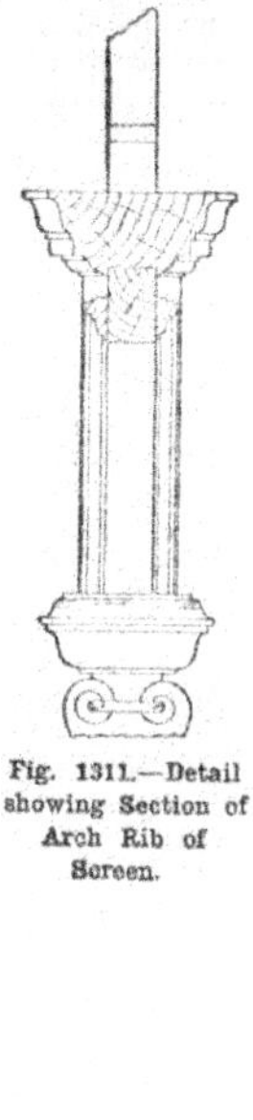

Fig. 1311.—Detail showing Section of Arch Rib of Screen.

Fig. 1312.—Panel on Shaft of Screen.

at angles. The flat member of the moulding is enriched by circular sinkings. The part of the shaft between the cap and the base is sunk and moulded to form a panel on each face; the return face having a panel with a semicircular head (Fig. 1312). That portion of the shaft above the lower cap is part square and fluted, and part shaped in square form with a scroll-shaped and carved base; and immediately below the intermediate cap it is scrolled, and has an ornamental carved swag falling from the volutes (see Fig. 1308). The return face of the intermediate shaft is shown in Fig. 1311 to be sunk moulded, the sinking being carried down and stopped, diminishing to one-third the distance of its length. The intermediate cap previously mentioned is formed by mitering round the shaft a bold moulding as shown. Springing from the latter cap is a shaped truss, moulded on the edge, the faces being sunk and incised. The truss supports a moulded and denticulated cap, which in turn carries the shaped rib forming the centre arch (see Fig. 1315). The outer pillars are constructed similarly to those already described, but on the face are two-thirds of the thickness of the centre ones. The rib forming the semicircular arch at the side opening is 2 in. thick, square on the edge, and with a face moulding planted on each side (see Fig. 1311). A small key block is formed at the crown (Fig. 1311), over which runs a double moulding cut between the square fluted shafts. Each end of the moulding has a projection forming an internal and external mitre, which in turn supports the shaped panel piece with square vertical bars (see Fig. 1316). A bold moulding is fixed under the head lining of the opening, forming a cornice which is broken round the square part of the shafts.

Room Screen.

The screen illustrated by Fig. 1314 is preferably made in mahogany or of the same wood as the furniture in the room where the screen may be used. Fig. 1315 shows a design less elaborate in the fretwork panel, but more varied in the glazing bar. The screen may be two- or three-fold. The screen measures 5 ft. 1 in. from the base to the top of the fret panel, and about 2 ft. 3 in. in width. Prepare two lengths for the stiles, 4 ft. 9 in. by 1¾ in. by 1 in.; two lengths, each 2 ft. 3 in. by 3 in. by 1 in., for the middle and lower rail; and one length, 2 ft. 3 in. by 2 in. by 1 in., for the top rail, which has a groove $\frac{7}{16}$ in. by ½ in. deep. This groove secures the fretwork, which is inserted from the top after framing up. The panel is prepared from ½-in. stuff, reduced to $\frac{7}{16}$ in. Make a full-size drawing of the fretwork, and trace it on the wood. Use a centre-bit for removing the largest spaces, taking care not to cut into the lines, and as soon as the centre of the bit is felt through the wood, withdraw it and finish cutting the hole from the back. This method prevents tearing the edges of the hole. Finish with a keyhole saw, and cut outside the lines, trimming off with a file and glasspaper. Cramp the two stiles together, and square off the lengths of the mortices and their widths with the marking gauge. The rails are stump-tenoned 1¼ in. deep, the tenon on the bottom rail being cut as shown in Fig. 1316. The glazing bars are ⅝ in. by 1 in., and are fitted to the top and middle rails before these are framed to the stiles (see Fig. 1317). In the screen shown by Fig. 1315, the joints of the bars are mitered and glued, with the exception of the large square, which should be mitered at the corner joints B, and stump-tenoned as shown in Fig. 1318. A slip, of the section shown in Fig. 1319, is glued to the stile at A (Fig. 1314), and a similar slip is glued to the bars to keep the glass in position. Glass of a pale greenish-yellow colour, similar to that used in stained-glass windows, has a pleasing effect. Fig. 1320 shows an enlarged view of fretwork panel. The lower panel is made by stretching tapestry with a reversible design. It should be one that harmonises in colour with the upholstery in the room. Cut it so as to bring the pattern even in the frame, allowing enough for turning down and securing with tacks, working from opposite centres alternately, and finishing at the corners. Fix thin wood strips on each side with fine brads as shown in the section, Fig. 1321. Another way is to stretch fine canvas across and paste Japanese wallpaper on both sides.

Two small finials dowelled on the top of the stiles will complete the screen.

Drawing-room Three-fold Fire-screen.

In the three-fold screen shown by Fig. 1322 the panels are of clear, plain glass, bevelled on both sides. The leaves are 5 ft. 6 in., 5 ft., and 4 ft. 6 in. in height respectively, and 1 ft. 8 in. wide. The large leaf is shown much the plainest (see Figs. 1323 and 1324), and is a suggestion for those who may not care to execute the richer designs given on the other leaves in

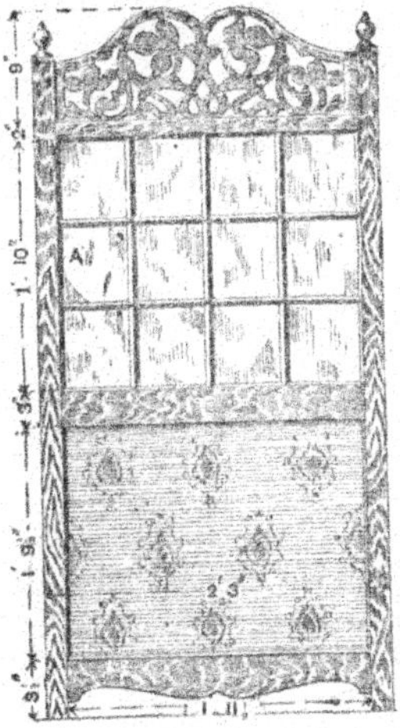

Fig. 1314.—Elevation of Room Screen.

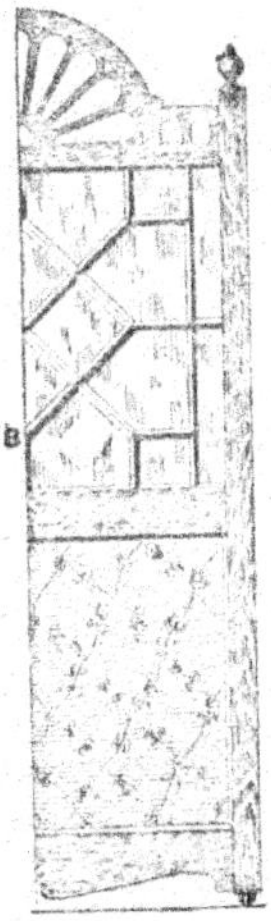

Fig. 1315.—Half Elevation giving Alternative Design for Room Screen.

Fig. 1316.—Joints in Screen Framing.

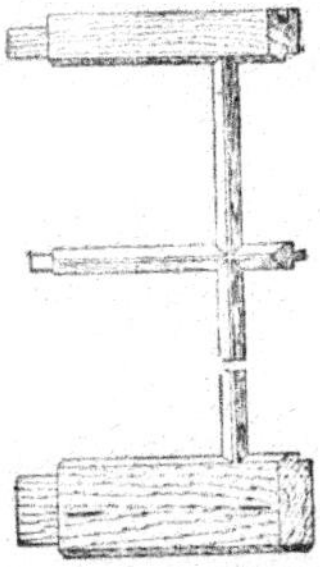

Fig. 1317.—Bars and Top and Middle Rails of Screen.

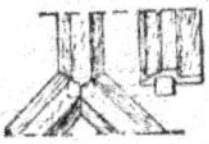

Fig. 1318.—Mitre and Joints for Screen Panels.

Fig. 1319.—Section showing Bar and Slip attached to Stile of Screen.

Fig. 1320.—Fretwork Panel for Screen.

Fig. 1321.—Method of Securing Tapestry to Screen Frame.

Fig. 1322 and in Fig. 1325. Polished walnut or mahogany could be used, or, possibly, enamelled mahogany might by preferred. The centre leaf represents the effect of a painting on the plain oval of the bottom panel. In Fig. 1323 the haunchings and tenons of the stiles are shown by dotted lines. As mentioned, another design for the top rails is given in Fig. 1325. The construction of the bottom part of the framing is indicated in Fig. 1326, which also shows the carving in the bottom rail. The carving in each case can be finished before the frames are glued up, but the bottom should be shaped with a fret saw before the screen is put together, and cleaned up afterwards. Fig. 1327 is a horizontal section through the centre of each leaf. The whole of the framing is prepared from 1-in. stuff, moulded with a small ovolo moulding, rebated, and mortised and tenoned together. The glass is fixed by beads cut and mitered into the rebate of the frames. The stiles are 2½ in. wide, the top rails are 7¾ in. wide, and the bottom rails are 10 in. wide.

Three- or Four-fold Screens.

Figs. 1328 and 1329 illustrate a folding screen with three leaves, each leaf being of the same pattern. If desired, another leaf may be added, making a four-fold screen. Other designs for the leaves are illustrated by Figs. 1330 and 1331. The framework of the screen may be of pine, painted and enamelled white or any light tint. If for use in rooms other than the drawing-room, mahogany or walnut, french-polished, will be more suitable. The four square panels in the upper part of the

Fig. 1322.—Front Elevation of Drawing-room Three-fold Fire-screen.

9

OTTOMAN AND MUSIC CABINET.

screen are of clear glass; the panels below are covered with some light-coloured fabric, such as damask, or wall-paper or Japanese paper. The dimensions of the leaves are: Height to top rail, exclusive of pediment, 6 ft.; width, 1 ft. 10 in.; pediment, 6 in. high and $\frac{3}{4}$ in. thick. The side posts, which are $1\frac{1}{8}$ in. square, are got out of what is termed $1\frac{1}{4}$-in. stuff. The top rail A (Fig. 1328) and the middle rail B are also $1\frac{1}{8}$ in. thick, but 1 in. wide on the face; the bottom rail is 3 in. wide. The ribs C which separate the glass panels are $\frac{3}{4}$ in. wide on the face, and, of course, are made level with the posts and rails. To avoid having to turn the side posts in a lathe, the turned ornaments on the top are fixed separately with a pin turned on the ornament. In putting the parts together, the posts are mortised to receive the tenons on the rails; and the cross rail and upright between the glass panels are halved where they cross each other. The shaped span-rails D are 9 in. high, and are made of very thin wood, or of three thicknesses of veneer glued together, the middle piece being the opposite way of the grain to the outside pieces. The frames are first glued together; then the glass panels are secured in position with beads $\frac{1}{4}$ in. thick, bradded to the

Fig. 1323.

Fig. 1324.

Figs. 1323 and 1324.—Carved Top of Fire-screen.

Fig. 1325.—Alternative Carved Top of Fire-screen.

Fig. 1327.—Part Horizontal Section of Fire-screen.

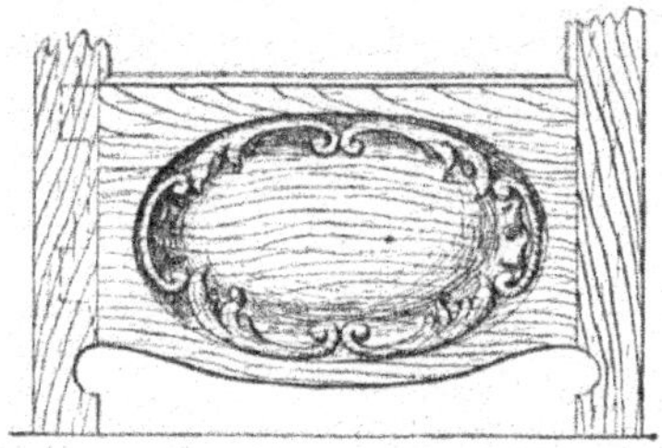

Fig. 1326.—Carved Lower Rail of Fire-screen.

posts and rails. This is clearly shown in the section (Fig. 1332).

Blind-frame, etc.—The lower panels E, to make a foundation for the covering, will require what is called a blind-frame. This is made of ½-in. stuff, the stiles and rails being about 3½ in. or 4 in. wide, and halved together. Calico or print stuff should be first tacked on the frame, and afterwards the outer covering. This frame also is secured with beads, a section of which is shown in Fig. 1333. In securing the span-rails D, a portion of the beads will have to be cut away to receive them. The beads on one side of the screws should be first fixed and mitered at the corners; then the

Fig. 1328.

Fig. 1329.

Figs. 1328 and 1329.—Elevation and Plan of Three-fold Screen.

Fig. 1332.

Fig. 1333.

Figs. 1332 and 1333.—Sections of Screen Framing, showing Beading of Panels.

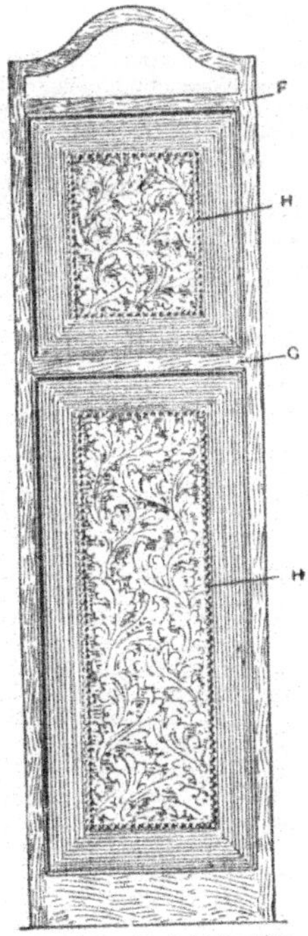

Fig. 1330.—First Alternative Design for Screen.

Fig. 1334.—Half Elevation of Screen Pediment (see Fig. 1328).

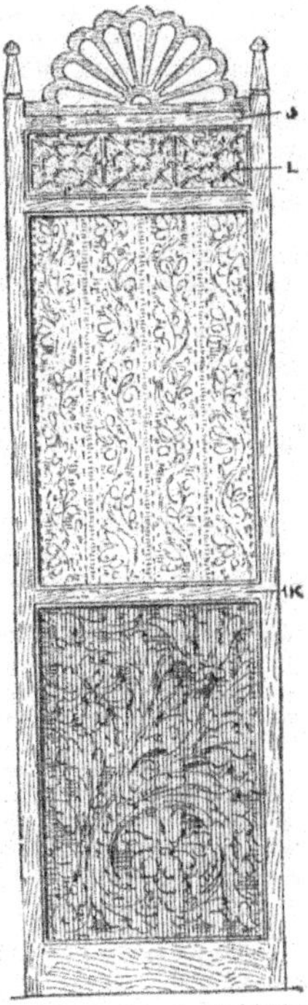

Fig. 1331.—Second Alternative Design for Screen.

Fig. 1335.—Panel for Screen (see Fig. 1331).

Fig. 1336.—Half Elevation of Screen Pediment (see Fig. 1331, above).

panels are put in, and next the beads to secure them. Should the screen be used for hiding any portion of a room, coloured "cathedral" glass may be substituted for the plain glass, or wood panels covered with plain silk or other suitable material may be used. An enlargement of half the pediment is shown in Fig. 1334. To make a clean job, the frame and beads should be enamelled or french-polished before the panels are finally fixed.

Hingeing Screen.—The leaves are joined together with 3-in. brass butt hinges, one being 6 in. from the top, another the same distance from the ground, and the third half-way between the two. The hingeing arrangement is shown in the plan (Fig. 1329). If the screen is to be draught-proof, bands of fabric or leather binding, about 2 in. wide, must be tacked to the edges where hinged; this is best done when the leaves are folded together.

Alternative Patterns of Screens.—A screen with leaves of the kind shown in Fig. 1330 is of the same dimensions as the one already described, with the following exceptions. The height is 6 ft. 3 in. at the sides, and 6 ft. 8 in. at the centre over all; the top rail F is 4 in. lower than the top of the side posts, and the rail G is 4 ft. from the floor. The upper and lower panels will require blind-frames. There are several ways of forming the margin at H. Close brass-head studs or nails may be inserted. Gimp, with a plain material for the borders, and a figured material for the centre of the panels, would look well; or tinted paper for the borders and figured wall-paper for the centre, with narrow gold beading separating the border and filling; or, as an alternative to the beading, a gilded line, about ¼ in. wide, of gold paint or leaf. The leaves are hinged in the same way as in the other screen. The dimensions with leaves as shown by Fig. 1331 are: Height to rail J, 5 ft. 10 in.; rail K, 2 ft. 9 in. from the floor; sight size of panel L (shown enlarged in Fig. 1335), 5 in.; height of pediment, 9 in.; terminals of side posts, 6 in. high. The terminals are shaped on each side as shown in the enlarged half view of the pediment in Fig. 1336. The ornament in the pediment is cut through with a fret-saw. The design of the panel L (see Figs. 1331 and 1335) is cut out of very thin wood or veneer, and glued on the face of the panel; the sunk portions are then roughened by tapping with a semi-blunted point of iron, thus giving a granulated effect. The panels may be covered with material as suggested for those in Fig. 1330. There is a great variety of suitable materials available, such as Lincrusta-Walton, Anaglypta, etc.

Window Screens.

The three window screens illustrated by Figs. 1337, 1338, and 1339 are designed to allow of expansion or contraction to suit the different sizes of windows without spoiling the general effect; while the side portions of Figs. 1337 and 1339 are also adaptable for bay windows. A suitable wood for making is pine, painted and enamelled white or some art shade to match the decoration of the room. If the scheme of decoration is dark, as a dining-room would be, the screen would look well in mahogany or walnut, finished with french polish. Before making the screen, it is advisable to prepare a full-size drawing of half the design. To get the right proportions, construct a scale as follows: Divide the design into as many parts as the desired height of the screen in inches —say 2 ft. 9 in. = 33 parts—calling 12 parts 1 ft. The various details may then be measured from the design, and transferred to the working drawing. The construction of the screen shown at Fig. 1337 will be described first. The framing of this should finish as thick as 1-in. stuff will allow when planed up. The side posts A are 1½ in. wide; bottom rail B, 2½ in. wide; top rail C and middle rail D, which is in three pieces, 1 in. wide; inner posts E, 1 in. wide. The cross rails B and C are tenoned into the side posts. The inner posts E are connected in the same way to the rails B and C, the three cross rails D being stump-tenoned into the uprights. The balusters are ⅝ in. square, and are sunk ¼ in. deep into the rails B and D. As an alternative, the balusters may be of brass cased tubing ½ in. in diameter. To facilitate cleaning, these should be made removable.

To do this, let the lower ends rest in holes bored ¼ in. deep in the bottom rail, and bore holes ½ in. deep in the rail above. Then, if the rods are cut to ½ in. longer than the space between the rails, they can be pushed up into the upper holes and allowed to drop into the lower ones, their removal thus being an easy matter. The shaped cut-through panels are of ½-in. stuff, and grooves ¼ in. deep are made in the posts and rails to receive them. In marking the tenons and mortices to the rails and posts, due regard must be paid to the grooving, as this will take away a part of the tenon (see Fig. 1340). A ¼-in. mortice chisel will be a suitable tool to use. If the screens are made of hardwood, it will be desirable to fix the panels with beading, as the polishing can be done better and more conveniently when the panels are out. A section showing this method of fixing the panels is given at Fig. 1341, while Fig. 1342 shows the rebating of the posts and rails to receive the panels and beads. The pediment is of ¾-in. stuff, screwed or nailed to the top rail; if of hardwood, it should be dowelled.

Alternative Designs.—The construction of Fig. 1338 is similar to that of Fig. 1337. The oblong panel in the centre is made separately. The side posts are 1½ in. wide; bottom rail, 2 in. wide; other rails, 1 in. wide. The stiles and rails of the centre framing are 1¼ in. wide, and are dovetailed together, the inner edges being grooved to receive the ½-in. stuff panel. The frame should be glued together with the panel inserted; and when dry, the grooves on the outer edges, to receive the four shaped panels surrounding it, should be made. The two short centre stiles and side rails are stump-tenoned into the frame. The ends of the pediment F (Fig. 1338) are housed $\frac{3}{16}$ in. deep into the side posts, and the lower edge is connected to the top rail by means of dowels. In the screen shown at Fig. 1339 the posts and rails are mortised and tenoned as described for the first screen. The outer and inner posts are 1¼ in. wide; bottom rail, 2¼ in. wide; rails G and H, 1 in. wide. The shaped balusters J are ⅝ in. thick, the ends being sunk ¼ in. deep. The side cut-through panels are of ½-in. stuff, inserted in grooves in the same way as those previously described. The centre portion of the screen has two small silk or muslin curtains, which may be drawn apart if desired. The supporting rod for the curtains may be of brass cased tubing of ⅜ in. diameter, and should be removable, so that the curtains may be taken off for cleaning. Bore holes in the inner posts to receive the rod, which should be ½ in. longer than the sight measure, making the hole in one side deeper than the other, as described for the brass balusters in Fig. 1337. Rings of ⅝ in. diameter are sewn to the curtain. The two side pediments are ⅝ in. thick, and may be fixed with screws driven from the under side of the top rails. The shaped span-rail over the curtain rail may be fixed in the same way, but it should also be housed $\frac{3}{16}$ in. deep into the inner posts.

Fixing Window Screens.—The screens may be fixed to the window frames with small brass glass-plates, or by means of dowels in the under edge of the bottom rails, and small brass sliding bolts fixed on the face of the top rails.

Carved Fire-screens.

Oak Screen with Cathedral Glass Panel.—The fire-screens about to be described are useful and ornamental. The screen shown in front and end elevations and plan by Figs. 1343, 1344, and 1345 is suitable for the dining-room, library, or study, and is constructed of oak, with a panel of tinted "cathedral" glass. The colour of the glass should be in harmony with the general colour scheme of the room in which the screen is placed. The measurements over all are 2 ft. 4 in. high, 2 ft. 3¾ in. wide, and 8½ in. across the feet. The material required is as follows: Two stiles 2 ft. 4 in. by 2 in. by 1 in.; top rail, 1 ft. 6½ in. by 3½ in. by 1 in.; bottom rail, 1 ft. 6½ in. by 1 in.; two beads, 1 ft. 6 in. by ⅜ in. by ¼ in.; two beads, 1 ft. 4½ in. by ⅜ in. by ¼ in.; and for the feet, two pieces 8½ in. by 2 in. by 1¼ in., and two pieces 6 in. by 2 in. by 1¼ in. The brackets, and the small fret attached to the middle of the bottom rail, are cut from one piece of ⅜-in.

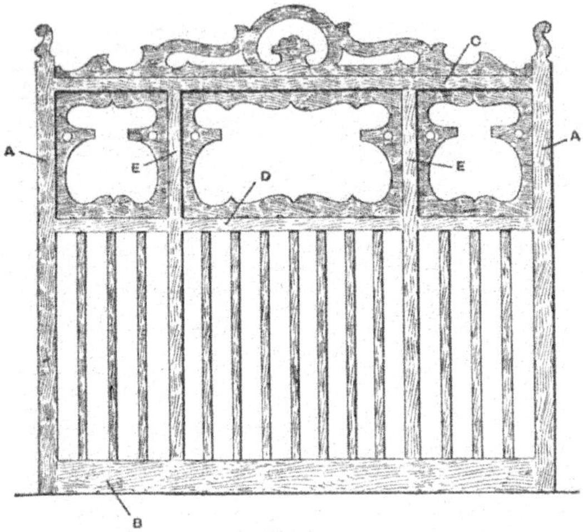

Fig. 1337.

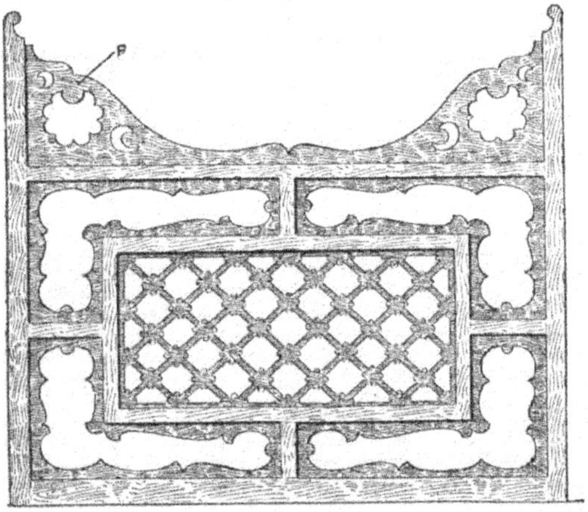

Fig. 1338.

Figs. 1337 and 1338.—Elevations of Window Screens.

stuff. After the timber is planed up square and true, the frame should be set out, and the pattern of the carving traced on. The carving should be done before the mortising; if the mortising were done first, the carver's gouges would be driven through into the mortices. The carving is a full $\frac{1}{8}$ in. in relief, and very simple in character, as shown by Fig. 1346. The frame is stop-rebated at the top and the bottom for the glass. A tenon on the ends of the stiles runs through the feet and is wedged. The feet are halved together as shown in Fig. 1347, which is an underneath view. The brackets and the small fret are cut with a band or bow saw, cleaned up with a spokeshave, files, and glass-paper, and fixed with dowels (see Figs. 1348 and 1349). Care should be exercised, when marking the pattern of the brackets on the board, to ensure that the straight grain will follow the general line of the bracket, as indicated in Fig. 1348. Fig. 1350 is an enlarged section taken at A (Fig. 1343), showing how the

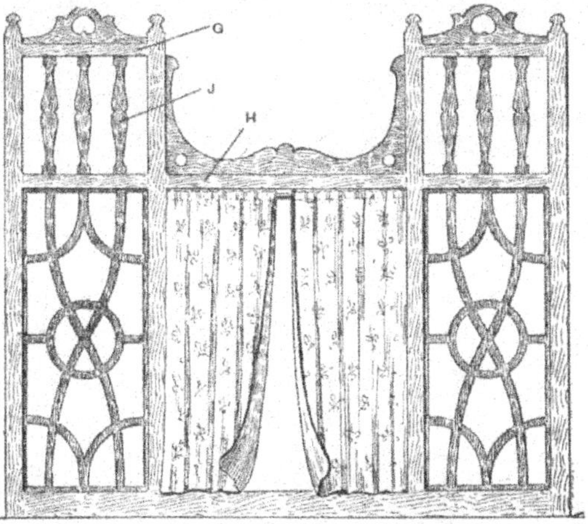

Fig. 1339.—Window Screen with Curtains.

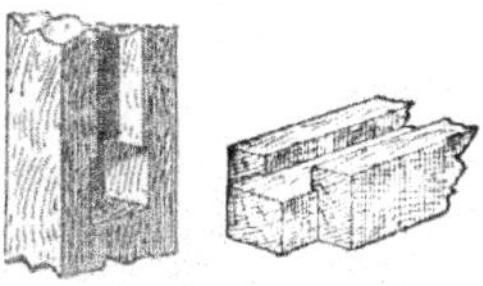

Fig. 1340.—Grooving Screen Rails and Posts for Panels.

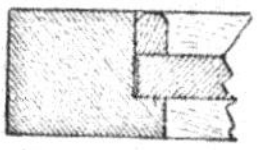

Fig. 1341.—Section showing Screen Panel secured with Beading.

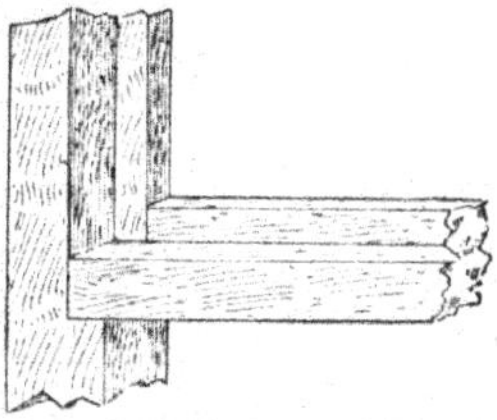

Fig. 1342.—Screen Post and Rail Rebated for Beaded Panels.

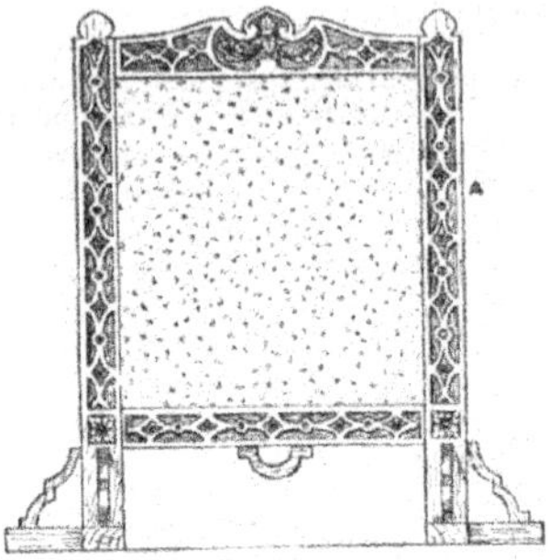

Fig. 1343.

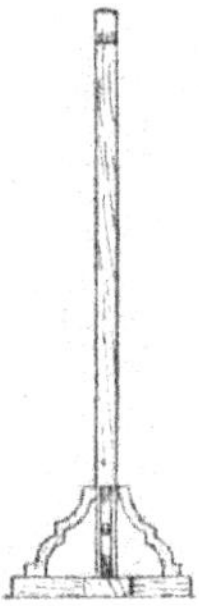

Fig. 1344.

Fig. 1345.

Figs. 1343 to 1345.—Elevations and Plan of Carved Fire-screen with Glass Panel.

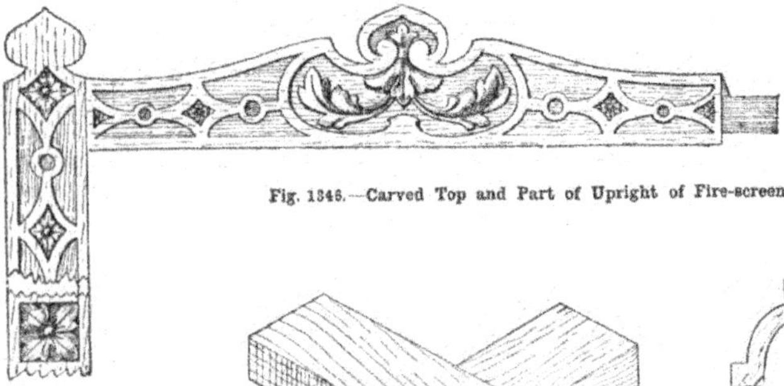

Fig. 1346.—Carved Top and Part of Upright of Fire-screen.

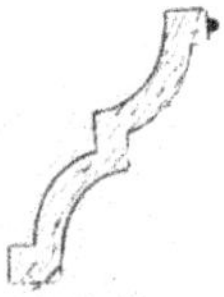

Fig. 1348.—Fire-screen Bracket.

Fig. 1350.—Section of Fire-screen Upright at A (Fig. 1343).

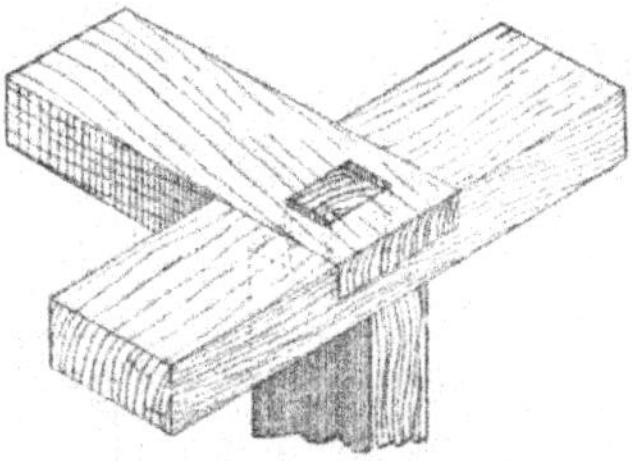

Fig. 1347.—Underneath View of Fire-screen Feet.

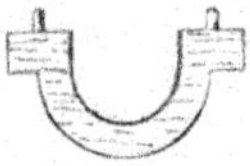

Fig. 1349.—Fret for Fire-screen.

Fig. 1352.

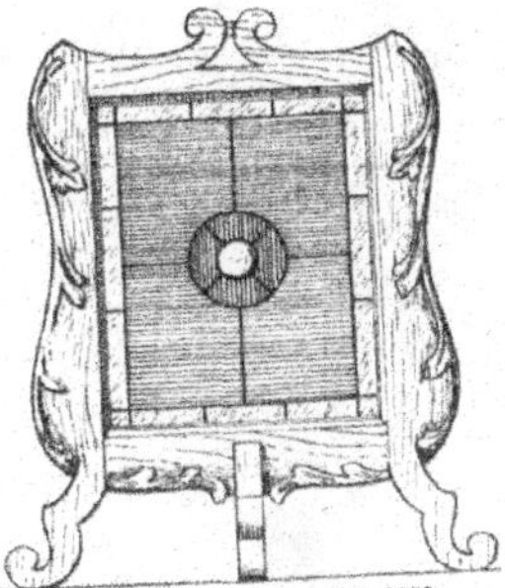

Fig. 1351.

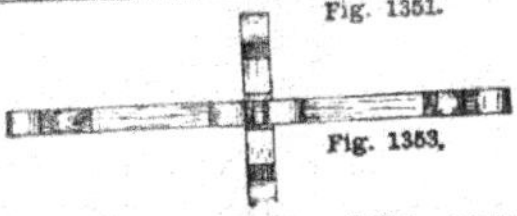

Fig. 1353.

Figs. 1351 to 1353.—Elevations and Plan of Fire-screen with Leaded Glass Panel.

Fig. 1354.—Section of Fire-screen Frame.

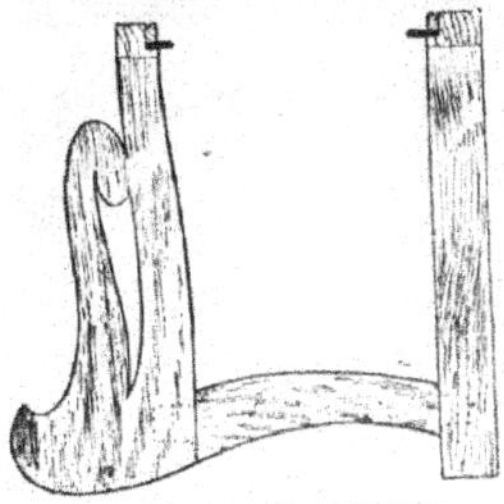

Fig. 1360.—Detail of Folding Fire-screen.

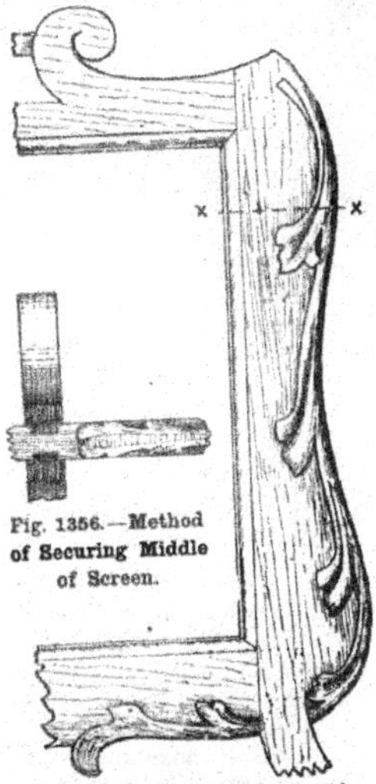

Fig. 1356.—Method of Securing Middle of Screen.

Fig. 1355.—Half Elevation of Fire-screen Frame.

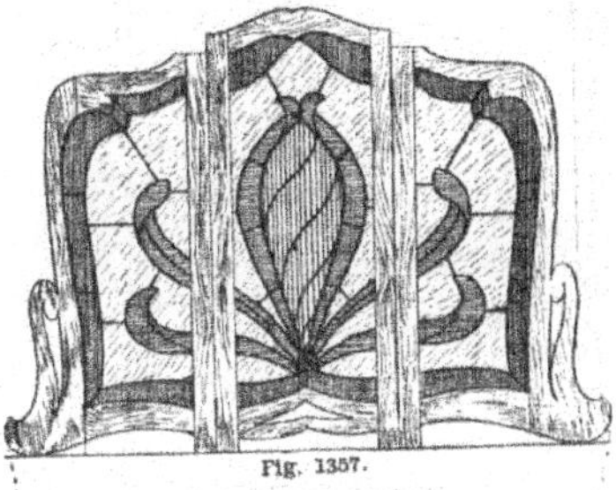

Fig. 1357.

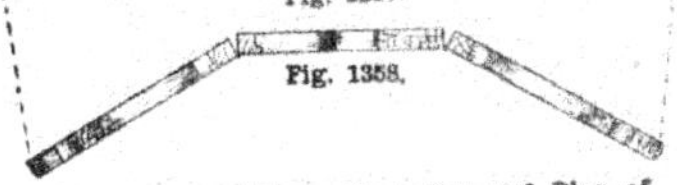

Fig. 1358.

Figs. 1357 and 1358.—Elevation and Plan of Folding Fire-screen.

Fig. 1359.—Method of Cutting Fire-screen Stiles

glass is secured with brads. When the screen is glued up and cleaned, it should be finished with No. 1½ glass-paper, and then fumigated and wax-polished.

Walnut Screen with Leaded Glass Panel.—Figs. 1351, 1352, and 1353 are respectively front and end elevations and plans of a drawing-room screen in polished walnut. It is 2 ft. 4 in. high, 2 ft. wide, and 9½ in. across the middle feet. This screen, with the exception of the beads, is prepared entirely from 1¼-in. walnut. The two stiles, which also form the outside feet, are 2 ft. 2¾ in. by 5½ in.; the top rail is 1 ft. 4½ in. by 4½ in.; and the bottom rail is 1 ft. 4½ in. by 4 in. by 1¼ in. The frame is moulded with a ⅜-in. ovolo, rebated, mitered, mortised and tenoned together as shown in Figs. 1354 and 1355, the latter figure also giving in detail the carved ornament. The cross feet of this screen are in the middle, and are dovetailed into the bottom rail as shown in Fig. 1356. The leaded-glass panel is made up of simple

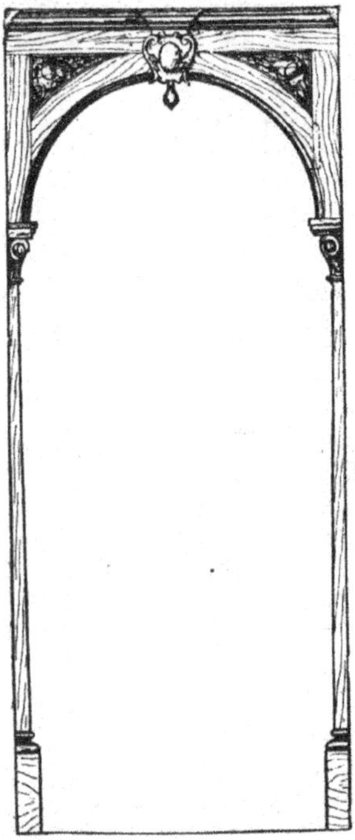

Fig. 1361.

Fig. 1362.

Fig. 1363.—Shield for Carved Arch.

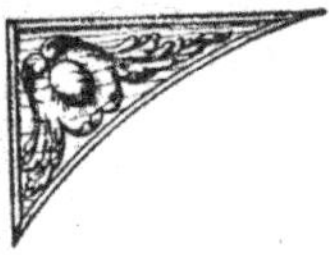

Fig. 1364.—Spandril for Carved Arch.

Figs. 1361 and 1362.—Elevation and Enlarged Vertical Section of Carved Arch for Corridor.

forms of coloured glass. A bulb of glass is suggested for the centre circle.

Mahogany Folding Screen.—Fig. 1357 shows an elevation of a three-fold drawing-

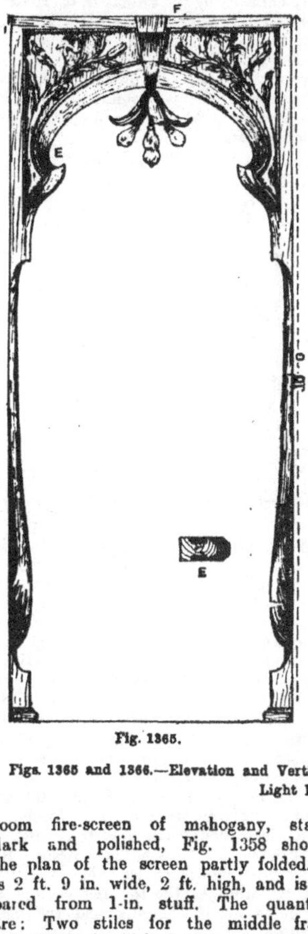

Fig. 1365.

Fig. 1366.

Figs. 1365 and 1366.—Elevation and Vertical Section of Carved Arch with Electric Light Pendant.

Fig. 1367.—Key-block for Carved Arch.

Fig. 1368.—Detail of Upper Part of Arch

room fire-screen of mahogany, stained dark and polished, Fig. 1358 showing the plan of the screen partly folded. It is 2 ft. 9 in. wide, 2 ft. high, and is prepared from 1-in. stuff. The quantities are: Two stiles for the middle frame, 1 ft. 10¾ in. by 1¼ in. by 1 in.; top rail, 9 in. by 2¼ in.; bottom rail, 9 in. by 2¾ in. For the outside frames, two hanging stiles, 1 ft. 10 in. by 1¼ in.; one piece for the shaped stiles of the outside frames (see Fig. 1359), 1 ft. 11 in. by 5 in. by 1 in.; two top rails, 7½ in. by 3 in.; two bottom rails, 7½ in. by 2 in. When the stuff is planed up, the frames should be set out, cut to the required shapes, and cleaned with spokeshave and glass-paper. The rebating should be done if possible on a vertical spindle. The beads for fixing the glass can be easily worked with a scratch tool, and cut to the curves with a fret saw. A template the full size of the glass should be carefully drawn before the leaded lights are

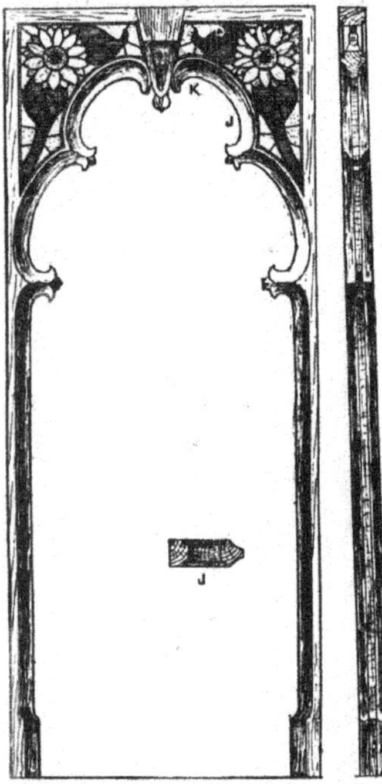

Fig. 1369. Fig. 1370.

Figs. 1369 and 1370.—Elevation and Vertical Section of Arch with Stained Glass Spandrils.

made, to insure the lines of the leaves following on easily and truly from the centre panel to the outside panels. An enlarged section and detail of the construction is given at Fig. 1360. This screen is hung with $1\frac{1}{2}$-in. brass butts.

Ornamental Arches for Bay Windows, Corridors, etc.

The openings of bay windows may be greatly enriched by the formation of arches, which can be designed in such a manner as not to exclude the light. Long narrow corridors may also be beautified with a few pretty arches. The designs about to be described will be found particularly useful and suitable for the decoration of restaurants and lounges.

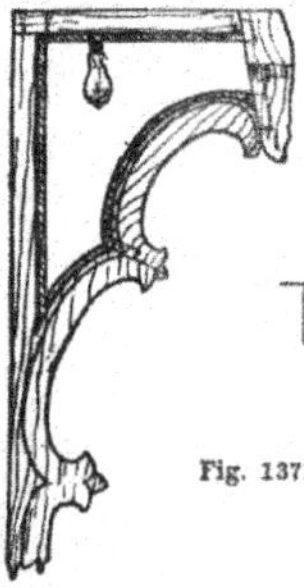

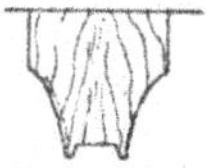

Fig. 1371.—Key-block for Arch.

Fig. 1372.—Detail of Upper Part of Arch.

Carved Arch for Corridor.—Fig 1361 would look well in fumed oak, and a good position for it would be the end of a corridor leading from the hall. It is a semicircular arch, prepared from 4-in. stuff, and is rebated to receive the carved spandrils, and tenoned into the key-block. Illustrations of the shield and spandril are presented by Figs. 1363 and 1364. The lower part of the key-block is covered by the carved shield (see Fig. 1362, which represents an enlarged vertical section taken through the middle of the arch), and the upper part has a cornice mitered round it. The construction is clearly shown in the illustration. The head runs through the key-block. A handsome design for the caps is shown in elevation, together with the stops at the top and bottom of the flutes on the pilaster and the base moulding.

Carved Arch with Electric Light Pendant.—A very pretty arch, with an electric light pendant, is shown by Figs. 1365 and 1366, Fig. 1367 representing a profile view of the key-block and Fig. 1368 a detail showing the construction, while E (Fig. 1365) is a horizontal section. The carving on the spandril represents a lily and foliage; and the pendant, which may be of armour-bright steel, copper, or brass, is in keeping with the carving. Fig. 1366 represents a vertical section taken at F (Fig. 1365) and an elevation of the intrados of the arch, illustrating the manner in which the chamfers are cut to leave a fancy edge.

Arch with Stained Glass Spandrils.—A novel idea for an arch is Fig. 1369, Fig. 1370 being a vertical section at K (Fig. 1369). This arch should be of 4-in. oak. Stained glass is used for the spandrils, and an electric light is fixed between the two glass panels on each side, ensuring a very pretty effect at night when the light is switched on. The position of these lights is shown in the illustration of the framework. A profile view of the key-block is given at Fig. 1371; a detail

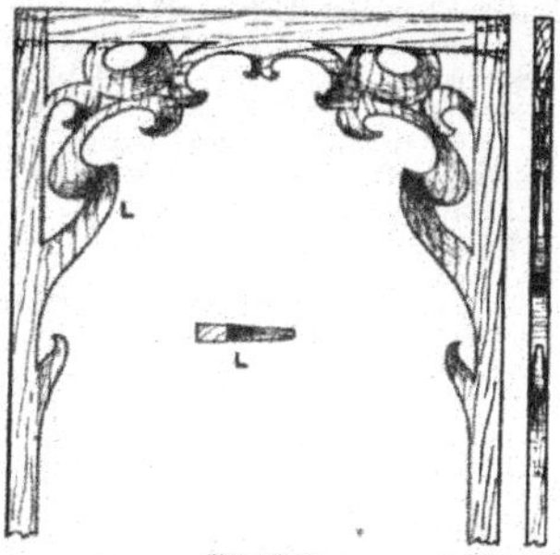

Fig. 1373. Fig. 1374.

Figs. 1373 and 1374.—Elevation and Vertical Section of Lightly Constructed Arch.

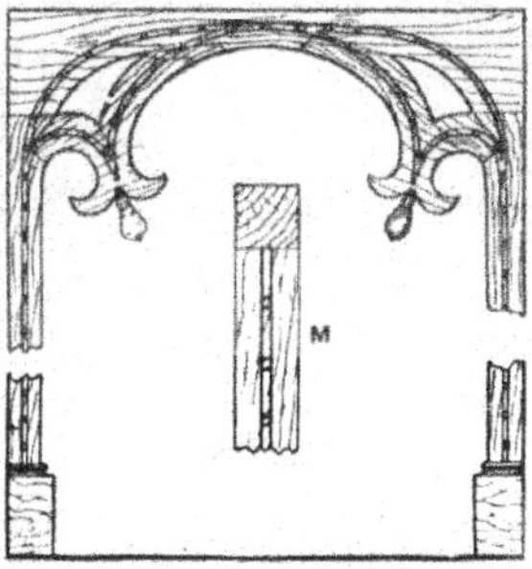

Fig. 1375.—Elevation of Severely Designed Arch.

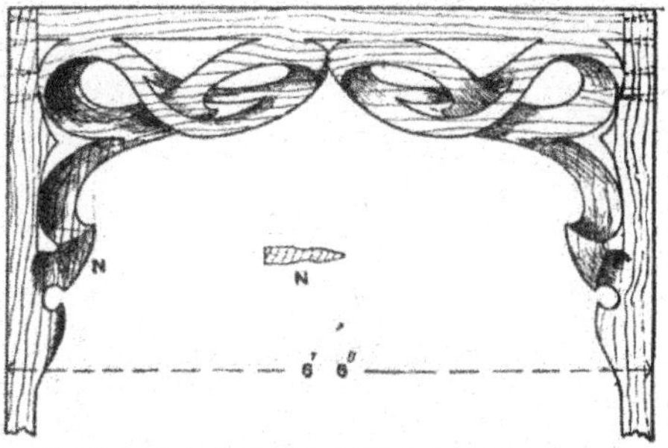

Fig. 1376.—Arch for Opening of Bay in Drawing-room.

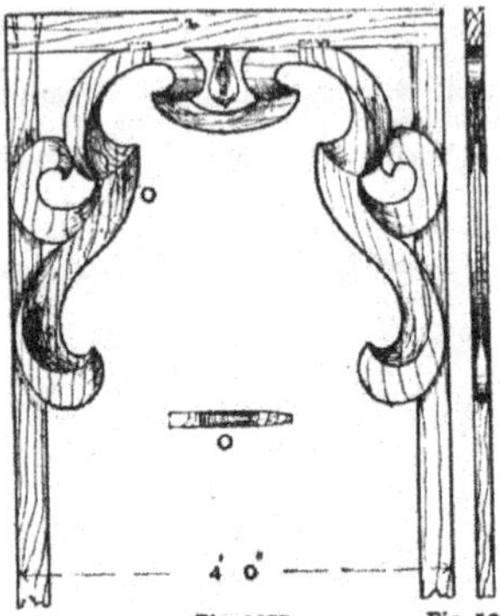

Fig. 1377. Fig. 1378.

Figs. 1377 and 1378.—Elevation and Vertical Section of another Lightly Constructed Arch.

showing the construction at Fig. 1372; Fig. 1369 includes a horizontal section at J.

Lightly Constructed Arch.—Fig. 1373 shows a design for a lighter and less expensive arch, to be prepared from 1½-in. stuff, with a horizontal section at L; Fig. 1374 being a vertical section through the centre of Fig. 1373.

Severely Designed Arches.—For an arch on somewhat severe lines Fig. 1375 will be found a useful design. This should be in 4-in. oak; M is an enlarged vertical section. Fig. 1376 represents a good design for the opening of a bay in a drawing-room, with a horizontal section at N.

Another Lightly Constructed Arch.—Another design for a light arch is reproduced at Fig. 1377, with a horizontal section at O, Fig. 1378 being a vertical section through the centre of Fig. 1377.

Arch for Restaurant or Lounge.—A very rich arch, for a restaurant or lounge, is illustrated at Fig. 1379, with an enlarged illustration of the carving (Fig. 1380), a detail elevation Q, and vertical sections R and S. This should be prepared from 3-in. stuff, which will allow good relief for the carving on both sides.

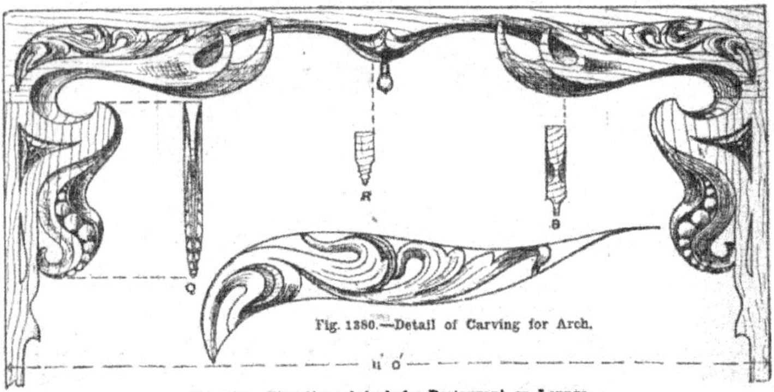

Fig. 1380.—Detail of Carving for Arch.

Fig. 1379.—Elevation of Arch for Restaurant or Lounge.

PLANT, FLOWER, AND LAMP STANDS.

Hardwood Palm Stand.

THE palm stand shown in general view by Fig. 1381, and in front elevation by Fig. 1382, may be made of almost any kind of hardwood. The legs are 1¼ in. square at the top, tapering to ¾ in. square at the bottom. The top is ⅝ in. thick, and should be moulded round the edge to the section shown in Fig. 1383. The shelf also is ⅝ in. thick, and should have a hollow worked round the edge, as shown at Fig. 1384. The corners of the shelf should be cut out on the slant, so as to fit round the legs.

Fretted Rails.—To set out the fretted rails under the top (see Fig. 1385), proceed as follows: Draw the centre line A B, and across this draw the lines C and D, with a space of 5 in. between them. Mark the portions of the legs as indicated by the dotted lines, and then, by drawing lines parallel to the insides of the legs and ¼ in. away, the sloping ends of the rail are obtained. From the point where D cuts A B, measure ½ in. up, this being the rise of the arc which forms the bottom edge of the rail. Join this point to E by a straight line, and bisect it, and continue the line obtained till it cuts A B prolonged; the point of intersection is the centre for the curve of the arc. Draw another arc ¾ in. above and parallel to this one. Next, from the line C, and on A B, mark a point 3½ in. down, and with this point as centre, and a radius of 2¾ in., draw the semicircle. Bisect the radius on line A B, and draw the horizontal line F; then, with centres G G, and a radius of 5⅛ in., draw the two arcs, and round off the corners as shown. The rail below the shelf (see Fig. 1386) can be set out in a similar way.

Top, Shelf, etc.—The top of the stand is secured to the legs by means of four short dowels, one in each leg (see section, Fig. 1387). Two housings, 5 in. long by ½ in. wide by ¼ in. deep, are cut on the tops of the legs to take the ends of the rails. The top, legs, and rails should be well glued together and blocked in the angles underneath as shown. A good method of securing the shelf and rails to the legs is shown at Figs. 1388 and 1389. The small blocks H are cut to the shape shown, and screwed to the under side of the shelf. The leg is notched out directly under the shelf, as shown in Fig. 1389. The back of the block should be cut a trifle short, to ensure the whole being brought up tight together when the screws are inserted in the legs. The shelf and rails should also be glued and blocked underneath.

Flower Stand with Shelf and Splayed Legs.

The flower stand illustrated by Fig. 1390 is suitable for a bay window, the height from the floor to the top of the stand being 3 ft. 3 in. Any of the hardwoods, thoroughly seasoned and free from defects, may be selected in preference to deal, pine, or whitewood. The legs splay outwards diagonally, so a full-size template must be made from thin stuff for marking them out economically on the plank. The original size will be 2½ in. square in section when cut by the band saw; but as the facets will stand the wrong way for the rails, the angles must be worked down to form facets going parallel to the sides of the stand. This will be more readily understood by referring to the dotted lines in Fig. 1391. The finished size should be 1⅝ in. from the top down to the lower shelf, where they are gradually rounded and tapered, widening out again at the feet. On reference to A in the section (Fig. 1392) it will be seen that a rail supports the lower shelf, there being another on the other side also; and in the fronts, facing the window and room, the shaped brackets take the place of the rails. Both the top and the lower side rails are stub-tenoned to the legs in the usual way; but the top front rails are dovetailed to the

sufficiently to clear the shelf and drop in its proper groove; by this method the top rails may all be mortised and tenoned as shown in Fig. 1394. The top is of stuff 1 in. thick, framed and rebated to receive tiles (see Figs. 1395 and 1396). A fillet runs across to support the tiles in the centre joint. The tiles are fixed with glazier's putty, and the rails are pinned to the legs with brass wire, the holes being stopped with either

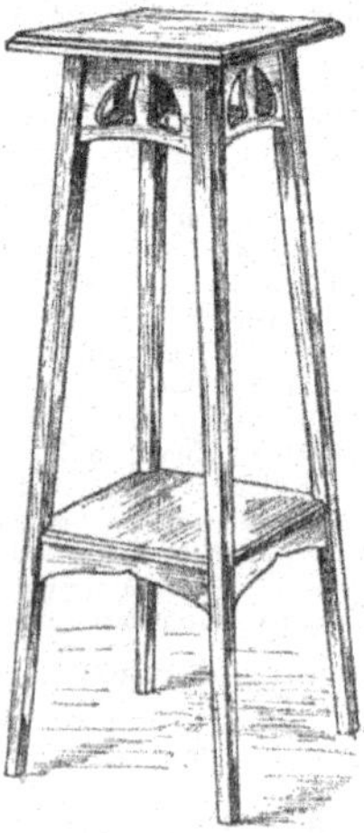

Fig. 1381.—**Hardwood Palm Stand.**

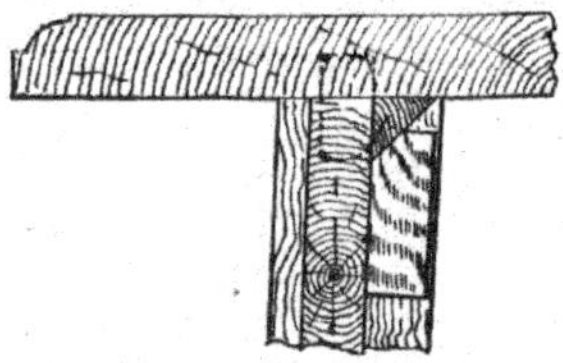

Fig. 1383.—Section through Top and Rail of Palm Stand.

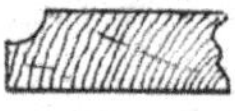

Fig. 1384.—Section of Palm Stand Shelf.

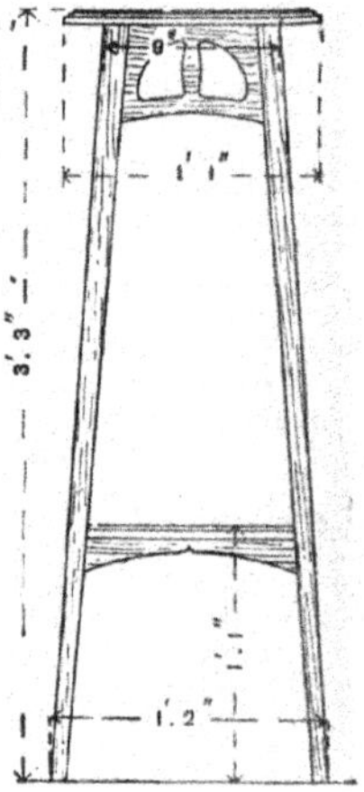

Fig. 1382.—Elevation of Hardwood Palm Stand.

legs as shown in Figs. 1391 and 1393, this method allowing a fretted panel to be placed in its groove in the shelf while the rail above is dropped to its position. The side shelves are continuous, or in one piece, screwed to the rails A (Fig. 1392), and grooved to fit over the shaped brackets, which are in turn haunched and tenoned to the legs; the mortices should alternate or break joint with the mortices on the opposite side of the leg. The fretted panels are housed about $\frac{1}{8}$ in. to the shelf and $\frac{1}{4}$ in. to the rails, or the rails may be grooved about $\frac{1}{2}$ in. deep and the panel pushed up in the groove

coloured hard stopping or wood-dust and glue. All screws should be secretly driven. The method of fixing the top is shown in Fig. 1392.

Lamp Pedestal or Fern Stand.

The stand illustrated by Fig. 1397 may be utilised as a lamp pedestal or as a fern stand; and if the top were made to hinge, the gallery could be silk-lined to form a lady's work-table. Of course, in the latter case, the height should not exceed 2 ft. 8 in., while the width may be conveniently increased by 3 in. or 4 in. For either of the first-named articles, the height should be from 3 ft. 3 in. to 3 ft. 6 in., and the legs 1½ in. square at the top, tapering to 1 in. square at the bottom. The legs are 7 in. apart where the shoulders of the top rails abut, and 1 ft. 2 in. apart at the base. Set out half-size on a board, and get the bevel for the shoulders of the rails and mortices. The mortices for the middle rails should be from ⅝-in. stuff. The legs may be slackened back off the tenons while the shelves are being fitted. Square off the dowel holes for the turned balusters in the shelf and top rails; remove, and bore holes in the shelf and rails; also counterbore and bore holes for screws at the angle shown in Fig. 1399 for securing the top, which is 1 ft. square by ¾ in. thick. These holes

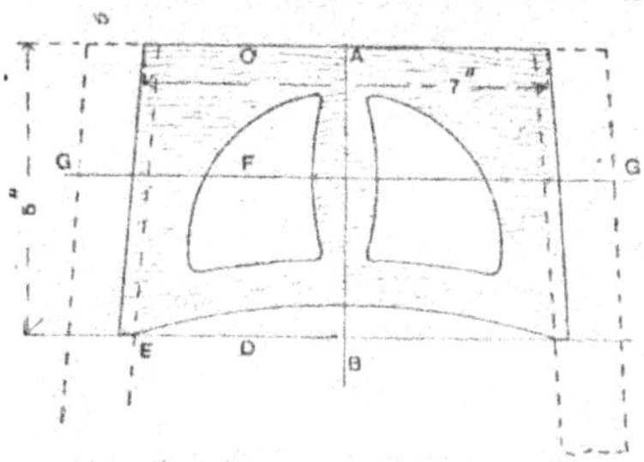

Fig. 1385.—Upper Rail of Hardwood Palm Stand.

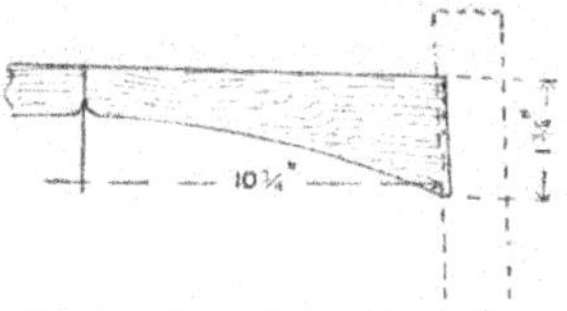

Fig. 1386.—Lower Rail of Palm Stand.

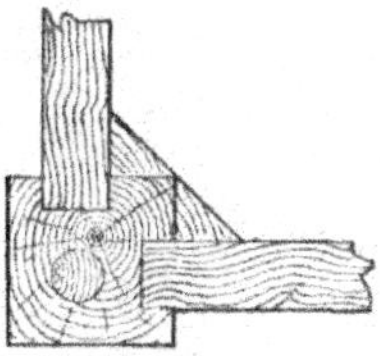
Fig. 1387.—Plan of Corner of Palm Stand, showing Dowel.

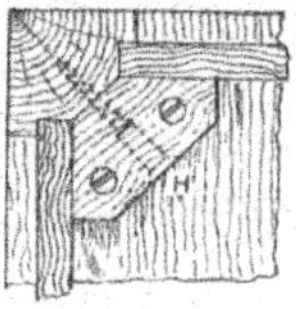

Fig. 1388.—Section through Palm Stand Leg and Rails under Shelf.

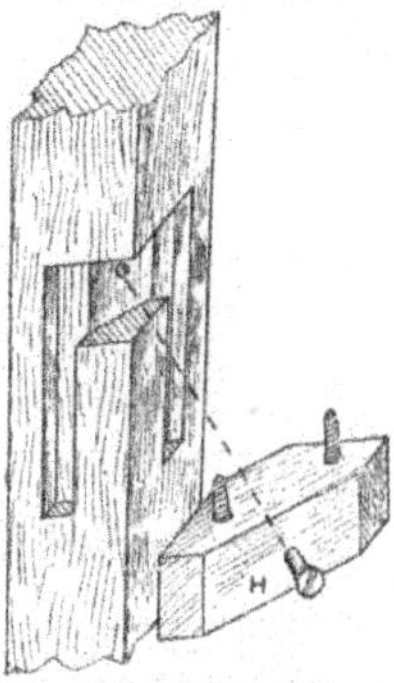

Fig. 1389.—Housing in Palm Stand Leg for Shelf Rails.

set off 10 in. from the top, and the mortices for the lower rails 1 ft. 6 in. from the bottom line. The rails are 1½ in. by 1⅛ in. wide. Cut the tenons and mortices and try together, making adjustments as necessary to the shoulders to obtain a fair bearing. The top rails are dropped in, being housed ¼ in. at the top edge of the rail to nothing at the lower edge on account of the splay of the legs (see Fig. 1398). The mortices are open-ended. The dimensions can now be obtained for the shelves, which are made should afterwards be stopped. To give a finish to the stand, beaded or reeded edges may be worked round the shelves and rails with a hand-beading tool (see Figs. 1400 to 1404). The work may be polished in parts, but care should be taken when jointing-up to wipe off the surplus glue before it has time to set properly. All the rails may be pinned from the inside, as shown in Fig. 1394 (p. 402). The turned balusters are tapered at their dowel ends, and are fitted in the second shelf first; the top rails

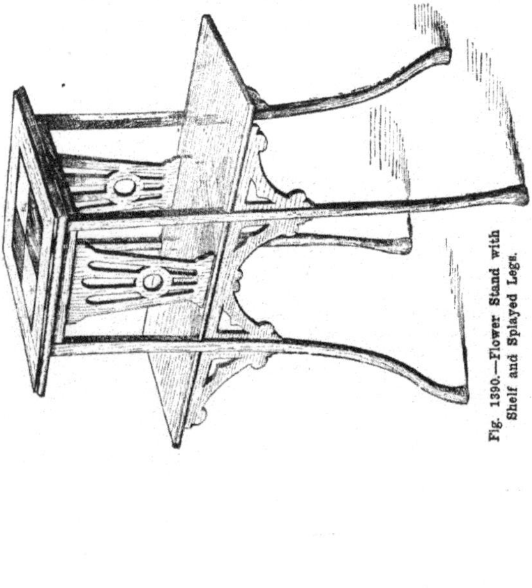

Fig. 1390.—Flower Stand with Shelf and Splayed Legs.

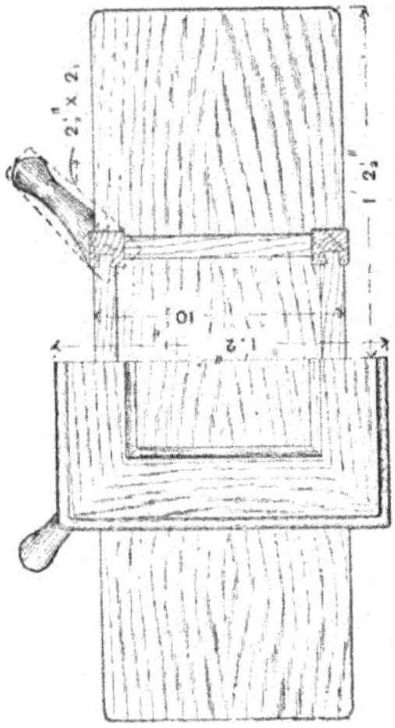

Fig. 1391.—Part Plan and Horizontal Section of Flower Stand.

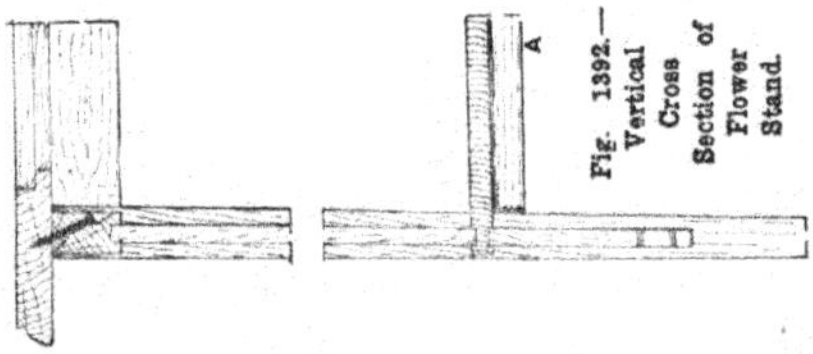

Fig. 1392.—Vertical Cross Section of Flower Stand.

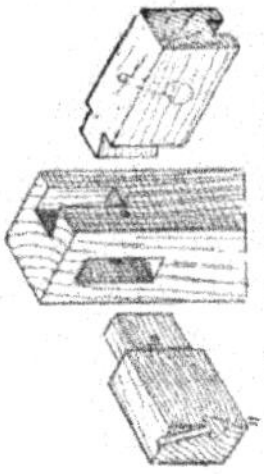

Fig. 1393.—Top Rails jointed to Leg of Flower Stand.

Fig. 1394.—Joint of Flower Stand Leg and Rails.

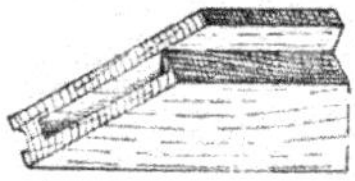

Fig. 1395.

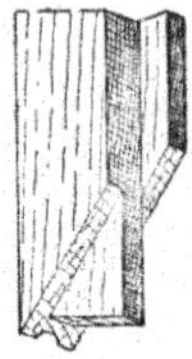

Fig. 1396.

Figs. 1395 and 1396.—Mitered and Rebated Frame of Flower Stand Top.

are then placed in position over them, and pinned on tenons, and finally the top is secured by screws driven up through the rails from outside as in Fig. 1399.

Fig. 1397.—Lamp Pedestal or Fern Stand.

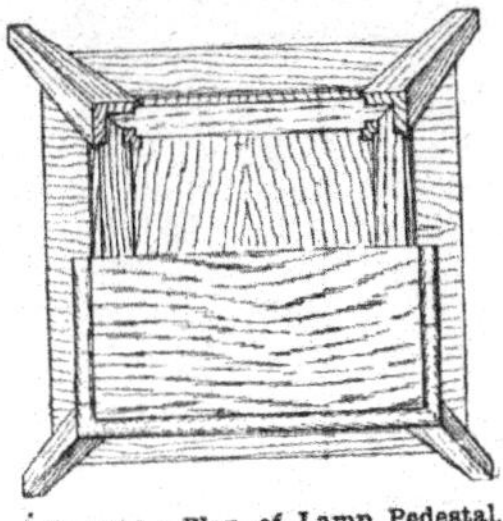

Fig. 1398.—Plan of Lamp Pedestal.

Fig. 1401.

Fig. 1402.

Fig. 1403.

Fig. 1404.

Figs. 1401 to 1404.—Mouldings for Lamp Pedestal Shelves, etc.

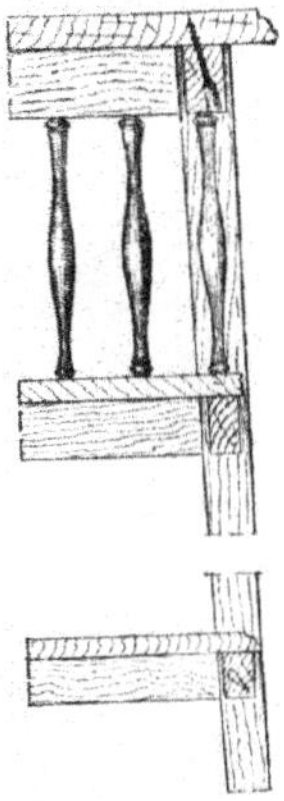

Fig. 1399.—Part Vertical Section of Lamp Pedestal.

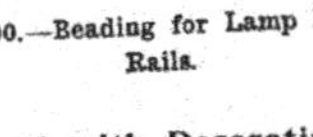

Fig. 1400.—Beading for Lamp Pedestal Rails.

Plant Stand with Decorative Panels.

The plant stand illustrated at Fig. 1405 may be made of almost any hardwood. The panel spaces lend themselves to a variety of treatment, such as carving, repoussé work, fretwork, or inlaying; or ornamental tiles may be used. A full-sized section should be prepared before starting the work, the leading dimensions being given in the illustrations. Fig. 1406 is an enlarged part vertical section through the centre. The posts are $1\frac{1}{4}$ in. square, and are mortised and stop-rebated on the insides (see Fig. 1407) to receive the rails and panels. The rails should have the edges rebated, the tenons cut on the ends, and be fitted to the posts, before being shaped. The two

Fig. 1405.—Plant Stand with Decorative Panels.

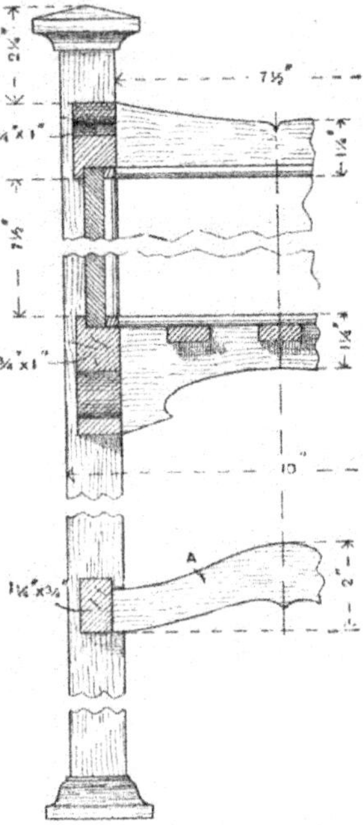

Fig. 1406.—Vertical Section of Plant Stand.

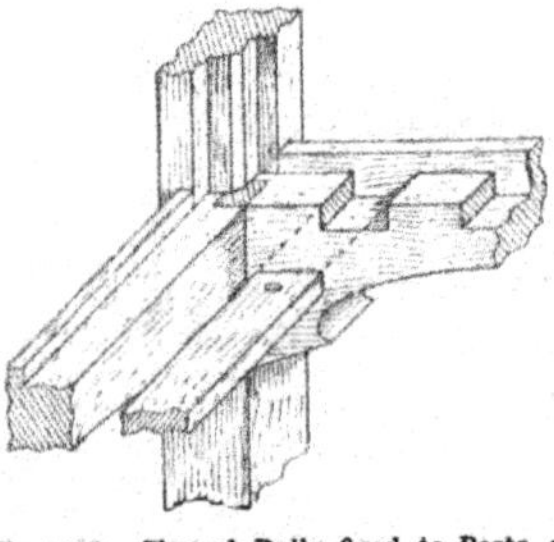
Fig. 1407.—Shaped Rails fixed to Posts of Plant Stand.

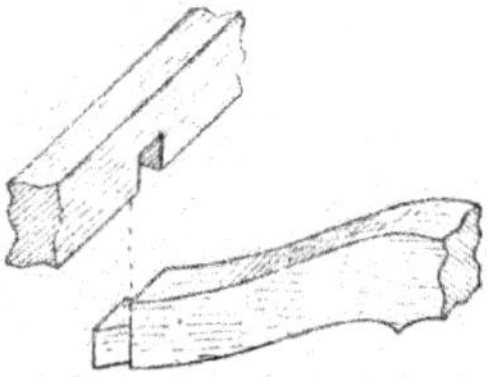
Fig. 1408.—Joint of Plant Stand's Lower Rails.

Fig 1409.—Alternative Shape for Top of Plant Stand Post.

Fig. 1410.—Carved Panel for Plant Stand.

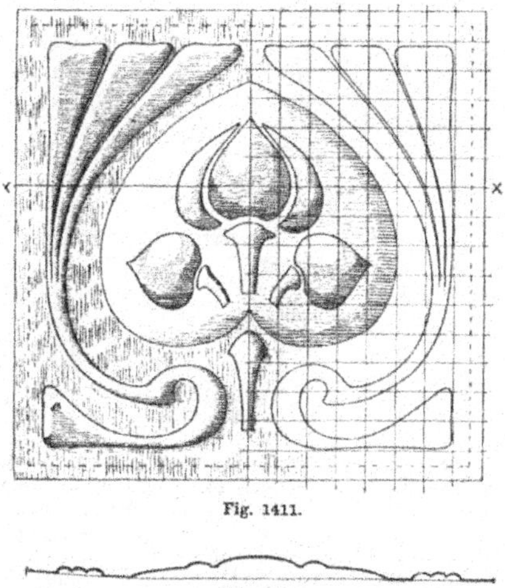

Fig. 1411.

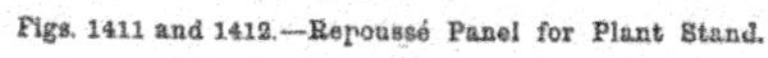

Fig. 1412.

Figs. 1411 and 1412.—Repoussé Panel for Plant Stand.

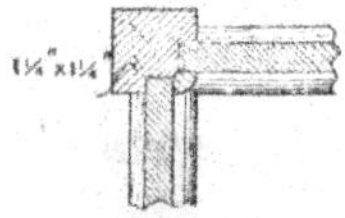

Fig. 1413.—Horizontal Section through Plant Stand's Post and Wooden Panels.

Fig. 1414.—Horizontal Section through Plant Stand's Post and Wooden Panels (Alternative Design).

lower rails are tenoned into the posts (see Fig. 1406), the top edges being 10 in. from the ground. The shaped cross rail A is connected to the lower rails by lap-dovetailing it into the under side of the rails, as shown at Fig. 1408. For supporting the tray on which the flower vase or pot is to stand, three laths, 9 in. by 1 in. by $\frac{3}{8}$ in., are notched into two of the rails, and screwed at each end (see Figs. 1406 and 1407). The moulded blocks for the feet are $2\frac{1}{2}$ in. square by 1 in. thick, and should be fixed with $\frac{3}{8}$-in. birch dowels about 2 in. long, the dowel holes being bored right through the feet. The capping pieces are obtained from blocks of the same size, and should be secured with dowels let into the blocks and posts about $\frac{3}{8}$ in. and 1 in. respectively. An alternative method of treating the tops of the posts is shown at Fig. 1409. If done in this way, the posts will have to be cut 1 in. longer. Fig. 1410 is a daffodil design for a carved wood panel. It will be advisable to have some of the flowers, or a good photograph of them, close at hand while carving. The lines ruled across the outlined half of the design are to be $\frac{1}{2}$ in. apart; this also applies to Fig. 1411. The

Fig. 1415.—Palm Stand or Jardinière with Shaped Legs.

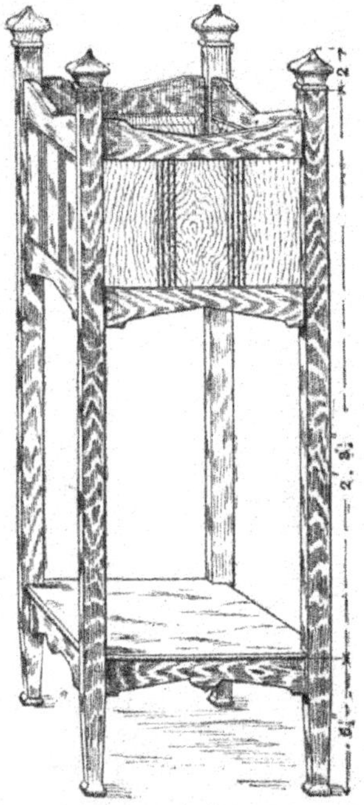

Fig. 1416.—Palm Stand or Jardinière with Plain Legs.

carving should be done in fairly bold relief, the ground being, say, $\frac{5}{16}$ in. deep. The flowers and bulbs should be the most prominent parts, and the leaves should be kept fairly flat. These panels should be secured with small beading (see Figs. 1406 and 1413). The horizontal beads are mitered at the angles, and are fixed before the angle beads, which may require scribing a little at the ends to make them fit nicely. Fig. 1411 is an alternative design for a repoussé panel. The section of the raised portions should follow the course of an ellipse rather than the arc of a circle. The

Fig. 1417.—Templates for Palm Stand Legs.

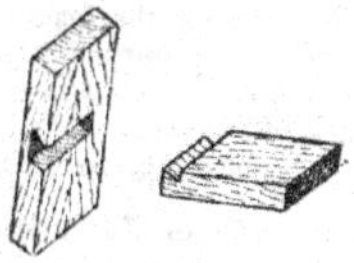

Fig. 1420.—Joint for Inner Bottom Rails of Palm Stand.

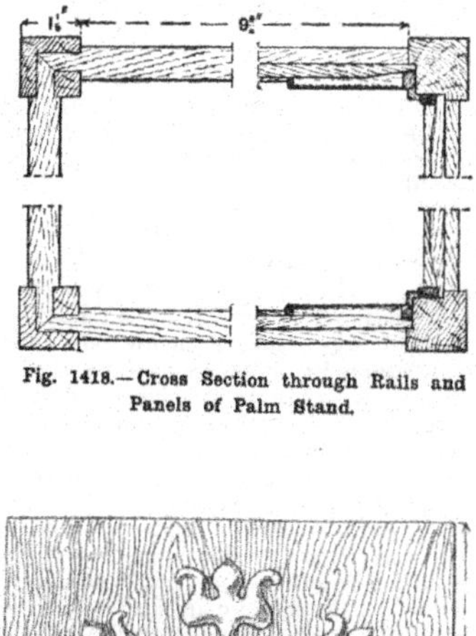

Fig. 1418.—Cross Section through Rails and Panels of Palm Stand.

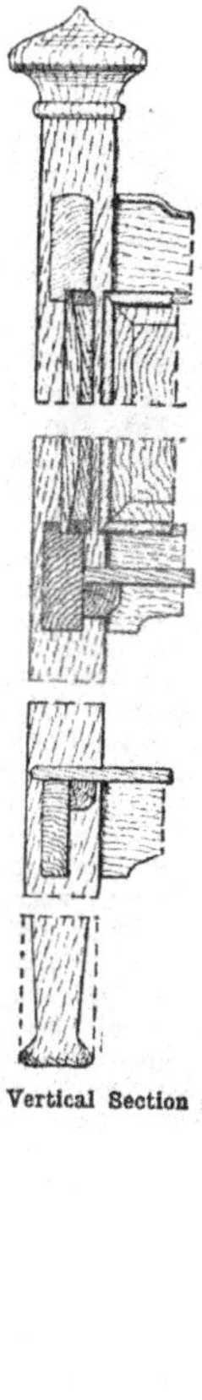

Fig. 1419.—Part Vertical Section of Palm Stand.

Fig. 1421.—Panel of Palm Stand.

copper plate should be backed up with a ¼-in. wood panel (see Fig. 1414), secured with beading in the same way as the carved wood panels. Fig. 1412 shows a section through the repoussé panel on the line x x (Fig. 1411). The metal tray or pan should be of such a size as to enable it to be conveniently withdrawn at any time. It should be provided on the inside with two drop rings for lifting purposes, and afterwards painted with two coats of enamel. The woodwork of the stand may be finished by staining, or polishing, as desired.

Palm Stands or Jardinières.

Figs. 1415 and 1416 show perspective views of palm stands constructed of hardwood. The stand shown by Fig. 1415 has ornamental panels which may be carved in wood, hammered in repoussé, or ornamented with gesso. The four panels are not necessarily alike, and the secondary panels in Fig. 1415 have heart-shaped ornament in low relief. In Fig. 1416, carved ornament is dispensed with, the panels being centre-beaded, but a lower shelf and curved rails are introduced and relieve the squareness. Well-seasoned American whitewood, or satin walnut, may be used, both being unusually free from knots and easy to work; also they take stain and polish readily. For Fig. 1415, prepare the material as follows: Four legs, 3 ft. long, 1⅜ in. square in the rough; the 11½-in. plain part is 1⅛ in. square; the thick part of the cabriole is 1¼-in. square section. Draw the curve for the legs on cardboard, and cut out with a sharp knife, thus making a plus-and-minus template (Fig. 1417). Trim the wood up square and set off the vertical distances, then fix the lower template (Fig. 1417) to the legs with a couple of drawing pins, and mark the curves on the four sides of each leg with a pencil, carefully working away the surplus wood. While finishing, try the upper template on the curves, this greatly ass.sting in getting them all alike. Where the bottom rails are mortised, the legs are 1 in. square, tapering to ⅞ in. and spreading to 1¼ in. for the feet. The finials are dowelled on, and are 1½ in. square in the largest part. Cut the mortices in the legs to receive stump-tenons from the rails; the tenons should only fit hand-tight, as driving is liable to split the legs; rebate the rails and legs on the inside to receive the panels, which are secured with bradded slips (see Figs. 1418 and 1419). The shelf is prepared from board 11 in. by ½ in., and rests on fillets secured to the rails (see Fig. 1419). Fig. 1420 shows the joint for the inner bottom rails. These are half-dovetailed to the outer rails, and stopped ⅛ in. from the top edge; use the glue thin and very hot when jointing up. In Fig. 1416 the legs are parallel except below the lower shelf, where they taper to $\frac{15}{16}$ in. square and swell out again for the feet. The finials are cut from separate pieces of wood, 2 in. high and 1¾ in. square, and are dowelled to the legs. The upper and middle rails are ¾ in. thick by 2 in. deep, and the bottom rails supporting the shelf are ½ in. thick by 2 in. deep, the top and bottom shelves being prepared from 11-in. by ½-in. and 12-in. by ½-in. stuff respectively. Fig. 1421 represents a panel design.

Columnar Palm Stand.

The palm stand shown in front elevation and sectional plan by Figs. 1422 and 1423 is of somewhat novel design, and may be made of any suitable hardwood. The top measures 10½ in. square over the mouldings, and is formed by attaching mouldings A (see Figs. 1424 and 1425), mitered at the corners, to a piece B, which is about ½ in. thick. Immediately below B, four moulded pieces C, mitered at the corners, are attached, and in order to secure a lighter appearance are cut away, as at D (Fig. 1422), to a curve which balances the contour of the moulding. Screwed to the pieces C are four columns E (Fig. 1424), fitted close up to which are angle blocks F, as shown in the inverted plan (Fig. 1425). The parts G (Fig. 1424) are short pieces of mitered moulding. The columns, which are 2 ft. 2 in. long and of 1½-in. square section at the largest part, are attached to the base as shown in Fig. 1426. The square pieces H have tongues along each edge, fitting in grooves in the mitered mouldings J. At each corner the pieces H are cut away (see Fig. 1427), and screws may be used to connect the columns firmly

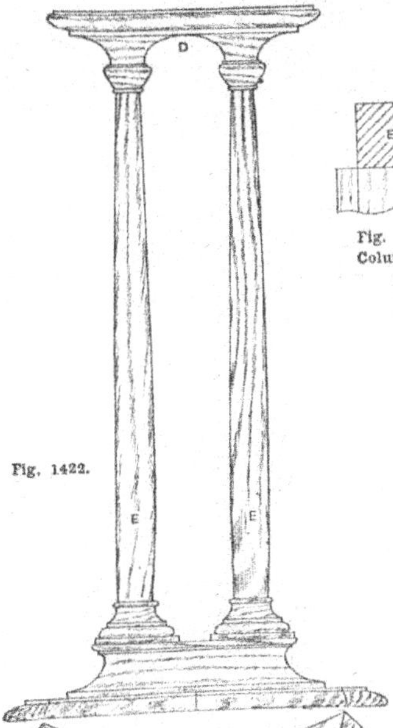

Fig. 1422.

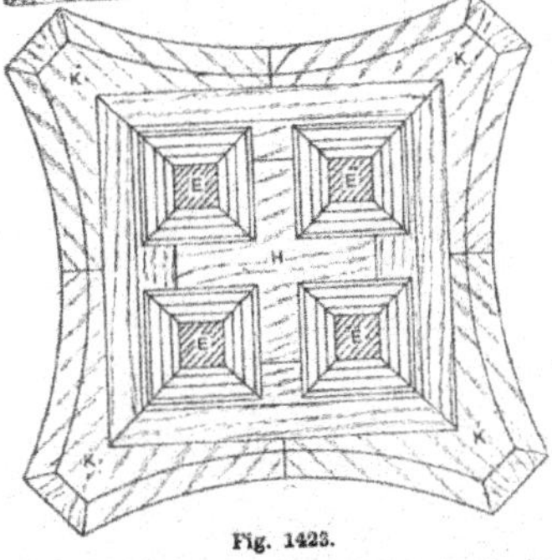

Fig. 1423.

Figs. 1422 and 1423.—Elevation and Horizontal Section of Columnar Palm Stand.

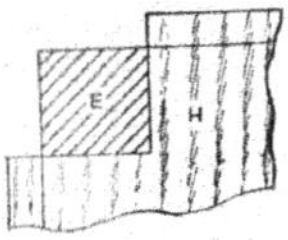

Fig. 1427.—Fixing Column of Stand to Base.

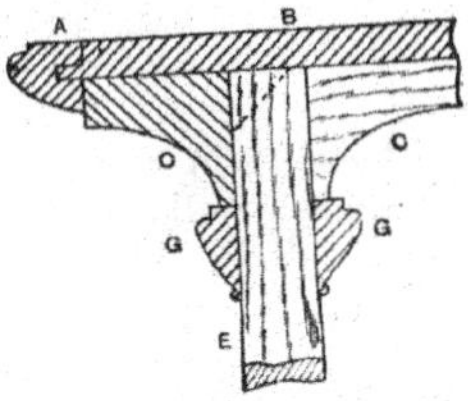

Fig. 1424.—Vertical Section of Top of Columnar Stand.

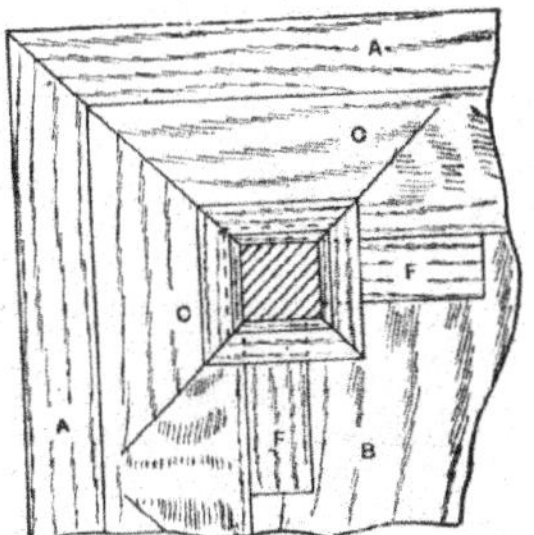

Fig. 1425.—Underneath View of Top of Columnar Stand.

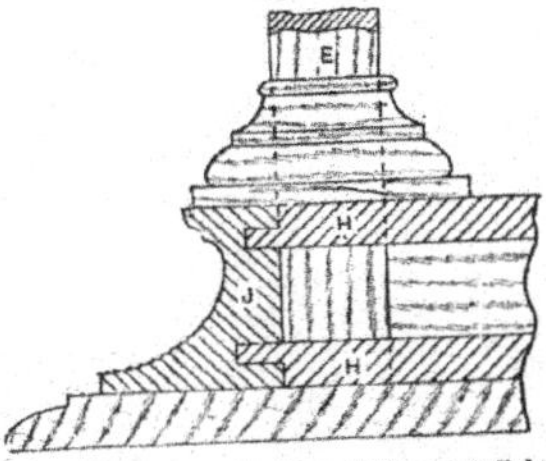

Fig. 1426.—Vertical Section of Base of Columnar Stand.

to these pieces. The four curved base pieces K (Fig. 1423) are fastened to the parts above with screws.

Table Plant Stand.

Two plant stands for the table are shown by Figs. 1428 and 1429. The stand shown in Fig. 1428 may be made in any wood to accord with the furniture of the room. If it is to be polished, use oak, mahogany, or walnut; if stained and varnished, pitch-pine; if enamelled, white or red deal, or kauri wood. The material should be thoroughly dry, as the article when finished

Fig. 1428.—Front Elevation of Table Plant Stand.

Fig. 1432.—Alternative Ornament for Top of Table Plant Stand.

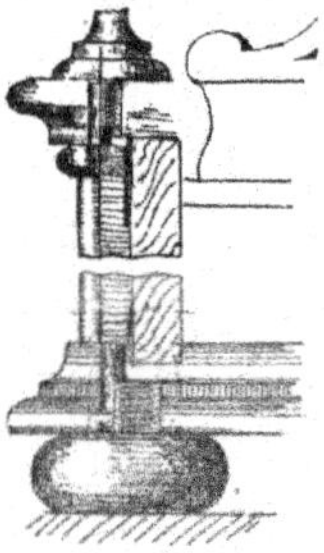

Fig. 1431.—Vertical Section through Corner of Table Stand.

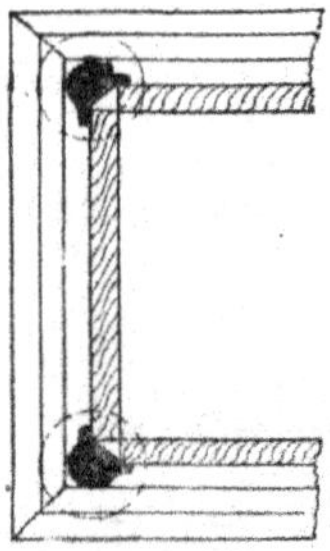

Fig. 1430.—Part Horizontal Section of Table Plant Stand (see Fig. 1428).

Fig. 1429.—Front Elevation of More Elaborate Table Plant Stand.

is to be used in a room which, of course, is generally warm. The base should be a flat piece of wood $\frac{3}{4}$ in. thick and about $7\frac{1}{2}$ in. square; on it should be planted an ogee moulding, carefully mitered at the corners. The corner pillars may be semicircular, as shown by the section (Fig. 1430), or semi-hexagonal if preferred, and should be dowelled into the base. Ball feet (Fig. 1431) would also fix these pillars. The top should be made from wood $1\frac{1}{8}$ in. or $1\frac{1}{4}$ in. wide and $\frac{3}{4}$ in. thick, moulded round the edge, and mitered at the corners. The corner knobs may have either a pin or a screw, which, passing through this top moulding, is fastened to the pillars. The 6-in. square tiles should be slid into place before the upper moulding is fixed together. The ornament on top of the upper moulding may be used or left off, according to the taste of the maker. It is cut out in fretwork and screwed to the top moulding with fine screws. Fig. 1432 gives an alternative pattern for this ornament. Fig. 1429 shows the elevation of a plant stand slightly more ornamental in design than the one before described. The base should be of thoroughly seasoned wood, cut out to the shape shown at Fig. 1433. It is to be $7\frac{1}{2}$ in. or 8 in. square, having the extended corners 1 in. by 1 in. in addition. The moulded pillars at the corners into which the tiles fit should be of the section shown at Fig. 1434, and should be screwed up from the bottom of the base. A groove $\frac{1}{8}$ in. deep should be cut in the base, as shown at Fig. 1435, to take the tiles. The top must be of the same shape and size as the base but $1\frac{1}{8}$ in. wide, and is grooved on the under side to take the 6-in. square tiles. The corner pillars may be dowelled into the top. The ornamental pillars at the extended corners are turned, and are fixed by pins or screws as shown at Fig. 1435. A small, plain moulding should be fitted on the under side of the upper moulding, and on the upper side of the base round the tiles, and, if it is wished, ornamental fretwork scrolls may be fixed to this stand with good effect. The tiles fitted in these stands should be quiet in pattern and colour; or really natural paintings of flowers would look well.

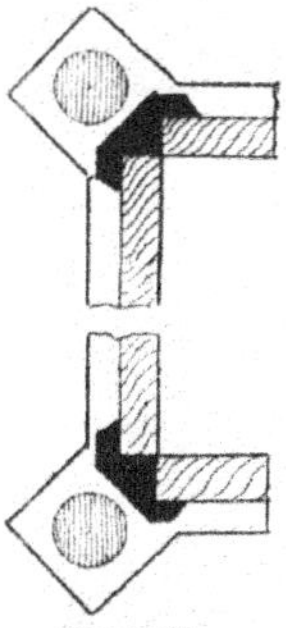

Fig. 1433.

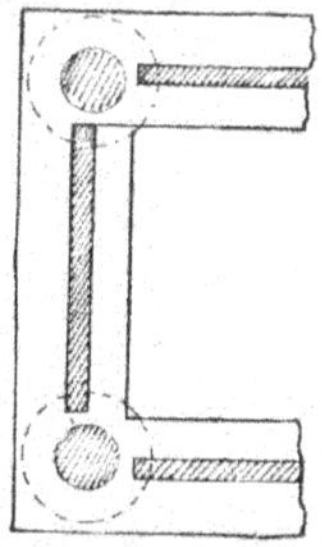

Fig. 1434.

Figs. 1433 and 1434.—Part Horizontal Sections of Table Plant Stand (see Fig. 1429).

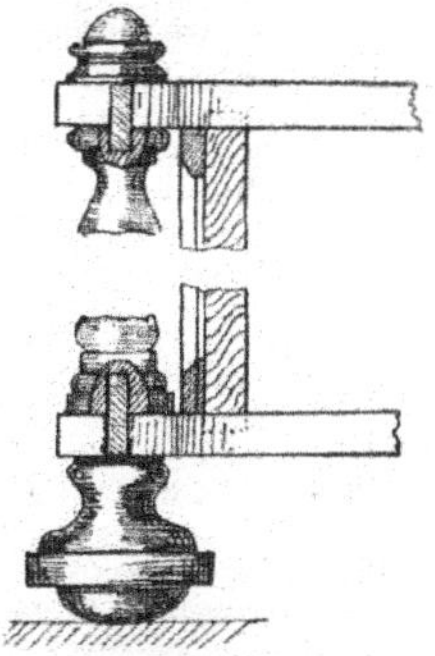

Fig. 1435.—Details of Corner of Plant Stand.

JEWEL BOXES.

Jewel Casket with Combination Lock.

THE casket illustrated by Fig. 1436 (p. 414) is intended for holding jewellery, private papers or correspondence, and personal nicknacks. Fig. 1436 shows a side elevation; Fig. 1437 part plan; Fig. 1438 an end elevation; Fig. 1439 part section through side of casket; Figs. 1440 and 1441 reduced vertical and horizontal sections; Fig. 1442 shows an enlarged detail of spoon moulding A (Fig. 1438); Fig. 1443 enlarged section of moulding B (Fig. 1438); Fig. 1444 lock plate; and Figs. 1445 and 1446 plan and section of lock in position. The casket is made of mahogany overlaid with fretted oak, and ordinary carved oak picture-frame moulding, combined with gold spoon picture moulding, is used in its ornamentation. It is fitted with a five-dial number lock of simple construction. As the casket cannot be opened until all the indicator hands are pointing to the correct figures, it is impossible for anyone not knowing the combination to open it, especially if it is well fitted without shake or play.

The Box.—The box measures 1 ft. 8 in. by 11¾ in. by 11½ in. high over all, the body, cover, and bottom being of ½-in. mahogany. The body is 1 ft. 5¾ in. long by 9¼ in. wide by 9 in. deep, and should be dovetailed together for greater strength, blind dovetails not being necessary, as the box ends will be covered by the fretwork overlay. The cover and bottom each measure 1 ft. 8 in. by 11¾ in., and must be well rounded on all edges, as shown in the elevation views and in the section of the box side. It is best to fix the bottom in position with fine screws from the inside, the heads being let into the sides flush. A piece of mahogany $\frac{5}{16}$ in. thick is required to fit nicely inside the box; its grain should cross that of the cover, to which it should be glued and screwed centrally each way, thus stiffening the cover and enabling the lock to catch the box front some little way down from the top edge.

Fretwork.—Some oak fretwood, of $\frac{3}{16}$-in. or ¼-in. thickness, will next be required for the fretted side, end, and cover pieces. The side and end pieces should be prepared the full depth of the box sides, and should be fitted in place with mitre joints before the pattern is marked out and cut. Cut out the cover piece 1 ft. 7⅜ in. by 11⅛ in., and well round the edges on one side only, as shown in the elevations and in the section of the side. Next mark out the fretwork designs on them all, these being enlarged from the drawing by an ordinary pantograph. Only half of the top is shown in the plan (Fig. 1437), the other half being the same with the exception of the centre where the lock-face comes. In the other half this space is filled in by another ornamental circle, making five on that side. The end pieces have a ⅜-in. space allowed all round the plates of the brass drop-handles; this should be marked out to suit the pattern of handles to be used; these, of course, should be bought beforehand, and may be of brass or gun-metal.

Mouldings.—After fret-cutting all the designs, the pieces should be secured in position with joiner's fine brads, punched in, and the holes filled in with putty; but screws should be used in the parts to be covered later by the mouldings. When they are all fixed, some ⅝-in. gold spoon moulding should have the back part forming the rebate cut away as indicated by the dotted line across the enlarged sectional view (Fig. 1442). After being mitered at the corners, this should be fixed to the top and bottom edges of the box all round, the top mouldings flush with the upper edges of the box, and the lower pieces in the angle formed by the junction of the bottom and sides. A ½-in. gold spoon moulding should also have a rebate cut away in a similar manner, and should be mitered and fixed on the outside of the cover, $\frac{3}{16}$ in. back from the edges of the fretted top, as shown in the plan and elevations. A slip of mahogany ¼ in. thick and ½ in. wide, having its outside edge well rounded, should next be mitered and fixed on the inside edges of the gold spoon moulding on the sides to protect it, as shown in the section of the side and in the elevations. Some 1¼-in. carved oak moulding, of any suitable pattern, is now required, and the backing of this is also cut away as shown by the dotted line in the enlarged sectional view of the oak moulding (see Fig. 1443). If the remaining carved front is now over ⅜ in. thick, it should be planed down to this thickness, care being taken not to injure the pattern. Then mitre-joint the ends, and brad in position next to the mahogany slips on the sides, and all round inside the spoon moulding on the cover. These mouldings, if nicely fitted, will give a handsome appearance to the jewel casket at very slight expense.

Cover Hinges.—The cover should next be hinged, butt-hinges, 1⅜ in. wide, being used; these will be of sufficient width to reach across the spoon moulding and the top edge of the box back, thus adding to the stability.

Combination Lock.—The five-dial lock (Figs. 1444 to 1446) is of simple design, and, if well fitted, very effective in working. It is a job for a metalworker, who can easily work from the following instructions. The main bolt is 5⅞ in. long by 3½ in. wide by ¼ in. thick. A mahogany pattern of it should be made slightly larger, to allow for contraction and filing; ⅛ in. longer and a good $\frac{1}{16}$ in. wider will be sufficient. It should be cast in good tough gun-metal, and after being cleaned should be finished all over with a fine file. The five brass rollers may also be cast and turned, or may be parted off a solid bar in the lathe. They should be 1¼ in. in diameter, ¼ in. thick, with $\frac{3}{16}$-in. centre holes. The dials and hands should be cut from stout sheet brass. The dials should be turned 1¼ in. diameter, with $\frac{3}{16}$-in. centre holes; and the hands should be ¾ in. long, with $\frac{3}{16}$-in. centre holes drilled in them. All these holes should be a tight fit for a piece of $\frac{3}{16}$-in. diameter brass wire. A few inches of brass tubing, to form bushes for the central pins, will be necessary. The wire should fit the tube exactly. If this size tubing cannot be conveniently obtained, the bushes may be bored and turned from a small brass rod in the lathe. Next file out the slots in the rollers to take the bolt lugs. They should be ¼ in. deep, and accurately fitted. When ready, the bolt and rollers should be assembled in their correct positions on the inside of the cover. The two outside holding lugs of the bolt are placed just level with the outside front edge of the cover stiffening piece, as shown in the view of the lock (Fig. 1446), the roller lugs being in position for opening the box—that is, slid into the roller slots. When all is ready, file off one end of the brass wire quite square, dip it into some colouring matter, and carefully mark off the centre holes of the rollers by passing down the wire and turning round in the holes. This done, mark the centre of each exactly, and carefully bore holes through the cover to fit the bush tubing tight, the holes to be bored through exactly square. Next plug these holes outside the cover with hardwood, and, with a centre-bit to fit them, let in the dial plates flush with the mahogany surface, leaving the oak fret cover with its 1¼-in. holes for dials to protect the brass hands when fixed. The dials should have the figures engraved or stamped on them with small steel stamps. Fix the dials in their recesses by hammering them home, after coating the backs with some shellac cement; then cut suitable lengths off the bush tubing, and drive them from inside the cover to meet the dial plates, taking care not to shift the plates, and using the same kind of cement as before. Next cut the centre pins to length,

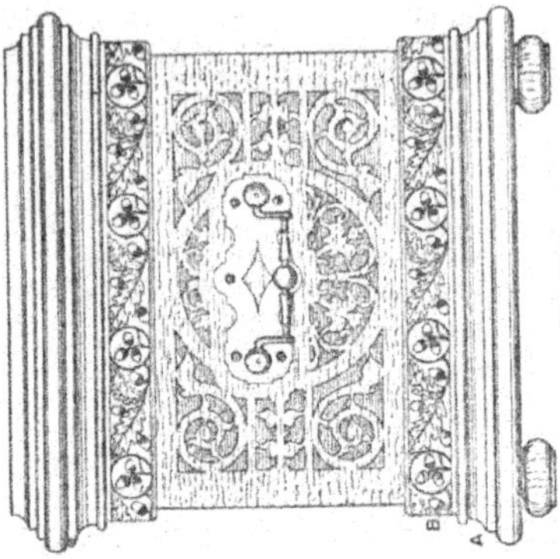

Fig. 1438.

Fig. 1436.

Fig. 1437.

Figs. 1436 to 1438.—Side Elevation, Part Plan, and End Elevation of Jewel Casket with Combination Lock.

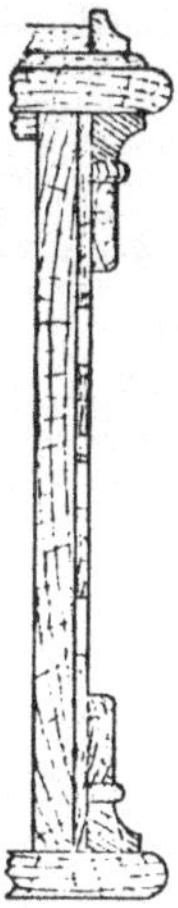

Fig. 1439.—Vertical Section through Side of Jewel Casket.

Fig. 1440

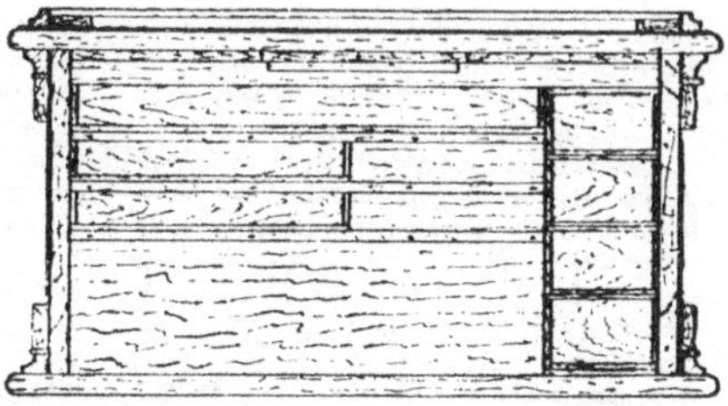

Fig. 1441.

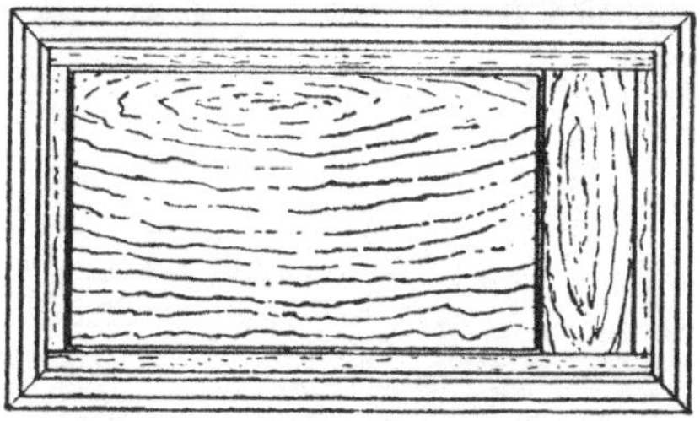

Figs. 1440 and 1441.—Vertical and Horizontal Sections of Jewel Casket.

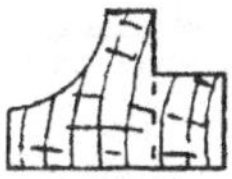

Fig. 1442.

Fig. 1443.

Figs. 1442 and 1443.—Sections of Mouldings on Jewel Casket (see A and B, Fig. 1438).

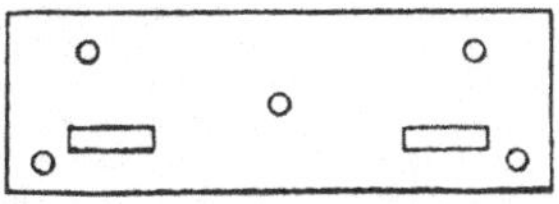

Fig. 1444.—Lock Plate of Jewel Casket.

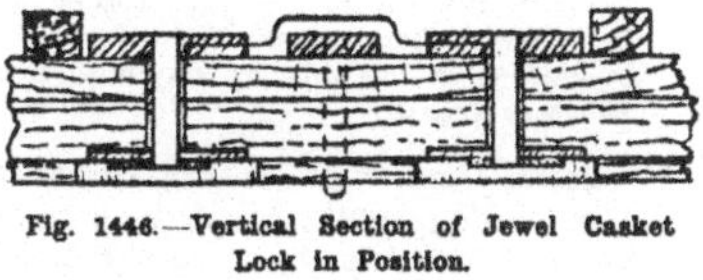

Fig. 1446.—Vertical Section of Jewel Casket Lock in Position.

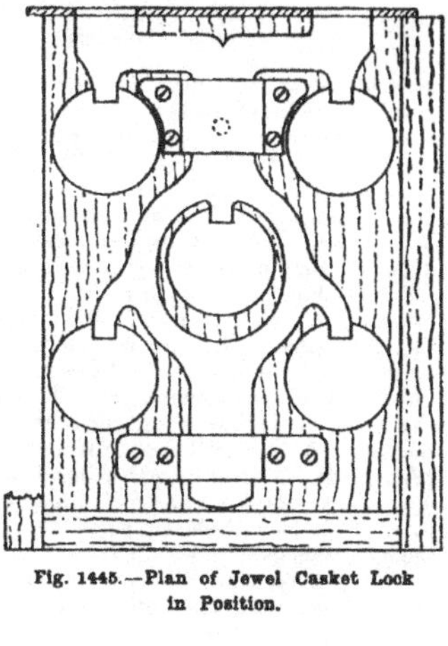

Fig. 1445.—Plan of Jewel Casket Lock in Position.

about 1½ in., and solder one into each roller flush with the ends; then pass the pins through the bushes into the cover, and, after filing to the correct length, solder on each hand in its allotted place when the bolt is in the opening position. Two brass hold-down clips are then bent to fit over the straight centre shank of the bolt as shown in the view of the lock (Fig. 1445), holes being drilled and the bolt screwed on. A hole ⅛ in. in diameter should be drilled in the bolt in the position indicated by the dotted circle, and a piece of brass wire should be soldered in this flush on the bottom, projecting on top about 1¼ in. to reach through the cover and project a little as shown in the section of the lock (Fig. 1446), and on the plan (Fig. 1445). This is for drawing back or shooting the bolt from outside. The hole in the cover to take this wire should allow of ¼-in. movement—just enough, in fact, to allow the bolt lugs to clear the rollers when the box is locked. A brass plate 4½ in. long by 1½ in. wide is required with holes cut $\frac{5}{16}$ in. down to take the two holding lugs of the bolt; it should be let in flush with the inside of the box front, and screwed in position. A view of this plate is also given (Fig. 1444). A strip of mahogany ⅜ in. by ½ in. should be got out and fixed round three sides of the lock as shown, so that a $\frac{3}{16}$-in. piece of mahogany screwed over all will just clear the lock clips and cover all in. It would be safer to make a private note of the exact positions of the indicator hands before closing the box, otherwise a failure of memory may necessitate forcing the box open. Should it ever become necessary to force the box, it would be best to break open the bottom and afterwards make a new one.

Completing Jewel Casket.—The box may be fitted inside with four drawers at one end as shown in the vertical section (Fig. 1440), the remainder of the box being partly occupied by a top till filling the remaining space, and two smaller half-length tills to slide to and fro on the runners. Thus a considerable space is left below for larger articles. The small end drawers have countersunk drop-handles in their fronts, and to enable the drawers to clear the $\frac{5}{16}$-in. thick till runners, pieces ⅜ in. thick must be fitted the full depth of the box and the total width of the drawers. These interior fittings need not be of mahogany; some cheaper and lighter material may be used instead. The four mahogany feet are turned 1⅝ in. in diameter and ¾ in. long, with a ⅜-in. pin ⅜ in. long on each. Holes are to be bored to fit these, and some thick glue will secure them in place. The box may be left with the material in a natural state, or finished in the ordinary way with french polish as desired.

Fig. 1446.—Detail of Bead, etc., of Jewel Case with Secret Drawers.

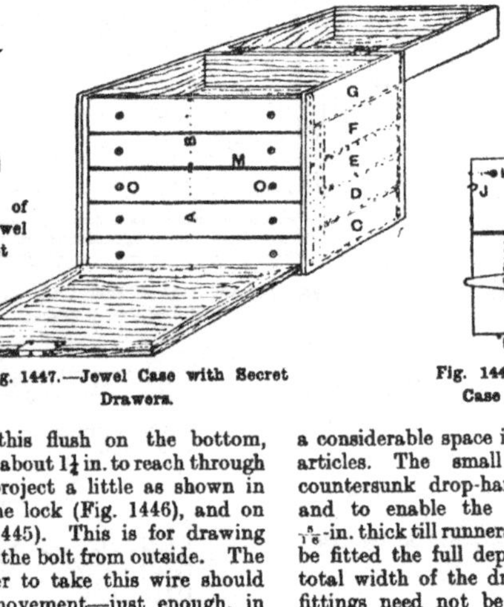

Fig. 1447.—Jewel Case with Secret Drawers.

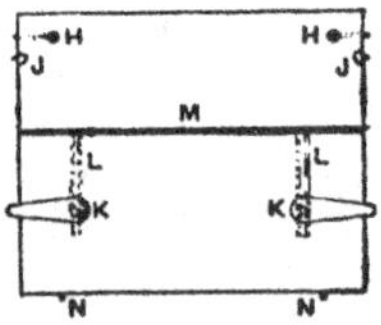

Fig. 1449.—Mechanism of Jewel Case with Secret Drawers.

Jewel Case with Secret Drawers.

A jewel case with secret drawers, the size being about 12 in. by 10 in. by 8 in., is shown open by Fig. 1447. The carcase is put together with secret dovetail and mitered joints. The front or flap is mitre-clamped and veneered on the face; the four drawers which the case contains are all hidden. The front A B (Fig. 1447) is made in two parts, and represents the fronts of five drawers, A being made the height of drawers C D E, whilst B is the height of drawer F and tray G. The bottom of G is a fixture, as are also the divisions between drawers F E, E D, D C; the front A is made as shown, with two scratch beads at equal distances. The bead M, which divides the front, is loose; to it are fixed two steel forks, which fit into the square mortices (Fig. 1448); the two drawer knobs K K (Fig. 1449) have a small square spindle attached, over which the steel fork passes, and fixed on the end of the spindle is an iron tongue and a nut forming a turn-buckle. When the knob is turned so that the front is fixed, the fork K is dropped and fixes the front A, and, until lifted, the latter cannot be moved. Dowels N N are fitted into the bottom of the case; the front B is made to work on pivots J J and is fixed by springs H H (Fig. 1449). These springs are hidden by the silk lining of the tray, and, until released, the front A will not move; when the springs are released the front will fall on the bottom of tray G, giving access to the bead M. In a shallow case it will be necessary to form the movable knob at O O, or the forks L L will not draw out sufficiently to release the front A.

Another Box with Secret Drawer.

The box shown by Fig. 1450 is made of ½-in. stuff, mahogany for preference. Two pieces 9 in. long and 4½ in. wide are required for the sides, and one piece 4½ in. by 4 in. for the end; the other end, being the slide, is only ⅜ in. thick; the top and bottom are ¼ in., and the moulding is ½ in. deep. The joints at the ordinary end are mitered; at the other, ⅜ in. from the end, cut a groove ¼ in. wide and ¼ in. deep; a piece ⅛ in. thick and 4½ in. by 3½ in. must be firmly fixed into this, and the ends cut as shown in section (Fig. 1451). Now cut the ⅜-in. piece to slide into the end. Before fitting together, however, 1 in. from the bottom of the sides cut out a groove ⅛ in. wide and ⅛ in. deep to take the false bottom, and then fasten on the top and bottom, the moulding hiding the joints. A line is now made round the box 1 in. from the top edge, and the box is cut in two along this line with a tenon saw. Withdraw the slide, and cut out a piece from the top along the dotted line (Fig. 1451) on the inside ¼ in. deep, and fix a mitered piece into its place to correspond to the other joints. A piece will have to be taken off the ⅛-in. piece, as that fastened on the slide will hang over flush with the inside of the box, the slide apparently being ½ in. thick. The drawer is made of ⅛-in. stuff, and is ⅞ in. deep; it cannot be taken out without the box being opened. Divisions are made in the box if desired.

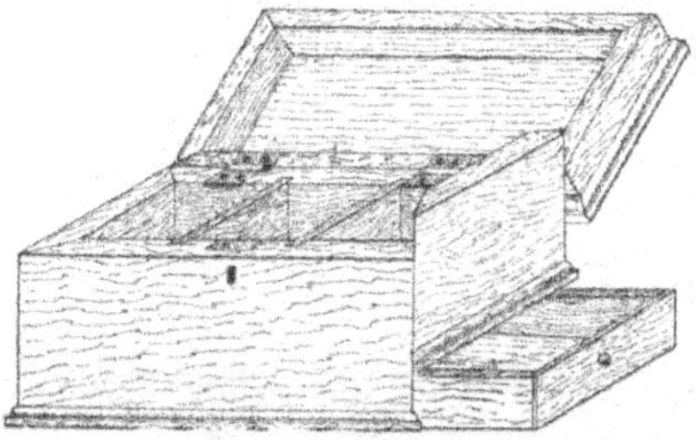

Fig. 1450.—Box with Secret Drawer.

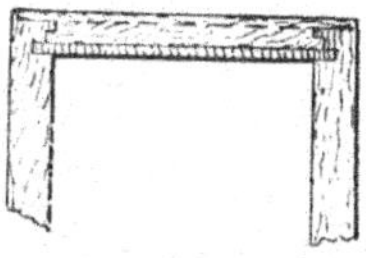

Fig. 1451.—Part Horizontal Section of Box with Secret Drawer.

Ornamental Workbox.

THE majority of workboxes are plain externally, and it is only when they are opened that the eye is relieved by the silk and plush with which the interior is decorated. Fig. 1452 is a part elevation, and except that if access can be had to a spindle machine a great deal of time and labour will be saved. The bottom part of the box is secret dovetailed; the mouldings on the upper part forming the lid being merely mitered and glued, and then one tier of mouldings screwed to the tier above. The bottom

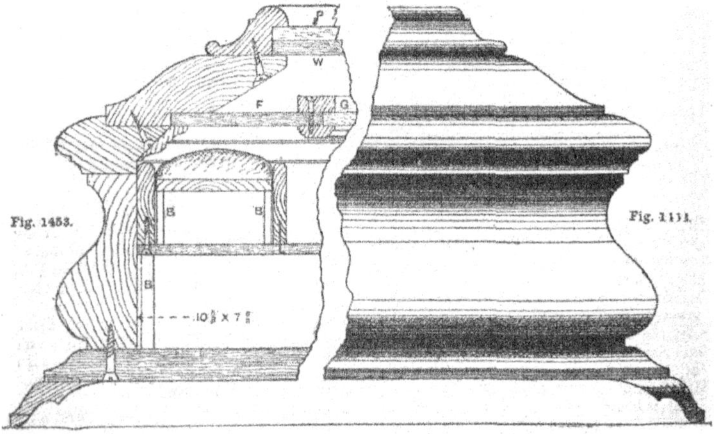

Figs. 1452 and 1453.—Part Elevation and Part Vertical Section of Ornamental Workbox.

Fig. 1453 a part vertical section through the centre of a box which is of good appearance outside as well as inside. The box is a mass of mouldings screwed and hinged together, mahogany of a good figure being very suitable for it. Nothing need be said on the running of the mouldings, is screwed on as shown, and the feet, which are cut out on two sides to the shape given, are glued and screwed to the bottom with a very light screw. On the top of the box, and held in place by the rebate of the top moulding, is a photographic view (P, Fig. 1453) mounted on glass, with a piece of thin

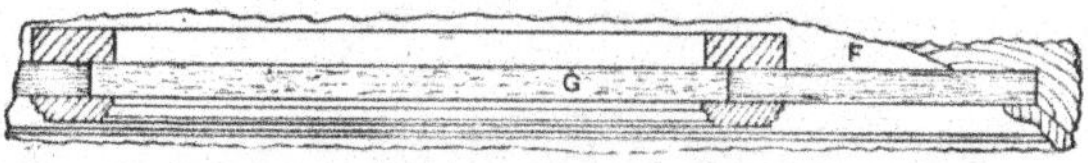

Fig. 1454.—Part Section through Workbox Lid.

Fig. 1455.—Part Plan of Inside of Workbox Lid.

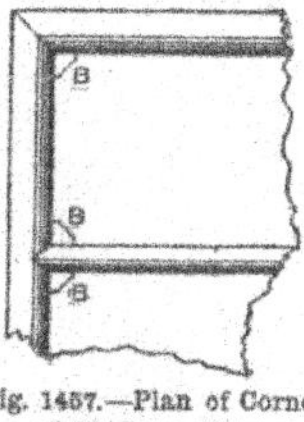

Fig. 1457.—Plan of Corner of Workbox Tray.

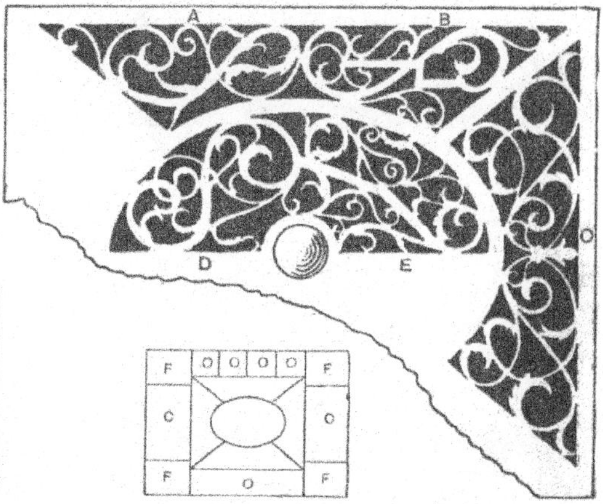

Fig. 1458.—Fret Cover for Workbox Lid.

Fig 1456.—Diagram of Workbox Tray.

Fig. 1459.—Fret Designs for Central Cover of Workbox.

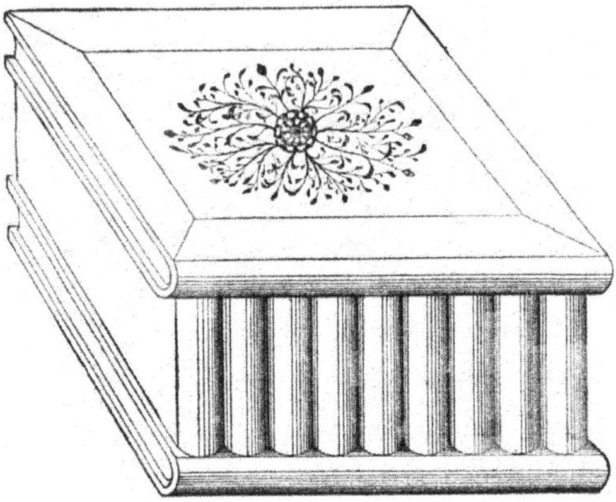

Fig. 1460.—Workbox with Secret Compartment.

Fig. 1461.—Vertical Cross Section of Workbox with Secret Compartment.

pine W below it. The view should be procured before proceeding to make the box, to get the right size of the top moulding. On the inside of the lid, resting in a rebate, and kept in place by a moulding, is a fret border F (Figs. 1453, 1454 and 1455), having a small piece of mirror plate G in the centre. A small piece of pine is glued to the back of the fret to form a rebate for the glass, and a small screw used to draw the mitered moulding hard up and bradded to the divisions. The frame of the tray is dovetailed, and the round on the top edge mitered. The divisions are rebated into and mitered to the frame and to each other. This is shown in Fig. 1457, which is a part plan of one of the corners of the tray. The middle part of the tray is an ellipse, with four diagonal pieces rebated and mitered to it. The ellipse is not essential, and either a circle or a square may be substituted. These central

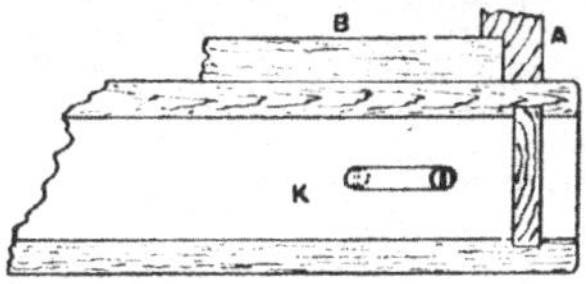

Fig. 1462.—Part Longitudinal Section of Bottom Book of Workbox.

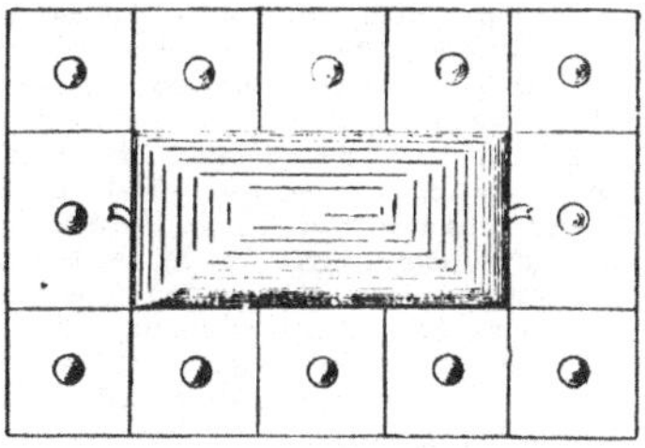

Fig. 1467.—Diagram of Workbox Tray.

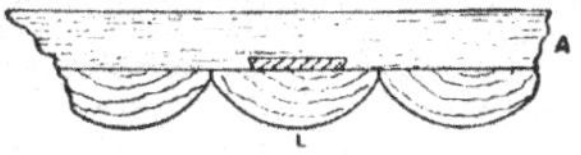

Fig. 1463.—Cross Section through Workbox Book Backs (L, Fig. 1461).

Fig. 1465. Fig. 1466.

Figs. 1465 and 1466.—Mouldings on Workbox Lid.

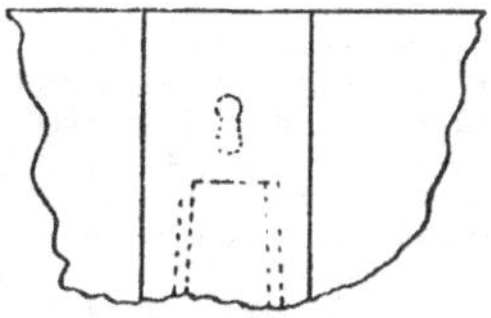

Fig. 1464.—Part Elevation of Workbox Book Backs.

to the face of the fret and glass. This is shown in Fig. 1454, which is a longitudinal section of the glass. Fig. 1455 is a part plan of the inside of the lid, and shows a running design for the fret border, which can be varied according to taste. Fig. 1456 is a line diagram of the tray, showing the divisions, the letters denoting where the space is intended to be covered by a fret F, cushion C, or left open as shown by the letter O. A section of the tray is shown in Fig. 1453, which shows the blocks B for supporting the cushion cover, as also the blocks which are glued to the corners of the box to support the tray. The bottom is screwed to the under edge of the outer portion of the tray pieces are kept low enough to allow the top of the fret to come in with the bottom edge of the round. Fig. 1458 is a design for the fretwork cover of the corner spaces. A, B, C, D, E, in Fig. 1459, are different fretwork designs for the centre of the tray. Holes are bored in the centre of the frets, and small turned knobs are glued in. The interior of the box and tray is covered with either velvet or plush. A piece of pink silk is stretched and glued to the top side of the fret border. This gives it a good appearance, and also hides the interior of the lid from view. The inside sizes are given in Fig. 1453, but the size can vary with requirements.

Workbox with Secret Compartment.

The box shown in Fig. 1460 is constructed in the imitation book form, and comprises the workbox proper and a secret receptacle for papers at the bottom. It is intended to be finished in hardwood, and polished. The construction is rather complicated, and requires careful study before it can be understood. The cross-section (Fig. 1461) shows it as plainly as is possible without giving sections of every detail, which is unnecessary. In Fig. 1461, A A is a box, secret dovetailed, with a yellow pine bottom B glued and bradded into a rebate. If it is determined not to cover the inside of the box with any material, the bottom should be mahogany. The front C and the two sides of the top should be secret dovetailed together at the front, and the sides fitted to the back D, which should be sunk parallel with the round in imitation of the binding of a book, as shown in Fig. 1460. It can then be lap dovetailed, so as to show side wood right into the margin. This, when glued up, forms the frame of the top book or lid of the workbox. A strip of yellow pine E, to which has been clamped a thin strip of mahogany, is then glued to the back. After this has been allowed to get hard and then flushed off on both sides, narrow strips, flush on the inside of the lid, are mitered at the front and butted against the back on the lower side, and screwed. These form the mock binding of the book. There are different methods which may be adopted to secure a good finish on the top. The finish shown in Fig. 1460 is composed of a central piece, to which a mitered border has been feathered and glued. The central portion, which may be a light-coloured wood, has been enriched by the addition of a floral design. The section (Fig. 1461), however, shows a raised and fielded frame of a different coloured wood, preferably dark, glued down to the top. After the top F has been cleaned all over, it should be glued down with thin hot glue to the top edge of the frame, and kept in close contact, either with hand-screws or thumb-screws, until the glue has thoroughly set. It may now be laid aside until the bottom book has been finished. The bottom book is a secret receptacle, and it is necessary to exercise great care in fitting the various parts together, so that it will remain a secret. The wood, in all such work as this, must be thoroughly dry to begin with; otherwise, when it begins to shrink, the manner of opening becomes plain to all who look at it. The back G moves along far enough to allow of one of the ends sliding out. The other end is secret dovetailed to the front H, similar to the top book, and lap dovetailed at the back to the piece K. The top and bottom of this book are grooved at one end to carry the sliding piece. Fig. 1462 is a part longitudinal section through one end of the bottom book, showing the end sliding in the groove, and the slot cut in the piece K for the screw to travel in for sliding the back. The sliding end is mitered to the front, and abuts against the back G. Before fixing either the top or bottom, two slots should be cut in the top at each end, to allow the bottom book to slide back on the screws, which are fixed to the bottom B through the slots. The slots and screws are shown dotted. The keyhole is covered with one of the books, or rather imitation book backs L. There is a dovetailed piece glued to the back of it, and a raggle cut in the box A. This raggle has a slight draw on it, so that the back L tightens as it comes up, and it is to allow this to drop down clear of the keyhole that the bottom book is required to slide back. Fig. 1463 is a cross-section through part of the workbox, showing the dovetail on the back L. Fig. 1464 is an elevation showing the position of keyhole and raggle for dovetailed piece. After the slots are cut in the top of the bottom book, it may be screwed to the top edge of the frame. The book backs L may now be glued on the middle, one being dovetailed. The screws can now be inserted into the bottom B, and then the bottom M can be screwed on. The top should now be fitted down to the top of the box and hinged, and the lock fitted. It should then be taken off and the mouldings planted in the inside of the lid. The corner moulding is fitted to blocks, which are glued to the corners. A small bevel plate mirror, kept in place by a small moulding, is used to relieve the surface. Figs. 1465 and 1466 are enlargements of the mouldings. The lid may now be re-hinged, and the lock got into good working order. The tray, which rests on small blocks at the corners, may next be proceeded with. Fig. 1467 is a line

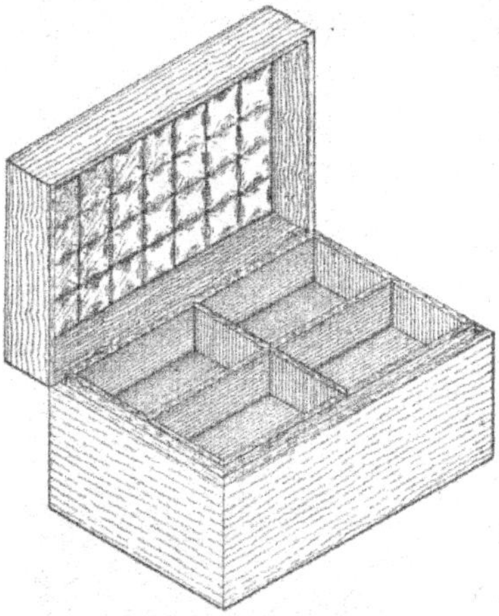

Fig. 1468.—Simple Workbox.

diagram of the tray, showing the divisions, which are raggled into one another. The spaces are covered with lids, the central one being padded and lifted out with a small piece of ribbon at each end.

Simple Workbox.

An ordinary form of workbox (Fig. 1468) is of rectangular shape, and has nothing in its construction which the cabinetmaker will not understand at a glance. Fig. 1469 is a vertical section showing the tray.

Inlaid Fancy Box with Secret Drawer.

Fig. 1470 shows an inlaid fancy workbox fitted with a secret drawer which is partially open. The outside dimensions of the box illustrated by Figs. 1471 and 1472, and not including the plinth, are :—Length, 7 in. ; width, 4 in. ; and height, 4 in. at the edges, rising to $4\frac{5}{8}$ in. at the crown. It is shown made in the solid, out of $\frac{7}{16}$-in. stuff ; and a suitable wood is Spanish mahogany, inlaid with satinwood, or lacquered brass might be used for the inlay with good effect. The secret drawer is concealed in the plinth, the square portion of which at one end of the box forms the drawer front, the moulding being cut off and fixed to the box. A false bottom, as shown in Fig. 1473, closes the receptacle, and the piece of the plinth on the drawer is mitered to fit the fixed parts.

Construction.—To construct the box, cut the sides and ends rather full to the lengths, having previously planed the wood

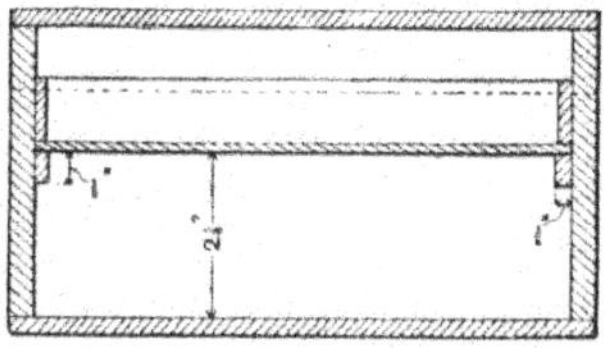

Fig. 1469.—Vertical Section of Simple Workbox.

true, and gauged to thickness. Then gauge to a width the sides up to the junction with the top, as shown in Fig. 1473, and the ends wide enough to reach the crown. The end piece at the right hand may, if desired, be smaller by the depth of the square plinth. Mark the mitres on the edges, square over the inside, and groove or rebate. Three methods of forming the angle joints are shown, the one at A (Fig. 1473) being the strongest, and that in Fig. 1475 the easiest mode. If the joint at B (Fig. 1473) is chosen, the lip may be bradded, the brads being driven in the seat of the inlay. Plough or saw a $\frac{1}{8}$-in. groove in the two sides and one end to receive the false bottom. Work a small tongue on the top edges of the sides as shown in Fig. 1473, and cut the ends to the sweep. Then make a rebated mitre on the inside of the ends on the top edge, as shown in Fig. 1476, and prepare the mitres at the angles. The mitre at the drawer end of the box must be stopped

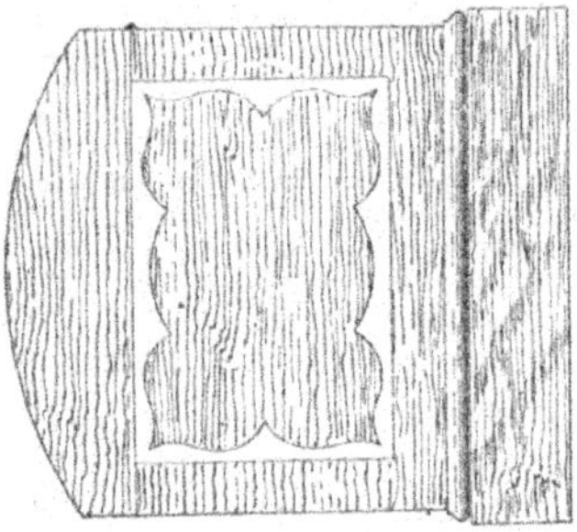

Fig. 1471.

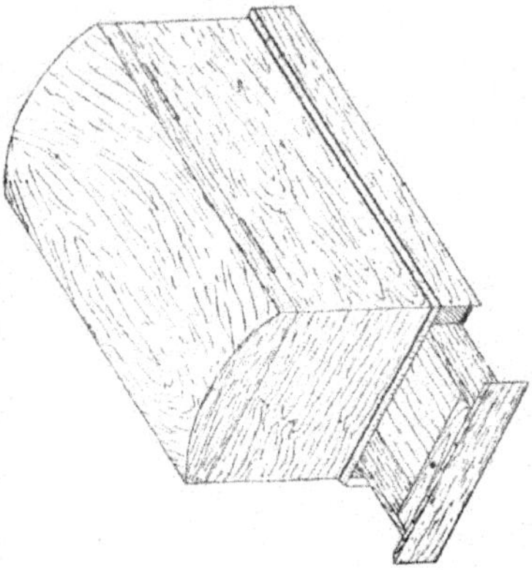

Fig. 1470.—Inlaid Fancy Box with Secret Drawer.

Fig. 1472.

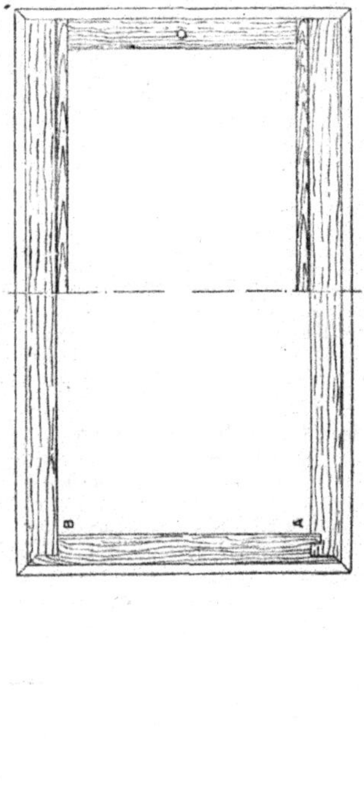

Fig. 1473.

Figs. 1471 to 1473.—End and Front Elevations and Half Horizontal Sections of Inlaid Fancy Box with Secret Drawer.

in line with the plinth, and the end brought out square, as shown in Fig. 1470. An easier, but less workmanlike, method would be to shoot the mitre right through, and afterwards fill in these pieces flush with the face of the box. When ready, glue and brad together temporary blocks, to which the hand-screws can be fixed. Next prepare the top, as shown in outline in Fig. 1477, 1/8 in. wider than the finished size. Shoot parallel and plough the grooves. Then mark the inside sweep with a template, and work it out with a round plane. Next set a bevel to an angle of 45°, and working off the top, shoot the ends to fit, trying it into the opening in the box. When this is fitted accurately, glue it in and fix with handscrews until dry, when the top can be cleaned off to the sweep of the ends, as shown by the dotted line in Fig. 1477.

False Bottom.—Next fit and slide in the false bottom. Bore a 1/8-in. hole in the middle of the right-hand end up through the thickness sufficiently to take the bolt, as shown in Figs. 1473 and 1474. To avoid making a separate illustration of the interior of this end, the bolt is shown in the section (Fig. 1474); but it is fixed inside the other end. Next brad on the true bottom, and clean off flush all round. At this stage gauge lines may be run round from the bottom edge for the straight sides of the inlay and for the joint of the lid. Also gauge the return lines on the ends and round the lid; use a sharp cutting gauge. Next make and fit in the drawer. This is too slight to be dovetailed, as the sides are of 1/8-in. stuff. They are simply glued and bradded on two 3/8-in. pieces, as shown in Figs. 1470 and 1474. The bottom may be made of tin-plate or two-ply veneer glued together crossways, as shown in Fig. 1473. The drawer should be fitted to slide rather tightly, so that no rattling occurs. If it moves stiffly, a little powdered French chalk rubbed all over the moving parts will correct this. Clean the drawer flush with the box, and prepare and fix the plinth. A small scratch tool will be suitable for working the moulding. The plinth should be mitered up all round, the solid end fixed first, then the sides, and lastly the drawer end. This piece must have the moulding cut off with a fine saw, and fixed on the box between the side pieces. Then the square

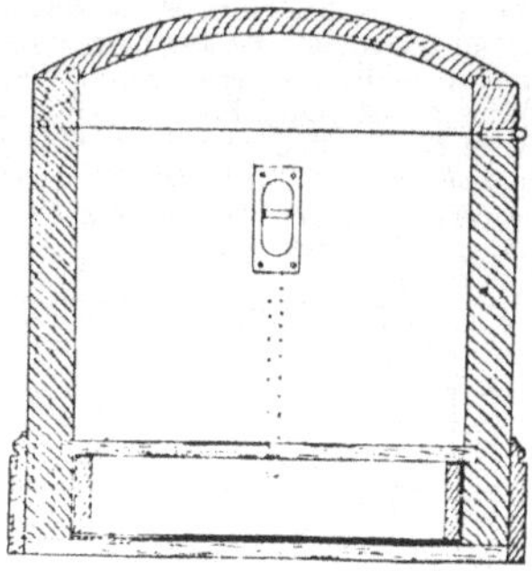

Fig. 1474.—Cross Section of Inlaid Fancy Box.

Fig. 1475.—Joint for Angles of Fancy Box.

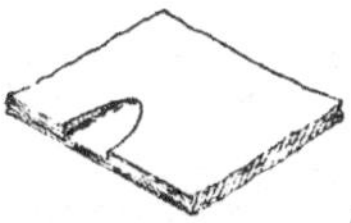

Fig. 1478.—Thumb Notch for Drawer of Box.

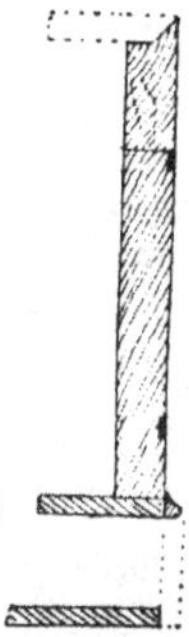

Fig. 1476.—Section of Drawer End of Box.

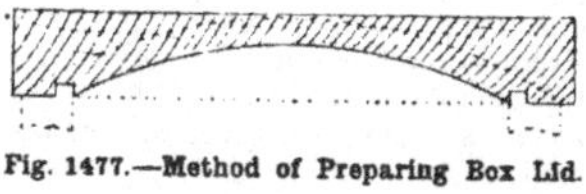

Fig. 1477.—Method of Preparing Box Lid.

part is fixed on the drawer front, the top edge of the drawer and the butt ends of the box being greased or oiled to prevent any glue sticking, which would prevent the drawer being withdrawn.

Completing Fancy Box.—Next cut the lid along the gauge lines with a fine saw, and clean up the edges. Fit the hinges, $1\frac{1}{4}$-in. cast brass butts, and again remove the lid previous to inlaying. Prepare paper patterns of the inlaying which are all in the nature of "repeats," and paste them on the veneer or metal, whichever is employed. Cut them out with a fret-saw, mitre up, and lay A piece of thin baize glued over the bottom will hide the thumb slot (see Fig. 1478), and it will be impossible to open the drawer without first opening the box.

Ladies' Hat- and Bonnet-Box.

A ladies' hat- and bonnet-box, such as is shown in front elevation by Fig. 1479, may be made from almost any kind of wood, and painted, stained, and varnished, clear varnished, or polished according to taste.

Fig. 1479.

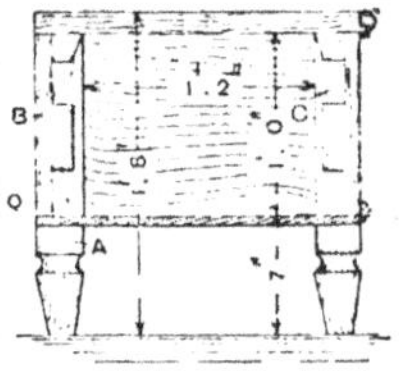

Fig. 1480.

Figs. 1479 and 1480.—Front Elevation and Cross Section of Ladies' Hat- and Bonnet-box.

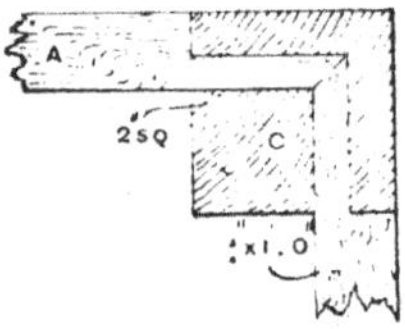

Fig. 1481.—Horizontal Section of Leg and Rails of Box.

Fig. 1482.—Section of Bottom Moulding of Box.

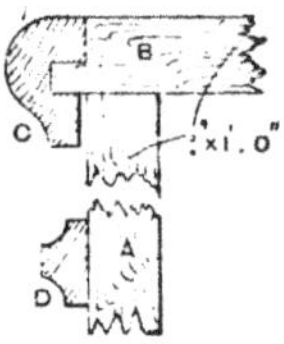

Fig. 1483.—Section of Top and Nosing of Box.

in position to the gauge lines. Mark round the outline with a hard, sharp pencil, having first chalked the surface of the box, and cut in the outline with gouges and chisels; or better, if any suitably shaped templates are at hand, run in the outline with a sharp penknife, chop up the core with chisels, and carefully remove. Level off the sinking with a router, fill in with clear glue, and rub in the veneer. If metal is employed, scratch the under side with a bradawl, and mix a little gold-size with the glue; clean off with hot water. Next fit in the bolt, rehang the lid, and fit in the lock, when the box will be ready for polishing.

For finishing in paint, dry yellow pine should be used, but for a stained and varnished finish white wood is preferable. For simply clear varnishing, pitchpine or Oregon pine might be employed; while for polishing, oak, mahogany, or walnut may be used. The legs A (Fig. 1480) are shaped and mortised, C showing the inside of the end rails, and the line of tenons mortised into the legs. The front and back rails B are framed into the legs at right angles. The thickness of the front and end rails is first gauged on to the face side and face edge of the leg, and from the lines the widths of the mortices should be set back, and the mortice gauge set accordingly. After the legs have been set out carefully in pairs, they are gauged and mortised to receive the front, back, and end rails A (Fig. 1481), all of which are fitted and mitered as shown, C being the leg. When gluing up the box, the front and back rails should be first glued into the legs, and allowed to dry; then the end rails should be glued and cramped together and the box squared, and the whole again allowed to dry. Next cut and fix on the bottom, and clean off and rub down with glass-paper, working in the direction of the grain if the wood is to be varnished. If the box is painted, it may be cross-rubbed. Next mitre on the moulding (Fig. 1482) to covet the joint, and plant on the moulding D (Fig. 1483), as shown in Fig. 1479. Hang the top with $2\frac{1}{2}$-in. brass butts, allowing the front and ends to project $5\frac{1}{16}$ in. beyond the outside of the box. Then on the under side of the lid mark carefully with a fine pencil the line of the outside of the box, disconnect the lid, work on the tongue, just leaving in the line; mitre on the nosing, and when the wood is dry, clean off and finish. The top is $\frac{3}{4}$ in. thick. The dimensions of the several pieces required are as follows:—Four legs, each 1 ft. 9 in., 2 in. square; two rails, each 3 ft. 6 in. by 1 ft. by $\frac{3}{4}$ in.; two end rails, each 1 ft. 6 in. by 1 ft. by $\frac{3}{4}$ in.; one bottom, 3 ft. 7 in. by 1 ft. 6 in. by $\frac{5}{8}$ in., or made up to the required width; one top, 3 ft. 7 in. by 1 ft. 7 in. by $\frac{3}{4}$ in.; 7 ft. of moulding to Fig. 1482; 7 ft. of nosing to C (Fig. 1483), and 9 ft. of moulding D. If the ends are treated in the same way as the front, additional moulding will be required. The box should be provided with a strong lock.

CURIO CASES AND TABLES.

Curio Case.

THE curio case illustrated in Figs. 1484 and 1485 has the front, sides, and shelf of glass. It stands 3 ft. high, the upper 1 ft. 6 in. being taken up by the case itself, and the remainder by its legs. The actual case has a width of 1 ft. 8 in. and a depth from back to front of 1 ft. 3 in. If, as is recommended, the wood employed is of a soft kind for ebonising, the legs (Fig. 1486) should be cut from 1-in. board. They are 2 ft. 11¾ in. long, and each will cut into a 4½-in. width, though, as usual, material may be saved by sawing two legs or more from one board.

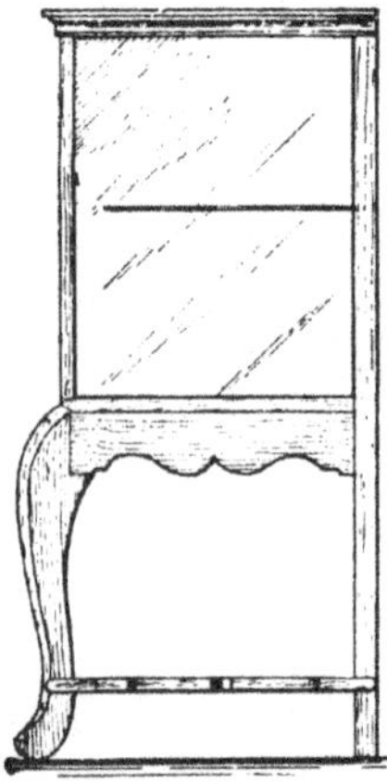

Fig. 1485.

Fig. 1484.

Figs. 1484 and 1485.—Front and End Elevations of Curio Case.

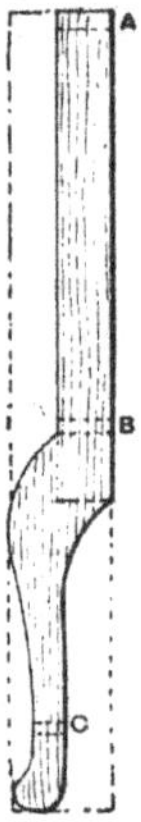

Fig. 1486.—Leg of Curio Case.

The hinder legs range parallel with the wall as is shown in section in Fig. 1487, and the front legs are set diagonally. In Fig. 1486 the dotted line at A indicates where the legs are held together by the false top; the lines at B show where they are grasped by the case bottom, and those at C where they are held by the ornamental diagonal braces. Fig. 1487 shows the under side of the false top and the adjacent pieces. The false top is of ¾-in. board, 1 ft. 8 in. long by 1 ft. 2¾ in. wide. Cuts at D and E are made for the tops of the legs, which are strongly screwed in place, thus bracing the upper part of the case firmly together. The bottom of the case is a board of the same dimensions as the false top, and in a similar way holds the legs together at their middles. Lower down, 3½ in. from the ground line, they are again held together by the diagonal braces (Fig. 1488). These are cut with the frame-saw from ½-in. stuff, and are, of course, halved where they cross in the centre. The boarding of the back G (Fig. 1487) is of ¼-in. stuff. The pieces lie horizontally, and are together 1 ft. 7 in. by 1 ft. 5 in., and their ends are fixed into rebates cut in the legs. The case is still further braced, as shown in Figs. 1484 and 1485, by the ornamental strips, which are of ½-in. board 3 in. wide, their ends being splayed to fit the diagonal legs to which they are screwed; the shorter strips are attached to the back legs by short pieces screwed to the legs behind them. The true top, indicated by the dotted lines H (Fig. 1487), is of ¼-in. board, 1 ft. 10 in. long by 1 ft. 4 in. wide, and overhangs the case by 1 in. at the front and ends. Its edges and front corners are rounded, and it is fixed in place by screws driven into it through the false top, the edges of which are hidden by strips of moulding J, ¾ in. deep by ½ in. wide.

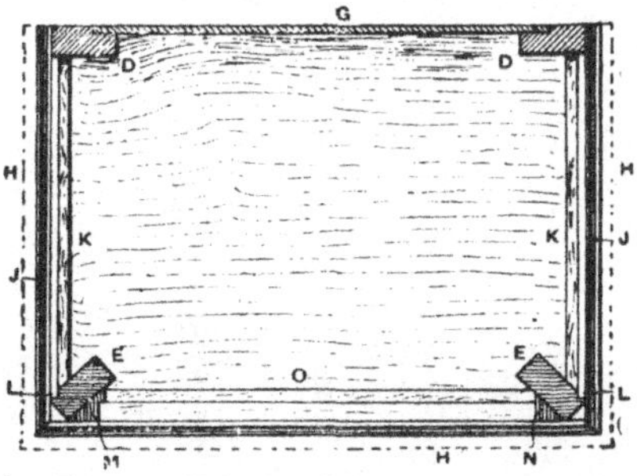

Fig. 1487.—Underneath View of False Top of Curio Case.

Glazing.—The glazing of the sides is effected by fixing strips K, ⅜ in. by ¼ in., ¼ in. from the edges of the false top, of the bottom, and of the back legs. In the front legs grooves L must be cut to receive the glass. Before puttying-in the glass it will be well to brush the wood with linseed oil; this makes the putty stick better. A convenient height for the glass shelf will be 9 in. from the bottom. Its back edge may rest on, and be puttied into, a strip of wood fixed for that purpose to the back. This shelf may be as long as the case will admit (1 ft. 7½ in.), and its breadth will be 1 ft. At the front its two ends will rest and be puttied into cuts made for them in the back angles of the diagonal legs.

Door.—The door of the curio case (see Fig. 1484) is hung and locked against two wooden triangular strips M and N screwed to the diagonal legs (see Fig. 1487). The width of both is 1 in. at sight, but M, to which the door locks, is ⅜ in. wider than N, to give room for a rebate in which the door may close. In like manner, for the support of the door, ⅜-in. strips O are fixed to the false top and to the bottom, ¾ in. from their front edges. The door is framed in eight strips of ⅜-in. board disposed in two layers. The four strips of the inner layer are 1¾ in. wide, the upright ones are 1 ft. 5¼ in. long, and the cross ones are 1 ft. 2½ in. long. In the outer layer the strips are 2 in. wide, and are, as regards the uprights, 1 ft. 1¼ in. long, and as regards the cross pieces, 1 ft. 5 in. long. Thus, when these are screwed together at the corners, the ¼-in. rebate formed for the glass is on the inner side. The illustrations are 1 in. to the foot, except Fig. 1487, which is 1½ in.

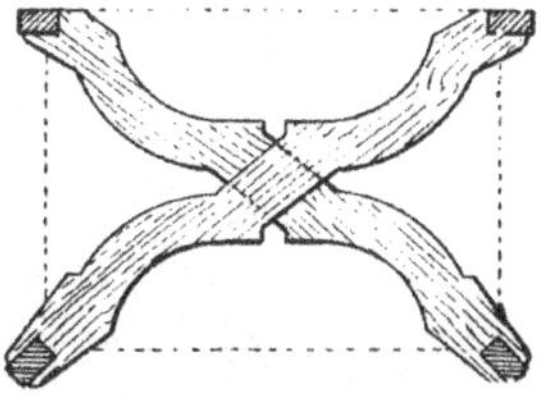
Fig. 1488.—Diagonal Braces for Curio Case Legs.

Sheraton Pattern Curio Table.

The curio table shown by Fig. 1489 is intended for the reception of small china ornaments, coins, medals, etc., the shelf underneath being useful for plants in antique vases. The extreme height, including the top, is 2 ft. 3¾ in.; the length of the top is 2 ft. 4½ in., and the width 1 ft. 7½ in. The following material, which should be of mahogany, will be required, all the figures given being finished sizes:—Top A: Two pieces 2 ft. 4½ in. by 3 in. by ¾ in.; and two pieces 1 ft. 5½ in. by 3 in. by ¾ in. Legs B: Four pieces 2 ft. 3 in. by 1¼ in. by 1¼ in. Side frames: Four pieces 2 ft. 2 in. by 1¼ in. by ¾ in.; four pieces 1 ft. 5 in. by 1¼ in. by ¾ in.; and eight pieces 6 in. by 1¼ in. by ¾ in. Bottom C: One piece 2 ft. 2 in. by 1 ft. 5 in. by ⅜ in. Shelf D: One piece 2 ft. 2 in. by 1 ft. 5 in. by ½ in. The exact dimensions of the shelf, however, should be determined by leg-to-leg measurements. The four frames which constitute the sides of the case are connected to the taper legs by means of mortice-and-tenon joints, and the bottom shelf holds the legs firmly in position, thus preventing the warping of the wood, which is very likely to happen unless great care is exercised in the selection of the timber. The top is made up of four pieces of wood jointed together to form a framework, the centre being glazed, preferably with a piece of bevelled plate. The sides of the case are also glazed, sheet-glass being used for these. To make the side frames, cut the sixteen pieces required, and plane them up to the dimensions given above. Take the 2-ft. 2-in. pieces, and on each end mark off 2 in. for the tenons. Divide the thickness of the wood into three, which gives ¼ in. as the thickness of the tenons. The depth should be arranged by cutting away the wood at the top of the rail which is on the upper part of the case, and from the bottom of the rail at the lower part, as shown in Fig. 1490. Open-end mortices are cut in the stiles of the same size as the tenons, which project through to a distance of ⅜ in. The two short sides or ends of the case are framed up in exactly the same way. Run a ⅜-in. satinwood banding along the bottom of each frame, making it flush with the edge; and all round the inside opening, at a distance of ¼ in. in, run a $\frac{3}{32}$-in. boxwood stringing. The arrangement of these inlays is shown in Fig. 1491.

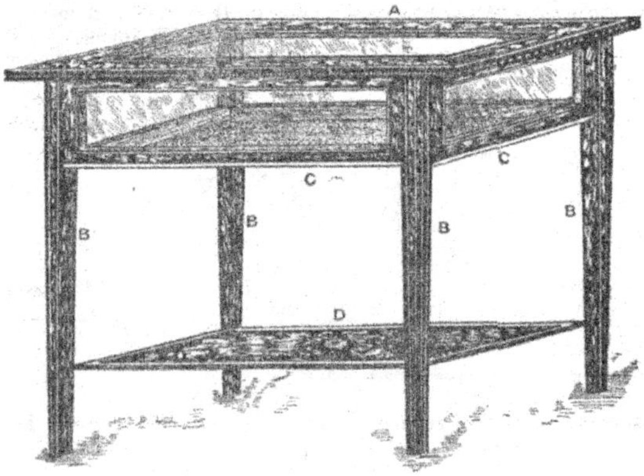

Fig. 1489.—Sheraton Pattern Curio Table.

Legs.—The legs may now be taken in

hand. Plane all four up to $1\frac{1}{4}$ in. square, and mark a line 6 in. from the top of each, this section being kept square to receive the side frames. The legs are tapered to $\frac{3}{4}$ in. square at the bottom. Mortices will have to be cut on the insides of the legs to receive the projecting tenons of the side frames, the tenons being recessed $\frac{1}{4}$ in. The two outside faces of each leg are inlaid with $\frac{3}{32}$-in. boxwood stringing, as shown in Fig. 1492. Glue the sides in position, cramp up, and set aside to dry.

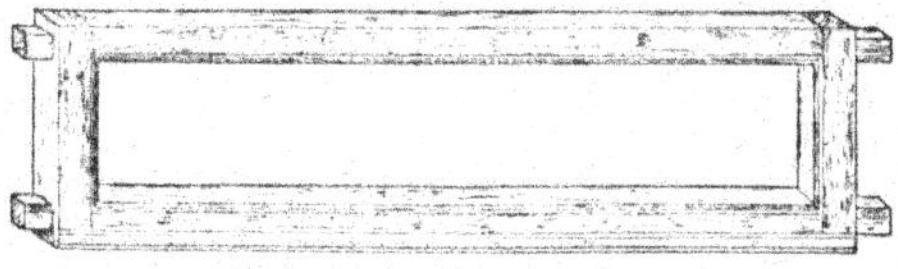

Fig. 1491.—Side Frame of Sheraton Pattern Curio Table.

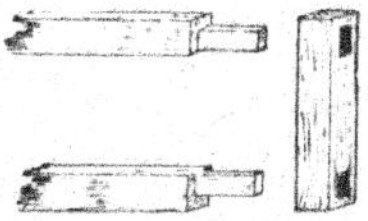

Fig. 1490.—Joint of Curio Table's Side Framing.

Fig. 1493.—Corner of Top of Curio Table.

Fig. 1492.—Leg of Curio Table.

Top or Lid.—Next cut the pieces for the top or lid, and plane up to finished sizes. The length of the short rails allows 2 in. on each end for the tenons, which should be 2 in. in breadth and $\frac{1}{4}$ in. in thickness; the inch reduction in breadth from that of the actual wood is made from the outside. Cut corresponding mortices in the long rails, and glue up the framework. When dry, the inlaying can be done. On the edge of the top a fancy banding can be used, such as oblique ebony and boxwood (see Fig. 1493), or any fancy banding, provided it is black and white. On the top of the lid, on the outside edge, run a $\frac{3}{4}$-in. satinwood banding, and on the inside edge run a $\frac{3}{32}$-in. boxwood stringing $\frac{1}{8}$ in. from the beading. In each of the four corners a 2-in. fan may be inlaid, this greatly improving the appearance of the finished article.

Lower Shelf.—To obtain the exact measurements of the lower shelf, mark on the leg the height at which it is desired to fix it, and carefully measure the distance between each inside corner, and in cutting allow $\frac{1}{8}$ in. less all round. Band the top edge of the shelf with $\frac{3}{4}$-in. satinwood banding, and let in a fan or shell in the centre. The shelf is fixed by means of brackets, the screwed ends of which are screwed into the leg, the plate being fixed to the under side of the shelf with short screws.

Glazing, etc.—A beading will have to be run all round the inside edge of the table top, to form a rebate for the glass, a small flat bead being also put all round each of the side frames. The glazing of the sides can be done with ordinary sheet-glass free from blemishes, that for the top being for preference a piece of bevelled plate. All the glass can be fixed by pinning a small flat bead round the inside of the apertures. The top is connected by means of two $2\frac{1}{2}$-in. brass butt hinges, a lock and key being fitted to the opposite side. The bottom of the show case, which should be lined with plush of a colour suitable to harmonise with the

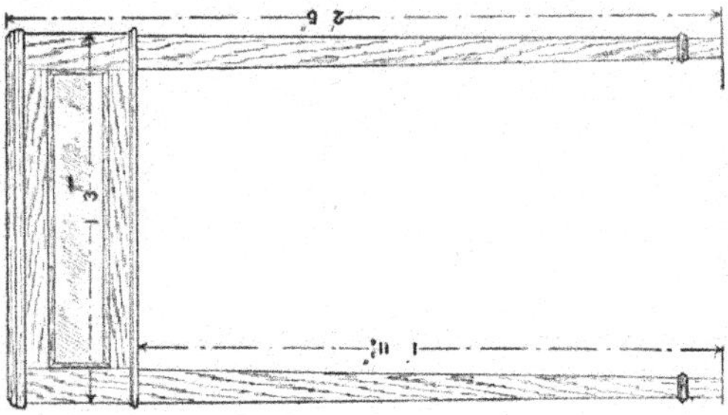

Fig. 1493.—End Elevation of Curio or Show-case Table.

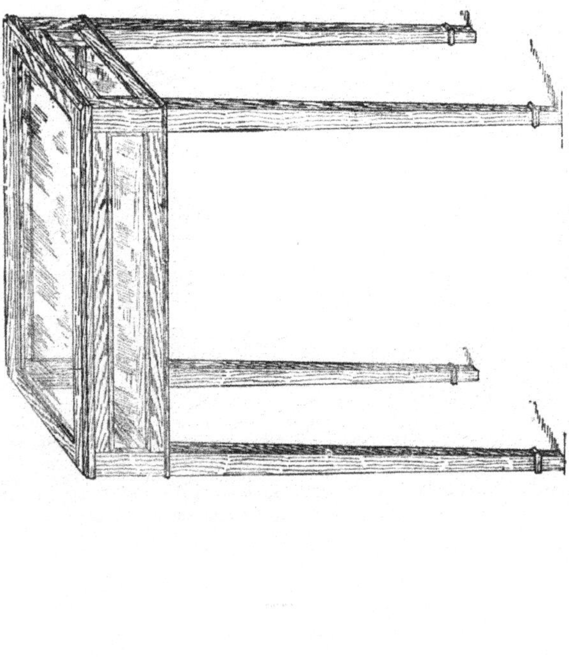

Fig. 1494.—Curio or Show-case Table.

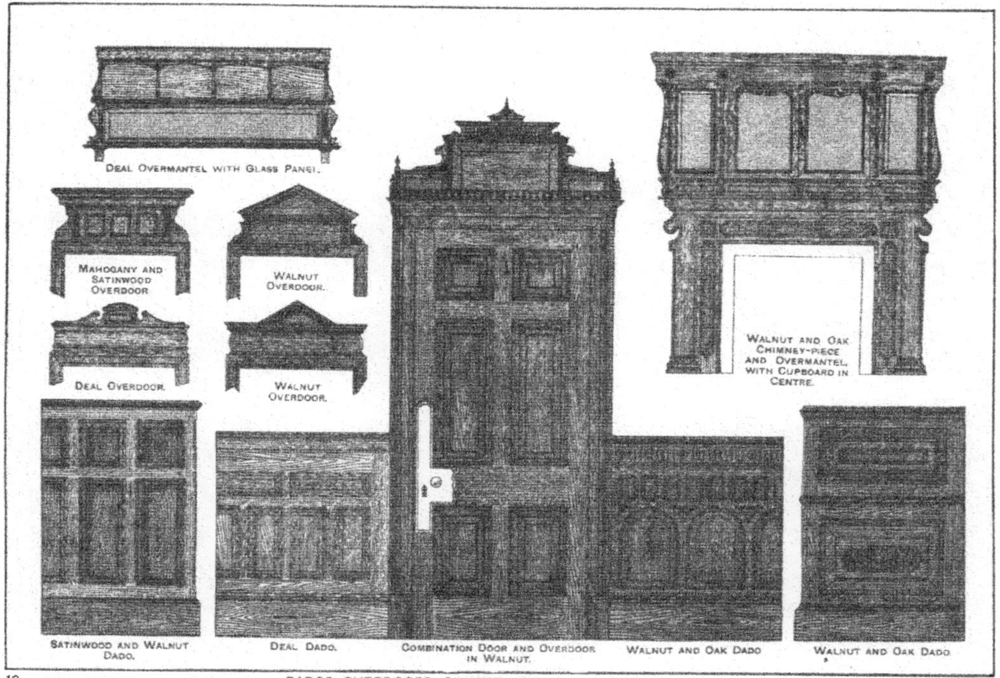

DADOS, OVERDOORS, CHIMNEY-PIECE AND OVERMANTELS.

surroundings, is secured from underneath by means of fine screws. The edges of the bottom should be slightly chamfered, to prevent them from showing.

Another Curio Table.

Fig. 1494 is a perspective view of a curio or show-case table, suitable for holding small articles of value, Fig. 1495 being an end elevation. Four legs are required, each $1\frac{3}{8}$ in. square at the top, tapering from the rail down to the foot, where it measures $\frac{7}{8}$ in. The top rails A (Fig. 1496) are $\frac{7}{8}$ in. wide by $\frac{7}{8}$ in. thick. The legs are mortised to receive the tenons, which are rebated on the top edges. The bottom rails B (Fig. 1496) should be $1\frac{1}{8}$ in. deep, showing $\frac{7}{8}$ in. deep on the face after the small moulding C is fixed on. The inside edges of the bottom rails are rebated, for the bottom, of pine $\frac{3}{8}$ in. thick. The inside corners of the legs are rebated at E (Fig. 1497) flush with the inside of the rails. The small moulding D (Fig. 1496) is glued and sprigged to the edges of the rails, and mitered at the corners. The glass is laid on the rebate thus formed, and beaded in from the inside. The moulding C (Fig. 1496) is glued along the bottom of the rails, crossing the legs, and is mitered at the corners. It may be secured by a few small sprigs. The leg near the bottom has a moulding housed into it, and glued and mitered (see Fig. 1498). After the table has been framed up, the bottom, $\frac{3}{8}$ in. thick, should be carefully fitted in the rebates on the bottom rails, and sprigged in. The lid is framed up like a door, the mortices being in the front and back rails, and the tenons on the short rails. Fig. 1499 shows a section of the lid. The moulding F is glued and mitered to the inside edge of the lid, and the glass beaded in from the inside. The ogee moulding is run on the outside edges after the lid is framed up. A pair of $1\frac{1}{2}$-in. brass butt hinges is required to hinge the lid to the top rail. The table may be of mahogany and french-polished. The bottom, on which the curios will be laid, should be covered with plush or velvet of a suitable colour.

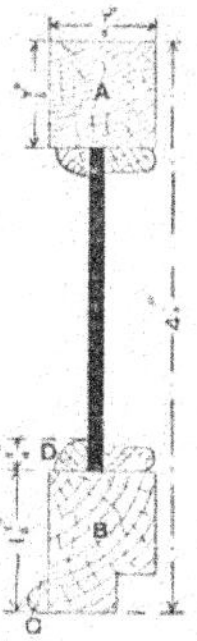

Fig. 1496.—Enlarged Section of Rails of Curio Table.

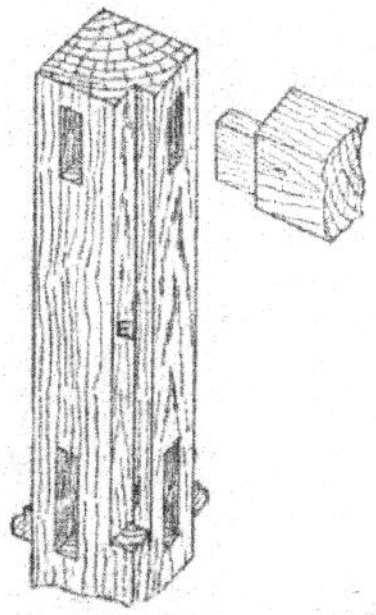

Fig. 1497.—Joint in Curio Table.

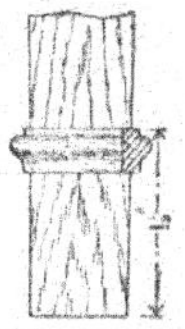
Fig. 1498.—Foot of Curio Table Leg.

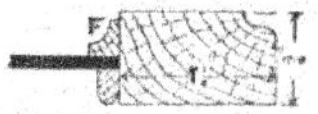

Fig. 1499.—Part Section of Curio Table Lid.

SEATS, COSY CORNERS, AND SETTEES.

Indoor Recess Seat.

FIGS. 1500 to 1508 show a design with dimensions and constructional details for a seat in a recess. The seat is made of oak or pitchpine, and the construction, as shown by the details, is extremely simple, and the seat is substantial without appear-

Fig. 1500.

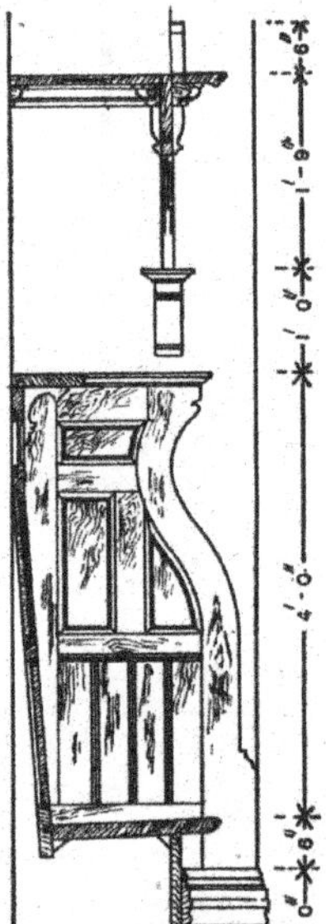

Fig. 1501.

Figs. 1500 and 1501.—Elevation and Vertical Section of Indoor Recess Seat.

ing heavy. The height of the canopy is optional, and can be arranged according to the height of room in which the seat is fitted. The seat is hollowed out as shown in panels were inserted in the back. Figs. 1500 and 1501 are reproduced to the scale of ½ in. to 1 ft., and the remaining figures to the scale of 2 in. to 1 ft. This fitment affords

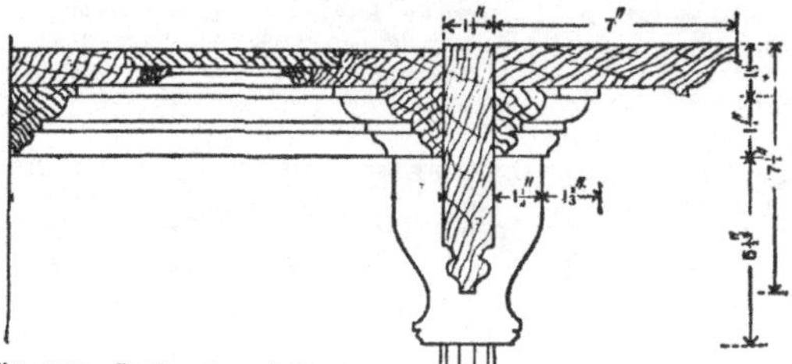

Fig. 1502.—Section through Canopy of Indoor Recess Seat.

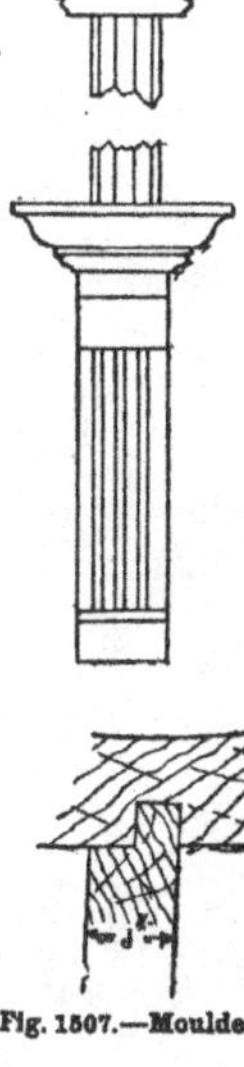

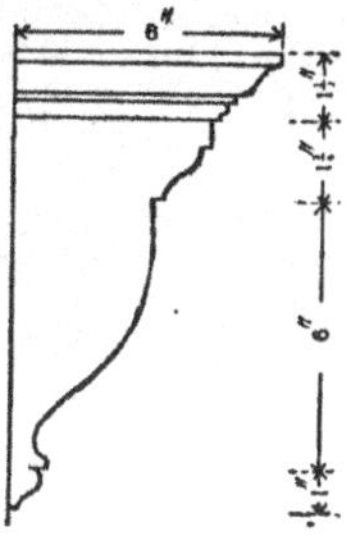

Fig. 1503.—Bracket of Recess Seat Canopy.

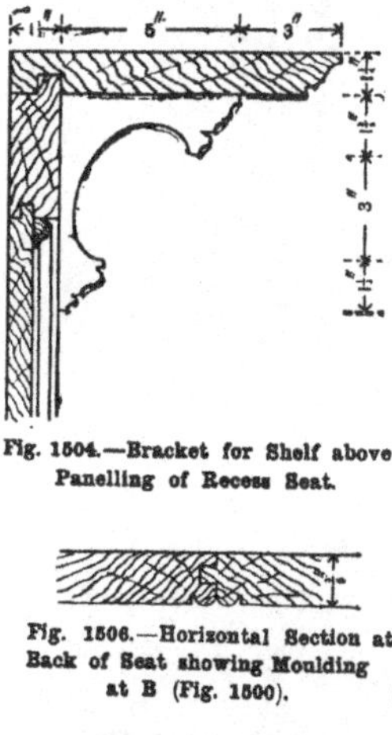

Fig. 1504.—Bracket for Shelf above Panelling of Recess Seat.

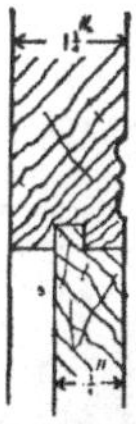

Fig. 1505.—Vertical Section at Back of Seat showing Moulding at A (Fig. 1500).

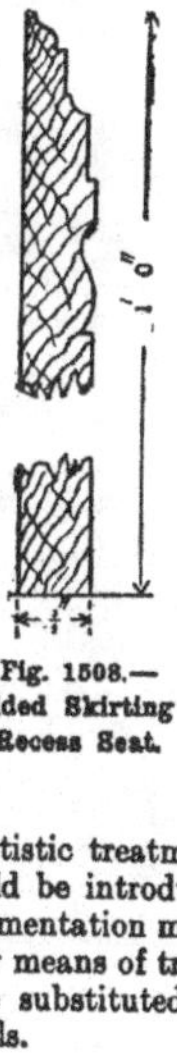

Fig. 1508.—Moulded Skirting of Recess Seat.

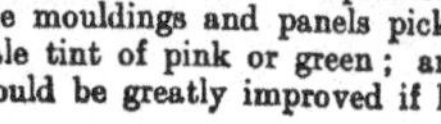

Fig. 1506.—Horizontal Section at Back of Seat showing Moulding at B (Fig. 1500).

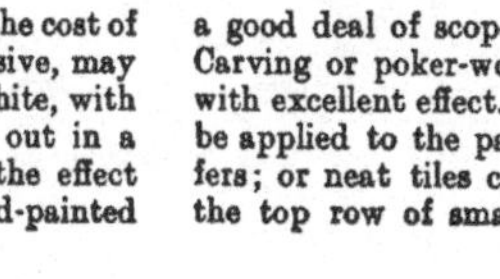

Fig. 1507.—Moulded End of Recess Seat.

the detail (Fig. 1507). The work, if the cost of constructing it in oak is too expensive, may be made in red deal and painted white, with the mouldings and panels picked out in a pale tint of pink or green; and the effect would be greatly improved if hand-painted a good deal of scope for artistic treatment. Carving or poker-work could be introduced with excellent effect. Ornamentation might be applied to the panels by means of transfers; or neat tiles could be substituted for the top row of small panels.

Ingle Nook for Drawing-room or Boudoir.

A general view of an ingle nook suitable for a drawing-room or boudoir is presented by Fig. 1509. Vertical section, half elevation, and half horizontal section are shown by Figs. 1510 to 1512. An enlarged detail of the plinth is illustrated by Fig. 1513; an enlarged section through the cornice by Fig. 1514; and an enlarged section through the end panelling by Fig. 1515.

Cosy Corner Settee.

A settee of the form shown by Fig. 1516, commonly described as a cosy corner, is particularly adaptable to large rooms. It will look well if made in mahogany, upholstered in plain crimson velvet, or

Fig. 1509.—Ingle Nook for Drawing-room or Boudoir.

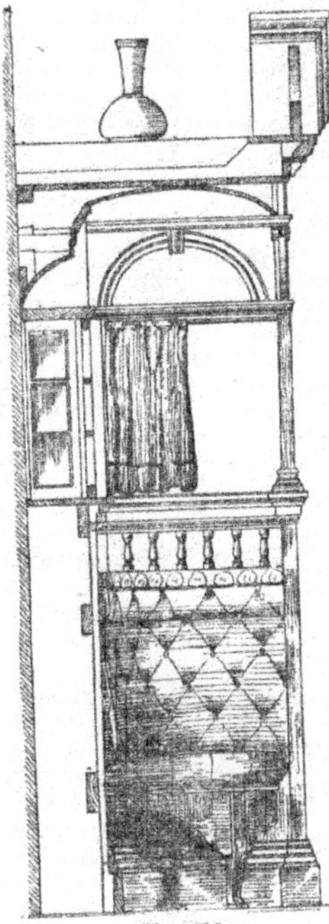

Fig. 1510.

Fig. 1511.

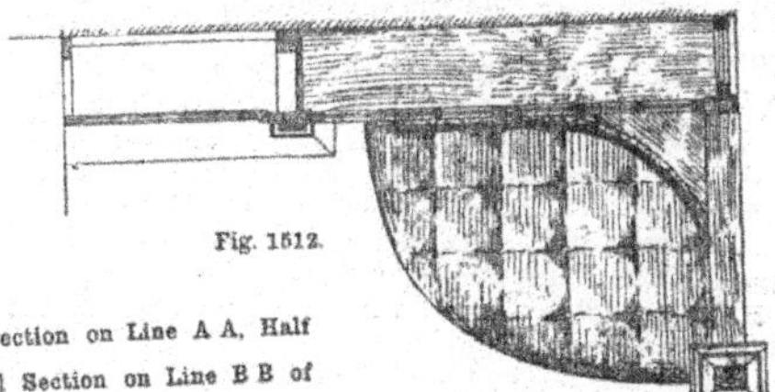

Fig. 1512.

Figs. 1510 to 1512.—Vertical Section on Line A A, Half Elevation, and Half Horizontal Section on Line B B of Ingle Nook.

the frame may be entirely of birch and white enamelled. Walnut with blue velvet, and oak with green velvet, are also suitable. The overall dimensions are 4 ft. by 4 ft., by 4 ft. 6 in. high. Fig. 1517 shows the framework, and gives the necessary dimensions. Supposing mahogany to be the material chosen, all the woodwork that shows should be made in that wood, but birch may be used for the remainder. The various parts should be got out to the following finished sizes :—One corner upright, 4 ft. 3 in. ; two back uprights, 3 ft. 2 in. ; two front uprights, 1 ft. 10 in. ; one leg, 11 in. ; two top uprights, 1 ft. 2 in. ; two end arm rails, 1 ft. 6 in.—all 1½ in. square ; two end

Fig. 1513.—Section of Ingle Nook Plinth.

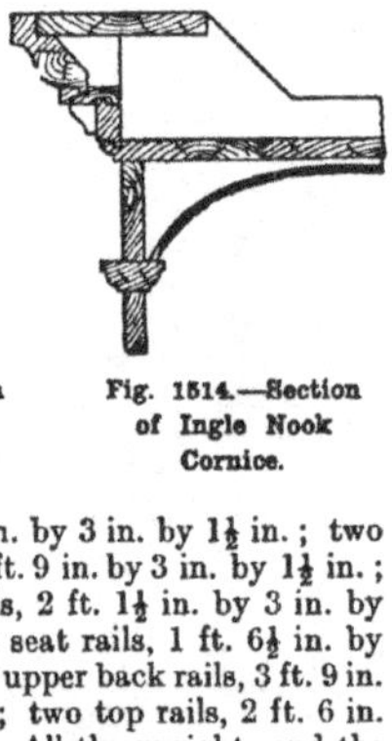

Fig. 1514.—Section of Ingle Nook Cornice.

seat rails, 1 ft. 6 in. by 3 in. by 1½ in. ; two back seat rails, 3 ft. 9 in. by 3 in. by 1½ in. ; two front seat rails, 2 ft. 1½ in. by 3 in. by 1½ in. ; four cross seat rails, 1 ft. 6½ in. by 2 in. by 1 in. ; two upper back rails, 3 ft. 9 in. by 2 in. by 1½ in. ; two top rails, 2 ft. 6 in. by 3½ in. by 1 in. All the uprights and the arm rails should be of solid mahogany ; the front and end seat rails, and the upper back and top rails, will do in birch faced with ¼-in. mahogany ; and the cross and back seat rails may be of solid birch. It is important that the ends of the rails shall be squared true. The dowelled joint is employed in putting the framework together, as in Fig. 1518, which shows the joint of the seat rail and upright.

End Frames.—The two end frames should be made first. These must have the three ornamental pieces, of 3-in. by ¾-in. mahogany, put in at the same time, in the manner shown by Fig. 1519. These end frames are then connected to the back corner upright and front leg by the six long rails, which must be put together in one operation. Fig. 1520 shows the pattern of the top rails, which should be cut before being dowelled in place ; they are fixed flush with the top ends of the corner and short top uprights, and ⅛ in. from the front side. The cross rails are intended to give strength to the frame, and support the upholstered seat ; they are let ½ in. into the front and back rails, being glued in and nailed from the top edge. To make it still more rigid, braces of hardwood may be fitted and glued and screwed in place at the corners (see Fig. 1521).

Brackets.—The next thing will be to make eight brackets to the patterns shown by Fig. 1522 ; these are fitted under the

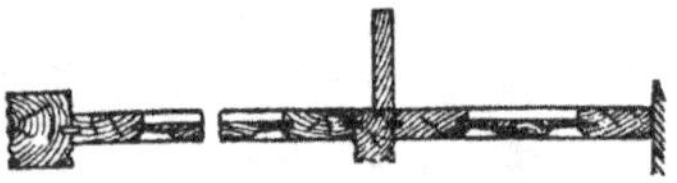

Fig. 1515.—Section through End Panelling of Ingle Nook.

seat rails, and are kept ⅛ in. back from the front. They are of ¾-in. mahogany, 5 in. by 5 in., with the grain running as in the diagram. Besides improving the appearance, they add to the strength of the settee if neatly fitted, sprigged, and well glued in place. Fig. 1523 is the pattern for the upper brackets, which are also of ¾-in. material. These may be glued and lightly sprigged, and small blocks fixed behind.

Shelf for Top.—The shelf for the top is of 1-in. board (see Fig. 1524), mitre-jointed as shown in Fig. 1525, and afterwards moulded on the under front edge to a flat ogee (see Fig. 1526) ; it is screwed flush with the back of the uprights. The front uprights are crowned with pieces 3½ in. square by 1 in. thick, which are moulded all round on the under side and dowelled to the uprights (see Fig. 1527). The ornaments of the back uprights are 3 in. by 2½ in. in section, and of vertical grain. To make these, two blocks are required 3½ in. long by 3 in. square, ¾ in. of the length being turned to a pin for fixing (see Fig. 1528). The ornament itself will require to be carved to the shape.

Completing the Woodwork.—The work is now ready for polishing. To do this, it is first necessary to clean it up with glasspaper; and the sharp corners of the uprights and rails should be slightly rubbed down. After polishing, ascertain that the six legs are quite level. This is easily done by stretching a string tightly over the extreme ends of two back legs, and comparing with the others. Square rim castors are then fitted to the legs, and the woodwork portion is complete.

Upholstering Cosy Corner.—The upholstering of the settee is perfectly straightforward. A frame for the seat and two back frames are required; these are made in sound deal, of 3 in. by 1 in. section. For the seat frame will be required: Two back rails, one 3 ft. 3½ in., and the other 3 ft. 6½ in.; two front rails, one 3 ft. 3½ in., and the other 1 ft. 10½ in.; one cross rail, 1 ft. 2 in. long; and two end rails, each 1 ft. 8 in. These are the finished sizes when the ends have been squared up ready for jointing. The 3-ft. 3½-in. back and front rails must first be dowelled to the left-hand end rail and the other back rail, with the cross rail about midway between; then the other front and end rails may be joined on. For the back frames the following are required: Four rails, 3 ft. 3 in.; four stiles, 2 ft.; and two stiles, 1 ft. 6 in. The rails are dowelled to the long stiles, with the short ones midway. The seat frame may now be fitted in place, allowing about ¼ in. play at the ends and the back corner, and about ⅛ in. from the surface of the front rails; it should rest on the back rails about ½ in. When in its exact position, it can be temporarily fixed with a few nails; then four holes should be bored with a ½-in. dowel bit, at points about 6 in. from the ends and from the front corner, and ¾ in. from the front edge. They should go right through the frame, and about 1 in. deep into the seat rails. The frame may then be removed, and dowels, slightly pointed, glued in the seat rails, to stand up about ¾ in. These will keep the upholstered seat in place. Pieces of wood of 1 in. square section are screwed on the upper side, flush with the front edges, as shown in Fig. 1529, which is a plan of the seat frame, showing the webbing on the under side and the

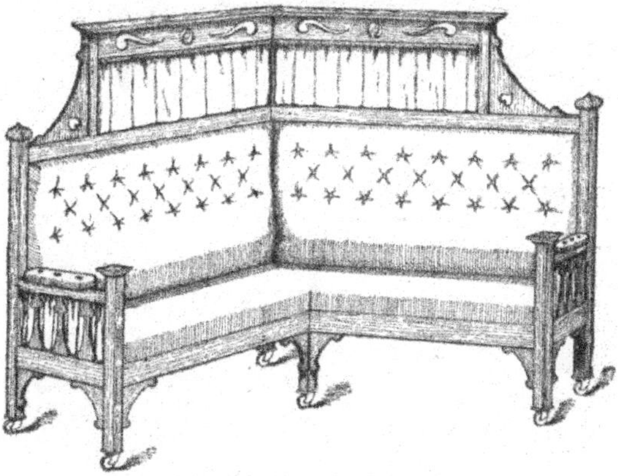

Fig. 1516.—Cosy Corner Settee.

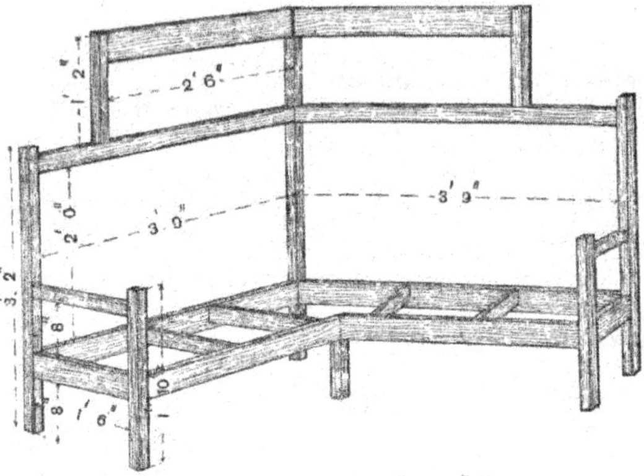

Fig. 1517.—Framework of Cosy Corner Settee.

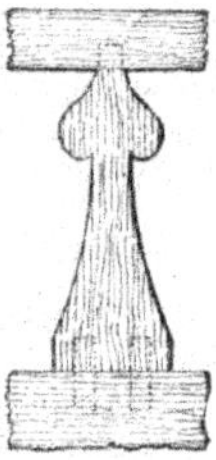

Fig. 1519.—Ornamental Piece in End Frame of Cosy Corner Settee.

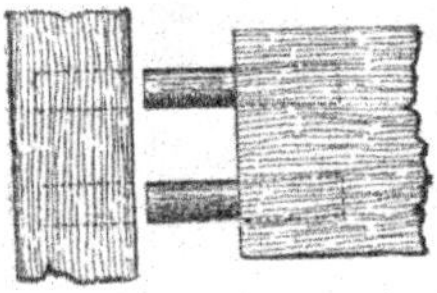

Fig. 1518.—Joint of Rail to Leg of Cosy Corner Settee.

Fig. 1520.—Part of Top Rail of Cosy Corner Settee.

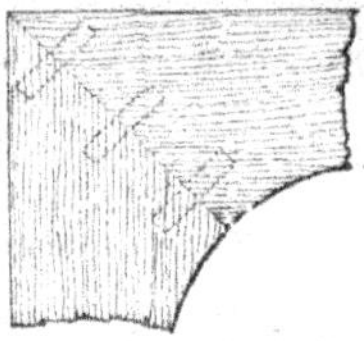

Fig. 1525.—Joint of Shelf in Cosy Corner Settee.

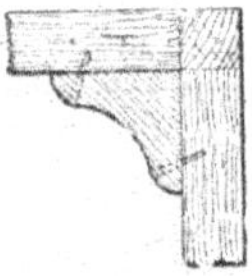

Fig. 1521.—Fixing Brace to Seat Rails of Cosy Corner Settee.

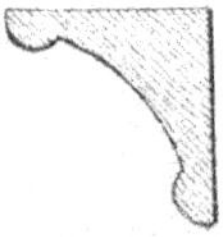

Fig. 1522.—Bracket for Legs of Cosy Corner Settee.

Fig. 1524.—Half Plan of Top Shelf of Cosy Corner Settee.

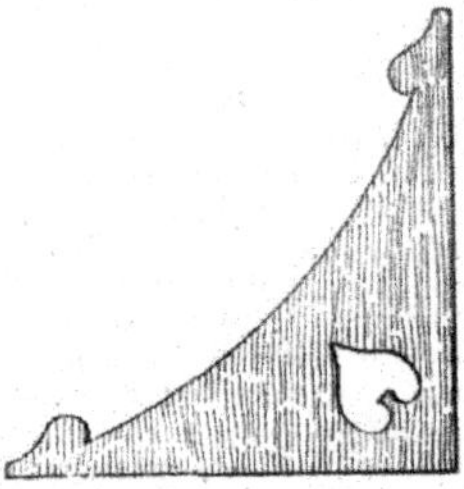

Fig. 1523.—Back Bracket of Cosy Corner Settee.

Fig. 1526.—Section of Shelf of Cosy Corner Settee.

Fig. 1527.

Fig. 1528.

Figs. 1527 and 1528.—Top Ornaments of Front and Back Uprights of Cosy Corner Settee.

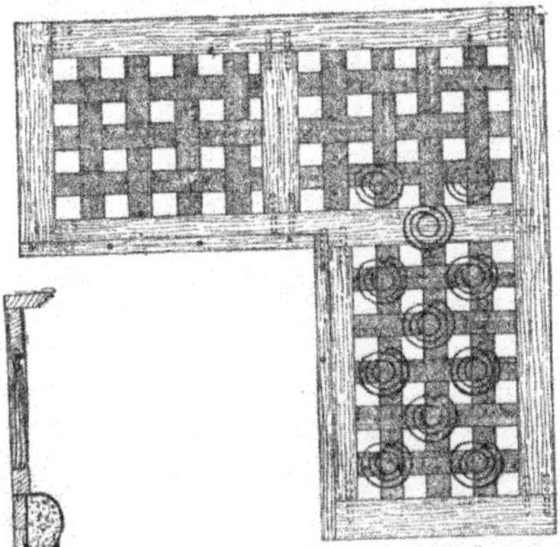

Fig. 1529.—Plan of Seat Frame of Cosy Corner Settee.

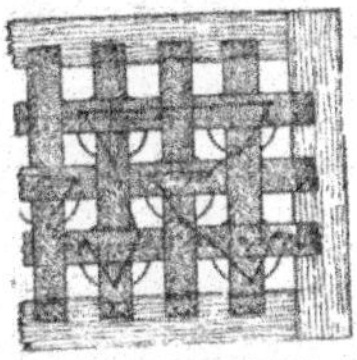

Fig. 1530.—Part Underneath View of Seat Frame of Cosy Corner Settee.

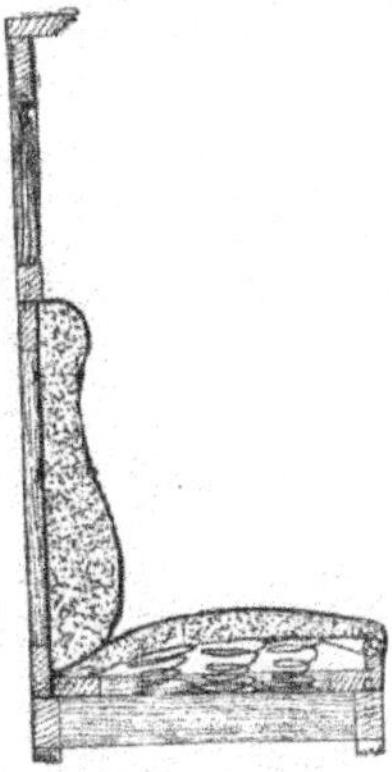

Fig. 1534.—Section through Seat and Back of Cosy Corner Settee.

Fig. 1532.—Method of Fixing Backs of Cosy Corner Settee.

Fig. 1535.—Plan of Arm Pad.

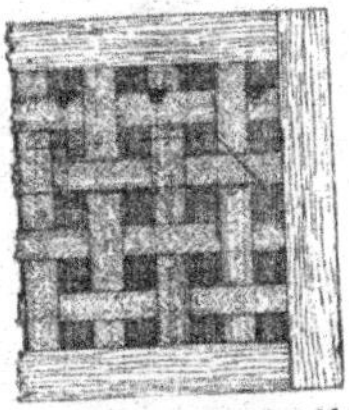

Fig. 1533.—Part Outside View of Back of Cosy Corner Settee.

Fig. 1531.—Section of Seat of Cosy Corner Settee.

Fig. 1536.

Fig. 1537.

Figs. 1536 and 1537.—Part Side Elevation and Section of Arm Pad.

position of the springs. The webs should be stretched as tight as possible, and secured with $\frac{5}{8}$-in. large-headed tacks. The springs are stitched in as shown in Fig. 1530, each spring being fixed with a knot at three points. They are then laced down evenly with strong twine, first crosswise, then lengthwise. This is done by partly driving a tack into the back rail directly behind the springs, tying on the string and diiving home. The string is twice slip-knotted to the top coil of the springs, which can then be drawn down to the required level, the twine being secured with a tack at the front edge (see Fig. 1531). They are treated lengthwise in the same manner. To pad the front of the seat, a strip of canvas about 6 in. wide must first be tacked along, then a

Fig. 1538.—Drawing-room Settee.

Fig. 1539.—Front Elevation of Drawing-room Settee.

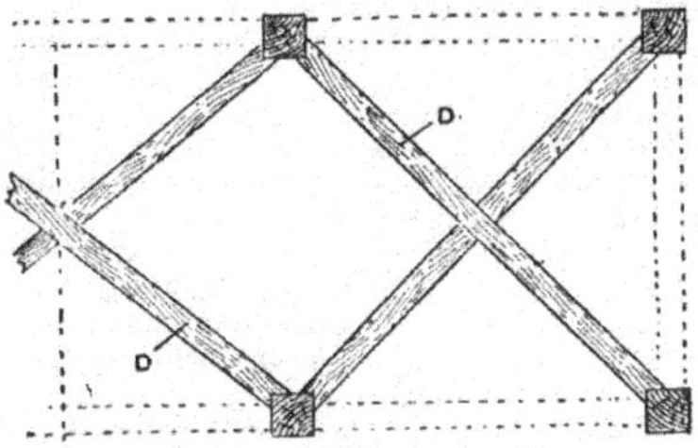

Fig. 1541.—Part Plan of Cross Rails of Drawing-room Settee.

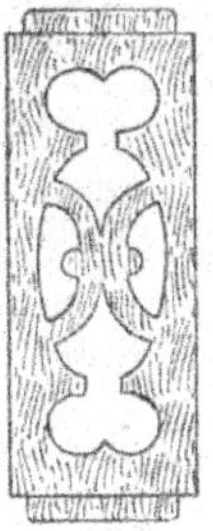

Fig. 1542.

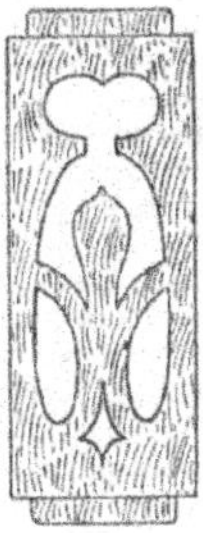

Fig. 1543.

Figs. 1542 and 1543.—Balusters of Settee.

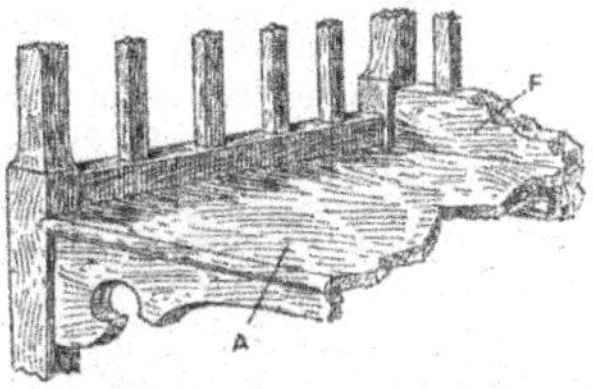

Fig. 1544.—Inside View of Settee Seat and End.

Fig. 1545.—Fixing Inner Legs of Settee.

Fig. 1546.—Jointing Cross Rail to Leg of Settee.

Fig. 1547.—Cross Rail Screwed to Leg of Settee.

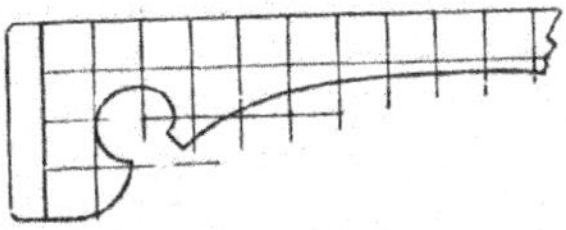

Fig. 1548.—Diagram of Settee Span-rail.

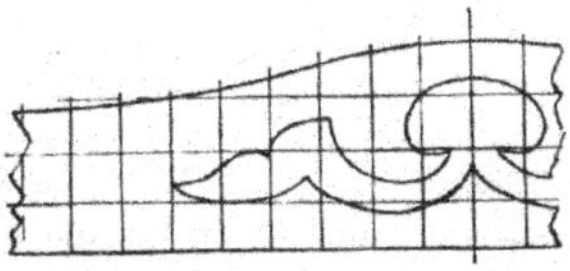

Fig. 1549.—Diagram of Settee Top Rail.

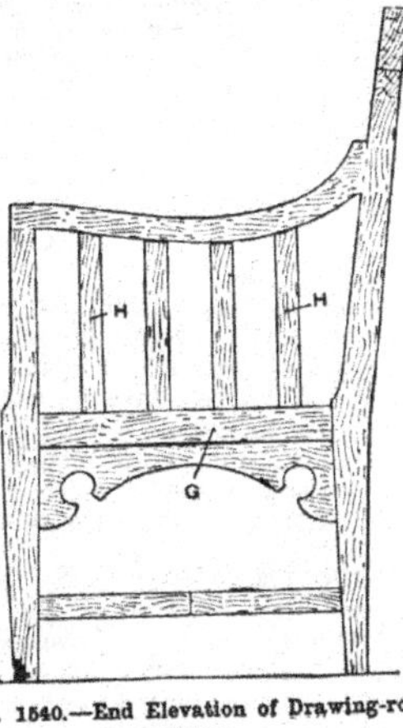

Fig. 1540.—End Elevation of Drawing-room Settee.

string should be stretched along the top side on tacks at each corner and midway, to keep in place the stuffing, which is packed under it. Hair or fibre is generally used for the stuffing, partly because it is better for stitching than flock. It should be put on to be about 1 in. higher when the canvas is tacked over, and may then be stitched to give it firmness and shape. A good strong canvas must be put over the springs and padding, stretched tight, and stitched to the springs. The stuffing, which may be of hair or wool flock, can now be distributed evenly over it. Hair has the advantage of lightness, and keeps soft much longer, but it is more expensive. There are cheaper qualities, but a medium wool flock is preferable; the cotton flock is mostly used for cheap work, and is very heavy. A covering of "scrim" (a light canvas) or calico is then put over, and any unevenness in the shape put right with a regulator or a steel knitting wire. A sheet of wadding will further modify any slight lumpiness in the appearance, after which the final velvet covering may be put on. This will require to be joined at the mitre, and is tacked on the under front edge first. Width may be given to the velvet, if necessary, by stitching black linen on the back edge.* It must be free from any looseness or wrinkles. A black linen cover tacked on the under side completes the seat. The backs may now be taken in hand, the frames being first fitted to allow about $\frac{3}{16}$ in. all round for the stuff. They are kept in place by $\frac{1}{2}$-in. dowels let into the upper back rails, four to each, to stand out about $\frac{3}{4}$ in., these fitting into corresponding holes in the frame (see Fig. 1532). They are webbed on the front side, and covered with canvas; then strings are stretched on tacks from corner to corner, to keep the stuffing on. This must be kept $1\frac{1}{4}$ in. from the lower edge, and should swell to about $4\frac{1}{2}$ in. high till about half-way up, then hollow to $2\frac{1}{2}$ in.; to rise again 3 in. high for the top $4\frac{1}{2}$ in.; this is when the scrim covering is on. They will require some regulating before the wadding and velvet are put on, and when this is done they are ready for buttoning. The buttoning is necessary to keep the stuffing in place and preserve the shape, besides improving the appearance. There are three rows of buttons, the first being $4\frac{1}{2}$ in. from the top, and about 6 in. apart from each other. Another row is put 7 in. lower, directly under, and the others between. The quickest way is first to mark them out with a touch of chalk; then the needle and stitching twine, having a knot tied on the end, must be passed through from the back, coming out at the mark; it is passed through the button, entering again about $\frac{3}{8}$ in. from the first piercing, and so on, along each line; they should not be drawn very tight. It then has the appearance shown by the two lower lines in Fig. 1533. When the last button is put in, a slip-knot is made with the free end and the last loop, a leather tuft put under, and the string drawn tight and secured; and so on, back to the first button (see upper line of Fig. 1533). When the outside back of black linen is put on, the backs are slipped in place and screwed to the back edge of the seat frame (see Fig. 1534). Two arm pads are required, for which two pieces of wood must be got out 1 ft. by $2\frac{1}{4}$ in. by $\frac{3}{4}$ in. They are stuffed about 2 in. above the wood, with a scrim covering, and stitched all round, as shown in plan by Fig. 1535 and in part side view by Fig. 1536. A little more stuffing must be added, finished off with wadding, then the velvet. Before it is tacked down, the buttoning must be done. Three buttons to each arm will be enough, and they are put in by securing the string with a tack at the side, passing the needle up through and out at the top centre, through the button, to the other side, drawing tight and fixing with a tack (see Fig. 1537). The velvet is tacked to the under side of the wood, the pads being then secured with screws driven through the arm rails from underneath. A thin brass rod is fitted under the arm rails behind the ornamental pieces to take curtains of silk plush, and others under the top back rail.

Drawing-room Settee.

In constructing a settee as illustrated in Fig. 1538 upholstering is avoided, as the seat is detached, being known as a squab seat. The settee will look well if made of mahogany, french-polished, or it may be painted and finished with white enamel. In the latter case, close-grained birch for the legs is recommended, and kauri pine or birch for the other parts. The front elevation (Fig. 1539), end elevation (Fig. 1540), and half plan (Fig. 1541) are drawn to a scale of 1 in. to the foot, so that the measurements of the various parts can be taken from the drawings. However, to avoid mistakes, the extreme length is 5 ft.; height from floor to wood seat A (Fig. 1539), 1 ft. 2 in.; height of back at sides, 2 ft. 9 in.; outside measurement from front to back of seat, 1 ft. 8 in.; thickest parts of legs, 2 in. square, tapering to 1⅜ in. square at the bottoms. The back leans 3 in. out of the perpendicular, and the perforated balusters (see enlargements Figs. 1542 and 1543) and the straight or shaped balusters B and C (Fig. 1539), which are alternative patterns, are of 1-in. stuff, planed up as thick as they will carry. The shaped span-rails below the seat are of 1½-in. stuff, finishing when planed up about 1¼ in. full, and standing back $\frac{1}{16}$ in. from the faces of the legs. The cross rails D (Figs. 1539 and 1541) are of 1-in. stuff and 1⅜ in. wide. The seat A (Fig. 1539) should be got out of 1-in. stuff, and when planed up it will finish about ⅞ in. thick. The top rail E (Fig. 1539) and the seat rail F are of 1½-in. stuff. It will be seen that the portions of the front and the back legs above the seat are reduced in thickness, say to about 1¼ in. at the top, and tapering to 1½ in. near the seat. The seat ends with the arms should be first taken in hand, the end rail G (Fig. 1540) being tenoned into the front and the back legs, and also the shaped span-rail beneath. Next the back parts of the arms are tenoned into the back legs, the front parts of the arms having the mortices, and the top of the front legs the tenons. The end balusters H are tenoned into the seat rail as shown in Fig. 1544. The front shaped span-rails below the seat are cut out in one length, and the outer ends J (Fig. 1539) are tenoned into the end legs. The inner legs are cut away, forming a continuous mortice, as shown in Fig. 1545. The inner legs can be further strengthened by driving screws through the back parts of the legs and into the span-rails. The cross rails D (Figs. 1539 and 1541) are halved where they cross. In connecting the rails with the legs, the latter are cut away as in Fig. 1546, thus forming a recess to receive the butt ends of the rails. When finally putting the framing together, due consideration must be given to getting the cross rails into position. The latter are further secured by slanting screws driven from the under side of the rails as shown in Fig. 1547. As an alternative to having the rails D (Fig. 1539) crossing, they can be carried straight from leg to leg. The seat A (Fig. 1544) is fitted against the back rail F and between the ends. To support the seat at the back and the ends, strips of wood about 1½ in. wide by 1 in. thick are screwed to the back rail and the insides of the ends. The front edge of the seat is rounded. The lower ends of the balusters in the back are tenoned into the back rail F (Fig. 1539), and the upper ends into top rail E. In fixing the top rail E, the mortices are cut in the latter, and the tenons on the top ends of the back legs. Fig. 1548 is an enlarged view (set out in squares for reproduction) of the shaped span-rails, and Fig. 1549 is an enlargement of the centre portion of the top rail with the cut-through ornament. The loose or squab seat should be about 2½ in. thick. Horsehair is about the best material for stuffing; but should this be too expensive, cheaper materials are fibre or flocks. The covering can be of tapestry or rep, the design and colouring to accord with the rest of the furniture.

Farmhouse Settle.

A farmhouse settle is shown in elevation and sections by Figs. 1550 to 1552. A list of the pieces of wood necessary will be useful. Two ends are required, 6 ft. by 2 ft. 1¾ in., by 2 in. thick; the seat is 8 ft. 10 in. by 1 ft. 9 in. by 1¼ in.; the top is 9 ft. 5 in. by 1 ft. 0½ in. by 1¼ in.; the moulding is 14 ft. 6 in. by 2¼ in. by 1⅛ in.; the back is 4 ft. 6 in. by 6 in. by 1¾ in., with four pieces each 4 ft. 6 in. by 8 in. by 1¾ in.; two pieces each 4 ft. 6 in. by 4½ in. by 1¾ in.; sixteen pieces each 2 ft. 3 in. by 6½ in. by 1 in.; sixteen pieces 7 in. by 6½ in. by 1 in.; seventeen pieces 1 ft. 6 in. by 6½ in. by 1 in.;

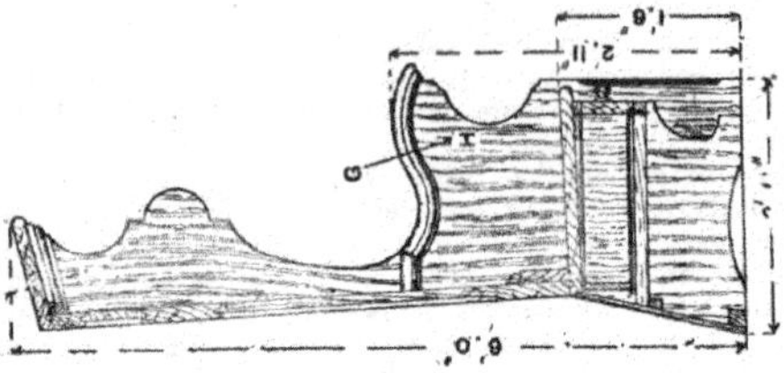

Fig. 1552.

Figs. 1550 to 1552.—Front Elevation, Horizontal Section, and Vertical Cross Section of Farmhouse Settle.

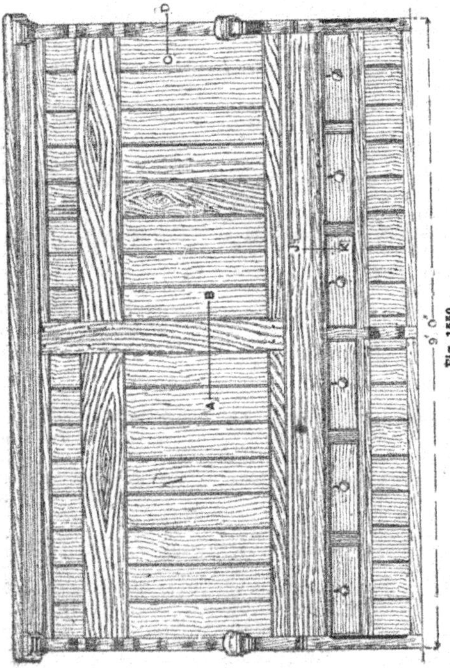

Fig. 1550.

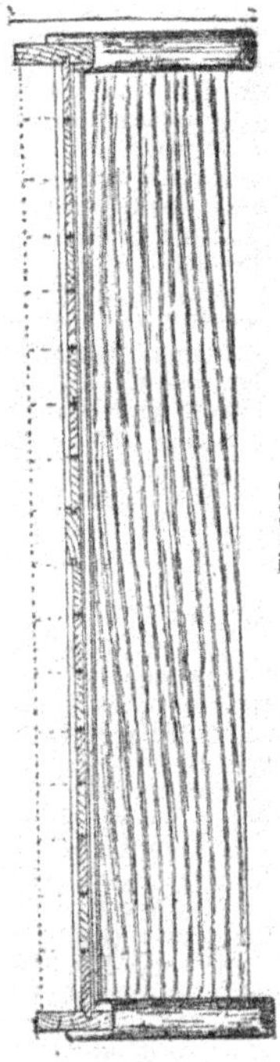

Fig. 1551.

one piece 9 ft. by 2 in. by 2 in.; one piece 9 ft. by 1½ in. by 1¼ in.; and one piece 8 ft. 10 in. by 6 in. by 1 in. For the drawers, have two pieces each 4 ft. 6 in. by 1⅛ in. by 1¼ in.; two pieces 4 ft. 6 in. by 2 in. by 1¼ in.; four pieces 1 ft. 9 in. by 3 in. by 1¼ in.; four pieces 1 ft. 9 in. by 3 in. by 1 in.; four pieces 8 in. by 2 in. by 1¼ in.; one piece 1 ft. 7 in. by 1 ft. 10 in. by 2 in.; six pieces 1 ft. 5 in. by 5¼ in. by 1⅛ in.; twelve pieces 1 ft. 7 in. by 5¼ in. by ½ in.; six pieces 1 ft. 5 in. by 4½ in. by ½ in.; six pieces 1 ft. 5 in. by 1 ft. 6 in. by ½ in.; two pieces 1 ft. 9 in. by 4½ in. by 3¾ in.; and two pieces 10 in. by 2½ in. by 1¼ in.; with six drawer knobs and six drawer locks. The above is a complete list of the pieces required. Fig. 1553 is a section on A B (Fig. 1550); Fig. 1554 is a section on C D (Fig. 1550); Fig. 1555 a section of the top; Fig. 1556 a section on G H (Fig. 1552); and Fig. 1557 a section on J K (Fig. 1550). Large settles are a common feature in farmhouse kitchens, but the above, which is 9 ft. 6 in. long over all, may be reduced in size by altering the dimensions to suit any room.

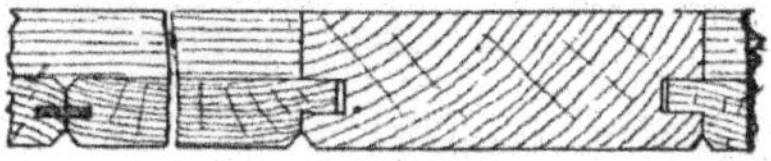

Fig. 1553.

Fig. 1554.

Figs. 1553 and 1554.—Horizontal Sections of Settle Back on Lines A B and C D (Fig. 1550) respectively.

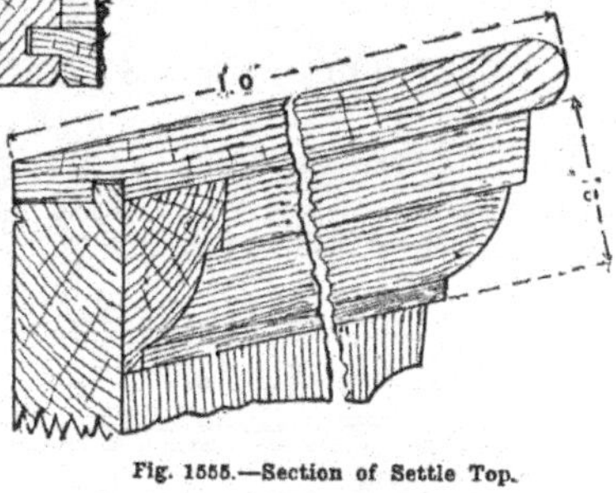

Fig. 1555.—Section of Settle Top.

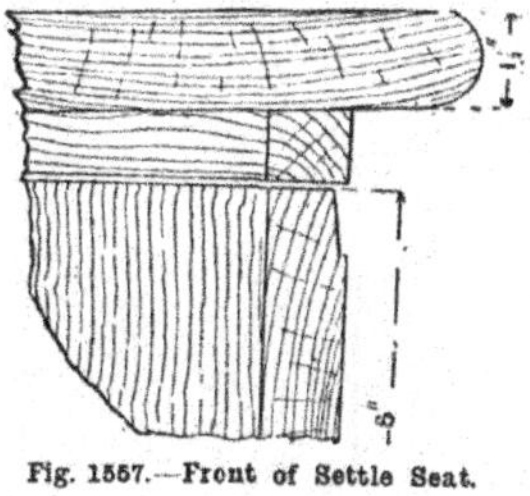

Fig. 1557.—Front of Settle Seat.

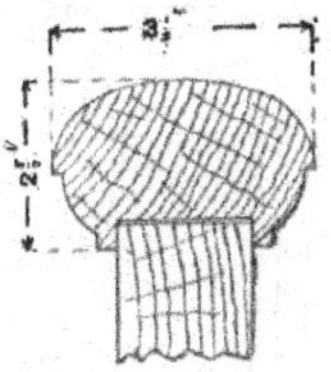

Fig. 1556.—Cross Section of Settle Arm.

MIRROR FRAMES AND PICTURE FRAMES.

Wall Mirror.

The mirror illustrated in Figs. 1558 and 1559 is suitable for hanging over a mantelpiece in a parlour or drawing-room. It will look well if painted or enamelled white and made of pine. The extreme length is 3 ft. 10 in., and the height is 3 ft. 4 in.

Fig. 1558.

Fig. 1559.

Figs. 1558 and 1559.—Front and Side Elevations of Wall Mirror.

Fig. 1560.—Part Section of Wall Mirror Frame.

Fig. 1561.—Side Elevation of Mirror Cornice.

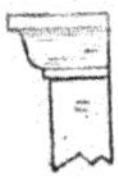

Fig. 1562.—Side Elevation of Mirror Pediment.

The foundation-frame is of 1¼-in. stuff, the side stiles being 2 ft. 11½ in. by 7½ in. wide. The top rail is 11 in. wide and 3 ft. 2 in. long, this allowing for tenons 4 in. long, and the bottom rail is of the same length and 9 in. in width. The rails are tenoned into the stiles. Fig. 1560 is a section of a stile, A being the inside edge of the frame, which is cut to shape with a fret- or bow-saw. The strips of ½-in. stuff B are nailed and glued on the face, and form the rebate for glass, the width at the narrow part

Fig. 1563.—Front Elevation of Fretted Oak Frame for Oil Painting.

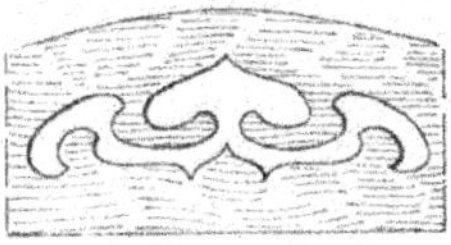

Fig. 1564.—Fretwork of Frame shown Enlarged.

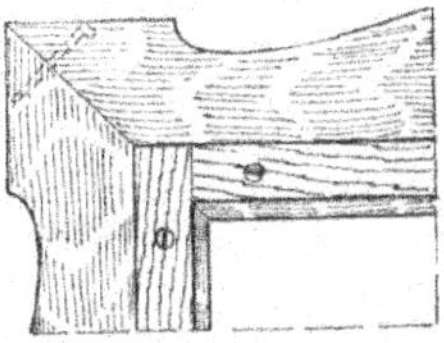

Fig. 1565.—Back Corner of Fretted Frame (see Fig. 1563).

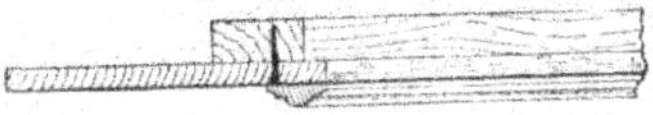

Fig. 1566.—Part Horizontal Section of Fretted Frame (see Fig. 1563).

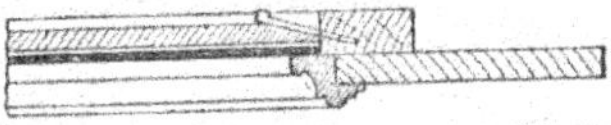

Fig. 1568.—Part Horizontal Section of Fretted Frame (see Fig. 1567).

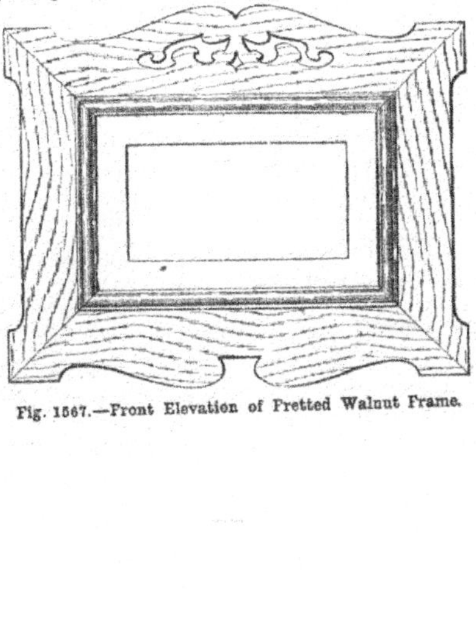

Fig. 1567.—Front Elevation of Fretted Walnut Frame.

Fig. 1569.—Front Elevation of Mirror with Side Brackets.

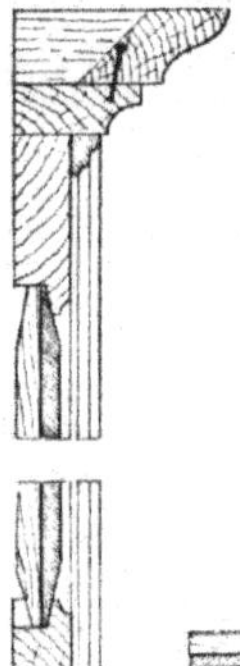

Fig. 1570.—Enlarged Vertical Section of Mirror.

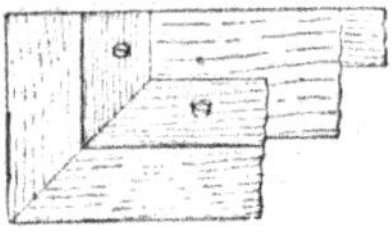

Fig. 1571.—Part Plan of Mirror Cornice.

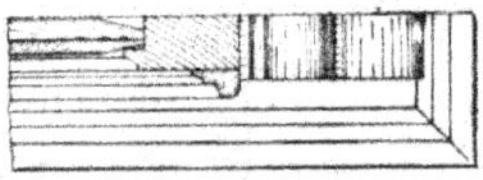

Fig. 1572.—Part Horizontal Section of Mirror Frame above Brackets.

being $2\frac{5}{8}$ in. A small hollow is worked on the inside edge. At each end of the strips are pieces $5\frac{1}{2}$ in. by $1\frac{1}{8}$ in., jointed and glued to the outer edges to support the mouldings C where they break outwards at the corners. The mouldings are worked in suitable lengths with rebate and hollow planes, and are then fixed with glue and nails to the facing B. The cornice and side pediment mouldings (see Fig. 1561) are of 1-in. stuff $1\frac{1}{2}$ in. wide; they are worked and mitered round the ends of the frame. The side pediments are fixed on the face of the frame, and the return ends D (Fig. 1558) are carved with chisels and gouges, as these cannot be mitered and yet keep the fillets vertical. The moulding E on the centre of the pediment is 1 in. wide and $\frac{5}{8}$ in. thick, and is planted on the face of the frame and mitered round the ends (see Fig. 1562). The curved lines of the scrolls are carved with a small gouge or parting tool. To make a good job, the silvered glass should have a $1\frac{1}{4}$-in. bevel. The glass is fixed by small wedge-shaped blocks about 2 in. long, and placed at intervals as in Fig. 1560. These blocks may be used as a support for the $\frac{1}{2}$-in. back, which is secured with small screws driven on the slant into the frame. For fixing to the wall, brass plates —that is, brass plates with three holes for screws—are screwed to the back of the frame, and the wall is drilled and plugged to receive screws which pass through the plates.

Frames for Oil Paintings.

The frames about to be described afford some variety and originality not usually associated with those made entirely from stock mouldings, which are in many cases but crude, over-coloured, and gaudy productions. The wood used should be selected with care, as the finish is either brown or art green stain, and french polish. Therefore those woods having a beautiful natural grain will give the best results when finished. The frame shown in Fig. 1563 is suitable for an oil painting or any highly coloured picture. There is, of course, a limit to the size in which a frame of this design will look best; probably any size over

2 ft. 6 in. wide will appear the reverse of pleasing. In the present case the proportionate sizes are as follows: Size of picture accommodated, 1 ft. 6 in. by 1 ft.; border moulding, $1\frac{1}{8}$ in. wide; outer fretted boards, 5 in. wide by $\frac{1}{2}$ in. thick. The stuff for the outer frame is first planed up smooth, and cut off in lengths 1 in. longer than finished

Fig. 1573.—Mirror with Candle Sconces.

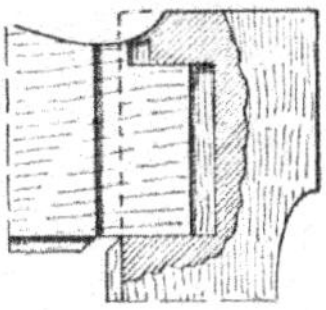

Fig. 1574.—Joint of Mirror Rails to Stiles.

Fig. 1577.—Quarter Plan showing Method of Striking Out Saucer of Candle Sconce.

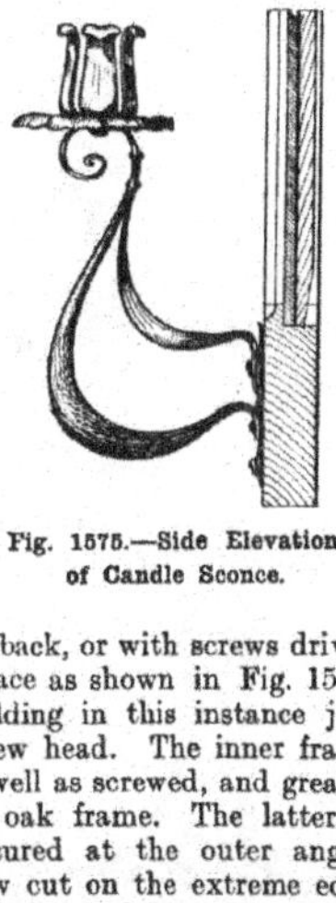

Fig. 1575.—Side Elevation of Candle Sconce.

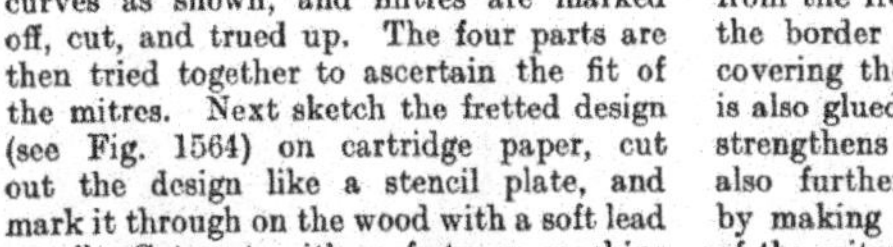

Fig. 1576.—Development of Scroll in Candle Sconce.

size. The outer edges are shaped to the curves as shown, and mitres are marked off, cut, and trued up. The four parts are then tried together to ascertain the fit of the mitres. Next sketch the fretted design (see Fig. 1564) on cartridge paper, cut out the design like a stencil plate, and mark it through on the wood with a soft lead pencil. Cut out with a fret-saw machine for preference. Finish off with a file and glass-paper. A sharp, well-defined outline is essential with this class of design. The mitres can now be glued and cramped up till set. Next fit up a frame of soft wood from stuff $1\frac{1}{2}$ in. wide by $\frac{7}{8}$ in. thick, the joints at the angles being halved and glued together. This frame is made $\frac{3}{8}$ in. larger than the inside edge of the oak frame, so as to form a rebate for the glass and picture as shown in Fig. 1565. It is secured with screws from the back, or with screws driven from the front face as shown in Fig. 1566, the border moulding in this instance just covering the screw head. The inner frame is also glued as well as screwed, and greatly strengthens the oak frame. The latter is also further secured at the outer angles by making a saw cut on the extreme edge of the mitres, running it down about $1\frac{1}{2}$ in., and then inserting a glued slip or feather. The nail shown by dotted lines in Fig. 1565 is only recommended when soft woods,

Fig. 1578.—Ornamental Frame with Trellis Pattern.

Fig. 1579.

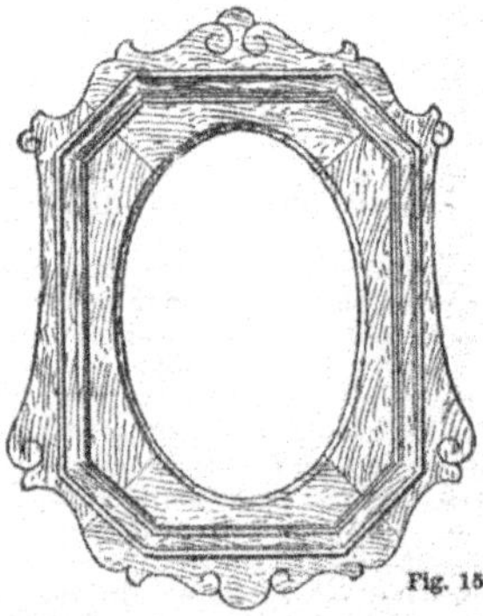

Fig. 1580.

Figs. 1579 and 1580.—Ornamental Frames with Mouldings and Oval Openings.

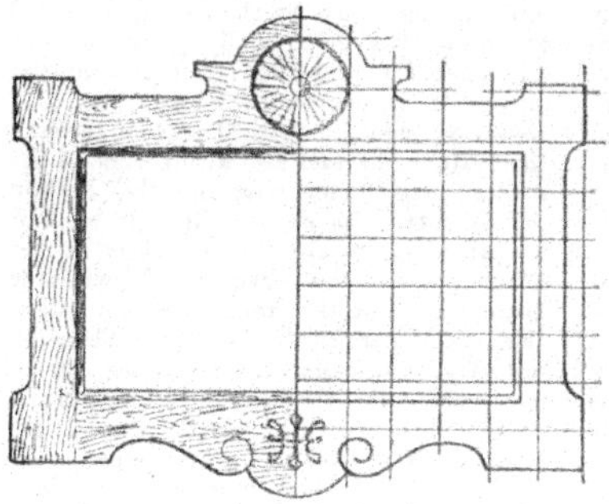

Fig. 1581.—Rectangular Frame with Circular Ornament in Pediment.

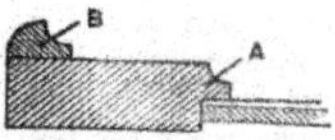

Fig. 1583.—Section of Frame shown by Fig. 1578.

Fig. 1584.—Portion of Carving on Frame shown by Fig. 1578.

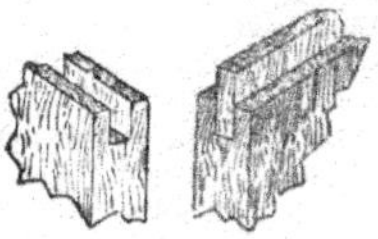

Fig. 1585.—Mitered Joint used in Frames (see Figs. 1579 and 1580).

Fig. 1586.

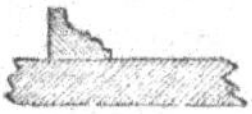

Fig. 1587.

Figs. 1586 and 1587.—Sections of Outer Mouldings of Frames shown by Figs. 1579 and 1580 respectively.

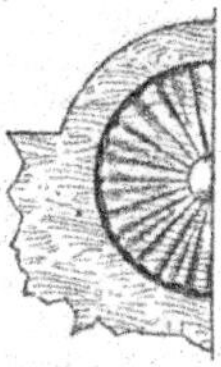

Fig. 1588.—Carved Pediment of Frame shown by Fig. 1581.

Fig. 1590.—Fretted Pediment of Frame shown by Fig. 1582.

Fig 1582.—Three-opening Frame with Fretted Pediment.

Fig. 1589.—Section of Carved Pediment (see Figs. 1581 and 1588).

Fig. 1591.—Shelf for Three-opening Frame.

Fig. 1592.—Bracket for Three-opening Frame.

such as basswood or satin walnut, are used. Fig. 1567 gives the design of a frame on similar lines, but with several distinctive features as compared with Fig. 1563. The top rail only is fretted, while the lower rail and uprights are of independent design. Also a bolection moulding is used for the inner border. This is shown in the enlarged sectional view (Fig. 1568).

Mirror with Side Brackets.

Fig. 1569 gives a front elevation of a mirror frame or overmantel of unique design. Any hardwood is suitable for its construction. A good size for the bevel-edge mirror is 1 ft. 2 in. wide by 1 ft. 6 in. high. The top and bottom rails are 4 in. wide by $\frac{7}{8}$ in. thick, the stiles are $3\frac{3}{4}$ in. wide at the top, and diminish below the curves to a width of $2\frac{1}{2}$ in. The rails and stiles have mortice-and-tenon joints, the rebates for the glass, etc., being worked out of the solid. The border moulding is worked, mitered, and glued on, and then two pieces are wrought and moulded and screwed to the bottom rail to form the base. Two shaped and carved brackets are also secured to the stiles and base, giving a pleasing effect to the lower part of the frame. On the top rail, a batten 2 in. wide by $\frac{3}{4}$ in. thick, with a mould on the front edge, and returned at the ends, is next secured, and above this a cornice mould is attached as shown in the enlarged vertical section (Fig. 1570) and part plan (Fig. 1571). Fig. 1572 is a horizontal detail section just above the carved bracket.

Mirror with Candle Sconces.

Fig. 1573 is a design for a mirror frame with wrought-metal candle sconces. The frame is simple in character, the decorative features being the sconces and the sunk ornament in the top rail. The rails and stiles are bevelled on the front inner edge, and also rebated. The thickness of the stuff used can vary from $\frac{3}{4}$ in. to 1 in., according to the size of the frame. The rails and stiles are mortised and stump-tenoned together. An enlarged view of the tenon entering the mortice in the stile is shown at Fig. 1574, which is a rear view, also showing the set-back for the moulded edges of the rebate to mitre together properly. The stiles should be left square on the top ends, as shown by the dotted lines, till the frame has been finally glued up and set, when the projections can be trimmed off to suit the curve on the rail. Fig. 1575 shows an enlarged vertical section through the bottom rail, and also a side view of the sconce. The inner edge of the bottom rail is curved as shown. It will be best to carry the rebate for the glass at the lowest point of the curve, parallel with the bottom edge of the frame, and not follow the inner curve of the rail. Thus an ordinary rectangular piece of glass can be used, and save the trouble of shaping it to fit the curve. The candle sconces can be made from sheet brass or copper. The latter metal is more suitable for an oak frame; the sheet metal should be about $\frac{1}{16}$ in. thick. The front scroll is bifurcated, its development being shown by Fig. 1576. After the metal has been marked out and cut to the shape given in Fig. 1576, it should be polished and then bent round to form a scroll. Then hammer at the back with the ball-pene of a hammer, laying the scroll on a pitch block or a block of hardwood or lead, so as to give the scrolls the rounded or repoussé appearance shown in Figs. 1573 and 1575. The front scroll is supported by a shorter one riveted at the upper end, and also to the back plate. The latter is of ivy-leaf form, and is attached to the frame with three small round-headed screws. The method of marking out the saucer is shown in quarter plan by Fig. 1577. The candle holder is cut from a piece of flat sheet, and consists of four or five petals projecting at equal distances round a $\frac{3}{4}$-in. diameter circle struck on the sheet metal. When the surplus metal has been cut and filed away round the petals, the latter are bent up to form the holder as seen in Figs. 1573 and 1575. The method of marking out on the flat is similar to that shown in Fig. 1577. The saucer and holder are fixed to the scroll either with a small copper rivet or a brass screw and nut.

More Ornamental Picture Frames.

Examples of cabinetwork frames are shown by Figs. 1578 to 1582. They may be made in almost any kind of wood—mahogany stained dark, American walnut, oak fumigated or stained in imitation of old oak, or American ash stained green, each being finished with french polish. The dimensions will be regulated by the size of the pictures or engravings to be framed. In making the frame illustrated by Fig. 1578, first get out the top and bottom and side-pieces of ¾-in. or 1-in. stuff, with the top and bottom rails tenoned into the side stiles, and work a hollow moulding on the inside edges, as shown at A in Fig. 1583. The trellis pattern is drawn on the wood, and then cut through with a fret-saw; the openings are cleaned up with a file and glass-paper. The ornamental raised border is got out of ⅜-in. stuff, the joins being made wherever the small scrolls come together. Tracings should be made of the several portions, and pasted on the ⅜-in. stuff, and then cut out with a bow- or fret-saw. The next operation is to carve them to the section shown at B (Fig. 1583). The terminal scrolls are rounded on the face as shown in Fig. 1584, with the addition of a small leaf where shown in Fig. 1578. In first-class work it is customary to carve and finish off the ornamental border before finally gluing it to the frame, as then the frame and the carving can be polished separately, thus ensuring a clean job without any sticky corners in the polishing. For convenience of carving, the sections are temporarily glued to a board, paper being inserted where glued, to enable the carving to be stripped off afterwards with a knife. The finished carving is glued to the frame, and when set the outside edges are cleaned off and glass-papered and then polished. The picture may be fixed in the usual way with thin back-boarding, and with brown paper pasted on to keep out the dust. The frame illustrated by Fig. 1579 is made of ¾-in. stuff, mitered at the corners; the edges are grooved, and a loose tongue inserted, as shown in Fig. 1585, and then glued. A hollow and a rebate are worked round the oval, as in Fig. 1583. A section of the outer moulding is shown in Fig. 1586. The small shaped pediment is of ¾-in. stuff, and may be fitted after the frame has been put together; small nails driven in from the top will secure it. The frame shown by Fig. 1580 is mitered together in the same way as Fig. 1579; the thickness is the same, but the shaped portions surrounding the moulding form part of the frame. To secure the moulding to the frame, the pieces may be screwed from the back. A section of this moulding is shown in Fig. 1587. The design given in Fig. 1581 is for a long frame, which is constructed in the same way as Fig. 1579, the section of moulding surrounding the picture also being the same. Half the circular ornament in the pediment is shown enlarged in Fig. 1588, and a section of the petals in Fig. 1589. The frame shown in Fig. 1582 is suitable for small engravings or photographs, and the shelves provide accommodation for small ornaments. The thickness of the wood is ¾ in. or 1 in., and to prevent the frame from warping it should be made like a drawing-board, by clamping the end-pieces as indicated by the grain of the wood in Fig. 1582. The openings for the pictures will have to be cut out with a fret-saw, and the hollow moulding and rebate carved with a gouge and chisel. The ornament in the pediment (shown enlarged in Fig. 1590) is cut through with a fret-saw. The top shelf may be of ½-in. stuff about 3 in. wide, and shaped as in the half-plan (Fig. 1591). Three small brackets, shaped as in Fig. 1592, will be required to support the shelves. The lower shelf is semi-circular in plan. The shelves are fixed with screws driven in from the back of the frame. The scrolls and ornamental lines at the bottom of this frame may be carved with a small gouge or parting tool.

Horseshoe Mirror with Glove-box.

The mirror illustrated by Fig. 1593 is so designed that hat and coat pegs may, if desired, be screwed to the upright bars. It is made preferably of good sound dry walnut, and is finished by polishing. Full details are shown in Figs. 1594 to 1599. The upright and cross-bars of the framework are dovetail-halved together. The ends of each bar are cut circular, and relieved by carving as shown in Fig. 1597. The mirror frame is made from a block of wood, cut to the requisite shape and then worked out to the section shown in Fig. 1596. After the representative nails have been driven in, the monotony of the sunk part is relieved by being slightly scored. The mirror plate has bevelled edges, and may be obtained ready cut and bevelled at a glass merchant's. It is kept in position by the backing board fitting flush over the glass and on to the edges of the frame, to which it is screwed (see Fig. 1596). The whole is fixed to the bar frame by screws passing through the cross-bars. The stud nails (Fig. 1599) are cut out of solid pieces of brass, a hole being drilled through the centre to receive a brass-headed nail. Another method of making them would be to cut them out of wood and glue in position, afterwards covering them with gold leaf. As may be seen in Fig. 1593, these nails are placed at the parts where the bars intersect each other. Below the bottom cross-bar and between the two upright bars is fixed a box, suitable for the reception of gloves, brushes, etc. It may be seen from the sectional plan and sectional elevation (Figs. 1594 and 1595) that the front and ends are mitered and tongued together, and the ends grooved into the back piece, which is of sufficient length and width to fit into a rebate in the adjoining bars, to which it is secured by glue and screws. The bottom is rebated in, and the box front finished with a narrow, deep-chamfered moulding. The lid is left plain, with the exception of a deep chamfer round the face edges. The front panel of the box may, if desired, be embellished with carving. The mirror is fixed to the wall with four strong eyeplates screwed to the cross-bars (see Fig. 1593). The wall should be properly plugged to receive the nails or screws. All the necessary dimensions may be obtained from the illustrations, in which the front elevation is reproduced to a scale of 1 in. to the foot, the remainder being reproduced to a scale of 4 in. to the foot.

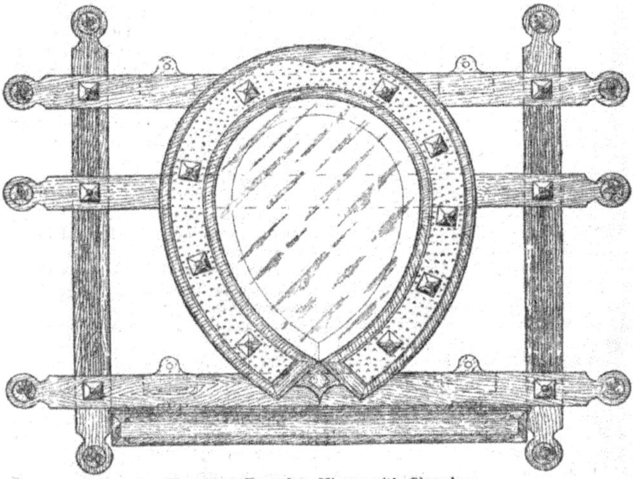

Fig. 1593.—Horseshoe Mirror with Glove-box.

Combined Mirror and Picture Frame.

The combined mirror and picture frame shown by Figs. 1600 to 1602 is of an ornamental character, intended to be constructed in oak or walnut. Fig. 1603 is a section taken on the line A A (Fig. 1600), and Fig. 1604 is a section taken at B B. The frame is 1 ft. 11 in. wide and 1 ft. 8 in. high from the top of the shelf to the top of the cornice mould. Most of the framing is 1 in. on the face by ⅞ in. thick, the widths of the top and middle rails only being increased to allow for fixing the cornice and necking moulds. The sizes of other parts are as follows: Shelf, 2 ft. 11 in. by 5 in. by ⅝ in. (giving 6 in. projection at each end over the frame); pediment, 1 ft. 11 in. by 2½ in. by ¾ in.; raking moulds to pediment, 7½ in. by 1 in. by ⅝ in.; cornice mould 1⅞ in. by 1¼ in.; necking mould, ⅞ in. by ½ in.; side brackets above shelf, 10 in. by 5½ in. by ⅝ in.; side brackets below shelf, 5½ in. by 5½ in. by ⅝ in.; brackets supporting the shelf, 5½ in. by 4⅞ in. by ⅝ in.; shaped spandrel piece between under shelf 1 ft. 9½ in. by 5½ in. by ⅝ in.; turned pendants to stiles of frame, 2⅞ in. by 1 in. in diameter; small spindles in frieze, 2 in. (between shoulders) by ⅝ in. in diameter, and with ⅜-in. pins ½ in. long. The central mirror is of bevelled plate, 1 ft. by 8½ in., with a ¼-in. backboard. The two pictures should be mounted on cardboard, fixed in ¾-in. gilt frames (without glass), backed

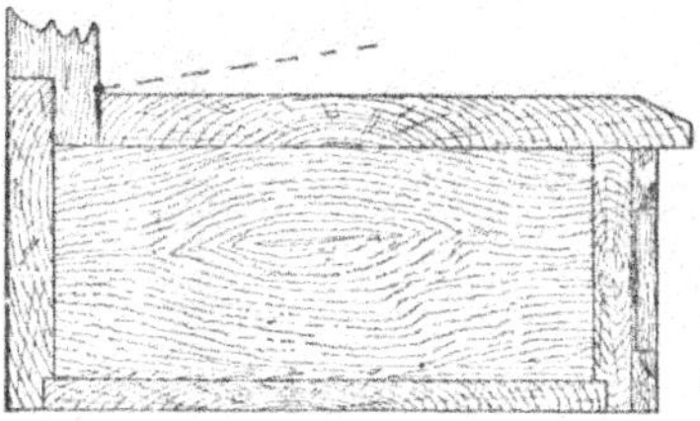

Fig. 1594.—Vertical Cross Section through Glove-box of Mirror.

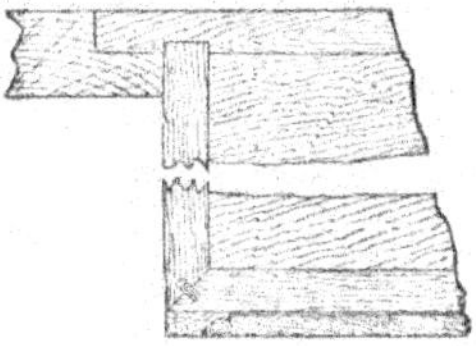

Fig. 1595.—Part Horizontal Section of Glove-box of Mirror.

Fig. 1596.—Section of Horseshoe Mirror Frame.

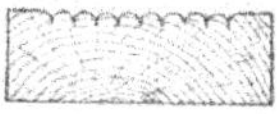

Fig. 1598.—Cross Section of Vertical Bar of Mirror.

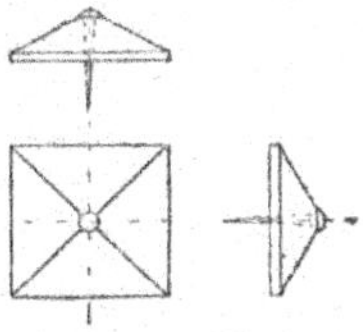

Fig. 1599.—Stud Nail for Horseshoe Mirror.

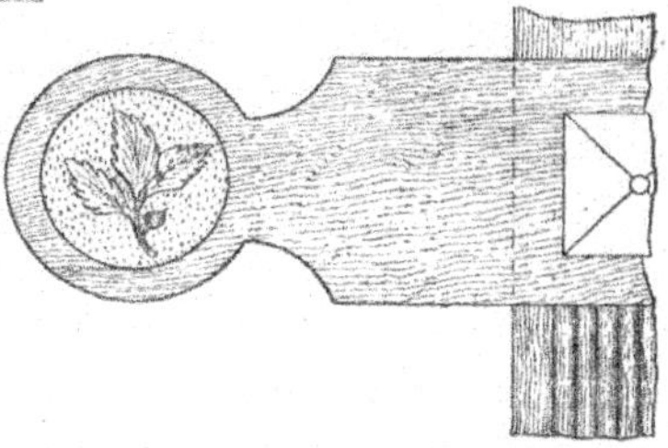

Fig. 1597.—End of Bar in Horseshoe Mirror.

with thin board, and then placed bodily in the main frame with a sheet of glass over all, to protect the gilt as well as the picture. The construction is as follows: The material is cut out, planed to size, and marked out accurately for joists, rebates, and middle rails. The spindles (Fig. 1605) are glued into these first, by using a round stick, hollowed one side to hold the glue. Dip the stick into the glue, hold sloping with the hollowed edge upwards, and carefully and quickly insert into the holes, with-

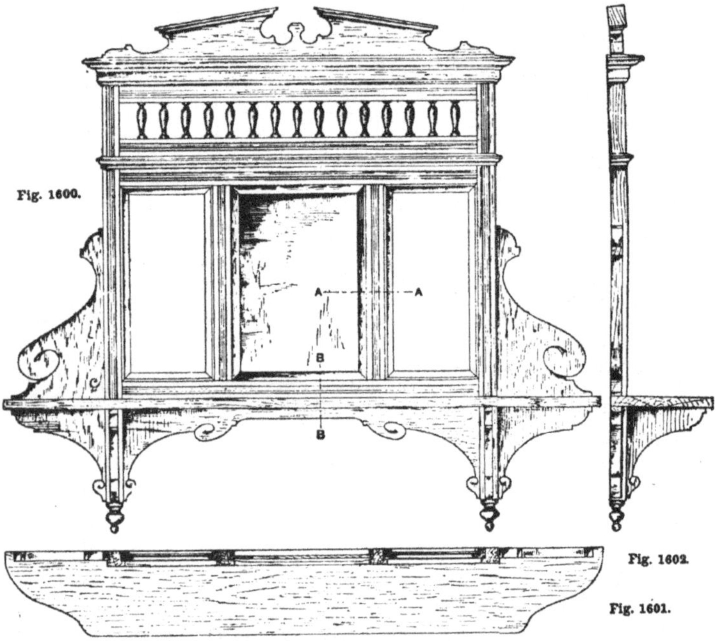

Fig. 1600.

Fig. 1602.

Fig. 1601.

Figs. 1600 to 1602.—Front Elevation, Horizontal Section, and Side Elevation of Combined Mirror and Picture Frame.

housings, and boring for spindles. Mortices are cut first (so as not to pass through to the seen edges), the reeding being done next, either with a reeding plane or a hand reeder. The tenons are then formed, the rebates and housings worked for the glass brackets and shelf, and the holes bored for the spindles. The frame is now cleaned and glued up, commencing with the top out dropping any glue on the work. Gently knock the spindles into one piece, turn bodily over, and insert into the other piece, and cramp up square. Allow this to set, and during this time prepare other parts, such as moulds, brackets, pediment, etc. Then glue up the whole frame, cramp and set aside. Complete the remainder of the parts whilst setting. Next clean off the frame, fit the shelf, and screw from the under side to the bottom rail. Then fit and fix all the brackets and spandrel, mitre the cornice and necking moulds (Figs.

Fig. 1603.—Section of Mirror Frame on Line A A (Fig. 1600).

Fig. 1605.—Turned Spindle.

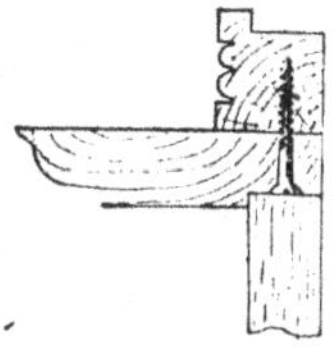

Fig. 1604.—Section of Mirror Frame and Shelf on Line B B (Fig. 1600).

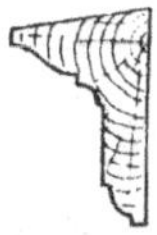

Fig. 1606.—Cornice of Combined Mirror and Picture Frame.

Fig. 1607.—Necking Mould for Frame.

Fig. 1609.—Joint in Frame at C (Fig. 1600).

Fig. 1608.—Turned Pendant for Frame.

1606 and 1607), fixing with glue only. Finally dowel on the pediment and fix the turned pendants (Fig. 1608). The joint at C (Fig. 1600) is shown enlarged at Fig. 1609. The frame must now be polished. If a first-class job is desired, instead of fixing all parts, simply fit the shelf, brackets, spandrel, pediment, and pendants in position, take apart, polish, and, when quite hard, carefully fix together, using a soft cloth and avoid handling as much as possible. Any slight marks caused by the warmth of the hand can be spirited off when complete. Now fix in the picture frames with their glass, then the mirror and backboard, and finish off by backing the part of the frame containing the mirror and pictures with strong paper, damped, stretched, and glued on. The whole can be hung by means of mirror plates and screws.

Hanging Mirrors with Candle Sconces.

Hanging mirrors of the kind illustrated by Figs. 1610 and 1611 are suitable for placing on walls that are insufficiently lighted, the combination of bevelled glass and lighted candles insuring a glistening and bright effect. They may be made of any of the usual hardwoods, french-polished; or of pine, painted and enamelled white or any light shade of colour to match the furniture or decoration of the room. The brass sconces are easily obtainable, and should be in hand before making the frames. The mirror frame shown in Fig. 1610 may be about 2 ft. 6 in. long, and proportionately high. To insure correct proportions, a full-size drawing of half the design should be prepared, the length being decided upon first. Half the design should then be set out to the same number of perpendicular spaces as shown at Fig. 1610; next the horizontal lines, the same distance apart, to form the squares. The shaped portion of the design in each square can then be filled in. The dimensions of the various pieces of wood can be obtained from the full-size drawing. The thickness of framing should be about $\frac{7}{8}$ in. The top and bottom rails are tenoned, and the side pieces mortised to receive them, as shown in Fig. 1612. The mouldings A (Fig. 1610), and those forming the rectangle, are about 1 in. wide by $\frac{3}{8}$ in. thick, with a hollow worked on each edge (see section, Fig. 1613). The angle ribs B (see also Fig. 1614) which support the mouldings should finish $\frac{1}{2}$ in. thick—this will allow a $\frac{1}{4}$-in. rebate for the glass. The angle ribs are fitted to the top, bottom, and sides of the frame, as shown in Fig. 1614, and are secured with glue and a small screw driven between the intersection. To protect the silvering of the glass plates, a back must be provided, fitting flush with the back of the frame. The ribs must therefore be less in width than the thickness of the frame (see Fig. 1613, which shows a section of the back C and of the ribs B). When the frame is glued together, and the ribs are fixed and glued, the outer shaped edges may be cut with the bow saw and cleaned up with a file and glass-paper. Next the face of the frame should be levelled with the plane, and glass-papered. The face mouldings may now be planted on as follows: Fit the top, bottom, and sides, mitering them at the corners, and temporarily secure them with screws driven from the

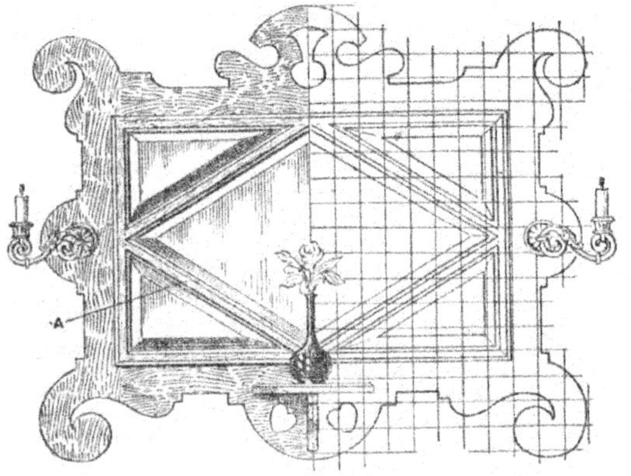

Fig. 1610.

Fig. 1611.

Figs. 1610 and 1611.—Front and Side Elevation of Mirror with Candle Sconces and Diamond Centre.

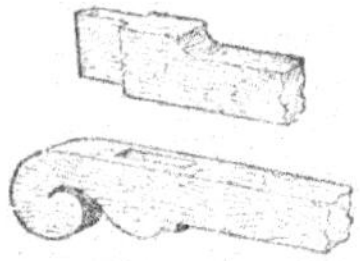

Fig. 1612.—Corner Joint of Frame.

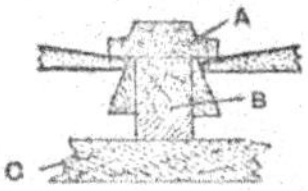

Fig. 1613.—Section of Face Moulding, Ribs, and Back of Frame.

Fig. 1616.—Plan of Mirror Shelf.

Fig. 1615.—Face Moulding of Mirror Frame.

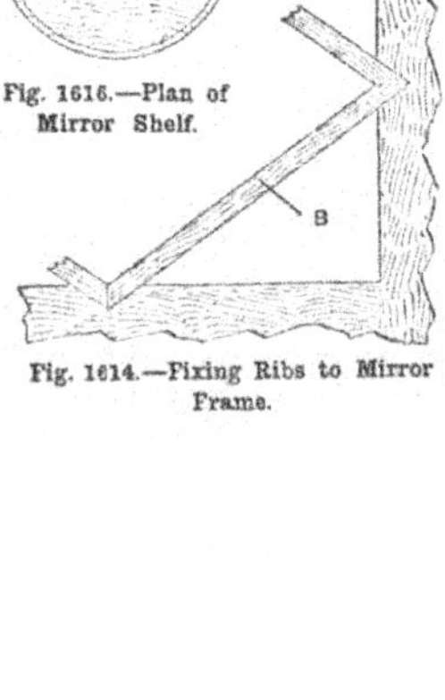

Fig. 1614.—Fixing Ribs to Mirror Frame.

under side of the frame. Next fit the angle pieces, making the joints and mitres as indicated in the enlargement at Fig. 1615. The intersection of the diamond centre is the most difficult part of the job, hence the full-size drawing (showing the relative lines of the frame, ribs, and facing) should be very carefully prepared. The mitres, being small, can be pared to shape with a chisel. The shelf, which is semicircular in plan (see Fig. 1616), is supported by a bracket (see end elevation, Fig. 1611), and fixed with screws from the back of the frame. The ornaments in the pediment and under the shelf are cut through. To obtain the best effect, bevelled glass should be used; but to lessen the cost, the bevelling may be dispensed with. The sconces should be fixed after the frame is polished. To suspend the mirror on the wall, brass plates are screwed to the back of the frame.

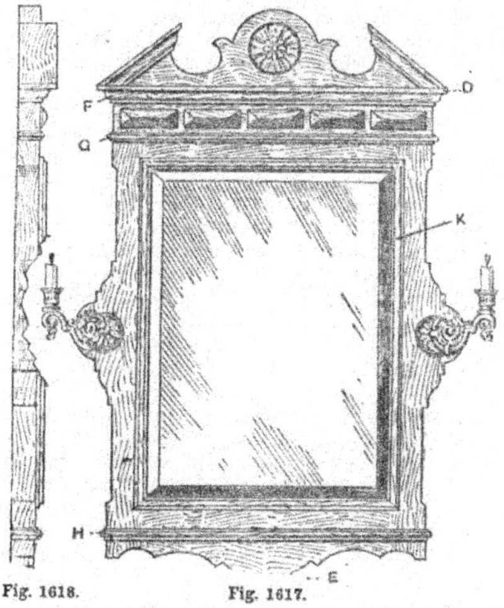

Fig. 1618. Fig. 1617.

Figs. 1617 and 1618.—Front and Side Elevations of Mirror with Candle Sconces and Circular Ornament in Pediment.

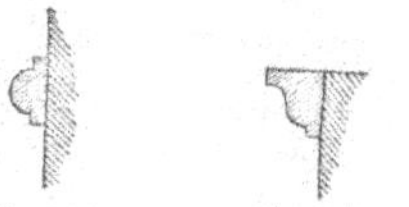

Fig. 1619. Fig. 1620.

Figs. 1619 and 1620.—Sections of Mouldings for Hanging Mirror.

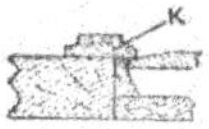

Fig. 1622.—Section of Mirror Frame and Face Moulding.

Fig. 1621.—Section of Mirror Tablet.

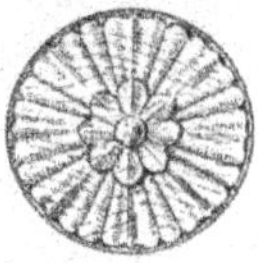

Fig. 1623.—Circular Ornament in Mirror Pediment.

Second Example.—The hanging mirror shown by Figs. 1617 and 1618 may be made to any size from 2 ft. high upwards. To insure the same proportions of width to length, half the design should be spaced out in squares as shown in Fig. 1610. The full-size drawing can then be prepared as already described. In height the frame extends from D to E (Fig. 1617), and the mouldings F, G, and H are planted on the face of the frame, and returned at the sides as shown in the end view (Fig. 1618), the corners being mitered. A section of the mouldings G and H is shown at Fig. 1619. The shaped part of the pediment is made separate from the frame; and the angle side mouldings are glued on the face (see section, Fig. 1620). The five tablets on the frieze are about $\frac{3}{8}$ in. thick at the highest point, and taper down to $\frac{1}{8}$ in. at the edges (see section, Fig. 1621). The mouldings K (Fig. 1617; see also section, Fig. 1622) surrounding the glass are planted on the face and a back is inserted as for the other mirror. The circular ornament in the pediment, shown enlarged at Fig. 1623, is carved with suitable gouges and chisels.

Third Example.—The dimensions of the mirror shown at Fig. 1624 are the same as for that shown at Fig. 1617. The frame, with nine openings, extends from L to M (Fig. 1624), and the face mouldings are planted on (see section, Fig. 1626). In mitering the various pieces of moulding, begin by fitting the top, bottom, and sides O; next the two inside uprights P, then the two short rails in each outer division, and last the two short rails next the panel. The mouldings may be nailed on, if for painting; but for polished hardwood they must be glued, and held down with wooden handscrews or iron clamps, or fixed with screws driven through from the back of the frame. If clamps are not available, the mouldings may be held down by a wrapping of webbing or strips of linen, this being allowed to remain on till the glue has set. The pediment moulding is planted on the face of the frame, mitered at the corners and returned at the sides (see end view, Fig. 1625), the mouldings R (Figs. 1624 and 1627) being treated in the same way. The pediment and shaped base are separate from the frame, and are fixed with dowels, the designs in each being cut with a fret saw. The centre panel on which the sconces are fixed, shown enlarged at Fig. 1628, is ¾ in. thick. The ornamental lines are hollow in section, and are formed with a fine gouge. The method of fixing the hanging mirrors to the wall is the same in each case.

Inlaid Frame.

Figs. 1629 and 1630 show half front and full sectional elevations of an inlaid frame made in mahogany and constructed to hold three photos—a large one at A (4½ in. by 6¼ in.) and two smaller ones at B (2 in. by 2⅞ in.). The grounds of all inlays are silver grey, the brackets and uprights being banded outside with mahogany. The photos should have a matt surface, and be mounted on a paper board as near as possible the colour of the grey veneer. When making an inlaid article, all inlays should where possible be laid first to allow them to sink properly and to avoid waiting, always remembering, when doing so, to use pale, clean glue, spreading it evenly over the surface to be laid on and not over the veneer. A few fine cabinet-makers' headless pins to keep the veneer in place, and a piece of clean paper to cover the inlay (to prevent it from sticking to the caul), should be at hand; the handscrews opened ready to the size required; and a well-heated caul, which should be applied quickly and evenly to the inlay. Cut out roughly to size the following pieces of wood for veneering. One piece of whitewood for the frieze, about 1 ft. 4 in. long and 1¾ in. wide, planed to 1 in.; one piece of whitewood for the pediment, about 1 ft. long and 3 in. wide and 1 in. thick; four pieces of soft mahogany for the brackets (the grain running horizontal), ⅜ in. thick; two pieces of soft mahogany for the uprights C, about 1 ft. 3 in. long and 1 in. thick (these will be 1 in. square when finished); two pieces of soft mahogany for the end uprights, about 6 in. long and ½ in. thick (these will be ½ in. square when finished). With a fine plane-iron, tooth the face side of all the pieces and coat with thin glue, and when dry, the inlays may be held down by means of a hot caul and handscrews. After the inlays have stood in the handscrews about twelve hours, they may be taken out and set aside to sink until required, when the paper must be toothed off, and the inlay scraped and finally glass-papered. The centre frame is made of whitewood, the top and two side rails being 1 in. by ¾ in., and mortised and tenoned together, the bottom rail being 1 in. wider to take the base D, which is 1 ft. 1¼ in. by 10¾ in. (see Fig. 1630). The two small frames E are of ⅝-in. by ⅜-in. stuff (see Fig. 1631), and are put together in the same manner as the centre, with the exception that the rails and stiles are all the same width (4⅞ in. by 4 in.). When made, tooth and size the face sides for veneering with mahogany. Cut out sufficient strips of mahogany veneer for all three frames, allowing enough to lap over each side and ends. Put all frames in hand at once, as the work requires following up. Get ready a caul and the necessary tools, etc., for veneering, and start with the stiles, first gluing the veneer down roughly to size on all, and letting it lap over both ends and sides. About half an hour after laying, take up the first one done and pare the veneer close to the wood where the mitres will be. Now mark the mitre, taking care to strike it on the bottom rail of the centre frame 1 in. from the bottom edge. Place a piece of about twelve hours; then with the aid of a chisel, rasp, and plane, trim the veneer round the edges of the frame. Next scratch the grooves in the centre frame, $\frac{3}{16}$ in. from each edge, and the depth of veneer to take a box line. Let the line be a shade above the surface rather than below. Take care to allow the 1 in. at the bottom as before. The small frames are also lined $\frac{1}{8}$ in. from each edge all round. Gauges should always

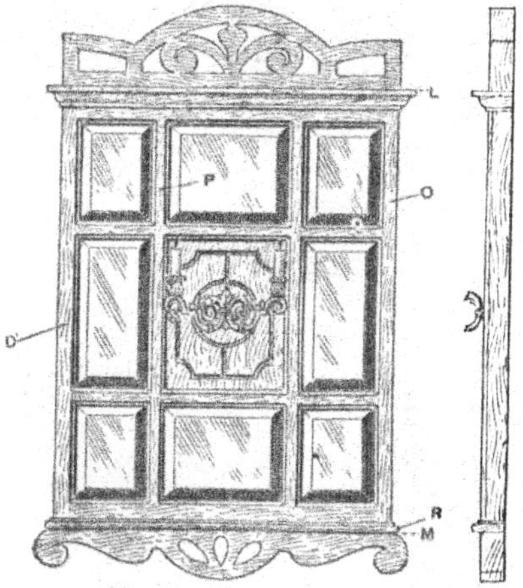

Fig. 1624. Fig. 1625.

Figs. 1624 and 1625.—Front and Side Elevations of Mirror Frame with Centre Panel and Candle Sconces.

Fig. 1626.—Section of Mirror Frame, etc.

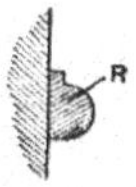

Fig. 1627.—Section of Mirror Moulding.

Fig. 1628.—Centre Panel for Mirror Frame.

hardwood with a straightedge exactly on the line, press it firmly, and cut the mitre with a dovetail saw, keeping the saw well up to the strip of wood. Then, with a chisel, take off the waste piece of veneer, which will come off easily, as the glue is not dry, and clean the glue off the frame. When the mitres are cut on the stile pieces the rails may be veneered. The mitres of these must be cut before laying, letting the veneer lap over each side as before. When gluing, put a pin at each end to keep the mitres well up. When the veneering is completed, allow the work to stand for be kept for this work, as different sizes are often used. Turn round the point of an ordinary marking gauge and file it to cut a groove the size of the line to be inlaid, so that it fits tight enough to be pressed in with a hammer. When the lining is done, put the work aside and polish the mouldings; this should be done before fixing. The cornice moulds, dentils, architrave, and bottom mould F ($2\frac{1}{2}$ in. by $1\frac{3}{4}$ in.) are all returned in the solid, as also are the moulds G on the wings. Now true up and scrape the uprights G and H, plane the frieze to a width, and scrape the inlay side. Scrape

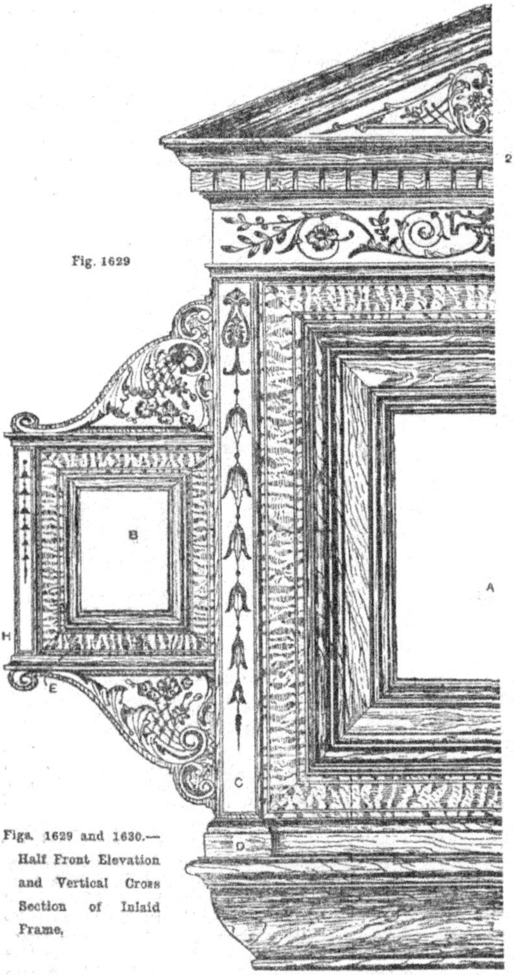

Fig. 1629

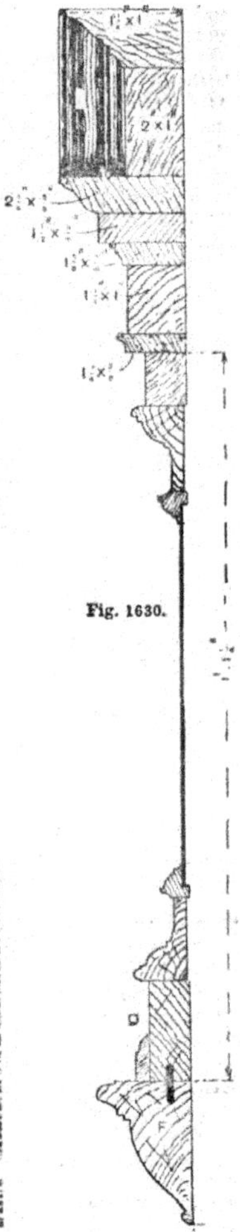

Fig. 1630.

Figs. 1629 and 1630.—Half Front Elevation and Vertical Cross Section of Inlaid Frame.

the piece for the pediment and the four bracket pieces before cutting to shape. Tooth and scrape the three frames, and all may be finely glass-papered. Then cut the brackets to shape, and finish. Bichromate all mahogany parts, and when dry finely paper again and polish, using a white polish for the grey inlays. Polish and oil especially must be kept a certain distance from the edges that are to be glued. When all the parts are polished, dowel in two places, and glue the uprights C to the centre frame. Mark a line across the back of the uprights always where brackets, etc., are dowelled to the other side, or they may be drilled out when boring for the other dowels. The base D should next be mitered and fixed with glue (see Fig. 1632), and when dry, saw off the ends of the uprights. Plane the bottom edge, dowel the mould F to the centre frame in three places, and fix with glue. Glue and screw the architrave to the centre frame and the upright C, and mitre and glue the frieze and cornice moulds. Dowel in two places, and glue the wing frames E to the upright C, keeping them central from the top of the base D and the bottom of the architrave. Dowel in two places and glue the end uprights to the frame and neatly saw off the ends. Glue and screw the four moulds G to the frames E, then dowel each of the brackets in two places to the uprights C, as this is the end grain of the brackets, and glue to uprights and moulds G (see Fig. 1631). A fine pin may be driven through the scroll end of the bracket into the mould G, which is $\frac{5}{16}$ in. deep, to keep the bracket in place, or it may be skew-screwed at the back. Marking out on paper the exact size of pediment, take out the thickness of top mould from the top sides, and cut the paper to shape; place it over the inlaid piece for the pediment, mark round, and cut the wood to the line; plane the edges, cut the centre mitre of the top mould, and glue to the top edge of the pediment, allowing the ends to lap a little. When dry, saw off the ends, plane the bottom edge level, dowel in two places, and glue to the top of the cornice. Mitre and fix the various moulds with glue. Fig. 1633 shows a sectional plan of the end upright with its base moulding. After a final touch up with polish the frame is completed. The ornamental moulding inside the centre frame is 1$\frac{1}{2}$ in. by 1 in., the glass being held in position with $\frac{3}{8}$-in. by $\frac{5}{16}$-in. beading. The outer glasses are held with beading $\frac{3}{8}$ in. deep (see Fig. 1631).

Oval Mirror.

A mirror frame as shown in Fig. 1634 may be made of hardwood, such as mahogany, walnut, oak, or ash, stained green and

Fig. 1631.

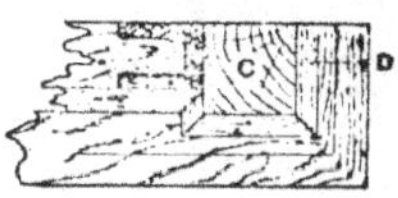

Fig. 1632.—Horizontal Section through Upright of Inlaid Frame.

Fig. 1633.—Horizontal Section through End Upright of Inlaid Frame.

Fig. 1631.—Vertical Cross Section through Side Glass of Inlaid Frame.

french-polished, or of pine, painted and enamelled. An effective size would be about 2 ft. high by 1 ft. 6 in. wide, with the framing about 2$\frac{3}{4}$ in. wide. The oval may be set out with the aid of a piece of string, two nails or pins, and a pencil as follows: Draw the centre lines through the height and width. Next, from a point at the extreme side of the oval, describe an arc with a radius equal to half the height of the oval. Where the arc cuts the perpendicular centre line, place the two nails. Now tie the string, which should be of sufficient length to reach the top or bottom of the oval, round the nails. The oval may be drawn by making the pencil travel right round the figure, and the inner line by shortening the string the width of the framing. The under portion of the frame is made by lapping two thicknesses of $\frac{1}{4}$-in. stuff in sections, the joints of one layer being placed in the middle of the other layer, as in Fig. 1635. These sections are ¼ in. less in width than the sight size of the frame, thus allowing a ¼-in. rebate for the glass (see section, Fig. 1636). The framing having been planed up true, the face portion may be planted on. Of course, the various sections will require gluing together, and the addition of screws, as in Fig. 1635, will help to secure them. A hollow is worked round the outer and inner edges of the frame, and pateras and flutes are carved on the face, as shown in Fig. 1634, and in the enlarged detail of the frame (Fig. 1637). The glass is secured with small wedges about 1½ in. long, placed at intervals of about 2 in., and a thin pine back A (Fig. 1636) is fitted in at the back of the frame and resting on the wedges. The back is then secured by driving thin screws in a slanting direction into the frame (as clearly shown in the sectional view, Fig. 1636). Two or more brass hanging plates, with three holes for screws, will be required for fixing the mirror to the wall.

Fig. 1634.—Oval Mirror.

Fig. 1635.—Two Thicknesses Lapped to form Frame.

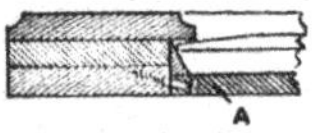

Fig. 1636.—Part Section of Oval Frame.

Fig. 1637.—Enlarged Detail of Oval Frame.

SHAVING CABINETS AND MIRRORS.

Simple Shaving Mirror.

OAK or any other hardwood may be used for making the frame of the shaving mirror illustrated in Figs. 1638 and 1639, but it should not be less than ¾ in. thick. The joints at the corners are made with hardwood dowels. The top and bottom pieces

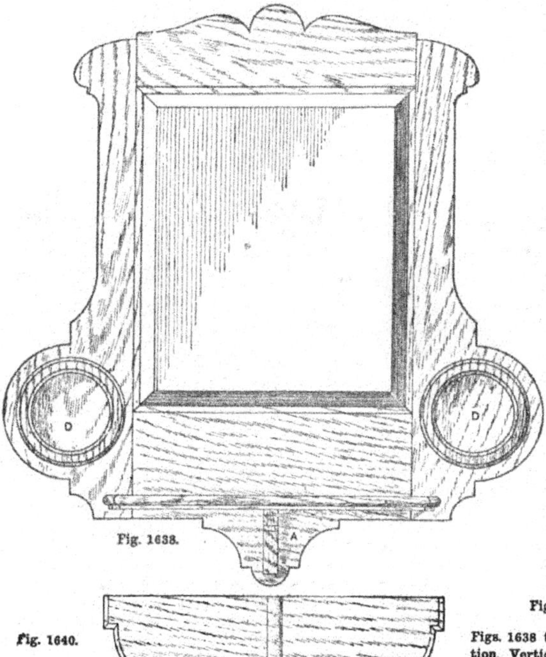

Fig. 1638.

Fig. 1639.

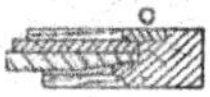

Fig. 1641.—Section of Shaving Mirror's Side Piece.

Fig. 1642.—Section of Circular Moulding of Shaving Mirror.

Fig. 1640.

Figs. 1638 to 1640.—Front Elevation, Vertical Section, and Shelf Plan (underneath) of Simple Shaving Mirror.

may be shaped, and the rebate at the back and the cavetto moulding at the front made before framing together; but at first it will be best to defer these operations on the side pieces. As the cavetto moulding is shaped from the solid, the corners must be mitered back. The frame, being dowelled together, but not glued, may be faced over with the plane (back and front) and (Fig. 1639), which is connected to the shelf by means of a housing joint. The mirror, which need not necessarily have bevelled edges, may be fixed in place with a thin wooden back as in framing a picture, but it is desirable to protect the back of the mirror from possible injury by interposing a few thicknesses of blotting paper. A rather neater method of fixing is shown

Fig. 1643.—Shaving and Toilet Cabinet.

the shape of the side pieces marked out. The pieces are more conveniently shaped if the frame is taken apart, but flats should be left at the top and bottom to form abutments for cramping when the frame is glued up. The flat abutments are, of course, shaped afterwards. The projecting piece A (Fig. 1638) at the bottom may be cut out separately and glued on. The shelf (Fig. 1640) may be attached either by wooden dowels or by screws inserted from the back, as also may the small bracket B in Fig. 1641, which is a horizontal section through one of the side pieces. Two rebates are shown, one to receive the mirror, and another to receive a narrow strip C, which projects over the wooden backing and is screwed down in place. The circular pieces D (Fig. 1638) are provided in order that two candle brackets (not shown), such as are fixed to pianos, may be screwed to the frame. These brackets are a very desirable adjunct to a mirror intended to be used when shaving. Fig. 1642 is a section

Fig. 1644.

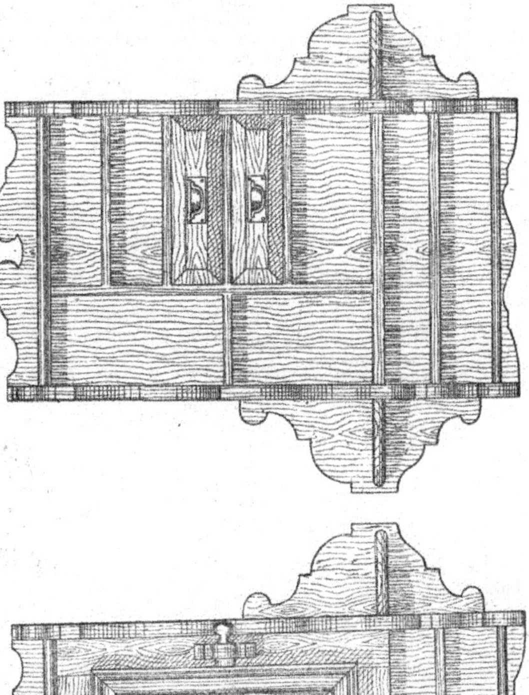

Fig. 1645.

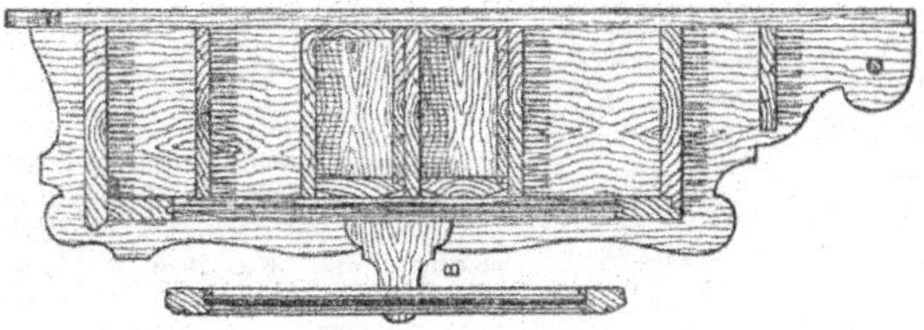

Fig. 1646.

Figs. 1644 to 1646.—Half Front Elevation, Front Elevation with Door and Mirror Removed, and Vertical Cross Section of Shaving and Toilet Cabinet.

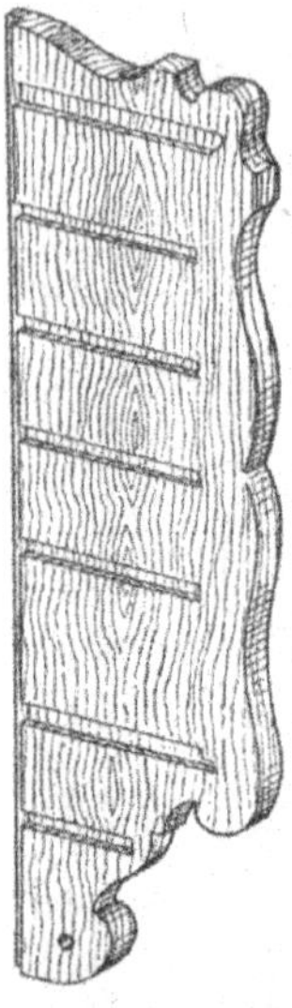

Fig. 1647.—Side of Shaving and Toilet Cabinet.

Fig. 1648.—Detail of Framing for Mirror.

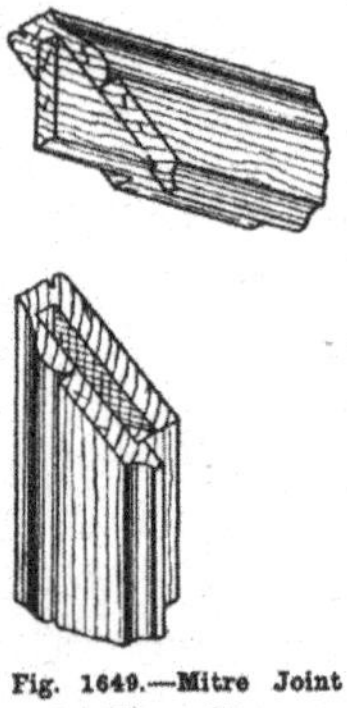

Fig. 1649.—Mitre Joint for Mirror Framework.

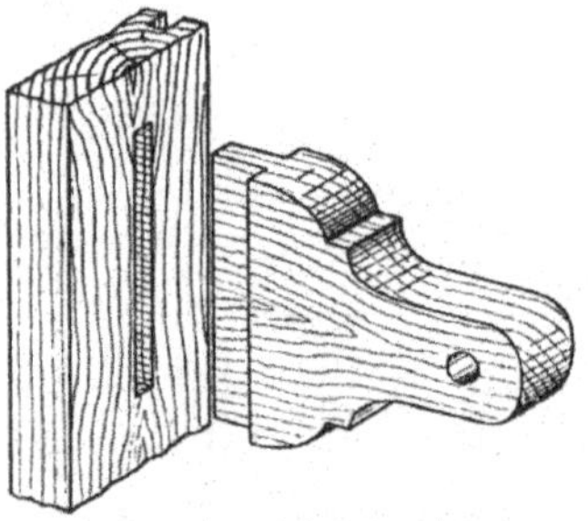

Fig. 1650.—Joint of Bracket (B, Fig. 1643) to Door Stile of Cabinet.

of the small turned moulding round D. For attachment to the wall, two eye-plates (not shown) may be fixed on the sides of the frame.

Shaving and Toilet Cabinet.

The cabinet, of which a general view is presented by Fig. 1643, is a straightforward piece of work. Fig. 1644 is a part front elevation, Fig. 1645 a front elevation with door and mirror removed, and Fig. 1646 is a vertical cross section, showing clearly the constructional details. An inner view of a side piece showing the grooves for shelves, etc., is presented by Fig. 1647. Fig. 1648 shows a detail of the mirror framing; Fig. 1649 the mitered joint of the mirror framing; and Fig. 1650 the connection of the bracket B (Figs. 1643 and 1646) to the door stile.

Shaving Stand with Mirror.

The shaving stand illustrated in front and side elevations by Figs. 1651 and 1652 may be constructed of any wood to agree with the furniture of the room in which it is to be placed. The stand is divided into two portions, the cupboard or table and the mirror and its supports. Four legs are required, each measuring 3 ft. 3 in. long by 1½ in. square. They are tapered towards the bottom 1 ft. from the end, the width and breadth being reduced to 1 in. as far as 3 in. from the end ; they are then curved out to the full thickness (see Fig. 1653). If preferred, the legs may be left straight or tapering all along. The back and sides fit into grooves worked in the legs, and the rails at the top and bottom and between the drawer and cupboard are tenoned into them. The grooves, which are cut ½ in. from the outer edges, are ¼ in. in depth and width, and extend 2 ft. 4¼ in. from a point ½ in. distant from the top of each leg. The front legs have grooves on the rear face only. The mortices for the side rails are ⅝ in. deep, ½ in. wide, and ¾ in. broad, the width being the horizontal measurement. Those at the top of the legs are flush with the end, and are therefore open. The lower mortices are 2 ft. 3¾ in. below those above, and all are ⅜ in. from the outer edge. Those for the back

Fig 1652.

Fig. 1651.

Figs. 1651 and 1652.—Front and Side Elevations of Shaving Stand with Mirror.

and front rails are similarly placed; but they are $1\frac{1}{8}$ in. deep. The two mortices being on the same level, they open into each other. Those taking the drawer rail are 6 in. below the ones above. Fig. 1654 shows the upper and lower mortices of the left front leg. The rails are all $\frac{3}{4}$ in. thick, and, with the exception of that beneath the drawer, which is $\frac{5}{8}$ in., are $1\frac{1}{4}$ in. wide. The three front and two back rails are 1 ft. 5 in. long, and the side rails $11\frac{1}{4}$ in. The tenons have an outer shoulder of $\frac{1}{8}$ in. and an inner shoulder of $\frac{5}{8}$ in.; the drawer rail, of course, has no inner shoulder. The back and front tenons are 1 in. long and the side tenons $\frac{5}{8}$ in. All the side and back rails are grooved to take the side and back boards; these grooves form a continuation of the

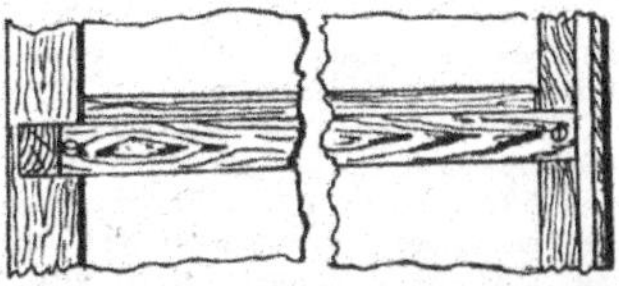

Fig. 1655.—Drawer Rail of Shaving Stand.

Fig. 1653.—Foot of Shaving Stand Leg.

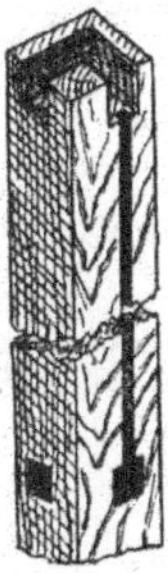

Fig. 1654.—Mortice in Shaving Stand Leg.

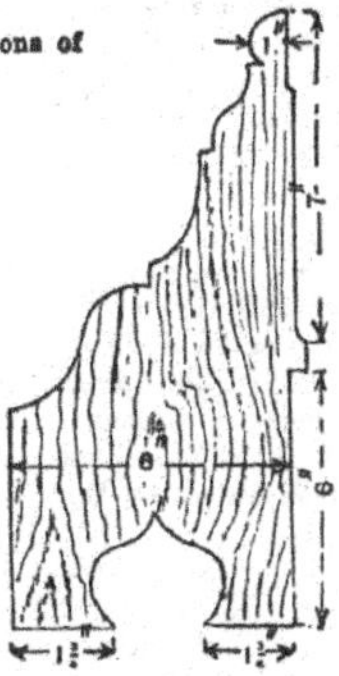

Fig. 1656.—Mirror Support for Shaving Stand.

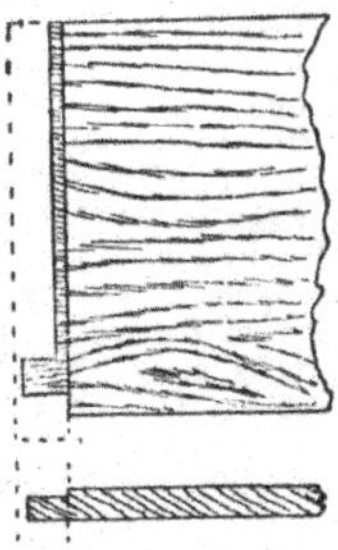

Fig. 1657.—End of Shelf of Shaving Stand.

leg grooves, and extend $\frac{1}{4}$ in. over the tenons. The back board measures 2 ft. $4\frac{1}{4}$ in. by 1 ft. $3\frac{1}{2}$ in. by $\frac{3}{8}$ in.; the side boards 2 ft. $4\frac{1}{4}$ in. by $10\frac{1}{2}$ in. by $\frac{3}{8}$ in. All are rebated $\frac{1}{8}$ in. on the outside. The various parts are glued together. Two side rails, measuring 1 ft. $0\frac{1}{8}$ in. by $\frac{3}{4}$ in. by $\frac{1}{2}$ in., are wanted for the drawer to slide on. A strip of $\frac{1}{2}$-in. stuff 1 ft. long by $1\frac{1}{4}$ in. wide is nailed, $\frac{5}{8}$ in. from the front end, to what is to be the outer $\frac{3}{4}$-in. face. Fig. 1655 shows these details. The rail is screwed to the back and front legs. The top of the stand is 2 ft. by 1 ft. 3 in. by $\frac{3}{4}$ in., the edges of which may be worked or left square as preferred. It is fixed by screws inserted through the top rails. The cupboard bottom consists of a piece of $\frac{1}{2}$-in. board 1 ft. $4\frac{1}{2}$ in. by $11\frac{1}{4}$ in., having rectangular pieces $\frac{3}{4}$ in. square cut from its corners to fit the legs. It rests upon and is glued to the lower rails. Two $\frac{3}{4}$-in. square rails, 11 in. long, are secured to the under surface of the top just above the drawer side rails, to prevent the drawer tilting when it is drawn out.

Cupboard Door.—The cupboard door is made up of two thicknesses of $\frac{3}{8}$-in. material, each 1 ft. 9 in. by 1 ft. 3 in., the grain of the inner piece running across the length, and that of the outer piece with the length. These are glued together, and a projecting panel of $\frac{3}{8}$-in. stuff (see Fig. 1651) is glued to the outside. It should measure 4 in. less than the door each way, and the edges should be bevelled. Two butt-hinges and a drop handle will complete the door.

Drawer.—For the drawer, saw out a piece of material 1 ft. 3 in. by 6 in. by $\frac{1}{2}$ in., and dovetail to the ends two pieces (the sides) 11 in. by 6 in. by $\frac{3}{8}$ in. The back is 1 ft. $3\frac{1}{4}$ in. by $5\frac{1}{4}$ in. by $\frac{3}{8}$ in., and is nailed and glued between the sides. The bottom measures 1 ft. $3\frac{1}{4}$ in. by $10\frac{1}{8}$ in. by $\frac{3}{8}$ in., and is glued to fillets $\frac{3}{8}$ in. in section fastened to the front and two sides of the drawer flush with the lower margins; it is also nailed to the back edge. To cover the dovetailing, a piece of $\frac{1}{4}$-in. material is glued over the front, and to this a panel is secured (see Fig. 1651).

Mirror and Frame.—The mirror and frame are supported by two uprights (Fig. 1656), cut from $\frac{3}{4}$-in. stuff, and united with a shelf and a back board. The illustration gives all necessary measurements. The back board is 1 ft. 11 in. by 6 in. by $\frac{3}{4}$ in. As the back of the mirror will not be seen, this back board may be screwed to the uprights in the recess made for it. The shelf, which is $4\frac{3}{4}$ in. wide by $\frac{1}{2}$ in. thick, is housed in the uprights to a depth of $\frac{1}{4}$ in., except at one point, where $\frac{1}{2}$ in., at a distance of $\frac{1}{4}$ in., from the front edge, is housed to $\frac{1}{2}$ in. depth (see Fig. 1657, in which the dotted lines indicate the upright). The length, including the tenon, is $11\frac{1}{4}$ in. One or two dowels or double-pointed nails make a firm union with the back. The mirror support is either dowelled or screwed to the top of the stand, the back board being 1 in. from the rear edge. The mirror, which may be bevelled, measures 1 ft. by 8 in., and if plain will cost about a shilling. It is framed with $1\frac{1}{4}$-in. picture moulding, the sides being 1 ft. 2 in. long and the ends 10 in. Cut the ends at an angle of 45° in the mitre block, and glue and nail the lengths together. When the moulding has set, insert the mirror; this is backed with thin wood or stout pasteboard, sprigged in place to keep the glass in contact with the rebates. A piece of $\frac{1}{4}$-in. material 1 ft. $1\frac{1}{2}$ in. by $9\frac{1}{2}$ in. is then screwed to the back of the frame. Finally, fix a pair of ordinary glass swivels to the frame and uprights, screwing the pins to the middle of the side lengths of the framing, and the sockets to the back of the uprights at the top (see Figs. 1652 and 1653).

OVERDOORS.

Decorative Overdoor.

ILLUSTRATIONS of a decorative overdoor are presented by Figs. 1658 to 1670. This fitment is intended as a decorative piece of furniture for fitting over a doorway. If it is to be fixed in a drawing-room, it should be made from whitewood, and finished with enamel, of a colour harmonising with the general tone of the room. For a dining-room it is best if made from mahogany or walnut, and polished. Choose good, well-seasoned, and straight-grained timber. The sides of the supporting brackets are slightly panelled and scored. The panels are obtained by cutting away the surface wood, as clearly shown in the section (Fig. 1662). The top edge of each bracket B should be first marked square and set off with a dip of ⅛ in. to the back edge, thus giving a better support for the fitment. The brackets are fixed to the architrave moulding M with a screw of sufficient length to enter the door-frame post F, as shown in Fig. 1668. The shelf S has moulded front and end edges, and is screwed to the bracket. Two corner pillars C, with half pillars against the wall W, are tenoned into the shelf. The pillars should have square tops and bottoms, with rounded shafts, or can be left square. The bottom ornamental pieces, shown in Fig. 1659, are cut from ⅜-in. stuff and grooved ⅛ in. deep into the pillars, and bradded from the back side to the shelf. This forms a protection for ornaments, etc. D is the door. Before cutting the centre trefoil ornament, a piece of coarse canvas should be stretched over the back of the board and fixed with glue. When the glue is dry the ornament can be cut, and will be much stronger by this method. The top rails T R are dovetailed into the corner pillars, as shown. The wall should be plugged as at P in Fig. 1664, to secure a brass strap plate round the half pillars, and along the end rails, fixed with No. 8 screws. This plate should be fixed in position before the front rail is put in. The rail moulding R is mitered at the corners and screwed to the pillars, the rails being bradded to it. A canopy piece, cut from ½-in. stuff, is bradded on top of the rail moulding and surrounded with a finishing moulding grooved on its top edges. The canopy is fixed by the ends of this moulding, being well secured, for which purpose the ends may be tenoned into the rail moulding. The top trefoil ornaments are cut with a fret-saw, and glued and bradded in position. To make these, get two pieces of wood $\frac{3}{16}$ in. thick, glue together, with the grain running at right angles to each other, and when dry cut to shape; this prevents warping. The over-all dimensions across the fitment depend upon the width of the door.

Overdoors with Carved Pediments.

The overdoor shown in front and end elevations by Figs. 1671 and 1672, and in plan by Fig. 1673, is very easily constructed, and may be of walnut or mahogany if it is to be polished; but if it is to be painted. American pine or whitewood may be used, It is 3 ft. 8 in. wide, 12¾ in. high, and 2¾ in. from back to front. Only five pieces of timber are required for building up this overdoor—namely, one piece for the back,

Fig. 1658.

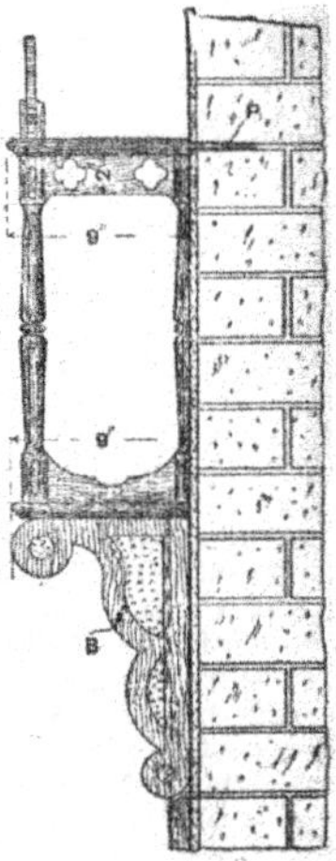

Fig. 1659.

Figs. 1658 and 1659.—Front and Side Elevations of Decorative Overdoor

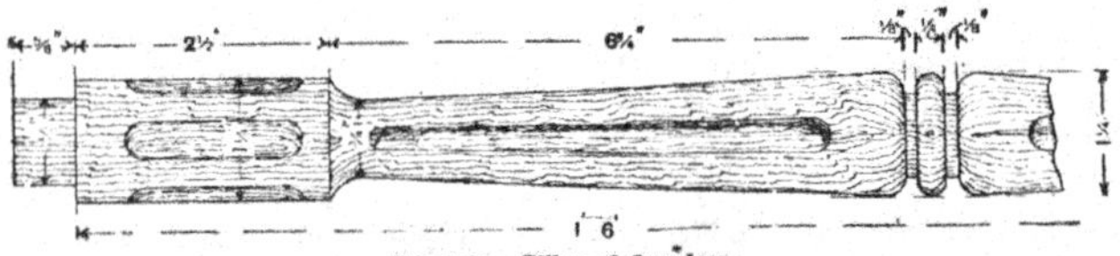

Fig. 1661.—Pillar of Overdoor.

Fig. 1662.—Horizontal Section of Overdoor Bracket.

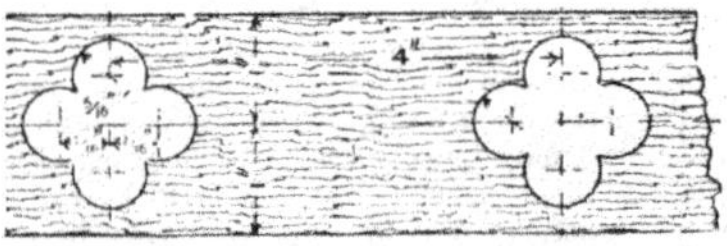

Fig. 1660.—Part of Overdoor Top Rail.

Fig. 1663.—Front of Overdoor Shelf.

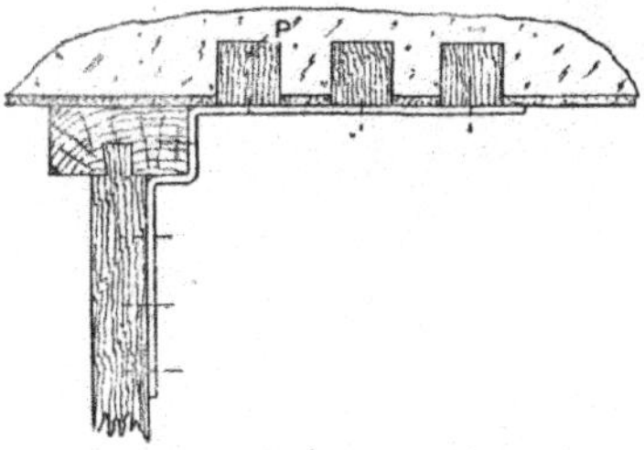

Fig. 1664.—Brass Strap and Wall Plugs for Overdoor.

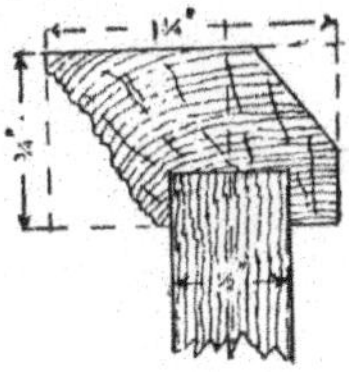

Fig. 1667.—Moulding for Top of Overdoor Pediment.

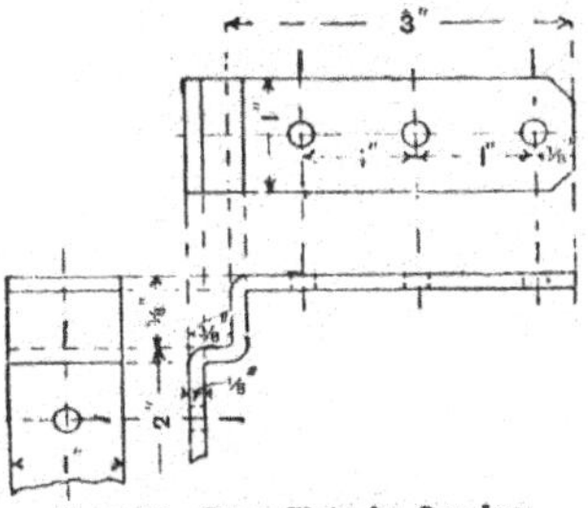

Fig 1665.—Strap Plate for Overdoor.

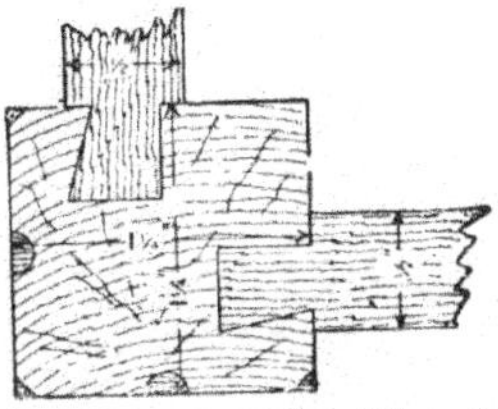

Fig. 1666.—Section of Overdoor Pillar and Rail Ends.

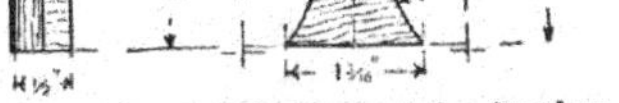

Fig. 1670.—Trefoil Ornament for Overdoor.

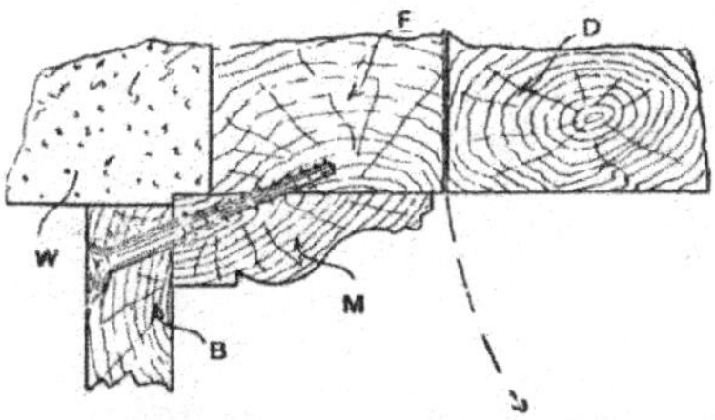

Fig. 1668.—Fixing Bracket of Overdoor.

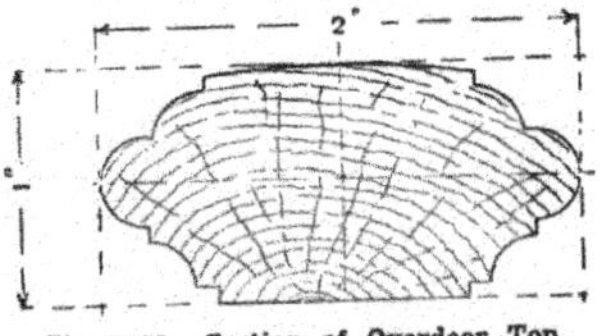

Fig. 1669.—Section of Overdoor Top Moulding.

Fig. 1671.

Fig. 1672.

Fig. 1673.

Figs. 1671 to 1673.—Front and End Elevations, and Plan of Overdoor with Carved Pediment.

Fig. 1678.—Section of Overdoor Pediment at B (Fig. 1677).

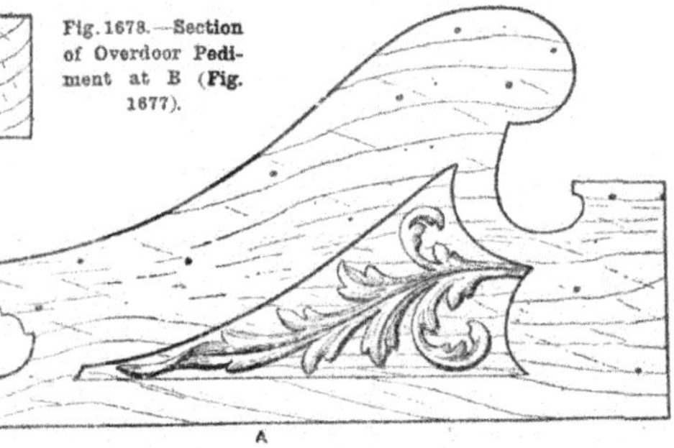

Fig. 1675.—Carving of Overdoor Pediment.

Fig. 1674.—Enlarged Vertical Section of Overdoor Pediment.

Fig. 1676.—Section of Carved Part of Overdoor Pediment at A (Fig. 1675).

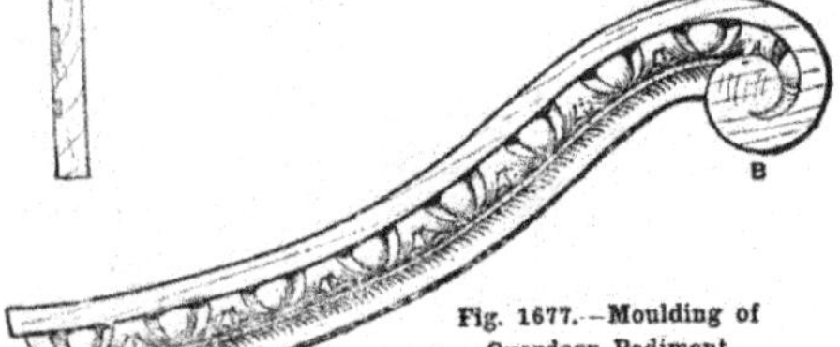

Fig. 1677.—Moulding of Overdoor Pediment.

Fig. 1679.—Carved Bracket of Overdoor.

3 ft. 8 in. by 12¾ in. by ¾ in.; two pieces for scrolls and egg-and-tongue moulding, 1 ft. 8¼ in. by 3½ in. by 2 in.; one piece for bracket, 7 in. by 7 in. by 1¼ in.; and one piece for centre shelf or capping, 5¼ in. by 2¾ in. by 1¼ in. The ¾-in. back covers the whole except the small piece of capping, as is clearly shown in the enlarged section Fig. 1674, which is taken through the centre of Fig. 1671. The back should be cut out with a bow or bandsaw. An enlarged illustration of half the back is given at Fig. 1675, showing the holes through which the screws are driven to secure the scrolls and the bracket, also an enlarged detail of the carving, which is about ¼ in. relief, and should be freely cut. Fig. 1676 represents a section taken at A (Fig. 1675). The pediment is now cut out and cleaned up. The mouldings are worked up to the scrolls with a vertical spindle, and finished with gouges; or, if the spindle is not available, the mouldings may be cut altogether with gouges, and the ovolo moulding carved as illustrated by Fig. 1677. A section taken at B (Fig. 1677) is reproduced at Fig. 1678. The bracket is then cut to shape and carved. A suitable design for this is enlarged at Fig. 1679. The border turns out into a scroll over the groundwork on each side. The pediment and bracket may now be screwed on to the back, and the back cleaned up with a spokeshave and gouges to the shape of the scrolls and mouldings. The capping is then screwed or bradded on to the bracket and back in the centre, and the overdoor polished or painted. Figs. 1680 and 1681 represent front and end elevations of an up-to-date overdoor, Fig. 1682 being the plan and Fig. 1683 a section through the centre. The ¾-in. back runs behind all except the shaped pieces C (Fig. 1680), which are bradded on after the

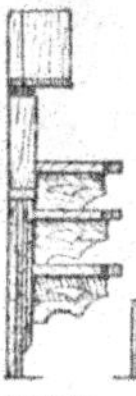

Fig. 1681.

Fig. 1680.

Fig. 1683.

Fig. 1682.

Figs. 1680 to 1683.—Front and End Elevations, Plan, and Vertical Section of Up-to-date Overdoor.

Fig. 1684.—Carving of Overdoor.

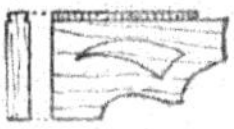

Fig. 1685.—Side Elevation of Overdoor Bracket.

back is cut out with a saw, cleaned up and carved. An enlarged illustration of the carving is given at Fig. 1684. It is a ¼-in. relief, cut into the ¾-in. back. If the carving presents any difficulty, it may be left out, and another small shelf can be fixed in the centre. The shelves are made lighter in appearance by being perforated as shown in the plan of the overdoor (Fig. 1682). The brackets (see Fig. 1685) are shaped, dovetailed into the shelves, and screwed on to the ⅜-in. piece, which is shaped with a bandsaw, and moulded with a small gouge, or with a machine. If hardwood is used, this ⅜-in. shelf may be prepared in two pieces and jointed in the centre, as illustrated in Fig. 1680; but if whitewood is used, it may be prepared in one piece. The section through the middle of Fig. 1680 (see Fig. 1683) shows the mouldings, etc., screwed on from the back.

Plain Overdoors.—Four plain designs are presented by Figs. 1686 to 1689, the first being suitable for execution in mahogany and satinwood, the second in deal, and the third and fourth in walnut.

Fig. 1686

Fig. 1688.

Fig. 1687.

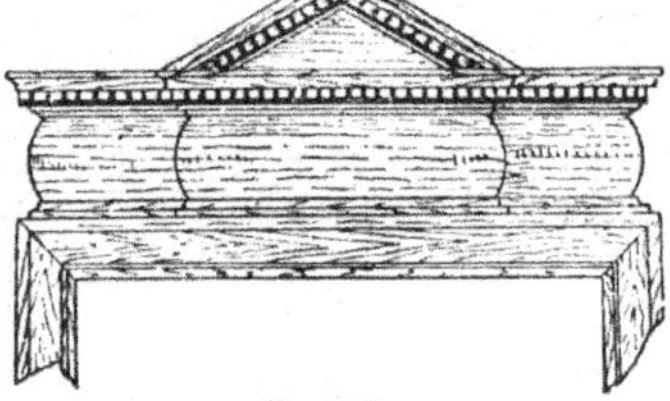

Fig. 1689.

Figs. 1686 to 1689.—Plain Overdoors.

PRESSES.

Cabinet Trousers Press.

FIGS. 1690 to 1693 show a trousers press which will accommodate several pairs at one time. The press is operated by two ⅞-in. diameter square thread joiner's bench screws. In the carcase are fitted two drawers of equal size, and at each end drop leaves are attached, which are supported with hinged brackets. If desired, the press could be made in a dwarf or table form; that is, the drawers and drop leaves being dispensed with, the posts need only be 4 in. to 6 in. below the top of the carcase, which would simplify the construction considerably. The most suitable wood is one of the hardwoods, such as oak, teak, or ash, and next in favour comes pitchpine. Fig. 1690 is a front elevation of the cabinet with the leading dimensions. The following are some of the principal sizes of material. The posts are 3 ft. 7 in. long by 1¾ in. by 2 in. in section, and tapered at their lower extremities, as shown in Figs. 1690 to 1692. All the sizes given are to be taken as the finished sizes of material, therefore allowance must be added for cutting and planing. The front and back rails are 1 ft. 6 in. long between the shoulders, and 1½ in. by 1¾ in. in section; the bottom rails are 1¼ in. thick by 1¾ in wide. Fig. 1691 is a vertical cross section of the press near the centre, but showing the carcase with the drawers removed. The side rails are 1 ft. 2 in. between the shoulders and of the same section as the front rails. Prepare the cross pieces A (Figs. 1690 and 1691), which are 1 ft. 2 in. long from the faces of the shoulders, 3 in. deep by 2¾ in. thick at the centre, and diminishing to 2¼ in. thick at each end. Lay the posts on the bench, and mark the positions for the mortices which are to receive the tenons of the various rails. Commence with the lower mortice at 1 ft. 6 in. from the floor end of the posts. Next measure up a distance of 1 ft. 4½ in. for the top face of the top rail; then midway between these mortices, set out a mortice (on the front posts only) for the reception of the parting rail between the drawers. This rail is ⅞ in. thick by 1¾ in. wide. The measurement having been marked off on one post first, the remaining posts can be brought close together, and the lines produced across their faces with a try square. The rails can then be turned over, and the mortices for the end rails set out in the same way. This will economise time and give true results, providing the posts are planed up true and square. Then cut the mortices, shape the tenons as shown at Figs. 1694, 1695, and 1696, and fix them all together temporarily.

Top of Carcase.—Next prepare the top of the carcase from boards 6½ in. wide by 1¼ in. thick, or 1 in. if hardwood is used. The boards can be grooved and tongued, jointed or dowelled, as preferred, the ends being clamped in the usual way. At the centre under side of the carcase top a batten, 2 in. wide by 1¼ in. thick, is screwed to the boards and also notched to the front and back top rails as shown at Fig. 1697.

Nut and Screw.—Next bore the holes in the top cross pieces A for the reception of the nut and screw; a 1-in. hole should be bored for a ⅞-in. diameter screw, while the hole for the nut should be of such diameter that it will stand driving in. If the nut is provided with ribs as shown in Fig. 1696, corresponding grooves must be cut in the cross pieces to receive them. The nut is kept in position and prevented from dropping out (in the event of the wood shrinking) by a light metal plate fixed with two screws. The square thread screws are 10 in. long, the lower ends being shouldered down, as indicated in Fig. 1698, to receive a metal plate $\frac{1}{8}$ in. thick and of similar shape to that shown in the plan at Fig. 1695. The part projecting through this plate is also shouldered down to receive a suitable washer, and finally this end is riveted over. The reduced part on which the metal plate fits should be about $\frac{1}{32}$ in. longer than the thickness of the plate. To ensure the screw working freely after the washer has been riveted over, the screws with the plates attached and drilled for the

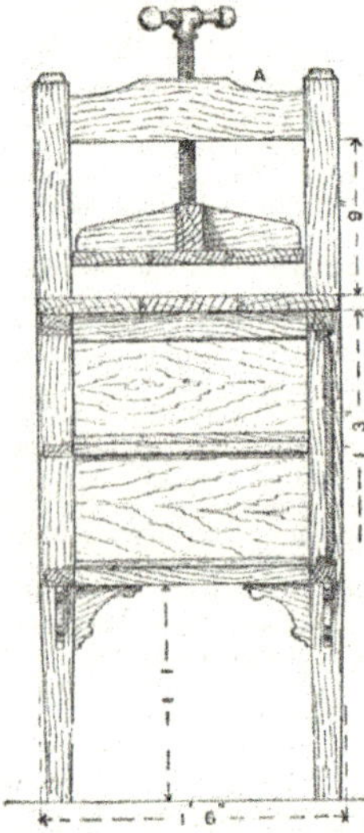

Fig. 1691.

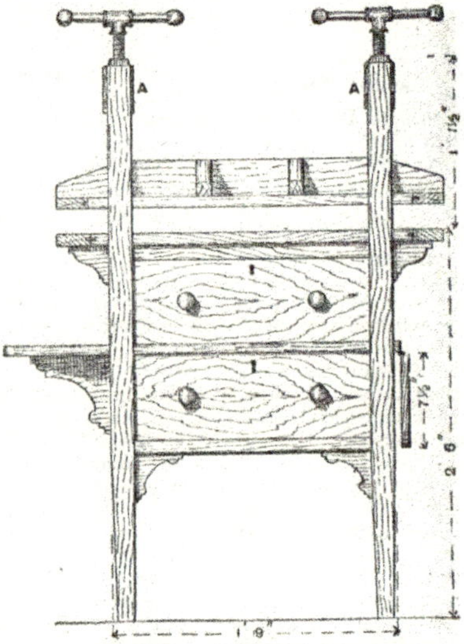

Fig. 1690.

Figs. 1690 to 1692.—Front Elevation, Vertical Section, and Part Plan and Horizontal Section of Cabinet Trousers Press.

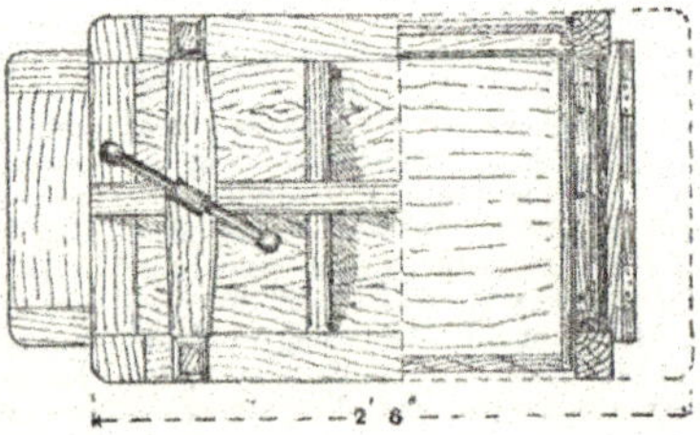

Fig. 1692.

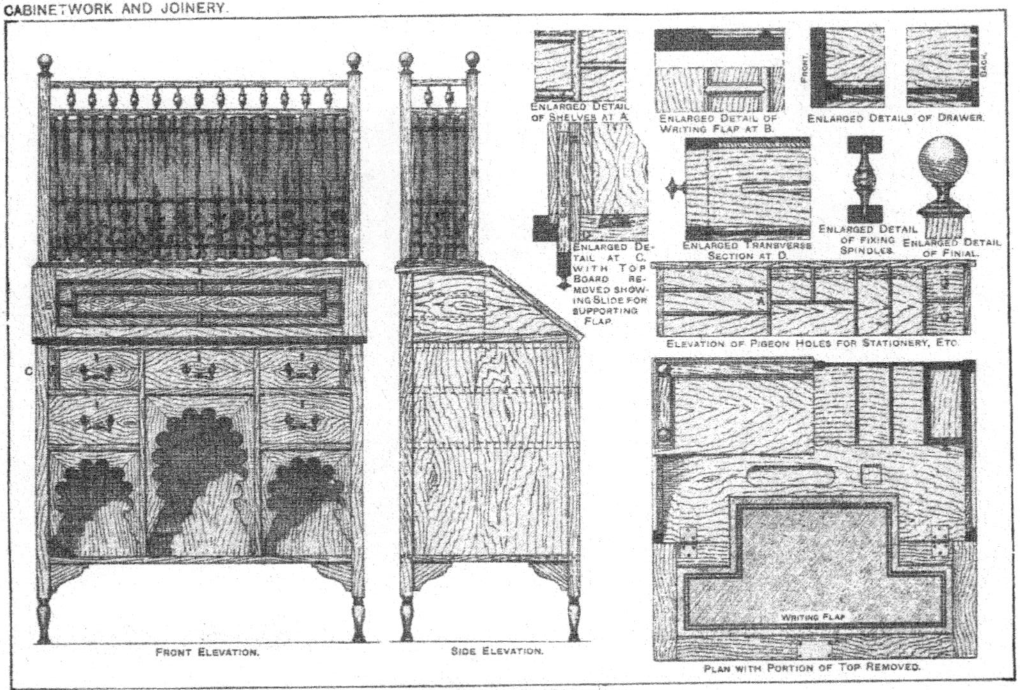

WRITING BUREAU.

wood screws must be in the cross pieces before they are finally framed to the posts. Then the drawer runners and guide fillets and stops can be fixed. Also, the carcase ends and backing, cut from stuff $\frac{3}{8}$ in. thick, can be fitted and blocked in, as shown in Fig. 1692.

Drop Leaves.—The drop leaves are $7\frac{1}{2}$ in. by 1 ft. 3 in. by $\frac{3}{4}$ in. thick, and are hinged to the fillet which supports the top drawer. These fillets project slightly beyond the outer faces of the posts. The brackets which support the leaves are cut with the grain running vertically, and short dowels, formed on each end nearest to the carcase, fit into holes bored in the top projecting

Fig. 1693.—General View of Cabinet Trousers Press.

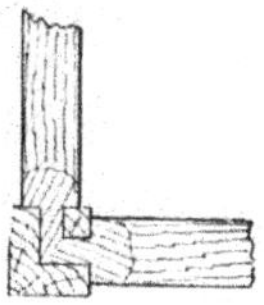

Fig. 1694.—Rails of Press Jointed to Posts.

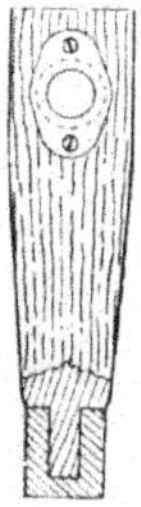

Fig. 1695.—Section of Cross-piece at Joint to Post of Press.

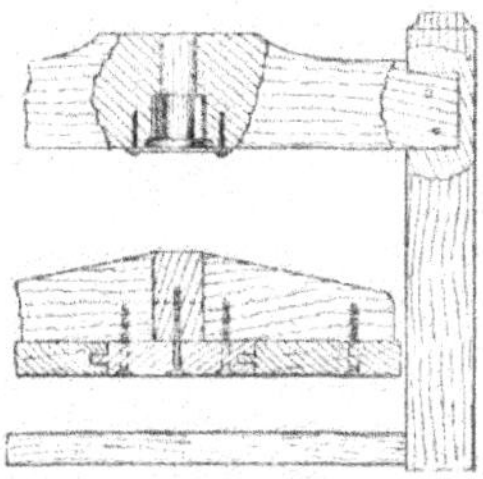

Fig. 1696.—Rising Top and Cross-piece of Press.

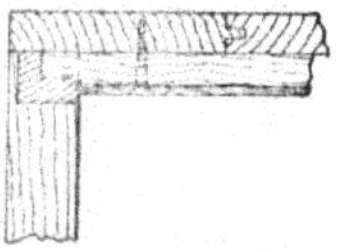

Fig. 1697.—Detail of Carcase Top of Press.

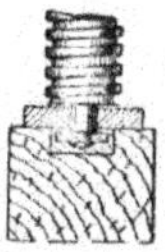

Fig. 1698.—Attachment of Screw to Press.

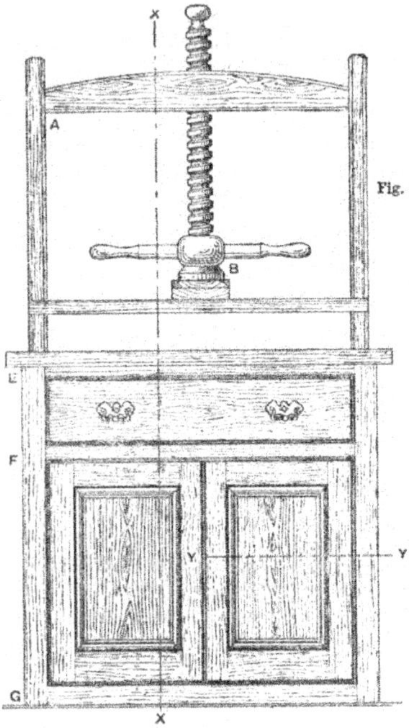

Fig. 1699.

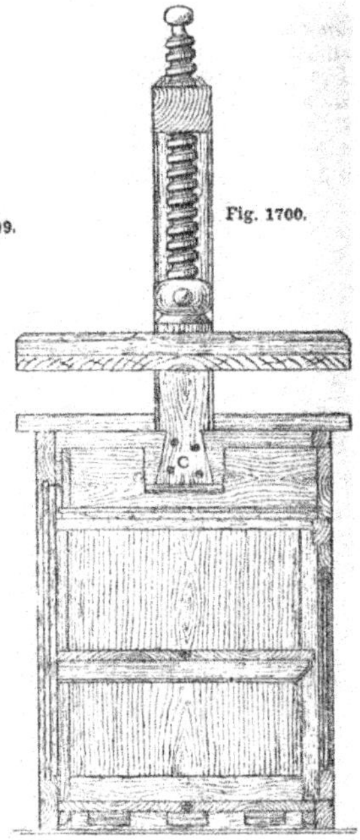

Fig. 1700.

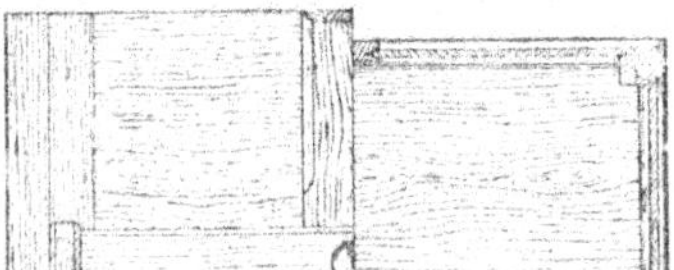

Fig. 1701.

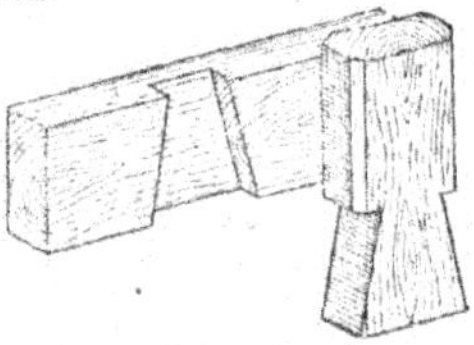

Figs. 1699 to 1701.—Front Elevation, Vertical Section **no Line X X, and Half Plan and Horizontal Section on Line Y Y of Linen Press.**

Fig. 1702.—Lap Dovetail Joint of Linen **Press (see C, Fig. 1700).**

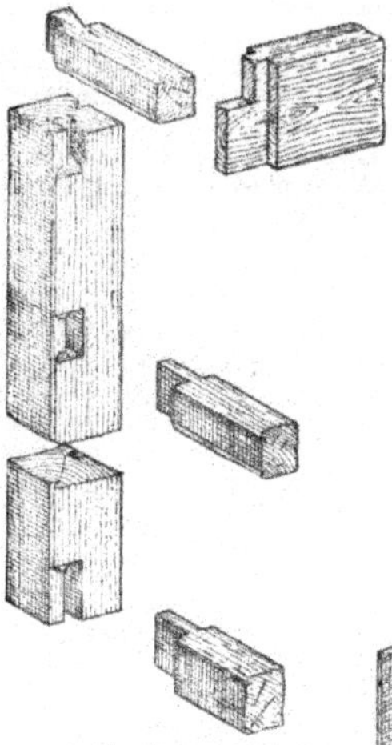

Fig. 1704.—Joints in Linen Press at E, F, and G (Fig. 1699).

Fig. 1703.—General View of Linen Press.

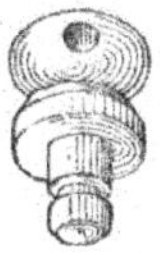

Fig. 1706.—Lower End of Linen Press Screw (B, Fig. 1699).

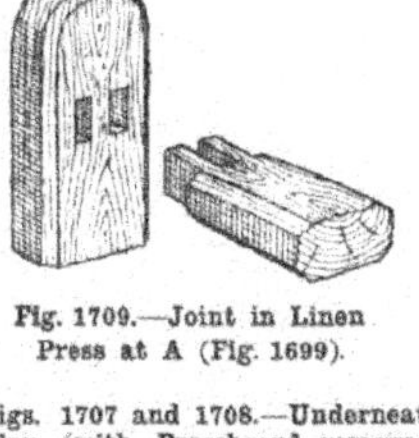

Fig. 1709.—Joint in Linen Press at A (Fig. 1699).

Fig. 1707.

Fig. 1708

Figs. 1707 and 1708.—Underneath Plan (with Pressboard removed) and Elevation at Lower End of Linen Press Screw.

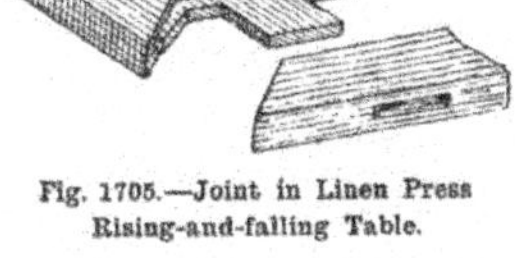

Fig. 1705.—Joint in Linen Press Rising-and-falling Table.

fillet and to the top edge of the lower rail respectively. Next fit and brad the small shaped angle brackets under the lower rails and projecting ends of the top. For the drawers, the fronts are cut from stuff $\frac{3}{4}$ in. thick, and the sides and back are of $\frac{1}{2}$-in. stuff, while the bottoms are $\frac{3}{8}$ in. thick. Knobs of wood or earthenware, or drop handles and locks, can be fitted as desired.

Rise-and-Fall Top.—The rise-and-fall top of the press is made up similar to the top of the carcase. Three boards are planed up true both on the faces and the edges, and are either grooved and tongued or secured with dowels. The two ends are clamped and wedged, then the ribs are secured to the rising top to stiffen it when in use, and also to distribute the pressure of the screws more uniformly over the top of the press. The central rib is 2 ft. 6 in. long by 3 in. deep and $1\frac{1}{4}$ in. thick, and is notched to fit tightly over the four cross ribs. All the ribs are secured with countersunk screws driven from the under face of the press (see Fig. 1696). The screws are attached to the rising top by the oval plates already mentioned. Therefore it follows that the top is raised and lowered with the screws, the pressure being taken first on the oval plates, which should be preferably of sheet brass. A little french chalk can be used on the screws, and will be cleaner than oil or any other kind of lubricant. The woodwork can be stained and sized and varnished, or filled up and polished according to taste.

Linen Press.

The linen press shown by Figs. 1699 to 1701 is complete with cupboard and drawer, the last-named being made with dovetail joints. Fig. 1702 shows the lap dovetail joint at C (Fig. 1700). Fig. 1703 presents a general view of the press. Three joints —respectively at E, F, and G—are shown in Fig. 1704. Fig. 1705 shows the joint at D (Fig. 1703) in the rising-and-falling table. The lower end of the screw at B (Fig. 1699) is shown separately by Fig. 1706. The fixing of the screw to the rising-and-falling table is shown by Figs. 1707 and 1708, the former figure being an underneath view with the pressboard J (Fig. 1708) removed. Finally, Fig. 1709 shows in detail the mortice and tenon joint at A (Fig. 1699).

CHURCH FURNITURE AND FITMENTS.

Notice Board.

In the notice board shown by Fig. 1710, the size of the panel clear of framing is 2 ft. 3 in. by 2 ft. 3 in. Fig. 1711 shows a section on line A B (Fig. 1710). The frame is made of $2\frac{3}{4}$-in. by $2\frac{3}{4}$-in. stuff, framed at the top and bottom joints as shown in Figs. 1712 and 1713, and is grooved on the inner edge to receive the board or panel, which should finish $\frac{3}{4}$ in. thick. The grooves should not run through from end to end, but should be stopped at the halvings on the two side pieces, and between the mortices on the top piece. In putting together the framing, the bottom and side pieces should

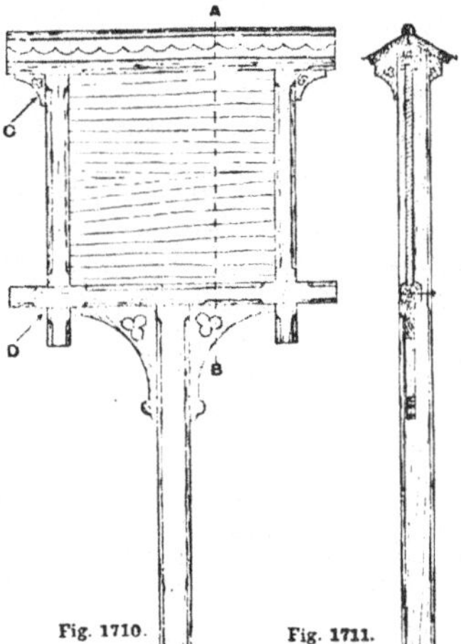

Fig. 1710. Fig. 1711.

Figs. 1710 and 1711.—Front Elevation and Vertical Section of Notice Board.

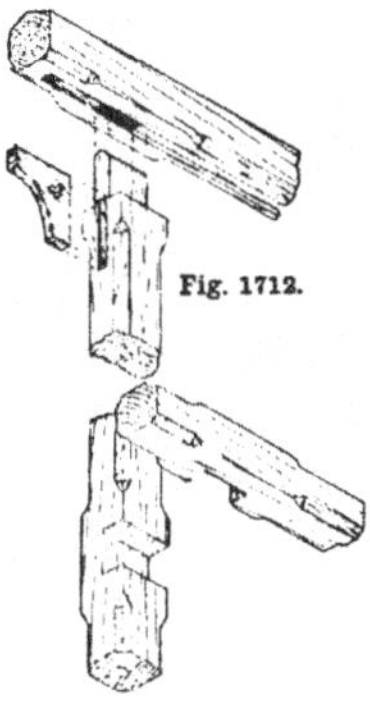
Fig. 1712. Fig. 1713.

Figs. 1712 and 1713.—Joints in Notice Board Framing at C and D (Fig. 1710).

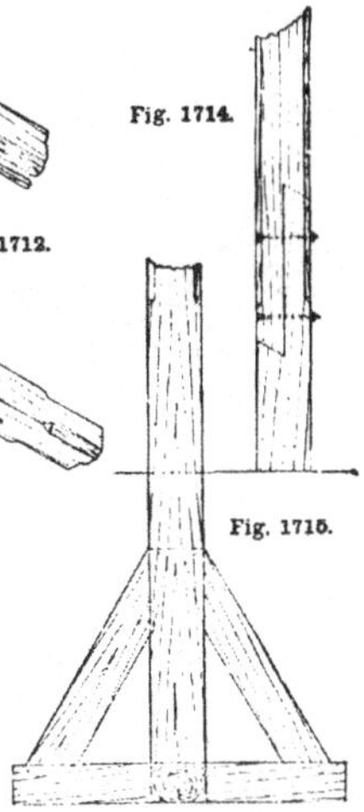
Fig. 1714. Fig. 1715.

Figs. 1714 and 1715.—Scarfed Joint in Notice Board Post and Underground Strutting.

Fig. 1716.—Ornamental Notice Board.

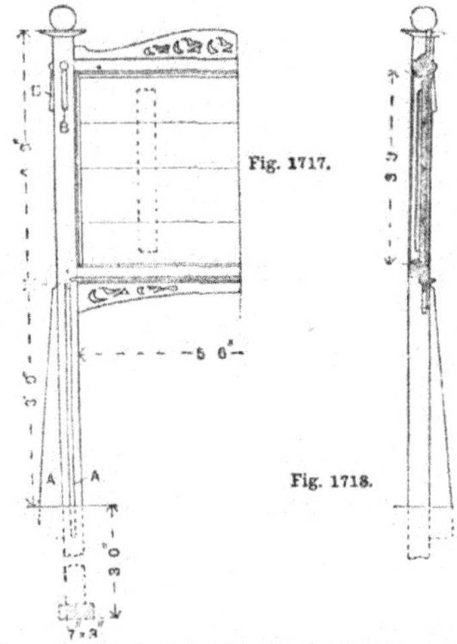

Figs. 1717 and 1718.—Half Front Elevation and Vertical Section of Notice Board.

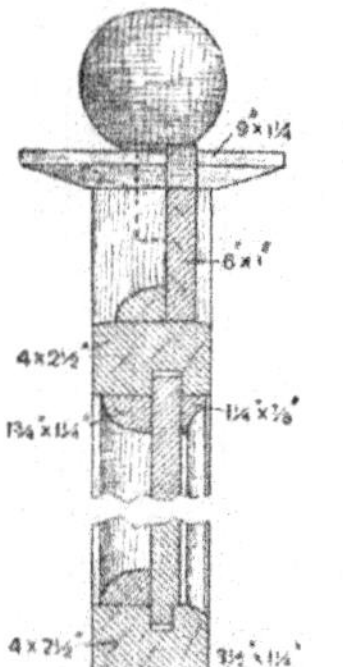

Fig. 1719.—Enlarged Vertical Section of Notice Board.

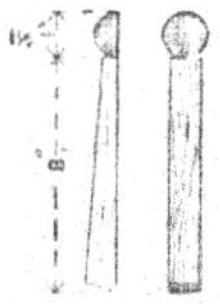

Fig. 1722.—Details of Ornament on Notice Board Post (see B, Fig. 1717).

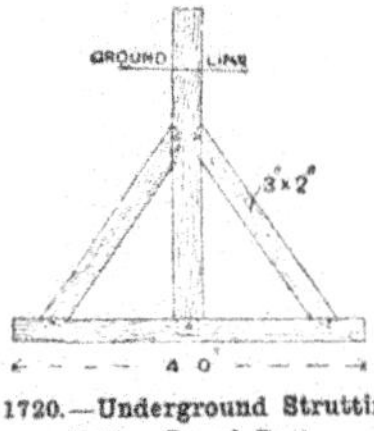

Fig. 1720.—Underground Strutting of Notice Board Post.

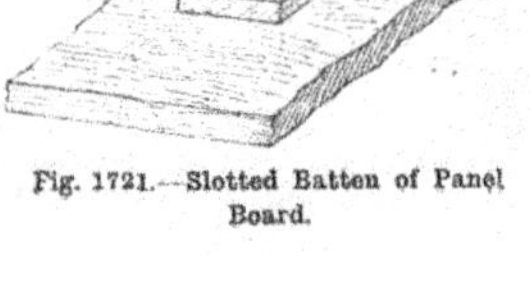

Fig. 1721.—Slotted Batten of Panel Board.

be knocked together, the panel being inserted in the grooves, and the head knocked on and wedged and pinned. The panel is then free in the grooves to expand or contract. The framed panel is mounted on a 4-in. by 4-in. upright post, which runs up under the capping, being cut away and fitted to the back of the framing as shown in Fig. 1711. The capping is formed of two pieces of 7-in. by 1-in. stuff, mitered and screwed to the top of the framing, which is bevelled on each side to receive it, and is further supported with brackets, which should be housed in ¼ in. to the face of the framing, which is surmounted by a roll over which the lead or zinc covering is dressed. The two brackets under the framing should be got out of 1¼-in. board, and housed ¼ in. into post and framing. The two small brackets under the head of the framing are ¾ in. thick, and should also be housed into the framing. If desired, the post could be scarfed as at Fig. 1714, and secured with two ½-in. flat-headed bolts. The board could then be removed for re-lettering, etc., without much trouble. The frame is secured to the post with two ½-in. flat-headed bolts through the top and bottom rails. If the frame is to be painted, it should be made from good yellow deal, and the panel from yellow pine carefully jointed with secret screws. The post should be of English oak, and the parts that go under ground (see Fig. 1715) should be given two or three coats of hot carbolineum. All joints, before being put together, should be carefully painted with good red-lead paint. If, however, the frame is to be varnished, teak would be found to be a handsome and durable material. The lead or zinc covering of the capping is turned up at the eaves to form a small gutter to prevent the wet finding its way between the covering and the capping. The capping piece is cut as shown in Fig. 1710, and soldered to the side pieces, which are secured to the capping under the capping piece. The height from the ground to the under side of the framing is 4 ft., but, of course, this may be varied to suit the position of the notice board. Figs. 1710 and 1711 are reproduced to the scale of ½ in. to 1 ft., and Figs. 1712 to 1715 to the scale of ¾ in. to 1 ft.

Ornamental Notice Boards.

First Example.—Figs. 1716 to 1724 illustrate a design for an ornamental notice board which is intended to be used for a permanent notice. Either good red deal, or yellow pine, would be suitable for the construction. The principal dimensions are given on the illustrations, which show clearly also the details of construction. The posts for Fig. 1716 are 4 in. by 4 in., and about 11 ft. long, 3 ft. of this length being below the ground. The rails and posts are stub-tenoned and mortised together, and draw-bore pinned. The inside edges of the rails, and the portion of the posts between them, are grooved as seen in the enlarged section at Fig. 1719. The groove in the lower rail is not so wide as that in the posts and top rail respectively. The bottom edge of the panel board is therefore rebated as shown, thus forming a shoulder, as a preventive against rain getting into the groove. The rails are chamfered on the top edges, to throw off the water. The panel is formed of about four boards, grooved and tongued together. Two battens, each 2 ft. 9 in.

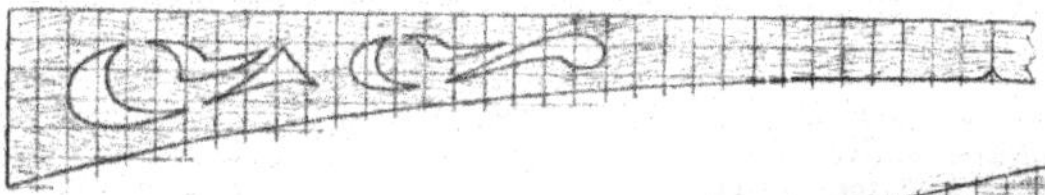

Fig. 1723.

Fig. 1724.

Figs. 1723 and 1724.—Pierced Rails of Notice Board (see Fig. 1717).

by 3½ in. by 1 in., are screwed to the back as indicated by the dotted lines in Fig. 1717, and are slotted as seen at Fig. 1721. This allows the screws to move freely, and prevents the joints being drawn apart in the event of any contraction of the boards. Two pieces, each 9 in. square and 1¼ in. thick, bevelled as shown, form the caps for the tops of the posts. The surmounting balls are 5 in. in diameter, and are dowelled through the caps and into the tops of the posts. The best method would be to make the balls and dowels in one piece while turning them in the lathe. The pierced rails are shown to an enlarged scale at Figs. 1723 and 1724; the lines ruled across being, in the actual work, 1 in. apart. The pieces A A (Fig. 1717) are 4½ in. wide at the bottom, tapering to 1 in. at the top. They are glued and nailed in position as indicated. The pieces B B (Fig. 1717; for enlarged detail, see Fig. 1722) are also glued and nailed to the posts, and, besides being a decorative feature, they also cover the pins used for holding the joints tightly together. A scotia moulding (see enlarged section at Fig. 1719) is nailed to the posts and top rail as seen. It is mitered at the corners, and the lower ends are cut on the slant to fit against the chamfer on the bottom rail. The feet of the posts are treated as shown in Fig. 1720. A moulded piece, 3½ in. by 1¼ in., is screwed to the under side of the cross rail as shown at C (Fig. 1719). The ends of the moulded nosing are returned, and are cut to fit round the posts. The ends of the pierced rails are housed into the posts, and skew-nailed to the cross rails. A plain moulding is mitered round the back of the panel as seen, and a piece is also nailed on the top and to the pierced rail. A couple of braces may be used at the back of the board, to add to its rigidity, and they need not detract in any way from the appearance of the design.

Second Example.—The smaller board shown by Figs. 1725 to 1730 is for the reception of bills, etc. Its general construction differs very little from that illustrated by Fig. 1716, and only a short description need be given. The posts are 3 in. thick and 3 in. wide at the top, tapering to 4½ in. at the ground line. They can easily be cut out of a 9-in. plank by sawing through its length obliquely and reversing the ends. The posts are about 9 ft. 3 in. long, of which length 2 ft. 9 in. should be buried in cement concrete, well rammed. If the board is in a much exposed position, where greater rigidity is required, the feet of the posts may require to be treated as shown in Fig. 1720. The cross rails, moulded and chamfered as shown, are secured to the posts by stub-tenons and mortices drawbore pinned (see Fig. 1728). The cross rails and posts are grooved to receive the panel, and two holes are bored in the bottom rail (as shown by the dotted lines in Fig. 1727) to allow of the escape of any water that may find its way into the groove. The panel is made up of three wide boards, each 1 in. thick, which are grooved and tongued together. The top rail is in one piece, and the moulded capping is nailed on as shown in section at Fig. 1727. The capping is splayed on the top edge, and the ends are cut to fit round the posts as shown at Fig. 1728. A moulding is nailed to the frame round the front of the panel as seen, and a bead secured round the back. Two blocks are carved to the scroll pattern shown at Fig. 1730. The ground is sunk about ¼ in. deep, and is matted as shown. The blocks are slightly housed into the posts, and well screwed into the positions indicated at Fig. 1725. An enlarged detail of the turned caps is given at Fig. 1729. A ¾-in. dowel is turned on the ends and sunk well into the tops of the posts. All the parts to go below ground are well tarred. The mortices and tenons, and all joints, are well coated with red-lead paint before being put together. The woodwork is carefully knotted and stopped, then painted with three coats of plain colour, and varnished. The board shown by Fig. 1716 would look well finished in white, whilst that illustrated by Fig. 1725 may be oak-grained.

Gothic Notice Board.

The church notice board shown at Fig. 1731 is of Gothic design, and is about 6 ft. wide by 5 ft. 4 in. high without the standards. The board is suitable for attachment to the wall of the church, or may be mounted on standards as illustrated. The standards are 6 in. by 4 in. in section, tenoned and wedged to cross pieces and braced at the base, which is sunk and well rammed into the ground a distance of 3 ft. 6 in. The sill of the board is about 4 ft. from the ground line, the posts being reduced to receive it (see Fig. 1732, which is a section taken at A, Fig. 1731); the remaining portion of the post is carried up 3 ft. 2 in. farther to support the back of the board. Well-seasoned straight-grained stuff, 1¼ in. thick, free from knots and other defects, should be used for the panelling. The joints are grooved and tongued, and secured with battens at the back, these being notched to the sill and jointed to a rail connecting the top ends of the standards at the back. Sloping battens, 3 in. by 2 in., are fixed to this top rail, to form a support for the upper portion of the

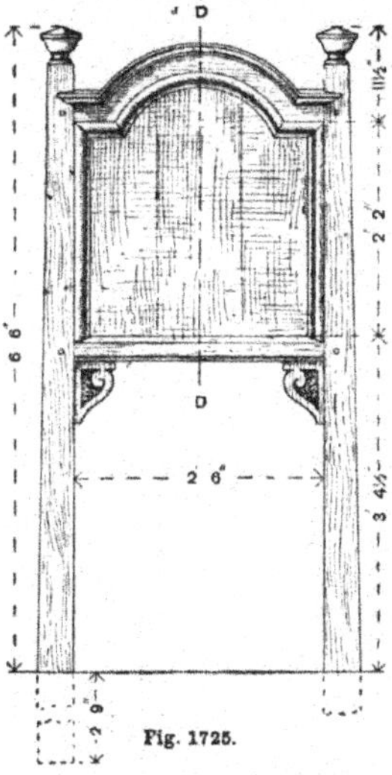

Fig. 1725.

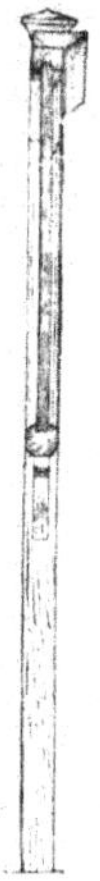

Fig. 1726.

Figs. 1725 and 1726.—Front Elevation and Vertical Section of Smaller Ornamental Notice Board.

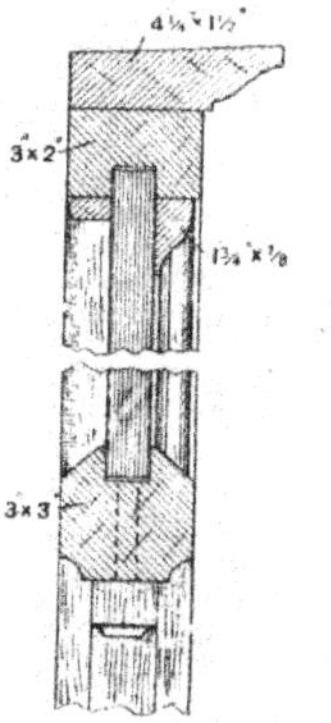

Fig. 1727.—Enlarged Vertical Section of Notice Board.

Fig. 1729.—Turned Cap of Notice Board.

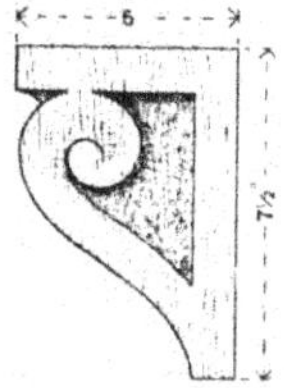

Fig. 1730.—Carved Bracket of Notice Board.

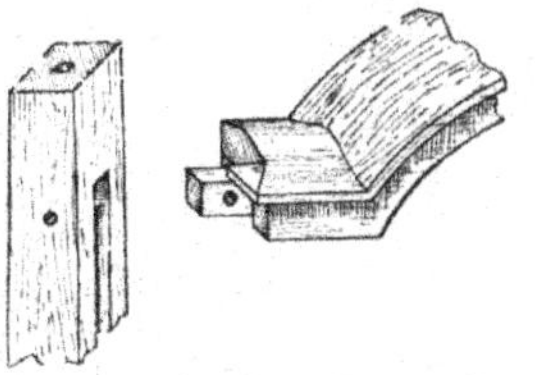

Fig. 1728.—Joint at Top of Notice Board.

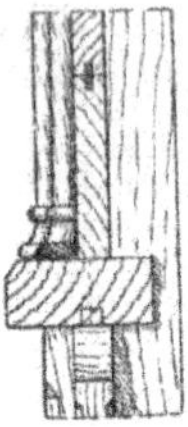

Fig. 1732.—Section of Gothic Notice Board at A (Fig. 1731).

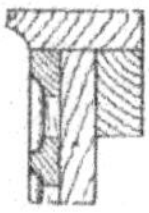

Fig. 1733.

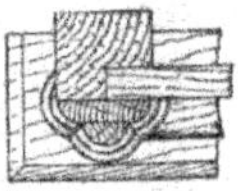

Fig. 1734.

Figs. 1733 and 1734.—Sections of Gothic Notice Board at B and C (Fig. 1731).

Fig. 1731.—Front Elevation of Gothic Notice Board.

notice board (see Fig. 1733, which is a section taken at B, Fig. 1731). The capitals are turned up solid (and carved if desired), and then sawn in halves, after which they are painted, and bradded to the front face of the panels. The columns are made in two parts (see section, Fig. 1734), the bases also being turned up separately in three pieces, and cut and mitered; the joints should be coated with red- or white-lead paint before bradding. The cusped ornaments and trefoil centre are also painted and bradded on.

Three-panel Notice Board for Chapel.

Fig. 1735 shows a notice board more suitable for a chapel. The top and bottom panels are for the name of the chapel and pastor respectively, the centre panel being for the usual notices. This board is 4 ft. 10 in. wide by 5 ft. 6 in. high, including top and bottom horizontal rails; the triangular pediment adds another 1 ft. 9 in. to the height. The stiles are 6 ft. high, and

Fig. 1735.—Front Elevation of Three-panel Notice Board.

Fig. 1736.—Section of Moulding on Notice Board (see D, Fig. 1735).

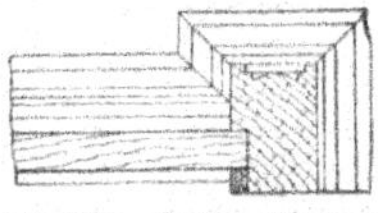

Fig. 1737.—Section through Stile and Panel of Notice Board.

Fig. 1738.—Section of Dentilled Moulding of Notice Board.

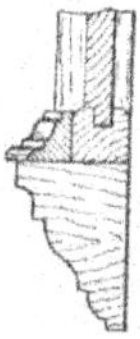

Fig. 1739.—Section through Bottom Rail and Bracket of Notice Board.

5½ in. by 4 in. in section at the lower end, being reduced at D to 5 in. by 3 in. (see section, Fig. 1736). The semblance of a raised panel is worked out of the solid on the face of the stiles, which are mortised to receive the rails and rebated for the panel boarding (see section, Fig. 1737). The cusped ornament is cut from ⅜-in. stuff, and bradded to the pediment and centre panel boards. A section of the dentilled moulding attached to the second rail is shown at Fig. 1738; the other mouldings are mitered at the angle and bradded to the rails and stiles. This notice board is intended to be fixed against the wall of the building, resting upon wood brackets built into the wall (see section, Fig. 1739).

Fig. 1740.—Front Elevation of Hymn Board.

Fig. 1743.—Part Vertical Section of Hymn Board (see Fig. 1740).

Fig. 1744.—Spring at Back of Hymn Board Door (see Figs. 1741 and 1742).

Fig. 1746.—Plan of End of Spring (see Figs. 1741 and 1742).

Hymn Boards.

Hymn boards are illustrated by Figs. 1740 to 1746. They may be executed in pitchpine, or, better, in oak. The board shown in Fig. 1740 should be constructed as follows: The pillars at the sides should be octagonal and plain, with the top and bottom finials slightly carved. The pillars are connected at the top and bottom by moulded pieces mortised into them. The quatrefoil ornament should be carefully and geometrically cut out, and the corner carved panels fitted as illustrated. The board on which the numbers are displayed is fixed into the centre of the ornament, and is made with moulded edge and divided into three spaces

Fig. 1741.

Fig. 1742.

Figs. 1741 and 1742.—Front Elevation and Vertical Section of Hymn Board with Gothic Cresting.

Fig. 1745.—Section of Hymn Board Back Box.

by moulded ribs, with flat back edge on which the printed or painted tickets rest (see Fig. 1741). A door is hinged to the bottom edge of the back of the panel and fastened at the top by a spring clip. At the back of the door (see Fig. 1742) fix three springs as shown; these will keep the printed cards flat against the moulded front. The tickets on which the names "Hymn," "Chant," "Anthem," etc., and the numbers are printed or painted should be of stiff card. The door being opened, the cards can be altered as required, and when the door is closed and fastened with the spring it will keep them in the required position. If preferred, the open spaces in the ornamental portion of the board may be filled in with wood covered with cloth of red or blue colour, and thus a rich effect will be produced. Fig. 1741 shows a different design. The pillars at the ends are square, and the mouldings, top and bottom, are carried through and returned at the ends. A carved cresting of Gothic design crowns the top moulding, and a cresting of simpler character is also affixed to the bottom moulding. These ornamental portions may be altered or not used at all. The hymn board occupies the whole oblong space, and is divided into three compartments by moulded ribs as before described. The back is closed in by a flat board, as shown in Figs. 1744 and 1745, the ends being open so that the cards containing the names or numbers may be passed into position from either end. These cards are kept in position by a spring, as shown in Fig. 1745, with a long slot hole at each end, as shown in Fig. 1746, and commanded in the centre by a knob. On pulling this knob back the spring will come flat and allow the cards to pass. On releasing the knob, sufficient pressure is given by the spring to keep the cards in place. The ornamental portions of these boards may be altered to suit the ornamental work of the choir seats or other woodwork near which they are to be used. For instance, a wrought-iron or a brass cresting may be substituted for the carving on Fig. 1741, but this must be left to the taste of the maker.

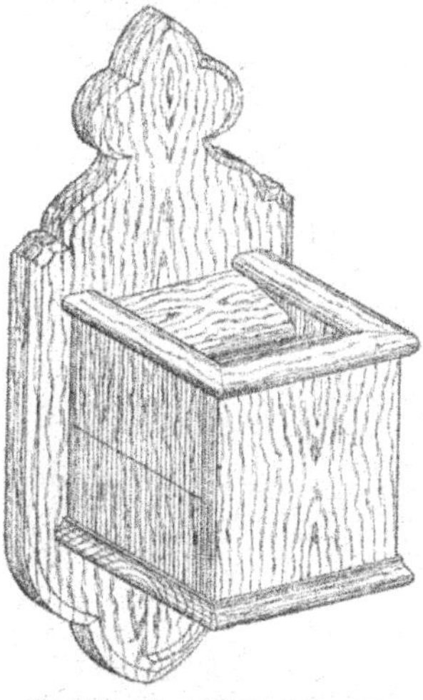

Fig. 1747.—General View of Alms Box.

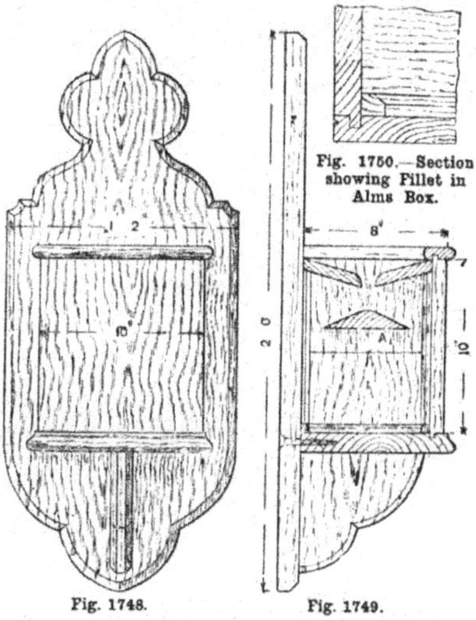

Fig. 1748. Fig. 1749.

Figs. 1748 and 1749.—Front Elevation and Vertical Section of Alms Box.

Fig. 1750.—Section showing Fillet in Alms Box.

Alms Box.

Fig. 1747 is a general view of an alms box for a church, Fig. 1748 being a front elevation and Fig. 1749 a vertical section. If the top is in two pieces on the incline, with an opening between them as shown, the box allows two or three persons to drop in their

Fig. 1751.—Front Elevation of Vestry Cupboard for Books and Robes.

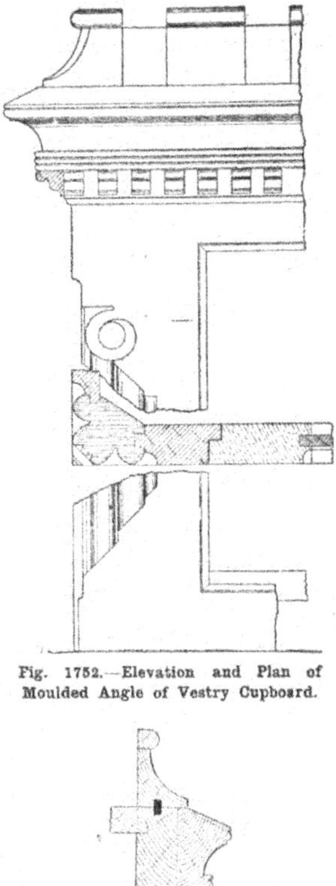

Fig. 1752.—Elevation and Plan of Moulded Angle of Vestry Cupboard.

Fig. 1754.—Inside View of Vestry Cupboard for Books and Robes.

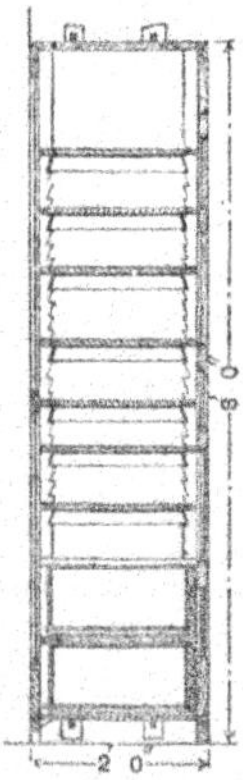

Fig. 1755.—Section through Drawers of Vestry Cupboard.

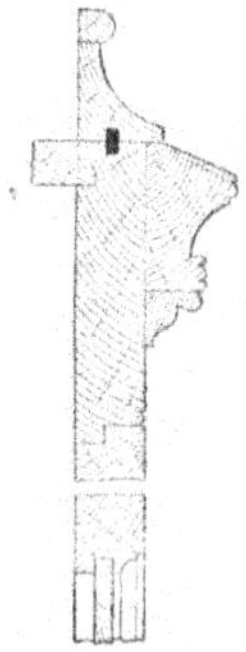

Fig. 1753.—Section through Vestry Cupboard Cornice.

offerings at the same time. The piece A (Fig. 1749) prevents anyone pushing in a rod with sticky stuff on the end to remove a coin. The lower part of the left-hand side is hinged for opening, and this door will require a lock, which may be screwed on inside, or a padlock with fancy staples may be used. The sides, front, and bottom should be grooved and tongued together, and the several parts secured to the back with screws. To obviate screwing through the front and into the side a chamfered fillet may be glued in the internal angles, as shown at Figs. 1749 and 1750. Oak is a suitable wood, but the box may be constructed of any other wood that matches the church furniture. The leading dimensions are figured in Figs. 1748 and 1749, and the thickness of the back, bottom, and bracket may be about 1 in., and of the other parts ¾ in.

Fig. 1759.—Capping of Chancel Rails.

Fig. 1760.—Mould on Edge of Framing of Chancel Rails.

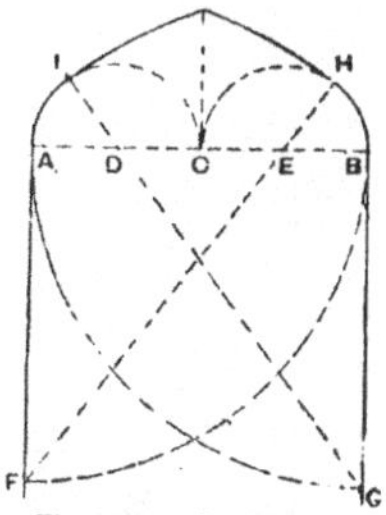

Fig. 1761.—Setting Out Four-centered Arches.

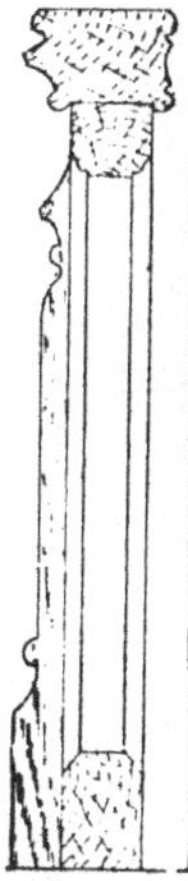

Fig. 1758.

Fig. 1756.

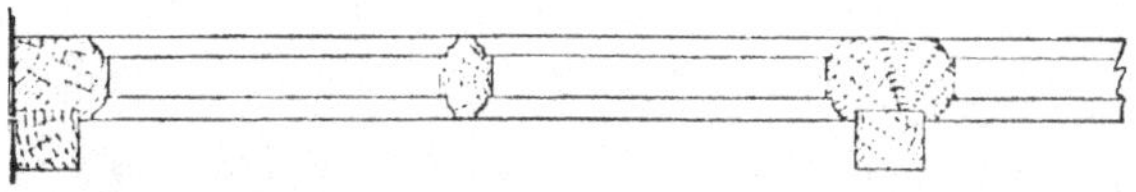

Fig. 1757.

Figs. 1756 to 1758.—Front Elevation, Horizontal Section, and Vertical Section of Chancel Rails.

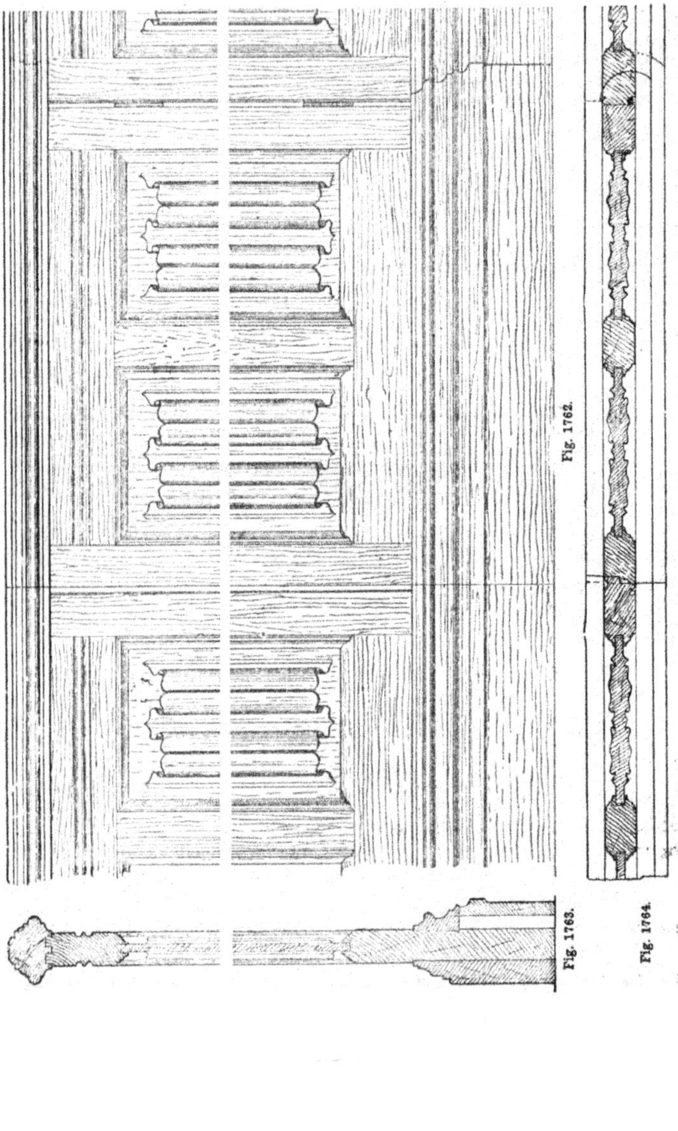

Figs. 1762 to 1764.—Front Elevation, Vertical Section, and Horizontal Section of Dwarf Chancel Screen.

Vestry Cupboard for Books and Robes.

The cupboard shown by Fig. 1751 is for the accommodation of books and clergy robes in the vestry of a church, and can, of course, be adapted for a variety of purposes and situations. Fig. 1751 gives the elevation of the front, with doors hung folded. The front is framed up 1½ in. thick, with angle posts. The moulding shown by Fig. 1752 is shaped to the section front 3 in. by 3 in., and is tongued on to the frame and sides, and stopped at the top and bottom

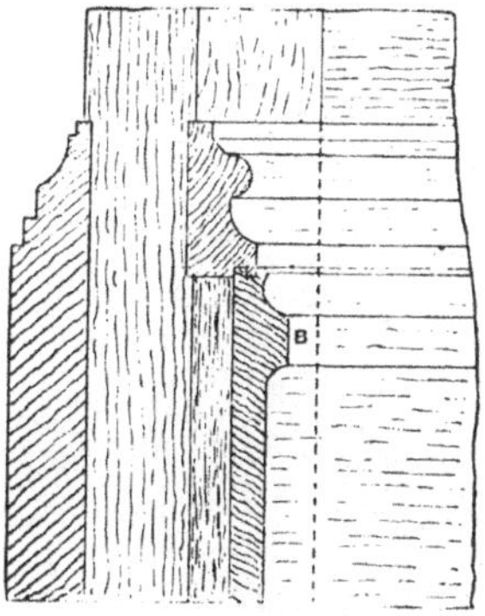

Fig. 1765.—Part of Striking Edge of Door in Dwarf Chancel Screen.

of the cupboard frame. The top moulding (Fig. 1753) is worked the full length required, cut back out of the solid, mitered round, and fixed with dowels to the top of the case. Fig. 1754 shows the elevation of the inside of the cupboard, with the upright standard keyed through top and bottom. On the one side, notched fillets are fixed for shelving, with two 9-in. drawers at the bottom. On the other side there is a single shelf with brass dress-hooks fixed for hanging surplices. The back is framed up with 1-in. pine framing and panels. The doors are hinged in four leaves, and filled in with V-jointed diagonal panels, fitted with brass flush bolts on the edges at top and bottom. Fig. 1755 shows a section through the drawer side of the cupboard, the handles of the drawers being turned or sunk on the drawer fronts.

Chancel Rails.

The chancel—that part of the choir where the altar stands—was in earlier times enclosed by lattices or crossbars, but is now generally enclosed by rails of a design suited to the architecture of the church. Of the design about to be noted, Fig. 1756

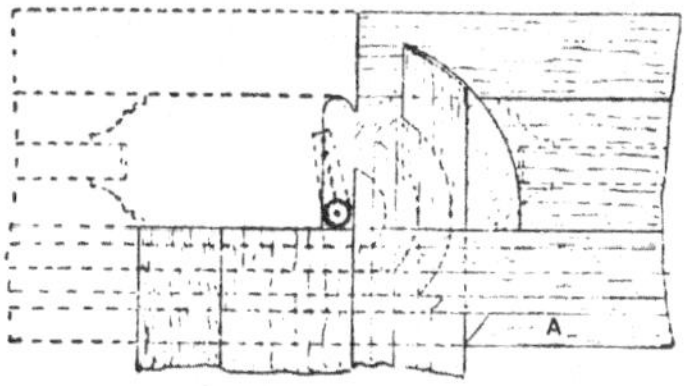

Fig. 1766.—Plan of Joints at Hanging Stile in Dwarf Screen.

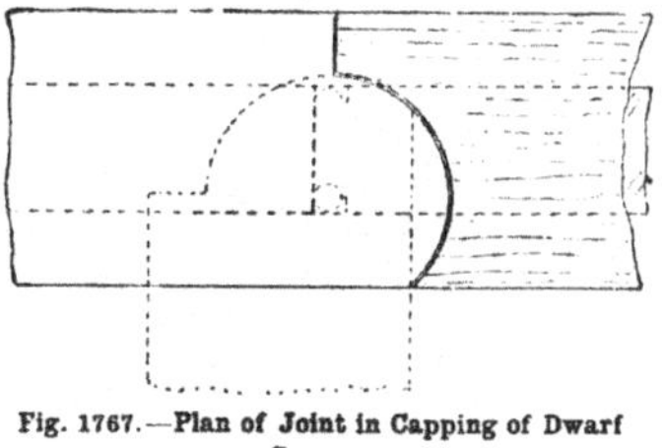

Fig. 1767.—Plan of Joint in Capping of Dwarf Screen.

is part elevation, Fig. 1757 horizontal section, and Fig. 1758 vertical section. The rail here shown should preferably be made of oak. The width of the opening of the chancel should be divided into a convenient number of bays, and the centre bay be made to open as shown in Fig. 1757. This bay should be hung with three 4-in. strong brass butt hinges, and should be provided with a small brass bolt on the inside to hold it in its place when closed up. The rail is 26 in. high from the floor to the top of the capping, and the width between the pilasters—that is, between the bays—is 23½ in. The capping, which is moulded

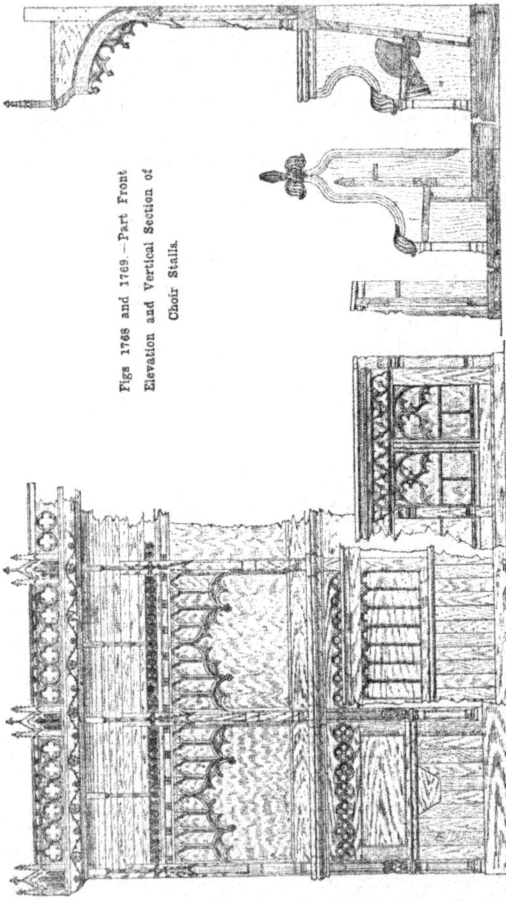

Figs 1768 and 1769.—Part Front Elevation and Vertical Section of Choir Stalls.

Fig. 1769.

Fig. 1768.

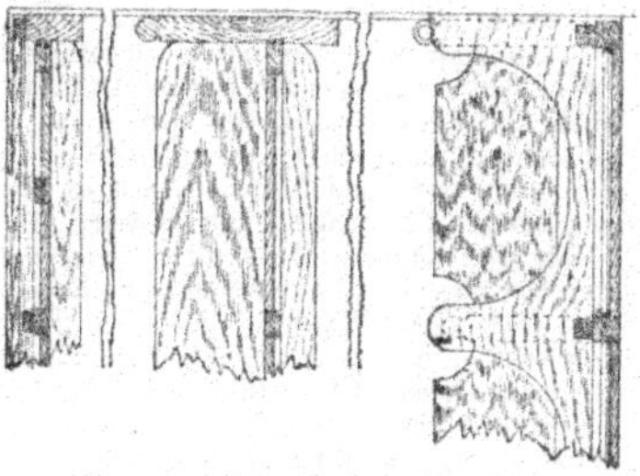

Fig. 1770.—Part Plan of Choir Stalls.

Fig. 1771.—Section through Front Framing of Choir Stalls.

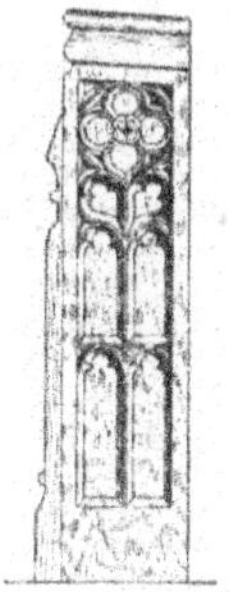

Fig. 1772.—Details of Front Framing Ends of Choir Stalls.

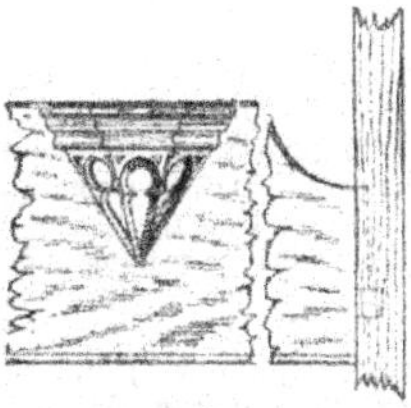

Fig. 1776. Fig. 1777.

Figs. 1776 and 1777.—Elevation and Vertical Section of Miserere Seat when turned up.

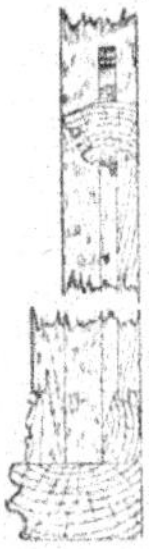

Fig. 1778.—Section of Miserere Capping, etc.

Fig. 1775.—Mould on Seat Ends of Choir Stalls.

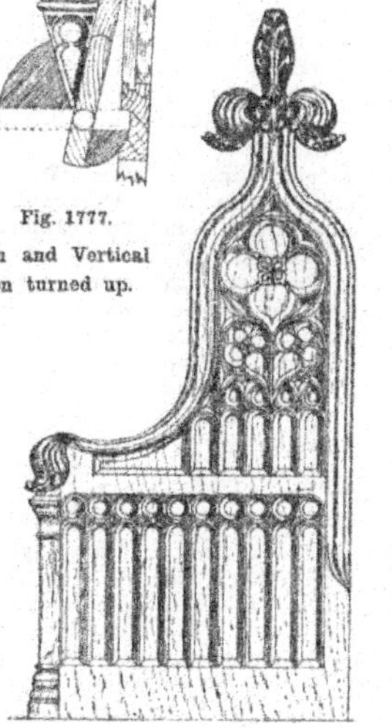

Fig. 1773. Fig. 1774.

Figs. 1773 and 1774.—Design and Alternative Design for Seat Ends to Choir Stalls.

on both edges and grooved on the under side for the top rail of the framing, is 4½ in. by 3 in., as shown by the enlarged detail (Fig. 1759). The shaped pilasters shown in section (Fig. 1758) are 2 in. wide and 1¾ in. thick. The framing is sunk ¼ in. deep to receive these. The framing is of 2½-in. stuff, and consists of 3½-in. by 2½-in. bottom rails, 6½-in. by 2½-in. shaped top rails, 3-in. by 2½-in. wall stiles, 3¾-in. by 2½-in. muntins, 2¼-in. by 2½-in. hanging stiles, and 1⅝-in. by 2½-in. bars. The bottom rails are chamfered, and the mould on the stiles and muntin is continued down to cut on to the chamfer. The mould on the edges of the framing is shown on enlarged detail (Fig. 1760). The top rails are shaped in the manner known as the four-centered arch, and the spandrils filled in on both sides of the framing, with very deeply sunk carving of flowers, fruit, wheat, acorns, leaves, etc., and no two panels should be alike. The method shown at Fig. 1761 of setting out the four-centered arches will be found sufficiently simple. The width A B is divided into four equal parts as A D C E B. With radius D C describe arc A C. With radius E C describe arc B C. From B, with radius A B, describe A G. From A, with radius A B, describe B F. From G through D draw a line to cut the arc at I. From F draw a line through E to cut the arc at H. With radii F H and G I complete the drawing. The framing should be fixed to the floor with strong brass screws through the bottom rail, their heads being sunk sufficiently deep to allow the screw-hole to be filled up with a pellet of oak. The capping should be dowelled on to the framing. The work should be left in its bare state, or just stained to suit any adjoining work.

Dwarf Chancel Screen.

The dwarf screen here described is suitable for a church in the fifteenth-century style, and should be executed in oak. The elevation (Fig. 1762) shows the outside of a portion of the screen, with a door 2 ft. 5 in. wide. The capping and plinth extend across continuously, and the method of fixing them is shown in Fig. 1763, which is a vertical section through the centre of a panel in the framing. Fig. 1764 shows the framing in section, and the capping and plinths in plan, with their respective joints. The framing is of 2-in. stuff, moulded in the solid, with cavetto and sunk fillets. The mitres are scribed, the bottom rail is chamfered, and the top rail worked with a double flute and fillet, stopped at the shoulders. The panels are carved; the outer plinth is made of two pieces, the upper piece being sunk in the rail and screwed from the back side. When this has been done, the back or inner plinth is fixed in its sinking by screws inserted from the front. Tapered dovetailed keys, fitting into corresponding grooves in the skirting, are fixed to the face of the rail at short intervals for the purpose of fixing the lower members of the face plinth. The skirting is driven on and secured with screws at the bottom edge, and should be fixed dry, in order that it may be removed for scribing to the floor. A filling-in piece to match the plinth should be inserted at the ends, as shown in Fig. 1765. The capping is double-tongued and glued on, the tongues and grooves being stopped ½ in. from the ends, in the doorway and on the door. It is fixed after the door has been hung. The hinge is sunk flush with the outside bead, and the door opens out at right angles. Figs. 1765, 1766 and 1767 (p. 497) illustrate enlarged details of the joints in the plinth and capping. Fig. 1766 is a plan of the plinth with the door open to its fullest extent, the dotted lines indicating its position when closed. The intersections of the various members give the points from which the radial paths of each are drawn. For marking the ends of the plinth, moulds of thin material are used, struck from the common centre of the hinge. The position of the moulds on the fixed piece A is found by first hanging the door, then securing the piece A temporarily in place, opening the door at right angles, and sliding the door piece against the fixed piece, and scribing its profile with the marking awl. The respective edge moulds should now be applied and the superfluous stuff cut away with gouges, until the moulds fit neatly when held against the door, and work easily when the latter is turned. Recesses will have to be cut, to the same sweep, in the bottom end of the hanging stile of the frame, as indicated by the dotted lines in Fig. 1766. In the case of the chief piece B (Fig. 1765) this path is cut right through the stile and into the plinth behind. This recess is shown by a full line in Fig. 1766, and is confined to the piece B, the flat below requiring a much smaller recess. The door piece may next be placed in position, and its end scribed with the compass to fit the recesses in the fixed piece. The compass is kept wide enough to allow for the sweep at the back, and the material grained portion indicates the fixed part of the capping; the mould for this is made of thicker stuff, and cut accurately square at the end. The cut on the moulding is marked with a half pencil round the curve, the position at which to apply the mould being found by opening the door (see Fig. 1767). The fixed piece is cut first, and then the other fitted to it. Figs. 1762 to 1764 are drawn to the scale of 1½ in. to the foot, and Figs. 1765 to 1767, 3 in. to the foot.

Fig. 1780.—End Elevation of Choir Stalls Canopy.

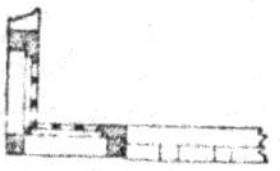

Fig. 1781.—Part Plan of Corner of Choir Stalls Gallery.

Fig. 1782.—Cresting and Top Rail of Choir Stalls.

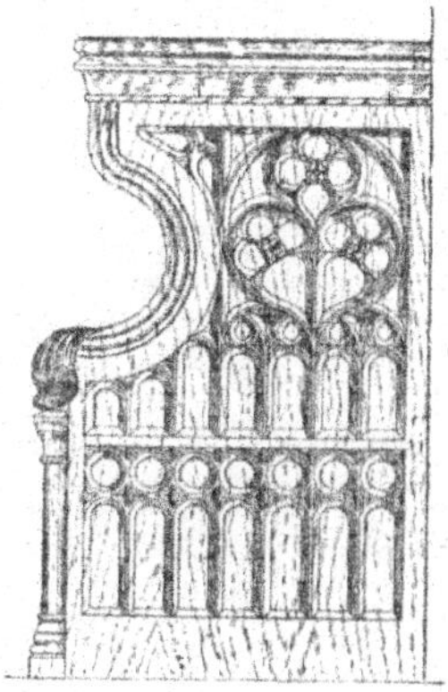

Fig. 1779.—Miserere End of Choir Stalls.

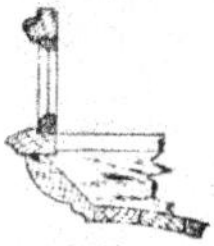

Fig. 1784.—Section of Cornice and Gallery of Choir Stalls.

Fig. 1783.—Canopy and Gallery of Choir Stalls.

cut away to the marks, the end being, however, shaped to the plan curves drawn with the moulds. The piece is cut off at first rather longer than required, to ensure a good fit. When the end is fitted so that it works easily, and fits close when shut, the opposite end is marked and cut off to a bevel, slightly longer outside, and fixed. The inner joints should be slightly bevelled, as shown in Fig. 1764. The remaining pieces of the plinth are then fixed at the striking end. A similar joint is made in the capping, although, in this case, a pair of moulds should be made, as shown in Fig. 1767. The Fig. 1765 shows the door wide open, with a portion of the fixed plinths, part of the bottom rail, and the hanging stile.

Choir Stalls.

The choir stalls illustrated by Figs. 1768 to 1784 conform to the Perpendicular style of Gothic architecture, and include front framing, an ordinary bench or seat, and at the back a row of misereres or subsellia, as they are variously called. Fig. 1768 represents a part front elevation, Fig. 1769 a section, and Fig. 1770 a part plan, and on referring to these it will be seen that the front framing and first seat are elevated 3 in. above the church floor, and the misereres 6 in. This elevated floor is composed of 1¼-in. narrow grooved and tongued boards laid on 3-in. joists placed 12 in. apart, with an oak curb mitered round. The bench ends are tenoned into these curbs. The front framing is 2 in. thick, and is 2 ft. 7 in. high from the raised floor to the top of the capping. It consists of an 8-in. by 2½-in. moulded top rail, with a trefoil design sunk and moulded as shown in Figs. 1768 and 1771. The framing is divided into panels by 3-in. muntins, every alternate muntin being moulded on its face, while the other muntins are left plain. A shaped pilaster or buttress projects in front. A geometrical moulded pattern is sunk in the faces of the panels, which are 1¼ in. thick. The bottom rail, which is 4½ in. by 2 in., has a moulded top edge. The ends to the framing are 8½ in. wide and 3 in. thick, and are shaped as shown by Fig. 1772, with carved sunk panels on the outer face, the front edge being shaped as a buttress. The capping or desk board is 7 in. by 1¼ in., and its front edge is moulded. The book-board is 4 in. by 1¼ in.; this also has a moulded edge, and is supported on shaped brackets at a height of 1 ft. 8½ in. above the raised floor.

Seats.—The seats are 2 ft. 10½ in. high to the tops of the backs, which consist of 2-in. framing, having a 4½-in. by 2-in. moulded top rail, 5-in. by 2-in. bottom rail, 3-in. by 2-in. muntin, with 1¼-in. panels sunk and carved to the design shown in Fig. 1768. The seat, 13 in. wide by 1¼ in. thick, is tongued into the bottom rail. The front of the seat, down to the floor, is filled in with 3-in. by 1-in. matchboarding nailed to 1¼-in. by 1-in. fillets nailed to seat and floor. The seat-ends are 4 ft. 3 in. high above the raised floor, and are 2 ft. wide by 3 in. thick. The edges are moulded (see Figs. 1773, 1774, and 1775), the moulds being stopped by the carving in front and diminished out on the back edge, and gradually diminished on the front edge, as in Fig. 1773. The front edge beneath the carved boss or carved head is shaped as an engaged pillar. The seat-ends are finished with a carved finial terminating in a poppy-head. The outer faces of the seat-ends are divided into sunk, moulded, and carved panels, as shown in Figs. 1773 and 1774. The sinking and carving of the ends is varied in design, and two designs may be made alternate, as shown by Figs. 1773 and 1774; but the profile should be kept to one pattern. The carved heads may be of grotesque design, or may represent notabilities connected with the Church. Only one seat is shown here between the front framing and the misereres, but there may be as many rows as the size of the choir demands. Each seat, however, should rise 3 in. higher than its front neighbour.

Seat Bracket or Rest.—The bracket or rest under the seat should be in one solid piece. It is carved and moulded, and is 11 in. by 9 in. by 5½ in. In old work these brackets were carved very grotesquely, one notable example being preserved in Wells Cathedral, where the bracket is formed by a carved representation of a cat playing a fiddle.

Misereres, or Subsellias.—The misereres are each 2 ft. 7 in. long, with a seat 1 ft. 2½ in. wide by 1½ in. thick. Fig. 1776 shows an elevation of the under side of miserere seat when turned up, and Fig. 1777 a section of seat and back framing. This seat is hung on pivots, or on solid pins formed on the seat, and has a bracket on the under side, and this bracket, when the seat is turned up, forms a rest for the occupant to lean upon when in a standing position. The ends and divisions of the misereres have a sinking ½ in. deep on their inner surfaces, to allow the seats to open back; the seats, when down, rest in the sinking. The backs of the misereres recline, and are composed of a top rail 8 in. by 2 in., with sunk and moulded quatrefoils, and bottom rail 5½ in. by 2 in., with its face hollowed out to receive the seat (Fig. 1777). Panels, 1¼ in. thick, are tongued into the framing. The framing, ends, and divisions are surmounted by a solid top 4 in. thick, with a shaped and moulded edge, as in Figs. 1769, 1770, and 1778. The miserere ends are sunk, moulded, and carved as shown in Fig. 1779, and may be varied alternately. The opening beneath the seats is matchboarded to the floor, boards being nailed to 1½-in. by 1-in. fillets.

Wall Lining, etc.—Above the misereres or subsellias is a panelled wall lining, 8 ft. 10 in. high above the floor, with a canopy and gallery (see Figs. 1780 and 1781). This lining consists of 4-in. by 3-in. posts 5 ft. 2 in. long, 4-in. by 2-in. chamfered bottom rails, and 4-in. by 3½-in. moulded top rails with carved ornaments at intervals in the sunk moulding. Over the top rail is a carved cresting, 3 in. high (see Fig. 1782). Tongued into the framing are 1¼-in. panels, with 1-in. thick tracery work in front of them. On the face of the posts are shaped and moulded buttresses, 2¼ in. thick, with carved finials.

Canopy, Cornice, Gallery, etc.—The canopy is supported by ribs ⅞ in. thick, with their edges perforated, moulded, and traceried, and terminating in a 1½-in. square carved drop (see Figs. 1768, 1769, and 1783). The end ribs have, in addition, a moulded sinking on their outer faces (see Fig. 1780). These ribs are covered with 1-in. grooved and tongued narrow boards, slightly V-jointed, these boards being secured to 2-in. shaped deal bracket-pieces tenoned into posts (see Fig. 1769). These brackets also support the carved cornice mould and gallery. Fig. 1784 shows a section of cornice and gallery. The cornice is 5½ in. by 1½ in., with flowers and stems carved in relief upon it (see Fig. 1783). Above this is a 3-in. by 2-in. mould, surmounted by the gallery. The gallery is composed of a 5½-in. by 1-in. board pierced with moulded quatrefoils, with a 2-in. by 2-in. crenellated capping. At intervals a group of three carved finials, each group consisting of two 1¼-in. by 1¼-in. outer posts, with a ⅞-in. sunk, shaped, pierced, and moulded panel, with a carved finial in the centre (see Figs. 1768, 1780, 1781, and 1783). The stalls should be executed in oak, not polished or varnished, but left in its natural state.

Communion Rails.

The panels and balusters of the oak communion rail shown in elevation by Fig. 1785 are enriched with carvings of the vine and wheat, symbolical of the wine and bread. The rail is 17 ft. 6 in. long, and 2 ft. 6½ in. high through the middle of the gate, which rises 2½ in. higher than the handrail on each side, while the greatest thickness, measured across the handrail above the balusters, is 9¼ in. The rail breaks back over the carved panels 1½ in. on each side, as shown in the plan (Fig. 1786). The handrail is prepared from 4-in. stuff, cut to shape with a band-saw, moulded as much as possible with a vertical spindle, and the mitres finished with carver's gouges. The top rail of the gate is also worked in the same manner. The bottom rail is from 2½-in. oak, 8 in. wide below the balusters, and 5½ in. wide below the panels. An enlarged detail of one length of the rail is given at Fig. 1787. The panels are from 2½-in. oak, perforated, and stump-tenoned into the top and bottom rails as shown by dotted lines. The balusters are 7½ in. square across the thickest parts, and are richly ornamented with wheat. Fig. 1788 represents a section taken at A (Fig. 1787), showing more clearly the bold treatment of the carving. A section taken at B (Fig. 1787) is reproduced in Fig. 1789, in which it will be noticed that the panels are set forward so that they project within about 1 in. of the front edge of the bottom rail. Fig. 1790 represents a part section at C (Fig. 1787). The tenons go right through the rail, and are wedged. The gate is constructed in a similar way to the rail, the top being in one piece. The letters I H S are introduced in the top of this panel, and they are also stump-tenoned into the top rail. The gate is hung with brass parliament or shutter hinges, one hinge being on the bottom rail and the other on the baluster (see Fig. 1791). The centre of the knuckles of the hinges must be on a line with the greatest projection of the mouldings. The hanging baluster of the gate is cut through at right angles, but the shutting baluster is cut sufficiently bevelled to allow the gate to swing clear. When closed, the gate is held fast by a small automatic ball-catch. The communion rail is "screw-slotted" to secure it in position at each end. Figs. 1792, 1793, and 1794 show this effectual and secret method of fixing. Fig. 1792 represents a piece of deal with a hole bored large enough to take the head of a stout screw, and a slot cut to receive the plain part of the screw, let in flush with and nailed to the wall; then the

Fig. 1785.

Fig. 1786.

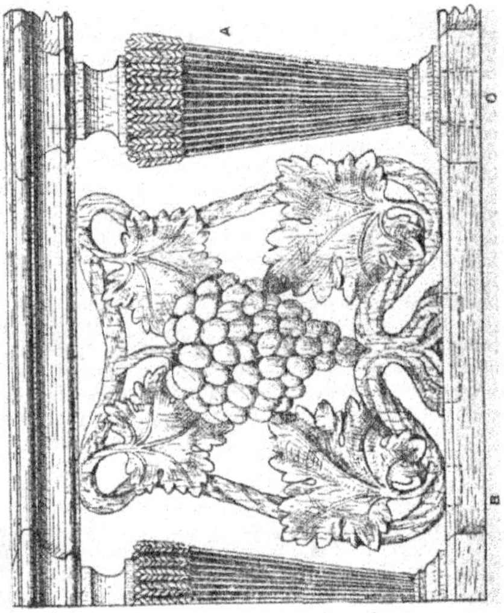

Fig. 1787.—Detail of Carving in Communion Rails.

Figs. 1785 and 1786.—Front Elevation and Plan of Communion Rails.

screws (see Fig. 1793) that have been driven into the end of the communion rail are slipped into the circular hole and driven down the slot, as in Fig. 1794, until the bottom rail touches hard on the floor. The bottom rail drops on to iron dowels that are leaded into the stone floor. The communion rail is fumed and wax-polished.

Simpler Design.—Fig. 1795 shows a much simpler design for inexpensive communion rails. Pitchpine, as shown, is very suitable for this class of work. A crimson cord is used instead of a gate. The extreme length, including the cord, is 18 ft. 3 in., and the height 2 ft. 4 in. The handrail is 4 in. by 3 in., and moulded as shown in the enlarged detail (Fig. 1796), which also includes a section of the bottom rail, 3 in. by $2\frac{1}{2}$ in., and section and front elevation of the balusters, which are prepared from 1-in. stuff, tenoned through the bottom rail, and wedged. The rail is secured to the wall at the ends in the same way as Fig. 1785, and the bottom rail is nailed to a wood floor. The rail should be prepared for varnishing, with two coats of size, the second being put on after the first has been rubbed down with glass-paper, then finished with elastic oak varnish. Fig. 1797 represents a plan of the inside ends of the rails, showing how the cord is held in position with brass rings on each end, and hooks fixed to the rails, from which the cord is easily detached.

Lectern.

Before constructing the lectern shown by Figs. 1798 to 1808, a very careful selection of oak should be made. This class of work calls for specially good craftsmanship, as church furniture of this description occupies a prominent position, in full view of all comers. Unfortunately, good dry and sound English oak is difficult to obtain; but carefully selected Riga oak makes an efficient substitute; and, when possible, the timber should be cut out to the sizes required and left for a few weeks before the work is put in hand. Fig. 1798 shows a front elevation of the lectern, and Fig. 1799 a side view. The heights given are those that are generally found suitable. An error of construction that is often committed is that of giving too much slope to the desk, causing the book to overlap the bottom edge and throwing it out of shape. A slighter slope, as shown, is much preferable. The base is formed of two chamfered pieces of $3\frac{1}{2}$-in. by 3-in. stuff, halved, and secured with a coach screw, and let into the column. In Fig. 1801 the dotted lines indicate the desk, and the

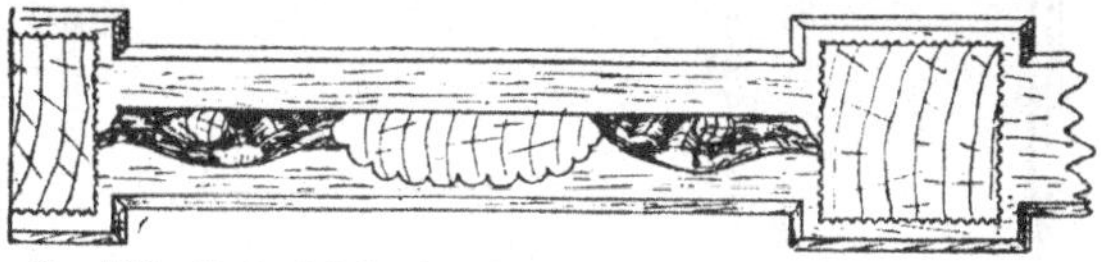

Fig. 1788.—Horizontal Section through Communion Rails at level of A (Fig. 1787).

Fig. 1789.

Fig. 1790.

Figs. 1789 and 1790.—Part Sections of Base of Communion Rails at B and C (Fig. 1787).

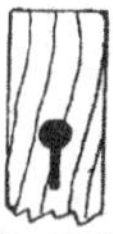

Fig. 1792.

Fig. 1793.

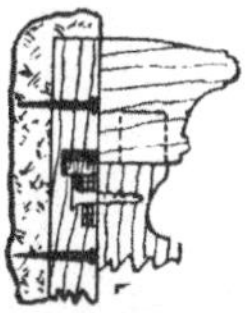

Fig. 1794.

Figs. 1792 to 1794.—Details of fixing Ends of Communion Rails.

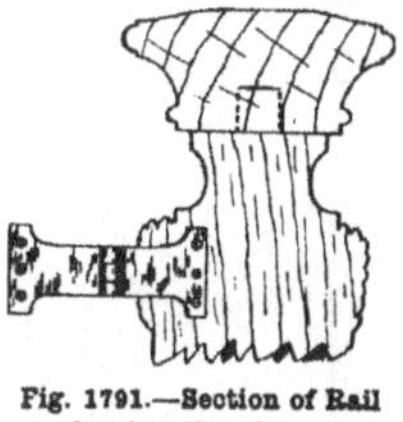

Fig. 1791.—Section of Rail showing Top Hinge.

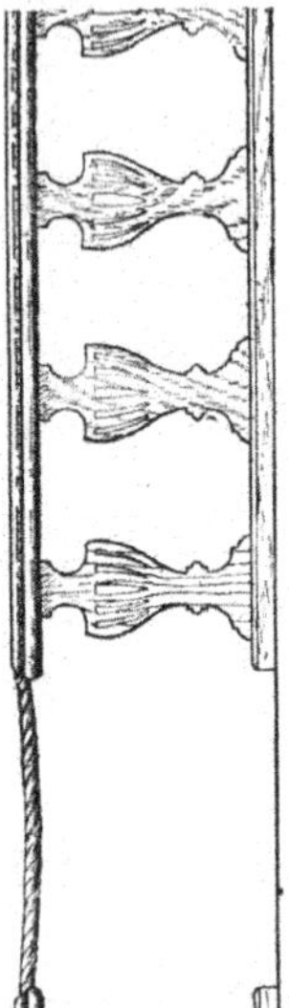

Fig. 1795.—Front Elevation of Simpler Design of Communion Rails.

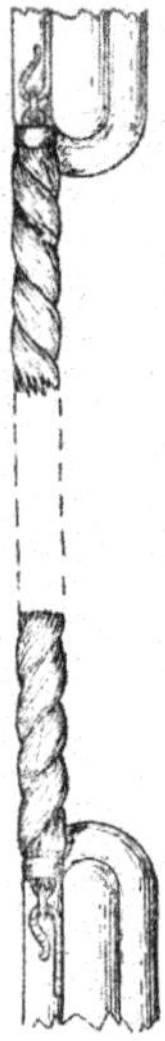

Fig. 1797.—Part Plan (Enlarged) of Communion Rails showing Connection of Cord.

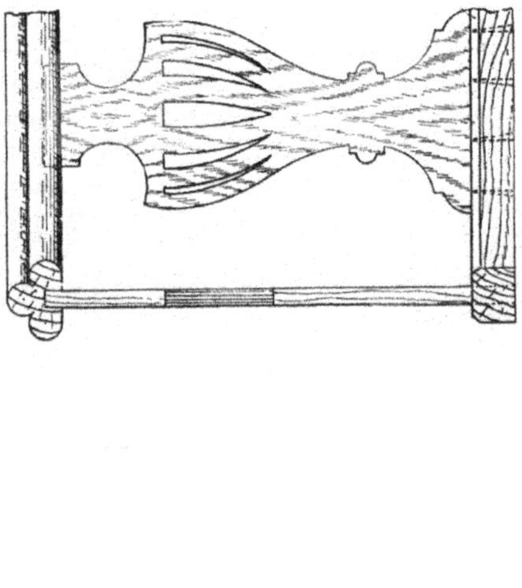

Fig. 1796.—Enlarged Detail of Communion Rail.

brackets above the cap. The octagonal column is in one piece, running from the base up to the under side of the desk. The bottom brackets are cut from 2-in. stuff, shaped, chamfered, and housed into the base and the octagonal column. Cusped brackets, with open panels of the same thickness, are also fitted underneath the desk. Figs. 1800 and 1801 show various plans, and Figs. 1802 and 1803 several details, while Fig. 1804 represents an elevation of the tracery panel in front of the desk. The desk is framed up as shown in Fig. 1805, and is cut out of 1½-in. stuff, with brackets fitted below as before described. Cutting the front tracery panel right through is much easier than cutting it out of the solid. A thin board screwed in from the back as shown in Fig. 1805 gives the appearance of a solid tracery panel. A cheaper method can be adapted for this lectern by making up the moulded octagonal cap and base in sectional pieces with mitered angles as shown by the hatched portions and sectional plan (see Figs. 1806 to 1808). This would do away with cutting the octagonal cap and base in the solid; and if it is decided to make them up as described, the central column should be in one piece and cut back to receive the moulded and mitered cap and base as shown. The finish of the wood is a matter of taste. It could be left clean, or beeswaxed, polished, oiled, or fumed, as required.

Revolving Lectern Top.

There will now be given details of a double desk lectern to surmount a similar pedestal stand to that just described. Figs. 1809 and 1810 show respectively a side elevation and a front elevation. The desk for receiving the book is 1 ft. 10 in. by 1 ft. 6 in. Alternative methods are shown for the moulded stop at the end of the sloping book-board. The arrangement of the fittings forming the revolving top is shown in Fig. 1811, and consists of a central pivot riveted to an octagonal-shaped plate, the latter being secured to the top of the column by means of screws, while a second plate is fixed to the bottom board of the desk as shown. The lower pivoted plate should be fixed slightly above the level of the carved brackets, so that the bottom board of the lectern will just clear them. To make this central pivot rigid, it can be run through a block of wood A, about 4 in. by 4 in. by 5 in., which is securely screwed from the board underneath, the pivot being finished on top with a large washer and screw. Fig. 1812 shows the detail of the end tracery panel, which can be cut out of ¾-in. stuff, and pierced right through. Fig. 1813 shows an enlarged detail through tracery panel on line X X (Fig. 1812), and Fig. 1814 an enlarged detail through line Y Y (Fig. 1812). A ½-in. backboard is screwed to the tracery, a small moulding being planted on to stop the joint. The panel can therefore be taken out at any time, in order that the fittings to the revolving top may be attended to. Good dry English oak should be used, but carefully selected American wainscot oak would make a good substitute.

Alternative Design for Lectern.

Figs. 1815 and 1816 show side and front elevations of another lectern, Fig. 1817 being a section on line A B (Fig. 1816), and Fig. 1818 an enlarged section on line C D (Fig. 1816). The tracery panels are 1½ in. thick, and after being cut out should have the back panels inserted as shown in detail. This is a much easier method than carving the tracery from the solid. The central shaft is quatrefoil in shape, and should run from the under side of the book-rest to the floor below. The mouldings to the base may be either solid or mitered at the angles, and planted on; but the lower part of the base must be in the solid, as the other parts of the lectern will be built up from this. The wood must be chosen very carefully. It must be perfectly dry, and free from all knots and shakes. English oak or teak would be very suitable.

Lectern of Substantial Construction.

The elevation (Fig. 1819) and the section (Fig. 1820) of the lectern about to be noted are reproduced to the scale of 1 in. to 1 ft. Fig. 1820 shows the general construction of the framing, which should be of the following dimensions :—The top, 1½ in. thick, moulded on all edges, and mitre-clamped at ends; the side standards, 2½ in. thick; and the

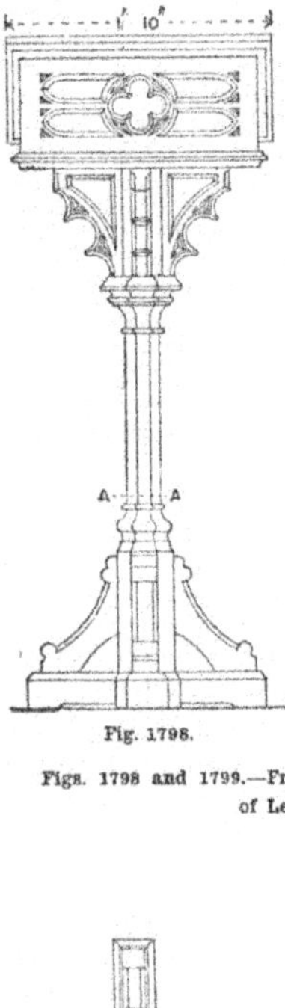

Fig. 1798.

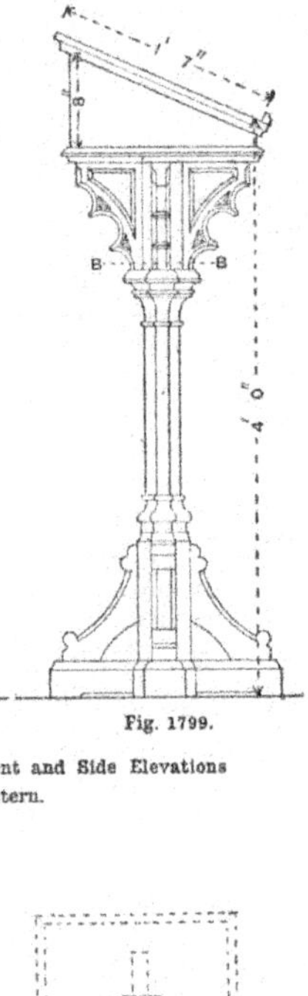

Fig. 1799.

Figs. 1798 and 1799.—Front and Side Elevations of Lectern.

Fig. 1800.—Horizontal Section of Lectern at A A (Fig. 1798).

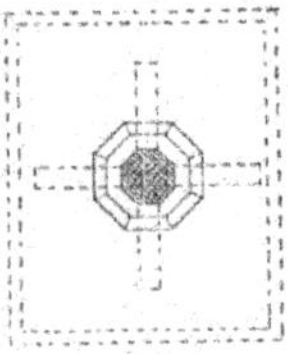

Fig. 1801.—Horizontal Section of Lectern at B B (Fig. 1799).

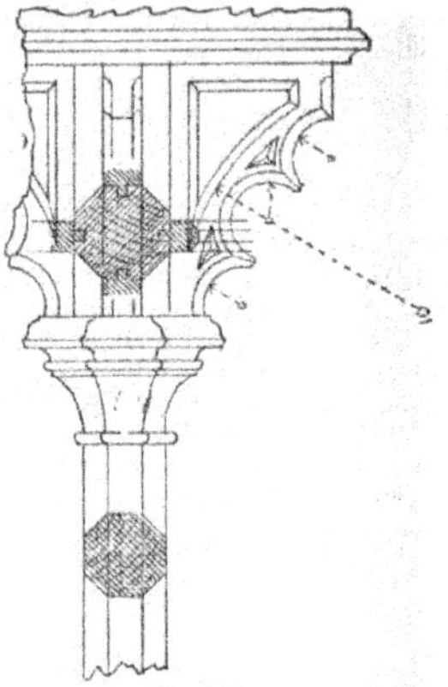

Fig. 1802.

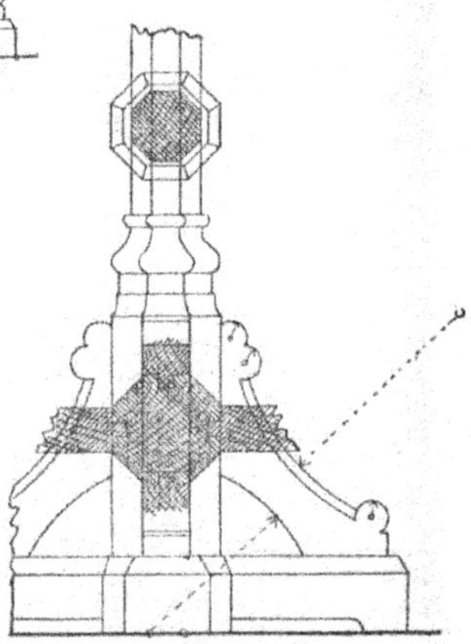

Fig. 1803.

Figs. 1802 and 1803.—Enlarged Detail Front Elevation and Sections of Lectern without Desk.

shaped brackets at the top, in front of the side standards, 2 in. thick; the main filling to the front elevation, $2\frac{1}{4}$ in. thick, the upper part having two pointed arch shapings cut into it, and trefoil filling pieces, $\frac{3}{8}$ in. thick, tongued to the soffit. The main pilasters are $3\frac{1}{2}$ in. by $2\frac{3}{4}$ in. in section, cut and fitted at back to the main standards. The smaller pilasters may be worked out of 3-in. by 3-in. stuff. The stiles and mullion to the lower portion are worked from 3-in. by 3-in.

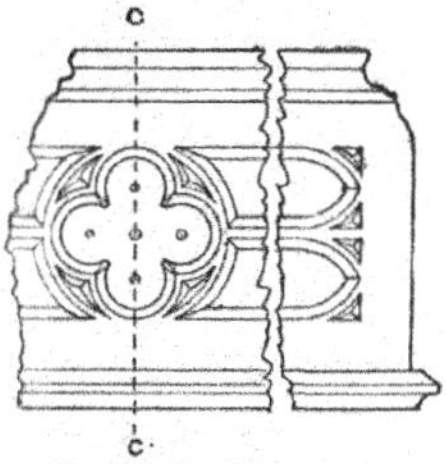

Fig. 1804.—Tracery Panel in Desk of Lectern.

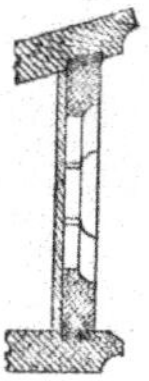

Fig. 1805.—Vertical Section of Tracery Panel in Lectern on Line C C (Fig. 1804).

stuff, notched out to receive the pilasters. The small rose-shaped enrichments to the front elevation may be sunk from the face to a depth of $\frac{3}{8}$ in. The perforations to the two lower panels give a rich and solid appearance to the lower portion. To obtain the best effect, the mouldings should be deeply undercut. In the illustrations, the base is formed by a projection in the curb to the platform adjoining, but this portion of the design could easily be altered to suit special requirements and circumstances. The wood used should be well-seasoned, carefully selected oak or pitchpine.

Seat to Lectern or Reading-Desk.

The side elevation (Fig. 1821) and the front elevation (Fig. 1822) of the seat to lectern or reading-desk are reproduced to the scale of $\frac{3}{4}$ in. to 1 ft. Fig. 1822 shows the general construction. The dotted lines

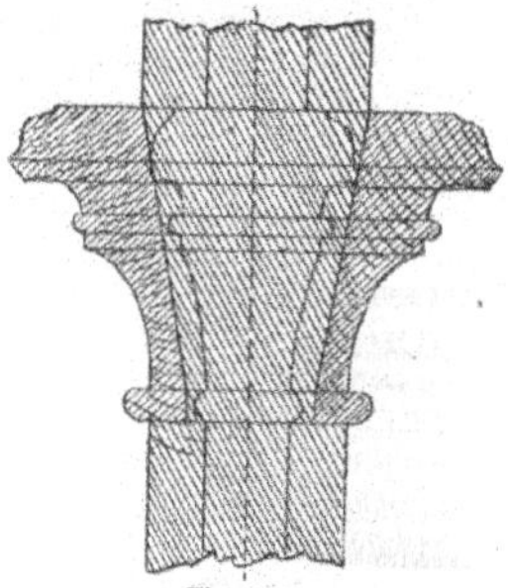

Fig. 1806.

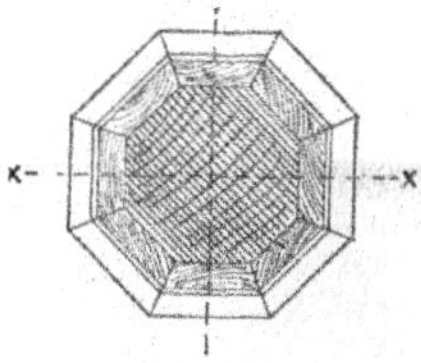

Fig. 1808.

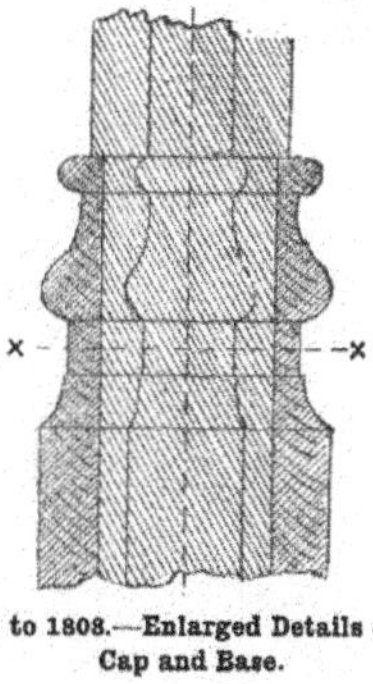

Fig. 1807

Figs. 1806 to 1808.—Enlarged Details of Lectern Cap and Base.

indicate the position of the seat, the seat back, and the V-jointed boarding under. The seat itself, which is 1½ in. thick, is tongued in at the back to the seat-back. The two ends are housed into the standards, and the front edge is moulded. The seat-back is 1¾ in. thick, housed in at the ends to the side standards, moulded on the top edge, and tongued and moulded on the bottom edge, while the shaped perforations shown in the front elevation are sunk in from the solid. The ½-in. matched and V-jointed boarding under the seat gives a substantial appearance to the lower part. The side standards are worked from 2-in. stuff, carefully jointed, cross-tongued, and

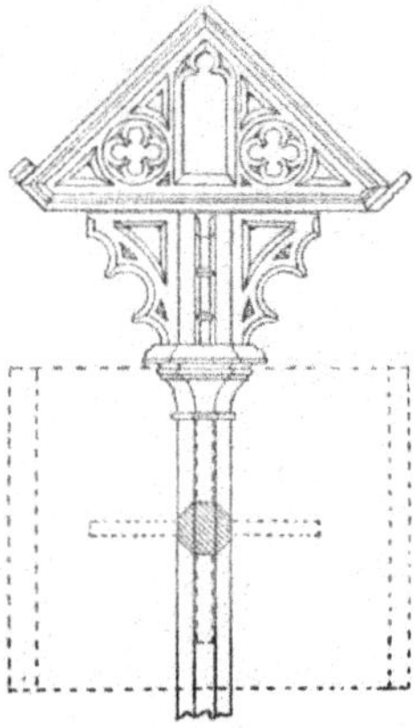

Fig. 1809.

Fig. 1810.

Figs. 1809 and 1810.—Side and Front Elevations of Revolving Lectern Top.

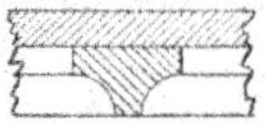

Fig. 1813.—Part Horizontal Section of Lectern Tracery Panel on Line X X (Fig. 1812).

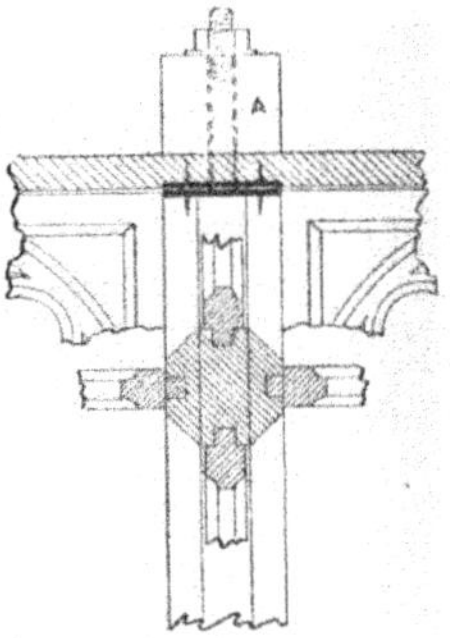

Fig. 1811.—Fixing of Pivot of Revolving Lectern Top.

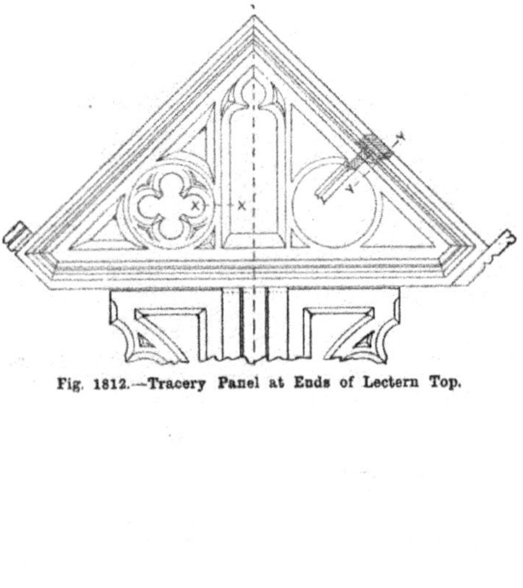

Fig. 1812.—Tracery Panel at Ends of Lectern Top.

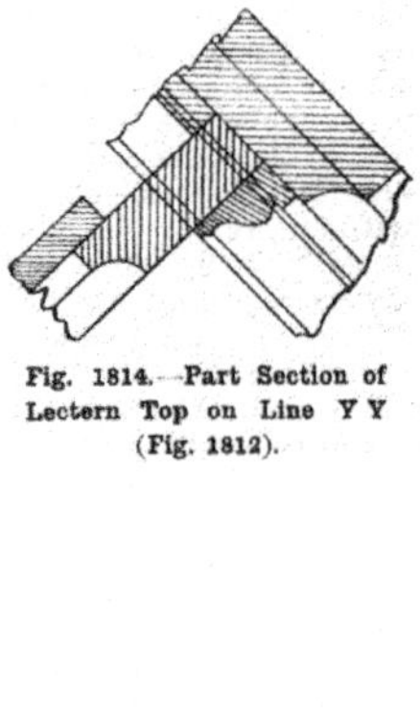

Fig. 1814.—Part Section of Lectern Top on Line Y Y (Fig. 1812).

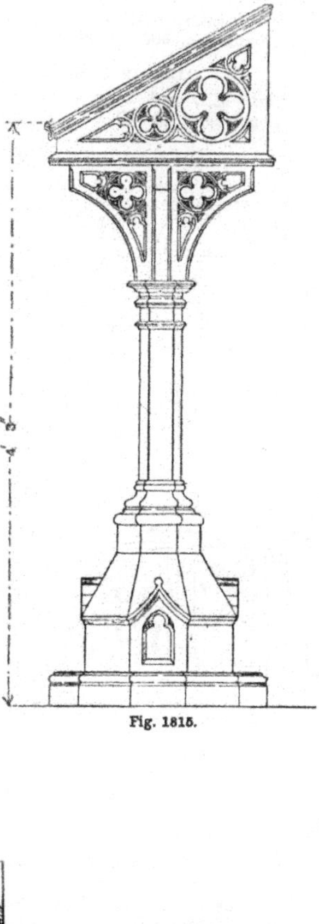

Fig. 1815.

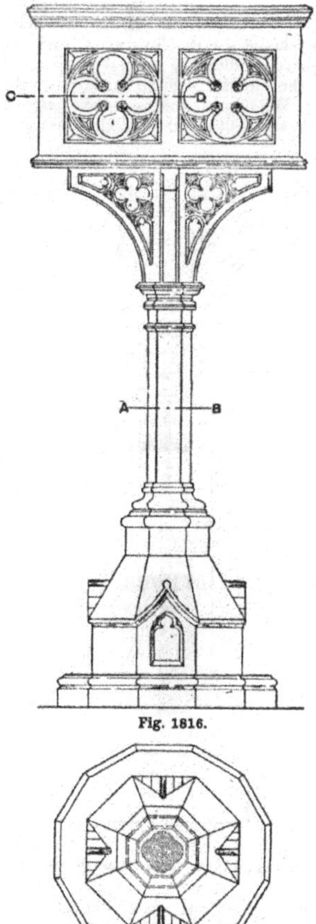

Fig. 1816.

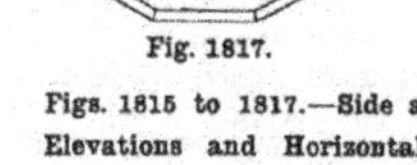

Fig. 1817.

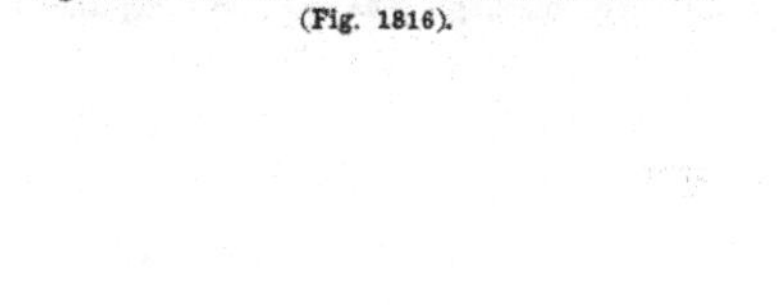

Fig. 1818.—Section of Lectern Desk on Line C D (Fig. 1816).

Figs. 1815 to 1817.—Side and Front Elevations and Horizontal Section on Line A B of Alternative Design of Lectern.

glued up. The perforations to the lower portion should be carefully sunk, and the trefoil-headed filling-in pieces tongued in before the circular column is fixed in the centre ; the shaped sinkings above may be sunk to the section shown, this giving a very pleasing effect. After the standards have been notched in front, the circular column, having square cap and base, should be carefully fitted and fixed. The standards on each side are tongued to a 3½-in. by 2¼-in. chamfered bottom rail, which, in its turn, may be screwed to the floor. The material used, and the general finish of the work, should accord with those specified for the lectern itself as described in the previous paragraph (pp. 507 and 509).

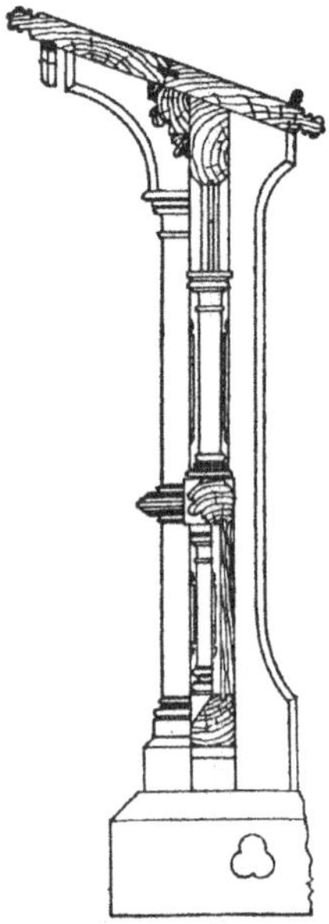
Fig. 1820.

Fig. 1819.

Figs. 1819 and 1820.—Front Elevation and Vertical Section of Lectern of Substantial Construction.

Litany Desk.

Figs. 1823 and 1824 show front and sectional elevations of a litany desk, Fig. 1825 showing the plan. This desk is usually placed in the middle of the chancel, and close to the entrance steps. The base is formed of 3-in. by 3-in. oak, with chamfered edges and returned angles. At the back portion a kneeling-board C, 8 in. by 5 in. by 1 in., is screwed to the base. On this board can be placed a small carpet or cushion for kneeling. The sides forming the desk are of 1½-in. stuff (see Fig. 1824), cut, shaped, and chamfered, and tenoned and mortised into the base and secured with oak pegs. The panelled front (see Fig. 1823) is made up as follows:—The main framing forming the panel is 1½ in. thick, housed into the sides, base, etc., as shown. This part of the panelling is chamfered and grooved to receive the 1-in. tracery panel. Fig. 1824 represents a sectional elevation of the desk, showing the main panel tongued into the base of the desk. A moulded rail is fitted up under the top of the desk, giving a neat finish to the panelled front. Fig. 1826 is an enlarged diagram showin g setting-out lines for front tracery panels; Fig. 1827 is an enlarged plan through front tracery panel on line A B (Fig. 1823).

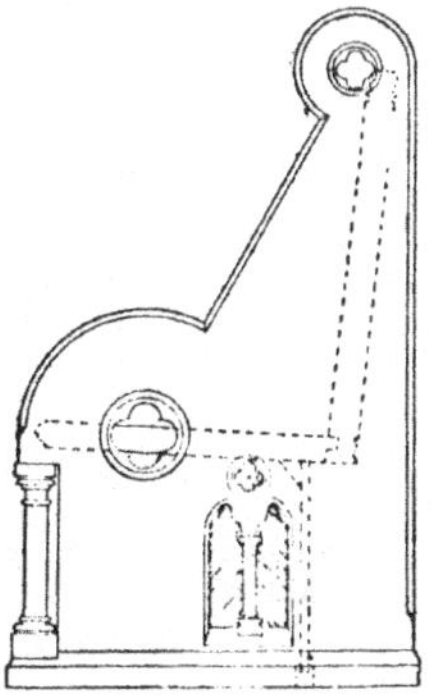

Fig. 1821.

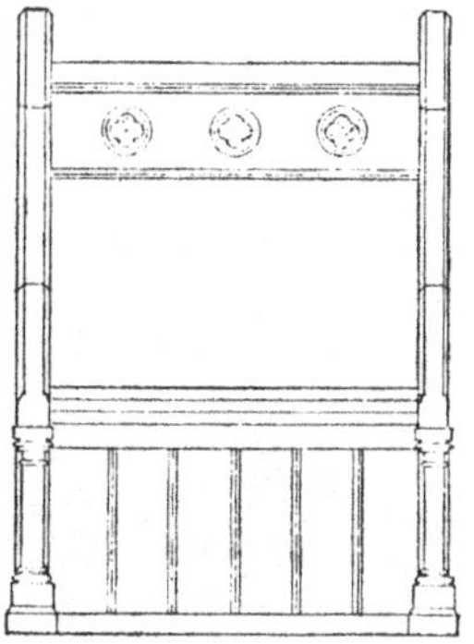

Fig. 1822.

Figs. 1821 and 1822.—Side and Front Elevations of Seat to Lectern.

Choir Book-rest.

The choir book-rest shown in Figs. 1828 and 1829 is drawn to the scale of ¾ in. to 1 ft. Fig. 1828, which represents a section taken through the centre of one bay, shows the general construction of the rest. The book board, ¾ in. thick and 7¼ in. deep, has a clear depth of 6¼ in.—sufficient to support any ordinary-sized book—and should be housed in at the ends to the end standards (finished size, 8 in. by 2½ in.). The 1½-in. thick skeleton framing under the book board is supported by shaped brackets at ground level, and housed into standards at ends. The three-quarter pilasters on the front, and the shaped brackets, should be carefully fitted and screwed, the heads of screws, of course, being let in and pelleted in grain. The shaped hatchings to the trefoil filling pieces at the top, and the panels at the bottom, as shown in elevation, indicate perforations; and if these are carefully worked, they greatly increase the rich appearance of the finish. The pilasters at the foot of the end standards are circular, and may have their bases circular and formed in the same piece of stuff, and the whole length need simply have a straight joint, with the standard at the back; or, instead, the pilasters may have their bases square, the lowest portion, including the chamfer, being formed solid in the standard, and the pilaster carefully fitted into the notching prepared for it. Fig. 1829 represents two bays only; the total length of book-rest will, of course, vary according to the number of choristers to be accommodated; but in any case it will be found that two bays will give ample width for each person. The materials used and the style of finishing of this book-rest should match those described in

Fig. 1823.

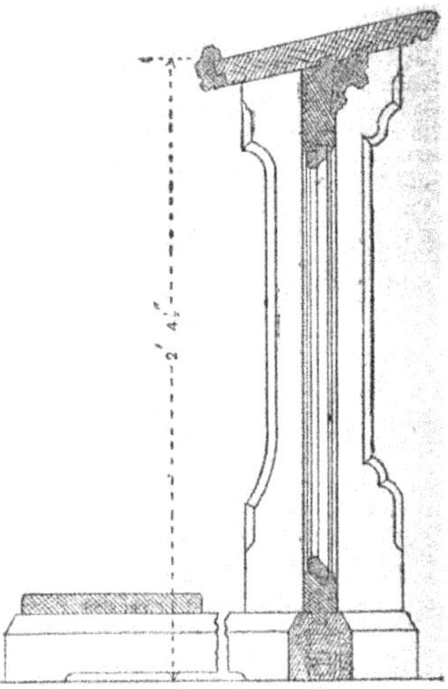

Fig. 1824.

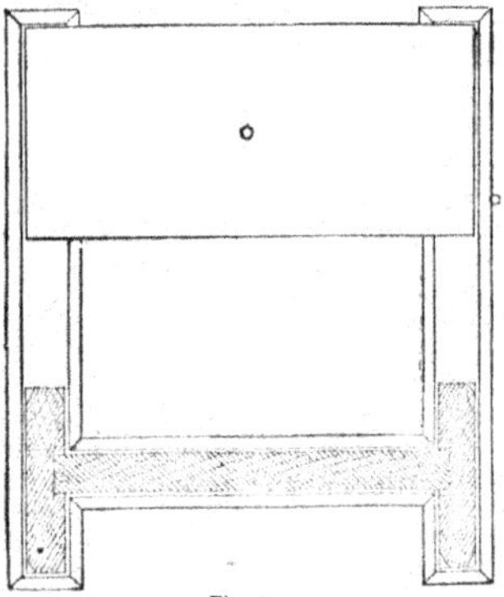

Fig. 1825.

Figs. 1823 to 1825.—Front Elevation, Vertical Section, and Plan of Litany Desk.

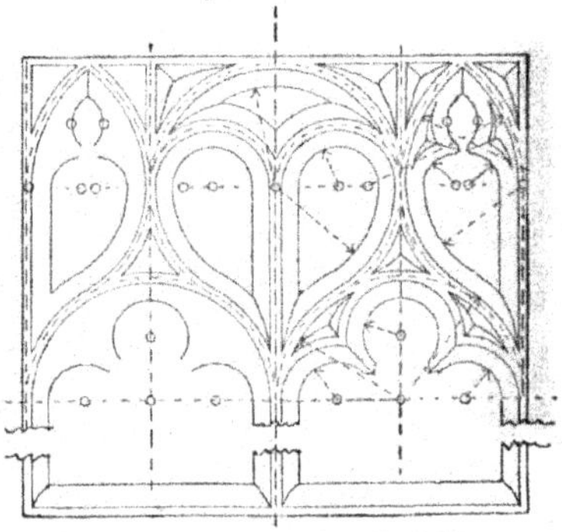

Fig. 1826.—Front Tracery Panel of Litany Desk.

the specification for the lectern illustrated by Figs. 1819 and 1820 (p. 512).

Pulpit.

Figs. 1830 and 1831 represent front and side elevations of a pulpit, standing partly on the floor of the church and partly on that of the choir. The measurement from the choir floor to the top of the pulpit is 5 ft. 10 in., the internal width and depth is 3 ft. 6 in., and the measurement from the choir floor to the floor of the pulpit 2 ft. 10 in. The pulpit is provided with a movable reading board, which can be adjusted to any height to suit the convenience of the preacher, and with a seat, which is necessary if the pulpit is intended for a chapel, but may be dispensed with for a church. Two small shelves are shown (see plan, Fig. 1832), which are convenient for standing a glass or books on. From the choir floor the pulpit is reached by a short flight of stairs, the lowest step having rounded tread and riser at both ends (see Fig. 1832). Two 4-in. by 4-in. turned newels, 4 ft. 10 in. high, with octagon-shaped terminals, and a ball as finial, support the strings and handrail as shown in Figs. 1830 and 1832. Under the strings of the stair is shown an open perforated spandril of $1\frac{1}{4}$-in. stuff. The strings are made of $1\frac{1}{4}$-in. stuff, with a perforated and panelled piece of framing $1\frac{1}{4}$ in. thick on the face. The handrail is of a roll pattern, out of 3-in. by 3-in. stuff. Around the front and on one side of the base of the pulpit is fixed a 6-in. by $\frac{3}{4}$-in. plinth to $\frac{7}{8}$-in. V-jointed matchboards, which cover up the 4-in. by 3-in. framing that supports the choir floor. The choir floor is reached by one step, and the nosing of this floor is carried round the front and side of the pulpit. Above this nosing is a $2\frac{1}{2}$-in. by $1\frac{1}{2}$-in. curb, fastened to a plinth composed of 8-in. by $\frac{3}{4}$-in. and 3-in. by $\frac{3}{4}$-in. moulds (see detail, Fig. 1834). This plinth is fastened to the lower portion of the pulpit, which consists of $1\frac{1}{4}$-in. framing, secured to and supported by 4-in. by 2-in. rough framed and braced scantlings, shown in section (Fig. 1833). The panels in this lower portion are $\frac{1}{2}$ in. thick, screwed to the framing from the back, and incised to a pattern as shown on elevation (Fig. 1830). Attached to this lower framing is an 8-in. by $\frac{1}{2}$-in. shaped and perforated corbel table, crowned with a mould made up of $3\frac{1}{2}$-in. by $\frac{3}{4}$-in. and 3-in. by $2\frac{1}{2}$-in. moulds, shown on enlarged detail (Fig. 1835). The upper portion of the pulpit, which commences above this mould, consists of 2-in. framing, with 1-in. perforated panels, $\frac{1}{2}$-in. panels being

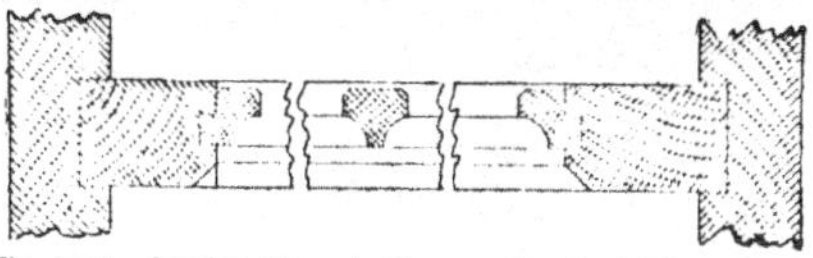

Fig. 1827.—Section through Tracery Panel of Litany Desk.

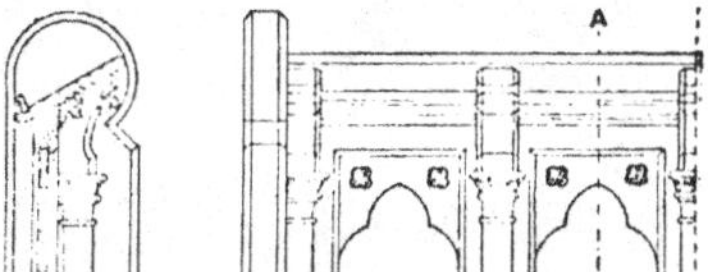

Fig. 1828. Fig. 1829.

Figs. 1828 and 1829.—Vertical Section and Part Elevation of Choir Book-rest.

screwed to the back, as shown by section (Fig. 1833) and enlarged detail (Fig. 1836). This top framing is surmounted by a 3½-in. by 3-in. deeply carved cornice. Fig. 1836 grooved and tongued boards, laid on 4-in. by 2-in. bearers. The back of the pulpit forms the seat back; it is fixed reclining, and consists of 1½-in. framing with 1-in.

Fig. 1830.—Front Elevation of Pulpit (from Church Floor).

shows an enlarged detail of the cornice, and Fig. 1837 displays the carving. Under the carved cornice is a 6-in. by ½-in. dog-tooth moulding (see Figs. 1830 and 1836). The floor of the pulpit consists of 1¼-in. V- and flush one side panels. The top rail is moulded and the middle rail ploughed to receive a 14-in. by 1½-in. round-edged seat supported at one end by a shaped bracket (see section, Fig. 1833). It will be noticed that the design of the centre panels is different from that of the side panels (see Fig. 1830). All rails have a return bead worked on the lower edge, and a deep chamfer on the top edge. The stiles are stop-chamfered; for detail of stop, see Fig. 1836. Around the centre rail of the upper framing is a 1¼-in. by ⅞-in. roll moulding, tongued into the rail (Fig. 1833).

Fig. 1831.—Side Elevation of Pulpit (above Church Floor).

Pulpit in Perpendicular Style.

The ancient wooden pulpits in England were usually polygonal, with their panels richly adorned with feathering, tracery, and mouldings. The pulpits of Continental churches are usually very large, and elaborately adorned with carved ornaments. Figs. 1838, 1839, and 1840 show, respectively, elevation, plan, and section of a pulpit in the Perpendicular style of Gothic architecture. The pulpit is octagonal in shape, and the portion above the pulpit floor, which is supported by eight trusses, projects over the lower portion. The floor of the pulpit is 4 ft. from the base, and is approached by a flight of five steps. These steps are 1¼ in. thick, with 1¼-in. risers, and are 2 ft. 3 in. wide between the strings. The strings are 11 in. by 2 in., with lower edge double moulded as shown by the enlarged detail (Fig. 1841), and with a moulded sinking to the outer face, as shown on elevation (Fig. 1838). The strings are housed and tenoned into two 7-in. by 7-in. bottom newels, each 5 ft. 7 in high above the floor, to the top of the finial. Each of these two newels has moulded sinkings, with trefoil heads on three faces and a moulded and carved finial, as shown on elevation (Fig. 1838), and by the enlarged detail (Fig. 1842). The two top newels are each 5 in. by 5 in., and reach from the base line to 13 in. above the capping. The lower portion of these newels has a moulded sinking, with trefoil head on one face only. The tops of these newels are 10 in. long, by 8 in. by 8 in. They are dowelled to the lower portion, and have a moulded quatrefoil sinking on each face, (Fig. 1843), and two trusses, each 6 in. wide by 3 in. thick, not shaped or moulded, but with square edges. These two last, being under the steps, are not seen. Four trusses are connected together at their tops with two 4½-in. by 3-in. deal floor joists, halved together at their intersections; and the other four trusses are connected to each other with two similar joists, 4½ in. lower down than the previous joists, but halved together at their intersections in a similar manner. Upon these two last-named joists

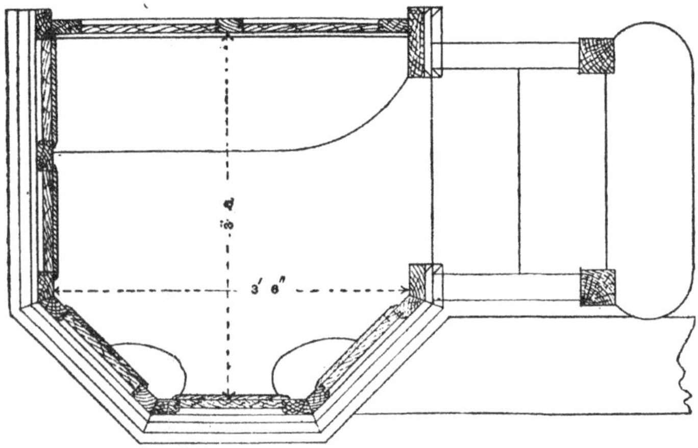

Fig. 1832.—Horizontal Section of Pulpit.

and a castellated mould at the crown. It will be noticed on the plan that these two newels are cut away between the capping and stringing course to form an abutment for the framing of the top portion of the pulpit, but above the capping and below the springing these newels are square. The handrail is 3½ in. by 3½ in., and is tenoned and housed into the newels (see Fig. 1841). Between the handrail and the 5-in. by 2-in. capping is a traceried panel (see Figs. 1838 and 1841). Forming the base of the pulpit are six shaped trusses, each 14 in. wide by 3 in thick, with double moulded edges, as shown on enlarged detail are laid 4½-in. by 3-in. packing pieces, to form a bearing level with the two first-mentioned joists to receive the 1¼-in. oak floorboards laid in narrow widths. The space between the trusses is filled in with 2-in. moulded framing, with raised carved panels, as shown in Figs. 1838 and 1844. This framing is supported by a 4-in. by 3-in. chamfered curb, tenoned into trusses. Around the base is a plinth. 1 ft. 2½ in. wide, made up of three moulds, the lowest of which is 7¾ in. by 2½ in., the middle mould being 3½ in. by 1¾ in., and the top mould 3¼ in. by 1½ in. The soffit of the lower part of the pulpit is formed of 7½-in.

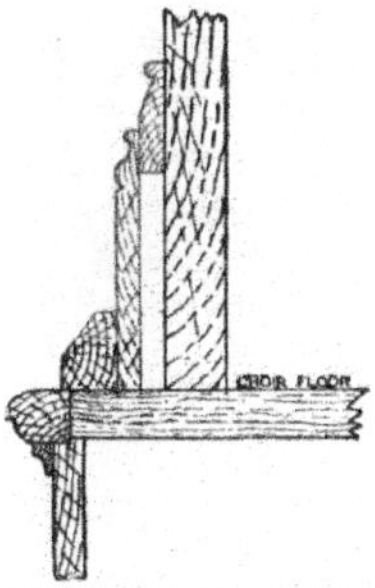

Fig. 1834.—Section of Curb and Base Moulds of Pulpit on Church Floor.

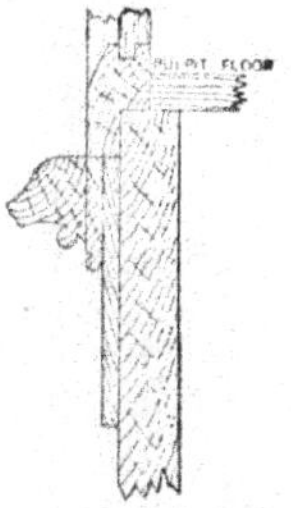

Fig. 1835.—Section of Moulds and Framing at Level of Pulpit Floor.

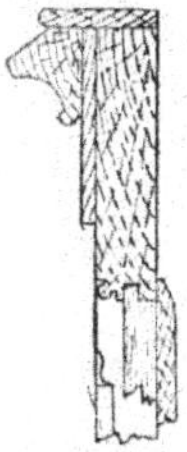

Fig. 1836.—Section of Cornice and Upper Framing of Pulpit.

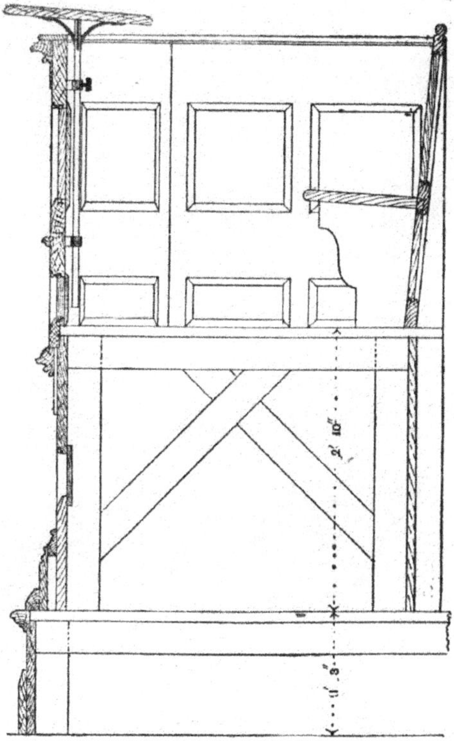

Fig. 1833.—Vertical Section of Pulpit.

Fig. 1837.—Carving in Cornice Mould of Pulpit.

by 2-in. plank, supported by the tops of the trusses, and rebated to receive the floor-boards. The upper surface of this plank also forms a mitered margin to the floor round the inside of the pulpit. In order to break the joint between the soffit and the lower framing, a 2-in. by 2-in. mould, as shown by Fig. 1845, is employed. The top portion consists of a 4-in. thick moulded and rebated skeleton frame, composed of 5-in. by 2¼-in. bevelled and rebated curb or string, 4-in. by 3½-in. moulded and rebated top rail, and 4-in. by 3-in. double moulded and rebated angle-bars. The top rail and string are

Fig. 1838.—Elevation of Pulpit in Perpendicular Style.

Fig. 1842.—Carved Finials to Pulpit.

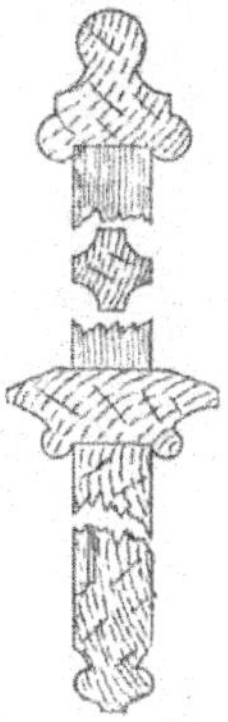

Fig. 1841.—Section of String, Capping, Handrail, and Panel of Pulpit.

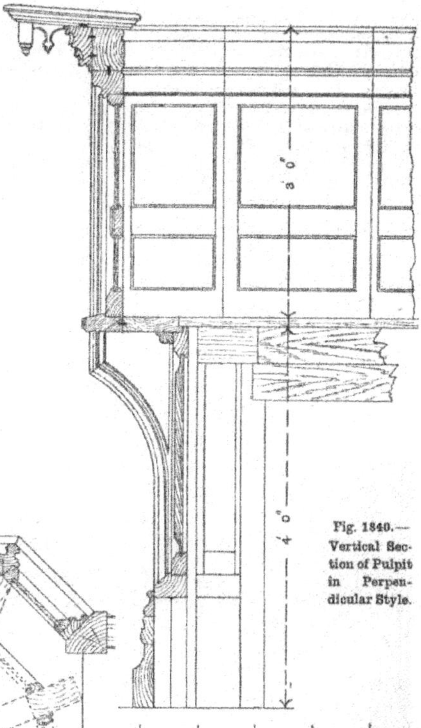

Fig. 1840.—Vertical Section of Pulpit in Perpendicular Style.

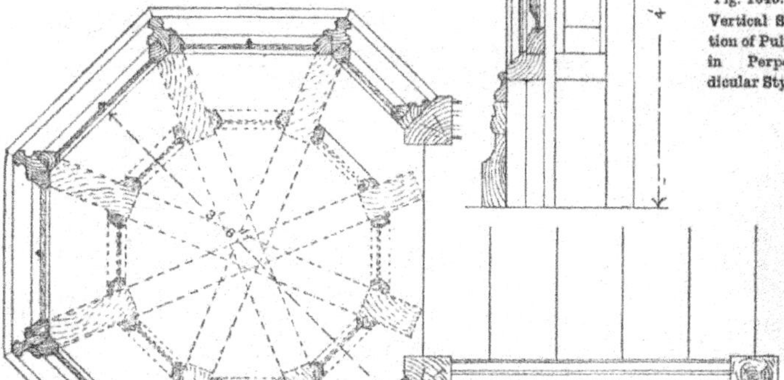

Fig. 1839.—Plan of Pulpit in Perpendicular Style.

mitered at the angles of the octagon ; each short length of top rail is not moulded throughout its length, but is left in its square state at each end, and the angle-bars are butt-jointed to this square portion, while the mould on the angle-bars is carried forward on the square ends of the top rail until the different members of the mould intersect with the corresponding members worked on the top rail, in the same manner that a mason joins the mullion of a window to a moulded head. In each bay of this skeleton framing is fixed 2-in. framed and moulded panelling, consisting of 2-in. by $1\frac{3}{4}$-in. moulded stiles and top rails, $3\frac{3}{4}$-in. by 2-in. chamfered bottom rails, $4\frac{1}{4}$-in. by $1\frac{5}{8}$-in. middle rail. The middle rail has a moulded traceried sinking on its face, and is intersected by a

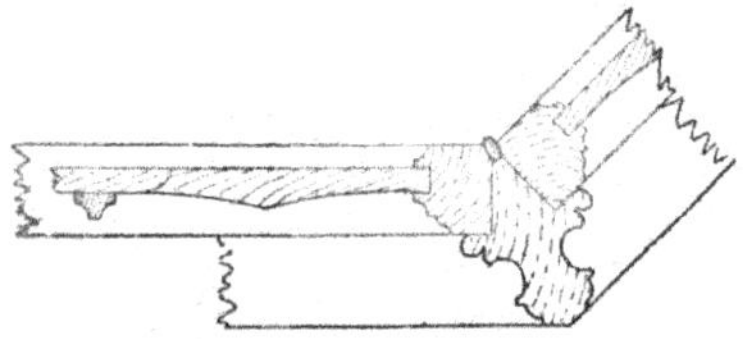

Fig. 1843.—Section of Top Panels, Framing, and Moulds on Edges of Trusses of Pulpit.

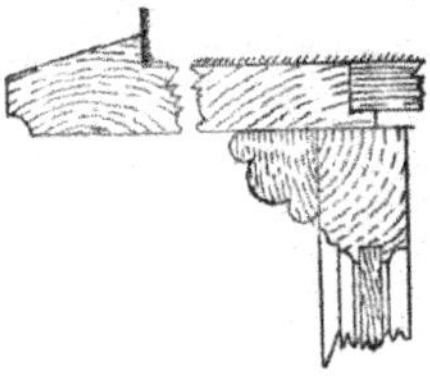

Fig. 1845.—Soffit Mould and Lower Framing of Pulpit.

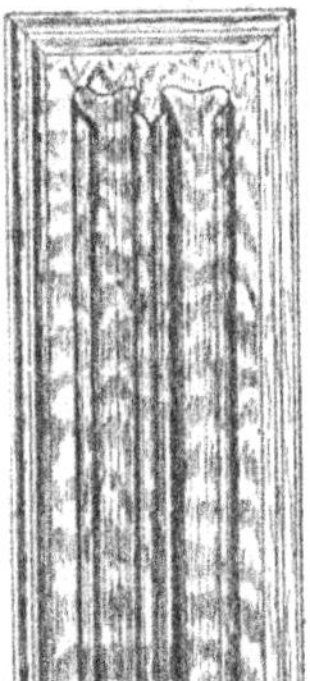

Fig. 1844.—Framed Panelling in Lower Portion of Pulpit.

Fig. 1846.—Framed Panelling in Top Portion of Pulpit.

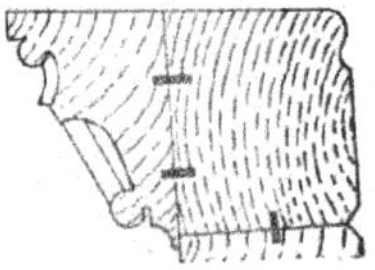

Fig. 1847.

Fig. 1848.

Figs. 1847 and 1848.—Capping of Pulpit.

Fig. 1849.—Pulpit Book-board.

Fig. 1850.—Side Elevation and Part Section of Pulpit on Circular Columns.

Fig. 1853.—Part Vertical Section of Pulpit at C (Fig. 1851).

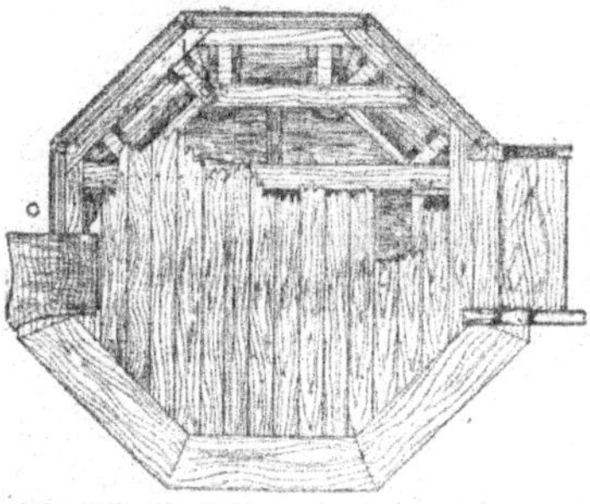

Fig. 1851.—Part Plan and Horizontal Section of Pulpit at A (Fig. 1850).

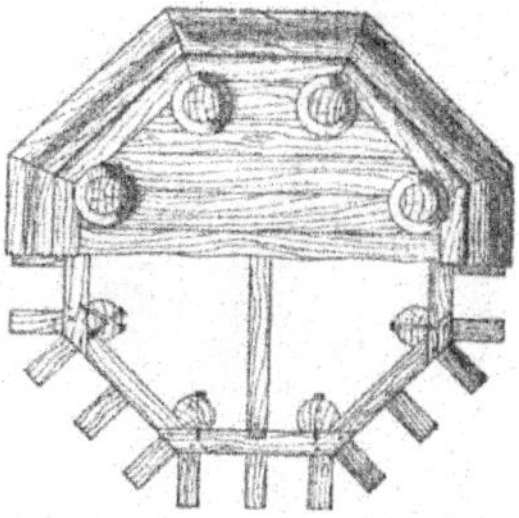

Fig. 1852.—Half Plan of Pulpit Base and Half Horizontal Section at B (Fig. 1850).

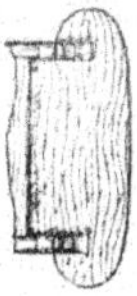

Fig. 1854.—Plan of Pulpit Bottom Step.

Fig. 1855.—Section of Pulpit Cornice.

Fig. 1856.—Section of Stiles of Pulpit (see A, Fig. 1850).

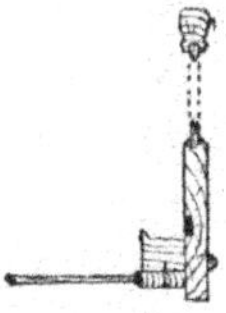

Fig. 1857.—Section of Pulpit Handrail, String, and Soffit.

Fig. 1858.—Carved Capital of Pulpit.

60.

Fig. 1859.

Figs. 1859 and 1860.—Part Elevation and Vertical Section of Reredos.

¾-in. muntin. The moulded traceried heads to the tops of each bay are worked out of a board in one piece, ½ in. thick, and carved to the pattern as in the enlarged elevation (Fig. 1846). The upper panel to each bay is raised in the centre, as in Figs. 1843 and 1846. The lower panel in each bay has a ⅜-in. sinking, with trefoil heads.

Capping.—The capping (Fig. 1847) is here shown built up in two pieces, but may be built up in a greater number if desired. It is 7½ in. wide by 5½ in. thick, made up of a 5½-in. by 3½-in. front portion and 5-in. by 4-in. back portion, tongued together. The front portion has a deep cove worked in it, and at intervals in this cove are raised and carved pateræ, while above these are curved dentils (see Figs. 1838, 1847, and 1848). These dentils should not be planted on, but carved out of the solid material.

Book-board.—The book-board is 18 in. long, 17 in. wide, and 2 in. thick. Underneath it, at the two front corners, are square drops with octagonal terminals. Between these drops is a shaped board, with carved cusps and moulded sinkings on the face. It is moulded on the shaped edges as shown by Fig. 1849. The book-board should be fixed at only a slight inclination. The elevation (Fig. 1838) is parallel to the stairs, and is not a front elevation. It will be noticed that the panels adjoining the newels are single, whereas all the other bays are double. The pulpit should be made of oak left in its natural state.

Pulpit on Circular Columns.

The pulpit with staircase represented by Fig. 1850 is octagonal in shape, and is supported by circular columns set at each corner of an octagonal base. Pine is used throughout, except for the columns and for the plain backs behind the fretwork panels, the columns and panels being prepared from teak, the dark wood contrasting well with the lighter. From the main floor to the top of the cornice is 7 ft. 9 in., and to the top of the pulpit floor 4 ft. 6 in., while the extreme width of the cornice is 6 ft. 11 in. and that of the base 5 ft. 9 in. The fretwork cornice, panels, and balusters make a rich and inexpensive decoration, while the newels are uncommon in form, and are well suited to the lines of the other decorative parts. The top newels are prepared from 11-in. by 2½-in. stuff, and the bottom from 7-in by 2½-in., and the balusters from 7-in. by ⅝-in. Bracketed carriage pieces are used for the construction of the stairs (see Fig. 1850). The strings are of 11-in. by 1½-in. pine. Fig. 1851 represents a part plan and section at A (Fig. 1850), with part of the 4-in. by 1-in. grooved and tongued floor-boards removed to show in plan the timbers used for building up the heavily moulded

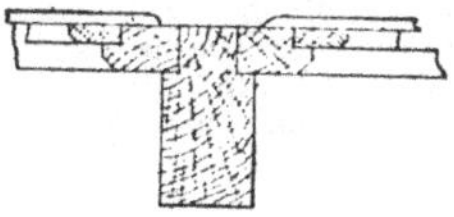

Fig. 1861.—Horizontal Section through Buttress and Framing of Reredos.

Fig. 1862. — Section through Top Rail of Reredos Outer Wings.

Fig. 1864.—Carved Finial of Reredos.

Fig. 1863.—Section of Moulded Capping to Centre Part of Reredos.

portion of the pulpit above the columns. Fig. 1852 shows a section at B (Fig. 1850), and a half-plan of the moulded base. The frame and brackets keep the moulding in position. The frame is from 9-in. by 3-in. stuff, tenoned and pinned together, with brackets nailed on. The bottom parts of the columns are bolted to the frame. Another view of this is given in the part section (Fig. 1853) taken at C (Fig. 1851), together with a section of another octagonal frame of 4-in. by 2-in. stuff, halved together and pinned, and resting on the columns, and to which the top parts of the columns are bolted. This frame also supports brackets, which are tenoned into it and into the 5-in. by 4-in.

joists of the floor above. These joists are unusually large, on account of the two brackets at each corner being tenoned into them. A section of the reading desk with a small shelf underneath is also given in Fig. 1853, as well as a section through the panelling. Fig. 1854 shows a plan of the shaped bottom step and newels, while Fig. 1855 represents an enlarged section of the cornice mouldings and detail of the perforated member. Fig. 1856 shows an enlarged section of the stiles at A (Fig. 1850), and Fig. 1857 gives sections of the string, handrail, and soffit, which is plain panelled. Fig. 1858 shows more clearly the detail of the carved caps of the columns. The pulpit looks well with the lighter wood sized and varnished, and the teak dull polished.

Reredos.

A reredos, dorsel, or lardrose, has been defined as "the wall or screen at the back of an altar, seat, etc." Formerly, the reredos was usually ornamented with panelling, etc., especially behind an altar, and sometimes was enriched with a profusion of niches, buttresses, pinnacles, statues, and other decorations, which were often painted with brilliant colours. A reredos of this kind not infrequently extended across the whole breadth of the church, and was sometimes carried up nearly to the ceiling, as at St. Albans Abbey. In village churches they were generally very simple in design. The reredos of which Fig. 1859 shows a part elevation should preferably be made of oak. Fig. 1860 shows a section. The reredos illustrated is 11 ft. 6 in. wide and 12 ft. high to the top of the finial, but could be modified to suit any width or height by altering the number or size of the panels. The two centre buttresses are 10 ft. 4 in. high, the two end buttresses 8 ft. high. These are all got out of 8-in. by 4-in. stuff, and are rebated out on the back edges to receive the 2-in. framing as shown on the enlarged detail (Fig. 1861). All four buttresses have sunk and chamfered panels on the face, and are shaped as shown on section (Fig. 1860). The two centre buttresses have a necking mould mitered round, and all four buttresses have a small moulded capping, with small carved cross or ornament to form a finial. The 2-in. framing consists of $3\frac{1}{2}$-in. by 2-in. stiles, $3\frac{1}{2}$-in. by 2-in. muntins, 9-in. by 2-in. bottom and top rails, $5\frac{1}{2}$-in. by 2-in. intermediate rails. This framing is chamfered at the front and rebated out at the back to receive the 1-in. perforated and chamfered panelling. This 1-in. panelling is backed by $\frac{1}{2}$-in. boards, which are screwed to the 2-in. framing, and secure the 1-in. panelling. The stiles of the framing are stop-chamfered as shown on Fig. 1862. All the under edges of the rails have a return bead worked on them, and all the top edges have a deep chamfer worked on. Around the base of the reredos is mitered a 9-in. by $1\frac{1}{2}$-in. moulded skirting, but in between the buttresses on the two outer wings of the reredos is fixed a 5-in. by 4-in. mould, and fixed on this at intervals are 5-in. by 2-in. moulded blocks, which give a castellated appearance. Details of these are shown in Fig. 1862. Fig. 1863 shows a detail of mould 4 in. by 2 in. fixed on the top edge of centre part of reredos. Fig. 1864 is an enlarged detail of the carved finial to the centre part.

SHIP FURNITURE AND FITMENTS.

Saloon Framing.

THE saloon side-framing about to be described is intended to be employed with hung windows. Fig. 1865 shows an elevation and Fig. 1866 a section through the middle frame and pediment, and shows the beam sole curving down to the top of the truss T. The three frames shown in Fig. 1865 are made separate and jointed with a slip-feather, the small frame having a bead on each stile to break the joint. The abacus of the truss is carried all round as shown, having a carved moulding below it. The cornice is formed by a dentil and an ogee moulding. The saloon cornice is formed by a cove resting against the beam side, and attached to a ground at the top. A moulding is planted on after the ceiling panel is up, and covers the deficiency. Thus the panel at any time can be taken down without interfering with the saloon cornice. This cove can be covered with Japanese paper, or it can be painted white. Apart from the designing of the framing, a great deal of the beauty depends upon the contrast of the colours in which the side and ceiling are finished. If this framing were dark-coloured, the flat portions on each side of the cornices, as well as the saloon cornice and ceiling, could be finished in white and relieved with gold. The colours in the stencils should be variegated as much as possible, to give the saloon a pleasing and gay effect, thus taking away the idea of its being a study in black and white.

Ceiling Panels.—Figs. 1867 and 1868 show two simple methods of dealing with the ceiling panels, and are intended to be used where the beam space is broken up with mock fore and aft beams. Fig. 1867 has a circular piece in the centre, to which the straight pieces are scribed. This circular piece can be used to carry either an electric or a swinging oil lamp. Fig. 1869 is a method which can be used on a panel which is not divided, as Figs. 1867 and 1868, though, owing to the labour connected with the circular mouldings, it is more expensive. These are some of the methods of working patterns with surface mouldings, some varieties of which are shown by Figs. 1870 to 1876.

Fittings for an Officer's Cabin.

The bed front shown in elevation in Fig. 1877 is one which may be applied to either the chief engineer's or chief officer's rooms, the finish of the inferior officer's furniture being generally of a plainer description. There are two features of special note in the bed here illustrated. These are a sliding writing flap and a swing washstand. The amount of space that can be allotted to officers' cabins, even on board the finest vessels afloat, is of necessity strictly limited; and therefore it is usually found expedient to render compact and portable the various articles required in the cabin. An additional amount of ornamentation is here introduced, which makes the bed a more pleasing feature in the room. Fig. 1878 is a vertical section on the line A B, and Fig. 1879 a section on the line C D. Fig. 1880 is a plan showing the sliding flap and also the top of the basin stand, Fig. 1881 being a section of part of the stand showing the method adopted to support it. The leeboard A and drawer stretchers B, as

Fig. 1866.

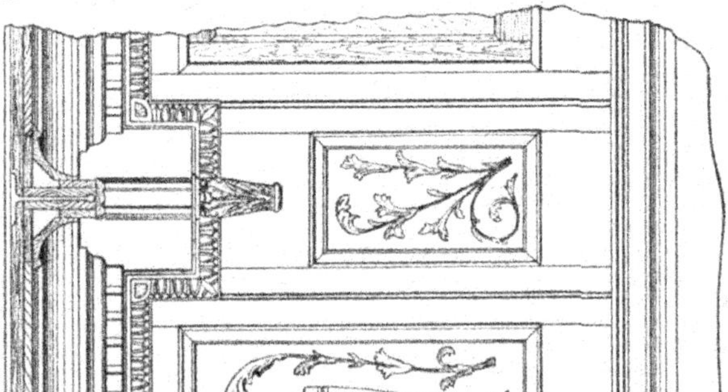

Fig. 1865.

Figs. 1865 and 1866.—Elevation and Vertical Section of Ship Saloon Framing.

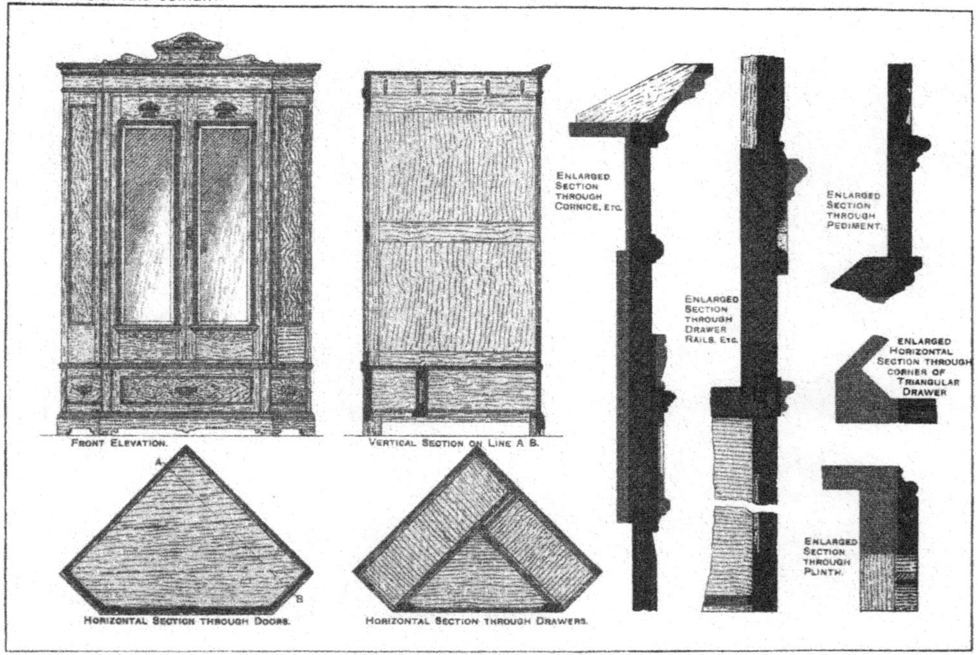

CORNER WARDROBE.

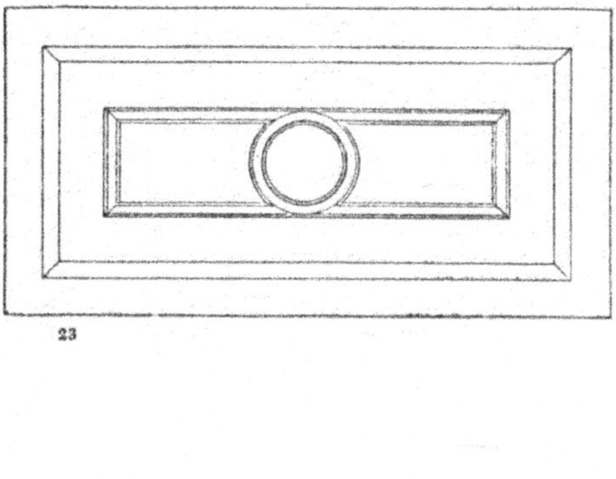

Fig. 1867.

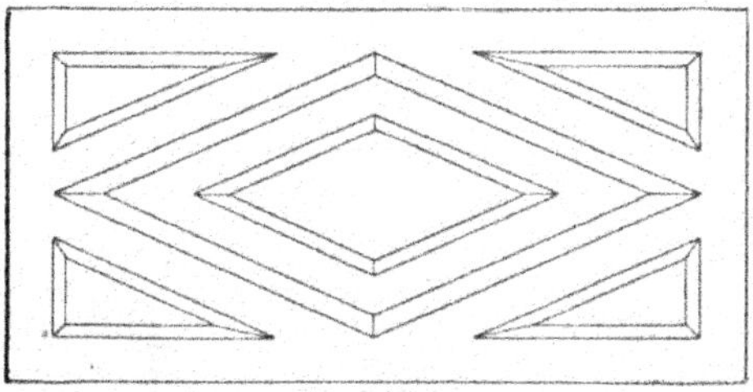

Fig. 1868.

Figs. 1867 and 1868.—Ceiling Panels for Ship Saloon.

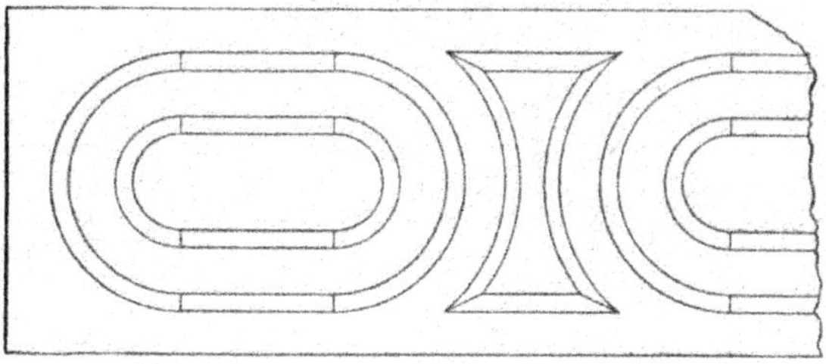

Fig. 1869.—Ceiling Panels for Ship Saloon.

Fig. 1870.

Fig. 1871.

Fig. 1872.

Fig. 1873.

Fig. 1874.

Fig. 1875.

Fig. 1876.

Figs. 1870 to 1876.—Sections of Surface Mouldings for Ship Saloon Ceiling.

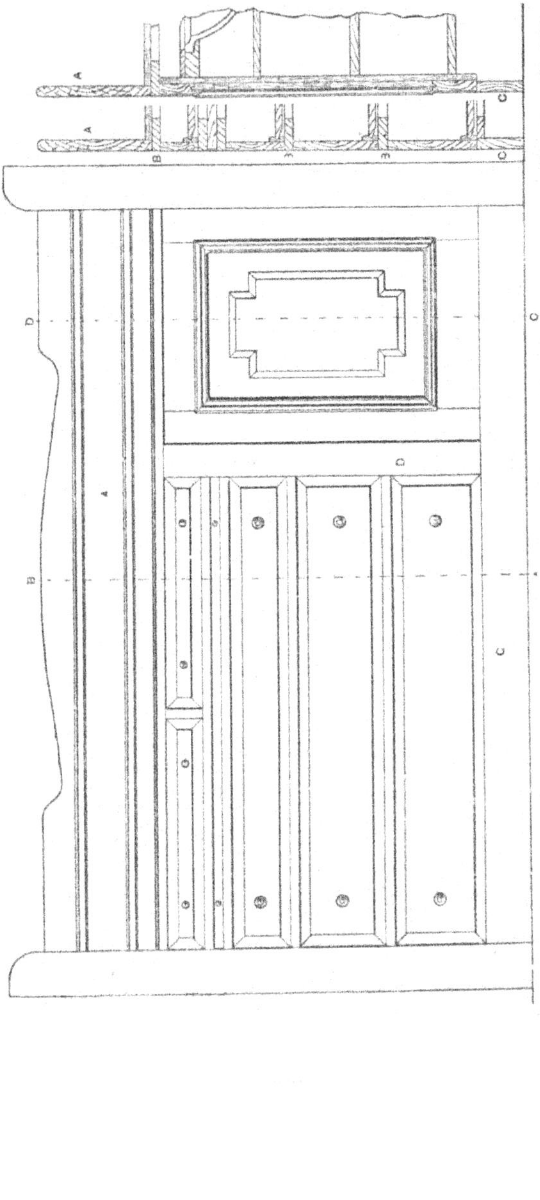

Fig. 1877. Fig. 1878. Fig. 1879.

Figs. 1877 to 1879.—Front Elevation and Vertical Sections on Lines A B and C D respectively of Bed for Officer's Cabin.

well as the bottom rail C, are tenoned to the left-hand haffit; the muntin D is tenoned to the bottom rail and top stretcher, and mortised on the edge to receive the short stretchers, the rest of the rails being fixed to the right-hand haffit as shown in Fig. 1877. The short stretchers are also tenoned to the yellow pine gables at each end. The short muntin between the two top drawers is raggle-dovetailed to the two stretchers. The drawers are dovetailed together in the usual manner, but the sides and front, instead of being grooved to receive the bottom, have grooved pieces glued and bradded to them. The writing flap, as shown at Fig. 1880, is about 15 in. broad, but the cross-ends, which are mitered at the front, are allowed to go right back and act as levers when the flap is drawn out for the purpose of writing upon. Pieces are screwed to the cross-ends at such a position that they stop the flap from coming out too far, and also prevent it going back further than to show the $\frac{1}{8}$-in. sinkage all round. An examination of the plan (Fig. 1880) and the sections (Figs. 1878 and 1879) will show the method on which the swing basin stand is constructed. Two gables are required, one of which is screwed to the back of the door, and the other one well screwed to the edge of it, with the addition of three or four iron angles in the inside. The shelves are raggled to the gables, the uppermost shelf being cut to receive the basin, which is covered round the edges with a broad coping. On the flying gable, as shown at Figs. 1880 and 1881, is fixed an iron rod, having a cross-head on the upper portion of it, travelling on the top of two iron quadrants. This supports the stand and takes the strain off the door. The door is planted with a good moulding, and a thin fielded panel of the shape shown in Fig. 1877 is glued and bradded to the plain panel. Two reeds are run on the face of the leeboard, and the top edge is rounded. However well fitted a cabin may be, there will be little comfort for its occupant unless it is watertight and well ventilated.

Ship's Sideboard.

Details of a ship's sideboard with storm doors are shown by Figs. 1882 to 1887.

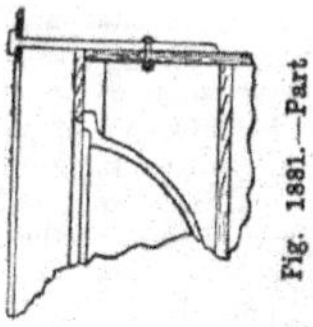

Fig. 1881.—Part Vertical Section of Washstand for Officer's Cabin.

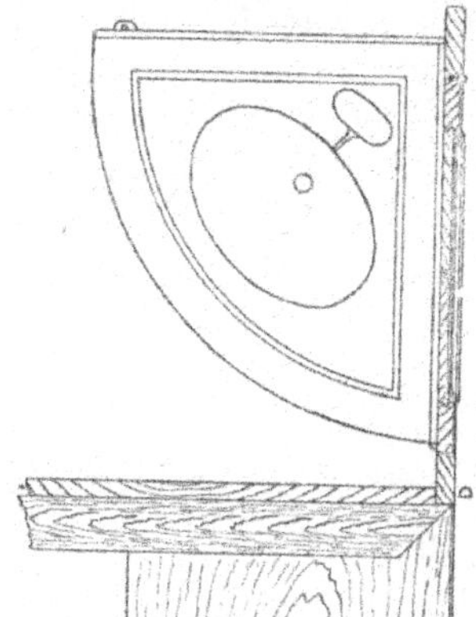

Fig. 1880.—Horizontal Section of Bed for Officer's Cabin, showing Writing Flap and Plan of Basin Stand.

This is a style of sideboard which is rarely used, owing, probably, to the extra labour and expense incurred in its construction. The usual height of a sideboard is about 3 ft., and the length to suit requirements and the space available. Fig. 1882 is a vertical section through one of the doors, Fig. 1883 a horizontal section or plan, and Fig. 1884 a front elevation. The sideboard is made in several parts. The base is formed into these upright pieces, and carry the top, which is sometimes a marble slab instead of hardwood. The top part is dowelled and screwed to the shelf carcases. Fig. 1882 shows the shape of the drawer front, and is kept inside the flush of the gables and centre-piece. The height of the doors is also seen in Fig. 1882. Fig. 1883 shows the position of the drawer carcases, and also the two outside gables which are dowelled to the

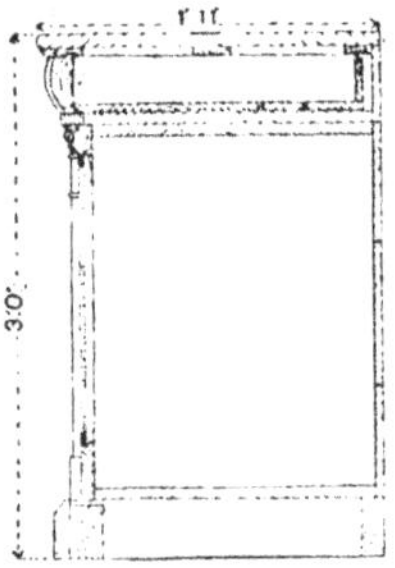

Fig. 1882.

Fig. 1884.

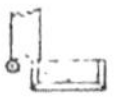

Fig. 1885.—Hinged Pilaster of Ship's Sideboard.

Fig. 1886.

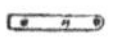

Fig. 1887.

Figs. 1886 and 1887.—Brass for Fixing on Sideboard Door.

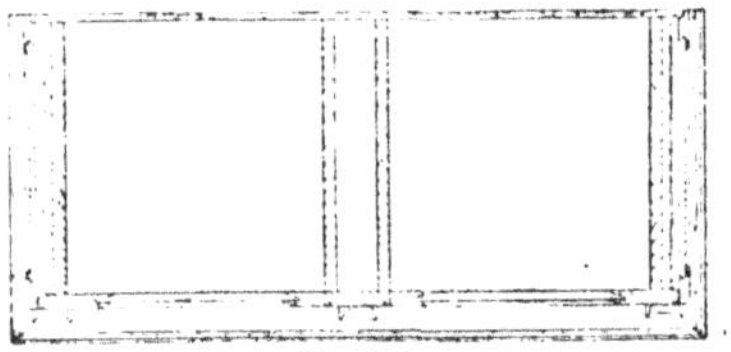

Fig. 1883.

Figs. 1882 to 1884.—Vertical Section, Horizontal Section, and Elevation of Ship's Sideboard.

of pieces of pine dovetailed together, having hardwood clamped to the top side as shown, and the hardwood base mitered at the corners and screwed to it. The two drawer carcases, shown in plan, are of pine dovetailed together, and blocked and screwed to the base. The top part is formed of two gables raggled into a pine sole, having hardwood clamped to the face edge and both ends. A centre drawer division is also raggled into this sole. Stretchers, the front one of which is clamped with hardwood, are dovetailed base and top part. The space which is left between the outside gables and the pine carcase must be fully the thickness of the door; and if bolection or raised mouldings are used, allowance must be made for them. The two outside pilasters with blocks are hinged to stand clear of this space, and the method of hingeing is shown in Fig. 1885. The method of sliding the door is clearly seen in Fig. 1883. A piece of brass, with a pin on it, is screwed to the top and bottom rails of the door; and the pin, being the centre of motion, must be kept half the thickness of the door from the edge of the stile. The

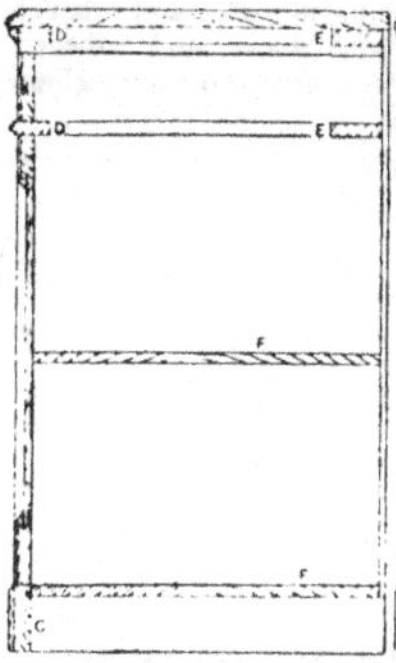

Fig. 1889.

Fig. 1888.

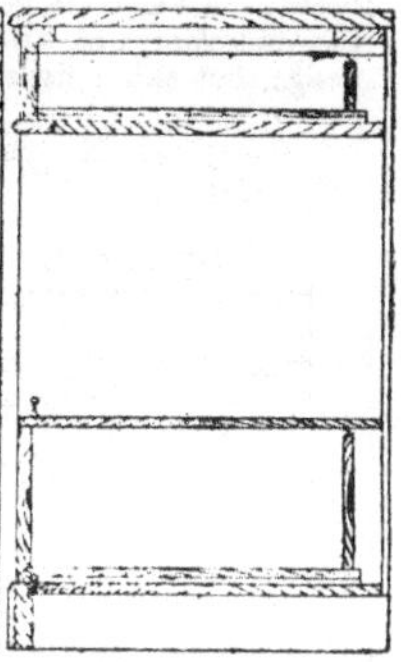

Fig. 1890.

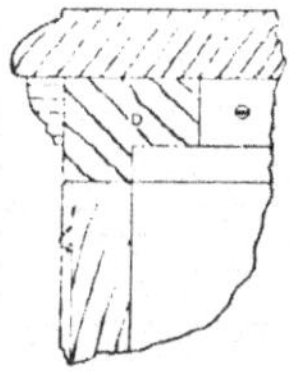

Fig. 1892.—Section through Front of Sideboard Top.

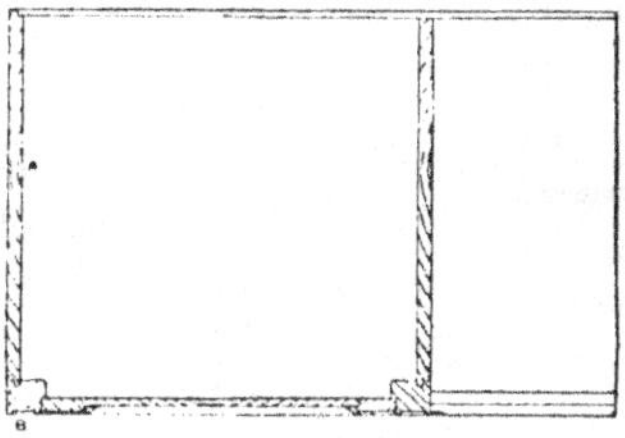

Fig. 1891.

Figs. 1888 to 1891.—Half Front Elevation, Vertical Sections, and Half Horizontal Section of Ship's Second-class Sideboard.

Fig. 1894.

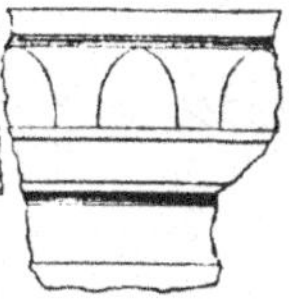

Fig. 1895.

Figs. 1894 and 1895.—Shelf in Centre Part of Sideboard.

Fig. 1893.—Part Section of Sideboard Door.

shape of this brass is shown in Figs. 1886 and 1887. Pieces of brass, with a slot in them to correspond with the pin, are screwed to the base and top part, between the outside gable and the pine carcase. A flush ring may be sunk into the edge of the door stile, to allow of the door being easily halved out from the recess. The pilasters, when closed, are kept in place with small circular spring pins. It will be seen from the plan and elevation (Fig. 1884) that the door stile, which is covered by the pilaster, is made broader to show the same margin outside. The pilaster is half-checked and screwed to the capital and base block. The end gables are framed to show the same as the door. The small gables on the top part of the sideboard could be of the same shape as the drawer fronts, and would therefore give it a heavier appearance.

Ship's Second-class Sideboard.

The second-class saloon sideboards on board a steamer are generally plainer than the first-class, and are therefore easier to design, but still a little taste is required to make a plain and comparatively cheap article look presentable. Fig. 1888 is a front elevation of half of a sideboard, which is divided into three parts. On each side are lockers, closed in with doors, and between them is an open space with a drawer below,

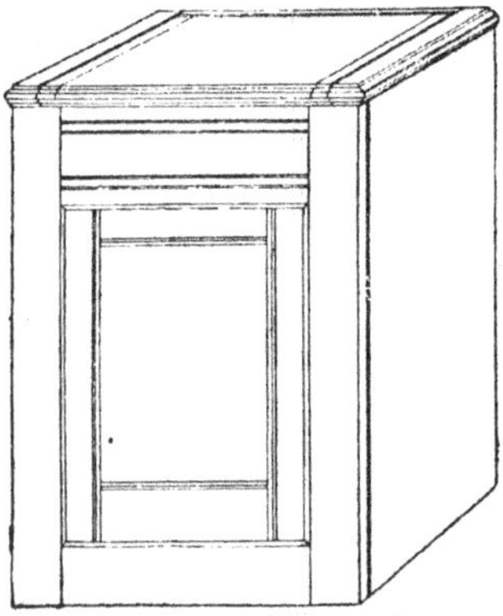

Fig. 1896.—Ship's Washstand.

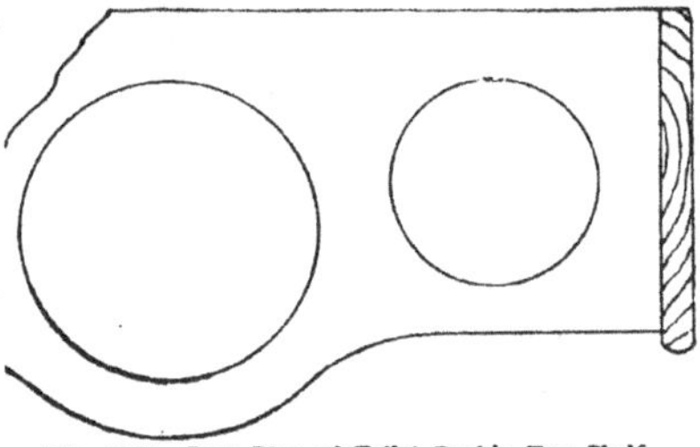

Fig. 1900.—Part Plan of Toilet Rack's Top Shelf.

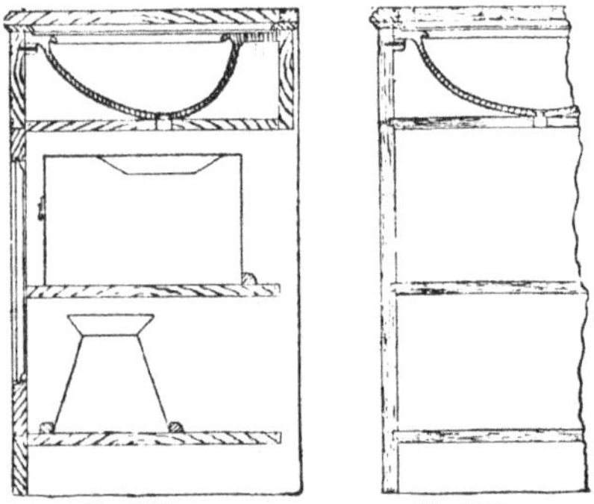

Fig. 1897. Fig. 1898.

Figs. 1897 and 1898.—Vertical Sections of Ship's Washstand.

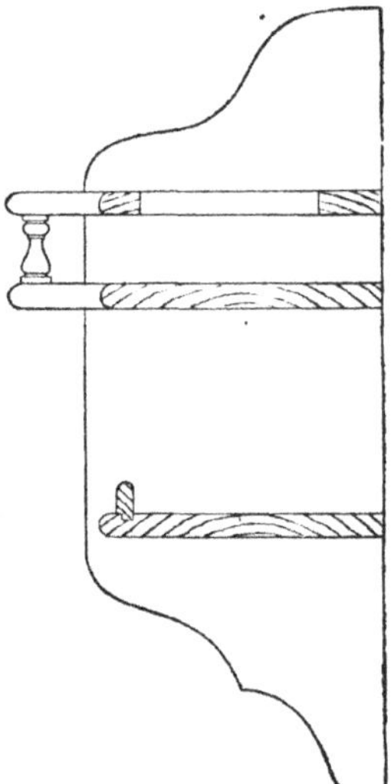

Fig. 1899.—Vertical Section through Toilet Rack.

while along the top are placed three shallow drawers. Fig. 1889 is a vertical section through one of the lockers, and shows the general construction. The shelves F are raggle-dovetailed to the gables, the bottom shelf being also feathered and grooved to the plinth. A small baluster rail is fixed to the top of the middle shelf to prevent articles rolling off, as shown in Fig. 1890, which is a vertical section through the central part of the sideboard showing the central space and upper drawer space and drawer. It will be seen by Fig. 1891 that the gables A are flush with the outside of the posts B, which are rebated to receive the doors. The plinth C (Fig. 1889) and the drawer stretchers D are tenoned to the outside posts. The central posts (see Fig. 1891) are tenoned to the plinth, and are carried right up, the lower drawer stretcher being carried over the face of them. The upper drawer stretcher, abutting against the posts with a square shoulder, is dovetailed down to them. The back stretchers E (Fig. 1889), which are of yellow pine, are dovetailed to the gables, though not brought through to the face. Fig. 1892 is an enlarged section of part of the top, showing the general finish more clearly. The top drawer stretcher D is made thick to carry the moulding, but is checked inside to reduce the weight. Fig. 1893 is a section through the door top rail, showing the flush and bead panel and moulding on the face. Fig. 1894 shows an enlarged section of part of the middle shelf and drawer front, and also shows the baluster rail, a vertical section being given at Fig. 1895.

Ship's Washstand and Toilet Rack.

Figs. 1896 to 1898 show a ship's washstand made in the example illustrated, of cypress, and stained a rich mahogany colour. The front is framed up out of ⅞-in. wood, not allowing the mortices to come through the stiles; the rails are advanced to mitre the bead, which is run on the inside edge. The front, after having been cleaned off, is stripped on both edges with the trying plane, and then grooved to receive the gables. The gables are squared on the top end, and raggled to receive the shelves as shown in Fig. 1898, and then a feather is wrought on the face edge to fit the groove on the front. They are then glued together, and the shelves inserted. A piece of yellow pine, rebated on the top edge to receive the basin (which in this case is a square Atlantic rimmed one), is fixed to the back edge of the top shelf (Fig. 1897), and to a rebate in the gables. A fillet is also screwed to the gables and front, to carry the front and sides of the basin. After the basin is fitted and the discharge hole cut, the cope should be mitered and screwed down on top of the carcase. The flap is contained between two haffits and a back rail, which are mortised and tenoned together, and screwed down to the cope, the holes being dowelled up. Small cross ends are also mortised and tenoned to the flap. The basin plug is attached to the under side of the flap with a brass chain. Two mouldings are run on the top rail of the front to break up the broad surface. As will be seen in Figs. 1896 and 1897, the door stiles and rails have a bead run on them, close to the inside edge. This is a very simple way of abolishing the moulding, and it looks very well. The receiver immediately below the basin, and the water jug on the bottom shelf, are made of zinc, and kept in place by fillets nailed to the shelf. Fig. 1899 is a vertical section through another necessary item in the furnishing of a room—namely, a toilet rack. These racks are generally made of teak or mahogany. The bottom shelf is for comb and brush, and the two upper ones are for a water bottle and two tumblers. Fig. 1900 is a part plan of the top shelf, showing the shape and arrangement of the holes. The middle shelf is solid, and is the same shape as the top one. As the portion outside the hole for the decanter is very weak, a baluster is inserted to stiffen it. The shelves are raggled to the gables, and glued. The small feather on the bottom shelf is also glued in.

Ships' Life-seats.

There have been many discussions as to the best means of saving life in the case of an accident on board a vessel, more especially a passenger steamer. With seats that are lashed to the deck by small rope, all that is necessary is to cut the lashing and lower the seat overboard, there being no fear of its capsizing. A life-seat comprises an ordinary sparred deck-seat with the addition of watertight tanks, usually made of copper, and a life-line, drawn through lashing eyes, is carried all round the seat. front rails R. The legs are turned with a round pin at the bottom, and are rebated on the face side to receive the front rail. Bearers B (Fig. 1901) are carried across the

Fig. 1901.

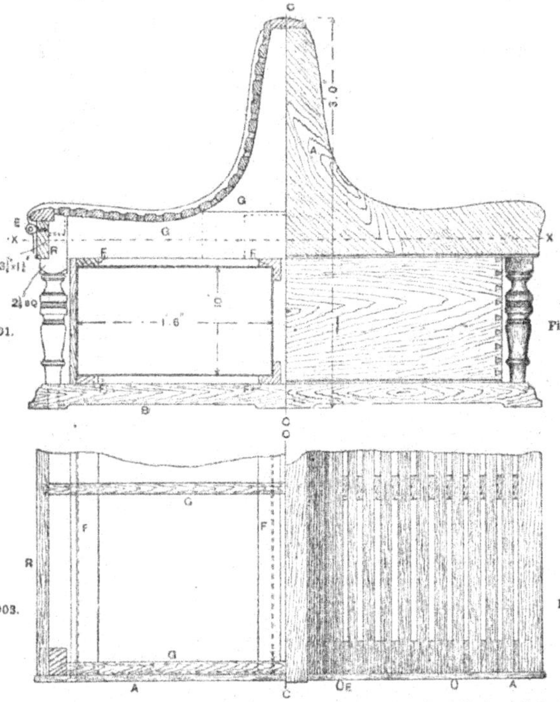

Fig. 1902.

Fig. 1903.

Fig. 1904.

Figs. 1901 to 1904.—Half Vertical Section, Half End Elevation, Part Horizontal Section, and Part Plan of Ship's Life-seat.

The seats should not exceed 8 ft. in length, as beyond that they are unwieldy. The seat shown in Figs. 1901 and 1902 is made up of grounds G (Figs. 1901 and 1903), the end ones being mortised into the end legs, and the intermediate brackets or grounds being raggle-dovetailed to the whole width of the seat, and have holes bored at each end to correspond with the pins at the bottom ends of the legs. The bearers are also shaped out on the under side as shown. The front rails are screwed to the legs, and the holes dowelled up. After the grounds, legs, front rails, and bearers have been put together, the frames can be turned upside down and the fillets F (Figs. 1901 and 1903) for receiving the tanks can be screwed on. The frame would then be turned on its feet again, the sparring fixed in place, and the beads for covering up the grounds bradded on. The ends would then be flushed, and the facing A (Figs. 1902, 1903, and 1904) fixed on. This facing has a bead on the under edge. A box is made for covering the tanks, and the length of the longest piece is the length of the seat outside the ½-in. facings at each end. The length of the end pieces is the size over the outside fillets, plus the thickness of the two long pieces. A bead is run on the bottom edge of the end pieces, and the pins are divided so as to allow the bead to run through to the front. The pins are put on the sides, so that the ends can be taken off to ship the tanks, as shown in Fig. 1901. The top cope C (Figs. 1901 and 1903) is fixed to the top ends of the grounds and allowed to project a little over the end facings and rounded off. The lashing eyes E (Figs. 1901 and 1904) for carrying the life lines are fixed at about 14-in. centres. A ring-and-plate is screwed to the facings about one-third of the height from the top, and another is fixed to the deck, and a thin lashing carried from one to the other. Fig. 1903 is a part section on the line X X (Figs. 1901 and 1902).

Ornamental Cover for a Ship's Ventilator.

In the case of large steamships in which the saloons, either first or second class, are scattered all over the deck, it sometimes occurs that a ventilator passes through some of them, and it is difficult to design a cover in keeping with the remainder of the wood in the saloon. Fig. 1905 is an illustration of a ventilator cover and the framing of the saloon. The sofa seat is carried along the athwartship bulkhead, and above the seat is the side framing, the design and construction of which are simple and yet effective. It is formed by planting mouldings of different curves on to a jointed panel which is set into a frame. In the centre of these mouldings a square- or diamond-shaped fielded panel is fixed, the square panel being carved and the others left plain. In the corners of the frame containing the carved panel a carved leaf ornament is nailed on. The pilaster is fluted and has a carved capital and a moulded block below. The sofa breast is plain; the cant on which it sits is covered

Fig. 1905.—Ornamental Ventilator Cover in Ship's Saloon.

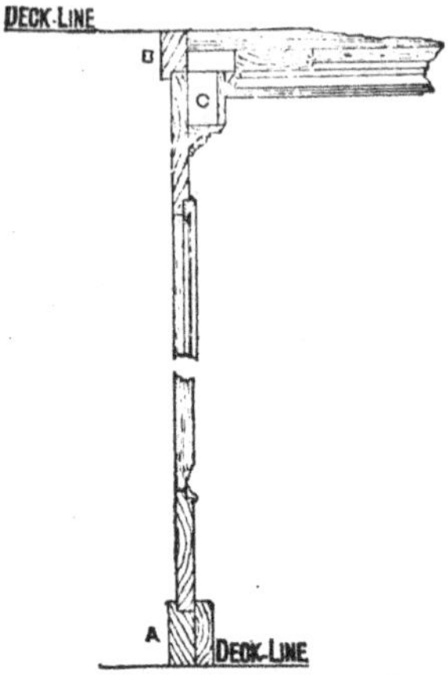

Fig. 1906.—Vertical Section of Ship's Ventilator Cover.

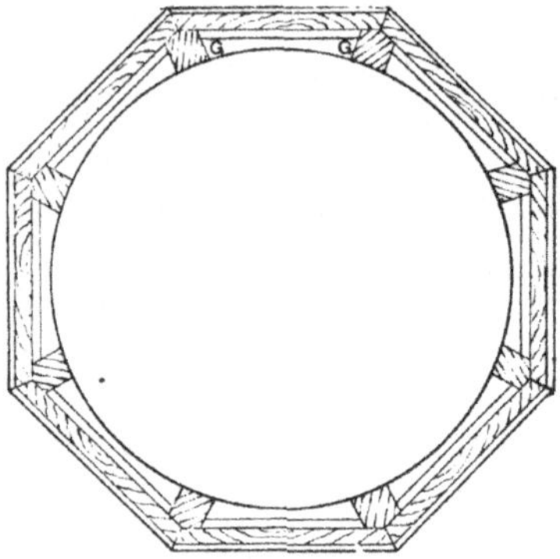

Fig. 1908.—Horizontal Section of Ship's Ventilator Cover.

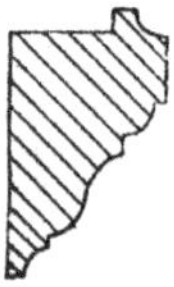

Fig. 1910.—Moulding for Fielded Panel.

Fig. 1912.—Section through Glass Panel.

Fig. 1909.—Section of Fielded Panel in Ventilator Cover.

Fig. 1911.—Section of Moulded Panel.

Fig. 1907.—Belt Rail of Ventilator Cover.

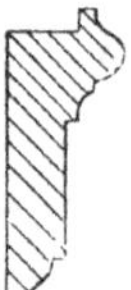

Fig. 1913.—Moulding to Secure Glass.

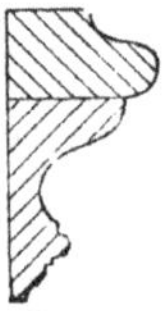

Fig. 1914.

Fig. 1915.

Figs. 1914 and 1915.—Sections of Dado Mouldings.

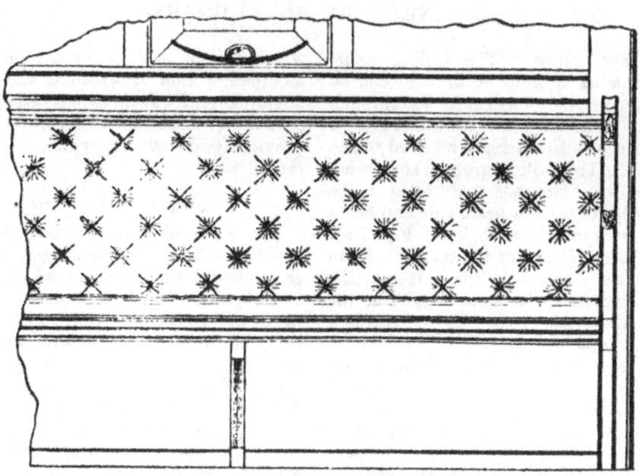

Fig. 1916.—Front Elevation of Ship's Saloon Sofa.

Fig. 1919.

Figs. 1919 and 1920.—Alternative Designs for Sofa Balusters.

Fig. 1920.

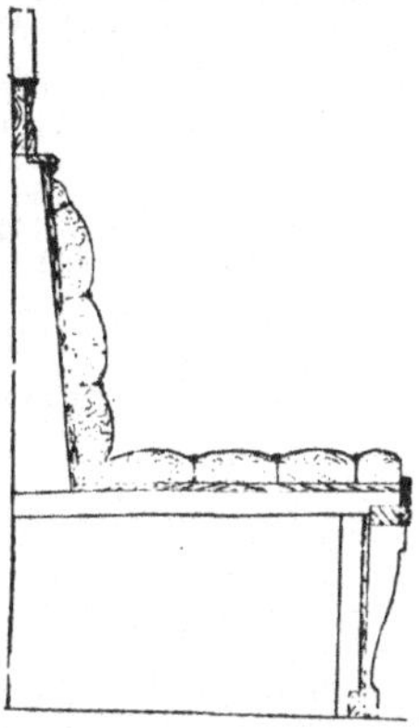

Fig. 1917.—Vertical Section of Sofa.

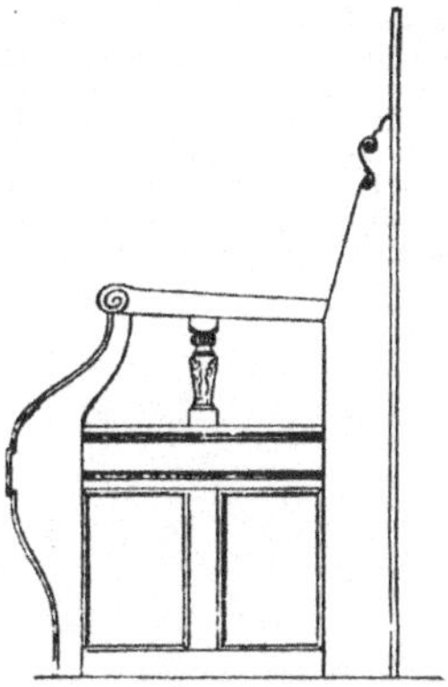

Fig. 1918.—End Elevation of Sofa.

by a polished base. The ceiling panel is framed with a muntin in the centre. Fig. 1906 represents a vertical section through the cover. A denotes the cant screwed to the lower deck, B the stringer screwed to the deck above. The ceiling ground and frame are also shown. Grounds C are fixed to the cover in order to carry a beam side and beam sole with neck moulding below. Fig. 1907 represents the belt rail of the frame. Dado mouldings are planted on the belt rail, the top one being kept the same height as the moulding above the sofa. Below the belt the panels are shown fielded, but above the belt on every alternate frame is a mirror. On the rest of the frames are ornamental panels as shown. Fig. 1908 is the plan or horizontal section showing the method of fixing the cover. G G denote the grounds fixed to the cant below and the stringer above. The cover is shown as if solid, which is sometimes the case in cheap work, mouldings being planted on to form imitation frames. The frames are screwed at the top and bottom to the stringer and cant, and to the ground at the back of the dado mouldings, which are then nailed on and the holes puttied. Fig. 1909 is an enlarged section of part of a bottom rail showing a design for a fielded panel; an enlargement of the moulding is shown at Fig. 1910. Fig. 1911 is another moulded panel with square stiles. Fig. 1912 is a section through one of the glass panels. The frame is rebated to receive the moulding on the face and the wood panel on the back. The back is fixed in, after which strips of felt are secured to it in order to form a soft bed for the glass, which is kept in place by the moulding, an enlarged section of which is shown at Fig. 1913. Figs. 1914 and 1915 represent enlarged sections of dado mouldings other than that previously shown, but which are very commonly used.

Ship's Saloon Sofa.

The style of sofa tends to restrict the design for a saloon finish, but it must be remembered that a finish which looks well is arrived at by the use of as little wood as possible. The general construction and design of a good type of sofa seat is shown by Figs. 1916 to 1920. Brackets are framed up and fixed to the groundwork at the back and to the coaming at the bottom. The seat and back are then covered up by ½-in. feather-and-grooved boards. Sometimes holes are cut in the seat, and lids fitted to them, so that the space below may be utilised as a steward's locker. The ship's skin is then framed up and the mounting put on. The design of the framing in this case is left to individual taste and requirements. The sofa breast is solid, and is fixed to the bearers at the back and the coaming. The pieces are butted on the vertical legs of the bearers, and the joint covered by a truss as shown. The sofa finishes at a doorway, and an elbow is required to stop all mouldings, etc., connected with the sofa. The end elevation (Fig. 1918) gives the shape of the elbow; this is framed up and a single turned and carved baluster is placed in the centre of the space above the cushions. Figs. 1919 and 1920 are alternative designs for the baluster.

MISCELLANEOUS EXAMPLES OF FURNITURE.

Newspaper Rack.

THE newspaper rack and book-holder, of which Figs. 1921 and 1922 show respectively front elevation and vertical section, Fig. 1923 giving a general view, may be made in oak or mahogany. The two wings are made to fall to an angle as shown, forming receptacles for papers or music, and when not in use may be folded back against the triangular centre-piece. The shelf underneath may be used as a convenient place for keeping books. For the two shaped ends, two pieces 2 ft. 3 in. by 11 in. by ¾ in. are required, the shape of the ends and the fretted design being shown by Fig. 1922. Grooves are cut at A A, 5½ in. long by ¼ in. deep, to receive the sides of the bookshelf, which is ⅝ in. thick; the shelf is housed into the sides and screwed from the outside. Round-headed brass screws may be used, two in each side, or ordinary screws sunk below the flush and the holes plugged with wood. The joint of the shelf is glued and bradded. Fig. 1924 shows the method of fixing the top shelf to the shaped ends. It is 7 in. wide by ⅝ in. thick, bevelled on the edges, and lap-dovetailed to the ends. A ⅜-in. piece B (Fig. 1925), 4½ in. wide, is glued to the top of the shelf, ⅛ in. from the ends. This forms the base of the triangle. For the two sides C, two pieces of ⅜-in. stuff, 1 ft. 1 in. wide and the same length as the base, are planed up for the sides of the triangle, the edges being bevelled where they meet at the top. The bottom edges are fitted to B, and glued, the joint being secured with a few brads. The ends of the triangle are left open, and the space is utilised for the reception of prints, stationery, etc. The two hinged wings are framed up, a mortice-and-tenon joint being used. The stiles and rails are 2 in. by ⅝ in., with a ½-in. rebate on the outside edges. There are five shaped straps, ⅜ in. thick; these have a small tenon on each end, and are fitted to the rails. The shaped strap is shown in detail by Fig. 1926. The wings, when framed together, are 2 ft. in length by 1 ft. 2¼ in. wide, and are hung with brass butt hinges as shown in Fig. 1927. It will be seen that the edge of the top shelf stops the wing and keeps it at the proper angle.

Lady's Workstand.

The lady's workstand shown in side and end elevation by Figs. 1928 and 1929 can be made in polished mahogany, walnut, or rosewood, and finished with glass lids and copper mountings. Fig. 1930 shows a plan of the lids. The box should be the part made first, and has sides and ends dovetailed together and grooved for the bottom, as shown in Figs. 1931 and 1932. The two lids have glass panels carried in a framing of stuff 1 in. wide by ½ in. thick, mitered and keyed and glued together, and rebated for the panel. A small bolection moulding finishes the face of the framing, after which the panel is inserted and secured by moulded glazing fillets. A dividing piece 2 in. wide is screwed to each side, and placed between the lids; they close on it. The lids are hinged to the box ends with 1¼-in. butt hinges, and furnished with a couple of small knobs. The box part is mounted on four legs, 1 in. square at the top, tapering to ¾ in. at the bottom, and cut to the shape shown in Figs. 1928 and 1929. The upper portion of each leg has the inside corner cut away to form a recess, in which the box is fitted (see Fig. 1933), and then screwed from the inner side, the dovetails in Fig. 1933, the side of the box stands ¼ in. back from the face of the leg. This admits of the side being panelled as illustrated. Glue on a chamfered moulding 1 in. wide by ¼ in. thick, and on the top part of the

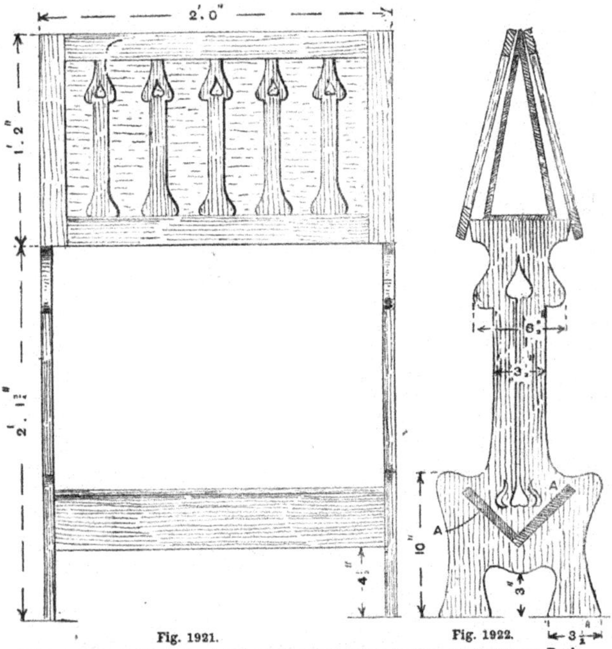

Figs. 1921 and 1922.—Front Elevation and Vertical Section of Newspaper Rack.

of the box thus being hidden. At 7 in. from the ground level, tenon to the legs a rail, 1 in. wide by ⅜ in. thick, and on this place a shelf ¼ in. thick. Fig. 1934 shows a section of the rail and shelf; the latter has a projecting rounded edge, and is cut into each leg on the corner, as shown in Fig. 1935. Glue the shelf to the rail, and strengthen the angle-joint with a number of glued blocks, as illustrated in Fig. 1934. As shown box the moulding projects upwards ½ in. to form a folding rebate in which the lids fall and fit (see Fig. 1932). At 3¼ in. from the bottom of the box (inside), two fillets are glued to support a loose tray, shown in plan in Fig. 1936. This tray is of stuff $\frac{3}{16}$ in. thick. The bottom is first cut and cleaned to size, and the edging, 1 in deep, with mitered corners, is glued and pinned to it. Divisions are formed for cottons, buttons,

Fig. 1923.—General View of Newspaper Rack.

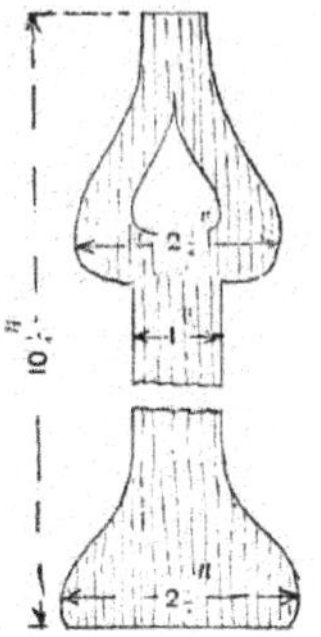

Fig. 1926.—Shaped Strap for Newspaper Rack.

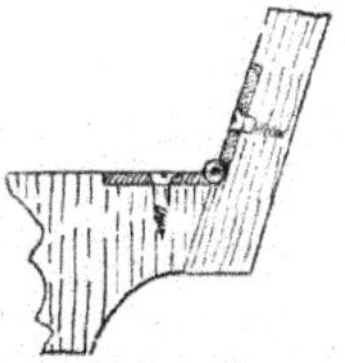

Fig. 1927.—Hinged Joint for Wings of Newspaper Rack.

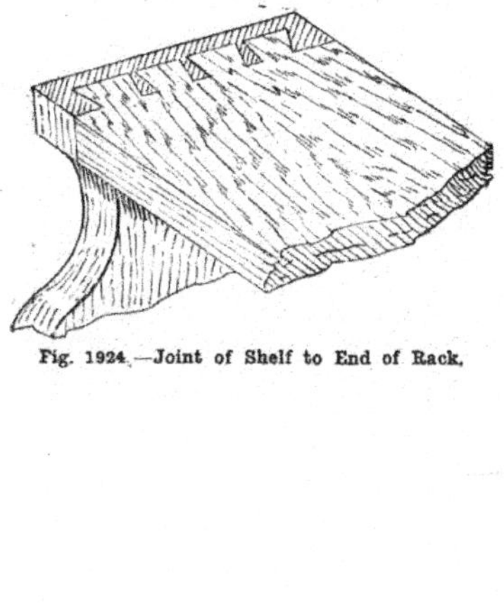

Fig. 1924.—Joint of Shelf to End of Rack.

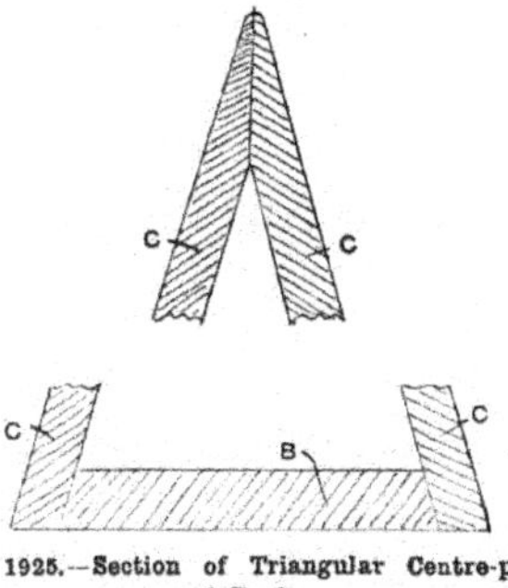

Fig. 1925.—Section of Triangular Centre-piece of Rack.

etc., by fixing strips, ⅛ in. thick, across the tray. They are fixed to each other and the tray sides by V-jointing, and need only fit tight. Line the interior of the box with pale blue silk, fixed with stiff paste applied cold. The tray can also be lined both inside and out in a similar manner. If this is done, the better plan will be to line the interior before the strips are fixed, and to cover each of the strips separately and then

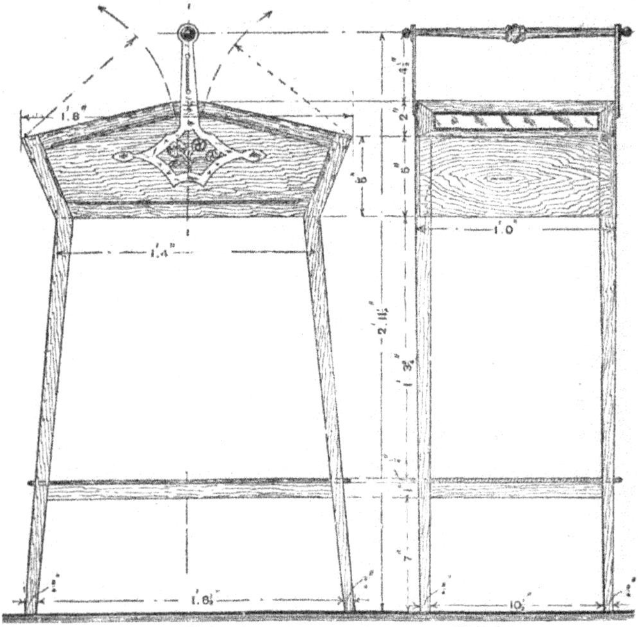

Fig. 1928. Fig. 1929.

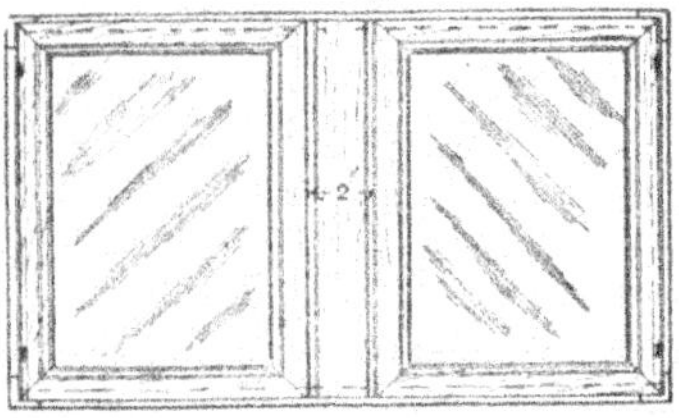

Fig. 1930.

Figs. 1928 to 1930.—Side and End Elevations and Lid Plan of Lady's Workstand.

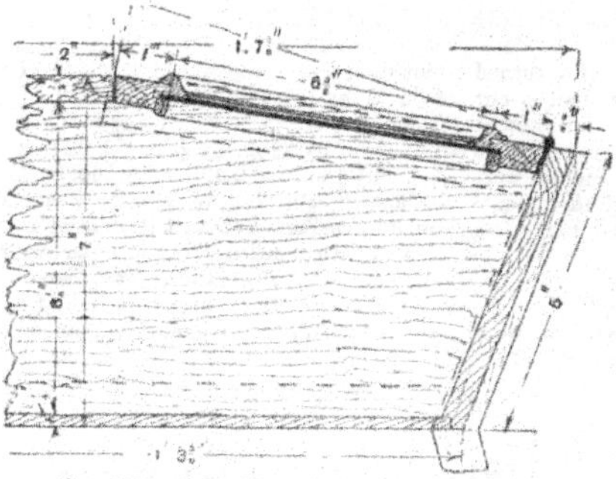

Fig. 1931.—Part Cross Section of Workstand.

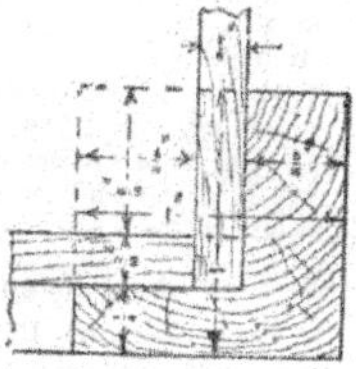

Fig. 1933.—Section through Top of Workstand Leg.

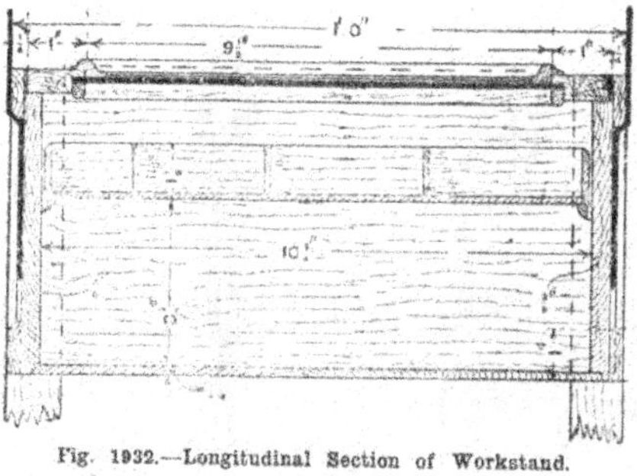

Fig. 1932.—Longitudinal Section of Workstand.

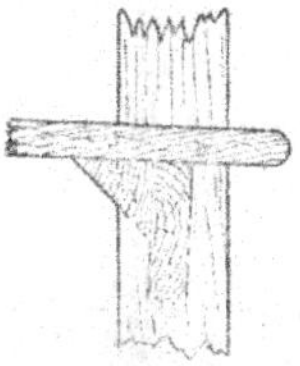

Fig. 1934.—Section of Workstand Rail and Shelf.

Fig. 1936.—Plan of Loose Tray for Workstand.

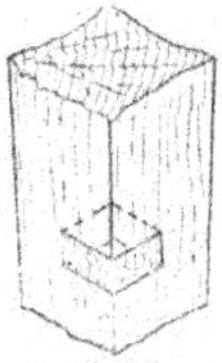

Fig. 1935.—Leg of Workstand cut to receive Shelf.

fix them. The handle consists of a turned length, carried in two fancy supports, cut from stiff polished sheet copper, and screwed to the sides. It is cranked to fit the chamfer moulding, and two turned knobs, with screwed ends, are passed through the mounting into the end of the turned handle to secure it.

Dinner-gong Stand.

Figs. 1937 and 1938 illustrate, in front and side elevation, a dinner-gong stand made in polished walnut, oak, or mahogany, the choice of wood depending on the furniture with which it is to be placed. The following quantities are required :— For pillar feet, 2 ft. by 4 in. by 1 in., and 1 ft. 4½ in. by 2 in. by ½ in.; pillars 4 ft. by 2 in. by 2 in.; capital moulding, 2 ft. by ¾ in. by ⅜ in.; pillar and scroll supports, 1 ft. 6 in. by 2½ in. by ½ in.; bottom rail, 1 ft. 2 in. by 1¾ in. by ⅝ in.; bottom rail mouldings, 3 ft. 8 in. by 5/16 in. by 3/16 in.; top rail, 1 ft. 2 in. by 5⅝ in. by 1½ in.; top rail mouldings, 5 ft. 3 in. by 9/16 in. by 9/16 in.; moulding above top rail, 4 ft. by ¾ in. by ¾ in.; for the entablature, 1 ft. 6 in. by 3½ in. by ⅞ in.; 1 ft. 9 in. by 4¾ in. by ½ in.; 1 ft. 2 in. by 2¾ in. by ¾ in.; 1 ft. 8 in. by 3¾ in. by ⅞ in., and 2 ft. 6 in. by ⅝ in. by ⅛ in.; drumstick, 1 ft. 2 in. by ¾ in. by ¾ in. The pillars (see Figs. 1939 to 1941) are of square section, with panelled sides above the capitals, as shown in Fig. 1939. The capital is carved in low relief, the cap moulding being worked separately, and glued and pinned on; Fig. 1942 is a section of this moulding. The shaft is fluted on each side as in Fig. 1940, and the base of the pillar as in Fig. 1941. The feet are made of 4-in. by 1-in. stuff, and finished as shown in Fig. 1943, with a moulding which is returned at the ends. The pillars are tenoned and wedged into the feet, a detail of this being given by Figs. 1943 and 1944. The feet are further supported by two scrolls (see Fig. 1938), made of ½-in. stuff, with panelled sides, and fixed with dowels and screws. The bottom rail is of 1¾-in. by ⅝-in. stuff, tenoned and secret-wedged into each pillar; the top and

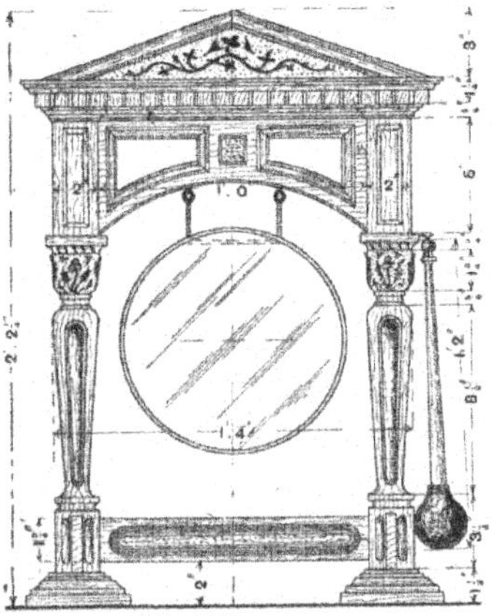

Fig. 1937.

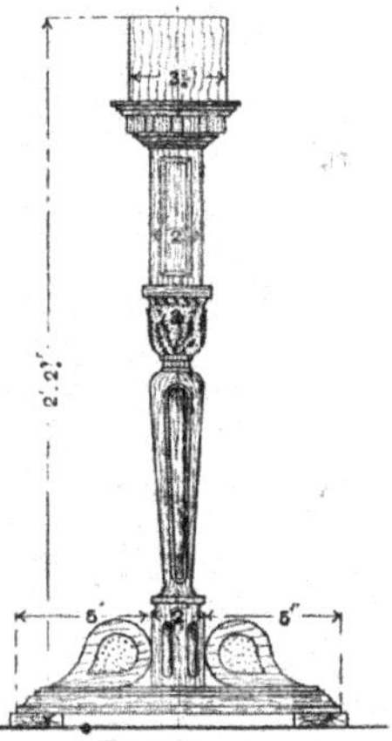

Fig. 1938.

Figs. 1937 and 1938.—Front and Side Elevations of Dinner-gong Stand.

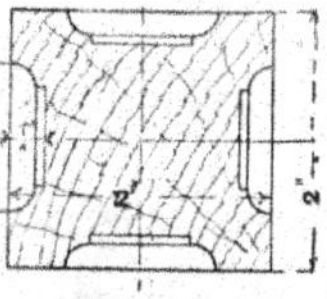

Fig. 1939.

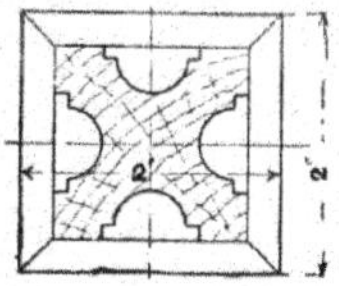

Fig. 1940.

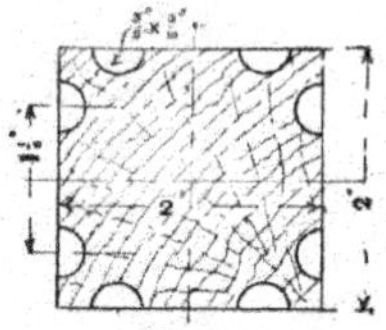

Fig. 1941.

Figs. 1939 to 1941.—Cross Sections of Dinner-gong Pillar.

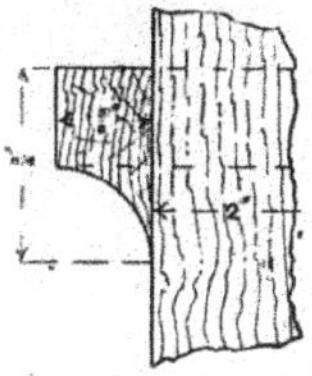

Fig. 1942.—Section of Dinner-gong Pillar and Capital Moulding.

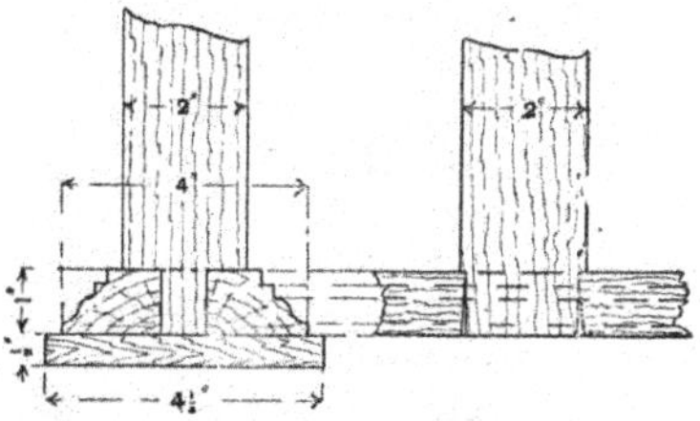

Fig. 1943. Fig. 1944.

Figs. 1943 and 1944.—Foot of Dinner-gong Pillar.

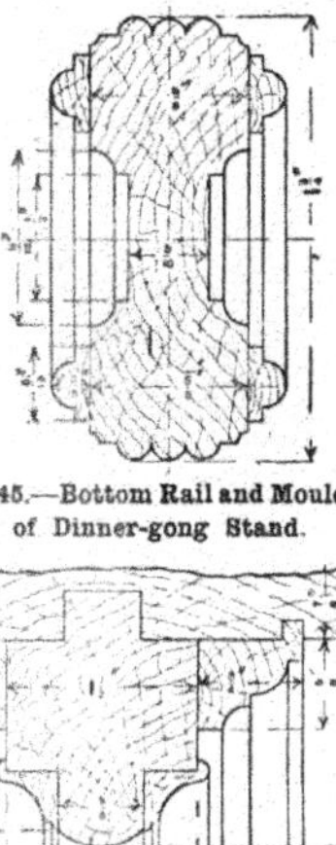

Fig. 1945.—Bottom Rail and Mouldings of Dinner-gong Stand.

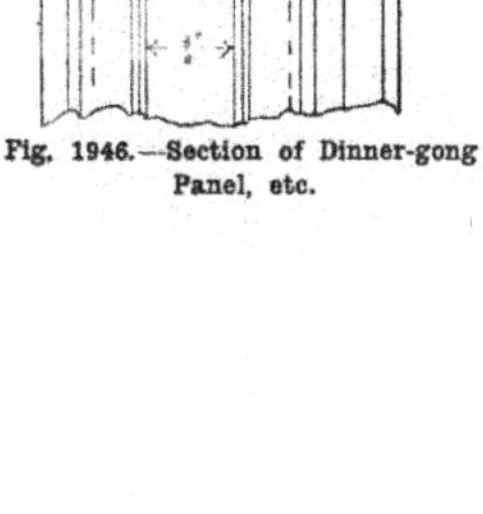

Fig. 1946.—Section of Dinner-gong Panel, etc.

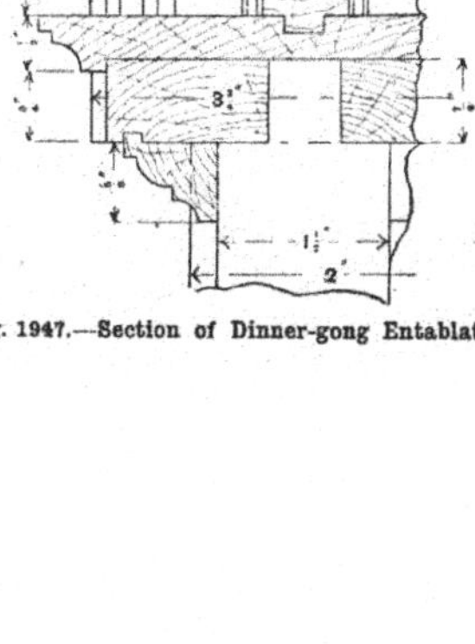

Fig. 1947.—Section of Dinner-gong Entablature.

bottom faces of the rail are beaded, and the sides panelled about $\frac{3}{16}$ in. deep; an independent ovolo moulding $\frac{5}{16}$ in. by $\frac{3}{16}$ in. surrounds the panel. The circular ends of the moulding are worked in the solid, and and stiles, and the rebating ploughed out. A section of the panel and moulding is given by Fig. 1946. A piece of stuff $3\frac{1}{2}$ in. by $\frac{7}{8}$ in. with squared edges is fitted over the top rail, and the dentil piece glued on.

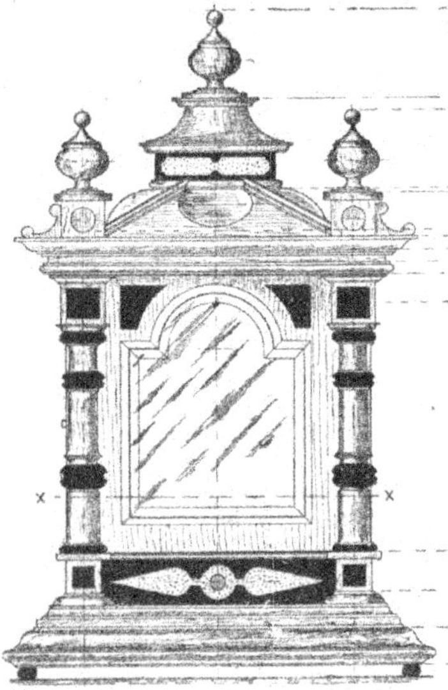

Fig. 1948.

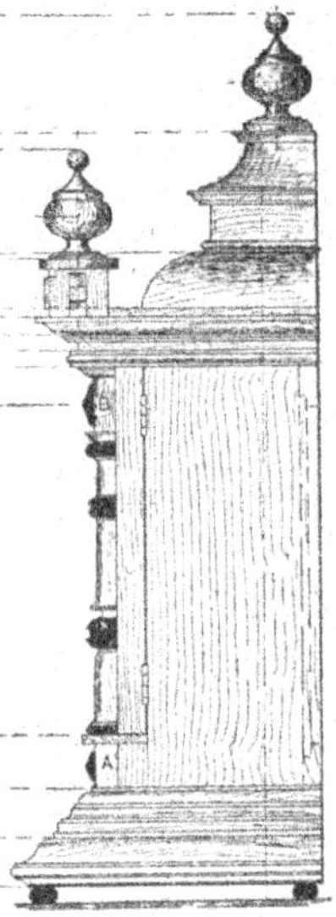

Fig. 1949.

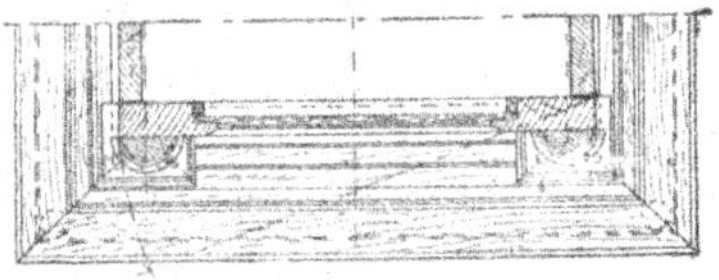

Fig. 1950.

Figs. 1948 to 1950.—Front and Side Elevations and Plan of Clock Case.

jointed to the straight lengths. Fig. 1945 is a sectional view of the rail with these mouldings on both sides. The top rail is cut from $5\frac{5}{8}$-in. by $1\frac{1}{2}$-in. stuff, and finished with open panels as in Fig. 1937. The rebates for the mouldings round the openings can be cut out with a router and chisel, or the rail may be framed together with rails Underneath this piece a moulding $\frac{3}{4}$ in. by $\frac{5}{8}$ in. is fixed to hide the joint, as shown in Fig. 1947. The lower part of the entablature is got out from a piece of stuff $4\frac{3}{4}$ in. by $\frac{1}{2}$ in., grooved along the centre to receive the tongue of the carved frieze panel. This panel is $\frac{3}{8}$ in. thick and of conventional design, carved in low relief; it is kept in position by the fixing of the cornice moulding. Fig. 1947 gives a detail section of the full depth of the entablature. The cornice moulding is worked from $3\frac{3}{4}$-in. by $\frac{7}{8}$-in. stuff, and is grooved over the frieze panel; the ends fit square into the lower part, to which they are screwed. The gong can be purchased from a furnishing ironmonger, and is suspended from two plated edges by a twisted silk cord. The beater (see Fig. 1937) is turned from $\frac{3}{4}$-in. square stuff, the ball being bound with cloth and covered with washleather.

Clock Case.

Figs. 1948 to 1950 represent front elevation, side elevation, and horizontal section respectively of a clock case made in walnut, and finished dull, with ebony mountings, which are shown black. The carcase is $11\frac{1}{4}$ in. high, $10\frac{1}{2}$ in. wide, and $4\frac{1}{2}$ in. from front to back. It is made of stuff $\frac{1}{2}$ in. thick, and the back is framed together to receive a door A (Fig. 1951), giving access to the clock movements. This door is made of rails and stiles $1\frac{1}{4}$ in. wide by $\frac{1}{2}$ in. thick, mitered and keyed together, and rebated for a panel $\frac{1}{4}$ in. thick, as shown in Fig. 1952. The carcase is grooved and glued into the foundation mouldings (see Fig. 1951). The mouldings are $2\frac{1}{8}$ in. thick, and project $2\frac{1}{4}$ in. beyond each side (see Fig. 1948) and the same distance at the front (see Fig. 1949). After setting out the mouldings, glue and dowel the piece on a backing of deal, the under side of which is rebated $\frac{1}{2}$ in. on by $\frac{3}{8}$ in. deep to take the bottom filling of the carcase, as in Fig. 1951. The foundation is mitered and keyed at each corner, and measures over all 1 ft. 3 in. long by $8\frac{1}{4}$ in. wide across the back. A bar 2 in. wide by about $\frac{1}{2}$ in. thick is screwed to tie the ends. Four ebony feet turned with dowel ends raise the whole $\frac{1}{2}$ in. The front door of the case is framed from stuff $\frac{3}{4}$ in. thick (see Fig. 1951). Cut the top rail semicircular, and chamfer the inside face corners of the rails and stiles, working a $\frac{1}{4}$-in. rebate on the inside to take the bevel plate glass, which is fixed with small nosing fillets pinned to the framing. The door is ornamented with ebony mountings and a column (Fig. 1953) half-round in section, at each side of the glass opening. Each column has rectangular bases A (Fig.

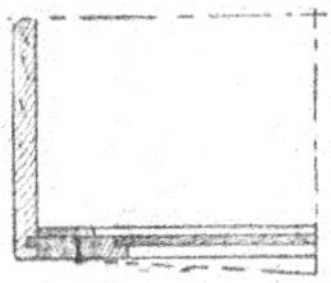

Fig. 1952.—Part Section of Back Door and Corner of Clock Case.

Fig. 1953.—Column of Clock Case.

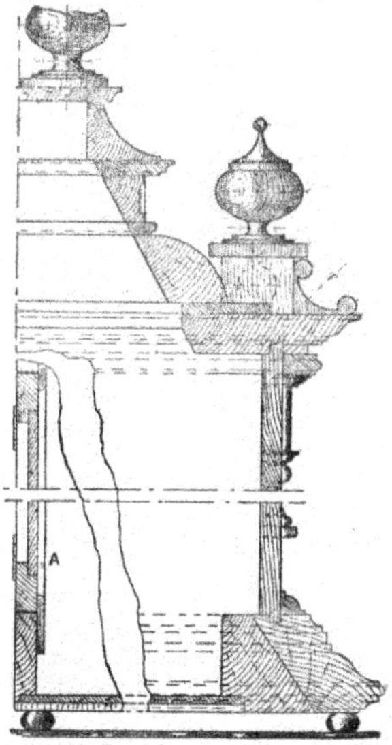

Fig. 1951.—Cross Section of Clock Case.

Fig. 1954. Fig. 1955. Fig. 1956.

Figs. 1954 to 1956.—Front and Side Elevations and Vertical Section of Grandfather Clock Case.

Fig. 1957.—Horizontal Section of Grandfather Clock Case at A (Fig. 1955).

Fig. 1959.—Upper Moulding of Clock Case Base.

Fig. 1960.—Front Elevation of Board for Supporting Clock Movement.

Fig. 1958.—Horizontal Section through Clock Case Base.

Fig. 1962.—Plan of Bottom Frame of Top Case.

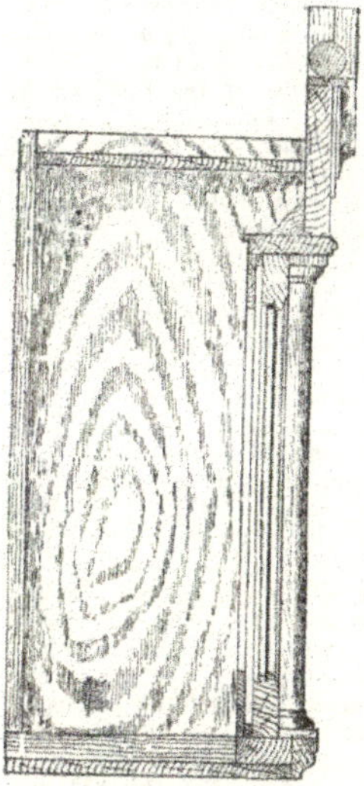

Fig. 1963.—Vertical Section through Centre of Top of Clock Case.

Fig. 1964.

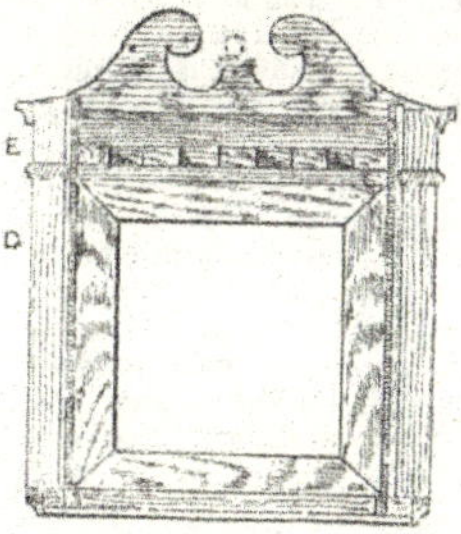

Fig. 1961.—Back View of Clock Case Top.

Fig. 1965.

Figs. 1964 and 1965.—Horizontal Sections through Clock Case at D and E (Fig. 1961).

1949) and capitals B, and the two columns can be turned as one piece, which is then carefully sawn down the centre. The portions shown in Fig. 1948 as ebony are stained in imitation. Below this door a rail, $1\frac{3}{4}$ in. deep, is housed into the carcase, and the base of the column is built over the joint. Between the base and mouldings fit an ebony veneer cut as shown in Fig. 1948, the panel below being pricked with a bradawl or carver's punch. The door hangs flush over the carcase sides, and folds under the cornice and over the rail. Use $1\frac{1}{4}$-in. brass butt hinges, and close the door with a spring knob. The cornice mouldings, illustrated in Fig. 1951, are $\frac{7}{8}$ in. thick, and overhang $1\frac{5}{8}$ in. Immediately below the cornice fix a small moulding to hide the joint, and increase the cornice to $1\frac{3}{8}$ in. in thickness. The moulded dome finish is built on the cornice mouldings, and each of the various members composing it is separately worked and glued to the preceding member, the amount of set-back being scaled from Fig. 1951. The lowest member is $1\frac{7}{16}$ in. high, worked on stuff $1\frac{3}{8}$ in. thick; the next member is $\frac{1}{4}$ in. thick, and has a small bead run along the face, and on this is fixed a panelled piece, 1 in. high, on which, at the front, is an ebony veneer (see Fig. 1948) showing a pricked panel, but at the sides it is left plain. This is covered by a moulded piece $\frac{5}{16}$ in. thick, with an overhang of $\frac{7}{16}$ in., and on this a cavetto moulding is placed, capped by a small overhanging piece. The top cap is $\frac{3}{16}$ in. thick, and carries the turned finials. The back of the dome portion is filled with a deal board cut to fit between the various mouldings, and fixed with brads to small corner blocks glued inside the carcase. The frieze panel is $\frac{3}{8}$ in. thick, and receives a moulding. On each side of the frieze is placed a block $1\frac{1}{4}$ in. high by $1\frac{1}{2}$ in. square, which has a boss or patera carved in walnut or ebony at the front. Finials of a similar design to the dome finial are dowelled so the blocks, and in each case a tight fit is all that should hold them, so that they can be readily removed for packing purposes. The clock movements are carried on a rail $\frac{5}{8}$ in. thick, fixed on blocks glued and screwed to the carcase. For the dial face house two uprights into the carcase top and bottom, and glue them to the sides. These pieces are not shown, as their position must be governed by the style of face and movement fitted. Finally, the deal parts are stained walnut.

Grandfather Clock Case.

The grandfather clock case shown in front and end elevations by Figs. 1954 and 1955 is of good proportions, and is fairly simple in construction. Fig. 1956 is a vertical section. It is 6 ft. 9 in. high, 1 ft. $8\frac{1}{2}$ in. wide, and 10 in. from back to front, and would look well in polished mahogany or fumed oak. The top case is constructed apart from the body, and is made to slide on easily from the front, thus enabling it to be removed without much trouble when the clock is out of repair. The first part to be built up is the middle case, or body, as most of the other framing is fitted to this. The pieces required for this portion are: Two for the sides, 3 ft. 6 in. by 7 in. by $\frac{1}{2}$ in.; two for the stiles, 3 ft. 6 in. by $1\frac{3}{4}$ in. by $\frac{1}{2}$ in.; one for the top rail, 1 ft. $0\frac{1}{2}$ in. by $7\frac{1}{2}$ in. by $\frac{1}{2}$ in.; and one for the bottom rail, 1 ft. $0\frac{1}{2}$ in. by 5 in. by $\frac{1}{2}$ in. These are mortised and tenoned together, and fitted into the rebate on the front edge of the sides, on the front edge of which a bead is glued to break the joint; this is shown in the enlarged section (Fig. 1957) taken at A (Fig. 1955), together with the rebated edge which receives the back. Dry pine angle blocks are fixed inside to secure the joints.

Bottom Case or Base.—The pieces necessary for the bottom case, or base, are: One for the front, 1 ft. 9 in. by 1 ft. $6\frac{3}{4}$ in. by $\frac{3}{4}$ in., and two for the ends, 1 ft. 9 in. by $9\frac{1}{2}$ in. by $\frac{3}{4}$ in., halved, mitered, and blocked together, as shown in section by Fig. 1958. The bottom moulding for the base—two pieces 10 in. by $3\frac{1}{2}$ in. by $\frac{3}{4}$ in., and one piece 1 ft. $8\frac{1}{2}$ in. by $3\frac{1}{2}$ in. by $\frac{3}{4}$ in.—is next cut to shape, mitered round, and fixed with screws driven from inside. The top moulding of the base is prepared from 1-in. stuff, two pieces being $9\frac{1}{4}$ in. long and one piece 1 ft. $6\frac{1}{4}$ in. long by 4 in. wide, and is mitered and fitted into the grooved top edge. The section given at Fig. 1959 shows how it is blocked and screwed to the body. Fig. 1956 is a section of the clock case taken at B (Fig.

1954), and an end view of the interior of the top with the door, pediment, etc., removed; a piece of ¾-in. stuff, 1 ft. 1 in. by 3¾ in., is screwed on the top of the ends of the main framing to support the movement. A front elevation of the board is given by Fig. 1960, which also shows the moulding that supports the top case, mitered round, blocked, and screwed from inside.

Top Case.—The top case is made up of several frames constructed in various ways. Fig. 1961 is a back view of the case, and Fig. 1962 a plan of the bottom frame. For the bottom frame, which should be put together first, one piece 1 ft. 4½ in. by 1⅝ in., and two pieces 8½ in. by 1⅝ in. by 1 in. are required; these are mortised, tenoned, and wedged together, and moulded on the outside edge with an ogee 1¼ in. by ¾ in. The frame fits over the square of the moulding C (Fig. 1960). Then the two ends of the case are stop-chamfered to the height of the columns, rebated to receive the ½-in. back, and grooved inside for ⅜-in. whitewood top and outside for the small moulding that rests on the top of the columns. The ends are screwed on to the bottom frame from underneath. Fig. 1963 is an enlarged vertical section through the centre of Fig. 1961, giving in detail the base and capital of the column. Fig. 1964 is an enlarged horizontal section taken at D (Fig. 1961), showing a section of the ¼-in. thick frame (which presses close to the face of the clock), the door frame, columns, and half-columns, and the 1-ft. 2½-in. by 1¾-in. by ½-in. pieces for the back of the half-columns. The columns are turned from 1-in. stuff, and the capitals and bases from 1¾-in. stuff, and are dowelled into the frames at the top and bottom. The door frame is 1 ft. 2½ in. square, the stiles and rails being 2 in. by ¾ in., moulded as shown in Fig. 1964, rebated to receive glass and beads, and mortised and tenoned in the usual way. The door is hung with 1½-in. brass butts about ⅛ in. forward of the front edge of the ends (see Fig. 1964). A ¼-in. by ⅜-in. slip is bradded round inside as a stop for the door, and the 2-in. by ¼-in. frame is mitered and bradded on to the slip from inside, making the case dust-proof. Fig. 1965 is a section taken at E (Fig. 1961), and shows the pediment mitered and halved to the end pieces, blocked and screwed from inside. The scrolls are also fixed with screws from the back, the ball in the centre being secured with a dowel. The ½-in. whitewood back is next fitted and bradded into the rebate. The door (shown in section at Fig. 1957) is moulded outside, rebated inside, fitted with a small cupboard lock and an escutcheon, and hung with 2-in. butts.

Child's Movable Table.

A table such as is shown in Fig. 1966 is suitable for a child learning to walk, and afterwards as a means to prevent the child getting to the fire or into other mischief. It will be found a source of pleasure when the child is able to move about by itself. It may be made of any light wood such as sound yellow pine for choice. The four legs A (Figs. 1967 to 1969) are 1½ in. by 1¼ in. by 1 ft. 8 in. long, and the eight rails B are 2 in. by ¾ in., two being 2 ft. 4 in. long, four 1 ft. 8 in. long, and two 1 ft. 2 in. long. These legs and rails are mortised and tenoned. The top C is of ¾-in. stuff 11½ in. wide, and is in two pieces C and C^1 (Fig. 1969), each 1 ft. 6 in. long. One (C^1) has a semicircular front, and is secured to the upper side rails with four 2-in. screws let in ⅛ in. to take plugs over the heads. A piece of American whitewood D (Fig. 1968), 1½ in. by ⅜ in. by 3 ft. 3 in. long, and with the upper edge rounded, is, after being steamed or boiled, bent round the front and secured with nine ¾-in. screws. If the wood shows signs of splintering, a backing of the same material, 1½ in. wide by ⅛ in. thick, is bent round with it, holes being bored through it large enough to let the heads of the screws pass through and secure the inside piece. Two fillets of hard wood E (Fig. 1970), ⅜ in. by ⅜ in., are secured to the side upper rail, and the movable part of the top is fitted on the under side with two runners F, 1 ft. 1½ in. long, so as to pass under the front piece 2 in. and hold it in place. Or two small cabin hooks and eyes may be attached to the under parts and will hold the two parts together. The hole in the top is 6½ in. by 6½ in., half being cut out of each piece, and the corners and edges being well rounded. Four small castors are attached to the lower parts of the legs. After the wood is cleaned off, two coats of size and one coat of oak varnish will make a good finish. For a young child procure a piece of duck or some strong material, 1 ft. 6 in. long and 7 in. wide, with an oval hole 4 in. by 3 in. cut out 3 in. from one end, as indicated in Figs. 1967 and 1968. This end is fastened to the front side of the hole by a small fillet at G (Fig. 1969), and three eyelets are worked in the other end so as to pass over three small hooks in the movable piece at H (Fig. 1969) and form a secure seat.

Café Tables and Chairs.

Table.—Oak is the favourite material for this class of furniture. The table (Fig. 1971) is 2 ft. 4 in. high, the top being 1 ft. 8 in. square; the dimensions may be varied if

Fig. 1966.—Child's Movable Table.

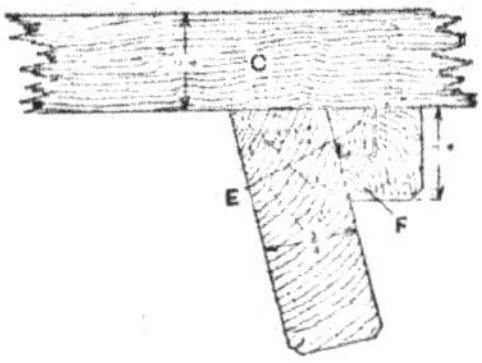

Fig. 1970.—Part Section of Table Top.

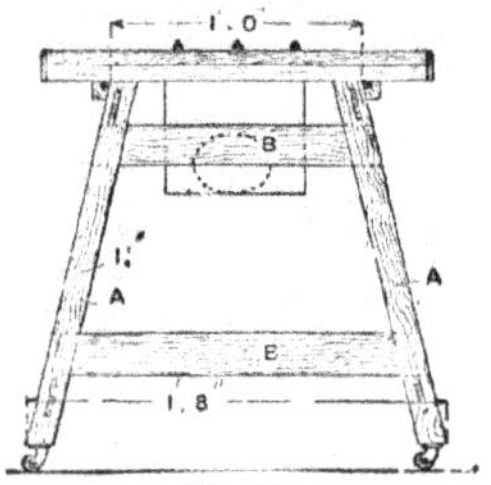

Fig. 1967.

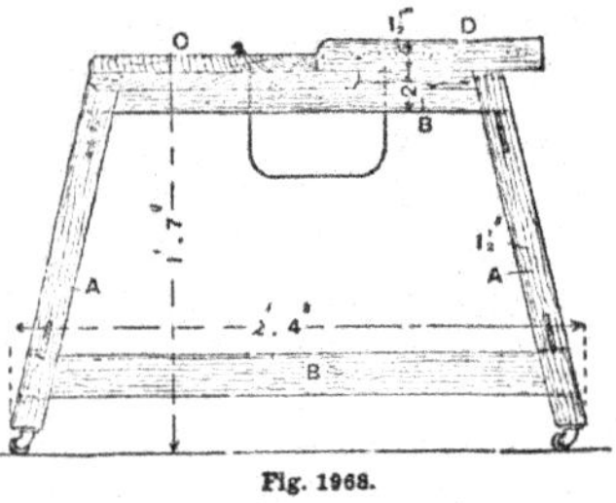

Fig. 1968.

Figs. 1967 to 1969.—End and Side Elevations and Plan of Child's Movable Table.

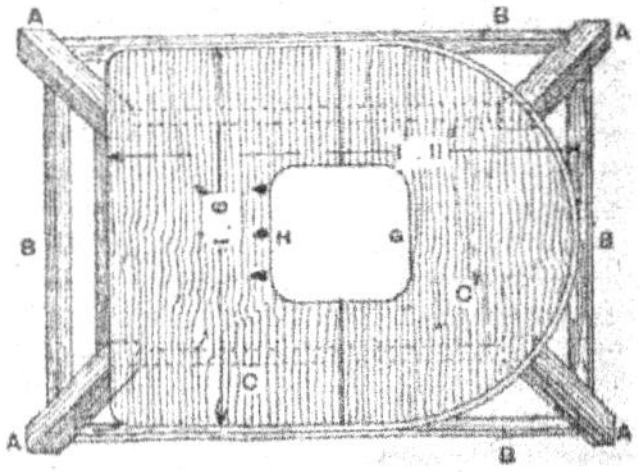

Fig. 1969.

desired—say 3 ft. by 2 ft.—without it being necessary to increase the strength of material. The legs are cut 2 ft. 4 in. long by $1\frac{3}{4}$ in. square; this allows $\frac{1}{2}$ in. at each end for working. The feet and top ends should be shaped in the lathe. The square portion at the top ends of the legs is 6 in. long; the rest of the leg is turned plain round as thick as the wood will allow, tapering slightly to the foot, which is turned to the full thickness. The two spindles supporting the shelf are 2 ft. long by 1 in. square; 11 in. of the centre must be left square, the ends being plain turned and having a pin $\frac{3}{4}$ in. in length by a full $\frac{1}{2}$ in. thick. The four span-rails are planed and squared to 1 ft. 2 in. long by 4 in. wide by 1 in. thick; the lower edge should be shaped as shown at Fig. 1972. They are dowel-jointed to the legs (see Fig. 1973). The shelf is 9 in. square by $\frac{3}{4}$ in. thick, the edges being shaped to a flat round; it is made to rest on the spindles, which are let into holes in the legs 8 in. from the floor, as shown at Fig. 1974, the square parts being cut to meet in the centre, and screwed to the shelf as shown in the underneath view (Fig. 1975). The top is 1 ft. 8 in. square by $\frac{5}{8}$ in. thick. A 1-ft. square is marked in the centre (see Fig. 1971), for covering with sixteen 3-in. square tiles (generally of plain green tint) cemented on with plaster-of-Paris mixed with thin glue to a consistency of thick cream; the tiles are surrounded with wood of the same thickness as the tiles, mitered at the corners, the edges being rounded (see section, Fig. 1976). To make the stand more rigid, it may be corner-blocked; then each rail is bored for two screws, for fixing the top. A quicker way than thumb-notching is to sink the surface with a 1-in. centre-bit, and then bore the screw holes, as in Fig. 1977. The top and stand having been screwed together, the table is complete.

Chair.—The café chair (Fig. 1978) is of a pattern to match the table, and combines lightness with strength. First get out a template for marking the back uprights on a board of $1\frac{1}{4}$-in. oak, 3 ft. long. The slight curve should begin just below the seat. When marking out on the board, keep the grain as straight as possible; but much labour will be saved by sending it to a saw-mill to be cut. For the front legs, pieces must be cut 1 ft. $6\frac{1}{2}$ in. long by $1\frac{1}{4}$ in. square, and for the front spindle 1 ft. $1\frac{1}{2}$ in. long by 1 in. square. All the plain spindles are of $\frac{7}{8}$-in. square stuff; the top side spindles are 1 ft. $0\frac{1}{2}$ in. long, the bottom 1 ft. $1\frac{1}{2}$ in., and the back 1 ft. The ends must taper slightly. The length of the spindles as given is the finished size; but $\frac{3}{4}$ in. is allowed over on the legs, the finished size of which is 1 ft. 5 in., with $\frac{3}{4}$ in. of pin at the two ends. The uprights must be cleaned up, and the top ends rounded over both ways (see plan, Fig. 1979). The top rails are cut 1 ft. $2\frac{1}{2}$ in. by $3\frac{1}{2}$ in. by $\frac{3}{4}$ in., with an oval-shaped hand hole, and the top edge rounded; they are mortised and tenoned to the uprights, this taking up 1 in. of each end, and the back spindles are let in 8 in. from the floor; at 1 ft. 6 in. there must be a space of 11 in. between the uprights. The exact length of the lower back rail can now be ascertained; this is of 2-in. by $\frac{3}{4}$-in. stuff, shaped to match the table rails; it is fixed 3 in. above the seat. Between the rails are two plain bars of 1-in. by $\frac{1}{2}$-in. section, and another 2 in. wide (shown enlarged at Fig. 1980), these being mortised and tenoned. The back frame is then glued up. The seat frame is 1 ft. 3 in. at the front, 1 ft. at the back, and 1 ft. 1 in. from back to front, and is of $2\frac{3}{4}$-in. by 1-in. stuff, rebated on the inside edge on the top side $\frac{3}{4}$ in. by $\frac{1}{4}$ in.; it is put together as shown at Fig. 1981, and is rounded on the front and sides. The back corners must be let into the uprights and secured with strong screws driven from behind, the heads being sunk below the surface and the holes afterwards plugged with wood. Now the legs should be connected by the front spindle, and let into the corners; then the side spindles should be inserted, connecting the back and front. The whole should be glued up in one operation. Both the chairs and the tables are now ready for staining and polishing. For the stuffed seats, a strong black canvas bottom cover should first be tacked in the rebate, stretched tight, and webbed, three each way, with No. 12 English web; this is covered with canvas, and a stuffing of hair put on, then calico. Cover finally with green leather, leather-cloth, or velvet, edged with copper studs.

Fig. 1971.—Café Table.

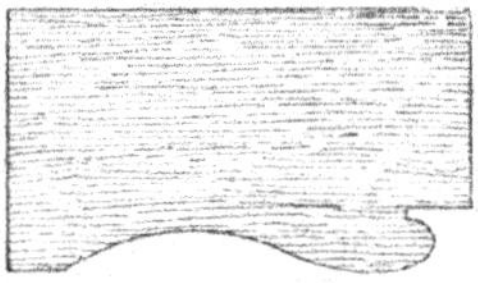

Fig. 1972.—Half Pattern of Café Table Rail.

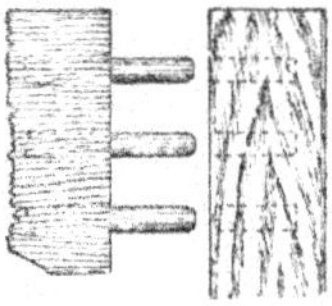

Fig. 1973.—Leg of Café Table dowelled to Rail.

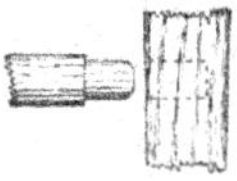

Fig. 1974.—Joint of Spindle to Leg of Café Table.

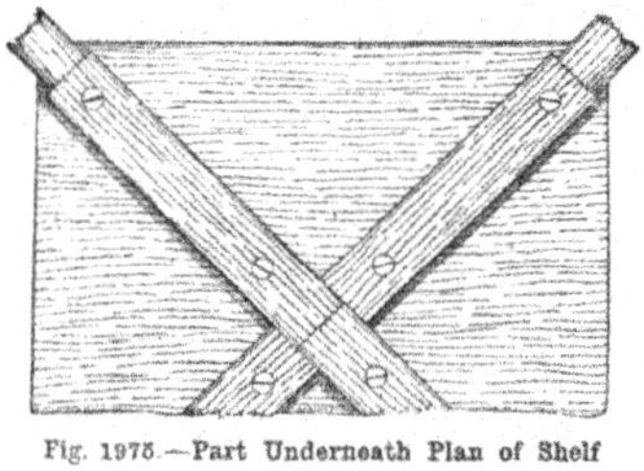

Fig. 1975.—Part Underneath Plan of Shelf and Spindles.

Fig. 1976.—Section of Top of Café Table.

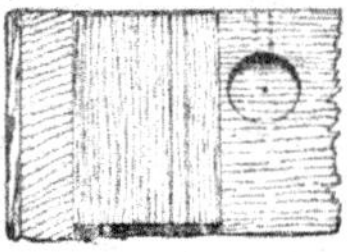

Fig. 1977.—Method of Screwing Café Table Top to Rails.

Triple Set of Telescopic Tea Tables.

Fig. 1982 gives a front view of a tea-table set which comprises three separate tables sliding one within the other, Fig. 1983 being a plan of the top. Fig. 1984 is a side view of the same set of tables, with a different top, of which Fig. 1985 is the plan. The rails between the legs across the front

Fig. 1978.—Café Chair.

of the two larger tables are lower than the side rails, to enable the inside tables to slide in and out. Figs. 1986 to 1989 show elevations and plans of the smaller tables, reproduced to the same scale as the large one. Figs. 1990 and 1991 show alternative designs for the sides and rails. The sliding principle can be applied to almost any square or rectangular table, as illustrated in elevation and plan at Figs. 1992 and 1993, where on the right-hand side a square table with double splayed legs is shown, and on the left-hand side a square table with curved splay legs. Tables with splayed legs are not so likely to be accidentally overturned as the others, but are liable to catch the foot sometimes. The legs are mortised to receive the barefaced tenons of the sides, with the shoulders of the tenons inside and the inner faces of the side pieces flush to receive the runners, which are screwed to the inner faces of the side pieces, but cut back to enable the hinged front piece of the large table to fold down. Fig. 1994 is a side view showing the flap raised and the smaller tables partly drawn out, and Fig.

Fig. 1980.—Centre Back Piece of Café Chair.

Fig. 1981.—Part Plan of Café Chair Seat Frame.

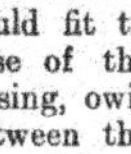

Fig. 1979.—Plan of Café Chair Upright.

1995 a front view. Fig. 1996 is a sectional view, showing the position of the hinged flaps and the runner of the large table, Fig. 1997 being a front view. The flap, when down, should fit tight between the legs; in the case of the splay-leg table it will require easing, owing to the double angle formed between the legs. The top of the large table may be shaped as desired, but the overhang of the tops of the smaller tables is necessarily limited. The rails and sides are mock-tenoned, as shown in Fig. 1998. The square table tops should have a flat ovolo or lamb's-tongue mould run round the edges, but for the shaped ones reeded edges and a flat chamfer on the top would be more suitable. Any of the usual hardwoods, polished, may be used for the construction of the tables. Birch with plenty of nice flower in the table tops would look extremely well, though unfortunately liable to show up stains. Figs. 1982 to 1993 are reproduced to a scale of 1 in. to 1 ft., and Figs. 1994 to 1998, 2 in. to 1 ft.

Reading-room Table.

For a reading-room or a waiting-room, a table of ordinary construction or design

Fig. 1982.

Fig. 1984.—Side View of Telescopic Tea Table with Different Top (see Fig. 1985).

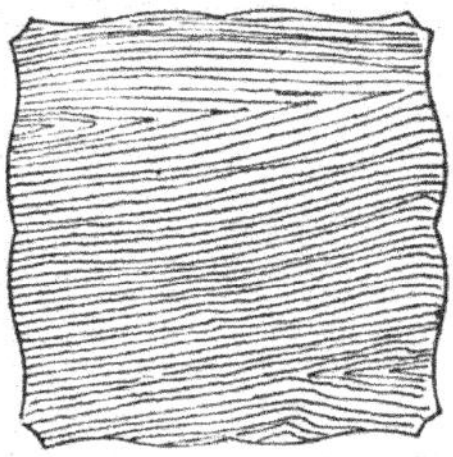

Fig. 1985.—Alternative Top for Table.

Fig. 1983.

Figs. 1982 and 1983.—Front Elevation and Plan of Telescopic Tea Tables.

Fig. 1986.

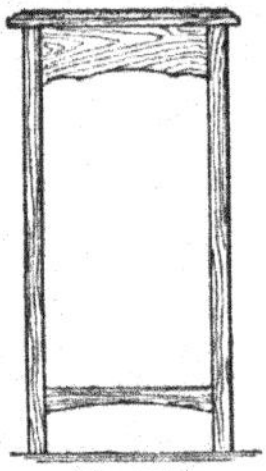

Fig. 1988.

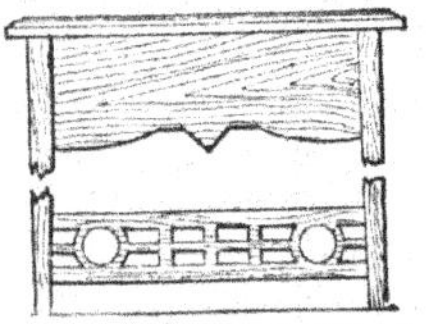

Fig. 1990.—Alternative Design for Tea Table Side and Rails.

Fig. 1989.

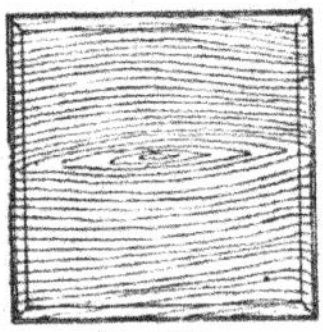

Fig. 1987.

Figs. 1986 to 1989.—Elevations and Plans of Smaller Tea Tables.

is unsuitable, as strength and durability, as well as a good appearance, are necessary. The table shown by Figs. 1999 to 2001 is 6 ft. long by 3 ft. wide, and 2 ft. 6 in. high; but if required, it may be enlarged to 9 ft. by one piece of timber. These joints will require four dowels, placed 3 in. from the ends and dividing the space between, gauging from the top side. If the top is composed of three boards, two should be

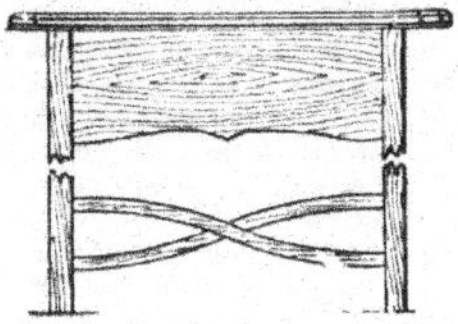

Fig. 1991.—Alternative Design for Tea Table Side and Rails.

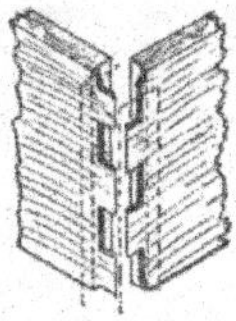

Fig. 1998.—Joint of Sides of Tea Table.

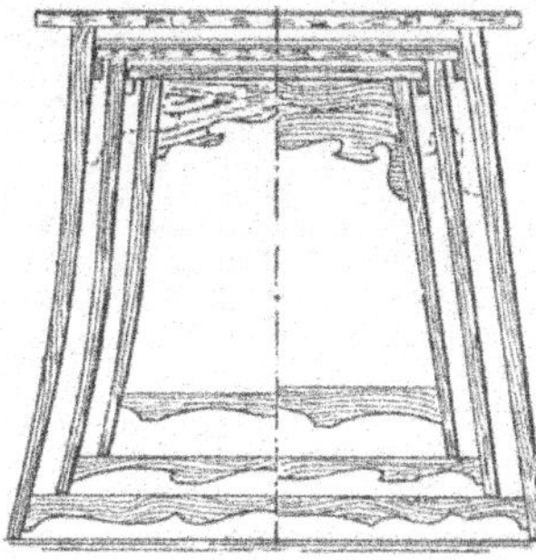

Fig. 1992.

Fig. 1994.

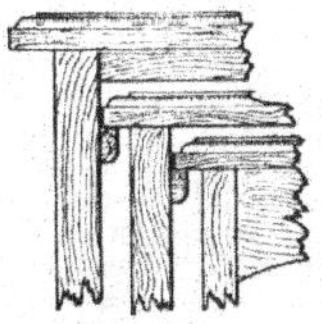

Fig. 1995.

Figs. 1994 and 1995.—Side and Front Elevations of Tea Tables Drawn Partly Out.

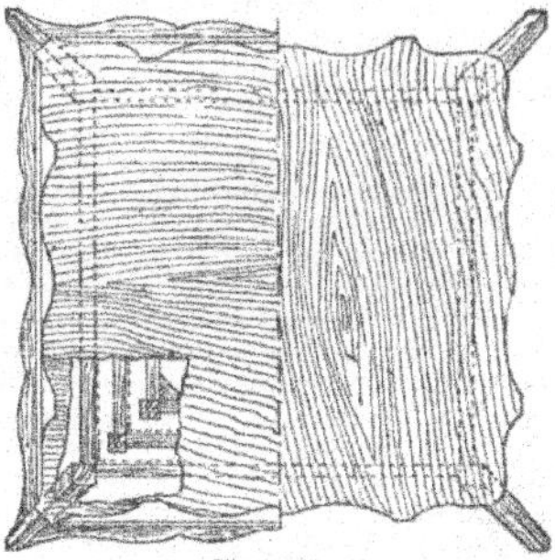

Fig. 1993.

Figs. 1992 and 1993.—Half Elevations and Half Plans of Splay-leg Telescopic Tables.

Fig. 1996.

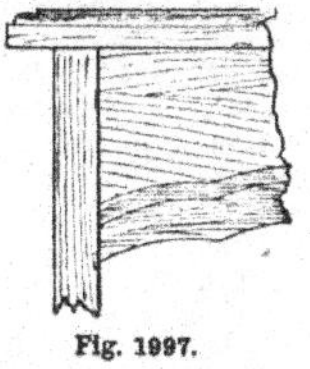

Fig. 1997.

Figs. 1996 and 1997.—Side and Front Elevations showing Hinged Flap, etc., of Tea Table.

4 ft. without any alteration in the construction, except perhaps an extra foot-rail. Oak is the best wood to use. For the top, three 1-in. boards 1 ft. wide, or four 9 in. wide, should be jointed lengthwise, and so put together that the grain of the wood may match as near as possible to represent jointed first, and the other added when the joint is set; but if of four boards, they may be jointed in twos, and the middle joint made afterwards. For the stand, of which Figs. 1999 and 2000 give the general design, four legs are required, of solid oak, 2 ft. 5 in. long by 3 in. square. About 3 in. of the

lower ends must be allowed for joining; there are 8 in. of plain square, and 10 in. of turning. Two base pieces are required at the ends, 2 ft. 4 in. long by 4 in. by 3 in.; the outer corners of these are rounded off, the top corners chamfered, and the under sides cut out 6½ in. from the ends and 1 in. deep, the under corners also being rounded off (see Fig. 2000). The legs are mortised and tenoned to the base pieces 2 in. from the ends, but should not be fixed till the two end rails are secured to the top ends of the legs. These are 2 ft. 8 in. by 4 in. by 1½ in.; the under corners are shaped to a flat ogee working 4 in. on the length and 1½ in. on the width, to show ½ in. of square at the lower edge and 2½ in. at the extreme ends, as in Fig. 2002, which also shows how the leg is cut away to leave a shoulder for the end and side rails, leaving 1½ in. square to be screwed to the inner side of the end rails only. When this is done the base joints can be glued, and wedges driven into the saw-kerfs of the tenons from underneath. The two long rails are 5 ft. 8 in. in length, and are made exactly the same as the end rails, but cut on the under edge in making the halved joints of the Oxford corners. When they are fitted to connect the legs, the stand will need joining at the base, and this is done by a rail which is also intended as a rest for the feet. It is 3 in. square in section, and the corners are chamfered and rounded off like the base pieces, the ends being cut to a shoulder on the under side by taking out 3½ in. by 1 in.; it is screwed from underneath. The next thing is to make a piece 2 ft. 8 in. by 3 in. by 1¾ in. as a support across the centre of the top, to be let into the rails, and then the stand is ready for the top The rails must be thumb-notched on the inside, about four on the long rails and three on the short ones, and the top should be laid face downwards on the bench or a pair of trestles; the stand is lifted on upside down, and placed evenly to allow 2 in. from the ends of the rails to the edges of the top; then the eight ends are bored with a ½-in. bit about ¾ in. deep, the holes being continued of a size to take screws 3 in. long, with which the top is secured, other screws being inserted where the rails are notched and through the centre support (see Fig. 2001). The moulding under the top is 2 in. by 1½ in. in section, and is fixed close against the ends of the rails, secured with screws through the notches. The table is now complete. The polishing process should be done without using any grain-filler or varnish, and should not be finished off too bright. When it is desired to take the table apart, the screws of the end rails only should be withdrawn; the top can then be lifted off with the long rails attached, and the stand turned up and the foot rail removed.

Invalid's Folding Chair.

The invalid's folding chair shown by Figs. 2003 to 2005 is strong, and is easily and cheaply constructed, the folding joints being simply swivelled on the small bolts which hold the frame together. The main thing to ensure the chair folding properly is to space and bore the bolt holes in each pair of the framing exactly alike. The front pair of wheels are 1 ft. in diameter, fitted with ¾-in. wired-on tyres. The back pair of wheels are 1 ft. 3 in. in diameter, also with wired-on tyres. The axles are 1 ft. 7 in. between the shoulders of the hub bearings. Probably for strength and durability the best wood for making the chair is sound, straight-grained, home-grown ash. The size of chair can be altered to meet special requirements, but the sizes here given are ample for the average adult. A side elevation of the chair is given in Fig. 2003. The arms A are bolted to the back legs B and the front legs C. The seat rails D and the foot rails E are respectively bolted to the front and back legs as shown in Figs. 2003 and 2004. The baluster rails are screwed on the back legs with round-headed 2½-in. screw nails. This makes a much stronger and stiffer job than tenoning the baluster rails into the back leg. Fig. 2005 is a side elevation of the chair folded together.

Turned Work.—If necessary the chair can be ornamented by turning. The pieces of wood required for the turned work, including the ends for cutting off, are: Two pieces 2 ft. 3½ in. long for the front legs (Fig. 2006), and two the same length for the foot rails (Fig. 2007); two pieces 1 ft. 7 in. for the arms (Fig. 2008); two pieces 1 ft. 6½ in. long for the seat rails (Fig. 2009); and two pieces 1 ft. 4 in. for the seat rail stretchers (Fig. 2010). These pieces are all dressed up to 1⅛ in. square before turning. For the baluster rails (Fig. 2011), two pieces of stuff 1 ft. 8½ in. by 1 in. are required, and five pieces 11 in. long by 1 in. for the balusters (Fig. 2012). The baluster

Bolt Holes, etc.—The bolt holes in the arms, seat rails, and foot rails are marked in the centres of the squares, and exactly 1 ft. 4 in. between the bolt-hole centres. The bolt holes in the front legs and back legs are marked 1 ft. 2 in. from the foot rail to the seat rail hole, and the arms are 10 in. from the seat. The holes are bored a tight fit for ½-in. bolts. The holes F in the seat rail (Fig. 2009) for the stretcher

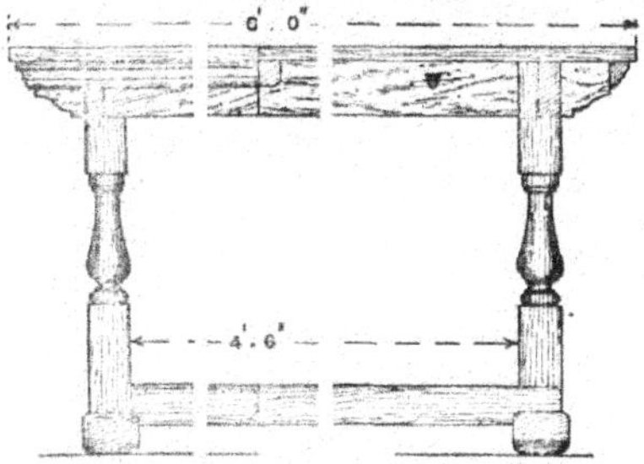

Fig. 1999.

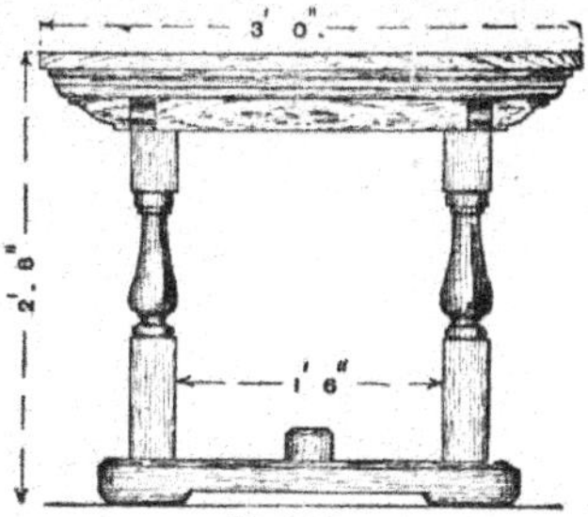

Fig. 2000.

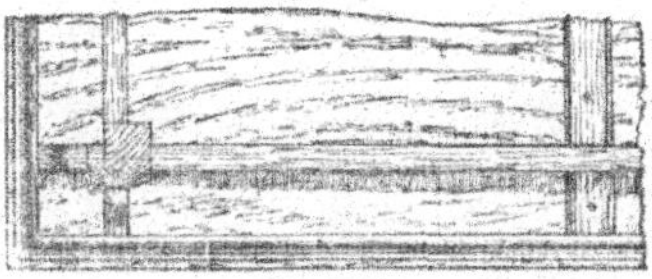

Fig. 2001.

Figs. 1999 to 2001.—Part Side Elevation and Part Section, End Elevation, and Part Underneath Top View of Reading-room Table.

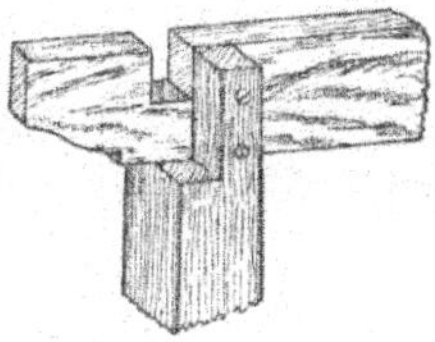

Fig. 2002.—Fixing End Rail of Table to Leg.

rails, as well as the arms, legs, etc., are planed before they are turned, to make the facets on the ends square and in line, so as to form good pivot joints. The balusters being finished in the lathe do not require planing. After they are planed, as a guide for the turner the squares are pencil lined, and the bolt holes are also marked and bored before beginning the lathe work. The squares at the ends (and front leg seat-joint) are left 1½ in. long, plus the allowance for cutting off in turning.

tenons are 1 ft. 1½ in. between centres, and they are bored ⅞ in. deep with a ⅝-in. centre-bit. The baluster rails are planed 1 in. square, the screw nail holes G (see dotted lines in Fig. 2011) are 1 ft. 4⅞ in. between centres, and bored at right angles to the holes for the balusters. These are equally spaced as shown, and bored with a $\frac{7}{16}$-in. bit, and ⅝ in. deep. The five balusters are turned 10 in. long with tight-fitting tenons and finished as shown in Fig. 2012. The seat rails and stretchers are, of course, left

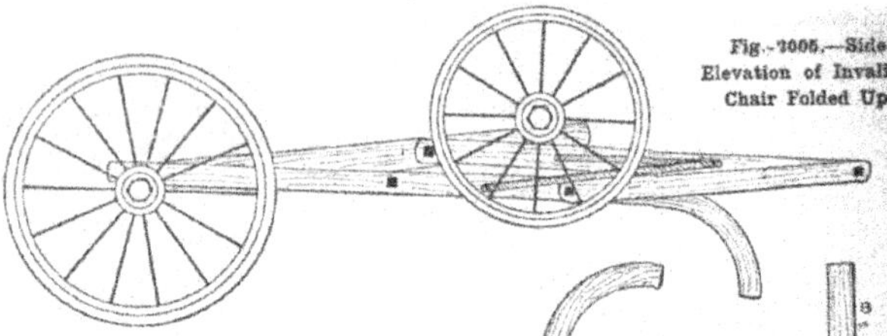

Fig. 2005.—Side Elevation of Invalid's Chair Folded Up.

Figs. 2003 and 2004. —Side Elevation and Half Front Elevation of Invalid's Folding Chair.

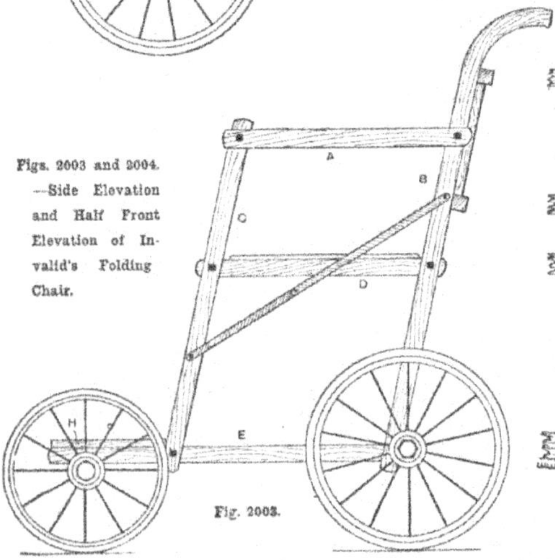

Fig. 2003.

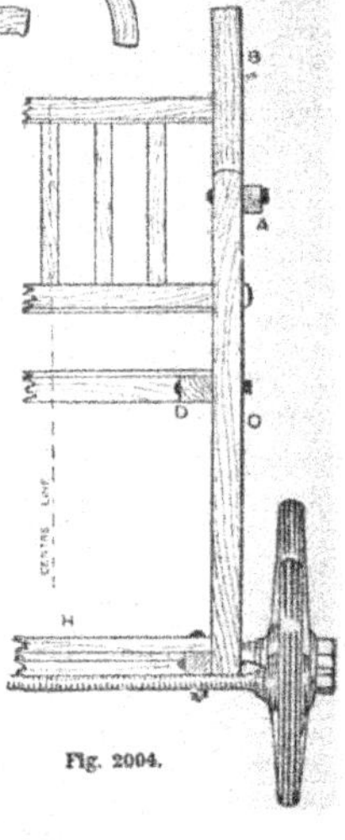

Fig. 2004.

Fig. 2006.—Front Leg of Invalid's Chair.

Fig. 2007.—Foot Rail of Invalid's Chair.

Fig. 2008.—Arm of Invalid's Chair.

Fig. 2010.—Stretcher for Seat Rails of Invalid's Chair.

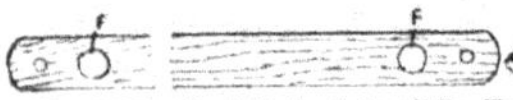

Fig. 2009.—Seat Rail of Invalid's Chair.

Fig. 2011.—Baluster Rail of Invalid's Chair.

unturned and square, excepting the ends, which are rounded off in the lathe. The stretchers (Fig. 2010) are 1 ft. 1½ in. long between the shoulders of the tenons, and the pins are turned a tight fit for the holes F (Fig. 2009). In turning the foot rail (Fig. 2007) the top part is left unturned and square to form the bed for the foot-board H (Fig. 2003). In finishing the turning, all the corners of the small flats or squares should have the sharpness taken off them.

Fig. 2012.—Baluster of Invalid's Chair.

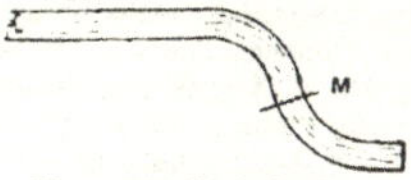

Fig. 2015.—Back Legs of Invalid's Chair made from Mailcart Shafts.

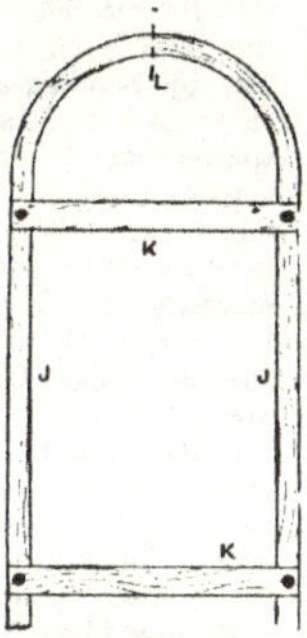

Fig. 2014.—Bending Back Legs of Invalid's Chair.

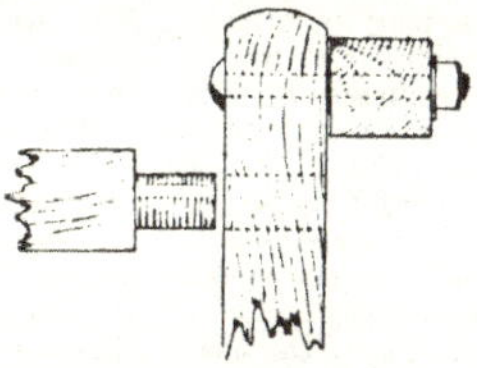
Fig. 2016.—Seat Rail Bolted to Back Leg.

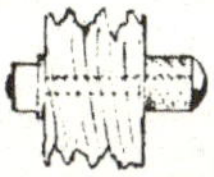
Fig. 2017.—Back Wheel Bolt of Invalid's Chair.

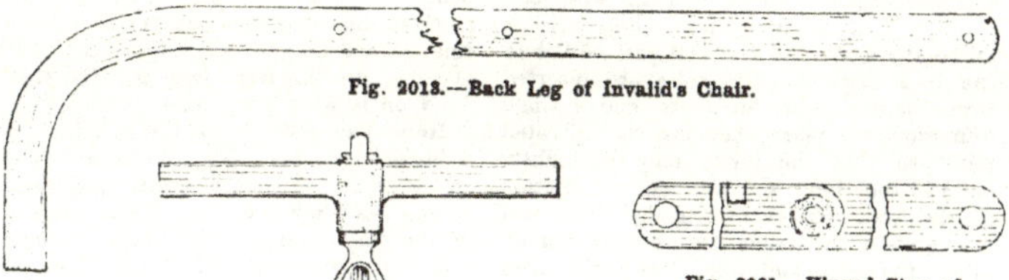
Fig. 2013.—Back Leg of Invalid's Chair.

Fig. 2021.—Hinged Stay of Invalid's Chair.

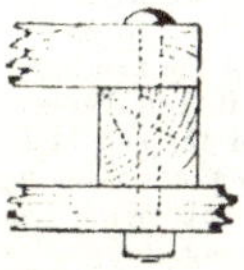
Fig. 2018.—Front Wheel Bolted to Foot Rail.

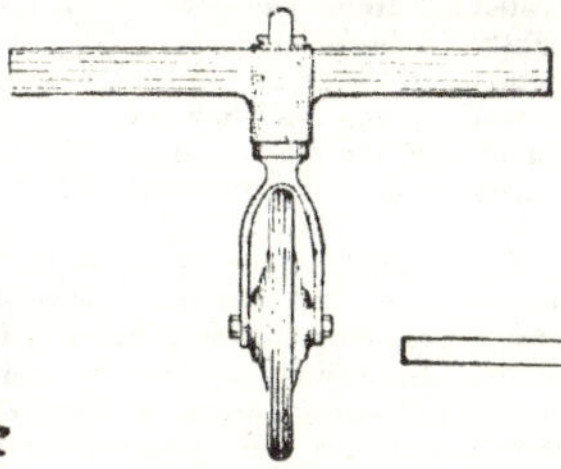
Fig. 2020.

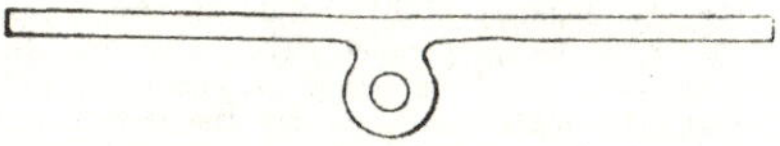
Fig. 2019.

Figs. 2019 and 2020.—Alternative Method of Fixing Back Wheel of Invalid's Chair.

Back Legs.—The back legs (Fig. 2013) are planed to 1⅛ in. square, and steamed and bent all in one piece to the shape shown at J (Fig. 2014) on moulds with a special appliance. After bending, slabs of wood K are nailed across as shown at Fig. 2014 to keep them in shape until thoroughly dry. The dotted line L shows where they are sawn across, and nothing but the cleanest straight-grained stuff is suitable for bending, and even then there is often much loss by breakages in bending. If desired, these back legs can be procured from bent-timber merchants. The back legs being thicker than the usual stock size for small trolly cars, there may be some difficulty in obtaining them. Therefore, a pair of ordinary mailcart shafts may be utilised by cutting off the top bends as shown by the line M (Fig. 2015). These mailcart shafts are of 1⅛-in. stuff, dressed and bent in breadths for two shafts. The back legs are cut to a reasonable length for wheeling (3 ft.), and the handle ends for about 6 in. should be nicely rounded with the spokeshave and smoothed with glasspaper.

Putting Chair Together.—To fit the chair together, give the stretcher tenons a touch of glue and knock them into the seat rails, and, having the edges of the rail and the stretcher square, drive a 1-in. wire nail through both to make fast. Fig. 2016 is an enlarged view of the seat rail bolted to the back leg. All the bolts are inserted from the inside of the joints, and a small iron washer is placed between each pivoted joint, so that the joints may be tightly bolted together without undue friction of the woodwork in folding up. The seat rails and the foot rails are then bolted on inside the back and front legs; the arms are bolted on the outside. The baluster rails and footboard can now be fixed. The balusters are knocked into the rails, measuring the distance at both ends to ensure the rails being parallel. The rails are screwed on the back legs 4 in. above the seat. To mark the holes for the screw-nails, insert the nails with the points projecting through the rails, and place in position on the back legs. Give each nail a tap to centre-pop the hole, then bore and screw on the rails. The footboard is dressed 8½ in. wide by ⅝ in. thick and 1 ft. 3¾ in. long, the front being rounded off on the top edge. It is nailed on the foot rails with half a dozen brass-headed stud nails. In fixing on the footboard, care should be taken to have it square with the outer edges of the rails, otherwise there may be trouble in folding the chair. A three-ply perforated seat cut to size and fixed on with small brass-headed stud nails finishes the woodwork, but before fixing the seat it is best to do the varnishing. The chair looks very well stained dark walnut colour, finished with two or three coats of copal varnish.

Wheels and Axles.—The wheels and axles can be obtained ready-made, but require painting and varnishing before fixing to the chair. The back wheels are bolted on the back legs as shown in Fig. 2017; the front wheels are bolted on underneath the foot rails, 3 in. from the front of the footboard, as in Fig. 2018. An alternative method of swivelling the rear wheel is shown in Figs. 2019 and 2020, Fig. 2019 being the plan of the socket, and Fig. 2020 an elevation of the wheel and socket. Should this method be adopted, a cross bar can be attached to the back legs and the socket fixed on it with ¼-in. bolts.

Iron Stay.—To keep the chair rigid, a hinged iron stay (Fig. 2021) is fastened to the front and back legs with snap-headed screws, as shown in Fig. 2003. One end of the stay forms a stop which fits into a notch in the other half of the hinge, and prevents it sagging downwards when straightened. The stay is made from $\frac{3}{16}$-in. by ⅝-in. iron, and the lengths of the two pieces are respectively 9 in. and 1 ft. 1 in. between the centre of the joint and the end holes. The shorter stay is fixed on the front leg (notched side upwards) 7½ in. above the foot-rail joint, the long end being screwed on about 1 ft. 7¼ in above the foot-rail joint of the back leg.

INDEX.

Printed by Cassell and Company, Limited, La Belle Sauvage, London, E.C.

Spotted something that isn't quite right?

Let us know and we'll fix it for future reprints.
And as a thank you for your diligence, we'll
send you a free Sothis book of your choice.

publisher@sothis.press

This page intentionally left blank

Made in United States
Orlando, FL
05 November 2026

90229678R00353